SUPERCONDUCTIVITY

SUPERCONDUCTIVITY

**S.L. KAKANI
SHUBHRA KAKANI**

ANSHAN LTD
11a Little Mount Sion,
Tunbridge Wells, Kent,
TN1 1YS

Co-published in the U. K. by

ANSHAN LTD, 11a Little Mount Sion, Tunbridge Wells, Kent, TN1 1YS
In 2009

Tel/Fax: +44(0)1892557767
e-mail: info@anshan.co.uk
Web Site: www.anshan.co.uk

© 2009 by New Age International (P) Ltd., Publishers

ISBN 978-1-848290-05-1

All rights reserved. No part of this book may be reproduced in any form, by photostat, microfilm, xerography, or any other means, or incorporated into any information retrieval system, electronic or mechanical, without the written permission of the copyright owner.

British Library Cataloguing in Publication Data
A Catalogue record for this book is available from the British Library

Preface

The field of superconductivity was born in 1911 when Onnes discovered this most exciting and amazing phenomenon. For a number of years thereafter, only a handful of experimentalists with access to liquid helium could study the known superconductors. The field of superconductivity grew, with important milestones being the propounding of microscopic observation of josephson effects, and the discovery in the late 1950s and early 1960s of new super conduction alloys that could be fabricated into composites capable of carrying high currents in high magnetic fields. After the exciting discovery of high temperature superconductivity in cuprates in 1986 by Bednorz and Muller a new era had begun. The initial skepticism about the discovery of Bednorz and Muller was soon dispelled when several other research groups confirmed the findings, and an enormous worldwide research effort began in quest of a room temperature superconductor. Within a year, the critical temperature (Tc) was raised to 90-93 K (above the boiling point of liquid nitrogen) with the discovery of yttrium compound, and then in 1988 it increased further to 120 K and 125 K with the successive discoveries: the bismuth and thallium cuprates. In 1993, the mercury cuprate went super conducting at 133 K when subjected to a pressure. We were past the halfway point to room temperature superconductivity. Indeed, the science fiction allusion of a room temperature super conducting system, which just a short time ago was considered ludicrous, appears today to be a distinct possible outcome of materials physics research. During this period, several new types of compounds were also found to exhibit superconductivity, such as the cubic perovskite BaKBiO, alkali and alkaline doped buckminsterfullerenes, organics, borocarbides, MgB_2, photo induced superconductors, etc. This generated renewed interest and an unprecedented amount of worldwide research on superconductors. As a consequence, superconductivity has grown into a huge field. Thousands of scientific papers have been published since Bednorz and Muller's pioneering paper on the cuprates in 1986. This poses a daunting challenge not only to the students entering this field as researchers but also to scientists with previous experience. Keeping this in view the present monograph Superconductivity is prepared.

The present monograph is spread over 10 chapters. For beginners and experimentalists, the monograph provides a readable introduction to the basics of superconductivity. For theorists, the monograph provides a brief description of the broad spectrum of experimental properties, theoretical concepts with all details, which theorists should learn, and provides a sound basis for students interested in studying super conducting theory at the microscopic level. A special chapter on the theory of high temperature superconductivity in cuprates is included.

Superconductivity is becoming increasingly accepted as a potential technology for the future with which one can address a number of global environmental problems. One of the most attractive of these applications is the construction of a global super conducting electrical power network, which should be able to increase the cost effectiveness of renewable energy sources such as wind and solar power. The current technology is developed to be applied to sources such as wind and solar power, and to enable reliable estimates of energy efficiency, including the cooling system and the total cost of the super conducting cable network system likely to be required.

Another promising candidate for the application of superconductivity is the MAGLEV train, which runs at speeds of up to 500 km/h. This speed can be more than doubled if the train is operated in a de-pressurized tube, providing us with the prospect of a new era of transporting passengers and goods at ultra high speeds. Because of its very low emissions, Maglev can contribute to an improved global environment, and high-level information technology for the first time can help to establish an energy saving society. Looking to this chapter 10 is devoted to applications of superconductivity. 'Glossary' of technical terms and useful Tables are provided.

In the preparation of this monograph we have received help from various publishers and authors. We are highly thankful to them for granting permission to produce figures, results etc., and we hope this monograph will have a significant influence on expediting the progress of future research in this field.

We are thankful to M/s New Age International Pvt. Ltd., Publishers and Anshan U.K. for their untiring efforts in bringing out the book within the shortest possible time period.

Suggestions for the improvement of the book are most welcome.

S.L. Kakani
Shubhra Kakani

Contents

Preface *(v)*

1 INTRODUCTION — 1

1.1 Introduction *1*
1.2 Classes of Superconductors *10*
1.2(*a*) Rare-Earth Transition Metal Borocarbides *12*
1.2(*b*) Sr_2RuO_4 *14*
1.2(*c*) Alkali Metal Doped C_{60} Fullerene Superconductors *14*
1.2(*d*) MgB_2 *14*
1.2(*e*) Quantum Spin-Ladder Superconductors *15*
1.2(*f*) LiV_2O_4 *16*
1.2(*g*) Heavy Fermion Superconductors (Superconductivity in *f*-electron systems) *16*
1.2(*h*) Itinerant-Electron-Ferromagnetic Superconductors *17*
1.2(*i*) Binary Alloys and Intermetallic Compounds *19*
1.2(*j*) Chevrel Phases *20*
1.2(*k*) Ternary Rare-Earth Rhodium Borides *22*
1.2(*l*) Organic Superconductors *23*
1.3 Potential Applications *25*
References 26

2 CHARACTERISTIC PROPERTIES OF SUPERCONDUCTING STATE — 29

2.1 Introduction *29*
2.2 Electrical Resistivity: Normal Metal *29*
2.3 The Superconducting State-Zero Resistivity i.e., Infinite Conductivity (No 'Individual Scattering') *35*
2.4 A.C. Resistivity *39*
2.5 The Wave Function of the Superconducting State *40*
2.6 Perfect Diamagnetism *41*
2.7 Ring Supercurrent *43*
2.8 Critical Magnetic Field *43*
2.9 The Thermodynamics of Superconductors *50*
2.10 Isotope Effect *58*
2.11 Acoustic Attenuation *58*

2.12 Mechanical Effects *58*
2.13 High Frequency Electromagnetic Properties *59*
2.14 Absence of Effects *59*
2.15 Characteristic Phenomenological Parameters *59*
2.16 Flux Quantization *62*
2.17 Magnetic Levitation *64*
2.18 Tunnelling Effects *66*
References 80

3 PHENOMENOLOGICAL THEORY 95

3.1 Introduction *95*
3.2 Two-Fluid Model *95*
3.3 London Theory *97*
3.4 Ginzburg–Landau (GL) Theory *99*
3.5 Type-I and Type-II Superconductors *102*
3.6 Lower Critical Field (B_{C_1}) *104*
3.7 Surface and Interface Effects in Superconductors *105*
References 108

4 CRITICAL CURRENTS OF TYPE-II SUPERCONDUCTORS 109

4.1 Introduction *109*
4.2 Mixed State *110*
4.3 Interaction between Vortices *111*
4.4 Abrikosov Lattice *113*
4.5 Anisotropic Type-II Superconductors *114*
4.6 Irreversible Properties: Metastable States *116*
4.7 Flux Flow *119*
4.8 Hysteresis Cycle *120*
4.9 Pinning of Flux Vortices *123*
4.10 Flux Creep *127*
References 130

5 MICROSCOPIC THEORY OF SUPERCONDUCTIVITY 138

5.1 Introduction *138*
5.2 Normal Metal *140*
5.3 Normal State Instability *142*
5.4 Origin of Attraction between Electrons Concept of Dielectric 'Constant' *143*
5.5 Dielectric Constant of a Gas of Electrons *144*
5.6 Motion of Ions in Metal *145*

5.7 Origin of the Attractive Interaction *147*
5.8 Microscopic (BCS) Theory *148*
5.9 Properties *160*
5.10 GL Theory and BCS Theory *163*
5.11 Strong Coupling Effects: Eliashberg Approach *164*
5.12 Optical Properties of Superconductors *167*
5.13 Microwave Properties *168*
References 169

6 HIGH TEMPERATURE SUPERCONDUCTING CUPRATES: GENERAL SURVEY — 179

6.1 Introduction *179*
6.2 Cuprates *179*
6.3 Layering Schemes for Cuprates *181*
6.4 Charge Carriers in Cuprates *183*
6.5 Crystal Structures *190*
6.6 Critical Temperatures *206*
6.7 Physical Properties of HTSC Cuprates *217*
6.8 Pseudogap *291*
6.9 Representative Phase Diagrams *295*
6.10 Electronic States of the HTSC Cuprates *323*
References 337

7 HUBBARD MODEL, ANDERSON LATTICE MODEL AND SUPERCONDUCTIVITY — 353

7.1 Introduction: Narrow Band Systems *353*
7.2 The Hybridized Systems *371*
7.3 Novel Mechanisms of Electron Pairing *374*
7.4 Coexistence of Antiferromagnetism and Superconductivity *387*
7.5 Conclusions *387*
References 390

8 THEORY OF HIGH TEMPERATURE SUPERCONDUCTIVITY IN CUPRATES — 392

8.1 Introduction *392*
8.2 Normal State Fermi Surface *395*
8.3 Experimental Results *397*
8.4 Resonating Valence Bond (RVB) Theory *397*
8.5 t-J Model *408*
8.6 Spin Fluctuation Mechanism *411*

- 8.7 Spin-Fermion Model *413*
- 8.8 Interlayer Pair-Tunneling (ILPT) Mechanism *416*
- 8.9 Collective Excitation Coupling to Quasi Particles *421*
- 8.10 Interlayer and Intralayer Effects in HTSC Cuprates *422*
- 8.11 Bipolaronic Model *424*
- 8.12 Lochon (Boson) Fermion Model *429*
- 8.13 Marginal Fermi Liquid (MFL) Model *440*
- 8.14 SO_2 and SO_5 Models *441*
- 8.15 Pseudogap Phenomenon in HTSC Cuprates *445*
- 8.16 Conclusion *449*

References 450

9 EMERGING SUPERCONDUCTORS   459

- 9.1 Introduction *459*
- 9.2 Intermetallic A 15 Superconductors *459*
- 9.3 Chevrel Phases *461*
- 9.4 Heavy Fermion and Ruthenate Superconductors *486*
- 9.5 Organic Superconductors *497*
- 9.6 Fullerene Superconductors *524*
- 9.7 MgB_2 *534*
- 9.8 Photo-Induced Superconductors *543*
- 9.9 New Discoveries since MgB_2 *543*
- 9.10 Future High-T_C Superconductors *545*

References 545

10 APPLICATIONS OF SUPERCONDUCTIVITY   555

- 10.1 Introduction *555*
- 10.2 Superconducting Magnets *556*
- 10.3 Electronic Applications of Superconductivity *603*
- 10.4 Superconducting Mixers *640*
- 10.5 Other Superconducting Devices *640*

References 642

Glossary   **648**

Appendix 1   **658**

Subject Index   **661**

Introduction 1

1.1 INTRODUCTION

A large number of metals and alloys when sufficiently cooled down to temperatures nearing 0 K, the *dc* electrical resistivity abruptly plunges from a finite value to one that is virtually zero and remains there upon further cooling. Materials that display this behaviour are called **superconductors**, and the temperature at which they attain **superconductivity** is called the critical temperature T_c. (We may note that T_c is also used to represent the curie temperature. However, both are totally different entities and should not be confused). The resistivity temperature behaviour for superconductive and non superconductive materials are contrasted in Fig. 1.1.

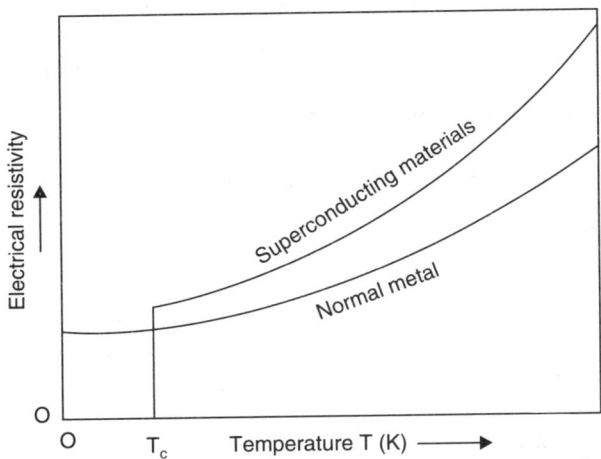

Fig. 1.1 Temperature dependence of the electrical resistivity for normally conducting and superconducting materials in the vicinity of 0 K

The critical temperature T_c varies from superconductor to superconductor but lies between less than 1 K and approximately 20 K for metals and metallic alloys. Recently, it has been demonstrated that some complex cuprate oxide ceramics have T_c in excess of 100 K. The transition from the normal to the superconducting state phase is often sharp and occurs within 10^{-2} to 10^{-4} K.

Super conductivity is a very old and exciting field discovered by H. Kammerlingh Onnes [1] in 1911. He showed that *dc* resistivity in mercury disappeared altogether at the critical temperature T_c ($\approx$ 4.2 K). His data are reproduced in Fig. 1.2.

Since its discovery in 1911, a great number of metals and alloys were found to exhibit this property. The values of T_c for some superconductors are given in Table 1.1.

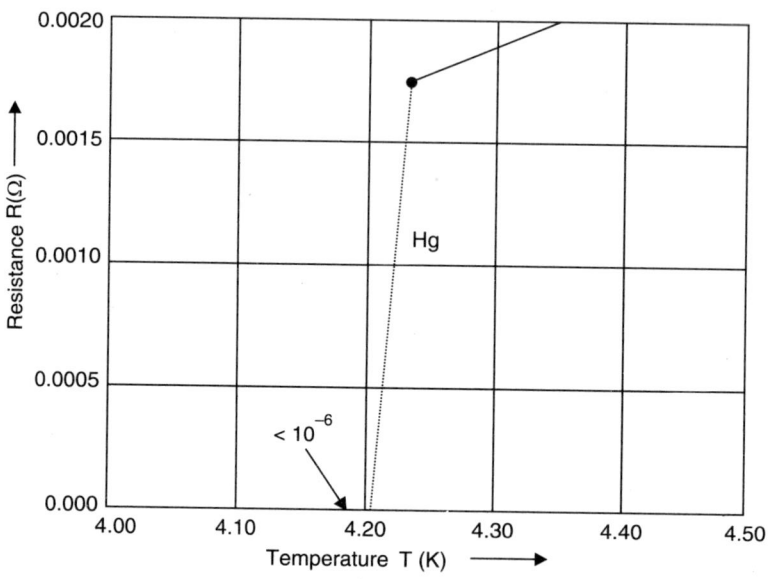

Fig. 1.2 Resistance versus temperature for mercury [1]

Table 1.1 Transition temperatures T_c for some superconductors

Superconductor	Critical Temperature T_c (°K)
Aluminium	1.196
Tin	3.72
Mercury	4.12
Lead	7.175
Thalium	2.4
Niobium	9.3
Thorium	1.4
Osmium	0.7
Nb_3Sn	18.05
$Nb_3(Al_{0.2}Ge_{0.2})$	20.05
Nb_3Al	17.5
V_3Ga	16.5
V_3Si	17.1
La_3In	10.4
InSb	1.9
NbN	16

...(Contd.)

Introduction

$PbMo_6S_8$	14.0
Nb-Ti alloy	10.2
Nb-Zr alloy	10.8
K_3C_{60}	19.3
Rb_3C_{60}	29.6
Rb_2CsC_{60}	31.3
Sr_2RuO_4	1.5
$CeCu_2Si_2$	0.7
UPt_3	0.5
URu_2Si_2	1.5
$U\,Ge_2$	0.75
U RhGe	0.25
$U\,Pd_2Al_3$	2.0
$U\,Ni_2Al_3$	1.2
$MgBr_2$	39
$Y\,Ba_2\,Cu_3\,O_7$	92
$Bi_2Sr_2Ca_2Cu_3O_{10}$	110
$Tl_2Ba_2Ca_2Cu_3O_{10}$	125
$HgBa_2Ca_2Cu_2O_8$	134

The sharpness of superconducting state transition depends on the state and purity of the sample, but in favourable situations it can occur within a temperature interval of less than 0.001 K. Fig. 1.3 illustrates schematically, two types of possible transitions. The sharp vertical discontinuity is indicative of that found for a single crystal of very pure element or one of a few well-annealed alloy compositions. The broad transition illustrated by broken lines is typical of the transition shape seen for materials which are inhomogeneous or contain unusual strain distributions. The temperature interval, over which the transition between the normal

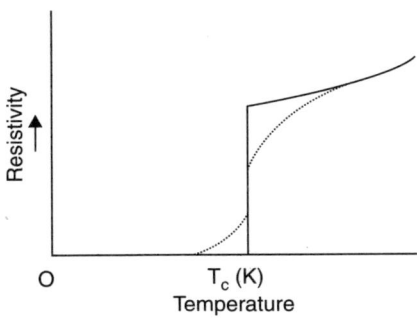

Fig. 1.3 Resistivity versus temperature for a pure and perfect lattice (solid line). Impure and/or imperfect lattice (dashed line)

and superconducting state takes place, may be of the order of as little as 2×10^{-5} K or several K in width, depending upon the material state. Present estimates of the resistivity of superconducting phase derived from the experiments on the lifetime of persistent currents (current induced in a superconducting ring) place it at less than 4×10^{-25} $\Omega - $m, while the lowest normal state resistivity observed in metals is of the order of 10^{-15} $\Omega - $m. Comparison of the resistivity of a superconductive body to heat of copper at room temperature reveals that the superconductive body is at least 10^{17} times less resistive. This is only an upper bound so that the resistivity of a superconductor for all practical purposes may in fact be essentially zero.

Zero resistance of a superconductor implied transmission of current at any distance with no losses, the production of large magnetic fields, or because a superconducting loop could carry current indefinitely storage of energy. These applications were not realized because, as was quickly discovered, the superconductors reverted to normal conductors at a relatively low current density, J_c, or in a relatively low magnetic field, called the critical field, B_c. The three material parameters, T_c, B_c and J_c, have become very important in the practical applications of superconductivity. Fig. 1.4 shows schematically the boundary in temperature-magnetic field-current density space separating normal and superconducting states. The position of this boundary will, of course, depend on the material. For temperature, magnetic field, and current density values lying between the origin and this boundary, the material will be superconductive, outside the boundary, conduction is normal.

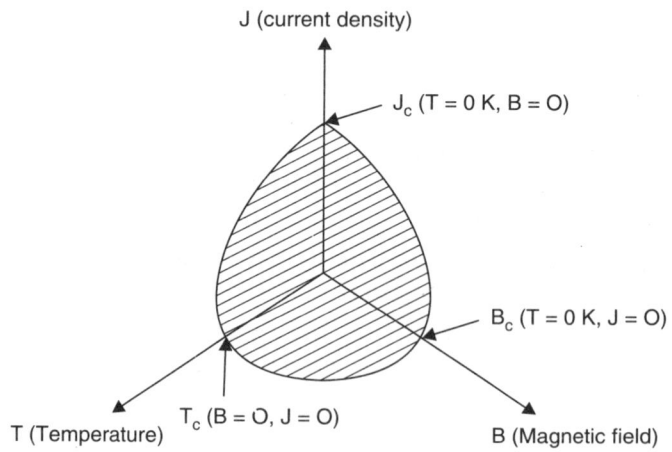

Fig. 1.4 Critical temperature, current density, and magnetic field boundary separating superconducting and normal conducting states

In 1916, Silsbee [2], hypothesized that the critical current for a superconducting wire was equal to that current which gave the critical field at the surface of the wire. The reason for this behaviour was not made clear until the discovery of the Meissner effect [3] in 1933.

The discovery and development, in the 1950s and 1960s, of superconductors which can remain superconducting at much higher fields and currents made practical the production of useful superconducting magnets. Abrikosov in 1957 [4] studied the behaviour of superconductors in an external magnetic field and discovered that one can distinguish two types of materials

Introduction

Table 1.2 Selected superconducting applications

Application	Advantage
• Generators with superconducting wires in rotors	• Life time cost savings upto 40%
• Energy storage rings	• Efficient, site independent, can revert from charging to discharging mode in less than 1 second. Can stabilize system.
• Power transmission lines	• Reduced resistive losses
• Magnets for magnetic resonance imaging (MRI)	• Superconducting magnets results in shorter exposure times and shorter images compared to conventional magnets. Largest current commercial application of superconductivity.
• Chip interconnects	• Lack of electrical resistance, reduces heat build up; permits dense-packing ..... rapid transmission of signal.
• SQUIDS (Superconducting quantum interference devices)	• Extremely sensitive to magnetic fields. Used for mineral exploration, antisubmarine warfare potential; development underway for use in medical diagnosis.
• Josephson junction switches	• Fast switching times, low power dissipation, dispersionless transmissions. Used in fast-sampling oscilloscope; potential computer logic elements.
• Josephson junction voltage standards	• Reliable, stable. Absolute voltage based on fundamental constants.
• Magnets for fusion devices	• Magnets to confine plasma.
• High energy physics	• High fields to guide beams, reduced energy consumption.
• Ship propulsion (motors)	• Smaller, quieter motor; elimination of gear box.
• Magnetohydrodynamic (MHD) power generator	• High fields interact with a plasma to generate electricity
• Magnets for MHD ship propulsion	• Quiet, more efficient, higher potential speeds
• Magnetic casting	• Eliminates contamination
• Magnetic separation	• Separates weakly magnetic materials
• Magnetic bearings	• Eliminates friction
• IR sensors	• Smaller packages
• Magnets for magnetically levitated trains (MAGLEV)	• Rapid and efficient mode of transportation

type-I and type-II superconductors. While the type-I expel magnetic flux completely from their interior, the type-II do it completely only at small fields, but partially in higher external fields. Thus due to the formation of the mixed-state, these materials can sustain superconductivity even in higher magnetic fields higher than 10 Tesla. Type-II superconductors are therefore the ones that are of interest for most large scale applications (Table 1.2). Such high-magnetic field large current carrying capability superconductors, which exhibits two critical fields B_{c_1} and B_{c_2}, are called "hard" or type-II superconductors, passes from the perfect diamagnetic state at low magnetic fields to a mixed state and finally to a sheath state before attaining the normal resistive state of the metal. We must note that a type-I superconductive body, as exemplified by many pure metals, exhibits perfect diamagnetism (the Meissner state) below T_c and excludes a magnetic field upto some critical field B_c, where upon it reverts to the normal state. Magnetic field-temperature dependence for type-I or 'soft' and type-II or "hard" superconductors are shown in Fig. 1.5.

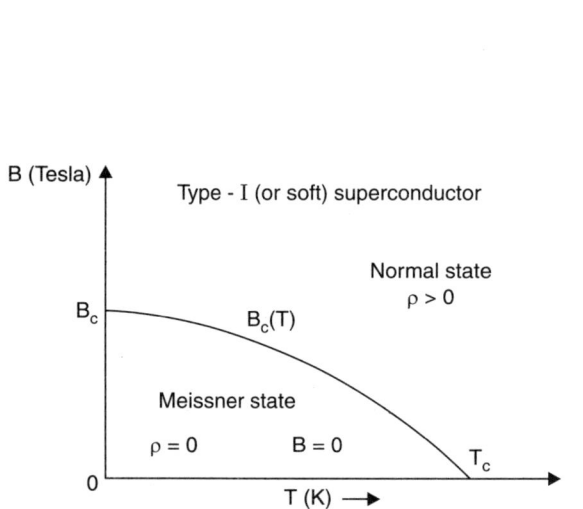

Fig. 1.5 (a) Magnetic-field-temperature dependence for type-I or 'soft' superconductors

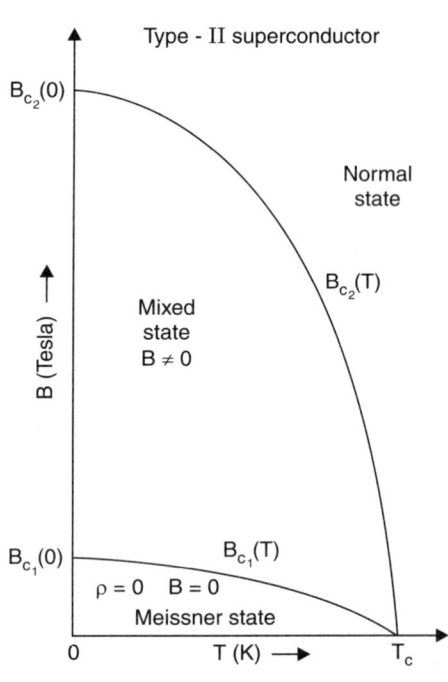

Fig. 1.5 (b) Schematic magnetization curve for type-II (or 'hard') superconductors

In 1950, Maxwell [5] and Reynolds et al. [6] independently discovered the isotope effect in superconductors. This experimental observation was an important key to the theoretical explanations of the mechanism of superconductivity. In the isotope effect, the critical temperature for many superconductors depends on the isotopic mass, indicating that lattice vibrations are involved in the superconductivity, and that the attractive coupling between electrons is through the lattice vibrations (i.e., phonon mediated). Thus the existence of isotope effect indicated that although superconductivity is an electronic phenomenon, it nevertheless depends in an important way on the vibrations of the crystal lattice in which the electrons move.

The discovery of Josephson effect [7] in 1962 opened up exciting potential for the use of superconductors in measurement science and in high speed electronic devices. According to Josephson [7], quantum tunnelling effects should occur when a supercurrent tunnels through an extremely thin layer (~ 10 Å) of an insulator. Josephson tunnelling of paired electrons through an insulating barrier is remarkable in that the tunnelling amplitude is that of an individual pair, despite the fact that the pairs comprise a correlated many body condensate.

Following the discovery of Meissner effect, F. and H. London in 1934 [8] proposed a simple two fluid model of superconductivity. The basic idea is that the electrons in a superconductor condense into a superfluid state, characterized by a single rigid 'wave function'

$$\psi(\mathbf{r}) = |\psi(\mathbf{r})| \exp[i\phi(\mathbf{r})]$$

Landau two fluid model explained the **Meissner effect** and predicted the *penetration depth* λ. Penetration depth (λ) is a characteristic length of penetration of the static magnetic flux into a superconductor. While the interior of a pure superconducting metal expels the magnetic flux and is therefore flux free (perfect diamagnetism), the static flux persists within a sheath of depth λ at the surface of the sample, its magnitude decreases exponentially towards the core of superconductor.

Ginzburg and Landau (GL) in 1950 [9], generalized the Landau theory to allow for both spatial and temperature variations in the amplitude of the "wave functions" by introducing the concept of complex temperature dependent order parameter $\psi(\mathbf{r}) = |\psi(\mathbf{r}) \exp i\phi(\mathbf{r})|$. GL's phenomenological theory (often called macroscopic theory) turned out to be surprisingly successful. One can obtain useful insight into the characteristic properties of most interesting superconducting materials by applying the results of GL theory.

In 1957, John Bardeen, Leon Cooper and Robert Schrieffer (BCS) [10] proposed the first successful microscopic theory of superconductivity. The basis of BCS theory is the interaction of a gas of conduction electrons with elastic waves of the crystal lattice. Ordinarily the electrons repel each other by the Coulomb force, but in the special case of a superconductor at sufficiently low temperatures there is a net attraction between two electrons that forms the so-called 'Cooper pairs'. Below critical temperature T_c, the attraction permits the formation of Cooper pairs that are pairs of electrons having opposite momenta and spin. This Cooper pairing or binding opens the familiar BCS gap in the electron energy spectrum. All Cooper pairs move in a single current motion, so a local perturbation, like an impurity, can not scatter an individual pair. Once this, collective highly coordinated state of 'coherent super-electrons' is set in motion, its flow is without any dissipation.

Weak coupling BCS theory of superconductivity explains most of the phenomena associated with it and provides the basis for our present understanding of superconductivity in 'conventional' low temperature superconductors, and to some extent plays a role of 'reference' theory in the on-going search or a correct description of superconductivity in the recently discovered HTSC cuprates, doped fullerenes, MgB_2, etc. The extension of BCS was presented by Eliashberg in 1960 [11]. MacMillan in 1968 [12] obtained an approximate solution of Eliashberg's equations.

Until 1986, the highest T_c observed for any superconductor was only 23.2 K in an alloy of niobium, aluminium and germanium. This meant that superconductors had to be cooled by liquid helium—an expensive and sometimes unreliable process. Consequently many potential applications of superconductors were not commercially viable. In addition, most scientists had come to regard superconductivity as a mature field with little possibility for any significant increase in critical temperature (T_c). All this suddenly changed with the discovery of K.A. Bednorz and J.G. Muller [13] of high temperature superconductivity in a new class of ceramic materials in 1986. More precisely, they found evidence for superconductivity around ~ 40 K in $La_{2-x} M_x CuO_4$ (M = Ba or Sr) ceramic. Following an initial disbelief of the naturally skeptic scientific community, their results were unambiguously confirmed and the era of high temperature superconductivity (HTSC) was ushered in. Bednorz and Muller's discovery, the result of several years of extensive investigations on metal oxides, some of which had earlier been shown to be superconducting. It is noteworthy that superconductivity in oxides had been known for many years. In fact a first metal oxide superconductor discovered was $Sr Ti O_3$ with T_c = 0.3 K. This oxide was demonstrated to be type-II superconductor. Substitution of Ca and Ba for Sr [14] raised the T_c to 0.5, the low critical temperatures limited general interest in these oxide materials.

According to Bednorz and Muller [13], their research was influenced by French work on La Ba CuO system [15]. However, the French scientists were not looking for superconductivity.

The end of 1986 and the begining of 1987 was marked by synthesis of rare-earth metal oxides with the discovery of the YBa CuO (YB CO) superconductor with a T_c of 93 K [16]. This was a significant breakthrough as it meant that for the first time the world has witnessed the existence of a superconductor with a T_c above that of liquid nitrogen (boiling point 77 K). Nitrogen is much more abundant than helium, much less expensive, and liquid nitrogen cryogenic systems are less complex than systems using helium refrigeration. One application which could benefit from nitrogen cooling is the development of hybrid microelectronic technology (semiconductor – superconductor devices)—both gallium arsenide and silicon can be tailored to perform better at liquid nitrogen temperatures.

The ease of making $Y Ba_2 Cu_3 O_7$ ceramics by mixing calcining and oxidizing the constituent powders permitted its investigation by many laboratories of the world. Only a year latter, early in 1988, Bismuth (Bi) and Tl cuprate oxides [17, 18] were discovered with T_c = 110 and 125 K respectively. These new HTSC containing Bi and Tl may have some advantages over ceramic superconductors containing rare-earths. Since the critical current density increases as T/T_c decreases, a T_c far above the opening temperature of liquid nitrogen (77 K) is advantageous. Moreover, the new materials are more stable than the rare-earth cuprate superconductors; they do not lose oxygen or react with water.

The maximum value of T_c has now increased to 133 K for mercury based cuprate $Hg Ba_2 Ca_2 Cu_3 O_{8+\delta}$ [19]. When this compound is subjected to high pressure ~ 30 G Pa, the onset of T_c increases to 164 K (more than half way to room temperature !) [20, 21]. While $Hg Ba_2 Ca_2 Cu_3 O_8$ cannot be used in applications of superconductivity at such high pressures, this striking result suggests that values of T_c in the neighbourhood of 160 K, or even higher, are attainable in cuprate oxides at atmospheric pressure. Several research groups have claimed even higher transition temperatures but none of them were reproducible or independently confirmed by other laboratories. The dramatic evolution of critical temperatures that have been observed since 1911 are illustrated in Fig. 1.6 where the maximum value of T_c is plotted versus date [22]. Approximately 100 different cuprate materials, many of which are superconducting, have been discovered since 1986.

High quality polycrystalline, single crystal and thin film specimens of HTSC cuprates have been prepared and investigated extensively to determine their fundamental, normal and superconducting state properties. Presently, two of the leading candidates for technological applications of superconductivity are the $Ln\,Ba_2\,Cu_2\,O_{7-\delta}$ and $Bi_2\,Sr_2\,Ca_2\,Cu_3\,O_{10}$ materials.

In addition to trying to develop new HTSC materials, researchers were also trying to fabricate materials with improved critical current densities (J_c). Current densities as high as $10^5 - 10^6$ A/cm^2 may be needed for applications such as magnets, motors, and electronic components.

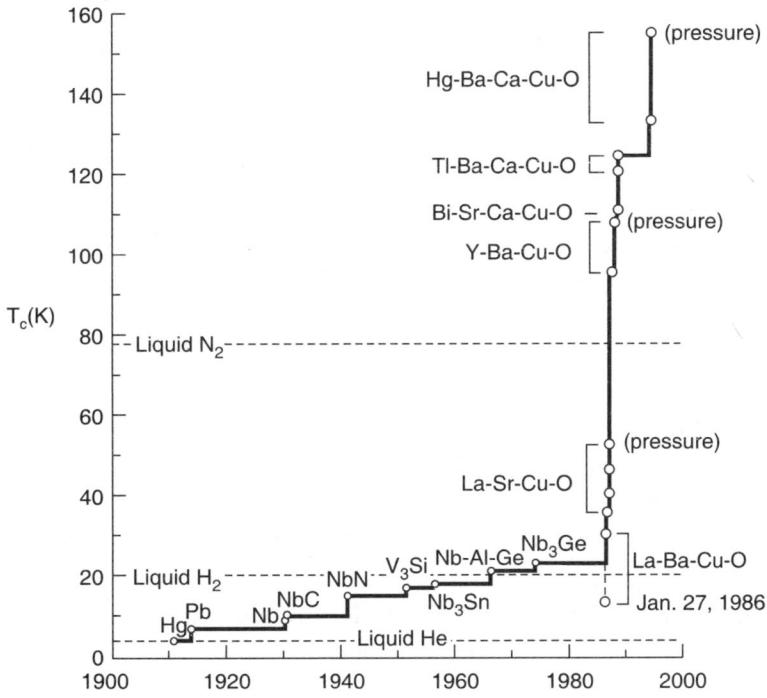

Fig. 1.6 Maximum superconducting critical temperature (T_c) versus date [22]

The HTSC are ceramics and have all the brittleness problems associated with non-superconducting ceramics. In addition J_c is not an intrinsic property of superconductors but is a function of the processing procedure. The rare-earth superconductors also have highly directional properties. Therefore, a crucial problem is to fabricate the material into a useful shape and still have sufficiently high J_c and mechanical strength for practical applications.

Single crystal films of YBCO have current densities above a million A/cm^2. However, results for bulk polycrystalline materials are orders of magnitude less. Recently, researchers have grown nonoriented polycrystalline thallium films with J_c in the millions [23]. Novel processing techniques such as explosive compaction, rapid solidification and laser ablation are currently being explored. Fortunately, these techniques yield values of J_c in high fields or in-plane grain oriented thin films of $Y Ba_2 Cu_3 O_{7-\delta}$ on flexible substrates at 64 K (pumped liquid nitrogen temperatures) [24].

All copper based HTSC possess the main properties of elemental superconductors, including zero resistance, Meissner effect, flux quantization and Josephson effect. In addition HTSC cuprates are characterised by

(*i*) highly anisotropic layered structure

(*ii*) short zero temperature coherence length ($\xi \approx 10 Å$)

(*iii*) transport properties in the normal state significantly different from those of normal metals, with strongly anisotropic resistivity, anomalous magnetic resistance, breakdown of Korringa law and anomalous Drude absorption

(*iv*) a variety of unusual and largely unexplained effects when the carrier concentration is not the one leading to the highest T_c (the so called underdoped compounds), with the opening of pseudogaps in the charge and spin excitations temperatures well above T_c [25]. There are a lot of studies for the pseudogap from both experimental and theoretical point of view. However, the complete understanding of the phenomenon remains to be obtained. We now probably know more experimentally about this class (HTSC cuprates) of materials than any other. But we still have a conudrum. The mechanism of pairing as well as the symmetry of the order parameter in HTSC cuprates are subjects of intense activity and discussion currently. We will discuss all these issues in chapter on high T_c cuprate superconductors.

In the light of tremendous progress that has been made in raising the transition temperature (T_c) of the copper oxide superconductors, it is natural to wonder how high the T_c can be pushed in other classes of materials. At present, the highest reported values of T_c for non-copper oxide bulk superconductivity are 33 K in electron doped $Cs_x Rb_y C_{60}$ [26], 30 K in $Ba_{1-x} K_x Bi O_3$ [27] and 39 K in $Mg B_2$ [28].

Hole-doped C_{60} was recently found [29] to be superconducting with a T_c as high as 117 K, although the nature of the experiment meant that the supercurrents were confined to the surface of the C_{60} crystal, rather than probing the bulk.

In the past two and half decades, a number of other noteworthy novel superconducting materials have been found. Although their T_c is low, but they exhibit several interesting properties. Broadly speaking they can be classified into following classes.

1.2 CLASSES OF SUPERCONDUCTORS

Many thousands of superconductors are known today, including metallic elements, alloys intermetallic compounds with non metallic compounds. The rigorous classification of these materials is often difficult and to a degree arbitrary. The cuprate based oxide superconductors discovered in 1986 [13], however, can be delimited relatively well. Because many of them have critical temperature $T_c > 77$ K, i.e., well above the generally liquid nitrogen temperature, they are generally referred to as high-temperature superconductors (HTSC).

Another relatively well defined class comprises organic superconductors, which were discovered in 1980 and currently have T_c values upto 13 K with the discovery of fullerene superconductors the highest T_c stabilized is about 33 K.

By far the largest class is composed of conventional or classical superconductors. It includes all superconducting elements and alloys, as well as many intermetallic compounds and metal-non metal compounds such as hydrides, sulfides, oxides, and nitrides. The maximum temperature for this group was 23 K (Nb$_3$ Ge), but with the discovery of Mg B$_2$, this has raised to 40 K. However, most members of the class have T_c < 10 K, so that helium cooling is required.

There are also "exotic superconductors" which includes all superconducting materials that can not be easily assigned to the conventional, organic, and HTSC. This class of superconductors include heavy fermion superconductors, Sr$_2$ Ru O$_4$, polymeric sulfur nitride, fullerenes, doped oxide superconductors, etc.

Some scientists also considers another class of superconductors, usually called as 'industrial superconductors', although these superconductors normally falls under the category of classical superconductors. Only in a few cases are pure metals suitable for applications; examples are Nb in SQUID technology and Pb in heat switches. Conventional industrial superconductors are generally alloys or stoichiometrically sharply defined compounds; all are of type-II. For specific technical requirements, they must be optimized in the following respects:

(*i*) High critical temperature : T_c ≥ 10 K for stable cooling with liquid ^{4}He (4.2 K).

(*ii*) High critical current density : J_c (4.2 K) ≥ 10^8 – 10^{10} A/m^2, especially for applications in power engineering and microelectronics.

(*iii*) High critical magnetic field: Bc$_2$ (4.2 K) ≥ 10 – 20 T for use in magnets.

(*iv*) Good mechanical properties: ductility, elasticity, and tensile strength of the above four points, the last is most critical. Superconducting materials display a tendency to become more brittle the higher their transition temperature (T_c), and this is also true of HTSC cuprates. Obviously, the processing of industrial superconductors thus entails complicated techniques.

Apart from niobium, the most important conventional industrial superconductors are NbTi 50 and the *A* 15 compound Nb$_3$ Sn. In addition the chevrel phase Pb Mo$_6$ S$_8$ and the hard material Nb N have some application potential for the future. Important data for these materials are summarized below. Nb Ti 50 stands out by virtue of its ductility; at present it is most widely used industrial superconductor (Table 1.3).

Table 1.3 Properties of some important industrial classical superconductors

System	T_c (K)	B_c ($T = O$), T	Uses
Nb	9.2	< 1	Microelectronics, SQUIDS, RF Cavities
Nb N	16	44	Microelectronics, Coating of fibers
Nb Ti 50	10	14	All round applications in multifilament wires
Nb$_3$ Sn	18	25	High field magnets
Pb Mo$_6$ S$_8$	15	60	Extremely high field magnets

1.2 (a) RARE-EARTH TRANSITION METAL BOROCARBIDES

Superconductivity was reported in a series of compounds with the formula RNi_2B_2C with a maximum T_c of 16.5 K for R = Lu [30, 31]. These materials have attracted a great deal of interest because they display both superconductivity and magnetic order and effects associated with the interplay of these two phenomena, similar to the tetragonal rare-earth rhodium borides having formula RRh_4B_4 and rhombohedral ternary molybdenum chalcogenides having formula RMo_6X_8 ($X = S, Se$). Investigations of superconducting and magnetic order and their interplay have been greatly faciliated by the availability of large single crystals of these materials [32] Recently, values of T_c that rival the T_c = 23.2 K value of intermetallic compound Nb_3Ge, the high T_c record holder for an intermetallic compound have been found in mixed phase materials of the composition $Y Pd_5 B_3 C$ (T_c = 23 K) [33] and $Th Pd_3 B_3 C$ (T_c = 21 K, $H_{C_2}(0) \approx 17$ K) [34].

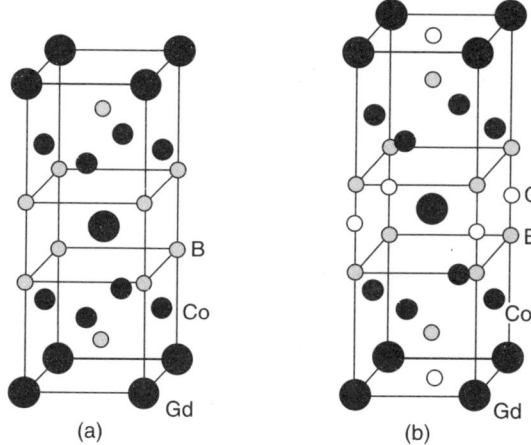

Fig. 1.7 $Gd Co_2 B_2$ has the $Th Cr_2 Si_2$ type structure where Gd resides on the Th, Co on the Cr and B on the Si sites, respectively

Table 1.4 Known R—T—B—C compounds with the $LuNi_2B_2C$-type structure. Compounds printed in bold face are superconductors

$CeCo_2B_2C$	$GdCo_2B_2C$	$LaRh_2B_2C$	**$ScNi_2B_2C$**	**$TmNi_2B_2C$**
$CeNi_2B_2C$	$GdNi_2B_2C$	$LuCo_2B_2C$	$SmNi_2B_2C$	UNi_2B_2C
$CePt_2B_2C$	$GdRh_2B_2C$	**$LuNi_2B_2C$**	$SmRh_2B_2C$	URh_2B_2C
$CeRh_2B_2C$	$HoCo_2B_2C$	$NdNiB_2C$	$TbNi_2B_2C$	YCo_2B_2C
$DyNi_2B_2C$	**$HoNi_2B_2C$**	$NdPt_2B_2C$	$TbRh_2B_2C$	**YNi_2B_2C**
$DyPt_2B_2C$	$HoRh_2B_2C$	$NdRh_2B_2C$	**$ThNi_2B_2C$**	**YPd_2B_2C**
$DyRh_2B_2C$	$LaIr_2B_2C$	$PrNi_2B_2C$	**$ThPd_2B_2C$**	YPt_2B_2C
$ErNi_2B_2C$	$LaNi_2B_2C$	**$PrPt_2B_2C$**	**$ThPt_2B_2C$**	**YRu_2B_2C**
$ErRh_2B_2C$	**$LaPt_2B_2C$**	$PrRh_2B_2C$	$ThRh_2B_2C$	$YbNi_2B_2C$

Introduction

Table 1.5 Borocarbide superconductors with $LuNi_2B_2C$-type structure: superconducting transition temperature T_c and magnetic ordering temperature T_n

Compound	T_c (K)	T_n (K)	Compound	T_c (K)	T_n (K)
$CeNi_2B_2C$	0.1	—	YRu_2B_2C	9.7	—
$DyNi_2B_2C$	6.2, 6.4	11	$ThPd_2B_2C$	14.5	—
$HoNi_2B_2C$	8, 7.5	5...8	YPd_2B_2C	23	—
$ErNi_2B_2C$	10.5	6.8			
$TmNi_2B_2C$	11	1.5	$LaPt_2B_2C$	10	—
$LuNi_2B_2C$	16.5	—	$PrPt_2B_2C$	6	—
YNi_2B_2C	15.5	—	YPt_2B_2C	10	—
$ScNi_2B_2C$	15	—	$ThPt_2B_2C$	6.5	
$ThNi_2B_2C$	8	—			

Table 1.6 Structural and magnetic properties of RCo_2B_2 and RCo_2B_2C phases with the $ThCr_2Si_2$- and $LuNi_2B_2C$-type structure, respectively.
F—ferromagnetic; A—antiferromagnetic; P—paramagnetic; imp—impurity phase; SR—spin reorientation; T_c—Curie temperature; T_n—Neel temperature

RCo_2B_2 RCo_2B_2C	a(Å)	c(Å)	z of the B site (4e)	Type of magnetic order	T_c (K)	T_n (K)
YCo_2B_2	3.5598	9.342	0.3780	P		
YCo_2B_2C				P (Pauli)		
$LaCo_2B_2$	3.6186	10.223	0.3750	P		
$LaCo_2B_2C$						
$PrCo_2B_2$	3.5985	9.951		F	19.5	
$PrCo_2B_2C$						
$NdCo_2B_2$	3.5920	9.8381	0.3750	F	32	
$NdCo_2B_2C$				A		≈ 3
$SmCo_2B_2$	3.5806	9.673		no (F imp)		
$SmCo_2B_2C$				A		≈ 6
$GdCo_2B_2$	3.575	9.561	0.3750	F	26	
$GdCo_2B_2C$	3.548	10.271		A		5.5
				A (helical)		
$TbCo_2B_2$	3.5670	9.4889	0.3750	?		
$TbCo_2B_2$				A		≈ 6

DyCo$_2$B$_2$	3.5548	9.331		A		9.3
DyCo$_2$B$_2$C				A		≈ 8
HoCo$_2$B$_2$	3.5517	9.251		A		8.5
HoCo$_2$B$_2$C	3.500	10.590		A SR at 1.46 K		5.4
ErCo$_2$B$_2$	3.5450	9.161		A		3.3
ErCo$_2$B$_2$C				A		≈ 4

1.2 (b) Sr$_2$RuO$_4$

This superconducting compound has the same structure as the La$_{2-x}$M$_x$CuO$_4$ (M = Ba, Sr, Ca, Na) HTSC cuprates [35], while the T_c of Sr$_2$RuO$_4$ is only ~ 1K, this compound is of considerable interest because it is the only layered perovskite superconductor without copper. The anisotropy of the superconducting properties of Sr$_2$RuO$_4$ is very large. It has been suggested that Sr$_2$RuO$_4$ may exhibit p-wave superconductivity.

1.2 (c) ALKALI METAL DOPED C$_{60}$ FULLERENE SUPERCONDUCTORS

Fullerenes are stable, cage like molecules that constitute the third form of pure carbon; the other two are diamond and graphite. The archetype fullerene is C$_{60}$, each molecule has the form of truncated icosahedron with 20 hexagonal faces and 12 pentagonal faces, like soccer ball. This is one of the 13 Archimedean solids characterized by having all their angles equal and all their faces regular polygons. C$_{60}$ crystallizes in a face centered cubic structure as shown in Fig. 1.8. One can dope such a molecular solid with alkaline metals and observe superconductivity T_c = 18 K, 30 K, 33 K and 40 K for K$_3$C$_{60}$, Rb$_3$C$_{60}$, RbCs$_2$C$_{60}$ and Cs$_3$C$_{60}$ respectively. The K-atoms in K$_3$C$_{60}$ occupy the octahedral sites in the cubic cell [36]. Hole doped C$_{60}$ (for C$_{60}$/CHBr$_3$ with 3 to 3.5 holes per C$_{60}$ molecule) was recently found to be superconducting with a T_c as high as 117 K (onset 117 K, rapid drop at 115 K), although the nature of experiment meant that the supercurrents were confined to the surface of the C$_{60}$ crystal, rather than probing the bulk [37]. By comparison, T_c's of graphite intercalated compounds with the same elements are only 0.5 K and 0.03 K for C$_8$K and C$_8$Rb respectively.

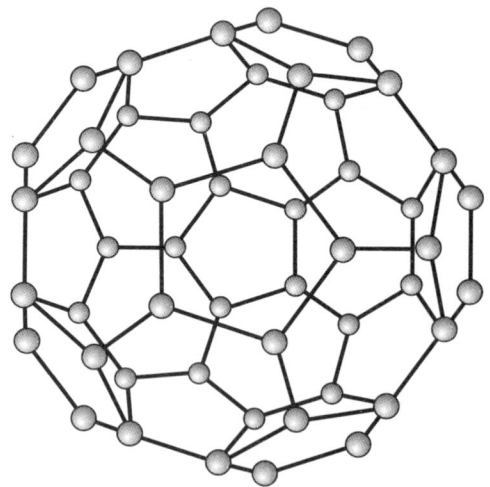

Fig. 1.8 The C$_{60}$ molecule

1.2 (d) MgB$_2$

The basic magnetic and electronic properties of most binary compounds have been well known for decades. The recent discovery of superconductivity at 39 K in the simple binary ceramic compound MgB$_2$ [38], probably highest yet determined for non-copper oxide bulk

Introduction

superconductor, was therefore surprising. Indeed, MgB_2 has been known and structurally characterized since the mid 1950s [39] and is readily available from chemical suppliers (it is commonly used a starting material for chemical metathesis reaction [40]. Crystal structure of MgB_2 is shown in Fig. 1.9.

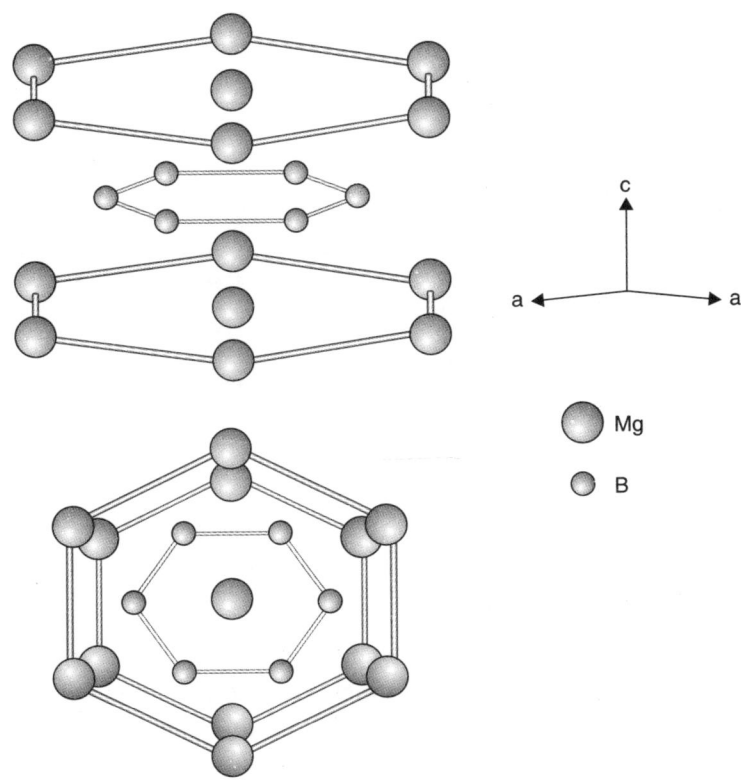

Fig. 1.9 Crystal structure of MgB_2. [38]

We note that MgB_2 is composed of two layers of boron and magnesium along the C-axis in the hexagonal lattice. The discovery of superconductivity in MgB_2 has generated excitement in condensed matter physics world: Does it push the limit of conventional superconductivity, or is the superconductivity based on a different mechanism?

1.2 (e) QUANTUM SPIN-LADDER SUPERCONDUCTORS

These materials have attracted much interest recently [41, 42]. Quantum spin-ladder materials consist of ladders made of anti ferromagnetic chains (AFM) of $S = 1/2$ spins coupled by interchain AFM bounds. $Sr Cu_2 O_3$ and $La Cu O_{2.5}$ are examples of 2-leg ladder materials whereas $Sr_2Cu_2O_5$ is example of a 3-leg material. Superconductivity has apparently been discovered in the ladder material $Sr_{0.4} Ca_{13.6} Cu_{24} O_{41.84}$ under pressure with $T_c \approx 12$ K at 3 G Pa [43]. Part of the interest in quantum spin ladder materials stems from the fact that they are simple model systems for theories of superconductivity based on magnetic pairing mechanisms. The crystal structure of $(Sr, Ca)_{14} Cu_{24} O_{41}$ (A) layer structured compound is shown in Fig. 1.10.

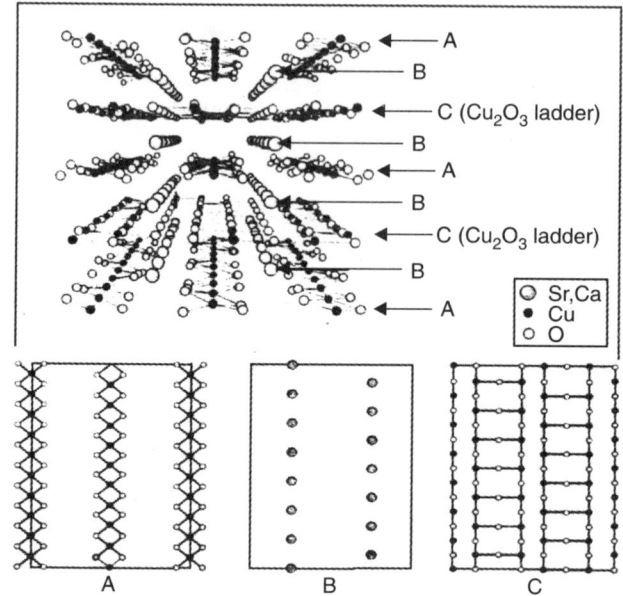

Fig. 1.10 The crystal structure of $(Sr,Ca)_{14}Cu_{24}O_{41}$. (*A*) Layers containing CuO_2 chains, (*B*) (Sr,Ca) layers, and (*C*) Cu_2O_3 layers containing two-leg ladders combine to form the three-dimensional compound

1.2 (f) Li $V_2 O_4$

The metallic transition metal oxide Li $V_2 O_4$ which has the FCC normal-spinel structure has been reported to exhibit a crossover with decreasing temperature from localized moment to heavy fermi liquid behaviour [44], similar to that which has been reported in strongly correlated *f*-electron materials [45]. The electronic specific heat coefficient at 1 K, $r \approx 0.42$ J/mole − K^2 is exceptionally large for a transition metal compound. No superconducting or magnetic order was observed in this system down to temperature as low as ~ 0.01 K. This behaviour can be contrasted with that of the isostructural compound Li $T_2 O_4$ which displays nearly T-independent Pauli paramagnetism and superconductivity with $T_c = 13.7$ [46].

1.2(g) HEAVY FERMION SUPERCONDUCTORS (SUPERCONDUCTIVITY IN *f*- ELECTRON SYSTEMS)

A small class of heavy fermion superconductors include one Ce based compound Ce $Cu_2 Si_2$ and few uranium based compounds U Be_3, U Pt_3, U $Ru_2 Si_2$, U $Ni_2 Al_3$ U $Pd_2 Al_3$, etc. [47, 48]. These compounds are characterized by enormous volumes of linear coefficient of the electronic specific heat ($C_e = \gamma T$) which can be as high as ~ 1 J mol^{-1} K^{-2}, and correspondingly large electron effective mass $m^* \sim 10^2 - 10^3 \, m_e$, where m_e is the free electron mass. The superconductivity in these systems, which are the typical strongly correlated electron systems (SCES), has not been explained from the microscopic point of view, mainly due to the complicated band structures and the strong correlation effect. A prominent puzzle is the coexistence of superconductivity and antiferromagnetic (AFM) ordering in most of these compounds. While for $CeCu_2Si_2$ there is evidence that both phenomena are in competition, they appear to coexist homogeneously in the U-based systems. UPt_3 with a hexagonal structure exhibit two superconducting critical temperatures (T_{c_1} = 0.58 K and T_{c_2} = 0.53 K) and three phases called A, B and C in superconducting phase diagram ($H - T$ diagram).

Recently a new class of heavy fermion superconductors $CeTIn_5$ (T = Rh, Ir and Co) has been discovered. While $CeRhIn_5$ [48a] shows pressure induced superconductivity. $CeIrIn_5$ [48b] and $CeCoIn_5$ [48c] show superconductivity at ambient pressure at 0.4 and 2.3 K respectively. The unique properties of $CeCoIn_5$ have attracted much attention in recent years. Various experiments such as specific heat [48d, 48e], thermal conductivity [48d], and $NMRT_1$ measurements [48f] have revealed that $CeCoIn_5$ is an unconventional superconductor with line nodes in the gap. From this, together with the suppression of the spin susceptibility below T_c [48f, 48g], this compound is identified as a d-wave superconductor. Recent rf penetration depth measurement [48h] and the flux line lattice imaging study by means of small angle neutron scattering [48i] also seem to be consistent with the existence of line nodes running along the c-axis. In unconventional superconductors, identification of the gap node structure is of fundamental importance in understanding the pairing mechanism.

The heavy fermion superconductors appear to exhibit an unconventional type of anisotropic superconductivity in which the superconducting energy gap $\Delta(\mathbf{K})$ vanishes at points or lines on the fermi surface and the electron pairing is mediated by anti ferromagnetic spin fluctuation. The origin of heavy fermion state is believed to be associated with the Kondo effect[1].

1.2 (h) ITINERANT-ELECTRON-FERROMAGNETIC SUPERCONDUCTORS

The intermetallic compound Y_9Co_7 has been shown to exhibit an interesting interplay between very weak ferromagnetism and some form of superconductivity *e.g.* the system shows ferromagnetism below 6 – 8 K, and at a lower temperature (~ 3 K) superconductivity sets in [49]. Unlike those ternary compounds in which magnetism originates from localized f-electron on rare-earth ions while superconductivity comes from d-electrons, the magnetism in Y_9Co_7 is considered to be itinerant, the conduction electrons from d bands contribute to both magnetism

---

1. Kondo effect. This is an abnormal, temperature dependent effect displayed in the thermal, electrical and magnetic properties of magnetic metals containing very small quantities of magnetic impurities. A striking example of the Kondo effect is the anomalous, logarithmic in the resistivity ρ with decreasing temperature. Other properties, such as heat capacity C_V, magnetic susceptibility χ, and thermoelectric power S, also display anomalous behaviour because of Kondo effect. This effect has been observed in a wide variety of dilute magnetic alloys.

Another common behaviour that is emerging for superconducting heavy fermion uranium compounds is the tendency of chemical substitutions to suppress both superconductivity and antiferromagnetism and induce local moment AFM or ferromagnetism (FM) with moment of the order of a μ_B.

The superconductivity in these systems, which are the typical strongly correlated electron systems (SCES), has not been explained from the microscopic point of view, mainly due to the complicated band structure and the strong correlation effect.

and superconductivity. Recently, Saxena et al. [50] reported the observation of superconductivity on the border of ferromagnetism in a pure system U Ge$_2$, which is known to be qualitatively similar to the classic d-electron ferromagnets. The superconductivity in U Ge$_2$ is reported below 1 K, in a limited pressure range on the border of ferromagnetism. Like Y$_9$ Co$_7$, superconductivity in U Ge$_2$ seems to arise from the same electrons that produce band magnetism.

Very recently ferromagnetic superconductivity have been observed in Zr Zn$_2$ [51] and U Rh Ge [52]. The superconductivity, like U Ge$_2$ and Y$_9$ Co$_7$ is confined to the ferromagnetic phase. Ferromagnetism and superconductivity are believed to arise due to the some band electrons. The persistence of ferromagnetic order within the superconducting phase has been ascertained by neutron scattering. The specific heat anomaly associated with the superconducting transition in these materials appears to be absent.

The ferromagnets Zr Zn$_2$ and U Rh Ge are superconducting at ambient pressure with superconducting critical temperatures T_c = 0.29 K and T_c = 0.25 K, respectively, Zr Zn$_2$ is ferromagnetic below curie temperature = 28.5 K with low temperature ordered moment of μ_S = 0.17 μ_B per formula unit, while for U Rh Ge curie temperature = 9.5 K and μ_S = 0.42 μ_B. The low Curie temperatures and small ordered moments indicate that compounds are close to a ferromagnetic quantum critical point.

The crystal structures of U Ge$_2$ and Zr Zn$_2$ are shown in Fig. 1.11 and 1.12 respectively.

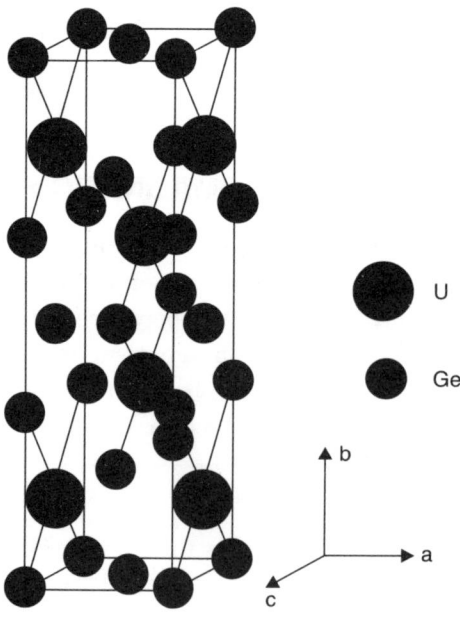

Fig. 1.11 The base centered orthorhombic C_{mmm} crystal structure of U Ge$_2$. The volume shown includes two crystal structures

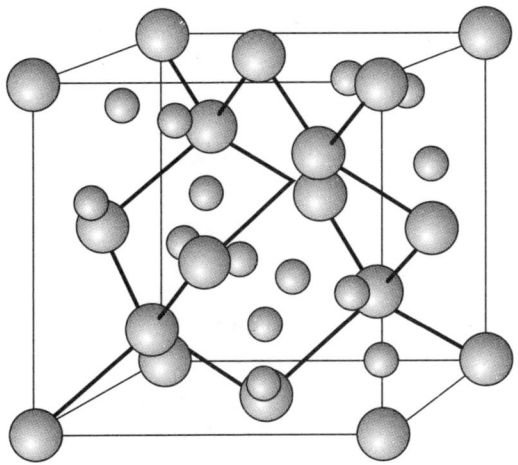

Fig. 1.12 Zr Zn$_2$ form a tetrahedrally coordinated diamond structure amid the zinc atoms

1.2 (i) BINARY ALLOYS AND INTERMETALLIC COMPOUNDS

The critical temperatures for most alloys and compounds are usually somewhat higher than in elemental metals (Table 1.7). Niobium is the metallic element with the highest T_c (= 9.3 K), but some alloys and intermetallic metallic compounds remain superconducting upto even higher temperature. For example, A-15 compound Nb$_3$ Ge has a T_c of about 23.2 K, which was the highest value ever observed until the discovery of high-temperature ceramic superconductors in 1986. Other cubic A-15 structure binary compounds with the values of T_c are Nb$_3$ Al (18.8 K), Nb$_3$ Ga (20.3 K), Nb$_3$ Si (~ 19 K), Nb$_3$ Sn (18.0 K), V$_3$ Ga (15.9 K) and V$_3$ Si (17.1 K). The structural arrangement of A-15 or β-tungsten is shown in Fig. 1.13.

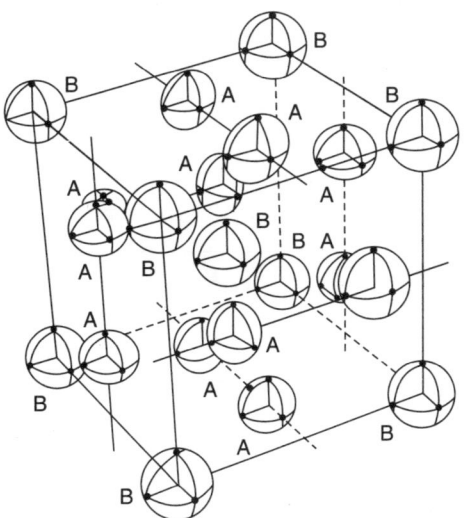

Fig. 1.13 Schematic structure of an A-15, *i.e.*, A$_3$B or β-tungsten compound

Table 1.7 The critical temperature (T_c) and critical magnetic field B_c/T of some A-15 compounds

Compound	T_c/K	B_c/T
V_3Al	9.6	—
V_3Ga	15.4	23
V_3Si	17.1	23
V_3Ge	7	—
V_3Sn	4.3	
Nb_3Al	18.9	33
Nb_3Ga	20.3	34
Nb_3Si	18.0	
Nb_3Ge	23	38
Nb_3Sn	18.3	24

B_{c_2} (T) values for A-15 superconducting compounds is very high. B_{c_2} (O), the upper critical field values as the temperature approaches 0 K for these A-15 compounds are also very high, e.g. 44 tesla for a pseudobinary A-15 compound with the composition $Nb_{79}(Al_{73}Ge_{27})_{21}$, 32 T for Nb_3Al, 39 T for Nb_3Ge, 23 T for Nb_3Sn, 21 T for V_3Ga and 25 T for V_3Si. The highest value of B_{c_2} (0) observed prior to the discovery of HTSC cuprates was about 60 T for the Chevrel phase compound $PbMo_6S_8$ [53].

1.2 (j) CHEVREL PHASES

In 1971, Chevrel et al. [54] discovered a series of ternary molybdenum chalcogenides having the general formula $M_xMo_6X_8$, where M represents any of a large (nearly 40) of metallic elements and rare-earths (RE) throughout the periodic table; x has values between 1 and 4, depending on the M element; and X is a chalcogens sulphur (S), selenium (Se) or tellurium (Te). The Chevrel phases are of great interest, largely because of their striking superconducting properties. Table 1.8 gives a selection of such compounds with their values of critical temperature T_c and critical field B_c.

Table 1.8 Critical temperature (T_c/K) and critical magnetic field (B_c/T) for few Chevrel phases

Compound	T_c/K	B_c/T
$SnMo_6S_8$	12	34
$PbMo_6S_8$	15	60
$LaMo_6S_8$	7	45
$SnMo_6Se_8$	4.8	—
$PbMo_6Se_8$	3.6	3.8
$LaMo_6Se_8$	11	5

Most of the ternary molybdenum chalcogenides crystallize in a structure in which the unit cell, that is, the repeating unit of the crystal structure, has the overall shape of rhombohedral angle close to 90°. Some of the ternary molybdenum chalcogenides display a slight distortion of the rhombohedral crystal structure at, or below, room temperature to triclinic structure, in which the three axes of the unit cell, and the three angles between them, are unequal.

The building blocks of the Chevrel phase crystal structure are the M-elements and $Mo_6 X_8$ molecular units or clusters. Each $Mo_6 X_8$ unit is a slightly deformed cube with x atoms at the corner (Fig. 1.14).

Many of the Chevrel-phase compounds (Table 1.4). The highest T_c (= 15 K) in the series is obtained in $Pb\, Mo_6 S_8$ with unusual high B_{c_2} value (= 60 T = 600 K Gauss).

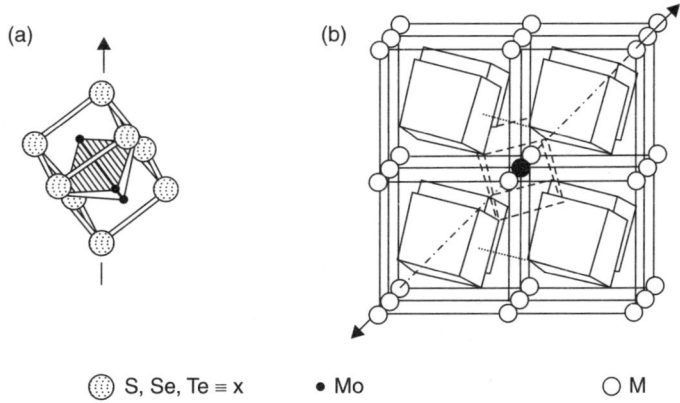

Fig. 1.14 Schematic structure of a Chevrel phase, $M_x Mo_6 X_8$: (*a*) One unit of $Mo_6 X_8$ (*b*) Stacking of 8 $Mo_6 X_8$ units into the rhombohedral unit cell.
We may note that the atom is centered at the origin

The large values of B_{c_2} as compared with $Nb_3 Sn$ and $Nb\, Ti$ make this material interesting for making superconducting wires. Critical currents (J_C) as high as ~ 3×10^{-5} A/cm^2 have been reported at 4.2 K and this provides an impetus for making wires out of these very brittle materials.

A number of Chevrel-phase compounds of the form $R\, Mo_6 X_8$, where R is a rare-earth element with a partially filled $4f$ electron shell and X is S or Se, display magnetic order at low temperatures in addition to superconductivity. It has been reported for the first time that antiferromagnetism of the rare-earth can coexist with superconductivity like in Gd, Tb, Dy, Er compounds where T_c is 1.4, 1.65, 2.1 and 1.85 K respectively. In Ho $Mo_6 S_8$, long range magnetic order produces ferromagnetism which destroys superconductivity. The phenomenon is named as 'reentrant' superconductivity. The material is superconducting only between two critical temperatures 2 K and 0.65 K. Below 0.65 K the material is ferromagnetic. The superconductivity in these systems is primarily with the mobile $4d$ electrons of Mo, while the magnetic order involves the localized 4f electrons of R atoms which occupy regular positions throughout the lattice. The coexistence of magnetism and superconductivity is possible because there is very small interaction between the rare-earth and the itinerant electrons which give rise to superconductivity and which are mainly Mo_6S_8 units.

1.2 (k) TERNARY RARE-EARTH RHODIUM BORIDES

The other class of rare-earth ternary compounds which exhibit superconductivity was first reported by Matthias in 1977 [55]. This series has the general formula $RE Rh_4 B_4$. The crystal structure of these compounds is isomorphic with the corresponding ternary RE cobalt borides. The primitive tetragonal crystal structure of $RE Rh_4 B_4$ with the unit cell in dashed-out line is shown in Fig. 1.15 [56]. The unit cell contains two formula units.

The compound for RE = Y, Er, Tm, and Lu exhibit superconductivity, whereas compounds for RE = Gd, Tb, Dy and Ho are magnetic and never exhibit superconductivity. The compound $E Rh_4 B_4$, a typical reentrant ferromagnetic superconductor, exhibits reentrant superconductivity. Superconductivity and antiferromagnetic order coexist in $RE Rh_4 B_4$ compounds (for RE = Nd, Sm, and Tm). Superconducting transition temperature (T_c), Curie temperature (Tm) or Neel temperature (T_N) for this series are given in Table 1.9.

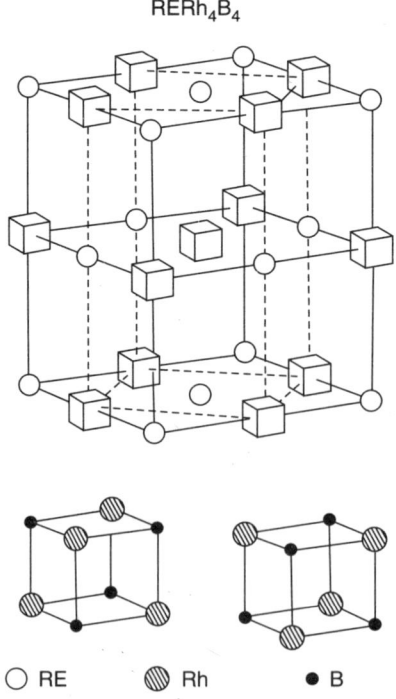

Fig. 1.15 The unit cell for ternary rhodium boride compounds. Dashed lines indicate the tetragonal unit cell. For clarity the cubes representing the $Rh_4 B_4$ clusters are not drawn to scale

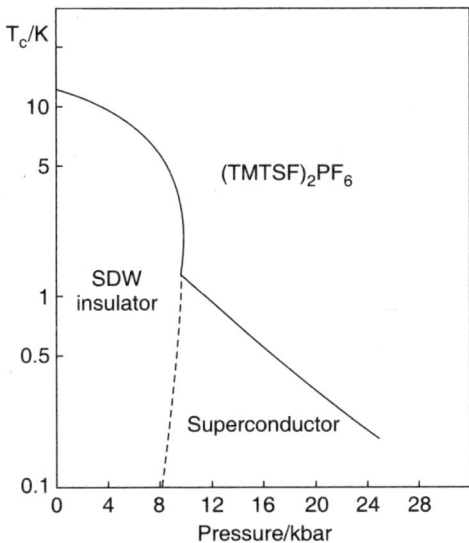

Fig. 1.17 Schematic electronic phase diagram of (TMTSF$_2$) PF$_6$ organic superconductor

1.3 POTENTIAL APPLICATIONS

Today, there is large range and diversity of potential applications of known superconductors as mentioned in the beginning of this chapter. Technological applications of superconductivity can be broadly divided into two major areas: (*i*) Superconducting electronics and (*ii*) Superconducting wires and tapes. The widespread use of recently discovered HTSC cuprates in technology has not yet been realized, steady and significant progress has been made towards this objective during the part two decades. Recent developments clearly indicate that HTSC cuprates will soon begin to have a significant impact on technology. The applications in superconducting electronics include : SQUIDS (Superconducting Interfere Devices), NMR Coils, wireless communications subsystems, MRI coils and NMR microscopes, and digital instruments. These are mainly thin film applications in ultrafast microelectronics or instrumentation. In the area of superconducting wires and tapes, applications that appear to be feasible, include : power transmission lines, motors and generators, transformers, current limiters, magnetic energy storage, magnetic separation, research magnet systems and current leads. Most of these applications are based on Josephson effect, a quantum phenomenon that enables the construction of the fastest nanoscopic switches, Josephson junctions and related device structures, SQUIDS. SQUIDS have been used to develop some of the most sensitive measurement instruments at the forefront of modern technology.

Conventional type-II superconductors (predominantly Nb compounds) due to their capacity to carry high transport currents with acceptably low energy dissipation have been used to construct commercial magnets producing fields in excess of 10^5 gauss. Wires made with these superconducting materials remain superconducting at 4.2 K even in magnetic fields considerably higher than the required performance peak of the magnet. Furthermore, for equal power and uppermost field the superconducting magnet is much smaller than its normal counterpart made out of copper wires.

Nevertheless, there is considerable work yet to be done, both in basic research and in the development of applications of HTSC cuprates. Although it is difficult to predict the future of HTSC science and technology, it should be pointed out that previous developments, such as the laser, required one or more decades of research and development before large scale applications emerged. All these issues are discussed in chapter 10.

REFERENCES

1. H. Kamerlingh Onnes, *Leiden Comm.* 120*b*, 122*b*, 124*c* (1911).
2. F.B. Silsbee and J. Wash, *Acad. Sci.* 6, 597 (1916).
3. W. Meissner and R.Ochsenfeld, *Naturewiess,* 21, 787 (1933).
4. A.A. Abrikosov, Zh. Eksperim, *i Teor. Fiz.* 32, 1442 (1957) [*Sov. Phys. JETP* 5, 1174 (1957)].
5. E. Maxwell, *Phys. Rev.* 78, 477 (1950).
6. C.A. Reynolds, B. Serin, W.H. Wright and L.B. Nesbitt, *Phys. Rev.* 78, 487 (1950).
7. B.D. Josephson, Phys. Lett. 1, 251 (1962); *Rev. Mod. Phys.* 36, 216 (1964); *Adv. Phys.* 14, 419 (1965).
8. F. and H. London, *Proc. Roy. Soc.* (London) A 149, 71 (1935).
9. V.L. Ginzburg and L.D. Landau, Zh. Eksperim, *i. Teor. Fiz.* 20, 1064 (1950).
10. J. Bardeen, L.N. Cooper and J.R. Schrieffer, Phys. Rev. 106, 162 (1957); 108, 1175 (1957).
11. G.M. Eliashberg, Zh.Eksperim, *i. Teor. Fiz.* 38, 966 (1960) [*Sov. Phys. JETP* 11, 696 (1960)].
12. W.L. McMillan, *Phys. Rev.* 167, 331 (1968).
13. J.G. Bednorz and K.A. Muller, Z. *Phys. B. Cond. Matt.* 64, 189 (1986), K.A. Muller and J.G. Bednorz, *Science* 237, 1133 (1187).
14. H.P.R. Frederiksee, et al., *Phys. Rev.* 16, 579 (1966).
15. C. Michel, et al., *Res. Bull.* 20, 667 (1985).
16. M.K. Wu. et al., *Phys. Rev. Lett.* 58, 908 (1987).
17. H. Maeda, et al., *Jpn. J. Appl. Phys.* 27, L209 (1988).
18. Z.Z. Sheng and A. H. Hermann, *Nature* 332, 138 (1988).
19. A. Schilling, et al., *Nature* 363, 56 (1993).
20. C.W. Chu, et al., *Nature* 363, 323 (1993).
21. L. Gao, et al., *Physica* C 235-240, 1493 (1994).
22. C.W. Chu in *Encyclopedia of Applied Physics*, Vol. 20, pp. 213-247, VCH Publishers, Inc. (1997).
23. D.S. Ginley, et al., *Appl. Phys. Lett.* 53, 406 (1988).
24. R.J. Cava et al., *Nature* 332, 814 (1988).
25. K.P. Sinha and S.L. Kakani, *Proceedings of National Academy of Sciences,* India Vol. LXXII, Section A, Part III, 153 (2002).
26. K. Tanigaki et al., *Nature* 352, 222 (1991).
27. R.C. Cava et al. *Nature* 332, 814 (1988).
28. R.C. Cava et al., *Nature* 332, 814 (1988), and preprint (2002).

29. R. Nagarayan et al., *Phys. Rev. Lett.* 72, 274 (1994).
30. R.J. Cava et al., *Nature* 367, 252 (1994).
31. B.K. Cho, *Phys. Rev.* B52, R 3844 (1995).
32. R.J. Cava et al., *Nature* 367, 146 (1994).
33. J.L. Sarro et al., *Physica C* 229, 65 (1994).
34. Y. Maneo. et al., *Natural* 372, 532 (1994).
36. S.L. Kakani, *'Superconductivity: Current Topics' Arihant Publishers,* Jaipur (2001), Ch. 4 and references cited there in.
37. J.H. Schon, et al., *Preprint* (2002).
38. J. Nagamatsu, et al., *Nature* 410, 63 (2001).
39. M.E. Jones and R.E. Marsh, *J. Am. Chem. Soc.* 76,1434 (1954).
40. E.G. Killian and R.B. Kaner, *Chem. Matter* 8,333 (1996).
41. M. Takano, *Physica* C 263, 468 (1996).
42. S. Maekawa, *Science* 273, 1515 (1996).
43. M. Uehara, et al., *J. Phys. Soc.* Japan 65,2764 (1996).
44. S. Kondo et al., *Phys. Rev. Lett.* 78, 3729 (1997).
45. M.B. Maple et al., *Phys. Rev. Lett.* 78,3729 (1997).
46. D.C. Johnston, *J. Low Temp. Phys.* 25,145 (1976).
47. K.P. Sinha and S.L. Kakani, *Magnetic Superconductors : Recent Advances,* Chapter 7 (and references citd therein), Nova Science Publishers, New York, USA (1989).
48. *Superconductivity: Key Problems,* Arihant Publishers, Jaipur (1996) (Chaper 4 and references cited therein).

    (a) H. Hegger, et al., *Phys. Rev. Lett.* 84, 4986 (2000).

    (b) C. Petrovic et al., *Europhys. Lett.* 53,354 (2001).

    (c) C. Petrovic et al., *J. Phys : Condens. Matter* 13, L337 (2001).

    (d) R. Movshovich et al., *Phys. Rev. Lett.* 86,5152 (2001).

    (e) S. Ikeda et al., *J. Phys, Soc. Japan* 70,2248 (2001).

    (f) Y. Kohori, et al., *Phys. Rev.* B 64,134526 (2001).

    (g) N.J. Curro, et al. *Phys. Rev.* B 64,180514 (2001).

    (h) E.E.M. Chia et al *Phys. Rev.* B 67, 014527 (2003).

    (i) M.R. Eskildsen et al., *Phys. Rev. Lett.* 90,187001 (2003).
49. A. Kolodziejez yk, et al., *J. Phys.* F 10, L-333-L 37 (1980).
50. S.S. Saxena et al. *Nature* 406, 587 (2000).
51. C. Pfleiderer et al., *Nature* 412, 58, (2001).
52. D. Aoki et al., *Nature* 412, 58, (2001).
53. P. Fulde (ed.), *Superconductivity of Transition Metals,* Springer Series in Solid State Sciences, Vol. 27 (1982).
54. B. Chevrel et al., *J. Solid State Chem.* 3, 515 (1971).

55. B.T. Matthias et al., *Proc. Natl. Acad. Sci*, USDA 24, 1334 (1977).
56. J.M. Vandenberg and B.T. Matthias, *Proc. Nat. Acad. Sci.* 74, 1336 (1977).
57. D. Jerome, et al, *J. Phys. Lett.* 41, L-95 (1980).
58. T. Ishiguro, K. Yamagi, G. Saito (Eds.), *Organic Superconductors,* 2nd Edition, Springer, Berlin (1998).
59. H. Urayama et al., *Chem. Lett.* 55 (1987).
60. K. Oshima, T. Mori, H. Inokuchi, H. Urayama, H. Yamochi, G. Saito, *Phys. Rev. B* 38, 938 (1988).
61. L.I. Buravov, N.D. Kushch, V.A. Merzhanov, M.V. Osherov, A.G. Khomenko, E.B. Yagubskii, *J. Phys. I* (France) 2,1257 (1992).

Characteristic Properties of Superconducting State 2

2.1 INTRODUCTION

In this chapter, we will describe superconducting state in more detail. To get a deeper insight into the physics of the superconducting state, it is useful to consider the origin of electrical resistivity in the normal metal and contrast it with the absence of resistivity in the superconductor. We introduce the basic concept of **superconducting wave function.** We then show that the **Meissner effect** is another distinct characteristic property of the superconducting state. We emphasize the difference in behaviours in external magnetic fields between type-I and type-II superconductors and then we briefly discuss critical currents in type-II superconducting materials. Then we present a short over view of other characteristic features of the superconducting state. The existence of the **energy gap** implies an **energy scale** which permits to understand the response of a superconductor to high frequency electromagnetic fields. At the end of the chapter, a brief discussion of flux quantization and Josephson effect is presented.

2.2 ELECTRICAL RESISTIVITY: NORMAL METAL

To understand what happens in a normal metal, it is useful to consider first the electrical resistivity of metallic conductors. Normal metal consists of a regular crystalline lattice of positively charged ions and a gas of free, non interacting conduction electrons that fill the inter-ionic space of the lattice. If we consider a typical case having one electron per ion, then we should have ~ 10^{23} electrons per cm^3. On account of this high concentration of free or conduction electrons, metals are good conductors. These conduction electrons can move nearly freely in the background field of the positive ions. As the electrons are of the opposite charge from the ions, the total charge is balanced and at equilibrium metal is electrically neutral.

Let us consider that an electric field is applied as an external perturbation to the gas of free electrons within the metal, the external force will accelerate the electrons and create a current flow of 'free' electrons. We know that ions are arranged in perfectly regular array, they do not scatter conduction electrons at $T = 0$. The scattering of electrons at $T = 0$ is really caused by deviations from the ideal periodic potential of the lattice, *i.e.,* imperfections in periodicity like dislocations, impurities etc.

We know that every real metal contains some impurities and imperfections and hence we observe some finite resistivity at very low temperatures. This resistivity, extrapolated to $T = 0$, is termed as residual resistivity, ρ_i. When we increase the temperature, the electrons also get scattered by thermal vibrations of the lattice (called phonons) and resistivity rises with temperature. This contribution to resistivity is termed as phonon resistivity, $\rho_p(T)$. Obviously, the temperature dependence of resistivity of a good metal, like silver, at low temperatures can be described by the following Matthieson's empirical relation

$$\rho(T) = \rho_i + \rho_p(T) \qquad \ldots(2.1)$$

Relation (2.1) provides a basis for understanding the resistivity of metals and alloys at low temperatures. Typically one atom percent of impurities leads to a residual resistivity of 1 $\mu\Omega$ cm. The intrinsic resistivity $\rho_p(T)$ due to electron-phonon interaction is a characteristic property of the host-metal. At sufficiently large temperatures the resistivity of metals is dominated by the phonon contribution $\rho_p(T)$ and a linear temperature dependence results for $\rho(T)$ above room temperature

$$\rho(T) = \rho(300 \text{ K}) + k_1 T \qquad \ldots(2.1(a))$$

At low temperatures, the intrinsic resistivity is typically proportional to T^5. Thus, the resistivity at very low temperatures is given by

$$\rho(T) = \rho_i + k_2 T^5 \qquad \ldots(2.1(b))$$

The constants k_1 and k_2 are characteristics of the metal in question. Fig. 2.1 shows the temperature dependence of the electrical resistivity of copper for selected values of the residual resistivity ratio (RRR), which is defined as

$$RRR = \frac{\rho(273 \text{ K})}{\rho(4 \text{ K})} \qquad \ldots(2.1(c))$$

RRR increases with increasing purity, reaching values above 10^4 in every pure metals. One can easily understand the difference between metals (conductors), insulators and semiconductors on the basis of band theory.

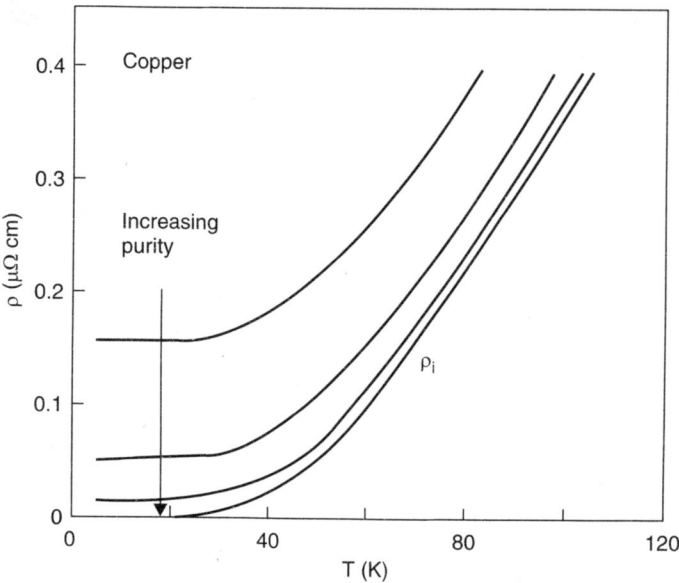

Fig. 2.1 Resistivity of copper for RRR values of 10, 30 and 100. At very low temperatures, the intrinstric resistivity $\rho_p(T)$ due to the electron-phonon interaction approaches zero

The Schrödinger equation for a free electron is

$$-\frac{\hbar^2}{2m} \nabla^2 \psi_k(\mathbf{r}) = E_k \psi_k(\mathbf{r}) \qquad ...(2.2)$$

where $\hbar = 1.0546 \times 10^{-34}$ J-s and m is the mass of the free electron. $\psi_k(\mathbf{r})$ is the wave function of the electron and E_k the corresponding eigenvalue of the energy. The use of periodic boundary conditions

$$\psi(x+L, y, z) = \psi(x, y+L, z) = \psi(x, y, z+L) = \psi(x, y, z) \qquad ...(2.3)$$

leads for an electron of momentum $\mathbf{p} = \hbar \mathbf{k}$ to the following wave function

$$\psi_K(\mathbf{r}) = \exp(i\,\mathbf{k}\cdot\mathbf{r}) \qquad ...(2.4)$$

where $K_x = 0, \pm\dfrac{2\pi}{L}, \pm\dfrac{4\pi}{L}, ..., K_y = 0, \pm\dfrac{2\pi}{L}, \pm\dfrac{4\pi}{L}, ..., K_z = 0, \pm\dfrac{2\pi}{L}, \pm\dfrac{4\pi}{L}, ...$ ...(2.5)

We must note that eqs. (2.3) and (2.5) are valid only for isotropic solids. The wave function (2.4) corresponds to plane waves propagating in the direction of the waves vector $\mathbf{k}$. Making use of eq. (2.4), eq. (2.1) leads to the following energy eigenvalues

$$E_k = \frac{\hbar^2}{2m} K^2 \qquad ...(2.6)$$

where $K^2 = K_x^2 + K_y^2 + K_z^2$. In a solid composed of N atoms each energy level splits into N levels. Since there are large number of atoms in a solid and hence it is justified to replace the discrete values of the wave number K by a continuous variable. One can easily represent the energy of free electrons in a one-dimensional lattice as a function of wave number as shown in Fig. 2.2.

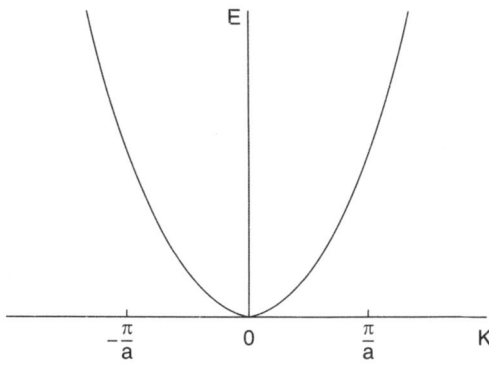

Fig. 2.2 Energy E of free electrons in a one-dimensional lattice versus wave number K; where a is the lattice constant

In the ground state, the electrons in atoms and solids occupy the lowest energy levels allowed by Pauli exclusion principle. Because of the two possible directions of the electron spin each energy level can be occupied by two electrons.

The Fermi-distribution law gives the probability that an energy state is occupied by the electrons in thermal equilibrium. We have the Fermi-Dirac distribution function $f(E)$ as

$$f(E) = \frac{1}{\exp\{(E-\mu)/k_B T\} + 1} \qquad ...(2.7)$$

where k_B is the Boltzmann constant ($= 1.38 \times 10^{-23}$ J/K), μ the chemical potential and T the temperature. In a fairly good approximation, one can easily replace μ by the Fermi energy E_F.

$$\mu = E_F \text{ for } T = 0$$

and $\qquad \mu \approx E_F \text{ for } T \neq 0 \qquad ...(2.8)$

We must remember that at $T = 0$, the Fermi-Dirac distribution function $f(E)$ is a step function with $f(E) = 1$ for $E < E_F$ and $f(E) = 0$ for $E > E_F$.

Now, we consider the density of states $D(E) = dN_\varepsilon(E)/dE$. The energy levels in the ground state can be represented by points in **k**-space. The occupied states are within a sphere of K_F, where the Fermi wave number K_F is defined by

$$E_F = \frac{\hbar^2}{2m} K_F^2 \qquad ...(2.9)$$

According to eq. (2.5) each energy state in an isotropic solid requires a volume $(2\pi/L)^3$ in **k**-space. Keeping in mind that each of these states can be occupied by two electrons having opposite spin directions, the number of electrons N_ε within the Fermi sphere of volume $(4/3)\pi K_F^3$ is

$$N_\varepsilon = \frac{V}{3\pi^2} K_F^3 \qquad ...(2.10)$$

where $V = L^3$ is the volume of the solid. Using eq. (2.10) the Fermi wave number, K_F can be expressed by the electron concentration $n_e = N_e/V$, and for the Fermi energy, E_F, we obtain

$$E_F = \frac{\hbar^2}{2m} \left(\frac{3\pi^2 N_e}{V} \right)^{2/3} = \frac{\hbar^2}{2m} (3\pi^2 n_e)^{2/3} \qquad ...(2.11)$$

The Fermi energy for a typical metal is of the order of electron volt.

Rearranging eq. (2.11), one obtains

$$N_e(E) = \frac{V}{3\pi^2} \left(\frac{2mE}{\hbar^2} \right)^{3/2} \qquad ...(2.12)$$

where $N_e(E)$ represents the number of states below the energy E. One obtains the resulting density of states $D(E)$ as

$$D(E) = \frac{dN_e(E)}{dE} = \frac{V}{2\pi^2} \left(\frac{2m}{\hbar^2} \right)^{3/2} E^{1/2} \qquad ...(2.13)$$

So far, we have neglected the periodically varying potential of the positive ions. Taking this into consideration, the solutions of Schrödinger's equation are Bloch functions

$$\psi(\mathbf{r}) = \exp(i\,\mathbf{K} \cdot \mathbf{r})\, u(\mathbf{r}) \qquad ...(2.14)$$

where the modulating amplitude $u(\mathbf{r}) = u(\mathbf{r} + \mathbf{X_L})$ reflects the periodicity of the lattice. In a one-dimensional lattice of spacing 'a' the Bragg condition is satisfied by electrons with wave vectors of $\pm \pi/a, \pm 2\pi/a, \pm 3\pi/a$ etc. These k values are boundaries of the Brillouin zones. Bragg scattering leads to interference of the electron waves propagating in opposite directions. From this interference, for example for $K = \pi/a$, one obtains the following two different standing waves:

$$\psi_+ \propto \left\{\exp\left(ix\frac{\pi}{a}\right) + \exp\left(-ix\frac{\pi}{a}\right)\right\} = 2\cos x\frac{\pi}{a}$$

$$\psi_- \propto \left\{\exp\left(ix\frac{\pi}{a}\right) - \exp\left(-ix\frac{\pi}{a}\right)\right\} = 2i\sin\left(x\frac{\pi}{a}\right) \qquad ...(2.15)$$

One finds that the potential energies for the wave functions ψ_+ and ψ_- are different, leading to an energy gap at the boundaries of the Brillouin zones. The energy of electrons in a periodic potential can be expressed by the free electron eq. (2.6) if the free electron mass is replaced by an effective mass m^*

$$E_K = \frac{\hbar^2}{2m^*}K^2 \qquad ...(2.16)$$

where
$$1/m^*(K) = \frac{1}{\hbar^2}\frac{d^2 E}{d K^2} \qquad ...(2.17)$$

The difference between m and m^* is a consequence of the electron-lattice interaction. In the vicinity of the boundaries of Brillouin zones m^* may deviate considerably from the free electron mass.

The band structure in a solid determines whether the solid is an insulator, or a conductor or a semiconductor. In insulators, the valence band is fully occupied by electrons, whereas the conduction band is completely empty. The energy gap E_g between valence and conduction band is typically larger than 2 eV for insulators. Therefore, thermal energy is not sufficient to excite electrons into the conduction band. As a consequence of the Pauli's exclusion principle and lack of empty states in the uppermost occupied band, the electrons can not be moved by an applied electric field.

Pure semiconductors are also insulators at $T = 0$. However, the energy gap E_g of semiconductors is only around 1 eV or less. The relatively small energy gap allows thermal excitation of a few electrons from the valence band into the condition band. In this process **holes** are created in the valence band which behave like positive electrons. Both the electrons in the conduction band and the holes in the valence band contribute to the electrical conductivity of semiconductors. The number of electron-hole pairs is proportional to $\exp(-E_g/2k_BT)$. Obviously, the number of electron-hole pairs increases with increasing temperature, leading to an enhanced intrinsic conductivity of semiconductors at higher temperatures.

In metals, the Fermi energy E_F is within the conduction band. Because of the empty energy states available in the conduction band, the electrons in a metal can be accelerated by an applied electric field $\mathbf{E_e}$ (Fig. 2.3). Fig. 2.3(a) shows the occupation of states for $\mathbf{E_e} = 0$.

Obviously, the same number of electrons move in the $+x$ and $-x$ directions and the net current flow is zero. On the other hand, for an electric field $\mathbf{E}_e$ applied in $-x$ direction (Fig. 2.3(b)), more electrons travel in the $+x$ direction and there is a net flow of current.

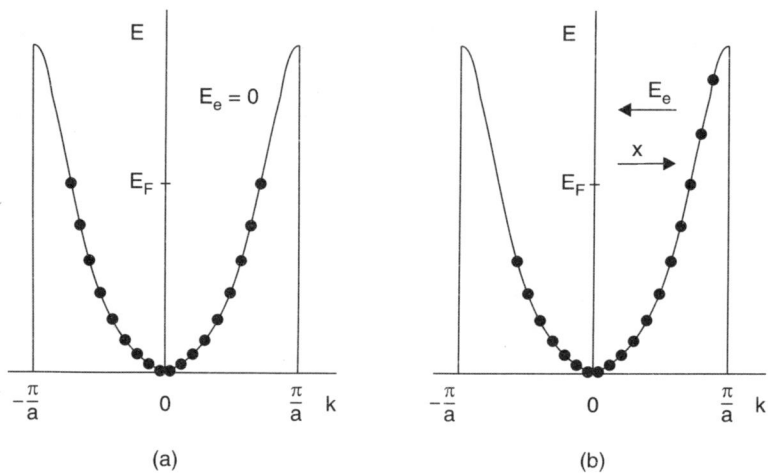

Fig. 2.3 (a) Occupation of energy states in a one-dimensional metal for $\mathbf{E}_e = 0$ and (b) an electric field $\mathbf{E}_e$ applied in $-x$ direction

Some characteristic quantities of the normal state for few materials are summarized in Table 2.1.

Table 2.1 Some characteristic quantities of the normal state for few materials

Material	n (10^{21}/cm^3)	Fermi velocity v_F (10^6(m/s))	Electron mean free path (characteristic distance) $l_e(nm)$	$\rho(100\ K)$ ($\mu\Omega$-cm)
Aluminium (Al)	180	2.0	130	0.3
Niobium (Nb)	56	1.4	29	3
La$_{2-x}$Sr$_x$CuO$_4$	5	0.1	~ 5	~ 100
YBa$_2$Cu$_3$O$_7$	7	0.1	~ 10	~ 60

Conduction electrons of maximum energy, E_F propagate with the Fermi velocity v_F related to the Fermi momentum $\mathbf{p}_F$ by

$$\mathbf{p}_F = m\mathbf{v}_F \qquad ...(2.18)$$

However,

$$E_F = \frac{1}{2} p_f\, v_f \qquad ...(2.19)$$

Moreover, Fermi wave vector $\mathbf{K}_F$ is always associated with a particle by de Broglie relation

$$\mathbf{p}_F = \hbar \mathbf{k}_F \qquad ...(2.20)$$

We have already mentioned in the beginning that the conduction electrons that propagate through the crystal with a characteristic Fermi velocity v_F are scattered by impurities or lattice imperfections. This gives rise to a resistivity. Between two scattering events, an electron covers on an average a *characteristic distance* (l_e) usually termed as the *electron mean free path*. According to elementary metal theory, the resistivity ρ_i of a metal is given by

$$\rho_i = \frac{m\, v_F}{ne^2\, l_e} \qquad \qquad ...(2.21)$$

where e and m represent the charge and mass of the electron. One can easily see that the conductivity in isotropic metals is equal to the inverse of the resistivity. In case of anisotropy; both quantities are tensors.

From eq. (2.21), we see that in the normal state of a given metal, the resistivity is inversely proportional to the electron mean free path. Obviously, the shorter the average distance between the scattering events the higher is the resistivity. We have already mentioned that the introduction of impurities into a metal reduces l_e and increases ρ_i.

2.3 THE SUPERCONDUCTING STATE-ZERO RESISTIVITY *i.e.*, INFINITE CONDUCTIVITY (NO 'INDIVIDUAL SCATTERING')

In section 2.2, we have read that the resistivity in the normal state of metals decreases continuously with the decrease in temperature and reaches a constant value ρ_o at low temperatures. In chapter 1, we have seen that the *dc* electrical resistivity in superconductors drops abruptly to an unmeasurably small value (less than 4×10^{-25} Ω–m), *i.e.* almost zero at the critical temperature, T_c (Fig. 2.4). This behaviour is remarkably different from the steadily decreasing resistance of non superconducting metals (Fig. 2.1).

In pure metals, the zero resistance state can be reached within a temperature range of 1 mK. In the case of impure metals the transition to the state may be considerably broadened. A transition width of ≈ 0.05 K was observed for impure tin. Above T_c the metal is in the normal state and the resistance is proportional to T^5. In many metals, the exponent is between 2 and 6, considerably different from the value of 5 predicted by Bloch theory. Matthias pointed out that T_c within the elements of a periodic table shows an interesting dependence on the number of valence electrons per atom (e/a) in the metal. A schematic plot of the dependence of T_c on e/a is given in Fig. 2.5. One can easily see that the transition metals with e/a less than or equal to 2 do not exhibit superconductivity. Further in this series, T_c is a maximum if e/a is close to an odd integer like 3, 5 or 7 and is minimum if it is an even integer like 4 or 6. Outside the transition metal series, one finds that in the *sp* metals like In, Sb and Pb, T_c monotonically increases with increase in e/a.

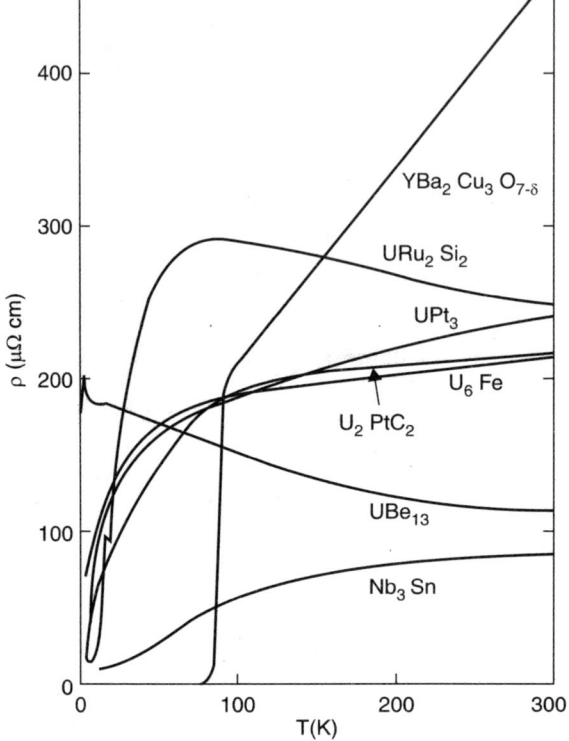

Fig. 2.4 The resistivity of several superconductors

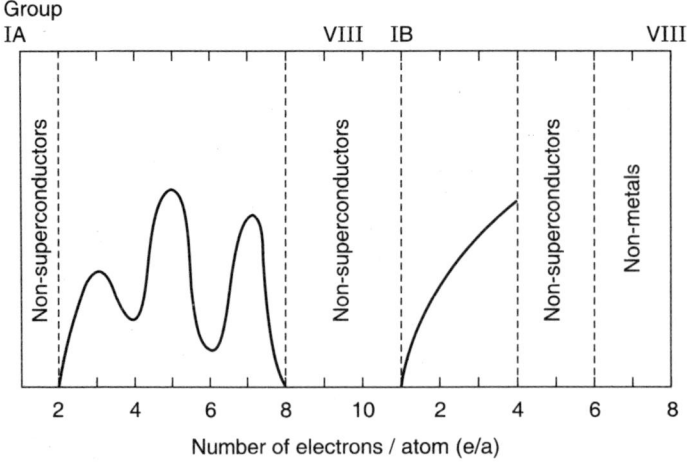

Fig. 2.5 T_c vs. number of electrons per atom for some metals

Fig. 2.6 shows resistance versus temperature for a single and a multiphase HTSC. Normal and superconducting metals are simple. Bi–, Tl, and Hg based HTSC cuprates are chemically complex materials in which there may exist several superconducting phases in one specimen. A two step transition reflects the presence of at least two superconducting phases. Generally,

the transition reflects the presence of at least two superconducting phases. Generally, the transition to the superconducting state, even in single-phase materials, is less sharp than in metallic temperature superconductors. The transition width ΔT_c for single-phase HTSCs is typically ≈ 1 K. In epitaxial $YBa_2Cu_3O_7$ films ΔT_c values as small as 0.3 K have been reported. Typically, above T_c a linear dependence of resistivity on temperature is observed.

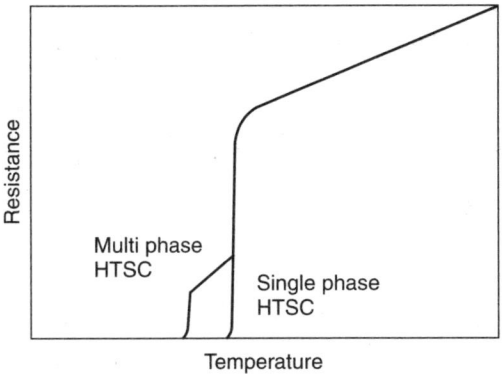

Fig 2.6 Resistance versus temperature curves of a single and a multi-phase high-temperature superconductor

Fig. 2.7 shows the resistance vs. temperature behaviour for a multifilamentary Ag/Bi-2212 wire of 1 mm diameter. Fig. 2.7 also illustrates some definitions of T_c often used in the literature.

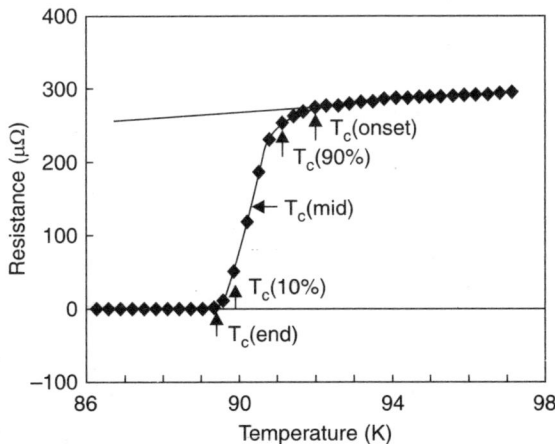

Fig. 2.7 Resistance versus temperature for a Ag/Bi-2212 multi core wire. The T_c - values resulting from different definitions of the critical temperature are marked

Because of the broadening of the superconducting transition in HTSC cuprates, the T_c-value can be defined in several ways. The first deviation from the linear behaviour in the normal state defines T_c (onset). Another possibility to define T_c is the end point of the superconducting transition, where the resistance has fallen to an unmeasurably small value. Generally, the determination of the onset or the end point of the transition is difficult. It is therefore more convenient to define T_c by the midpoint of the superconducting transition (T_c mid). The T_c - values corresponding to 10 and 90% of the normal state resistance can be used to define a transition width $\Delta T_c = T_c(90\%) - T_c(10\%)$. The width of the superconducting transition shown in Fig. 2.6 is about 1.2 K. Above T_c the electrical conductivity of the Ag/Bi-2212 multi-core wire is determined mainly by the silver matrix.

Several workers have performed experiments to establish upper limits for the resistivity in the superconducting state. An upper limit of 3.6×10^{-23} Ω-cm has been determined for the resistivity in low temperature type-I superconductors. As a consequence of flux creep effects, slightly decaying supercurrents have been observed in type-II superconductors. However, the creep rates are so small that the supercurrents in Nb_3 Sn, 3Nb-Zr and Nb powder would die out after times of more than 3×10^{92} years. Therefore, in any practical sense the supercurrents can be considered as persistent currents. Upper resistivity limits of 2×10^{-18} and 7×10^{-23} Ω-cm have been reported for YBaCuO at 77 K. As in conventional type-II superconductors flux creep effects have been observed. These values of the upper limit of the resistivity in the superconducting state have to be compared to the resistivity of annealed, very pure metals. The resistivity of aluminium with a RRR-value of 40000 is $\approx 10^{-10}$ Ω-cm at 4.2 K. The value is several orders of magnitude larger then the upper limits found for a possible resistivity in the superconducting state. For all practical purposes, one can thus treat the resistance to be essentially zero. Thus for a superconductor

$$\rho = 0 \quad \text{for all} \quad T < T_c$$

In the normal state of a metal, the conduction electrons behave like a gas of nearly free electrons that are scattered by lattice vibrations, lattice imperfections, etc. ..., all of which contribute to the resistivity. When we cool superconducting metal to below T_c, its dc resistance abruptly vanishes, *i.e.* its resistivity is zero. Now, question arises, what happened to the scattering of conduction electrons which contributed to the resistivity in the normal state? Why did it disappear? A satisfactory explanation of these puzzling questions are provided by BCS (Bardeen-Cooper-Schrieffer) theory of superconductivity [1].

When we cool a metal like Aluminium (Al) below T_c, the gas of the 'repulsive' individual electrons that characterizes the normal state transforms itself into a different type of fluid : a quantum fluid of highly correlated **pairs of electrons** (in the reciprocal **momentum space,** not in a **real space**). Below T_c a conduction electron of a given momentum and spin gets weakly coupled with another electron of exactly the opposite momentum and spin. These pairs are called **Cooper pairs.** The 'glue' is provided by the elastic waves of the lattice, called phonons. The 'distance' between the two electrons of the Cooper pair, called the **coherence length,** ξ. ξ has a large value 16000 Å in pure Al, 380 Å in pure Nb, while its value in superconductors only a few angstroms. Obviously, the behaviour of a fluid of correlated Copper pairs is different from the normal gas.

The electrons which forms the Cooper pairs have opposite momenta and opposite spins. Obviously, the net momentum of the Cooper pair is zero. By the de Broglie relation, eq. (2.20), the associated wave has an infinite wavelength (physically, the wavelength is of the order of the size of the sample). This reveals that superconductivity is a quantum phenomenon on a macroscopic scale. We must note that Cooper pairs can not be scattered by the usual scatters of individual waves, *i.e.*, there is no mechanism which could give rise to d.c. resistivity.

2.4 A.C. RESISTIVITY

A superconductor has no resistance for d.c. electric current, *i.e.* the d.c. electrical resistivity in the superconducting phase vanishes at T_c. This means that there is no voltage drop along the metal in the superconducting phase when a d.c. current is passed through it, and no power is generated by the passage of the current. Now, the question arises, what happens if the current is changing? If the current is changing an electric field is developed and some power is dissipated. One can understand the reason for this as some aspects of conduction electrons in superconductors.

Below T_c, the conduction electrons in a superconductor can be put into two classes, some behaving as '*superelectrons*' which can pass through the metal without resistance, *i.e.*, suffering no collisions, the remainder behaving as '*normal electrons*' which can be scattered and so experience just like conduction electrons in a normal metal. The fraction of superelectrons appears to decrease as the temperature is raised towards the T_c. At 0 K all conduction electrons, behave like superelectrons, but, if the temperature is raised, a few begin to behave as normal electrons, and further increases with the increase in temperature. Obviously, at T_c, all the electrons have become normal electrons and the metal loses its superconducting properties. Obviously, a superconductor below its T_c appears to be permeated by two electron fluids, one of normal electrons and other of superelectrons. The relative electron density in the two fluids of electrons depend on the temperature. Thermodynamic arguments based on the results of specific heat and similar measurements on superconductors seems to favour this 'two-fluid model'.

The current in a superconducting metal can in general be carried by both the normal and superelectrons. However, in the special case of constant d.c. current all the current in a superconductor metal is carried by the superconducting electrons. One can understand this by noting that, if the current is to remain constant, there must be no electric field in the metal, otherwise the superelectrons would be accelerated continuously in this field and the current would increase indefinitely. Obviously, if there is no field there is nothing to drive the normal electrons and so there is no normal current. This means that for a constant value of total current all the current is carried by the superelectrons. One can consider a superconducting metal just like two conductors in parallel, one having a normal resistance and the other zero resistance. One can say that the superconducting electrons *short circuit* the normal electrons. Let us put this another way, if we suddenly apply a voltage source, such as a battery, across a superconductor, the current tends to rise to infinity but is in fact limited by the internal resistance of the source. While the current is changing, the electric field must be present to accelerate the electrons. We must note that electrons have a small inertial mass and hence the supercurrent does not rise instantaneously but only at the rate which the electrons accelerate in the electric field. When an alternating field is applied, the supercurrent will therefore lag behind the field because of the inertia of the superelectrons. This means that the superelectrons present an *inductive impedance* and because of the presence of an electric field, some of the current will be carried by normal electrons. Obviously, the current is not, therefore, carried entirely by the superconducting electrons as in the d.c. case. We must note that, the normal electrons also have an inertial mass by their resulting *inductive reactance* is completely swamped by the resistance resulting from their being scattered in the metal. This means that one can represent the bulk properties of a superconducting metal by a perfect inductance in parallel with a resistance.

The fraction of the current diverted through the normal electrons dissipates power in the usual way. The mass of the electron is, however, extremely small, so the inductance due to their inertia is also extremely small. We must note that the inductance in henry of a typical superconductor due to the inertia of its superconducting electrons is only about 10^{-12} of its normal resistance in ohm, so at 1000 Hz, for e.g., only about 10^{-8} of the total current is carried by the normal electrons and there is only a minute dissipation of power. Obviously, this contrasts with the absolutely zero resistivity in the d.c. current case.

When the frequency of an applied electric field is sufficiently high, a superconducting metal responds in the same way as a normal metal. This is because, superconducting electrons are in a lower energy state than normal electrons, but if the frequency of the applied field is high enough, the photons of the electromagnetic field have enough energy to excite superconducting electrons into the higher states where they behave as normal electrons. This happens for frequencies greater than about 10^{11} Hz. Obviously, the behaviour of a superconductor at optical frequencies is therefore no different from that of a normal metal and there is, e.g., no change in the visual appearance of a superconductor as it is cooled below its T_c.

2.5 THE WAVE FUNCTION OF THE SUPERCONDUCTING STATE

In section 2.3, we have introduced the concept Cooper pairs, which are responsible for the superconducting state. The Cooper pair has twice the charge of a free-electron, $q = 2e$. The electrons are fermions; that is they are *indistinguishable* quantum particles subject to Pauli's exclusion principle.

Indistinguishability of the particles is defined by using the permutation symmetry. According to Pauli exclusion principle *no two electrons can occupy the same state*. Indistinguishable quantum particles not subject to Pauli's exclusion principle are called *bosons*. Bosons can occupy the same state multiply. Whether an elementary particle is boson or fermion is related to the magnitude of its spin angular momentum in units of $\hbar (= h/2\pi)$. Particles with integer spins are *bosons*, while those with half-integer spin are *fermions*. Fermions obey the Fermi-Dirac statistics and Pauli exclusion principle. Cooper pairs are quasi-bosons, obey the Bose-Einstein statistics and are allowed to be (all) in the same state. In contrast to the normal metal in which each electron has its own wave function, in a superconductor, all Cooper pairs are described by the following single wave function

$$\psi(\mathbf{r}) = \sqrt{n_s}(\mathbf{r}) \exp[i\phi(\mathbf{r})] \qquad ...(2.22)$$

where $n_s(\mathbf{r})$ can be considered as the number of Cooper pairs, *i.e. superconducting electrons*. We must note that $\psi(\mathbf{r})\psi^*(\mathbf{r}) = n_s(\mathbf{r})$ and $\phi(\mathbf{r})$ is a spatially varying phase. The gradient of the phase is related to the momentum of a particle by the de Broglie relation, $\mathbf{p} = \hbar\mathbf{k}$, or $\mathbf{v} = \dfrac{\hbar}{m}\nabla\phi$. Since all Cooper pairs are in the same state, the gradient of the phase becomes a macroscopic quantity, a quantity proportional to the current flowing in the superconductor.

2.6 PERFECT DIAMAGNETISM

In addition to resistanceless current transport, the superconducting state is characterised by perfect diamagnetism. The magnetic field behaviour of a superconductor is illustrated in Fig. 2.8. We have to distinguish between two different situations. In the first case the superconductor is cooled below T_c without an applied magnetic field (zero field cooled: ZFC, Fig. 2.8 left). Below T_c, a magnetic field is applied. Due to the time variation of the magnetic field $dB/dt \neq 0$, persistent screening currents are induced in the surface layer of the superconductor. These currents generate a flux density opposite to that of the applied magnetic field. Obviously, the magnetic flux density is therefore zero everywhere inside the superconductor. On the other hand, outside the superconducting sphere, the magnetic field is enhanced as a consequence of the superposition of the flux of the applied magnetic field and that of the screening currents flowing in the surface layer of the superconductor. We expect the similar behaviour for a perfect conductor with zero resistance. The superconductor is again unmagnetised when the applied magnetic field is removed.

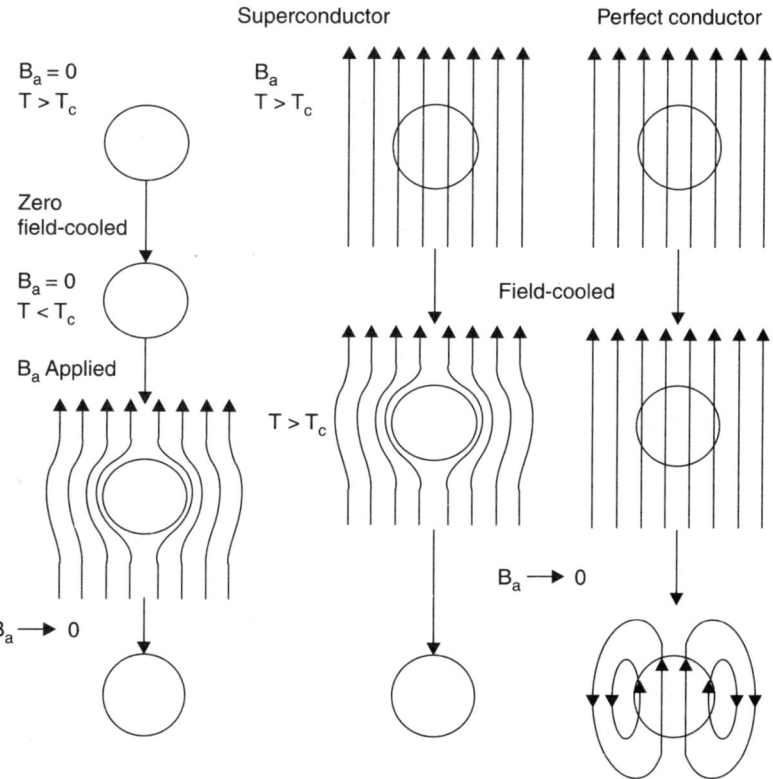

Fig. 2.8 The magnetic flux is excluded from the interior of a superconductor without field cooling (left) as well as with it (centre). In contrast to this behaviour, a magnetic flux would exist in the interior of a field-cooled perfect conductor (right)

In the second case, the superconductor is cooled below T_c while a magnetic field is applied (field cooled: FC, Fig. 2.8 centre). As soon as the temperature has fallen below T_c, the magnetic field is excluded from the interior of the superconductor. This remarkable behaviour of a superconductor is called the *Meissner-Ochsenfeld effect* [2] discovered in 1933. Consequently, we have

$$\mathbf{B} = \mu_0 \mathbf{H} + \mu_0 \mathbf{M} = 0 \qquad \ldots(2.23)$$

where **M** is the magnetization and $\mu_0 = 4\pi \times 10^{-7}$.

So the susceptibility χ is

$$\chi = \frac{M}{H} = -1 \qquad \ldots(2.24)$$

The superconductor therefore acts as an ideal or perfect diamagnet that excludes the flux from its interior by means of surface currents.

The Meissner effect can be demonstrated by a floating magnet as shown in Fig. 2.9. A small bar magnet above T_c simply rests on a superconductor dish. If temperature is lowered below T_c, the magnet will float as indicated. The gravitational force exerted on the magnet is balanced by the magnetic pressure due to the inhomogeneous magnetic field (B) surrounding the magnet, that is represented by the magnetic flux line as shown.

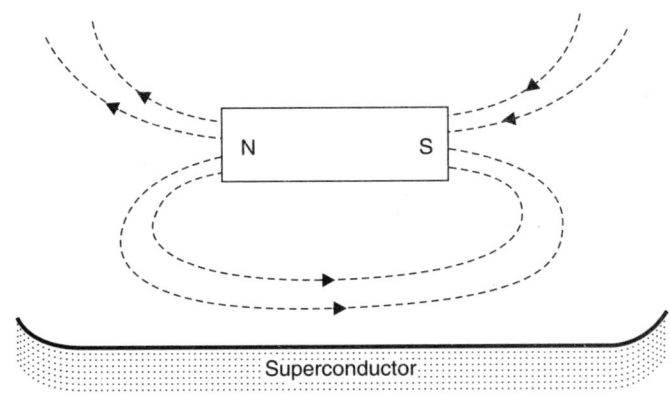

Fig. 2.9 A floating magnet

It is important to mention that Meissner effect is not a consequence of infinite conductivity, but of another intrinsic characteristic property of the superconducting state.

From Fig. 2.8, it is clear that in the normal state, at temperatures above T_c, the field lines pass through the metallic specimen. Upon cooling below T_c, a phase transition into the superconducting state takes place and the magnetic flux gets expelled out of the interior of the metallic sample. The expulsion of the magnetic flux implies that this new superconducting state is a true thermodynamic equilibrium state. One can easily explain this with the help of elementary formulae of electrodynamics. We can write Ohm's law $V = IR$ as

$$\mathbf{E} = \rho \mathbf{J} \qquad \ldots(2.25)$$

where **E** represents the electric field, ρ the resistivity and **J** the electric current density in the sample. Zero resistivity implies zero electric field. Now, taking Maxwell's equation

$$\nabla \times \mathbf{E} = -\frac{\partial \mathbf{B}}{\partial t} \qquad \ldots(2.26)$$

we have
$$\frac{\partial \mathbf{B}}{\partial t} = 0 \qquad \ldots(2.27)$$

Obviously, the magnetic induction B in the interior of the sample has to be constant as a function of time. The final state of the sample would have been different if it were cooled under an applied external field or if the field were applied after the sample has been cooled below T_c. In the former situation, the field would have remained within the sample, while in the latter it would have been zero. For the specimen to be in the same thermodynamic state, independent of the precise sequence that one uses in cooling or in applying the magnetic field, the superconducting metal always expels the field from its interior, and has $\mathbf{B} = 0$ in its interior. Thus the expulsion of the magnetic field ensures that the superconducting state is a true thermodynamic state.

2.7 RING SUPERCURRENT

Consider a ring-shaped superconductor. If a weak magnetic field **B** is applied along the ring axis and temperature is lowered below T_c, the field is expelled from the ring due to the Meissner effect. If the field is slowly reduced to zero, part of the magnetic flux lines may be trapped as shown in Fig. 2.10. It was observed that the magnetic moment so generated is maintained by a never-decaying supercurrent around the ring [3].

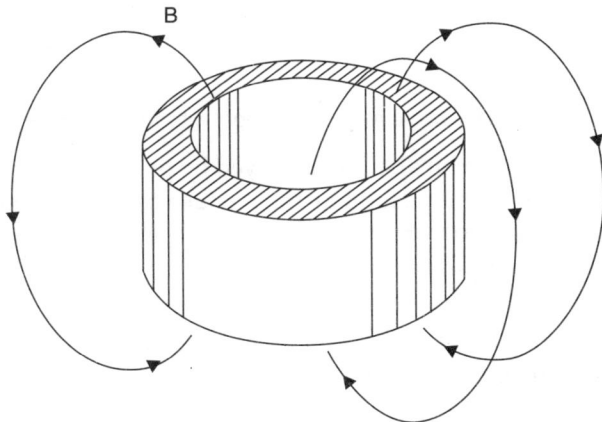

Fig. 2.10 A set of magnetic flux lines are trapped in the ring

2.8 CRITICAL MAGNETIC FIELD

One of the main characteristics of a superconductor is its magnetic behaviour arises from its vortex structure. The superconducting state can be destroyed by sufficiently large magnetic fields. The transition from superconducting state occurs at a certain magnetic field B_c. This critical thermodynamic field B_c is characteristic of the superconductor in question. The critical field increases as temperature is lowered and reaches a maximum value $B_c(0) \equiv B_o$ as $T \to 0$. The temperature dependence of the critical field B_c is typically well described by

$$B_c(T) = B_c(0)\left[1 - \left(\frac{T}{T_c}\right)^2\right] \qquad \ldots(2.28)$$

where $B_c(0)$ is the extrapolated value of B_o at $T = 0$. For pure elemental superconductors, the critical field B_0 is not very high. For example, the value of B_o for mercury (Hg), tin (Sn) and lead (Pb) are 411, 306 and 803 G (Gauss), respectively. The highest, about 2000 G, is exhibited by niobium (Nb). Fig 2.11 exhibits the temperature variation of the critical magnetic field $B_c(T)$ for some elemental superconductors.

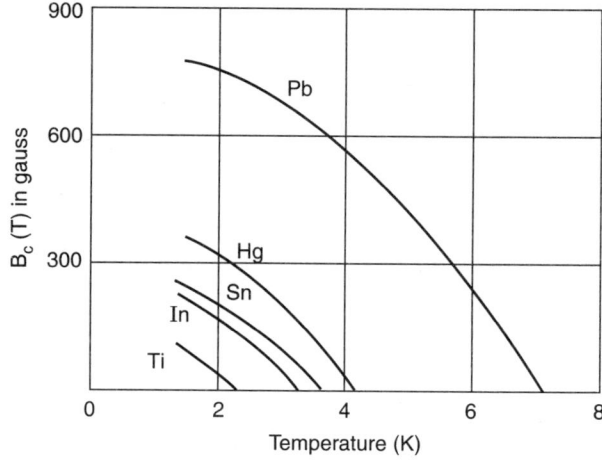

Fig. 2.11 Critical field B_c change with temperature

The superconducting state can also be destroyed by passing an excessive current through the material which creates a magnetic field at the surface of strength equal to or greater than $B_c(0)$. This limits the maximum current that the material can sustain and is an crucial problem for applications of superconducting materials.

Generally, both the transition temperature T_c and the critical field B_c depend on the crystal structure. For example the T_c values of hexagonal close-packed and face centered cubic lanthanum are 4.88 K and 6.00 K respectively. Some of the elements are superconducting only at high pressure phases.

2.8.1 Type-I Superconductors

These superconductors are characterised by perfect diamagnetism *i.e.*, these materials completely expel magnetic flux until they become completely normal. The screening currents flowing in the surface layer produce a magnetisation which cancels the applied magnetic field in the interior of the superconductor. These superconducting materials are usually termed as *Type-I* or *'soft'* or *'pure'* superconductors. With exception of V and Nb, all elemental superconductor and most of their alloys in the 'dilute limit' are Type-I superconductors. The strength of the applied magnetic field required to completely destroy the state of perfect diamagnetism in the interior of Type-I superconducting material is called the thermodynamic critical field B_c, whose variation with temperature is approximately parabolic and governed by relation (2.28) and shown in Fig. 2.12.

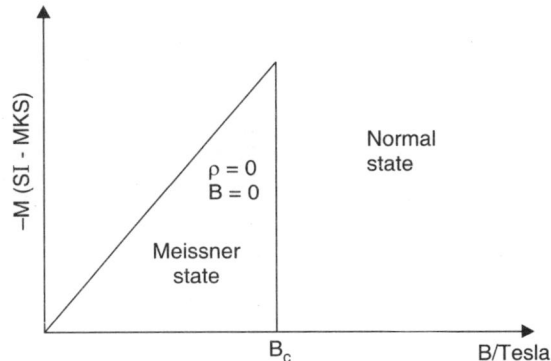

Fig 2.12 Variation of magnetization as a function of magnetic field for Type-I superconducting materials

2.8.2 Type-II Superconductors

There are superconducting materials for which the magnetization varies with the applied field in the manner shown in Fig. 2.13. These materials are called Type-II superconducting materials and are widely used in several applications *e.g.* magnets and energy applications. For these materials there are two critical fields : The lower B_{c_1} and the upper B_{c_2}. The magnetic flux is completely expelled only upto the field B_{c_1}. Obviously, in applied field smaller than B_{c_1}, the type-II superconductor behaves just like a type-I superconductor below B_c. For fields lying between B_{c_1} and B_{c_2} the magnetic flux partially penetrates the material although it is still in the superconducting state. Above B_{c_2} the material returns to the normal state.

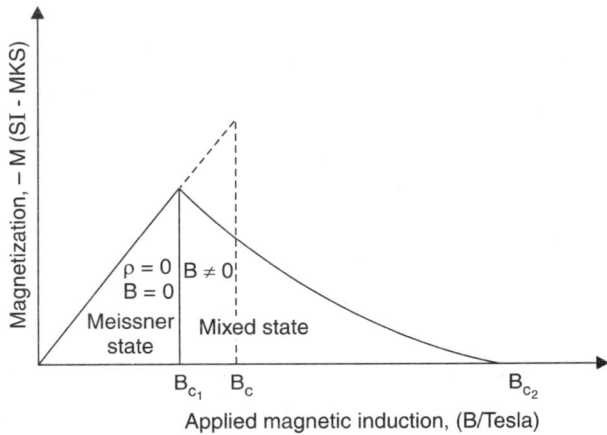

Fig 2.13 Magnetization curve for Type-II superconducting materials

Between B_{c_1} and B_{c_2} the superconductor is said to be in the *mixed state*. The maximum value observed for the upper B_{c_2} is about 800,000 gauss. The Meissner effect for the mixed state is partial. For all applied fields $B_{c_1} < B_c < B_{c_2}$, magnetic flux partially penetrates the superconducting specimen in the form of tiny microscopic field called *vortices* (Fig. 2.14).

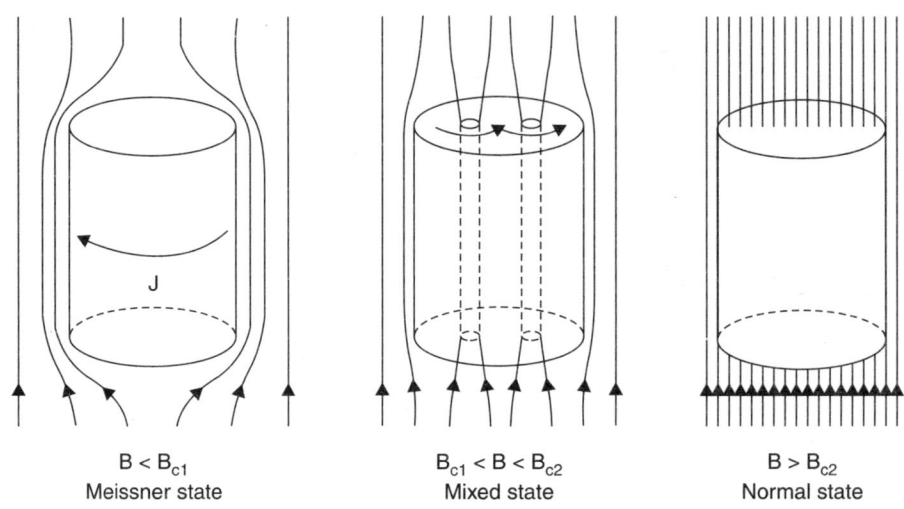

Fig. 2.14 Flux penetration in the mixed state for type-II superconductors

The diameter of a vortex in conventional superconductors is typically 100 nm. It consists of a normal core, in which the magnetic field is large, surrounded by a superconducting region in which flows a persistent supercurrent which maintains the field within the core.

Each vortex carries a magnetic flux given by

$$\phi_o = \frac{h}{2e} = 2.067 \times 10^{-15} \text{ Weber} \quad \ldots(2.29)$$

The magnetic induction B is directly related to the number of vortices per m², n as

$$B = n\,\phi_o \quad \ldots(2.30)$$

We must note that the creation of vortices keeps the magnetic energy smaller than the condensation energy, so the overall free energy of the mixed state in a type-II superconductor remains (thermodynamically) more favourable than the normal state even upto high magnetic fields.

The current displaces the vortices and this creates a non-desirable energy dissipation. The vortex in motion creates an electric field

$$\mathbf{E} = \frac{d\phi}{dt} \quad \ldots(2.30(a))$$

In the presence of field **E** (eq. 2.30(*a*)), the current **J** dissipates energy **E . J**. This energy dissipation is equivalent to resistivity. Theoretically critical currents of type-II superconductors are weak; still weaker for small values of B_{c_1} compared with B_{c_2} which is rather large. Now,

Characteristic Properties of Superconducting State

the question arises, how does one pass intense currents in without dissipation above B_{c1}? For this purpose, one has to prevent the motion of vortices so that critical current would not limited by B_{c1}. This is achieved by the so called *vortex pinning* or *flux pinning*. For this purpose, one has to create sites out of which the vortex cannot leave without large energy increase. For example, in conventional type-II superconductors one can find small inclusions of normal metal imbedded in the superconductor. The vortex will be pinned to such inclusion as it does not have to spend energy to destroy superconductivity in that inclusion. The typical size of efficient inclusions is the coherence length ξ, the diameter of the tube which is the normal state within the vortex. Obviously, inhomogeneities over distances of the order of the coherence length are responsible for the attainment of very high currents in some materials like Nb-Ti.

In HTSC cuprates where the ξ is very short ~ 10 Å, it is not obvious how to control the vortex pinning sites.

Since the supercurrent can flow in the mixed state through the superconducting regions between vortices type-II superconductors allow one to construct wires for high field magnets.

A type-I material can change to type-II on the substitution of some impurities. For example, Pb, a type-I superconductor with $B_{c1} \approx 600\ G$ at $4K$, when added to it 2 wt% indium, it becomes a type-II superconductor with $B_{c1} \approx 400\ G$ and $B_{c2} \approx 1000\ G$. On adding 20 wt% indium in type-I superconductor lead, type-II superconductor with $B_{c1} \approx 70\ G$ and $B_{c2} \approx 3600\ G$ is obtained. We must note that the area within the $M - B$ curve remains constant as the material goes from type-I to type-II on substitution of impurities.

2.8.3 Critical Current (J_c)

The existence of a critical magnetic field (B_c), beyond which superconductivity is disrupted, directly implies the existence of a critical current (J_c) (because of the intrinsic magnetic field generated by this current). Accordingly, each superconductor has a maximum transport current I_c that it can carry. This critical current is of central importance for industrial applications. It depends on the material properties and the geometry of the superconductor, the temperature, and the strength of any applied external magnetic field. Because the critical field B_c, B_{c1}, B_{c2} becomes smaller as the temperature rises and vanish for $T \geq T_c$, critical currents exhibit a corresponding temperature dependence.

Because type-I superconductors have relatively low critical fields, they also have little current carrying capacity. For example, a lead wire 1 mm in diameter has $I_c \approx 200A$ at 0 K. This value is easily derived, since the transport current flowing through the wire must generate at the surface of the wire a magnetic field with B_c (Pb) ≈ 50 mT.

Many type-II superconductors have very high (curve) critical fields, which imply very high critical currents. This gives rise to a new problem that does not exist for type-I superconductors. Due to the transport current, the fluxoids penetrating the superconductor in the mixed state experience a Lorentz force, perpendicular to the plane spanned by the transport current and the *fluoxid*. If the fluxoids are free to move, they begin to do so. This flux movement is associated with energy dissipation and leads to a finite resistance. Only if the fluxoids are pinned at traps can their movement be prevented or reduced. Traps include lattice defects and normally conducting illusions. Type-II superconductors with good flux pinning are said to be "hard". Conventional superconductors optimized in this way, such as Nb_3Sn or NbTi, achieve critical current densities (the critical current J_c divided by the conductor cross section) J_c of $10^9 - 10^{10}$ A/m² at $T = 4.2$ K.

In liquid-N_2 cooled HTSC cuprates, flux movement (sometimes referred to as *flux creep*) is more prominent because of the higher operating temperature and the fundamentally poorer pinning properties of HTSC cuprates, which result in increased thermal activation of the fluxoids away from their traps. The phenomenon is especially marked for bulk HTSC specimens, where the J_c ($T = 77$ K) values attainable at present are only 5×10^8 to 2×10^9 A/m², depending on the material, despite the high B_{c_2} values, a further drastic reduction occurs in an external field. In contrast, epitaxially grown HTSC films with thickness upto a few hundred nanometers exhibit J_c ($T = 77$ K) values upto almost 10^{11} A/m². The highly effective pinning mechanism in these thin layers has not yet been elucidated. Fig. 2.14(a) shows the critical current density J_c as a function of external magnetic field for some conventional and HTSC.

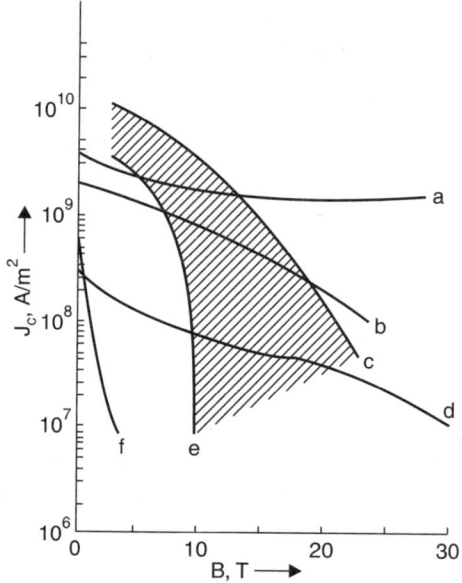

Fig. 2.14 (a) Critical current densities $J_c(B)$ for selected superconductors: (a) Bi-2212 tape at 4.2 K (B parallel to tape); (b) $PbMo_6S_8$ wire at 4.2 K; (c) Nb_3Sn at 4.2 K; (d) Y-123 specimen at 77 K (B parallel to CuO_2 planes); (e) NbTi-50 at 4.2 K; (f) Bi-2223 tape at 77 K (B parallel to tape); All commercial helium-cooled HTSC lies in the shaded area

Contrary to type-I superconductors where the critical currents are rather low, the J_c can be sufficiently high in many type-II superconducting materials to enable practical applications. For example, Nb-Ti alloys have $J_c \sim 10^6$ A/cm² at 4.2 K and are used for making wires. If B_{c_2} is very high (~ 14 T for NbTi, ~ 60 T for Chevrel phase), such wires can be used for construction of superconducting magnets that can and do produce magnetic field of tens of Teslas. HTSC cuprates are extreme type-II materials with an B_{c_2} of order 150 T. Values for the upper critical field B_{c_2} of selected superconductors are listed in Table 2.2.

Characteristic Properties of Superconducting State

Table 2.2 Upper critical fields B_{c2} of selected high-field superconductors

Material	T_c (K)	B_{c2} (T) T = 0
Nb Ti	9.5	13
Nb N	15	15
V_3 Ga	15	23
V_3 Si	17.1	23
Nb_3 Sn	13	23
Nb_3 Al	19.1	33
Nb_3 Ge	23.2	38
Pb Mo_6 S_8	15	60
Rb_3 C_{60}	29.6	57

The bulk of material goes normal when B_{app} reaches B_{c_2}, but the superconducting state can persist in a thin surface sheath for B_{app} upto the higher applied field value

$$B_{c_3} = 1.7 \, B_{c_3}$$

The thermodynamic critical field B_c is related to the depairing current density J_{depair} through the relation

$$J_{depair} = \alpha \, B_c / \mu_o \lambda$$

where the dimensionless coefficient α is of the order of unity.

The criterion for the type-I or type-II behaviour of superconducting materials depends on relative magnitudes of the two characteristic lengths for superconductors which are required by the BCS theory. One of these is the *London penetration depth* (λ_L), which describes the distance to which the magnetic flux penetrates inside the superconductor from the air-metal surface. λ_L depends on temperature and becomes infinite at T_c. The temperature dependence of λ_L is given by

$$\lambda_L(T) = \lambda_L(0) \left[1 - \left(\frac{T}{T_c} \right)^4 \right]^{-1/2} \qquad ...(2.31)$$

where $\lambda_L(0)$ is the penetration depth at zero temperature.

The other is the *coherence length*, ξ (the 'distance' between the two electrons of a Cooper pair), which is a measure of the distance r, within which the energy gap parameter, Δ, does not change drastically. If the superconductor is subjected to a spatially varying magnetic field, Δ is not uniform and is a function of position r. ξ denotes the distance over which $\Delta(\mathbf{r})$ does not change drastically. The ratio of the coherence length ξ and the penetration depth λ_L is called the *Ginzburg-Landau (GL) parameter* κ [4]

$$\kappa = \frac{\lambda_L}{\xi} \qquad ...(2.32)$$

In fact, the G_L theory predicts that depending on whether the parameter κ is smaller or larger than $1/\sqrt{2}$ the superconductor in question is type-I or type-II, respectively, *i.e.*,

$$\frac{\lambda_L}{\xi} < 0.71 \text{ Type I}$$

$$\frac{\lambda_L}{\xi} > 0.71 \text{ Type II} \qquad \ldots(2.33)$$

Table 2.3 gives for various superconductors the coherence length ξ and the *GL* parameter κ at $T = 0$. The coherence length λ_L is closely related to the mean free path of the electrons in a metal. Large values of the mean free path of the electrons lead to a large coherence length ξ. Most elements are therefore Type-I superconductors.

Table 2.3 Coherence length ξ and Ginzburg-Landau parameter κ for selected superconductors at $T = 0$

Material	T_c (K)	ξ (nm)	$\kappa = \lambda_L / \xi$
Al	1.18	550	0.03
In	3.41	360	0.11
Cd	0.52	760	0.14
Sn	3.72	180	0.23
Ta	4.5	93	0.38
Pb	7.2	82	0.48
Nb	9.25	39	1.28
NbTi	9.5	4	75
NB_3Sn	18	3	21.7
Nb_3Ge	23.2	3	30
Rb_3C_{60}	29.6	2	123.5
$YBa_2Cu_3O_{7-\delta}$	93	2	95

2.9 THE THERMODYNAMICS OF SUPERCONDUCTORS

In this section, we will try to gain deeper insights into the properties of the superconducting state by making use of the laws of thermodynamics.

(*i*) Entropy

In all superconductors, entropy decreases considerably upon cooling below T_c. As entropy is a measure of the disorder of a system and hence the observed decrease in entropy between the normal state and the superconducting state is more ordered than the normal state, *i.e.* the fraction of electrons that is thermally excited in the normal state becomes ordered in the superconducting state. The change in entropy is small, in

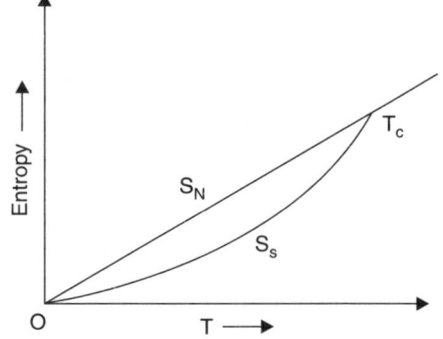

Fig. 2.16 Entropy versus temperature for zero applied field of aluminium in the normal and superconducting states

aluminium of the order of $10^{-4} k_B$ ($k_B \to$ Boltzmann constant). The small entropy change must mean that only a small fraction (of the order of 10^{-4}) of the conduction electrons participate in the transition to the superconducting state. Entropy (S) of aluminium in the normal (S_N) and superconducting state (S_s) as a function of temperature is plotted in Fig. 2.16.

(*ii*) **Heat Capacity or Specific Heat**

The specific heat C_v in a normal conductor at low temperatures consists of two contributions: (*a*) from electrons in the conduction band, C_{el} and (*b*) from the lattice, C_{ph}.

The lattice or phonon contribution C_{ph} to the specific heat can be well described by the Debye approximation

$$C_{ph} = 9N\, k_B \left(\frac{T}{\theta_D}\right)^3 \int_0^{x_D} \frac{x^4 \exp x}{(\exp x - 1)^2}\, dx \qquad \ldots(2.34)$$

where N is the number of atoms in the solid, k_B the Boltzmann constant, θ_D the Debye temperature and $x_D = \theta_D/T$. The Debye temperature θ_D is connected to the Debye frequency ω_D by

$$\theta_D = \hbar\, \omega_D/k_B \qquad \ldots(2.35)$$

well above the θ_D the specific heat C_{ph} reaches the classical value of $3R$, where $R = N k_B$ is the gas constant. On the other hand, the phonon contribution is proportional to T^3 for the temperatures below $0.1\theta_D$. At low temperatures C_{ph} is found from eq. (2.34) to be

$$C_{ph} = 234\, N\, k_B \left(\frac{T}{\theta_D}\right)^3 = \beta\, T^3 \qquad \ldots(2.36)$$

In addition to the phonons, the free electron gas contributes to the specific heat of a conductor. However, this contribution is typically less than 1% of the phonon specific heat at room temperature. As a consequence of Pauli exclusion principle thermal excitation is limited to those electrons that can reach empty energy states above E_F. Therefore, only a small fraction $k_B T/E_F$ of the free electrons contribute to the specific heat of a conductor in the normal state. The electronic contribution to the specific heat C_{el} can be written as

$$C_{el} = \frac{1}{3}\pi^2\, D(E_F)\, k_B^2\, T = \alpha\, T \qquad \ldots(2.37)$$

where $D(E_F)$ is the density of states at the Fermi energy. At low temperatures, $T \ll \theta_D$, the specific heat in the normal state can be expressed as

$$C_n = \alpha\, T + \beta\, T^3 \qquad \ldots(2.38)$$

Because of the rapidly declining phonon specific heat $C_{ph} = \beta T^3$, the elctron specific heat $C_{el} = \alpha T$ is the dominating contribution to the specific heat at sufficiently low temperatures. Eq. (2.38) can be rewritten as

$$\frac{C_n}{T} = \alpha + \beta\, T^2 \qquad \ldots(2.39)$$

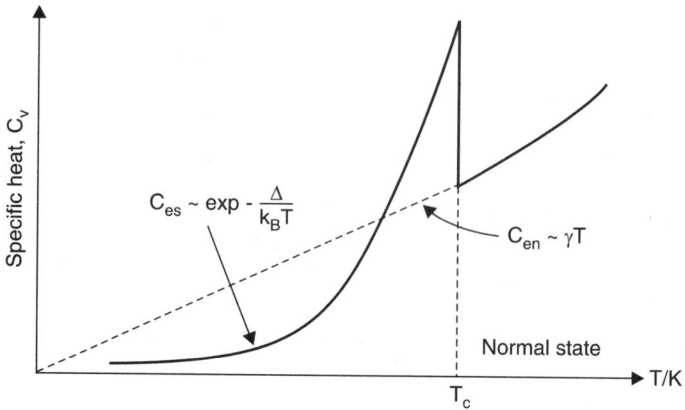

Fig. 2.15 Specific heat of a metal in the normal and the superconducting state in zero magnetic field $B(0)$. We note the characteristic jump at the transition temperature T_c

The transition to the superconducting state is accompanied by quite drastic changes in the thermodynamic equilibrium and thermal transport properties of the superconductor. In particular, the heat capacity of the superconductor changes at T_c in a characteristic way (Fig. 2.15). In zero magnetic field there appears a discontinuity at the transition temperature T_c. At temperatures immediately below T_c, the heat capacity is much larger than in the normal state so the sudden 'jump', together with the more rapid decrease with decreasing temperature, give rise to this characteristic shape.

The difference of the specific heats in the superconducting and normal state can be easily shown to be

$$C_s - C_n = \frac{2V}{\mu_o} B_c(0)^2 \left[\frac{3T^2}{T_c^2} - 1\right]\frac{T}{T_c^2} \qquad ...(2.40)$$

where V is the volume. Taking into consideration that $B_c(T_c) = 0$ the following expression for the jump in the specific heat at the transition temperature results:

$$C_s - C_n = \frac{VT_c}{\mu_o}\left(\frac{\partial B_c}{\partial T}\right)^2_{T=T_c} \qquad ...(2.41)$$

Eq. (2.41) is known as *Rutger's formula*. The BCS theory predicts for the jump in the electronic specific heat

$$\frac{C_{se} - C_{ne}}{C_{ne}} = \frac{C_{se} - \alpha T}{\alpha T} = 1.43 \qquad ...(2.42)$$

The normal state specific heat C_{ne} can be determined experimentally by applying a magnetic field $B > B_c$ to the superconductor. Eq. (2.40) indicates that the specific heats of the two states are equal for $T = T_c/\sqrt{3}$ (see Fig. 2.15). Below this temperature, the specific heat in the normal state is larger than that in the superconducting state.

(*iii*) Free Energy

We make use of Gibbs free energy to describe the thermodynamics of the normal and the superconducting state. The Gibbs free energy is given by

$$G = U - TS + pV - \mathbf{m} \cdot \mathbf{B} \qquad \ldots(2.43)$$

where U is the internal energy, S the entropy, p the pressure, V the volume, $\mathbf{m}$ the magnetic moment and $\mathbf{B}$ the applied field. The differential of the internal energy is

$$dU = Tds - pdV + \mathbf{B} \cdot d\mathbf{m} \qquad \ldots(2.44)$$

The resulting differential for the Gibbs free energy is

$$dG = -SdT + Vdp - \mathbf{m} \cdot d\mathbf{B} \qquad \ldots(2.45)$$

The independent variables are the temperature T, the pressure p and the magnetic field $\mathbf{B}$, which can be easily controlled in an experiment. Because the changes in volume are typically very small for solids, their specific heats at constant pressure C_p and at constant volume C_v are not very different. For the thermodynamically stable phase, the Gibbs free energy reaches a minimum.

At T_c the superconducting and normal phases coexist, and the Gibbs free energies of both are equal. The entropy S and specific heat C are related by the second law of thermodynamics

$$dS = \frac{C\,dT}{T} \qquad \ldots(2.45(a))$$

Using eqs. (2.37) and (2.38) the normal energy S_n can be derived:

$$S_n = \int_0^T \frac{T + \beta T^3}{T}\,dt = \alpha T + \frac{1}{3}\beta T^3 \qquad \ldots(2.46)$$

The Gibbs free energy at constant pressure can be found by integration of eq. (2.45). As a consequence of the third law of thermodynamics, the entropy S_0 at $T = 0$ is zero independent of the volume and the pressure. The resulting Gibbs free energy for the normal state is

$$G_n = -\frac{\alpha}{2}T^2 - \frac{1}{12}\beta T^4 \qquad \ldots(2.47)$$

In the presence of a homogeneous magnetic field, the term $\mathbf{m} \cdot d\mathbf{B}$ in eq. (2.45) contributes to the Gibbs free energy. In a long cylindrical rod, one can neglect demagnetization effects and the magnetic moment $\mathbf{m}$ is given by

$$\mathbf{m} = \mathbf{M}V \qquad \ldots(2.48)$$

where $\mathbf{M}$ is the magnetization and V the volume of the rod. We can express the magnetization as

$$\mathbf{M} = \chi \frac{\mathbf{B}}{\mu_0} = (\mu_r - 1)\frac{\mathbf{B}}{\mu_0} \qquad \ldots(2.49)$$

where μ_r is the relative permeability and χ is the magnetic susceptibility. Because μ_r is close to unity for non-ferromagnetic metals, the magnetic energy $\mathbf{m} \cdot d\mathbf{B}$ can be neglected for the normal state whereas it is important in the superconducting state. Because of perfect diamagnetism $\chi = -1$ in the superconducting state. For constant pressure and temperature, eq. (2.45) reduces to

$$dG = -\mathbf{m} \cdot d\mathbf{B} \qquad \ldots(2.50)$$

Using eq. (2.50) the field dependence of Gibbs free energy in the superconducting state can be easily derived. One obtains the following expression for the difference of the Gibbs free energies with and without applied magnetic field

$$G_s(T, B) - G_s(T, O) = -\int_0^B \mathbf{m} \cdot d\mathbf{B} \qquad \ldots(2.51)$$

where $\mathbf{m} = \chi \mathbf{B} V/\mu_o = -BV/\mu_o$. The resulting difference of the Gibbs free energies is

$$G_s(T, B) - G_s(T, O) = \frac{B^2 V}{2\mu_0} \qquad \ldots(2.52)$$

where $B \leq B_c(T)$ is the applied magnetic field. Obviously the magnetic energy is independent of the temperature. At a certain temperature $T < T_c$, the maximum possible energy is reached for $B = B_c(T)$. Furthermore, the Gibbs free energies of both the normal and superconducting states are equal at the phase transition. The Gibbs free energy of the normal state is considered to be independent of the applied field B. Neglecting the very small change in volume at the phase transition, the difference of the Gibbs free energies of the normal and the superconducting state at zero applied field can be expressed as

$$G_n(T) - G_s(T) = \frac{B_c^2(T) V}{2\mu_0} \qquad \ldots(2.53)$$

Fig. 2.17 shows the Gibbs free energies of both the superconducting and the normal state as functions of temperature. Using the expression $B_c(T) = B_c(0)[1 - (T/T_c)^2]$ together with eqs. (2.51) and (2.53), the Gibbs free energy difference in the presence of applied field B can be expressed as

$$G_n(T, B) - G_s(T, B) = \frac{V}{2\mu_o}[B_c^2(T) - B^2] \qquad \ldots(2.54)$$

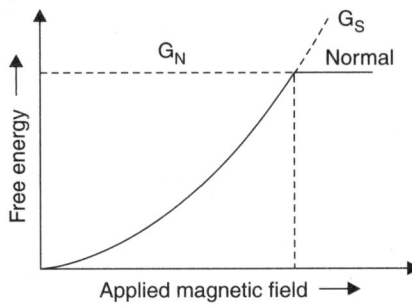

Fig. 2.17 Gibbs free energies of the superconductor and the normal states versus temperature for zero applied field

One can obtain the entropy difference $S_s(T) - S_n(T)$ by differentiation of the Gibbs free energy at constant applied field

$$S = -\frac{dG}{dt} \qquad ...(2.55)$$

Combining eqs. (2.54) and (2.55), one obtains the following expression for the entropy difference as

$$S_s(T) - S_n(T) = \frac{V B_c(T)}{\mu_o} \frac{dB_c}{dT} \qquad ...(2.56)$$

The specific heat C and the entropy S are connected by the expression

$$C = T \frac{\partial S}{\partial T} \qquad ...(2.57)$$

So far we have considered, the effects of temperature and magnetic field on the free energy of the superconducting state. However, T_c and B_c depend also on applied pressure. Negative pressure derivatives of the critical temperature dT_c/dp are related to a decreasing critical field B_c under applied pressure. At the transition from the normal to the superconducting state very small changes in the sample volume occur, which contribute to the Gibbs free energy. The application of hydrostatic pressure allows the study of the effects of a continuous variation of the lattice constants on the superconducting properties. Experimentally, it is observed that an applied pressure of 10 k bar causes a 10% reduction of T_c of indium whereas the T_c of (La, Ba)$_2$ CuO$_{4+\delta}$ is remarkably enhanced under applied pressure. A similar effect can be achieved by a chemical pressure. Large ions can be substituted by smaller ions and as a consequence, the lattice constants are reduced. The investigation of pressure effects may be a useful tool to search for new superconducting materials with high-T_c.

(iv) Thermal Conductivity

Thermal conductivity is one of most complicated kinetic phenomena occurring in solid matter. In the transport of thermal energy takes part directly the electron and phonon-subsystems. Assuming fulfilment of the Mathiessen's rule one can adopt that the total thermal conductivity κ is equal to the sum of two components: electron thermal conductivity and lattice thermal conductivity,

$$\kappa = \kappa_e + \kappa_{ph} \qquad ...(2.58)$$

In pure metals, the electronic component accounts for nearly all the heat conducted, while the lattice component, in most cases, is negligible. The electronic thermal conductivity is related to the electrical conductivity through the mean free path. In certain temperature regions, the value of the mean free path for both thermal and electrical conduction can be assumed to be the same; for these cases, the Wiedemann–Franz law is applicable

$$\kappa_e/\sigma T = L = \pi k_B^2/3e^2 \qquad ...(2.59)$$
$$= 2.45 \times 10^{-8} \text{ W } \Omega \text{ K}^{-2}$$

where σ is the electrical conductivity. The reciprocal of thermal conductivity, κ_e is the thermal resistivity W for metals. We can write analogously, $W = W_o + W_l(T)$, W_o is the term due to scattering by imperfections, and W_l due to scattering by lattice vibrations. Thus $W = 1/\kappa_e = W_o + W_o(T)$. In the low temperature region, the thermal conductivity is proportional to T. At higher temperatures, the thermal resistivity due to lattice vibrations exceeds W_o. In the case of alloys, W_o is much larger than in pure metals and a lattice thermal conductivity must also be included.

In superconductors at temperature below T_c, the electronic conduction is reduced; at sufficiently low temperatures, the thermal conductivity becomes entirely due to lattice waves and is similar to the form of the thermal conductivity of an insulating material. The thermal conductivity of HTSC cuprates remains a controversial transport coefficient. The temperature dependence of κ is similar in various HTSC compounds, namely κ rises at low temperature reaches a broad maximum, has a break at T_c and then increases or decreases at high temperatures depending on the materials.

Thermal conductivity is an interesting property and can be used in low temperature technologies for 'heat switches': thermal contact between two materials established by a superconducting material can be controlled by a magnetic field which switches off the superconductivity when heat transfer is needed, or alternatively preserves it (by turning off the magnetic field) when thermal separation is preferable.

(v) Energy Gap

The heat capacity in the superconducting state will be below T_c and varies as

$$C_v(T < T_c) \sim \exp\left(-\frac{\Delta}{k_B T}\right) \qquad \ldots(2.60)$$

$E_g = 2\Delta$ is a constant for a given material, called the *energy gap*. Near the critical temperature the energy gap is approximately

$$E_g(T) = 2\Delta(T) = 3.53\, k_B T_c (1 - T/T_c)^{1/2} \qquad \ldots(2.61)$$

Such a temperature dependence is characteristic of a system that has an energy gap in its spectrum of allowed energy states. The absorption of electromagnetic waves can be used to determine the energy gap of a superconductor. One can detect this gap by photoabsorption, quantum tunnelling, and other experiments. The energy gap $E_g(T)$ as determined from the tunnelling experiments is shown in Fig. 2.18. We note that the energy gap is zero at T_c and reaches a maximum $E_g(0)$ as the temperature lowered toward 0 K.

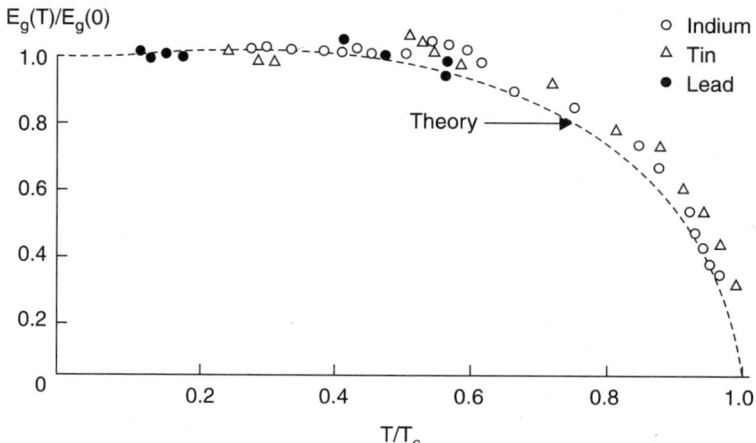

Fig. 2.18 The energy gap $E_g(T)$ versus temperature, as determined by tunnelling experiments [5]

There are materials with gapless superconductivity but most materials of interest to scientific community do have an energy gap. An important prediction of BCS theory is that the width of the energy gap $2\Delta(0)$ is closely connected to the transition temperature T_c

$$E_g(0) = 2\Delta(0) = 3.53\, k_B T_c \qquad \ldots(2.62)$$

or

$$\Delta(0) = 1.76\, k_B T_c$$

Fig. 2.19 shows the width of the energy gap $2\Delta(0)$ versus $k_B T$ for various superconducting elements. The solid line corresponds to the ratio $2\Delta(0)/k_B T_c = 3.5$ given by the BCS theory. Experimentally determined values of this ratio are typically between 3.2 and 4.6 for the superconducting elements, showing good agreement with the BCS theory. The manifestations of the energy gap in the low temperature heat capacity and in electromagnetic absorption provide strong confirmation of the BCS theory.

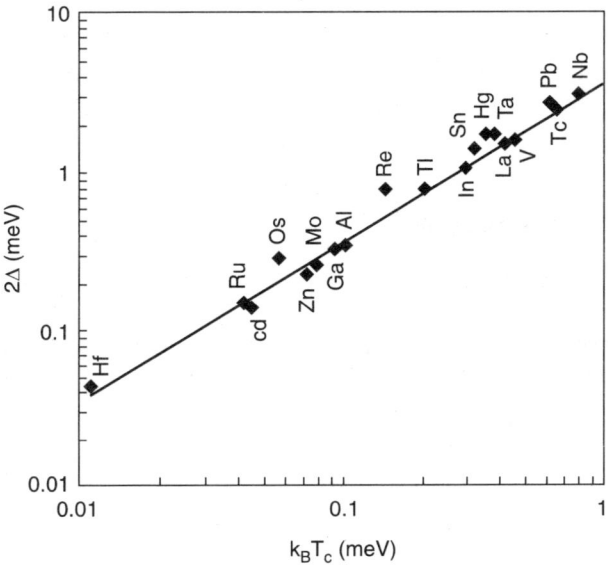

Fig. 2.19 Width of the energy gap $2\Delta(0)$ versus $k_B T_c$ for various superconducting elements. The solid corresponds to the ratio $2\Delta(0)/k_B T_c = 3.5$ given by the BCS theory

The gap in conventional superconducting materials with $T_c < 20$ K is of the order of 1 meV ($\because 1eV \sim 12000$ K) while in HTSC cuprates with T_c of ~ 100 K, $\Delta \sim$ several $(1-10)$ meV. We can easily see that these superconducting energy gaps are quite small as compared to semiconductor energy gaps with $E_g \sim 1.5$ eV in Ga As (~ 1.2 eV in Si at $T = 0$).

We must remember that the gap in the energy spectrum of semiconductors corresponds to the energy difference between the valence and the conduction band and is therefore on the scale of ~ 1 eV, whereas in the superconductor 2Δ corresponds to the energy needed to break a Cooper pair. Moreover, in semiconductors like Ga As, the electron-hole recombination across the gap releases photons of wavelength $\sim 1\,\mu$m, weakly coupled Cooper pairs in superconductors can be broken by less energetic individual photons of longer wavelengths ~ 0.1 to 1 mm (1 eV = 1.24 μ m).

2.10 ISOTOPE EFFECT

It has been observed that the critical temperature T_c of superconductors varies with isotopic mass. The relation valid for some simple metals, is given by

$$T_c \sim M^{-\alpha} \qquad ...(2.63)$$

where M is the atomic mass of the isotope and α is roughly 0.5. For example, for mercury, T_c varies from 4.185 K to 4.146 K as the isotopic mass M varies from 199.5 to 203.4. Thus, the existence of the isotope effect indicated that, although superconductivity is an electronic phenomenon, it nevertheless depends in an important way on the vibrations of the crystal lattice in which the electrons move. Fortunately, not until after the development of the BCS theory was it discovered that the situation is more complicated than it had appeared. For some superconductors the exponent of M, i.e. α is not $-1/2$, but near zero e.g., Ru and Zr, and for at least one it is positive.

The isotope effect, as studied by substitution of ^{18}O for ^{16}O in HTSC cuprate materials, is weak. The substitution of ^{18}O for ^{16}O in eq. (2.63) gives $\alpha \approx 0.02$ in YBCO and $\alpha \approx 0.15$ in LBCO. This has prompted the exploration of non-phonon electronic coupling mechanisms responsible for superconductivity in these materials. As stated above, a weak isotope effect is not conclusive.

2.11 ACOUSTIC ATTENUATION

A loss of intensity suffered by sound, radiation, etc. as it passes through a medium. It may be caused by absorption or scattering. When sound wave propagates through a metal, the microscopic electric fields due to the displacement of ions can impart energy to the electrons near the Fermi level, thereby removing energy from the wave. This is expressed by the attenuation coefficient, α, of acoustic waves. The ratio of α for superconducting and normal state is given by

$$\frac{\alpha_s}{\alpha_n} = \frac{2}{1 + \exp(\Delta/k_B T)} \qquad ...(2.64)$$

At low temperatures, this reduces to

$$\frac{\alpha_s}{\alpha_n} = 2[\exp(-\Delta/k_B T) \qquad ...(2.65)$$

2.12 MECHANICAL EFFECTS

When a superconducting material is mechanically stressed, it is found experimentally that both T_c and B_c are slightly altered. We can easily see that many of the mechanical properties of the superconducting and normal states are thermodynamically related to the free energies of these states, and the critical field strength depends on the difference in the free energies of the two states. There is an extremely small change in volume when a normal material becomes superconducting, and the thermal expansion coefficient and bulk modulus of elasticity must also be slightly different in the superconducting and normal states. However, the effects are extremely small and one can derive expressions for these effects by straightforward thermodynamic manipulation.

2.13 HIGH FREQUENCY ELECTROMAGNETIC PROPERTIES

The electrical and magnetic behaviour of superconductors at high frequencies differs from the zero frequency behaviour described earlier. In the radio frequency (up to about 10^8 Hz) and microwave frequency (from 10^8 to about 10^{11} Hz) regions of the electromagnetic spectrum, it is found that the superconductors do not have zero resistance to the flow of current. The resistance and the accompanying electrical energy loss are still much smaller than in the normal state, but they are not zero, and they increase with increasing frequency. On the other hand, in the optical region of spectrum (about 10^{16} Hz), the electromagnetic response in the superconducting state is indistinguishable from that in the normal state. One may confirm this simply by looking at a superconductor as it transforms from the normal state to the superconducting state, there is no change in its appearance. Clearly something interesting happens somewhere between 10^{11} Hz and 10^{15} Hz ($1e\ \nu \sim 10^{14}$ Hz). Obviously, the change in the frequency response occurs at $\nu \sim 10^{11}$ and $\sim 10^{12}$ Hz in the conventional and high T_c cuprate superconductors respectively. In this region (10^{11} Hz to 10^{12} Hz), depending on the material, the absorption of electromagnetic radiation by a superconductor rises quite sharply from a small value to the value characteristic of the normal state. This behaviour provides another clear indication of the presence of a gap in the electronic energy spectrum of a superconductor, rather like the gaps which occur in semiconductors. As stated earlier, in semiconductors gaps tend to be on the order of 1 eV, whereas the gaps in the superconductors are typically a thousand time smaller.

The sharp rise in the electromagnetic absorption occurs at the frequency for which the energy of a single photon (equal to Planck's constant times the frequency, $E = h\nu$) becomes just sufficient to produce an excitation of some sort (consisting, in fact, of two 'electrons') out of the superconducting state across the gap.

The HTSC cuprate materials exhibit light excitation on the T_c values. These materials are ceramic in nature, black in colour and *p* type in carriers. Though these materials are black in colour, they do not behave like black body. Their optical absorption spectra are continuous but consist of discrete bands corresponding to different energy levels. These are responsible for additional *photoinduced* charge carriers and giving rise to change in T_c. We may call it as *photodoping* and superconductors as *photoinduced* superconductors.

2.14 ABSENCE OF EFFECTS

We have seen that most of the electronic properties of a superconductor are profoundly affected by the transition to the superconducting state, while many other properties are changed very little if at all. These include the mechanical and elastic properties, tensile strength, sound velocity and density, among others.

2.15 CHARACTERISTIC PHENOMENOLOGICAL PARAMETERS

2.15.1 Penetration Depth (λ)

In our earlier description of Meissner effect, we mentioned that the superconductor expels a (weak) magnetic field B from its interior, *i.e.* $B = 0$ in the interior of a superconductor. The finer experiments reveal that the field B penetrates into the superconductor within a very thin surface layer. Consider the boundary of a semi-infinite slab. When the external field is applied parallel to the boundary, the B-field falls of exponentially

$$B(x) = B(0) \exp(-x/\lambda) \qquad \ldots(2.66)$$

as indicated in Fig. 2.20. Here λ, called the *penetration depth* of the magnetic field, is of the order of 500 Å in most superconductors at very low temperatures.

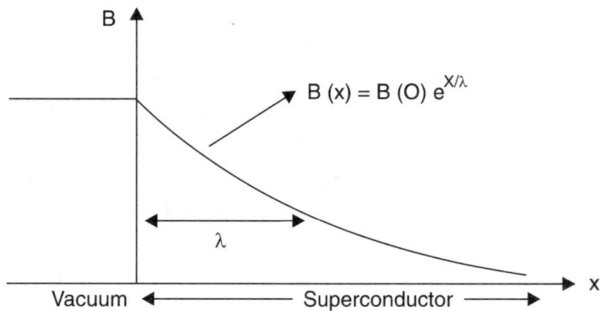

Fig. 2.20 Penetration of the magnetic field B into a superconductor slab

We must note that to cancel B at the surface of the superconductor, one requires currents on the surface which give rise to magnetization M so that in the interior of the superconductor $M + H = 0$. As the resistivity is zero these *surface currents* do not dissipate energy. These surface currents are also referred as *superfluid currents* or simply *supercurrents*.

The penetration depth (λ) is also the extension of the supercurrent around the aforementioned flux tubes (vortices) that penetrate into the sample in the mixed state (between B_{c1} and B_{c2}) of type-II superconductors.

If a superconducting film or filament is thinner than this, its properties are significantly different from those of the bulk material. In particular, the value of B_c increases as thickness decreases, and the special property of type-II superconductor arise from this.

Obviously, the presence of surface currents and associated magnetic fields will have a profound effect on the properties of thin film superconductors, or indeed of any superconductors whose dimensions are comparable to λ. The temperature dependence of λ is given by eq. (2.31).

2.15.2 Coherence Length (ξ)

In section 2.8.2, we introduced to the coherence length ξ as the distance between two electrons of the Cooper pair within the highly correlated coherent superconducting state. This was an elementary definition. Now, we re-define the coherence length more and more rigorously.

One can also define coherence length in another introductory way that ξ is a measure of the distance over which the gap parameter Δ, can vary, for instance in a spatially-varying magnetic field or near a superconductor-normal metal boundary. One can also define the intrinsic or BCS coherence length ξ_0, which is related to the Fermi velocity v_F and the 'energy gap' Δ, by the following relation

Characteristic Properties of Superconducting State

$$\xi_0 = \frac{\hbar v_F}{\pi \Delta} \qquad ...(2.67)$$

Here $v_F \sim 10^6$ m/s and $\Delta(0) \sim 10^{-3}$ eV. So $\xi_0 \sim 10^{-6}$ m. For λ_L or $\lambda \sim 10^{-8}$ m at 0 K, we have $\lambda/\xi_0 \sim 10^{-2}$ for pure metals. Obviously these behave as type-I superconductors. ξ is about 10 Å in the HTSC cuprates. Table 2.4 provides values of λ and ξ for few selected superconductors.

Table 2.4 Coherence length (ξ) and penetration depths (λ) perpendicular to (ξ_{ab}, λ_{ab}), and parallel to (ξ_c, λ_c) the c-axis for various anisotropic superconductors.

Anisotropic ratio $T = (m_c/m_{ab})^{1/2} = \dfrac{\xi_{ab}}{\xi_c} = \dfrac{\lambda_c}{\lambda_{ab}}$

Material	T_c (K)	ξ_{ab} (nm)	ξ_c (nm)	λ_{ab} (nm)	λ_c (nm)	T^a (ξ_{ab}/ξ_c, λ_c/λ_{ab})
UPt_3	0.46	–	–	782	702	–
YNi_2B_2C	15	6	5.5	–	–	–
k-$(ET)_2Cu[NCS]_2$	9	–	–	980	–	–
$(La_{1-x}Sr_x)_2CuO_4$	–	3.2	0.055	–	–	58
$YBa_2Cu_3O_{7-\delta}$	66	–	–	260	–	–
$Y Ba_2Cu_3O_{7-\delta}$	90	2.5	0.8	–	–	3.3
	89	3.4	0.7	36	125	5
$Tm Ba_2Cu_3O_{7-\delta}$	86	7.4	0.9	–	–	8.2
$Bi_2Sr_2CaCu_2O_8$	109		500			
$Bi_2Sr_2Ca_2Cu_3O_{10}$	109	2.9	0.09	–	–	31
$Tl_2Ba_2CaCu_2O_{8-\delta}$	100	–	–	182	–	–
$Tl_2Ba_2Ca_2Cu_3O_{10}$	100	–	–	175	2350	13.4
$HgBa_2Ca_2Cu_3O_{8+\delta}$	133	1.3	–	130	3500	27

When a small impurity is added to a metal λ increases very rapidly while ξ decreases. These effects are approximately represented by the following relations :

$$\lambda_{impurity} = \lambda \left(\frac{\xi_0}{l} + 1\right)^{1/2} \qquad ...(2.68)$$

$$\frac{1}{\xi_{impurity}} = \frac{1}{\xi_0} + \frac{1}{\alpha l} \qquad ...(2.69)$$

where $\lambda_{impurity}$ and $\xi_{impurity}$ are respectively the London penetration depth and coherence length for the impure metal while λ and ξ_0 are corresponding to pure metals, l is the mean free path and α is a constant of the order of unity. For pure superconductors $l \gg \xi_0$ while for impure superconductors $l < \xi_0$. Thus for impure superconductors, we have

$$\frac{\lambda_{impure}}{\xi_{impure}} = \lambda \left(\frac{\xi_0}{l} + 1\right)^{1/2} \left(\frac{1}{\xi_0} + \frac{1}{\alpha l}\right) \approx \frac{\lambda}{\alpha l}\left(\frac{\xi_0}{l}\right)^{1/2} \qquad ...(2.70)$$

Since $\xi_0 \gg l$ the ratio $\lambda_{impure}/\xi_{impure}$ is greater than 1 and the impure superconductor behaves as type-II superconductor.

2.15.3 Ginzburg-Landau Parameter (κ)

In section 2.8.2, we defined Ginzburg-Landau ratio κ as

$$\kappa = \frac{\lambda}{\xi} \qquad \ldots(2.71)$$

κ is an important parameter that characterizes the superconducting material. Close to T_c, κ is independent of temperature and it allows one to distinguish between type-I and type-II superconductors

$$\kappa = \frac{\lambda}{\xi} < 0.71 \text{ type-I}$$

$$\kappa = \frac{\lambda}{\xi} > 0.71 \text{ type-II}$$

In the latter case, the magnetic flux does penetrate the sample in the form of the cylindrical tubes called *vortices*. We must note that vortices have a radius λ and destroy superconductivity locally within a cylinder of radius ξ. It is energetically favourable for type-II superconductors to let the flux penetrate partially in the form of vortices.

2.16 FLUX QUANTIZATION

This refers to the fact that the magnetic flux threading a superconducting ring or solenoid cannot have an arbitrary value; it has to be a integral multiples of flux quantum $\phi_0 = h/2e$ = 2.0678×10^{-15} Weber. We introduced the wave function $\psi(r)$ (eq. 2.22) to describe the Cooper pairs. We know that in an isolated bulk superconductor, in the absence of an applied magnetic field the phase is same everywhere. This means that there is phase coherence in the whole sample. We must note that the absolute value of the phase has no physical meaning. We know that the gradient of the phase is related to the supercurrent. We now consider the postulate that all Cooper pairs described by eq. (2.22) and that phase coherence extends over the entire sample. In a superconducting ring or solenoid, the wave function ψ (eq. 2.22) has to go through an integral number of oscillations around the loop. For each oscillation of the wave function, the quantum of magnetic flux is $h/2e$.

Let us consider a ring of a wire whose diameter is much larger than 2λ. Obviously in the presence of an external magnetic field, the magnetic induction (**B**) and the current (**J**) are zero deep inside the superconducting wire at a distance greater than λ from the surface.

One can describe the superconducting state by a 'macroscopic' wave function :

$$\psi(\mathbf{r}) = |\psi(\mathbf{r})| \exp[i\,\phi(\mathbf{r})], \psi^*(\mathbf{r}) = |\psi(\mathbf{r})| \exp[-i\,\phi(\mathbf{r})] \qquad \ldots(2.72)$$

where $\phi(\mathbf{r})$ is the spatially varying phase of the wave function. We have the following relation between the momentum (and therefore the velocity) of Cooper pairs and the gradient of the phase as

Characteristic Properties of Superconducting State

$$\mathbf{V} = \frac{1}{m} \hbar \nabla \phi \qquad ...(2.73)$$

In the presence of an external magnetic field represented by a vector potential **A**, relation (2.73) reads as

$$\mathbf{V} = \frac{1}{m} (\hbar \nabla \phi - 2e\,\mathbf{A}) \qquad ...(2.74)$$

In fact, if we have a charge q and we suddenly apply an external magnetic field described by the vector potential **A**, one obtains an electric field

$$\mathbf{E} = -\frac{d\mathbf{A}}{dt} \qquad ...(2.75)$$

This electric field gives a momentum to the particle of $-q\,\mathbf{A}$. This explains the origin of the expression (2.74) for the velocity.

The electric current density is defined as

$$\mathbf{J} = 2e n_s \mathbf{V}_s \qquad ...(2.76)$$

and can be expressed as

$$\mathbf{J} = \frac{2e}{m} |\psi|^2 (\hbar \nabla \phi - 2e\,\mathbf{A}) \qquad ...(2.77)$$

Inside the sample, we have $\mathbf{J} = 0$

i.e., $\hbar \nabla \phi = 2e\,\mathbf{A}$ ...(2.78)

Let us now consider any path C around the interior of the ring deep inside the sample (Fig. 2.21).

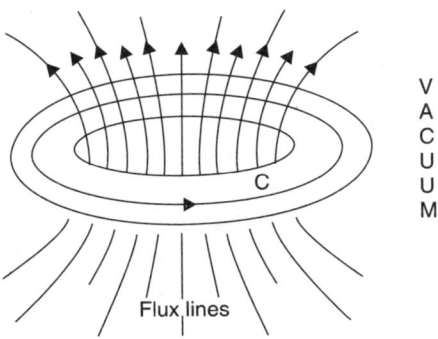

Fig. 2.21 Magnetic flux threading a superconducting ring

At any point, eq. (2.78) holds, so one finds

$$\oint \hbar \nabla \phi\, dl = \oint 2e\,\mathbf{A}\, dl \qquad ...(2.79)$$

As ϕ is a phase, one can increase it by any multiple of 2π without altering ψ. Now, if we go around the path C, the phase change is a multiple of 2π, i.e.

$$\oint \nabla \phi\, dl = 2\pi n \qquad ...(2.80)$$

where n is an integer. Since the circulation along the path of **A** is the flux enclosed by C, we have

$$\hbar\, 2\pi\, n = 2e\, \phi$$

This leads to $\phi = n\, h/2e$ ...(2.81)

i.e. the flux inside the ring is an integral number of the flux quantum, $h/2e$,

or
$$\phi_0 = \frac{h}{2e} = 2.0678 \times 10^{-15} \text{ Weber or tesla-m}^2.$$

This unit of flux is called a **fluxoid** or **fluxon.**

The flux threading through the ring is the sum of the flux ϕ_{ext} from external sources and the flux ϕ_{sc} from the superconducting currents which flow in the surface of the ring : $\phi = \phi_{ext} + \phi_{sc}$. As ϕ_{ext} is not quantized, it means that the supercurrents adjust themselves in order that ϕ assume a quantized value.

Superconducting devices can measure this tiny variation of magnetic flux, *i.e.* ϕ_0, which is exceedingly important in fundamental physics and in metrology and advanced instrumentation.

The first experimental demonstration of flux quantization was performed on superconducting cylinders, which have thin walls and were formed by evaporating a superconductor onto the surface of a glass fibre. The cylinder was cooled through its critical temperature (T_c) in an external field, and the magnetic moment associated with the trapped flux was measured by vibrating the sample and observing the induced voltage in a nearby coil.

The flux quantization in a superconducting loop has been used to produce intense magnetic fields. Fields of strength 5×10^5 A/m have been produced by flux compressor.

We must note that flux quantization is one of the most important evidence for the validity of the description of the superconducting state in terms of the macroscopic wave function.

2.17 MAGNETIC LEVITATION

One of the most fascinating demonstrations of superconductivity is the levitation of a superconducting particle over a magnet (or vice versa). Typically this is achieved by dropping a particle of one of the HTSC materials in a dish of liquid nitrogen, with a magnet underneath and watching the particle jump and however above the magnet when the temperature drops below T_c. In fact one can arrange the system so that the particle jumps out of the liquid nitrogen, warms above T_c, drops back onto the magnet, and then repeats the cycle. In order to understand the phenomenon, let us perform a simple analysis.

The repulsion of the particle from the magnet is caused by the flux exclusion from the interior of the material. For simplicity, we assume that the particle is spherical with radius R and that $R \gg \lambda_L$ so that initially we may neglect surface effects. We will also neglect the (typically very small) susceptibility of the sample in the normal state. Now, the difference in free energies (per unit volume) of the normal and superconducting states is

$$F_N - F_S = \frac{B_c^2 - B^2}{8\pi} \qquad \ldots(1)$$

where B is the average magnetic field inside the sample in the normal state and B_c is the critical magnetic field. We may note that the difference in free energies with no magnetic field applied is simply related to the maximum magnetic field that the superconductor can expel, and hence B_c is also known as the thermodynamic critical field (for both type I and type II superconductors). For a simple type-I system the two free energies become equal as B approaches B_c, and one gets a transition from the superconducting to normal state.

In order to produce a force on the sphere, of course one need a magnetic field gradient.

Let us assume that B decreases as $1/r$ in the vertical direction as shown in Fig. 2.21(a). We may note that this is a completely ad hoc assumption to simplify the calculation, and we are not addressing the question of the stability of the sphere parallel to the surface of the magnet. We may recall that we must satisfy $\nabla \cdot \mathbf{B} = 0$, and in fact directly over the centre of the magnet will be a point of unstable equilibrium. We may also exceed the critical field near the bottom of the sphere, and hence lose a portion of the energy given in eq. (2) in the region where the sphere is in the normal state[1]. The effective superconducting result thus will be reduced, and hence h_E will decrease. A similar result will occur in the opposite extreme, where $\lambda_L \gtrsim R$. The flux will penetrate into the sample, and the effective volume will again be reduced, reducing h_E. Obviously, very small particles will not levitate.

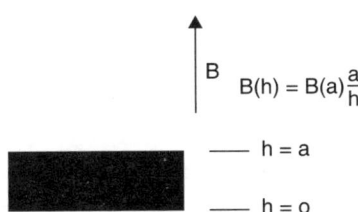

Fig. 2.21 (a)

Finally, if we raise the temperature towards, T_c, λ_L will increase, and eventually λ_L will become comparable to R. Hence we expect h_E to decrease with increasing temperature, and eventually the sphere will once again rest on the magnet.

The Meissner Oschsenfeld effect can be demonstrated strikingly by the levitation of a permanent magnet. If a small magnet is placed on a warm superconductor, which is then cooled, the magnet is raised up at the onset of superconductivity and remains in a levitated state [Fig. 2.21(b)].

---

[1] For a type II superconductor, one also get flux penetration above B_{c_1}, although the penetration will not be complete.

Fig. 2.21(b) A ring-shaped magnet is levitated above a liquid-nitrogen-cooled high-temperature superconductor by persistent supercurrents

Let h be the height of the sphere above the magnet and also assume that the average value of B may be taken at the centre of the sphere (*i.e.*, $h \gg R$). Now, the magnetic free energy is just

$$F_m = \frac{B^2(a)}{8\pi} \frac{a^2}{h^2} V \qquad \ldots(2)$$

where V is the volume of the sphere, a locates the surface of the magnet and $B(a)$ is the value of the field at the surface of the magnet. The gravitational potential energy on the other hand is just $mgh = \varrho V g h$. Minimizing the total free energy with respect to h then gives the equilibrium position h_E:

$$h_E = \left[\frac{B^2(a) a^2}{4\pi \rho g} \right]^{1/3} \qquad \ldots(3)$$

Obviously, the above result does not depend on the size of the particle, since both energies are directly proportional to V, and in fact the only materials-dependent parameter is the density ρ.

In the above calculation, h_E is independent of R is only valid in the regime that $h \gg R \gg \lambda_L$. If we let R increase, with h_E constant, then the bottom of the sphere will approach the magnet, and eventually it will sit on the magnet.

2.18 TUNNELLING EFFECTS

Tunnelling or barrier penetration is a quantum mechanical process which permits electrons to penetrate from one side to the other through an extremely thin potential barrier to electron flow (Fig. 2.22). The barrier would be a forbidden region if the electron were treated as a classical particle. A two-terminal electronic device in which such a barrier exists and primarily governs the transport characteristic (current-voltage curve) is called a tunnel junction.

Characteristic Properties of Superconducting State

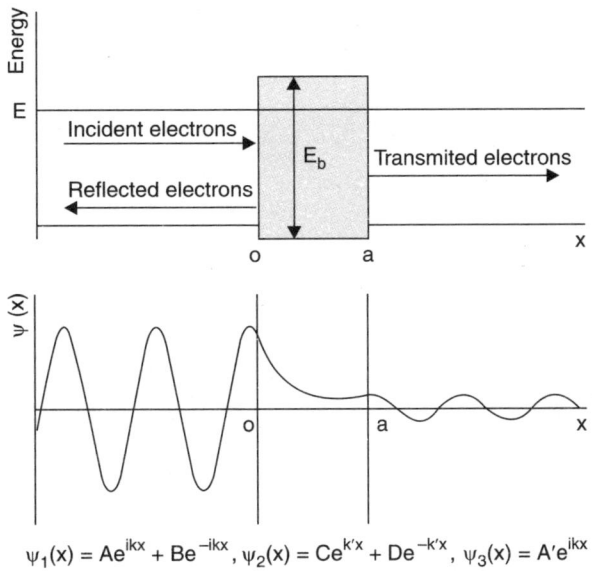

$\psi_1(x) = Ae^{ikx} + Be^{-ikx}$, $\psi_2(x) = Ce^{k'x} + De^{-k'x}$, $\psi_3(x) = A'e^{ikx}$

Fig. 2.22 Illustration of electron tunnelling through a sufficiently thin potential barrier. The electron wave functions (bottom) reveal that a few electrons are able to pass through the barrier

In 1957 L. Esaki [6] discovered the tunnel diode (also called the Esaki diode). This discovery demonstrated the first convincing evidence of electron tunnelling in solids. Tunnelling had been considered to be a possible electron transport mechanism between metal electrodes separated by either a narrow vacuum or a thin insulating film usually made of metal oxides (Fig. 2.23). I. Giaever [7] in 1960 demonstrated for the first time that if one or both of the metals were in a superconducting state, the current-voltage curve in such metal tunnel junctions revealed many details of that state [Fig. 2.24(a)].

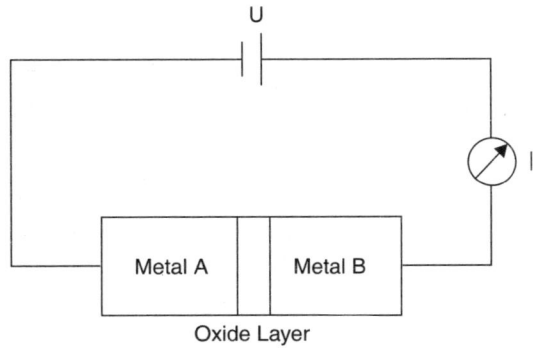

Fig 2.23 Tunnelling contact of two metals separated by a thin oxide layer

One can see from this figure that for normal metals and low voltages the I-V curve (current-voltage relation) of our sandwitch structure (N-I-N), often called tunnelling junction is ohmic, *i.e.* the current (I) is directly proportional to the applied field. When one of the metals becomes superconducting (N-I-S), the I-V curve changes into characteristic form as drawn in Fig. 2.24(b). From this figure, it is obvious that at absolute zero no current can flow until some characteristic voltage V_g is applied. V_g is directly proportional to the energy gap.

$$V_g = E_g/2e = \frac{\Delta}{e} \qquad \ldots(2.82)$$

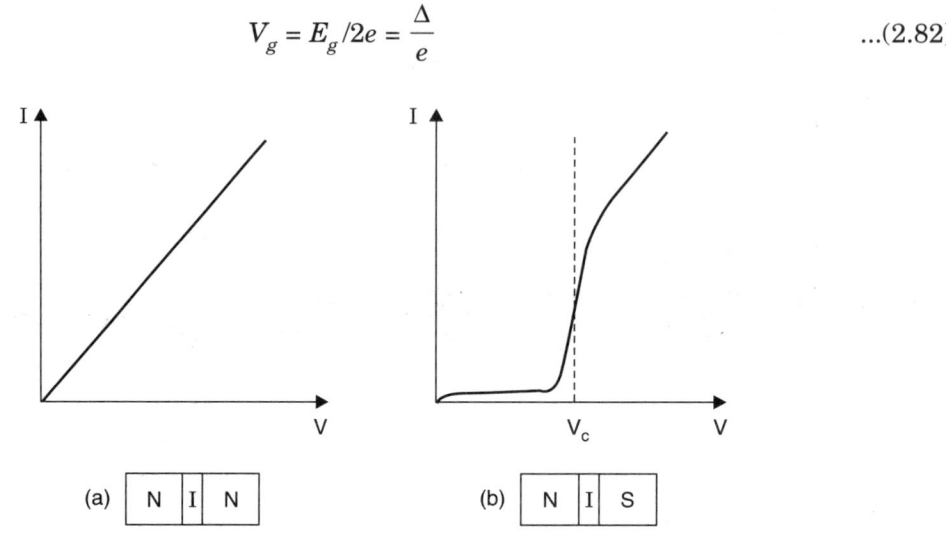

Fig. 2.24 Linear current-voltage (I-V) characteristics for two normal metals separated by a thin insulating barrier: N-I-N tunnelling. (b) The I-V characteristics when superconductor replaces one of the metal in Fig. 2.23: N-I-S tunnelling.

From eq. (2.82) it is obvious that at finite temperatures the thermal excitation of electrons allows the passage of a very small current through the barrier even at low voltages. Giaever's technique was sensitive enough to measure the most important feature of the BCS theory—the energy gap—which forms when the electrons condense into correlated bound pairs (so called Cooper pairs).

2.18.1 Josephson Effects

In 1962 B.D. Josephson [8] predicted theoretically that Cooper pair tunnelling through a very thin insulating layer (~ 2, nanometers) is possible. His theory predicted in addition to the Giaever current, the existence of supercurrent, arising from tunnelling of the bound electron pairs. Such a junction is called weak link. The effects of pair tunnelling include:

DC Josephson effect. A dc current flows across the junction resulting from Cooper pair tunnelling in the absence of any electric or magnetic field, *i.e.* at zero voltage across the insulating layer.

AC Josephson effect. Josephson also predicted that if a constant nonzero voltage V were maintained across the tunnel barrier an alternating supercurrent would flow through the barrier in addition to the dc current produced by the tunnelling of unpaired electrons. This effect has been utilized in a precision determination of $\hbar/e$. Further, an *rf* voltage applied with the dc voltage can then cause a dc current across the junction.

(i) dc Josephson Effect

To calculate the DC Josephson current flowing between two superconductors A and B separated by a thin oxide layer the time dependent Schrödinger equation has to be applied to the wave function of both Cooper pair systems. The resulting Schrödinger equations are

$$i\hbar \frac{\partial \psi_A}{\partial t} = E_A \psi_A + K_{co} \psi_B \qquad ...(2.83)$$

$$i\hbar \frac{\partial \psi_B}{\partial t} = E_B \psi_B + K_{co} \psi_A \qquad ...(2.84)$$

where K_{co} is the coupling constant describing the Cooper pair tunnelling through the insulating barrier. One can write the wave functions for the two systems as

$$\psi_A = n_{CA}^{1/2} e^{i\phi_A} \quad \text{and} \quad \psi_B = n_{CB}^{1/2} e^{i\phi_B} \qquad ...(2.85)$$

where n_{CA} and n_{CB} are the Cooper pair concentrations in the two superconductors. The phase difference between the two wave functions is

$$\delta = \phi_B - \phi_A \qquad ...(2.86)$$

Using eq. (2.85), and eqs. (2.83) and (2.84) leads to the following expressions

$$\frac{1}{2} n_{CA}^{-1/2} e^{i\phi_A} \frac{\partial n_{CA}}{\partial t} + i n_{CA}^{1/2} e^{i\phi_A} \frac{\partial \phi_A}{\partial t} = -\frac{i}{\hbar}(E_A n_{CA}^{1/2} e^{i\phi_A} + K_{co} \psi_B) \qquad ...(2.87)$$

$$\frac{1}{2} n_{CB}^{-1/2} e^{i\phi_B} \frac{\partial n_{CB}}{\partial t} + i n_{CB}^{1/2} e^{i\phi_B} \frac{\partial \phi_B}{\partial t} = -\frac{i}{\hbar}(E_B n_{CB}^{1/2} e^{i\phi_B} + K_{co} \psi_A) \qquad ...(2.88)$$

Multiplying eq. (2.87) by $n_{CA}^{1/2} \exp(-i\phi_A)$, eq. (2.88) by $n_{CB}^{1/2} \exp(-i\phi_B)$, and using eq. (2.86), one obtains

$$\frac{1}{2} \frac{\partial n_{CA}}{\partial t} + i n_{CA} \frac{\partial \phi_A}{\partial t} = -\frac{i}{\hbar}[n_{CA} E_A + K_{co}(n_{CA} n_{CB})^{1/2} e^{i\delta}] \qquad ...(2.89)$$

$$\frac{1}{2} \frac{\partial n_{CB}}{\partial t} + i n_{CB} \frac{\partial \phi_B}{\partial t} = -\frac{i}{\hbar}[n_{CB} E_B + K_{co}(n_{CA} n_{CB})^{1/2} e^{-i\delta}] \qquad ...(2.90)$$

One obtains the following expressions by separation of eqs. (2.89) and (2.90) into real and imaginary parts

$$\frac{\partial n_{CA}}{\partial t} = \frac{2 K_{co}}{\hbar}(n_{CA} n_{CB})^{1/2} \sin \delta = -\frac{\partial n_{CB}}{\partial t} \qquad ...(2.91)$$

$$\frac{\partial \phi_A}{\partial t} = -\frac{1}{\hbar}\left[E_A + K_{co}\left(\frac{n_{CB}}{n_{CA}}\right)^{1/2} \cos \delta\right]$$

$$\frac{\partial \phi_B}{\partial t} = -\frac{1}{\hbar}\left[E_B + K_{co}\left(\frac{n_{CA}}{n_{CB}}\right)^{1/2} \cos \delta\right] \qquad ...(2.92)$$

The resulting dc current I through the contact is

$$I = 2|e| \frac{\partial n_{CA}}{\partial t} V \qquad ...(2.93)$$

where V is the volume of superconductor A. Using eq. (2.91), one obtains the following expression for current through the contact for two identical superconductors.

$$I = \frac{4|e|K_{co}}{\hbar} n_c V \sin \delta = I_c \sin \delta \qquad ...(2.94)$$

where I_c is the critical current for the Josephson junction. We note that the Josephson dc current with no applied voltage through the contact is between $-I_c$ and I_c, depending on the phase difference δ of the two Cooper wave functions. We can see that this phase difference is time independent for zero voltage across the insulating layer. This is the dc Josephson effect.

If the injected current I_g through the junction is higher than the critical value I_c, the junction becomes resistive and there appears a voltage difference V across the junction. We can now define a characteristic voltage drop V_t across the junction:

$$V_t = RI_c \qquad ...(2.95)$$

where R is the resistance of the junction. Theoretical results yield

$$V_t = \frac{\pi \Delta}{2e} \qquad ...(2.96)$$

For metals most commonly used as electrodes, lead and niobium, V_t is of the order ~ 2.5 mV.

(ii) ac Josephson Effect

Let a dc voltage U be applied across the junction. One can do this because the junction is an insulator. This causes a time dependent phase difference between the two Cooper pair wave functions. For two identical superconductors result from eqs. (2.92)

$$\frac{d}{dt}(\phi_B - \phi_A) = \frac{1}{\hbar}(E_A - E_B) \qquad ...(2.97)$$

We know that an applied voltage U corresponds to an energy difference $E_A - E_B = 2eU$. The resulting phase difference between the Cooper pair wave functions vary as

$$\delta(t) = \delta(0) + \frac{2|e|U}{\hbar} t \qquad ...(2.98)$$

where

$$\nu_J = \frac{2|e|U}{\hbar} = \frac{U}{\phi_0} \qquad ...(2.99)$$

is the characteristic Josephson frequency. The ratio ν_J/U is given in terms of fundamental constants and corresponds to 484 M Hz/µV ($\lambda/U = 620$ µm/µV). Using eq. (2.94) and taking into consideration the time dependence of the phase difference $\delta(t)$ between the two Cooper pair wave functions, one obtains the following expression for the Josephson ac current I_{ac},

$$I_{ac} = I_c \sin[2\pi \nu_J t + \delta(0)] \qquad ...(2.100)$$

Characteristic Properties of Superconducting State

This shows that current oscillates with a frequency

$$\omega_J = \frac{2eU}{\hbar} \qquad \ldots(2.101)$$

This is the so-called *ac Josephson effect*. The relation (2.101) says that the tunnelling of a Cooper pair through a barrier is connected to the emission or absorption of a photon with an energy $E = h\nu_J = 2|e|U$. We must note that the ac Josephson current flows in addition to the current resulting from single electron tunnelling. By measuring the voltage and the frequency it is possible to obtain a very precise value of $e/\hbar$. The United States legal volt, V_{NBS} is now defined by eq. (2.99) through the assigned value given by eq. (2.102) at the National Bureau of Standards to an

$$2e/h = 483593420 \text{ MHz}/V_{NBS} \qquad \ldots(2.102)$$

accuracy of within a few parts in 10^8 using the ac Josephson effect; the standards of voltage of most other nations as well as the international volt are similarly defined and maintained.

There is also an *inverse ac Josephson effect* whereby the dc voltage is induced across the junction when an ac current is caused to flow through it, or when an electromagnetic field, as from microwaves, is incident on it.

(iii) Macroscopic Quantum Interference Effects

The macroscopic quantum interference effects in Josephson junctions are closely connected to the phase shifts of the cooper pair wave functions caused by an applied magnetic field. Fig. 2.25 illustrates the situation considered in the following calculations.

The magnetic field $B_y(x)$ is applied along the Y-direction. We can represent the vector potential **A** within the barrier as

$$A_z(x) = -x B_y(0) \text{ for } |x| \leq d/2 \qquad \ldots(2.103)$$

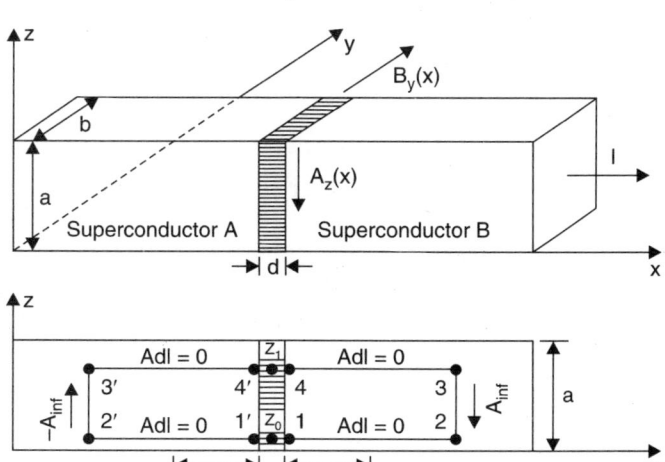

Fig. 2.25 S-I-S tunnel junction with magnetic field applied along the Y-direction (top). The integration path for the determination of the phase difference of the two Cooper pair wave functions at an arbitrary point z_1 (bottom).

The magnetic field $B_y(x)$ decays exponentially inside the two identical superconductors. The magnetic field is zero at a sufficiently large distance from the insulating layer and the vector potential attains a constant value $\mathbf{A}_{inf}$. Moreover, the screening currents are zero deep inside the superconductor.

One easily obtains the following condition for the phase shift

$$\oint \nabla \phi(\mathbf{r})\, dr = \frac{2\pi}{\phi_0} \oint \mathbf{A} \cdot \mathbf{dl} \qquad \ldots(2.104)$$

The phase difference between the two Cooper wave functions at a reference point z_0 in the barrier (Fig. 2.25 bottom) is obtained as

$$\phi_0 = \phi(1) - \phi(1') \qquad \ldots(2.105)$$

Now, the phase difference at an arbitrary point z_1 is given by

$$\phi(z_1) = \phi(4) - \phi(4') \qquad \ldots(2.106)$$

One can determine this phase difference by integration of the vector potential $A_z(x)$ along the paths $1 \to 4$ and $1' \to 4'$ as marked in Fig. 2.25. One obtains

$$\phi(z_1) = \phi_0 + \frac{2\pi}{\phi_0} \left\{ \int_1^4 \mathbf{A} \cdot \mathbf{dl} - \int_{1'}^{4'} \mathbf{A} \cdot \mathbf{dl} \right\} \qquad \ldots(2.107)$$

Since $\mathbf{A} = \mathbf{dl} = 0$ on the horizontal paths and hence only the vertical paths $2 \to 3$ and $2' \to 3'$ contribute to the line integrals in eq. (2.107). One obtains the resulting phase difference of the Cooper pair wave functions for two identical superconductors as

$$\phi(z_1) = \phi_0 + \frac{2\pi}{\phi_0}\, 2z\, A_{inf} \qquad \ldots(2.108)$$

where $z = z_1 - z_0$. Now, the total flux enclosed by the barrier as

$$\phi = \oint \mathbf{A} \cdot \mathbf{dl} = 2a\, A_{inf} \qquad \ldots(2.109)$$

One obtains the total flux approximately as

$$\phi = a(d + 2\lambda)\, B_y(0) \qquad \ldots(2.110)$$

where λ is the penetration depth and d the thickness of the insulating barrier.

Now, the phase difference of the two Cooper pair wave functions can be expressed as

$$\phi(z_1) = \phi_0 + \frac{2\pi\, \phi}{\phi_0}\, \frac{z}{a} \qquad \ldots(2.111)$$

We note that the applied field gives rise to a spatial variation of the phase difference $\phi(z)$ of the two cooper pair wave functions. Using eq. (2.111), one obtains the following expression for the dc current,

$$I = bJ_c \int \sin \phi(z)\, dz \qquad \ldots(2.112)$$

Integration of eq. (2.112) results in

$$I = I_c \sin \phi_0\, \frac{\sin \pi\phi/\phi_0}{\pi\phi/\phi_0} \qquad \ldots(2.113)$$

where $I_c = J_c ab$. The phase shift ϕ_0 in eq. (2.113) is not known. However, the maximum Josephson dc current I_{max} is reached for $\sin \phi_0 = 1$, corresponding to a phase difference of $\pi/2$, i.e.

$$I_{max} = I_c \left| \frac{\sin \pi \phi/\phi_0}{\pi \phi/\phi_0} \right| \qquad ...(2.114)$$

The interference pattern for the Josephson dc current is shown in Fig. 2.26. We can see that the interference pattern for the Josephson dc current is well known from the diffraction of light at a single rectangular slit. We note that the Josephson-Fraunhofer interference of two Cooper pair wave functions indicates that superconductivity is a macroscopic quantum phenomenon.

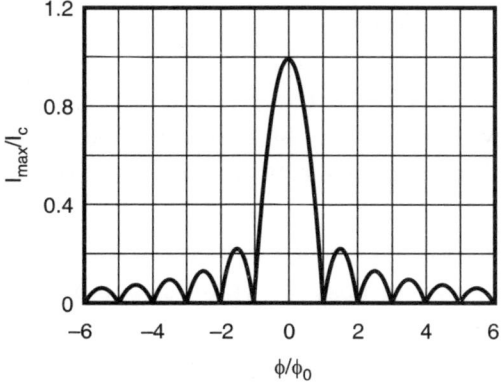

Fig. 2.26 Josephson–Fraunhofer diffraction pattern for the normalised maximum Josephson *dc* current I_{max}/I_c. Minima of the current appear at ϕ/ϕ_0 values of π, 2π, 3π, 4π. etc.

Now, we consider the quantum interference effects occuring in a loop of two identical *S-I-S* junctions. Let us consider the double-junction loop shown in Fig. 2.27.

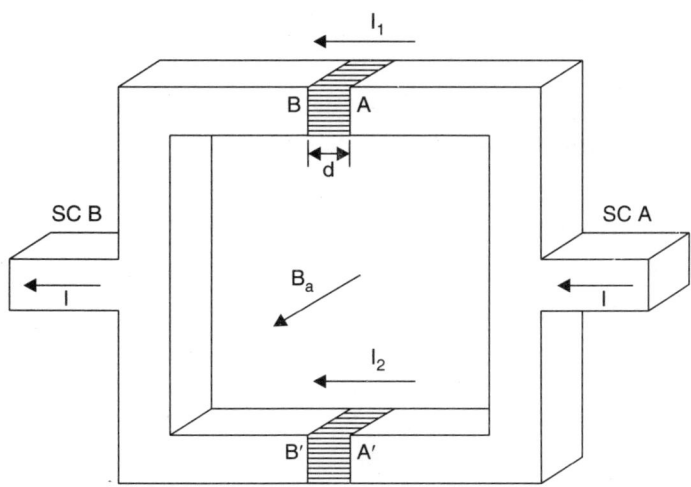

Fig. 2.27 Double contact loop obtained two identical S-I-S junctions

The same currents $I_1 = I_2$ flow in both junctions being identical. Assuming that the contribution to the currents to the magnetic field applied to the Josephson contacts can be neglected. We have to integrate similar to the calculation of the phase difference in single contact. The vector potential along a path which includes the whole area of the double contact loop.

We obtain the phase difference between the wave functions in the two junctions, analogously to eq. (2.111) as

$$\phi_{A'B'} = \phi_{AB} + 2\pi \phi_L/\phi_0 \qquad \ldots(2.115)$$

where ϕ_L is the magnetic flux within the loop. We must note that eq. (2.94) is valid for each of the two contacts and one obtains the total Josephson current as

$$I = I_c \left[\sin \phi_{AB} + \sin\left(\phi_{AB} + 2\pi \frac{\phi_L}{\phi_0}\right) \right] \qquad \ldots(2.116)$$

In eq. (2.116), we have neglected the phase difference within the sufficiently small individual contacts. Now, we can replace the undetermined phase difference ϕ_{AB} by $\phi_{AB} = \phi_0 + \pi \phi_L/\phi_0$ providing the following expression holds for the current flowing through the double contact

$$I = 2I_c \sin \phi_0 \cos(\pi \phi_L/\phi_0) \qquad \ldots(2.117)$$

We know that the maximum possible value of $\sin \phi_0 = 1$ and hence the maximum Josephson current is

$$I_{max} = 2I_c \mid \cos(\pi \phi_L/\phi_0) \mid \qquad \ldots(2.118)$$

One can obtain the maximum currents taking into account the phase difference within the individual loops by superposition of the single contact eq. (2.114) and the loop eq. (2.118) as

$$I_{max} = 2I_c \left| \frac{\sin \pi \dfrac{\phi_J}{\phi_0}}{\pi \dfrac{\phi_J}{\phi_0}} \cos \frac{\pi \phi_L}{\phi_0} \right| \qquad \ldots(2.119)$$

where ϕ_J is the flux in a single contact. The magnetic fluxes in the individual contacts and the loop are connected to the applied field B_a. The relations are

$$\phi_J = B_a A_J, \qquad \phi_L = B_a A_L \qquad \ldots(2.120)$$

where A_J and A_L are areas of the single contacts and the loop respectively. One can write eq. (2.119) by defining

$$B_J = \phi_0/A_J \text{ and } B_L = \phi_0/A_L \text{ as}$$

$$I_{max} = 2I_c \left| \frac{\sin \pi \dfrac{B_a}{B_J}}{\pi \dfrac{B_a}{B_J}} \cos \frac{\pi B_a}{B_L} \right| \qquad \ldots(2.121)$$

Fig. 2.28 shows the resulting diffraction patterns for ratios B_J/B_L of 4 and 10 obtained using eq. (2.121). We can see that the maxima of the rapid loop oscillations are limited by the slower current variations within the single contacts.

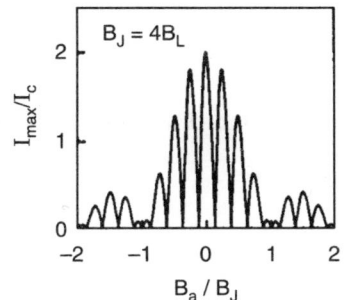
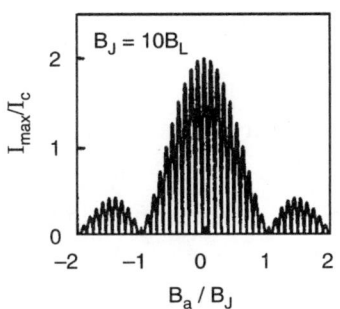

Fig. 2.28 Quantum interference effects in double loop Josephson junctions

(iv) Josephson Junction

The weak connections between superconductors through which the Josephson effects are realized are known as Josephson junctions. Historically, superconductor-insulator-superconductor (S-I-S) tunnel junctions have been used to study the Josephson effect, primarily because these are physical situations for which detailed calculations can be made. However, the Josephson effect is not necessarily a tunnelling phenomenon and the Josephson effect is indeed observed in other types of junctions, such as the superconductor-normal metal-superconductor junction. A particularly useful Josephson junction, the point contact, is formed by bringing a sharply pointed superconductor into contact with a *blunt* superconductor. The critical current of a point can be adjusted by changing the pressure of the contact. The low capacitance of the device makes it well suited for high-frequency applications. Thin-film microbridges form another-group of Josephson junctions. A practical Josephson junction can be represented as driven by a current source $I_c \sin \delta$ from equation

$$I = I_c \sin \delta$$

in parallel with a conductance G to account for quasiparticle tunnelling and a capacitor C to take into account displacement current, as shown in Fig. 2.29. The current flow in this equivalent circuit is governed by the following differential equation

$$I(t) = I_c \sin \delta + GV + C\frac{dV}{dt} \qquad ...(2.122)$$

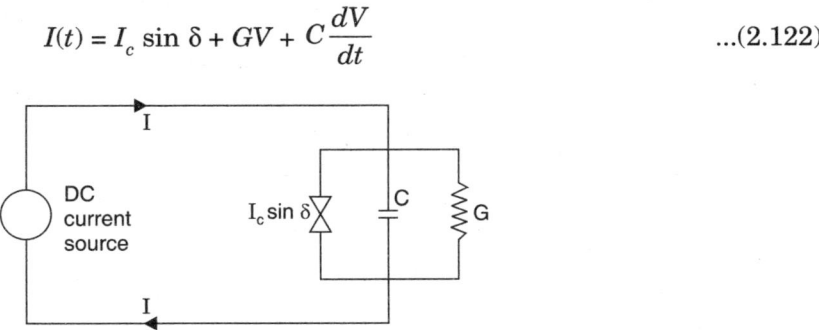

Fig 2.29 Parallel circuit representing a Josephson junction driven by the *dc* current source $I(t)$. The junction current source $I_c \sin \delta$ is in parallel with a capacitor C and a conductance G

with the help of the Josephson relation

$$\frac{d\delta}{dt} = \frac{2e}{\hbar} U$$

Equation (2.122) becomes

$$I = \frac{\hbar C}{2e} \frac{d^2\delta}{dt^2} + \frac{\hbar G}{2e} \frac{d\delta}{dt} + I_c \sin \delta \qquad ...(2.123)$$

One can define a critical voltage $U_c = I_c/G$, an associated Josephson frequency ω_c,

$$\omega_c = (2e/\hbar) U_c \qquad ...(2.124)$$

and a dimensionless admittance ratio β_c,

$$\beta_c = \omega_c C/G \qquad ...(2.125)$$

Solutions to eq. (2.213) are plotted in Fig. 2.30 for small ($\beta_c \ll 1$), medium ($\beta_c = 4$), and large ($\beta_c \gg 1$) values of admittance ratio. This figure is a plot of current vs the average voltage

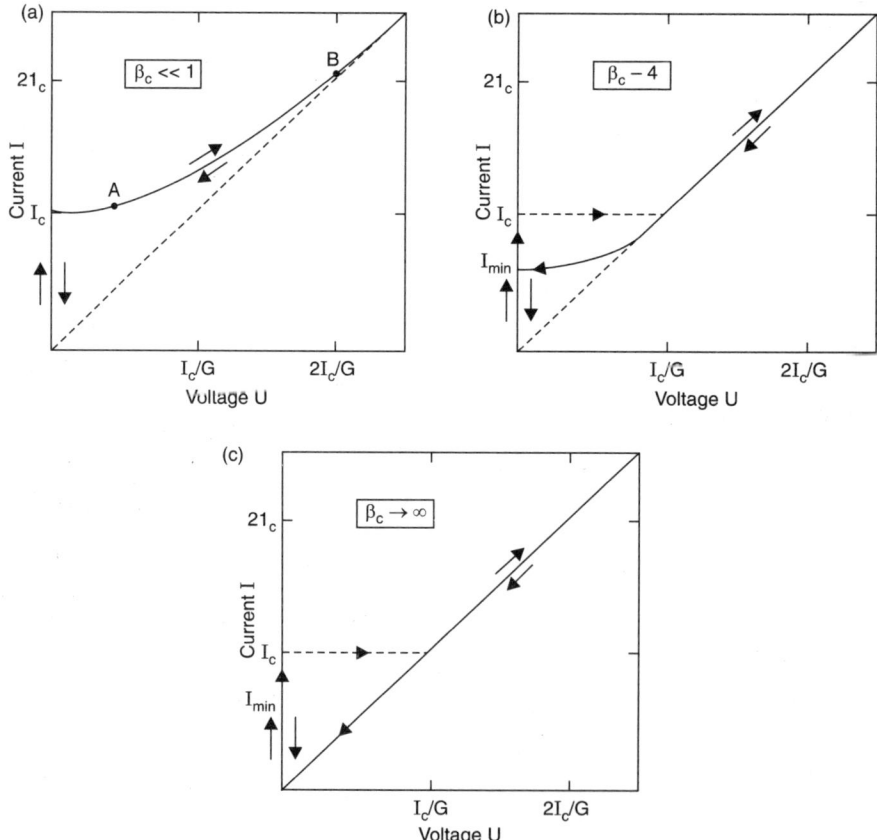

Fig. 2.30 Normalized current-voltage characteristics, *I* vs *U*, where *U* = <*U*> is the average voltage, for the Josephson junction circuit of Fig. 2.29 with (*a*) negligible capacitance, $\beta_c \ll 1$, (*b*) appreciable capacitance, β_c, and (*c*) dominating capacitance $\beta_c \Rightarrow \infty$, where $\beta_c = \omega_c C/G$

Characteristic Properties of Superconducting State

$U = <U>$, and Fig. 2.31 shows the oscillations of the voltage (U) at points A and B of Fig. 2.30(a). We may note the presence of hysteresis in the plots of Fig. 2.30.

Extremely small Josephson junctions called nano junctions exhibit new phenomena. The change in voltage due to applied magnetic field to a Josephson junction, $\Delta U = e/C$ arising from single electron tunnelling can become comparable with a typical junction voltage and produce a blockage of current flow (Coulomb blockade). Fluctuations can appear that produce a so-called coulomb staircase on an I vs U characteristic plot. A dc current biased ultra small Josephson junction can exhibit correlated tunnelling of cooper pairs leading to what are called *Bloch oscillations* at the frequency $\nu_B = I/2e$.

(v) Microbridge

The simplest microbridge is a short narrow contriction (length and width on the order of a few micrometers or smaller) in a superconducting film known as the Anderson-Dayem bridge. If the microbridge region is also thinner than the rest of the superconducting film, the resulting variable thickness microbridge has better performance in most device applications. If a narrow strip of superconducting film is overcoated along a few micrometers of its length with a normal metal, superconductivity is weakened beneath the normal metal, and the resulting microbridge is known as proximity-effect or Notary-Mecerean microbridge. Among the many other type of Josephson junctions are the superconductor oxide-normal metal superconductor junctions, and the so called SLUG junction, which consists of a drop of lead-tin solder solidified around a niobium wire. Few different types of Josephson junctions are shown in Fig. 2.32.

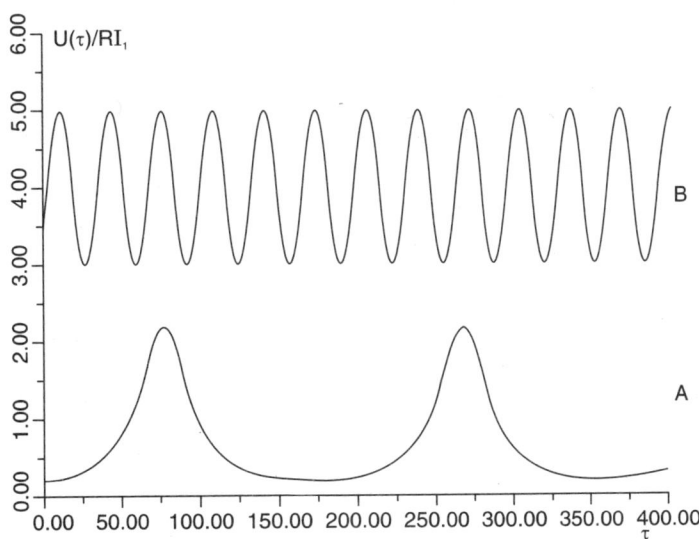

Fig. 2.31 Voltage oscillations across the Josephson junction of Fig. 2.29 for the negligible capacitance case $\beta_c \ll 1$, and small and large *dc* bias voltages as marked at points A and B, respectively of Fig. 2.30(a)

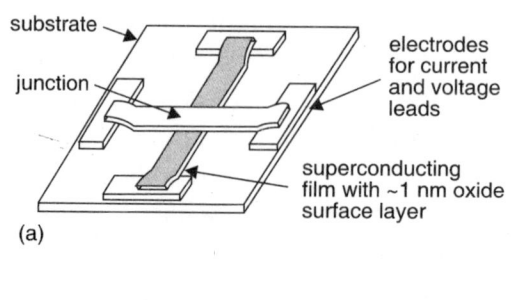
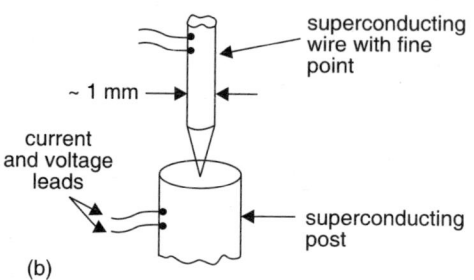
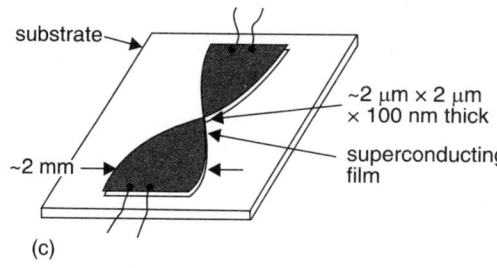

Fig. 2.32 Few different types of Josephson junctions : (a) Thin film tunnel junction (b) Point contact junction (c) Thin film weak link

The dc current-voltage characteristics of different types of Josephson junctions may differ, but all show a zero-voltage supercurrent, and constant-voltage steps can be induced in the dc characteristics at voltages given by

$$U = \frac{nh\,\upsilon}{2e}$$

(where υ is the frequency of the alternating voltage U).

when an ac voltage (U) is applied. A comparison of the dc characteristics of a microbridge and a tunnel junction is shown in Fig. 2.33.

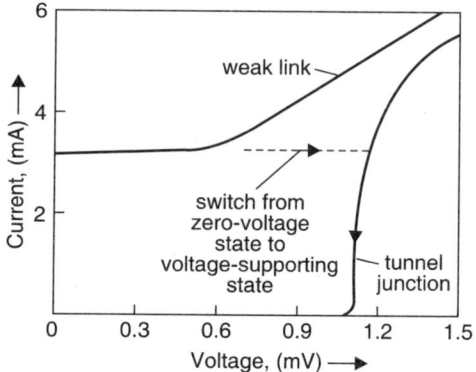

Fig. 2.33 DC current-voltage characteristics for a weak link and a tunnel junction

(vi) HTSC Josephson Junctions

There are wide variety of HTSC cuprates Josephson junctions. In fact, it has been proven relatively easy to fabricate HTSC cuprates devices which exhibit Josephson effects, with the technology ranging in sophistication from naturally occurring grain boundary weak links to all-epitaxial structures incorporating superconductors, insulators and deposited interlayers. However, it has been proven difficult to meet the requirements of some of the more demanding applications. For example, single flux quantum (SFQ) digital circuit require junctions with high resistance, high $I_c R_n$ products and $1 - \sigma I_c$ spreads less than 10%. No junction technology to date has consistently met these constraints. At present most widely used HTSC junctions are grain boundary weak links and SNS edge junctions with doped YBCO interlayers. The grain boundary devices have excellent characteristics but relatively poor I_c spreads, while the SNS edge junctions are more difficult to produce but appear to be viable candidate for fabrication of complex circuits.

Compared to semiconductor logic elements, the Josephson junction is superior, by orders of magnitude, in both speed and power consumption. Nevertheless, the cryogenic requirement and the concomitant thermal incompatibility with the semiconductors has restricted the incorporation of Josephson junctions into computers. With the discovery of HTSC cuprates, the situation may change. Josephson junctions based on YBCO are the fundamental building blocks for a variety of superconducting electronics applications operating at temperatures > 50 K. The properties of individual junctions fabricated in a variety of configurations are sufficiently close to ideal Josephson behaviour to meet application requirements. However, integration of junctions into multilayer circuits and demands on reproducibility of junction parameters with higher junction counts need have narrowed development efforts to a few promising configurations. Most of the HTSC cuprates circuit fabrication effort in industrial laboratories is based on edge SNS junctions which have been used for the most sophisticated and extendible digital circuit demonstrations. Further incremental improvements in the uniformity of these junctions to $1 - \sigma I_c$ spreads less than 10% will permit medium-scale integrated circuit fabrication. Researchers are also exploring higher risk alternative junction configurations intended to circumvent some of the limitations to junction uniformity that may exist for edge junctions.

Josephson junctions, and instruments incorporating Josephson junctions are used in other applications for metrology at dc and microwave frequencies, frequency metrology, magnetometry, detection and amplification of electromagnetic signals, and other superconducting electronics such as high-speed analog-to-digital converters and computers. A Josephson junction, like a vacuum tube or a transistor, is capable of switching states in as little as 6 ps and is the fastest switch known. Josephson junction circuits are capable of storing information. Moreover, because a Josephson junction is a superconducting device, its power of dissipation is extremely small, so that Josephson junction circuits can be packed together as tightly as fabrication techniques will permit.

The other applications of the Josephson effect include *voltage standards*, based on the AC effect, and the extremely sensitive magnetometers or 'SQUIDS' (Superconducting Quantum Interference Devices), which in principle only require the DC effect. In all such applications, the quantum mechanical phase of the Cooper electron pairs plays a key and vital role. These devices are described in detail in chapter 10.

REFERENCES

1. J. Bardeen, L.N. Cooper and J.R. Schrieffer, *Phys. Rev.* 108, 1175 (1957).
2. W. Meissner and R. Ochsenfeld, *Naturewises* 21, 787 (1933).
3. J. File and R.G. Mills, *Phys. Rev. Lett.* 10, 93 (1963).
4. V.L. Ginsburg and L.D. Landau, *Zh. Eksperim i Teor Fiz* 20, 1064 (1950).

4(a). N.C. Yeh, *Phys. Rev.* B 40, 4566 (1989) ; *ibid* 39, 9708 (1989).

5. I. Giaever and K. Megerle, *Phys. Rev.* 122, 1101 (1961).
6. L. Esaki, *Science* 183, 1149 (1974).
7. I. Giaever, *Science* 183, 1253 (1974).
8. B.D. Josephson, *Science* 184, 527 (1974).
9. P.J. Lee (Editor), *Engineering Superconductivity,* John Wiley and Sons, Inc. (2001).

Appendix 2.1—Magnetic Properties of Superconductors

1. Introduction

We have seen that one of the main characteristics of a superconductor is its magnetic behaviour, and this behaviour arises from its vortex structure. Therefore, it is important to understand the configurations and interactions of vortices and macroscopic properties that result from them. It is also important to understand the role of vortices in influencing electric current flow and their dependence on the sample shape.

2. Internal Field and Magnetization

We have the general expressions for the B and H fields as

$$B = \mu_0(H + M) = \mu H = \mu_0 H(1 + \chi) \qquad \ldots(1)$$

Relation (1) is valid both inside and outside a superconducting sample. Here M is the magnetization or magnetic moment per unit volume, χ is the dimensionless magnetic susceptibility, μ_0 is the permeability of free space where $M = 0$ and $\chi = 0$, and $\mu = \mu_0(1 + \chi)$ is the permeability of a medium. When an external magnetic field, $B_{app} = H_{app}/\mu_0$ is applied, the magnetization $M = 0$ in the free space outside the sample. Inside the sample, B and H fields are related to the magnetization through eq. (1), we have

$$B_{in} = \mu_0 H_{in}(1 + \chi) \qquad \ldots(2)$$

where
$$\chi = M/H_{in} \qquad \ldots(3)$$

is an anisotropic property of the medium. If the medium is anisotropic, then the susceptibility will have different values, χ_a, χ_b and χ_c along the three principal directions, with $\chi_a = \chi_b \neq \chi_c$ for axial symmetry.

For an ideal type-I superconductor $\chi = -1$, and if we assume that there are no demagnetization effect, we have

Characteristic Properties of Superconducting State

$$B_{in} = 0$$
$$M = -H_{in} = -B_{app}/\mu_0 \quad \ldots(4)$$

which are plotted in Fig. 1. For a type-II superconductor, which has two critical fields B_{c_1} and B_{c_2}, the analogous has the form

$$B_{in} = 0$$
$$0 \leq B_{app} \leq B_{c_1}$$
$$M = -H_{in} = -B_{app}/\mu_0 \quad \ldots 5(a)$$
$$B_{c_1} \leq B_{app} \leq B_{c_2}$$
$$\mu_0 M = -(B_{app} - B_{in}) \quad \ldots 5(b)$$

Fig. 2 shows the plots for B_{c_1} and B_{c_2}. In practice, the actual magnetization and internal field curves are more rounded than indicated by the figure.

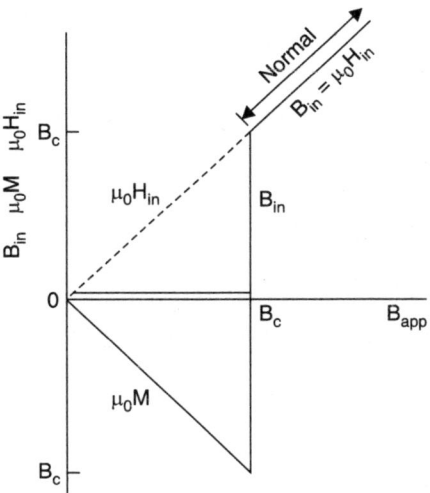

Fig. 1 Internal fields B_{in}, H_{in}, and magnetization M for an ideal type-I superconductor

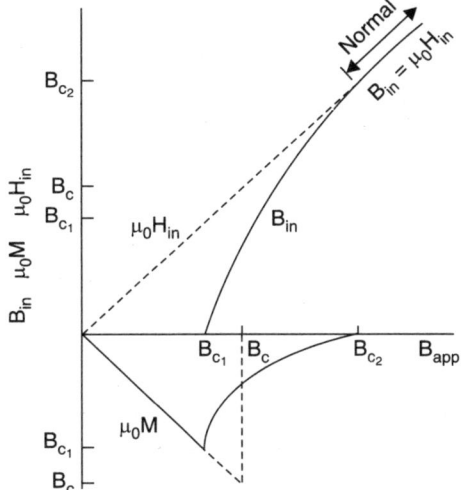

Fig. 2 Internal fields B_{in}, H_{in} and magnetization M for an ideal type-II superconductor

We have shown thermodynamic critical field, B_c in these figures, which is defined geometrically for a type-II materials by the equal area criterion corresponding to integrals

$$\int_{B_{c_1}}^{B_c} (B_{app} + \mu_0 M) dB_{app} = \mu_0 \int_{B_c}^{B_{c_2}} (-M) dB_{app} \quad \ldots(6)$$

(where M is a negative quantity), and energetically B_c involves the difference between Gibb's free energies of the normal and superconducting states,

$$G_n - G_s = B_c^2/2\mu_0 \quad \ldots(7)$$

where $B_c^2/2\mu_0$ is the condensation energy. We may note that all three quantities appearing in eq.(7) are temperature dependent and it is valid for both type I and type II superconductors.

The magnetization often exhibits hysteresis, *i.e.* it depends on the previous history of how the external field was applied. The sketches of a representative hysteresis loop exhibiting the coercive field B_{coer}, which is the value of the applied field that reduces the magnetization to zero, and the remnant magnetization M_{rem}, which is the magnitude of the magnetization when the applied field passes through zero is shown in Fig. 3. The temperature and field dependences of some typical low field and high-field hysteresis loops are shown in Fig. 4 and 5 respectively.

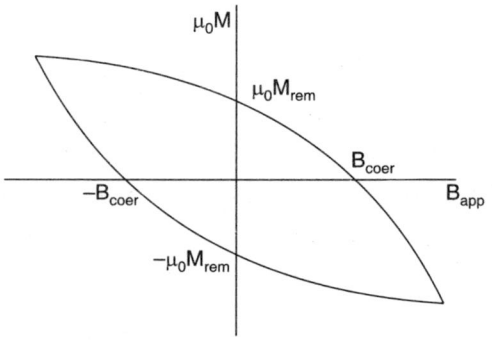

Fig. 3 Representative low-field hysteresis loop exhibiting the coercive field B_{coer} where the magnetization is zero, and the remnant magnetization M_{rem} that remains when the applied field has been reduced to zero

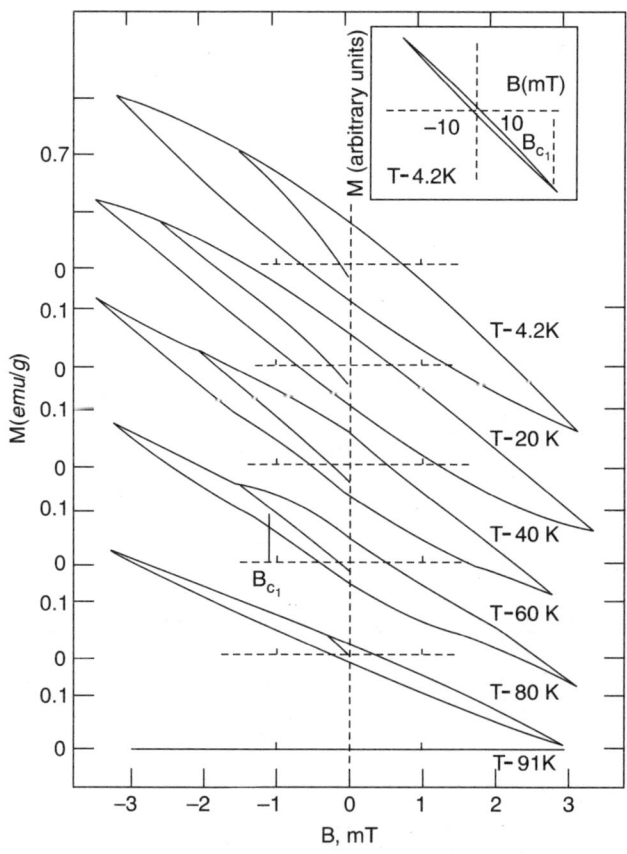

Fig. 4 Low-field hysteresis loops of $YBa_2Cu_3O_7$ cycled over the same field scan, -3 mT $\leq B_{app} \leq 3$ mT, for several temperatures. The loops gradually collapse as the temperature increases. The virgin curve for the initial rise in magnetization is also shown for each loop [1]

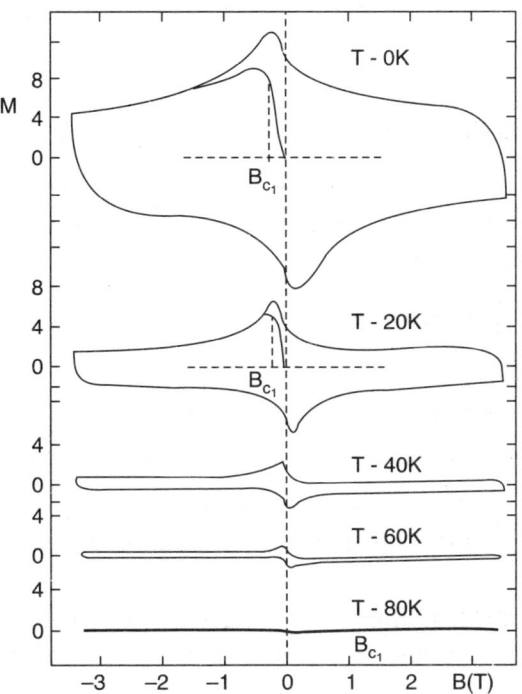

Fig. 5 High-field hysteresis loops of $YBa_2Cu_3O_7$ cycled over same field scan, $-3T \leq B_{app} \leq 3T$, for several temperatures. The loops gradually collapse as the temperature increases. The deviation of the virgin curve from linearity occurs near the lower critical field B_{c_1}, which increases as the temperature is lowered [1]

3. Critical Fields

For an isotropic type-II superconductor, the lower and upper critical fields are given by

$$B_{c_1} = \frac{\phi_0 \log \kappa}{4\pi \lambda^2} \qquad ...(8)$$

and

$$B_{c_2} = \frac{\phi_0}{2\pi \xi^2} \qquad ...(9)$$

where λ the penetration depth and ξ is the coherence length. In terms of thermodynamic critical field, B_c (given by eq. 10), B_{c_1} and B_{c_2} can be expressed as

$$B_c = \frac{\phi_0}{2\sqrt{2}\,\pi\lambda\xi} \qquad (10)$$

$$B_{c_1} = \frac{B_c \log \kappa}{\sqrt{2}\,\kappa} \qquad ...(11)$$

and

$$B_{c_2} = \sqrt{2}\,\kappa B_c \qquad ...(12)$$

$\kappa =$ is called the Ginzberg-Landau parameter ($\kappa = \lambda/\xi$). Fig. 2 shows the positions of B_c, B_{c_1} and B_{c_2}. The bulk of material goes normal when B_{app} reaches B_{c_2}, but the superconducting state can persist in a thin surface sheath for B_{app} up to the higher applied field value $B_{c_3} = 1.7\,B_{c_2}$.

The thermodynamic critical field B_c is related to the depairing current, J_{depair} through the relation

$$J_{depair} = \alpha B_c / \mu_0 \lambda \qquad ...(13)$$

where the dimensionless coefficient α is of the order of unity.

4. Vortices

The magnetic flux of an isolated, *i.e.* an individual vortex is determined by integrating its magnetic field over its area

$$\int \mathbf{B} \cdot d\mathbf{A} = \phi_0 \qquad ...(14)$$

and it equals the flux quantum ϕ_0,

$$\phi_0 = h/2e = 2.0678 \times 10^{-15} \text{ Tm}^2 \qquad ...(15)$$

This flux quantum ϕ_0 is associated with the Hall effect quantum of resistance R_H,

$$R_H = h/e^2 = 2\phi_0/e = 25.813 \, \Omega \qquad ...(16)$$

Some workers use $R_0 = h/4e^2 = 6.45 \, k\Omega$ as the *quantum of resistance* since the charge of a Cooper pair is $2e$.

For the high κ limit of HTSC cuprates, $\lambda \ll \xi$, the vortex magnetic field $B(r)$ and shielding current density $J_s(r)$ outside the core, $\lambda > \xi$, have radial dependences

$$B(r) = \frac{\phi_0}{2\pi \lambda^2} \kappa_0(r/\lambda) \qquad ...(17)$$

$$J_s(r) = \frac{\phi_0}{2\pi \mu_0 \lambda^3} \kappa_1(r/\lambda) \qquad ...(18)$$

where $\kappa_0(r/\lambda)$ and $\kappa_1(r/\lambda)$ are the zero order modified Bessel functions, respectively, with the properties that $\kappa_1(r) \gg \kappa_0(r)$ for small $r \ll \lambda$, and $\kappa_1(r) \approx \kappa_0(r)$ for large $r \gg \lambda$. Inside the core, $r \leq \xi$, we have $J_s(r) = 0$, and $B(r)$ may be approximated as having the constant value

$$B(r) = \frac{\phi_0}{2\pi \lambda^2} \kappa_0(\xi/\lambda) \qquad ...(19)$$

Near the core, eqs. (17) and (18) have the following asymptotic behaviours

$$B(r) = \frac{\phi_0}{2\pi \lambda^2} \log(1.123 \, \lambda/r) \quad (r \ll \lambda) \qquad ...(20)$$

and

$$J_s(r) = \frac{\phi_0}{2\pi \mu_0 \lambda^3} \lambda/r \quad (r \ll \lambda) \qquad ...(21)$$

and far away from the core

$$B(r) = \frac{\phi_0}{2\sqrt{2\pi} \lambda^2} \frac{\exp(-r/\lambda)}{\sqrt{r/\lambda}} \quad (r \ll \lambda) \qquad ...(22)$$

$$J_s(r) = \frac{\phi_0}{2\sqrt{2\pi} \mu_0 \lambda^3} \frac{\exp(-r/\lambda)}{\sqrt{r/\lambda}} \quad (r \ll \lambda) \qquad ...(23)$$

where the numerical factor 1.123 in eq. (20) is $2e^{-\gamma}$ and γ is the dimensionless Euler-Mascheroni constant.

Ordinarily, vortices arrange themselves in the hexadic pattern as shown in Fig. 6, separated by a distance d, where $\frac{\sqrt{3}}{2}d^2$ is the area occupied per vortex. Near the lower critical field $d \approx 2\lambda$ and near the upper critical field $d \approx 2\xi$. For large κ, the amount of flux ϕ_{core} in the core of an isolated vortex is much less than the flux quantum ϕ_0. Near the upper critical field B_{c2} the vortex cores almost overlap and the flux in each core approaches ϕ_0. The average field B_{in} inside the superconductor is expressed as

$$B_{in} = \frac{\phi_0}{\frac{\sqrt{3}}{2}d^2} = N_A \phi_0 \qquad ...(24)$$

where N_A is the number of vortices per unit area.

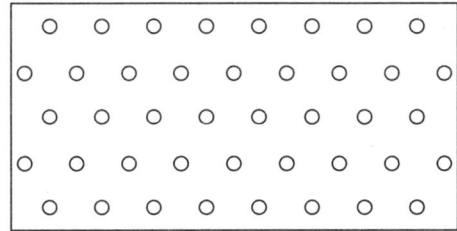

Fig. 6 Hexagonal vortex lattice [2]

5. Vortex Anisotropies

The characteristic length relationship of anisotropic superconductors is

$$\xi_a \lambda_a = \xi_b \lambda_b = \xi_c \lambda_c \qquad ...(25)$$

and the GL parameter in the ith principal direction is

$$\kappa_i = \frac{|\lambda_j \lambda_k|^{1/2}}{|\xi_j \xi_k|^{1/2}} \qquad ...(26)$$

By generalizing eqs. (11) and (12), one obtains the critical fields for ith direction respectively;

$$B_{c_1} = \frac{\log \kappa_i}{\sqrt{2}\,\kappa_i} B_c \qquad ...(27)$$

$$B_{c_2} = \sqrt{2}\,\kappa_i B_c \qquad ...(28)$$

Comparing eqs. (10) and (12), one obtains

$$B_c = \frac{\phi_0}{2\sqrt{2}\,\pi\,\xi_i \lambda_i} \qquad ...(29)$$

which is independent of the direction.

Axially symmetric superconductors with in-plane ($m_a^* = m_b^*$) and axial direction (m_c^*) effective masses are characterized by the anisotropic ratio T,

$$\Gamma^2 = m_c^*/m_{ab}^* = (\xi_{ab}/\xi_c)^2 = \left(\frac{\lambda_c}{\lambda_{ab}}\right)^2 \qquad ...(30)$$

where for the cuprates $m_c^* > m_{ab}^*$ and we have

$$\xi_c \ll \xi_{ab} \ll \lambda_{ab} \ll \lambda_c \qquad ...(31)$$

For the HTSC cuprates with the applied field in the ab plane (κ_{ab}) and along the c-direction (κ_c), respectively, we have

$$\kappa_{ab} = \frac{|\lambda_{ab}\lambda_c|^{1/2}}{|\xi_{ab}\xi_c|^{1/2}} \qquad ...(32)$$

$$\kappa_c = \lambda_{ab}/\xi_{ab} \qquad ...(33)$$

and eqs. (27) – (29) provide the critical fields for axial symmetry in the ab plane,

$$B_{c_1}(ab) = \frac{\phi_0 \log \kappa_{ab}}{4\pi \lambda_{ab}\lambda_c} \qquad ...(34)$$

$$B_{c_2}(ab) = \frac{\phi_0}{2\pi \xi_{ab}\xi_c} \qquad ...(35)$$

and along the c-direction

$$B_{c_1}(c) = \frac{\phi_0 \log \kappa_c}{4\pi \lambda_{ab}^2} \qquad ...(36)$$

$$B_{c_2}(c) = \frac{\phi_0}{2\pi \xi_{ab}^2} \qquad ...(37)$$

When the applied magnetic field is in the z-direction, along the c-axis, the vortex has axial symmetry, its cross-section is a circle, and it has a distance dependence

$$B_z(x,y) = \frac{\phi_0}{2\pi \lambda_{ab}^2} K_0[(x^2+y^2)^{1/2}/\lambda_{ab}] \qquad ...(38)$$

When the applied field is in the x-direction, along the a-axis, there is no longer axial symmetry, the cross-section is elliptical and the distance dependence is more complicated.

Vortex cross-sections for the two applied field directions is shown in Fig. 7. When the applied field is aligned at an oblique angle relative to the c direction, the expressions for the magnetic field and current density become very complicated, and neither the internal magnetic field nor the magnetization oriented in the same direction as B_{app}.

In the HTSC cuprates the coherence length ξ_c along the c-axis is less than the average spacing between layers of copper oxide planes. The Josephson coupling between layers is reported weak (theoretically). A vortex perpendicular to these layers is looked upon as a stacking of two-dimensional (2D) *pancake-shaped vortices* (Fig. 8).

Characteristic Properties of Superconducting State

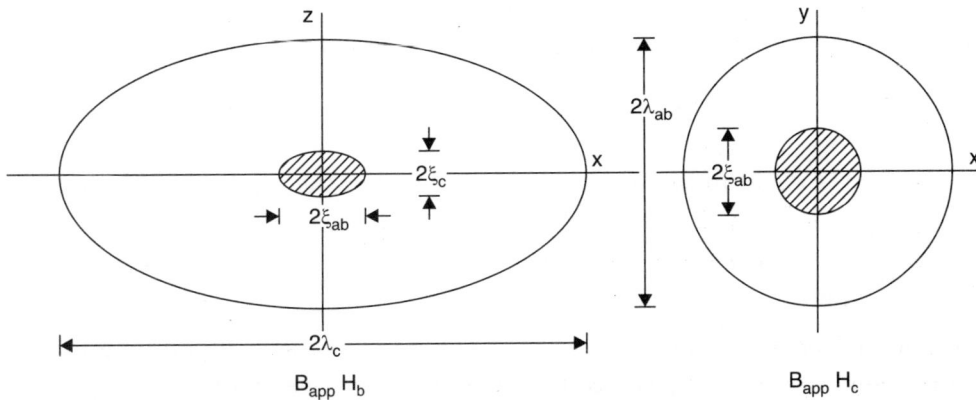

Fig. 7 Shape of the core (shaded) and the perimeter one penetration length from the center of a vortex for the applied magnetic field along b (left) and c (right) crystallographic directions respectively. The magnetic field is constant along each ellipse and along each circle. The figure is drawn for the condition $\lambda_c = 2\lambda_{ab} = 6\xi_{ab} = 12\xi_c$ [2]

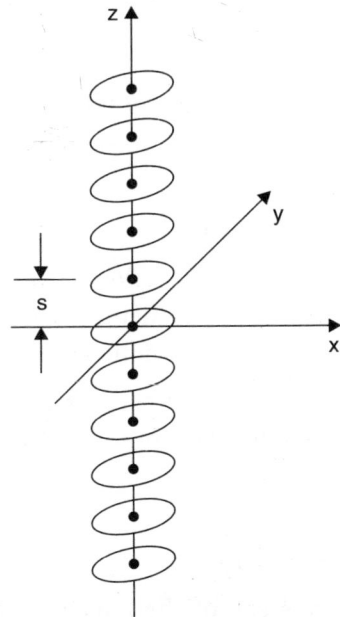

Fig. 8 Stack of two-dimensional pancake vortices along the c direction [3]

6. Individual Vortex Motion

The repulsive force per unit length F/L between two vortices arises from the Lorentz force interaction $\mathbf{F} = \mathbf{J}_1 \times \mathbf{B}_2$ between the current $\mathbf{J}_1$ from one vortex and the magnetic field $\mathbf{B}_2$ of the other

$$\mathbf{F}/L = \int \mathbf{J}_1 \times \mathbf{B}_2 \, r_2 \, dr_2 \, d\phi_2 \qquad \ldots(39)$$

For the hypothetical case of vortices that are far apart, $d \gg \lambda$, the current density of one vortex is fairly uniform in the neighbourhood of the other, and one can make the approximation

$$\mathbf{F}/L \sim \mathbf{J}(d) \times \int \mathbf{B_2}\, r_2\, dr_2\, d\phi_2 \qquad \ldots(40)$$

where the integral is equal to the flux quantum ϕ_0, and in the high κ approximation $\mathbf{J}_1(d)$ is given by eq. (23). For applied fields far above the lower critical field, $B_{app} \gg B_{c_1}$, nearest neighbour vortices overlap, i.e. are much close to each other than the penetration depth λ, and eq. (40) does not apply.

The Lorentz force required to depin a single vortex equals the pinning force, and the force per unit length needed to produce this depinning, F_p, was found to have the following temperature dependence

$$F_p(T) = F_{po}\,[1 - (T/T_c)]^n \qquad \ldots(41)$$

with F_{po} varying from 10^{-12} to 4×10^{-4} N/m, and n ranging from 1.5 to 3.5.

An isolated vortex ϕ_0 in a region of constant current density $\mathbf{J}$ has its steady-state motion governed by the equation

$$\mathbf{J} \times \phi_0 - \alpha n_s e(\mathbf{v} \times \phi_0) - \beta \mathbf{v} = 0 \qquad \ldots(42)$$

with the Lorentz force $\mathbf{J} \times \phi_0$ balanced by the two velocity-dependent forces, a dissipative force $\beta \mathbf{v}$ and the sideways-acting Magnus force $\alpha\, n_s\, e\, (\mathbf{v} \times \phi_0)$. The magnus coefficient α has different values for different theoretical models.

7. Transport Current in a Magnetic Field

Consider a transport current of uniform density J_x flowing along a superconducting wire located in a transverse magnetic field B_z. The superconductor is considered soft, i.e. the pinning forces are not strong enough to prevent flux motion so following three things can happen:

(i) The current exert the force $\mathbf{J} \times \phi_0$ on the vortices, causing them to move from one side of the wire to the other. The viscous drag $\beta \mathbf{v}$ limits this motion to a constant velocity v_ϕ and the Magnus force causes it to occur at an angle θ_ϕ defined by the ratio of the Magnus force to the drag force,

$$\tan \theta_\phi = \alpha\, n_s\, \phi_0 / \beta \qquad \ldots(43)$$

(ii) A magnetic field gradient is established across the sample in accordance with the Maxwell equation $\nabla \times \mathbf{B} = \mu_0 \mathbf{J}$, is given by

$$\frac{d}{dy} B_z = \mu_0\, J_x \qquad \ldots(44)$$

This is shown in Fig. 9.

Characteristic Properties of Superconducting State

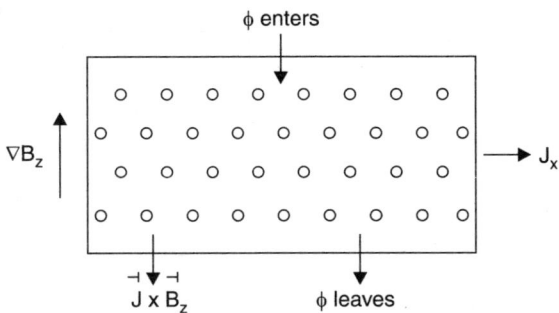

Fig. 9 Hexagonal lattice of vortices with a gradient $\nabla B_z = dB_z(y)/dy$ in the y direction due to the application of a transport current density J_x in addition to the magnetic field B_z. The Lorentz force $\mathbf{J} \times \mathbf{B}_z$ exerted by the current on a vortex is shown. In the absence of pinning forces, the current density causes the vortices to move downward at a constant velocity, with new ones entering at the top and old ones leaving at the bottom. Strong pinning forces can prevent this motion and provide dissipationless current flow [2]

(*iii*) By flux flow a magnetic field **B** moves across the sample at the constant speed v_ϕ and generates an electric field

$$\mathbf{E} = \mathbf{v}_\phi \times \mathbf{B} \qquad \ldots(45)$$

Perpendicular to both $\mathbf{v}_\phi$ and **B** that gives rise to the ohmic loss $\mathbf{J} \cdot \mathbf{E}$,

$$\mathbf{J} \cdot \mathbf{E} = \mathbf{J} \cdot (\mathbf{v}_\phi \times \mathbf{B}) \qquad \ldots(46)$$

and heat dissipation.

8. Magnetic Phase Diagram

A simplified phase diagram of the magnetic states of a type-II superconductor consists of a Meissner phase of perfect diamagnetism (absence of vortices) at the lowest temperatures, and

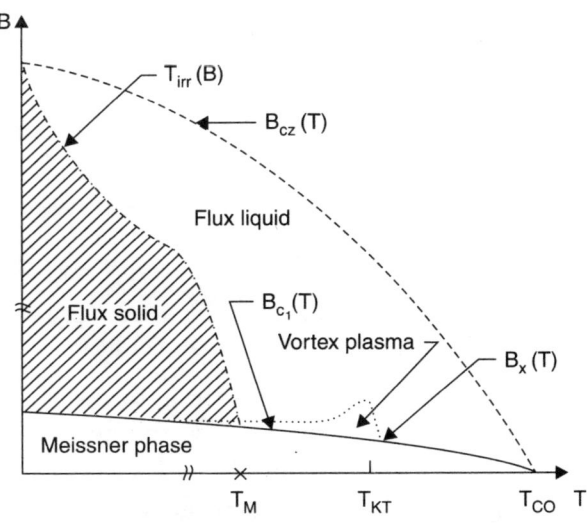

Fig. 10 Magnetic phase diagram showing the Meissner phase, the flux solid and flux liquid regions separated by the irreversibility line (T_{irr}), the plasma phase, the lower ($B_{c_1}(T)$) and upper ($B_{c_2}(T)$) critical field curves, and the melting (T_M) and Kosterlitz-Thouless (T_{KT}) temperatures [4]

a mixed (vortex lattice) phase at higher temperatures. The situation is actually much more complicated than this. Fig. 10 presents one of the many more realistic phase diagrams that have been proposed [4]. In addition to Meissner effect, this shows a flux solid phase with vortices pinned or otherwise held in place, and flux liquid phase where many vortices are unpinned or free to move, but with dissipation. These two phases are separated by a melting line that is also called the *irreversibility line* T_{irr}. Flux creep can occur to its left and flux flow to its right. In the narrow region called the *plasma phase* thermal fluctuations create short-lined positively and negatively oriented vortices that are called *intrinsic*.

9. Ellipsoids in Magnetic Fields

So far we have ignored demagnetization effects. We have implicitly treated the case of a long cylindrical superconductor in an external magnetic field applied along its axis.

When an ellipsoid with a susceptibility χ is placed in a uniform magnetic field $\mathbf{B}_{app}$ oriented along one of its principal directions, then its internal field $\mathbf{B}_{in}$ and $\mathbf{H}_{in}$ are parallel to the applied field. Their values can be obtained with the aid of the demagnetization expression

$$NB_{in} + (1-N)\mu_0 H_{in} = B_{app} \quad \ldots(47)$$

which relates the internal and applied fields, where N is the demagnetization factor that satisfies the normalization condition

$$N_a + N_b + N_c = 1 \quad \ldots(48)$$

for the three principal directions a, b, c. The largest value of N_i is for field alignment along the shortest principal axis, etc. For the common case of an ellipsoid of revolution with the c-direction selected as the symmetry axis, the semi major axes are $a = b \neq c$, and the demagnetization factors are

$$N_{\parallel} = N_c, \; N_{\perp} = N_a = N_b, \quad \ldots(49)$$

subject to the normalization condition

$$N_{\parallel} + 2N_{\perp} = 1 \quad \ldots(50)$$

Combining eqs. (2), (3) and (47), one obtains

$$B_{in} = B_{app} \frac{1+\chi}{1+\chi N} \quad \ldots(51)$$

$$H_{in} = \frac{B_{app}/\mu_0}{1+\chi N} \quad \ldots(52)$$

$$M = \frac{B_{app}}{\mu_0} \frac{\chi}{1+\chi N} \quad \ldots(53)$$

for the internal fields and the magnetization expressed in terms of the applied field. We must remember that χ is negative for a superconductor, so the denominators in these relations become small when χ approaches -1 and N approaches 1.

Experimentalists usually express the measured susceptibility χ_{exp} in terms of the applied field,

$$\chi_{exp} = \mu_0 M/B_{app} \quad \ldots(54)$$

and this is related to the true susceptibility χ ($= M/H_{in}$) as follows:

$$\chi_{exp} = \chi/(1 + N\chi), \chi = \chi_{exp}/(1 - N\chi_{exp}) \quad ...(55)$$

An oblate ellipsoid, i.e., one flattened in the a, b plane, has $c < a$ with $N_{\parallel} > N_{\perp}$ and Von Hipple [5] gives

$$N_{\parallel} = \frac{1}{\varepsilon^2} - \frac{[1-\varepsilon^2]^{1/2}}{\varepsilon^3} \sin^{-1}\varepsilon, c < a \quad ...(56)$$

where the oblate eccentricity ε is

$$\varepsilon = [1 - (c^2/a^2)]^{1/2}, c < a \quad ...(57)$$

For a prolate ellipsoid, i.e., one elongated along its symmetry axis so $c > a$ and $N_{\parallel} < N_{\perp}$, one finds

$$N_{\parallel} = \frac{1-\varepsilon^2}{\varepsilon^2}\left[\frac{1}{2\varepsilon}\log\left(\frac{1+\varepsilon}{1-\varepsilon}\right) - 1\right], c > a \quad ...(58)$$

where the prolate eccentricity ε is

$$\varepsilon = [1 - (a^2/c^2)]^{1/2}, c > a \quad ...(59)$$

Fig. 11 shows how the demagnetization factor depends on the c/a ratio.

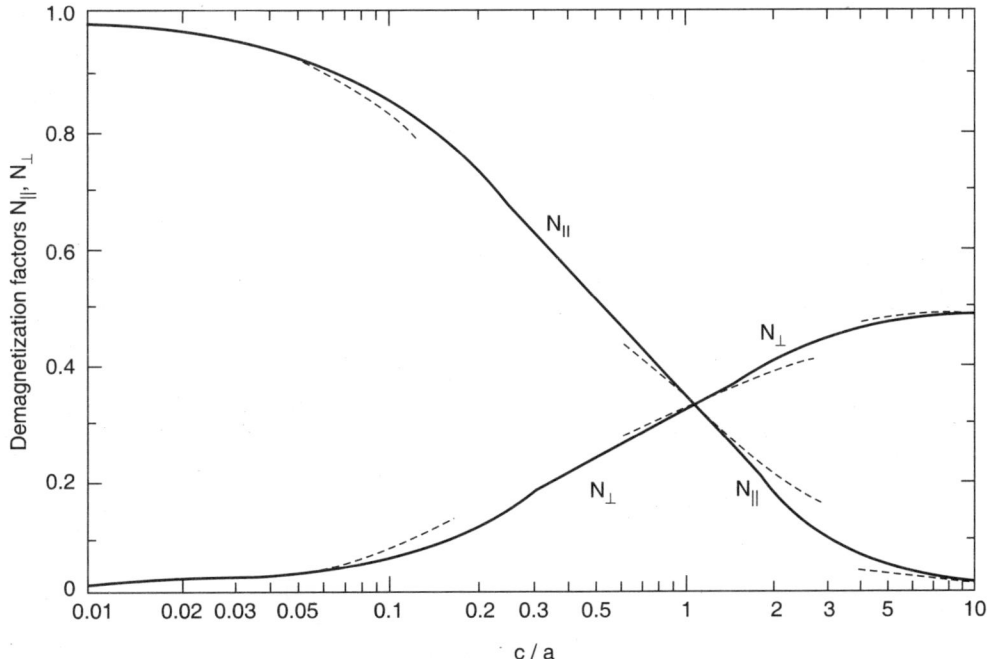

Fig. 11 Dependence on the ratio c/a of the demagnetization factors $N_{\perp} = N_x = N_y$ perpendicular to the axis, and $N_{\parallel} = N_z$ along the axis of an ellipsoid with semimajor axis $a = b \neq c$. [2]

10. Intermediate State of Type I Superconductor

Let us now consider a type I superconducting ellipsoid with a demagnetising factor N. It will exist in a true state excluding magnetic flux for applied fields less than $(1-N)B_c$ with the characteristics (viz eqs. (51) to (53))

$$B_{app} < (1-N)B_c \qquad \ldots(60)$$
$$B_{in} = 0 \qquad \ldots(61)$$
$$H_{in} = B_{app}/(1-N)\mu_0 \qquad \ldots(62)$$
$$\mu_0 M = -B_{app}/(1-N) \qquad \ldots(63)$$
$$\chi = -1 \qquad \ldots(64)$$

At higher applied fields, namely $(1-N)B_c \leq B_{app} \leq B_c$, the material is in the intermediate state in which it splits into domains of normal material with $\chi \approx 0$ embedded in pure superconducting regions with $\chi = -1$. The boundary separating normal from superconducting regions has the approximate thickness

$$d_{bound} \approx (\xi - \lambda) \qquad \ldots(65)$$

and the overall energy density per unit area of this boundary layer is

$$E_{bound} = (B_c^2/2\mu_0)\, d_{bound} \qquad \ldots(66)$$

The domains of normal material have dimensions and separations that are much greater than d_{bound}. The fields averaged over these regions of normal, boundary layer, and superconducting material in the intermediate state of an ellipsoid are given by

$$(1-N)B_c < B_{app} < B_c \qquad \ldots(67)$$
$$B_{in} = \frac{1}{N}[B_{app} - (1-N)B_c] \qquad \ldots(68)$$
$$H_{in} = B_c/\mu_0 \qquad \ldots(69)$$

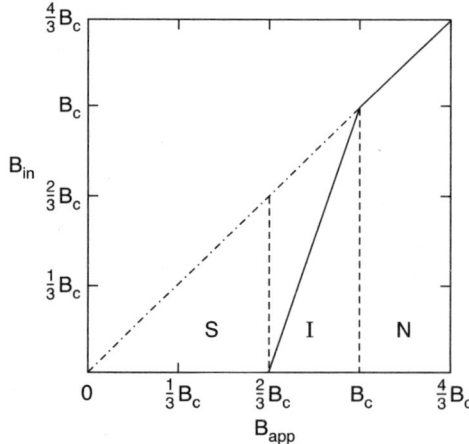

Fig. 12 Internal magnetic field B_{in} in the Meissner (S) and intermediate (I) state of a Type I superconducting sphere ($N = 1/3$) as a function of the applied field B_{app} (eqs. (61) and (68)). Solid lines represent the function being plotted; vertical dashed lines indicate the boundaries of the Meissner, intermediate, and normal regions [2]

$$\mu_0 M = -\frac{1}{N}(B_c - B_{app}) \quad \ldots(70)$$

$$\chi = -\frac{1}{N}\left(1 - \frac{B_{app}}{B_c}\right) \quad \ldots(71)$$

The internal field B_{in} in eqs. (61) and (68) is plotted versus the applied field B_{app}. For the case of a sphere ($N = 1/3$), these plots are shown in Fig. 12.

11. AC Susceptibility

An ac field $B_{ac}(t) = B_0 \cos \omega t$ applied to a superconductor causes the magnetization $M(t)$ to trace out a magnetic hysteresis loop during every cycle of the applied field. During the cycle $M(t)$ does not follow $B_{ac}(t)$ directly, but tends to lag, to distort in shape, and to shift in phase relative to the applied $B_{ac}(t)$, so it acquires an out-of-phase component that varies as $\sin \omega t$. To account for this, one can define in phase dispersion χ' and out-of-phase (quadrature) absorption χ'' susceptibilities [6]

$$\chi' = \frac{\mu_0}{\pi B_0} \int M(t) \cos \omega t \, d(\omega t) \quad \ldots(72)$$

$$\chi'' = \frac{\mu_0}{\pi B_0} \int M(t) \sin \omega t \, d(\omega t) \quad \ldots(73)$$

Fig. 13. shows the temperature dependences of χ' and χ'' determined for several values of the ac field amplitude applied at the frequency $\omega/2\pi = 73$ Hz with no dc field present. A simultaneously applied field affects the temperature dependence of χ' and χ''.

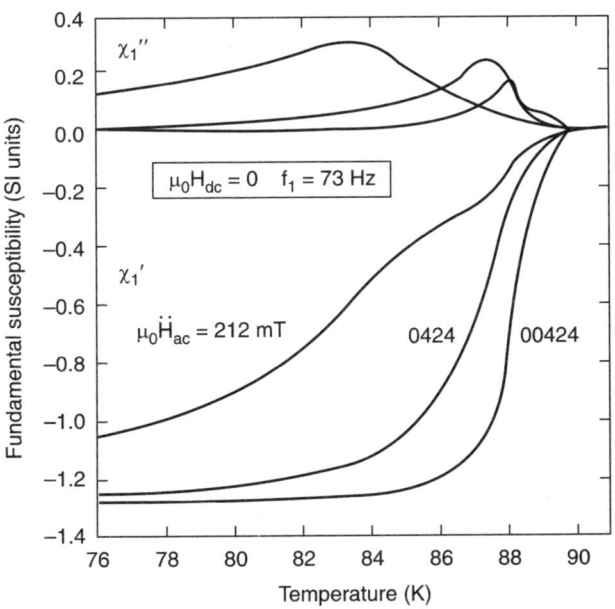

Fig. 13 Real (χ') and imaginary (χ'') components of the susceptibility of $YBa_2Cu_3O_{7-\delta}$ measured in the applied ac magnetic fields $\mu_0 H_{ac} = 0.0424, 0.424$ and 2.12 mT as a function of the temperature below T_c for the frequency 73 Hz. For this experiment no dc field was present, and the data were not corrected for the demagnetization factor [7]

REFERENCES

1. S. Senoussi et al., *J. Appl. Phys.* 63, 4176 (1988).
2. C.P. Poole, Jr. et. al, *Superconductivity*, Academic Press, New york [1995].
3. J.R. Chem, *Phys. Rev. B* 43, 7837 (1991).
4. N.C. Yeh, *Phys. Rev. B* 40, 4566 (1989) ; *ibid* 39, 9708 (1989).
5. A.R. Von Hipple, *Dielectrics and Waves*, MIT Press, Cambridge, MA (1954), pp.255.
6. Y. Matusmoto et al., *Physica* C 185, 1229 (1991).
7. T. Ishida and R.B. Goldfrab, *Phys. Rev. B* 41, 8937 (1990).

Phenomenological Theory 3

3.1 INTRODUCTION

We have read that superconductivity is characterized by three main phenomena : (i) the complete loss of dc resistance below T_c, (ii) the expulsion of magnetic flux from the interior (perfect diamagnetism), and (iii) flux quantization. Taken together, these phenomena imply that superconductivity is a remarkable example of quantum effects operating on a truly macroscopic scale. (Here, we have used the term "macroscopic" in the sense that macroscopic numbers of condensed electrons participate in a superfluid state that exhibits quantum effects over large distances. One should not confuse this with, for *e.g.*, macroscopic quantum tunnelling).

The principal phenomenological theories which have been constructed to explain the basic experimental properties are discussed in this chapter. These theories preceded the BCS microscopic theory [1] are no less useful for being phenomenological and incomplete. The original BCS theory was based on an idealized model, but nevertheless has been broadly successful in explaining the properties of real superconductors. BCS theory has been extended and elaborated to cover even more complex and realistic situations.

3.2 TWO-FLUID MODEL

Gorter and Casimir in 1934 [2] introduced one of the simplest phenomenological theory of superconductivity based on the assumption that in the superconducting state there are two components of conduction electrons "fluid" (hence the name given this theory, the "two-fluid model"). One, called the superfluid component, is an ordered condensed state with zero entropy, hence is incapable of transporting heat. It does not interact with the background crystal lattice, its imperfections, or the other conduction electron component and exhibits no resistance to flow. The other component, the normal component, is composed of electrons which behave exactly as they do in the normal state. It is further assumed that the superconducting transition is a reversible thermodynamic phase transition between two thermodynamically stable phases, the normal state and the superconducting state. The validity of this assumption is strongly supported by the existence of the Meissner Ochsenfeld effect and by other experimental evidence.

According to two-fluid model, in the superconducting state a fraction X of the electrons is in the normal phase and the remaining $(1-X)$ is in the condensed superfluid state. The free energy of the system as a function of X and T is written as

$$G(X, T) = X^{1/2} g_n(T) + (1-X) g_s(T) \qquad ...(3.1)$$

where
$$g_n(T) = -\frac{1}{2}\gamma T^2 \qquad ...(3.1(a))$$

and $$g_s(T) = -g_0 = -\frac{B_c^2(0)}{2}\mu_0 \qquad ...(3.1(b))$$

Here g_n and g_s are the contributions to the free energy in the normal and superconducting state respectively, g_s is assumed to be independent of temperature and denotes the condensation energy gained by the electrons in the superconducting phase. One can obtain the equilibrium fraction X at a given temperature by minimizing G with respect to X :

$$\left(\frac{\partial G}{\partial X}\right)_T = \frac{1}{2} X^{-1/2} g_n - g_s = 0$$

or
$$X = \left(\frac{g_n}{2g_s}\right)^2 = \frac{1}{16}\frac{\gamma^2 T^4}{g_0^2} \qquad ...(3.2)$$

Now, substituting the value of $\gamma = \frac{2B_c^2(0)}{T_c^2}\mu_0$ and g_0 from eq. (3.1b) in eq. (3.2), one obtains

$$X = \left(\frac{T}{T_c}\right)^4 \qquad ...(3.3)$$

Thus at $T = 0$, $X = 0$, hence all the electrons are superconducting while at $T = T_c$, $X = 1$, so all of them are in the normal state as expected.

Eq. (3.3) explains the observed dependence of penetration depth (λ) on T.

From a thermodynamic point of view, the superconducting phase appears below the T_c because the free energy of the superconducting phase becomes less than the free energy of the normal phase for all temperatures below T_c. The exclusion of magnetic flux by the superconductor in an applied field B increases the free energy per unit volume of superconductor by $\mu_0 B^2/2$ (in SI units); it costs energy to push the flux lines out of the superconducting region. When this increase in free energy becomes equal to the decrease in free energy associated with the normal-to-superconducting transition, it no longer pays the superconductor to remain superconducting, and the superconductor goes normal.

Hence, the superconducting condensation energy must equal $\mu_0 B_c^2/2$. One can verify this experimentally by comparing the results of critical field measurements with direct measurements of the heat capacity in the normal and superconducting states.

We have seen that the zero-field heat capacity shows a finite discontinuity at the T_c, not an infinite singularity. Obviously, this means that there is no latent heat at the transition, *i.e.*, the transition is 'second order', and therefore the entropy is the same in the superconducting and normal states at the T_c. At all but the lowest temperatures, the electronic heat capacity in the superconducting state is proportional to T^3. In the normal state, it is proportional to T, $C_n = \gamma T$, where γ is a constant. All of these facts may be combined to yield the results that the heat capacity in the superconducting state $C_s = 3\gamma T^3/T_c^2$ and that the entropy in the superconducting state is less than that in the normal state for all temperatures below the T_c, *i.e.*, the superconducting phase is more ordered than the normal state. There is thus a direct thermodynamic connection between the T^3 heat-capacity and the parabolic critical field curve ; one implies the other.

3.3 LONDON THEORY

Two years after the discovery of the Meissner effect, the first formulation of a macroscopic, but a phenomenological theory of superconductivity was proposed by F. and H. London [3, 4]. In the normal conducting state, the current density J and the applied electric field $\mathbf{E}$ are connected by Ohm's law $J = \sigma E$, where σ is the conductivity.

As a consequence of infinite conductivity in the superconducting state, Ohm's law has to be modified. In order to achieve it, they used two-fluid model. Of the total density n of electrons, there is a fraction n_s that behaves in an abnormal way and represents superconducting electrons. These (n_s) electrons are not scattered by either impurities or lattice vibrations, *i.e.*, phonons thus they do not contribute to the resistivity. Moreover these electrons are freely accelerated by an electric field. If $\mathbf{V}_s$ is their superfluid velocity, one can write the equation of motion as

$$m \frac{d\mathbf{V}_s}{dt} = e\mathbf{E} \qquad \qquad ...(3.4)$$

One can now define a superconducting current density

$$\mathbf{J}_s = n_s e \mathbf{V}_s \qquad \qquad ...(3.5)$$

which obeys the following equation

$$\frac{d\mathbf{J}_s}{dt} = \frac{n_s e^2}{m} \mathbf{E} \qquad \qquad ...(3.6)$$

In the normal phase, the current density in the steady state is given by

$$\mathbf{J}_n = \frac{n e^2 \tau}{m} \mathbf{E} = \sigma_n \mathbf{E} \qquad \qquad ...(3.7)$$

where τ is the relaxation time and σ_n is the conductivity in the normal state.

Taking curl of both sides of eq. (3.6), we have

$$\frac{d}{dt} \nabla \times \mathbf{J}_s = \frac{n_s e^2}{m} \nabla \times \mathbf{E} \qquad \qquad ...(3.8)$$

From Maxwell's equation, since the displacement $\mathbf{D}$ inside the superconductor vanishes, one obtains

$$\nabla \times \mathbf{E} = -\frac{\partial \mathbf{B}}{\partial t} \qquad \qquad ...(3.9)$$

and $$\nabla \times \mathbf{B} = \mu_0 \mathbf{J} \qquad \qquad ...(3.10)$$

From eqs. (3.8), (3.9) and (3.10), one obtains

$$\frac{d}{dt}(\nabla \times \nabla \times \mathbf{B}) = \mu_0 \frac{d}{dt}(\nabla \times \mathbf{J}_s)$$

$$= \mu_0 \frac{n_s e^2}{m}(\nabla \times \mathbf{E})$$

$$= \frac{\mu_0 n_s e^2}{m} \frac{d\mathbf{B}}{dt}$$

or
$$\frac{d}{dt}(\nabla\nabla \cdot B - \nabla^2 B) = -\frac{\mu_0 n_s e^2}{m}\frac{d\mathbf{B}}{dt}$$

or
$$\nabla^2 \mathbf{B} = \frac{1}{\lambda_L^2}\mathbf{B} \qquad \ldots(3.11)$$

where
$$\lambda_L = \left(\frac{m}{\mu_0 n_s e^2}\right)^{1/2} \qquad \ldots(3.12)$$

is a characteristic length scale known as the London penetration depth. This measures the extension of the penetration of the magnetic field inside the superconductor.

On integrating eq. (3.11) with time, one obtains

$$\nabla^2(\mathbf{B} - \mathbf{B}_0) = \frac{1}{\lambda_L^2}(\mathbf{B} - \mathbf{B}_0) \qquad \ldots(3.13)$$

where $\mathbf{B}_0$ denotes the magnetic flux at $t = 0$. We note that eq. (3.13) yields a particular solution $\mathbf{B} = \mathbf{B}_0$, i.e., the flux $\mathbf{B}_0$ within the metal can be frozen in when the metal makes a transition from the normal to the superconducting state in the presence of a magnetic field. This is contrary to the Meissner effect which requires that $B = 0$ in the superconducting phase. Obviously, a straightforward application of Maxwell's equations combined with the acceleration equation appropriate for zero resistance does not yield the correct description of the superconductor. As a consequence London proposed that in the case of superconductor, eq. (3.11) should be replaced by the equation

$$\nabla^2 \mathbf{B} = \frac{1}{\lambda_L^2}\mathbf{B} \qquad \ldots(3.14)$$

where $\mathbf{B}$ is an average of microscopic field $\mu_0 \mathbf{H}$.

Eq. (3.14) has a solution of the form:
$$B = B_0 \exp(-x/\lambda_L) \qquad \ldots(3.15)$$

for an external field applied parallel to the surface (located at $x = 0$). This shows that the field decays exponentially into the interior over the London penetration depth. Another consequence of the London theory is that the supercurrents which screen out the external field only flow within a thin layer, with a thickness of order λ_L, near the surface.

To obtain the estimate of order of magnitude of λ_L, substitute the values of m, μ_0, e and $n_s = 10^{28}/m^3$ in eq. (3.12), we obtain $\lambda_L = 532$ Å. We notice that this length is considerably larger than the interatomic distance.

It may be remarked that the decay of the magnetic field in real superconductors deviates considerably from the exponential law given by eq. (3.15). A more general definition of the penetration depth is

$$\frac{1}{\lambda_L} = \frac{1}{B(0)}\int_0^\infty B(x)\,dx \qquad \ldots(3.16)$$

Measured values of the penetration depth λ_L are typically a factor of too larger than the values predicted by the London theory [63].

Pippard [5] and Faber Pippard [6] proposed a series of experiments to demonstrate, that the supercurrent $\mathbf{J}_s$ responds non-locally to an applied vector potential $\mathbf{A}$, introducing the oncept of a "coherence" length ξ to characterise the distance over which this nonlocal response occurs. In a clean superconductor, for which the mean free path length l_e of the normal electrons is sufficiently long, the intrinsic coherence length ξ_0 varies inversely with T_c, i.e., $\xi_0 = a\, \hbar\, v_F / k_B T_c$, where a was determined to be about 0.15 for aluminium and tin [5]. In dirty or type-II superconductors, for which $l_e < \xi_0$, the total coherence length is given by

$$\frac{1}{\xi} = \frac{1}{\xi_0} + \frac{1}{l_e} \qquad \ldots(3.17)$$

where ξ_0 is the *BCS* coherence length. We are more concerned about dirty or type-II super conductors and hence we shall not need the non-local electrodynamics.

3.4 GINZBURG–LANDAU (GL) THEORY

Ginzburg and Landau [7], proposed in 1950 a highly, innovative phenomenological theory of superconductivity based on Landau's theory of phase transitions, which now bears their name. The London equation is not applicable to situations in which the number of superconducting electrons, n_s, varies; it does not link n_s with the applied field or current. Ginzburg and Landau generalized the Landau theory, to allow for both spatial and temperature variations in the amplitude of the "wave function" by introducing the concept of a complex, temperature-dependent order parameter $\psi(\mathbf{r}) = |\psi(\mathbf{r})| \exp[i\phi(\mathbf{r})]$. In the normal state $\psi(\mathbf{r})$ is zero. Then they assumed that sufficiently near T_c, where superconductivity is weak and $\psi(\mathbf{r})$ is small, the free energy of a superconductor can be expressed as a sum of a series of terms in increasing powers of $|\psi|^2$ and $|\nabla\psi|^2$; $|\psi|^2$ is interpreted as proportional to the super fluid density.

We get,
$$F_s(\mathbf{r}, T) = F_n(\mathbf{r}, T) + \alpha |\psi|^2 + \frac{\beta}{2} |\psi|^4$$
$$+ \frac{\hbar}{2m} (\nabla |\psi|)^2 + \frac{1}{2} |\psi|^2 m \mathbf{V}_s^2 + \frac{\mu_0}{2} \mathbf{B}^2(\mathbf{r}), \qquad \ldots(3.18)$$

where
$$\mathbf{V}_s = \frac{1}{m}(\hbar \nabla \phi - 2e\mathbf{A}) \qquad \ldots(3.19)$$

We can easily see that if the order parameter does not vary in space, one obtains back exactly to the London free energy and London equation by carrying out the minimization. If there is no magnetic field and the order parameter has no phase, we obtain the usual Landau theory. Obviously, the *GL* free energy is thus the way to introduce the London idea in the usual second order phase transition.

Equation (3.19) introduces, ab initio, two phenomenological parameters, α and β, into the free energy expression. The fourth term in eq. (3.19) is the energy associated with variations of ψ in space. We can see that it is written as if ψ represents a true quantum mechanical wave function; $A(\mathbf{r})$ is a vector potential at a point $\mathbf{r}$ and $\mathbf{B}$ is the microscopic magnetic field at the same point. We have $\mu_0 \mathbf{B} = \nabla \times \mathbf{A}$. From Landau theory, we have

$$\alpha = a(T - T_c)$$

and $\beta = \textit{positive}$ constant, independent of T. ...(3.20)

These phenomenological parameters can be determined by fitting the experimental results to the predictions of the *GL* theory.

Minimizing eq. (3.18) with respect to ψ and $\mathbf{A}$, Ginzburg and Landau derived two basic equations:

$$\alpha \psi + \beta |\psi|^2 \psi + \frac{1}{2m}(i\hbar \nabla - 2e\mathbf{A})\psi = 0 \qquad ...(3.21(a))$$

$$\mathbf{J} = \nabla \times \mathbf{B} = \frac{e}{m}[\psi(-i\hbar \nabla - 2e\mathbf{A})\psi + c.c] \qquad ...(3.21(b))$$

Although eqs. (3.21) resemble typical quantum mechanical wave equation, the quantity ψ must be considered as describing the *whole* condensate rather than a single charged particle; ψ is therefore normalized to the total number of superelectrons in the sample. Since a rapid spatial variation in ψ implies a corresponding large kinetic energy, changes in order parameter occur with a characteristic length ξ. Thus if ψ vanishes at some point, such as a normal-superconducting boundary, the order parameter requires a length ξ to rise to its value in the bulk sample. For this reason, ξ is sometimes termed as the **healing length.** In pure, *i.e.*, type-I superconductors, ξ is the order of 10^4 Å. The *GL* equations have also been derived from the microscopic theory. This derivation reveals that the effective charge and mass are twice the values for single electrons and leads to a precise definition of the order parameter as the wave function for the condensed bound electron pairs. Abrikosov in 1957 [8] reported a solution of the *GL* equations for the case $\lambda > \xi$ which constituted the first theoretical explanation of the mixed state in type-II superconductors.

3.4.1 Consequences of GL Equations

(*i*) **Thermodynamic critical field:** Let us consider the first *GL* eq. (3.21 *a*) with no magnetic field. One obtains for a homogeneous case,

$$|\psi_0|^2 = -\frac{\alpha}{\beta} \qquad ...(3.22)$$

No solution exists except for $T < T_c$ where $\alpha = \alpha(T - T_c)$ is negative. Obviously, superconductivity appears below T_c. Thus

$$F_s(T, O) - F_n(T, O) = \frac{-\alpha^2}{2\beta},$$

and from Gibb's potential equation

$$G_n(T, O) - G_s(T, O) = \frac{1}{2}\mu_0 H_c^2$$

where G_n and G_s are the Gibb's potentials of the normal and the superconducting state, we have

$$\mu_0 H_c^2 = \frac{\alpha^2}{\beta} \qquad ...(3.23)$$

(*ii*) **Magnetic penetration depth** (λ): Let us apply a small magnetic field and assume that we can neglect the variations of ψ. One obtains from the second *GL* equation (3.21 *b*)

$$\mathbf{J} = \nabla \times \mathbf{B} = -\frac{4e^2}{m} \mathbf{A} |\psi_0|^2$$

Taking

$$\frac{1}{\lambda^2} = 4e^2 \frac{|\psi_0|^2}{m} \mu_0 \qquad ...(3.24)$$

We obtain the London equation. Here λ is the penetration depth (put $n_s = 4|\psi_0|^2$ in London theory). It enables one to calculate the distribution of current and magnetic field :

$$\nabla \times \mathbf{B} = -\frac{1}{\lambda^2} \mathbf{A} \qquad ...(3.25)$$

We must remember that **B** is an average of microscopic field $\mu_0 \mathbf{H}$.

It is found that λ can vary from several thousands of angstroms down to less than a hundred. λ depends on temperature ; as $\psi_0 \to 0$ at T_c, $\lambda \to \infty$. We must note that at the transition temperature T_c, the magnetic field completely penetrates the sample.

(*iii*) **Coherence length:** We now consider *GL* equation (3.21*a*) in one dimensional case without external field :

$$-\frac{\hbar^2}{2m} \frac{d^2 \psi}{dx^2} + \alpha x + \beta |\psi|^2 \psi = 0$$

This equation defines the length scale :

$$\xi^2(T) = \frac{\hbar^2}{2m|\alpha|} \qquad ...(3.26)$$

We note that the solution of this equation depends only on $x/\xi(T)$.

The length ξ is called the *coherence* length and represents the length over which the order parameter $\psi(\mathbf{r})$ varies, when one introduces a perturbation at some point. We can see that this length also diverges when $T \to T_c$ as $\alpha \to 0$.

Thus both ξ and λ both diverge as $(T_c - T)^{-1/2}$ as the temperature approaches T_c from below.

(*iv*) **Ginzburg and Landau parameter** (κ): The properties of a superconductor depend sensitively on the ratio of the penetration depth (λ) to the coherence length (ξ), and it is convenient to define a dimensionless parameter,

$$\kappa = \frac{\lambda}{\xi} \qquad ...(3.27)$$

κ is called the Ginzburg-Landau parameter. It does not depend on the temperature and this is the only parameter that really appears in *GL* equations.

3.4.2 Applications of Ginzburg–Landau Equations

When we apply *GL* equations to a particular superconductor, the values of α and β can be obtained from experimental measurements.

Relation (3.23) is the first equation which enables one to calculate α and β from experimental data for a superconducting material. Combining eq. (3.23) with eq. (3.24), one obtains α and β as functions of H_c and λ :

$$\alpha = - \frac{4e^2}{m} \mu_0^2 H_c^2 \lambda^2, \qquad \ldots(3.28)$$

and

$$\beta = \left(\frac{4e^2}{m}\right)^2 \mu_0^3 H_c^2 \lambda^4 \qquad \ldots(3.29)$$

Now, if we measure the length in units of λ and the field in units of $H_c = B_c/\mu_0$, one can demonstrate that the free energy of *GL* depends only on κ. Obviously, κ completely characterizes a given superconductor phenomenologically.

Now, combining eq. (3.23) with eqs. (3.24) and (3.28), one obtains an important relationship

$$H_c(T)\, \lambda\,(T)\, \xi(T) = \text{constant} = \frac{\hbar}{2e\mu_0 \sqrt{2}} = \frac{\phi_0}{2\pi \mu_0 \sqrt{2}} \qquad \ldots(3.30)$$

3.5 TYPE-I AND TYPE-II SUPERCONDUCTORS

The phase transition from normal to superconducting state at T_c is always a second order phase transition in zero magnetic field.

For type-I superconductors, at a given temperature below T_c, the transition from superconducting state to a normal state is of the first order at B_c, contrary to the type-II superconductors which undergo second order phase transition at B_{c_2}. Now, we will characterize the difference in behaviour of type-I and type-II superconductors.

For a second order phase transition, the order parameter is small. Taking *GL* equation (3.21 *a*) and limiting our analysis to first order in ϕ, we have

$$\frac{1}{2m}(-i\hbar\nabla - 2e\mathbf{A})^2 \psi = -\alpha\psi \qquad \ldots(3.31)$$

where **A** is the vector potential of the applied magnetic field. The contribution due to the superfluid currents is given by *GL* second equation 3.21 (*b*) and since it is of the order of $|\psi|^2$ and hence one can neglect it. Eq. (3.31) describes the motion of a charged particle ($q = 2e$) in the magnetic field. The lowest eigenvalue of this Schrödinger equation is

$$E_0 = \frac{1}{2}\hbar\,\omega_c \qquad \ldots(3.32)$$

where $\omega_c = 2eB/m$ is the cyclotron frequency. One finds a nonzero solution for ψ in eq. (3.31) and hence the appearance of the superconducting state, if $B < B_{c_2}$ with

$$\frac{e\hbar B_{c_2}}{m} = -\alpha \qquad \qquad ...(3.33)$$

Using eq. (3.31), one obtains

$$B_{c_2}(T) = \frac{\phi_0}{2\pi} \frac{1}{\xi^2(T)} \qquad \qquad ...(3.34)$$

Combining eq. (3.34) with eq. (3.30), one obtains

$$B_{c_2} = \mu_0 H_{C_2} = \mu_0 \kappa \sqrt{2} H_c \qquad \qquad ...(3.35)$$

At this stage, one can distinguish two different situations :

(i) If $\kappa \equiv \lambda/\xi < 1/\sqrt{2}$, we have $B_{c_2} < B_c$.

Obviously, by decreasing the magnetic field the superconducting state appears at (and below) B_c with total expulsion of the flux. We have a type-I superconductor.

(ii) If $\kappa \equiv \lambda/\xi < 1/\sqrt{2}$, we have $B_{c_2} > B_c$. The superconducting state appears at and below B_{c_2}. As the flux expulsion is not complete, we have type-II superconductor.

The difference between the behaviour of type-I and type-II superconductors can be understood from the following analysis. To achieve $B = 0$ magnetic energy has to be considerable. Type-II superconductors minimize that energy by creating small normal cylinders called flux vortices or vortex lattice also called **flux line lattice** in the mixed state, *i.e.*, in the range $B_{c_1} < B < B_{c_2}$.

Each flux vortex has a normal core with a radius of order ξ, and contains exactly one flux quantum ϕ_0 within a region of circulating current with a radius of order λ. The creation of such a vortex costs energy

$$(F_s - F_n) \times \text{volume of the cylinder} = \mu_0 \frac{H_c^2}{2\xi^2} d ,$$

where d is the length of the vortex. The creation of such a normal cylinder permits the magnetic field to penetrate into the sample.

However, the magnetic field will penetrate inside of the superconductor only over the distance λ. Thus the gain in energy is $\mu_0 \frac{H^2}{2} \lambda d$

The energy balance is in favour of the creation of such flux vortex if $\lambda \gg \xi$ or $\kappa \gg 1$. Obviously, this explains why magnetic behaviour is different for $\kappa \gg 1$ (Type-II superconductivity) as compared with $\kappa \ll 1$ (type-I superconductivity).

Type-II superconductors are generally preferred over type-I for most bulk applications, such as wires and magnets, because the upper critical field B_{c_2} is usually much higher than the thermodynamic critical field B_c. However, somewhat surprisingly, an "ideal" type-II superconductor with a reversible M–B plot, would actually be quite useless for such applications. The reason is that each flux vortex experiences a "Lorentz" force per unit length $\mathbf{F}_L = \mathbf{J}_s \times \boldsymbol{\phi}_0$, where the flux ϕ_0 is directed along the vortex. In the absence of pinning, the vortices will thus move perpendicular to the current in a process known as *flux flow*. This leads to a finite longitudinal voltage, which can be thought of as arising either from Faraday's law due to the flux motion or from the fact that the phase gradient in the superconductor "slips" by 2π with the passage of each flux vortex. The combination of both a current and a voltage along the some direction implies that power is being dissipated, which is undesirable for most bulk applications. Bardeen and Stephen [9] were the first to discuss the specific mechanisms of power dissipation via the scattering of normal electrons.

Fortunately, nearly all real superconductors contains a variety of *defects*, or *pinning centers*, which tend to pin the flux vortices. As a result, no flux flow takes place until the applied current exceeds a certain critical current density J_c, above which the Lorentz force $\mathbf{F}_L$ exceeds a characteristic *pinning force* $\mathbf{F}_p$. The resistance is zero when the applied currents density J is less than J_c, but is non vanishing when $J > J_c$. All the relevant parameters, including B_{c_2}, B_{c_1} and J_c, are functions of temperature and can be plotted on a phase diagram. The study of HTSC cuprates has led to the development of additional important concepts, such as the *irreversibility line*.

3.6 LOWER CRITICAL FIELD (B_{c_1})

The lower critical field B_{c_1}, corresponds to the thermodynamic limit above which it is energetically favourable for vortices to penetrate into the superconductor. One can estimate B_{c_1}, by calculating the variation of the Gibb's energy which corresponds to the entry of n vortices per unit surface

$$\Delta G = n\, U - BH \qquad \ldots(3.36)$$

where U is the magnetic energy of the vortex,

$$U = \frac{\phi_0^2}{4\pi \lambda^2 \mu_0} \ln \frac{\lambda}{\xi} \qquad \ldots(3.37)$$

We may consider a vortex as a whirl of superfluid currents around a metal tube.

We have neglected the interaction between vortices in writing eq. (3.36) which, close to B_{c_1}, are rather far apart. Using $B = n\phi_0$, we have

$$\Delta G = B\left(\frac{U}{\phi_0} - H\right) \qquad \ldots(3.38)$$

The vortices will penetrate into the superconductor if this lowers the Gibbs energy. The condition is

$$B \geq B_{c_1} = \frac{U}{\phi_0} \qquad \ldots(3.39)$$

Using eq. (3.37), one finally obtains

$$B_{c_1} = \frac{\phi_0}{4\pi\lambda^2} \ln \frac{\lambda}{\xi} \qquad ...(3.40)$$

For $\kappa \gg 1$, the field B_{c_1} is small.

One obtains an interesting relation for the product $B_{c_1} B_{c_2}$ if one uses eq. (3.30);

$$B_{c_1} B_{c_2} = B_c^2 \cdot \ln \kappa \qquad ...(3.41)$$

As the value of the thermodynamic critical field does not vary very much in type-II superconductors between 0.1 T to 1 T, obviously this means that the higher B_{c_2} is, the lower becomes B_{c_1}.

In *HTSC* cuprates where B_{c_2} is very high (~ 100 T) due to the small coherence length, it means that B_{c_1} is very small, of the order of 10^{-2} Tesla.

3.7 SURFACE AND INTERFACE EFFECTS IN SUPERCONDUCTORS

3.7.1 Critical Field B_3

The solution of *GL* equations is rigorously valid only for an infinite sample. We have seen that below B_{c_2} the superconductivity appears within the volume of the sample. Let us consider now a situation where the sample is semi-infinite in the half-space $x > 0$ and where the field is applied parallel to the surface. We are interested in finding non-zero solution in this geometry and calculate the critical field. For this purpose, we have to solve the same linearized *GL* equation with an additional condition that there is no current perpendicular to the surface of the sample. The second *GL* equation yields the supercurrent so at the surface of the sample, one can impose

$$(-i\hbar\nabla - 2e\mathbf{A})_n \psi = 0 \qquad ...(3.42)$$

With the above boundary condition, one can show that the lowest-energy solution of the linearized *GL* equation (3.31) becomes

$$E_0 = 0.59 \times \frac{1}{2} \hbar \omega_c$$

Consequently the critical field at which superconductivity appears is

$$B_{c_3} = \frac{1}{0.59} B_{c_2} = 1.69 \, B_{c_2} \qquad ...(3.43)$$

Obviously, if the field B is parallel to the surface and has strength

$$B_{c_2} < B < B_{c_3} \qquad ...(3.44)$$

Then a superconducting layer of thickness $\sim \xi$ appears within the surface of the specimen. We must note that if one reduces the field to below B_{c_2}, the superconductivity appears in the volume of the sample.

In the *GL* approach, this effect is described by a boundary condition on ψ which implies that the superfluid current cannot escape from the sample. The appropriate boundary condition is a direct generalization of eq. (3.42). One can write for the interface

$$\frac{\partial \psi}{\partial n} - \frac{2ie}{\hbar} A \psi = \frac{1}{b} \psi \qquad \ldots(3.45)$$

where b corresponds to a length called the **extrapolation length** (Fig. 3.2). One finds several physical effects related to the existence of b, *e.g.*, the decrease of T_c of thin superconducting films deposited on the normal metal or disappearance of surface superconductivity below B_{c_3} under some specific conditions.

One obtains from microscopic theory for a superconductor-insulator interface

$$b \sim \frac{\xi_0^2}{a_0} \qquad \ldots(3.46)$$

At a superconductor-normal metal interface in the limit $\xi_0 \ll l$ (called clean limit), one obtains

$$b \sim \xi_n = \frac{\hbar v_F}{k_s T} \qquad \ldots(3.47)$$

while for dirty limit, *i.e.*, $\xi_0 \gg l$, one obtains

$$b \sim \xi_n = \sqrt{\frac{\hbar v_F l}{6\pi k_B T}} \qquad \ldots(3.48)$$

The coherence length in conventional superconductors is much larger than the interatomic distance ($\xi_0 \gg a_0$). For superconductors deposited on an insulator, b is very large and therefore the superconductivity is not affected.

When $\xi_0 \sim a_0$, *i.e.*, the coherence.

We can see that the existence of B_{c_3} in the aforementioned geometry has a direct consequence on the interpretation of various experiments. When the magnetic field is parallel to the length of the sample, one finds that the resistivity measurements will give B_{c_3} : the resistivity is zero below that field. The supercurrent passes within the superconducting layer near the surface. When one measures the magnetization, one obtains B_{c_2} as this experiment is sensitive to the state of the bulk of the sample and the flux is not being expelled until the magnetic field is reduced down to or below B_{c_2}.

Phenomenological Theory

The variation of B_{c_1}, B_{c_2} and B_{c_3} of type-II superconductor is shown in Fig. 3.1.

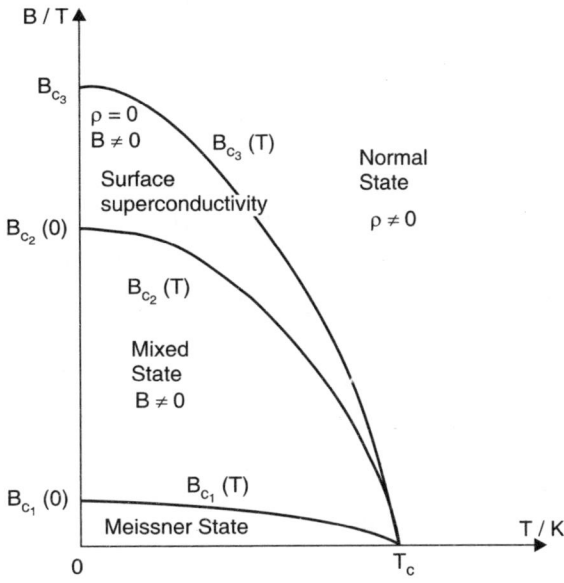

Figure 3.1 Variation of B_{c_1}, B_{c_2} and B_{c_3} of type-II superconductor with temperature

We must note that for certain type-I superconductors, the magnetic field B_{c_3} can be larger than B_c. In that particular case, one can observe surface superconductivity.

3.7.2 Proximity Effects

So far we have dealt with only individual superconductors. Let us consider the effects that result when a normal metal is deposited onto a superconductor (interface problem) the normal metal will alter the order parameter ψ close to the interface (Fig. 3.2) length is short, we can expect some important effects. HTSC cuprates provide a typical example. One can expect considerable proximity effects in films or in the bulk of new HTSC cuprates that contains some insulating grains.

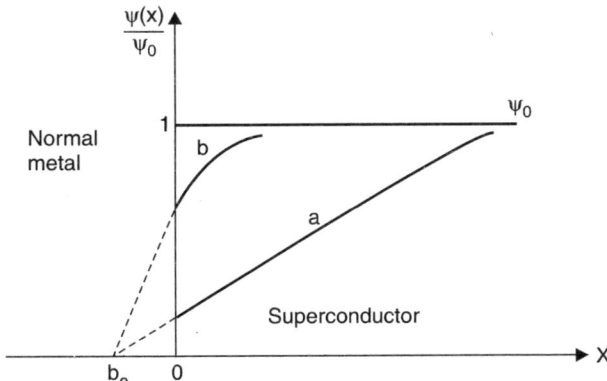

Figure 3.2 The order parameter at the interface between a normal metal and a superconductor: (a) near T_c and (b) at low temperatures

In the case of a superconductor normal metal interface, one can induce superconductivity in the normal metal within a sheath of thickness ξ_n. This length is termed as **normal coherence length.** We must note that the superconductivity is due to Cooper pairs which enter the normal metal.

REFERENCES

1. J.R. Schrieffer, *Theory of Superconductivity*, Benjamin/Cummings 1983.
2. C.J. Gorter and H.B.J. Casimir, *Physica* 1, 306 (1934).
3. F. London, Nature (London) 141, 643 (1938), *Superfluids*, I and II (Dover, New York, 1964).
4. F. and H. London, *Proc. Roy. Soc.* (London) A 149, 71 (1935).
5. A. B. Pippard, *Proc. Roy. Soc.* (London) A 216, 547 (1953).
6. T.E. Faber and A.B. Pippard, *Proc. Roy. Soc.* (London) A 231, 336 (1955).
7. V.L. Ginzburg and L.D. Landau, *Zh. Eksperim, i Teor Fiz.* 20, 1064 (1950).
8. A.A. Abrikosov, *Zh. Eksperim. i Teor. Fiz.* 32, 1442 (1957) [English Translation Sov. Phys. JETP 5, 1174 (1957)].
9. J.Bardeen and M.J. Stephan, *Phys. Rev.* 140, A 1197 (1965).

Critical Currents of Type-II Superconductors

4.1 INTRODUCTION

The critical currents of type-II superconductors are of considerable practical interest. We have mentioned previously that electromagnets capable of generating strong magnetic fields can be wound from wires of type-II superconductors, and clearly the more current that can be passed through the windings of such an electromagnet without resistance appearing the stronger will be the magnetic field that can be generated without heat being produced.

In chapter 2, we saw that, provided the specimen is considerably larger than the penetration depth, the critical current of a type-I superconductor is successfully predicted by Silsbee's hypothesis, *i.e.*, if the resistance is to remain zero, the total magnetic field strength at the surface, due to the current and applied magnetic field together, must not exceed B_c. The situation in type-II superconductors is, however, more complicated, because the state of the material changes at two field strengths, B_{c_1} and B_{c_2}, not at a single field strength B_c. In a magnetic field whose strength is less than B_{c_1} a type-II superconductor is in the completely superconducting state and behaves like a type-I superconductor, whereas in field strength greater than B_{c_1}, it goes into the *mixed state* (Fig. 4.1). This, so called, mixed state, also known as *Shubnikov phase* is required in most applications. The mixed state does not necessarily have zero resistance. Let us first describe the equilibrium state which is a triangular lattice of flux lines. We must note that no current can flow without dissipation in this ideal state, we discuss *metastable states* which are mainly caused by inhomogeneities in the material. These metastable states which lead to irreversible properties, are found to be necessary condition to

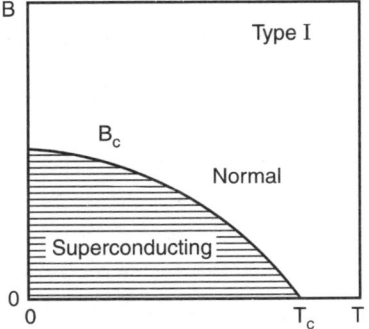 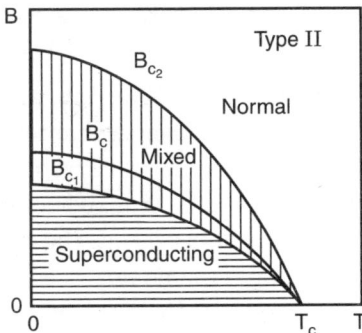

Fig. 4.1 Comparison of the phase diagrams for Type-I (left) and Type-II (right) superconductors

observe non-dissipative supercurrents in these materials. This leads to the concept of ciritical current and provides an opportunity for the study of different regimes obtained with a current flowing in the mixed state.

4.2 MIXED STATE

In a magnetic field whose strength is less than B_{c_1} a type-II superconductor is incompletely superconducting state and behaves like a type-I superconductor, whereas in field strengths greater than B_{c_1} it goes into the mixed state. Between B_{c_1} and B_{c_2} the flux partially penetrates into the sample in the form of flux tubes as shown in Fig. 4.2, called *Vortices*. We must note that the flux lines form a triangular lattice that one can directly observe.

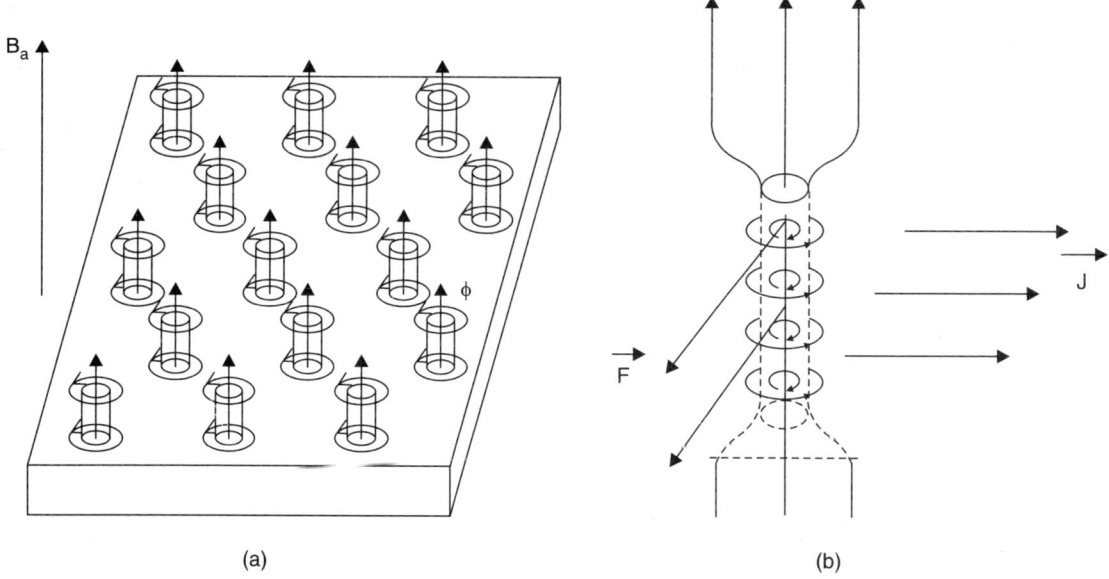

Fig. 4.2 (*a*) Type-II superconductor in the mixed state. Each vortex contains just a single fluxoid. Note that vortices form a hexagonal lattice.
(*b*) Lorentz force **F** on a flux line in the presence of the current **J**

This mixed state is a stable state and we have already seen the magnetization curve for an ideal type-II. Superconductor in chapter 2. However, what one requires for most applications is the highest possible transport current that can persist even in high magnetic fields. The maximum current that can be transported by type-I superconductors is very small; a relatively weak external magnetic field exceeds the critical field (B_c) thereby destroying superconductivity. Obviously, in most applications one uses type-II superconductors in which one can have high critical currents. However, the flow of current creates a magnetic field and vortices in the sample. We have already mentioned that the current flow in the mixed state is a rather complex problem.

Let us first obtain the ideal magnetization curve. We must note that if one makes a measurement of magnetization as a function of field one has difficulties in obtaining the predicted ideal behaviour. One will have to prepare samples carefully and thoroughly annealed. However, if one does cold working on such a 'perfect' sample, the behaviour becomes completely different (Fig. 4.3).

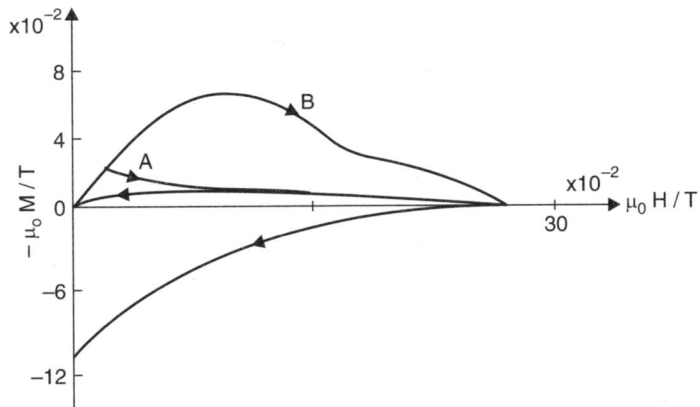

Fig. 4.3 Hysteresis in magnetization measurements: curve A represents the hysteresis cycle without annealing and curve B exhibit the same sample after cold working

Now, the magnetic field has difficulties to penetrate due to the inhomogeneties and defects. The magnetization at B_{c_1}, changes very little and hence it is difficult to measure this field B_{c_1}, the flux has difficulties to be expelled and stays trapped inside. One obtains, instead of a reversible magnetization curve a hysteresis cycle as in a permanent magnet.

We must note that it is this absence of reversibility which permits the existence of permanent magnets ; they are essentially in the *metastable state*. Similarly, we find that it is the hysteresis that permits one to have a current in the mixed state without dissipation. Obviously, if the system is perfect, the current will drive the vortices and cause dissipation. The possibility of having zero resistance in the mixed state is directly related to the possibility of hysteresis in the magnetization curve. *Bean model* or critical state model relate both effects. We must note that both effects are associated with the difficulties of having vortex motion within an inhomogeneous sample.

4.3 INTERACTION BETWEEN VORTICES

Let us calculate the energy of two vortices separated by the distance $\mathbf{r}_{12}$ where the materials have high values of κ. The vortices repel each other and the equilibrium condition for the vortex is that the total supercurrent vanishes at the core of the vortex.

In chapter 3, we have obtained an expression for the distribution $\mathbf{B}(\mathbf{r})$ of the field which minimizes the free energy as

$$\mathbf{B}(\mathbf{r}) + \lambda^2 \nabla \times \nabla \times \mathbf{B}(\mathbf{r}) = 0$$

This was modified as

$$\mathbf{B} + \lambda^2 \nabla \times \nabla \times \mathbf{B} = \frac{\phi_0}{\mu_0} \delta_2(\mathbf{r})$$

by two singularities which describe two vortices: one at the origin and the other at the point $\mathbf{r}_{12}$. We note that the equation is linear in B and hence one can show that the field created by the two vortices is just equal to the sum of the fields separately created by each vortex. As a result, one can easily show that the total energy U_t which is given by

$$U_v = \int d^3r \, \frac{\mu_0}{2} (B^2 + \lambda^2 J^2) \quad \text{(See ch. 3)}$$

is a sum of individual energies of the vortices, $2\,U_v$, plus an interaction energy term, i.e.,

$$U_t = 2U_v + U_{12} \qquad \qquad \text{...(4.1)}$$

where
$$U_{12} = \phi_0 \, B(\mathbf{r}_{12}) \qquad \qquad \text{...(4.2)}$$

Obviously, the interaction energy is equal to ϕ_0 multiplied by $B(\mathbf{r}_{12})$, the magnetic field created by one of the vortices and 'sensed' by the other. Since the vortices repel each other and hence this energy is positive. It is the magnetic pressure that forces them to enter the sample and compress themselves when the external field increases. We have the force exerted by vortex 1 on vortex 2 as

$$f_x = \frac{\partial U_{12}}{\partial x_2} = \phi_0 \, J_y \qquad \qquad \text{...(4.3)}$$

The force on vortex 2 is along the line that link the vortices and is proportional to the superfluid current created by vortex 1. We can see that this current is perpendicular to this very link and hence it is equivalent to the magnus force of a turbulence in hydrodynamics as illustrated in Fig. 4.2 (b). One finds that in an array of vortices the total force on one vortex vanishes because of the symmetry of the vortex lattice.

To study the thermodynamics of an ensemble of vortices in the sample, one will have to generalize the Gibb's energy.

$$\Delta G = B \left(\frac{U_v}{\phi_0} - H \right)$$

and take into account the repulsion between vortices

$$G_s(H) - G_s(0) = \frac{BH_v}{\phi_0} + \sum_{ij} U_{ij} - BH \qquad \qquad \text{...(4.4)}$$

Minimizing the Gibbs energy with respect to B, one obtains

$$B - B_{c_1} = \mu_0 \frac{\partial}{\partial B} \sum_{ij} U_{ij} \qquad \qquad \text{...(4.5)}$$

Using eq. (4.5), we can calculate the most stable lattice and the magnetization curve. One finds that the triangular lattice is the most stable one. In that case, one obtains the distance 'a' between two vortices given by

$$a = \left(\frac{2}{\sqrt{3}} \frac{\phi_0}{B}\right)^{1/2} \qquad ...(4.6)$$

Obviously, 'a' decreases with increasing magnetic field.

4.4 ABRIKOSOV LATTICE

Abrikosov predicted the penetration of flux in type-II superconductors through a lattice of flux lines. However, Abrikosov's analysis was for magnetic fields close to B_{c_2} where our presentation is not valid. Indeed, in that case vortices are very close and the order parameter is reduced everywhere. One has to use the full GL equations and calculate both the order parameter and the distribution of the field. Here we will restrict ourself to the results obtained by Abrikosov. The average value $<|\psi|^2>$ of the order parameter as a function of magnetic field vanishes linearly close to B_{c_2}.

$$<|\psi|^2> = |\psi_0|^2 \frac{1}{1.16} \frac{2\kappa^2}{2\kappa^2 - 1}\left(1 - \frac{B}{B_{c_2}}\right) \qquad ...(4.7)$$

One obtains the free energy as

$$F = \frac{B^2}{2\mu_0} - \frac{1}{2\mu_0} \frac{(B_{c_2} - B)^2}{1 + (2\kappa^2 - 1)1.16} \qquad ...(4.8)$$

we have
$$H = \frac{\partial F}{\partial B} \qquad ...(4.9)$$

and
$$M = \frac{B - \mu_0 H}{\mu_0} \qquad ...(4.10)$$

one finds
$$M = \frac{B_{c_2} - B}{1.16 \mu_0 (2\kappa^2 - 1)} \qquad ...(4.11)$$

For a given temperature, the magnetization vanishes at B_{c_2} as a linear function of the applied magnetic field. One can determine κ from the measurements of the slope of the magnetization vs. magnetic field close to B_{c_2}. This represents one of the few measurements of $\kappa = \lambda/\xi$. We know that κ is one of the fundamental quantities that characterize a given superconductor.

One can directly visualize the flux lattice by *magnetic decoration techniques*. One scatters finely divided ferromagnetic particles on the surface of a sample with a magnetic field perpendicular to the surface. These magnetic particles tend to accumulate on the flux line cores where they remain on warming up to room temperature. Subsequent examination under an electron microscope clearly shows the flux line lattice.

4.5 ANISOTROPIC TYPE-II SUPERCONDUCTORS

So far we have discussed the isotropic superconductors. The recently discovered HTSC cuprates are highly anisotropic. Now, we shall give here, the value of B_{c_1} and B_{c_2} for these anisotropic superconductors. As stated earlier, these HTSC cuprates have layered structures and hence one has to replace the electron mass by the *mass tensor* which has two principal values: m^c along the c-axis and m^{ab} in the ab-plane.

The mass along the c-axis is much larger than in the ab-plane. We introduce a small quantity,

$$\varepsilon = \sqrt{\frac{m^{ab}}{m^c}} \ll 1$$

This ratio is of the order 0.2 for $Y Ba_2Cu_3O_7$ HTSC cuprate. The *GL* equations introduce two coherence lengths:

$$\xi^c = \frac{h}{\sqrt{2m^c \alpha}}$$

and
$$\xi^{ab} = \frac{h}{\sqrt{2m^{ab} \alpha}} \qquad ...(4.12)$$

Obviously, $\xi^c \ll \xi^{ab}$ for the coherence lengths along the c-axis and in the ab-plane, respectively. We must note that the order parameter changes over a smaller length along the c-axis then in the ab-plane.

Let us introduce two penetration depths

$$\lambda^c = \sqrt{\frac{m^c}{\mu_0 \eta_s e^2}}$$

$$\lambda^{ab} = \sqrt{\frac{m^{ab}}{\mu_0 \eta_s e^2}} \qquad ...(4.13)$$

Obviously, in HTSC cuprates, we have

$$\varepsilon = \frac{\xi^c}{\xi^{ab}} = \frac{\lambda^{ab}}{\lambda^c} \qquad ...(4.14)$$

we can see that λ^{ab} is involved when the supercurrent shielding the field has to flow along the c-axis, whereas λ^{ab} when it flows in the ab-plane.

Now, by solving the anisotropic *GL* equations as a function of the angle θ (the angle between the c-axis and the magnetic field), one obtains

$$B_{c_2} = \frac{\phi_0}{2\pi (\xi^{ab})^2} (\cos^2 \theta + \varepsilon^2 \sin^2 \theta)^{-1/2} \qquad ...(4.15)$$

One finds the limiting values of B_{c_2} as

$$B_{c_2}^c = \frac{\phi_0}{2\pi (\xi^{ab})^2} \text{ for } B \parallel C \qquad ...(4.16)$$

$$B_{c_2}^{ab} = \frac{\phi_0}{2\pi \, \xi^{ab} \xi^c} \text{ for } B \perp C \qquad \ldots(4.17)$$

We note that the coherence lengths involved are the coherence lengths in the plane perpendicular to the magnetic field. The critical field parallel to the layer is much higher than the magnetic field perpendicular to it. We have

$$\frac{B_{c_2}^c}{B_{c_2}^{ab}} = \frac{\xi^c}{\xi^{ab}} = \varepsilon \qquad \ldots(4.18)$$

We must note that the triangles of the Abrikosov lattice are no longer equilateral except in the case where the field is parallel to the c-axis. One obtains isosceles triangles in all other cases. For example, if the field is applied along the b-axis, the ratio of the height of the triangle d_a to the distance d_c between vertices along the c-axis is

$$\frac{d_a}{d_c} \sim \frac{\lambda^c}{\lambda^{ab}}$$

one obtains,

$$B_{c_1}^c = \frac{\phi_0}{2\pi \, (\lambda^{ab})^2} \ln \frac{\lambda^{ab}}{\xi^{ab}} \text{ for } B \parallel C$$

$$B_{c_1}^{ab} = \frac{\phi_0}{2\pi \, \lambda^{ab} \lambda^c} \ln \left(\frac{\lambda^{ab} \lambda^c}{\xi^{ab} \xi^c} \right) \text{ for } B \parallel ab \qquad \ldots(4.19)$$

One has the inequality $B_{c_1}^{ab} \ll B_{c_1}^c$.

Obviously, the vortex lines exhibit a trend towards lying in the layer plane where they are formed at relatively small fields. One obtains, for the angle θ,

$$B_{c_1}^c (\theta) = \left[\frac{\cos^2 \theta}{(B_{c_1}^c)^2} + \frac{\sin^2 \theta}{(B_{c_1}^{ab})^2} \right] = 1 \qquad \ldots(4.20)$$

One can easily understand the qualitative behaviour of B_{c_2} within the *Lawrence-Doniach model*. If the field is parallel to the c-axis, $B_{c_2}^c$ is unaltered, *i.e.*, remains the same as in the *GL* model. For this particular orientation of the field, one finds that the electrons move within the layers and the phase of the order parameter does not change from one layer to another in HTSC cuprates; therefore the Josephson contribution to the energy does not enter the calculation.

However, the above is not true where the field is parallel to the ab plane, *i.e.*, some current will flow between the layers and the phase difference in the order parameter will introduce an energy variation between the layers. Close to T_c, one finds that the model is reduced to the anisotropic *GL* equation.

For the range of temperatures that satisfy the following condition

$$\frac{T_c - T}{T_c} > 2 \left(\frac{\xi_0}{s} \right)^2 . \qquad \ldots(4.20(a))$$

the magnetic field will penetrate between the CuO$_2$ layers of HTSC cuprates. As λ is large compared with the thickness of the superconducting layers, the magnetic field will uniformly penetrate without any cost in magnetic energy. Obviously, in this simple model, a thin superconducting film whose thickness is very small compared with λ could support an infinite parallel field without destruction of the superconducting state. This holds only because in

$$G_n(T, B_c) = G_n(T, 0)$$

we have neglected the decrease of the normal state Gibbs energy with magnetic field in a paramagnetic field. One can easily show that in this case of HTSC cuprates the critical field is given by what is called the *paramagnetic Clogston limit*. One obtains the field B_p as

$$B_p = \frac{\Delta}{\sqrt{2}\,\mu_B} \qquad \ldots(4.20(b))$$

More precisely, one obtains the upper critical field $B_{C_2}^{ab}$ as

$$B_{c_2}^{ab} = B_p \sqrt{\frac{T_c - T}{T_c}} \qquad \ldots(4.20(c))$$

for
$$\frac{T_c - T}{T_c} > 2\left(\frac{\xi_0}{s}\right)^2$$

This change of behaviour between the anisotropic *GL* model close to T_c and the two-dimensional (2D) behaviour at low temperatures, where the superconducting CuO$_2$ layers are more or less decoupled and behave as independent superconducting planes, can be very easily seen from the curve for $B_{c_2}(\theta)$, where θ is the angle between the c-axis and the magnetic field.

Close to T_c and for $\theta = \pi/2$, i.e., for a field close to the *ab*-plane, one obtains the smooth behaviour. If the layers are decoupled, one obtains a cusp behaviour for $B_C(0)$ around $\theta = \pi/2$, just one finds in the case of very thin films.

This permits one to determine the characteristic temperature T^* below which the *Lawrence-Doniach model* gives more appropriate description of cuprate layered HTSC compounds than the *GL* model. Let us estimate the temperature T^* for Bi$_2$ Sr$_2$ CaCu$_2$O$_{8-\delta}$ by taking $T_c = 85\ K$, $\xi_0 = 1$ Å and $s = 15$ Å. One obtains

$$T_c - T^* = 0.76\ \text{K}.$$

This shows that Bi – 2212 compound behaves very much as a 2D superconductor except when very close to T_c and hence one can approximate it as a superconductor-insulator multilayer structure with weak Josephson coupling between the superconducting planes.

4.6 IRREVERSIBLE PROPERTIES: METASTABLE STATES

We have mentioned that it is difficult to obtain the ideal magnetization curve. As soon as the material has defects, one obtains a hysteresis curve. This fact reveals that vortices have difficulties moving, either to enter or to leave the sample. This difficulty to move vortices permits the sample to sustain a current without dissipation. Indeed, one finds that a current drives vortex perpendicularly to the direction of the current. If the vortex moves, it means a local change of flux appears and through Maxwell's electromagnetic equations, an electric field appears. Obviously, when we have both an electric field and the current we have *dissipation*.

If the superconductor is an ideal superconductor, then any small current will create dissipation as there is nothing to prevent the motion of the lattice of vortices. The critical current would be the current which creates a field H_{C_1} at the surface of the sample, obviously, a very small current. Let d be the dimension of the sample, then we have

$$J \sim \frac{H_{C_1}}{d} \qquad \ldots(4.21)$$

Let us now consider a situation when we have a strong hysteresis, *i.e.*, the vortices have difficulties in moving. This means that a current will flow without moving them if the force exerted on the vortices is not strong enough to set them into motion. Thus a superconductor will support a current without dissipation. This means that the dissipation will occur only when the force exerted on the vortices is strong enough to overcome the barrier which prevents vortices to move, *i.e.*, there will be a critical current density J_c for the material. We must note that this J_C is of course very dependent on the irreversibility of the sample and zero in an ideal sample. Obviously, J_C can be increased by increasing the irreversibility.

4.6.1 Depairing Critical Current

Let us imagine a situation when one were able to completely prevent the motion of vortices. In such a situation, there is an intrinsic critical current that a superconductor can support which is called the depairing critical current. Obviously, it is the current which destroys pairs and thus superconductivity. One can understand this with the same reasoning that shows the existence of the critical magnetic field. The gain in energy per unit volume over the normal state, in the superconducting is $\frac{1}{2\mu_0} B_c^2$. If one puts a superconducting sample in a situation which increases its energy by $\frac{1}{2\mu_0} B_c^2$, superconducting sample will go back to the normal state. If one introduces currents in the superconducting state, then the energy increases by a value given by the kinetic energy of the current. In accordance with the London model, one obtains for the energy of the current :

$$n_s \frac{1}{2} m \mathbf{V}_s^2 = \frac{1}{2} \frac{m}{e^2 n_s} \mathbf{J}^2 = \frac{1}{2} \mu_0 \lambda^2 \mathbf{J}^2 \qquad \ldots(4.22)$$

The above energy has to be smaller than the energy gained in the superconducting state. Obviously, the current has to be smaller than J_{CL}, where

$$J_{CL} = H_c/\lambda = B_c/\mu_0 \lambda \qquad \ldots(4.23)$$

One can obtain the same depairing current by applying *GL* equations (except for a different numerical factor),

$$J_{CGL} = \frac{2\sqrt{2}}{3\sqrt{3}} \frac{H_c}{\lambda} = \frac{2\sqrt{2}}{3\sqrt{3}} \frac{B_c}{\mu_0 \lambda} \qquad \ldots(4.24)$$

We must note that one obtains the same relation for J_c by *GL* and *BCS* approaches except different numerical prefactors.

Obviously, the maximum current that can theoretically be sustained in a superconductor is of the order $B_c/\mu_0\lambda$.

We have in a conventional superconductor $B_C \sim 0.1\ T$ and $\lambda \sim 1000$ Å. Thus, we obtain

$$J_c \sim 10^{12}\ A - m^{-2} = 10^8\ A/cm^2.$$

This is the maximum critical current that one can expect in a conventional type-II superconductors. In real materials, the actual maximum current is $\sim 10^6 - 10^7$ A/cm^2 which is one or two orders of magnitude lower than the depairing current.

4.6.2 Different Regimes

We have already seen that an isolated vortex in a current **J** is subjected to a force per unit volume, often called the *Lorentz force*:

$$\mathbf{f} = \mathbf{J} \wedge \phi_0 \qquad \ldots(4.25)$$

Obviously, the lattice is subjected to a force density per unit volume,

$$\mathbf{F} = \mathbf{J} \wedge n\phi_0 = \mathbf{J} \wedge \mathbf{B} \qquad \ldots(4.26)$$

This force **F** tends to set in motion the lattice of flux lines. If the vortices can move freely, it is not possible to pass a current above B_{c_1} without energy dissipation. Obviously, to achieve a finite critical current, we have to pin the vortices, *i.e.*, to find mechanisms or geometries which prevent flux lines motion. Let F_p is the **average pinning force density** which prevents the lattice moving. For $F < F_p$, the lattice will not move and one has a non-dissipative current. For $F > F_p$ the lattice will move and one will have the flux flow regime. When $F = F_p$, one obtains the critical regime. Let us define J_c by

$$F_p = J_c B, \qquad \ldots(4.27)$$

then J_c is the critical current of the material at zero temperature. We must note that for $J < J_c$ there is no dissipation, whereas for $J > J_c$ there is dissipation. Now, we are interested in calculating F_p that we have introduced under the term of *average pinning force density*.

The pinning of vortices in conventional superconductors usually occurs at the inhomogeneities in the material: due to the local variation of ξ or λ the energy of the flux tube change accordingly and the vortex gets pinned to the energetically more favourable sites. From efficiency point of view, the inhomogeneities have to be of the order of ξ or λ, *i.e.*, 100 to 1000 Å in conventional superconductors. The contrary seems to be true for HTSC cuprates where $\xi \sim 10$ Å. The pinning problem in HTSC cuprates is *discussed latter*. Now, we describe briefly the essential features of different regimes, *e.g.*, (*i*) Flux flow (*ii*) hysteresis (*iii*) Flux creep and (*iv*) Taff. These regimes are illustrated in Fig. 4.4.

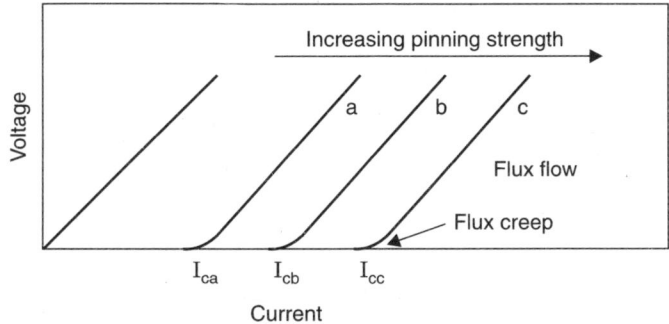

Fig. 4.4 Current-voltage characteristics for type-II superconducting two samples with different concentration of defects; critical currents are also different

4.7 FLUX FLOW

This is a dissipating regime where the flux line lattice moves and dissipation is nearly ohmic. When the current exceeds the critical current, i.e., $J > J_c$, flux flow is the dominant regime in conventional superconductors. The flux flow resistivity is linear in B.

Let us calculate the resistivity in the presence of a current larger than the critical current. We take the limit where the Lorentz force is large compared to the pinning force, so one can neglect the latter. We know that the motion of a flux line dissipates energy so we can describe this dissipation in terms of viscosity. Assuming that the vortex motion be damped by a force proportional to the velocity ($\mathbf{V}$), one can calculate the drift velocity of the vortex line as a function of this viscosity, η, by equating the Lorentz force to the friction force,

$$\mathbf{J}\phi_0 = \eta \mathbf{V} \qquad \ldots(4.28)$$

The motion of the lattice induces an electric field $\mathbf{E}$ parallel to $\mathbf{J}$, given by Maxwell equation

$$\mathbf{E} = \mathbf{B} \wedge \mathbf{V} \qquad \ldots(4.29)$$

Using these two relations, we obtain the resistance of the flux flow as

$$\rho_f = \frac{E}{J} = B\frac{\phi_0}{\eta} \qquad \ldots(4.30)$$

Here the viscocity is an unknown parameter. Experimentally, it is found that it does not depend on the current and does not vary very much with B. Obviously, we obtain an ohmic regime and the resistivity is linear in B. The microscopic calculation gives information about η and a simple interpretation of this ohmic regime as it leads to

$$\rho_f = \rho_n \frac{B}{B_{c_2}} \qquad \ldots(4.31)$$

Obviously, one obtains η in terms of resistivity in the normal state as

$$\eta = \frac{\phi_0 B_{c_2}}{\rho_n} \qquad \ldots(4.32)$$

We note that the resistivity in the flux flow regime is the same as that obtained for currents flowing inside the normal cores. In fact, if one considers that the vortex is a cylinder of normal metal with radius ξ then B/B_{c_2} represents the fraction of the normal metal. Flux flow resistivity as a function of field for different temperatures for a Nb–Ta sample is shown in Fig. 4.5.

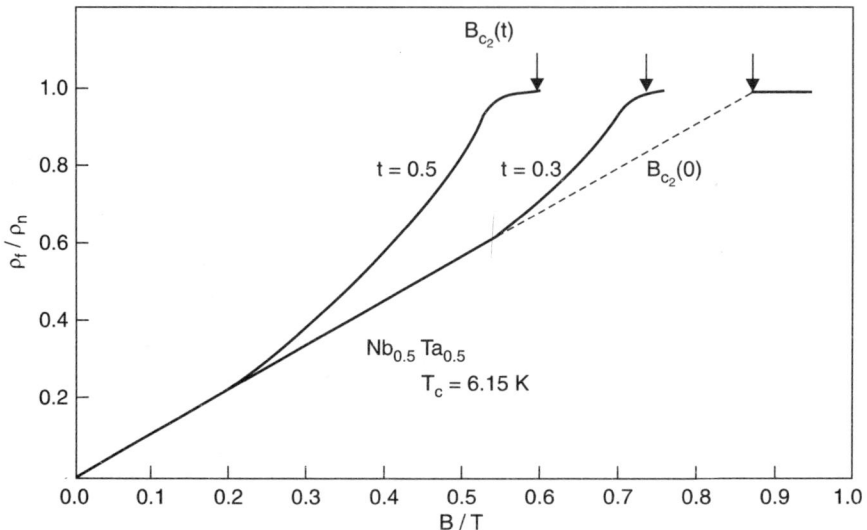

Fig. 4.5 Flux flow resistivity as a function of B for different values of $t = T/T_c$ for a Nb – Ta sample. The resistance increases as a function of the magnetic field and the temperature. The expected behaviour for $t = 0$ is shown by dotted line. Value of $B_{C_2}(t)$ is indicated by vertical pointers

4.8 HYSTERESIS CYCLE

We have already shown the magnetization curves as a function of the field in Fig. 4.3. When there are some inhomogeneities that perturb the vortex motion, one finds that the magnetization curves are irreversible.

Metallurgical defects are created by cold drawing of metals, which are usually the pinning centers in conventional superconductors. We see that the corresponding magnetization curve is strongly irreversible. We can carry out the sequence of annealing of the sample which reduces the number of defects and consequently the irreversibility of magnetization. We must note that the flux does not penetrate at B_{c_1}, it remains pinned to the surface so the magnetization curve only gradually deviates from the straight line of perfect diamagnetism. On the contrary, one finds that the magnetization always disappears at B_{c_2} independently of the irreversibility and the pinning forces always disappear at B_{c_2}. Even when the magnetic field is lowered below B_{c_2} the flux remains trapped within the specimen and one finds that it may exhibit paramagnetic ($B > \mu_0 H$), rather than diamagnetic, response.

The mechanism of pinning is not well understood but all the structural features of the size of the vortex are effective pinning centers. We can create them in a controlled manner either by cold working of the metal, or by precipitating a second phase within the superconductor, or by irradiating the sample. The pinning increases with the number of inhomogeneities in conventional superconductors.

Bean developed a simple model to interpret the irreversibility of the magnetization curves. Bean's model allows one to deduce the critical current of the sample. With the help of this model, one can estimate the critical current by measuring only the magnetization of the specimen. The force on the vortex is given by

$$\mathbf{f} = \mathbf{J}_{ext} \wedge \phi_0 \qquad ...(4.33)$$

Using eq. (4.26), one obtains the equation for the force per unit volume

$$\mathbf{F} = \nabla \times \mathbf{H} \wedge \mathbf{B} \qquad ...(4.34)$$

One can understand eq. (4.34) by considering one-dimensional case where H varies uniquely in one direction x:

$$F = \frac{dH}{dx} B \qquad ...(4.35)$$

Obviously, the force is related to the gradient of the field. Moreover, we find that it makes the field uniform in the sample.

Let us consider that F_p be the pinning force per unit volume. Equilibrium will not be achieved if $F > F_p$. The vortices will continue their motion until $F = F_p$ at all points. This physical regime is named as the *critical state*. The variation of the field in the interior of the sample is directly linked to the current that the sample can sustain without vortex motion, we have

$$F_p = J_c B = \frac{dH}{dx} B \qquad ...(4.36)$$

We must note that the Bean model, in its simplest form supposes that the effect of vortex pinning is to determine the maximum gradient of the magnetic field. The gradient is equal to the critical current, i.e., $J_c = \dfrac{dH}{dx}$.

To illustrate the Bean model, we consider the magnetization of a superconducting film of thickness d in the presence of a field parallel to its surface as shown in Fig. 4.6. Above B_{C_1}, the magnetic field will begin to penetrate into the film up to a depth Δ determined by the following critical condition

$$\frac{dH}{dx} = J_c$$

or
$$\Delta = \frac{H}{J_c} \qquad ...(4.37)$$

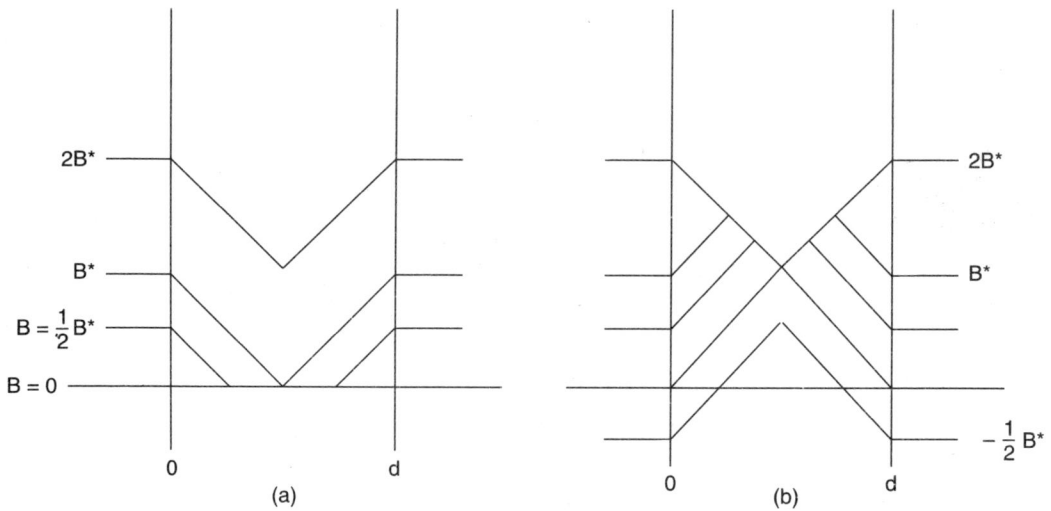

Fig. 4.6 Profile of the magnetic field in the interior of a superconducting sample (a) for increasing magnetic field and (b) for decreasing magnetic field

We find that above $B^* = \frac{1}{2} \mu_0 J_c d$ the magnetic field penetrates into the total depth of the film in accordance with the profile shown in Fig. 4.6 (a). We must note that B^* is the maximal external field which is completely screened in the middle of the superconductor. On reducing the applied field from $B_m \gg B^*$, one obtained the situation shown in Fig. 4.6 (b). When one brought the external magnetic field zero, a considerable quantity of flux still remains trapped in the interior of the sample. Unless one applies an inverse external field $\frac{1}{2} B^*$, the flux within the sample will not disappear. This causes hysteresis. If the external magnetic field is cycled with a maximum value $B_m < B^*$, we can easily calculate the energy dissipated during the cycle which is proportional to $(B_m)^3$. If we do cycling with $B_m \gg B^*$, we obtain the dissipated energy proportional to B_m. These losses due to hysteresis limit the potential use of type-II superconducting materials in ac applications.

It is assumed in Bean model that J_c is independent of B. However, J_c has to be zero at B_{c_2} where the pinning forces disappear. We have to take into consideration the following classical approximation,

$$J_c = \frac{\alpha}{B + B_0} \quad \quad ...(4.38)$$

where α and B_0 are adjustable parameters.

Bean model also help to measure the critical current. One can easily deduce $J_c(H)$ from the hysteresis cycle. The usual way is to take the difference $\Delta M(H)$, between the magnetizations measured in increasing and decreasing magnetic fields ($M \uparrow - M \downarrow$) as illustrated in Fig. 4.7. One can calculate the variation of magnetization using the Bean model as illustrated in Fig. 4.8. It is found to be proportional to the area shown in Fig. 4.8. Thus, one obtains

Critical Currents of Type-II Superconductors

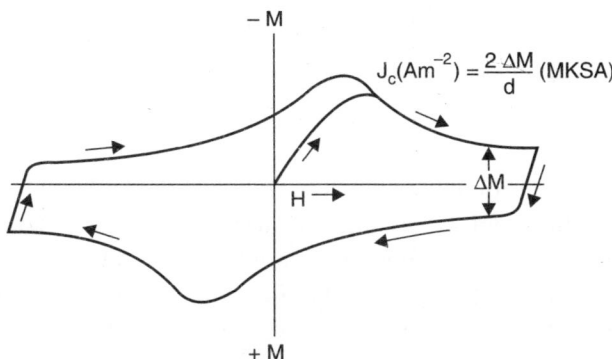

Fig. 4.7 Typical magnetization curve for type-II superconducting materials. The hysteresis ΔM in a given magnetic field H provides the measurement of the critical current J_c

$$J_c\,(A/m^2) = 2\,\frac{\Delta M}{d} \quad \text{(in MKSA)} \quad \ldots(4.39)$$

where d denotes the sample thickness, or the grain size for the case of polycrystalline ceramic specimens.

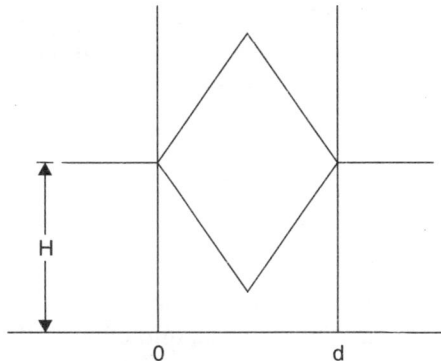

Fig. 4.8 Illustration of the modification of the profile of the magnetic field for a given external magnetic field for increasing and decreasing magnetic fields. One finds that the variation of the magnetization of the film is proportional to the area of the parallelopiped

4.9 PINNING OF FLUX VORTICES

We have mentioned earlier that all real superconductors contain a variety of defects, or **pinning centers,** which tend to pin the flux vortices. As a result, no flux flow takes place until the applied current exceeds a certain critical current density J_c, above which the Lorence force, **F** exceeds a characteristic pinning force $\mathbf{F}_P$. The pinning force F_P is related to the measured critical current by eq. (4.27):

$$F_P = J_c B$$

However, the above relation does not say anything about the mechanism of pinning nor how to increase the value of the critical current J_c. Moreover, J_c is not related to the pinning force of an individual vortex f_p. Obviously, to increase F_P, we have first to make the link with the pinning force on an individual vortex f_p and then study the mechanism which can pin an individual vortex line.

Let us assume random inhomogeneities in the bulk of the sample, which we call pinning points. One can describe these objects by an interaction potential with a vortex line. We must note that the pinning points are inhomogeneities which can either favour or inhibit the pair condensation responsible for superconductivity, i.e., which repel or attract flux lines. Let us first study the way these point interactions between flux lines and pinning points add to determine the pinning force density F_P. We write

$$F_P = N f_p \qquad \text{...(4.40)}$$

where N is the number of interactions between pinning points and vortices. However, this is not valid in general. Indeed, one must consider a rigid lattice of vortices and random inhomogeneities. One would have no pinning at all. The reason for this stems from the fact that the pinning forces are randomly oriented and stastically cancelled. We can also understand this result by considering the fact that the interaction energy in an infinite medium would be independent of the relative position of the rigid lattice of flux lines and the random array of pinning centers. As pinning effectively occurs an explication is needed. It is essential to pinning processes that the lattice be deformed. One may see that in this case, the total energy of the system is lowered by deformation of the lattice and pinning may occur if an energy increase is required to move the lattice with respect to the pinning array. One may also mention the opposite limit, i.e., if one has no lattice but a liquid of vortices. In that case, one have also $J_c = 0$ because it is needed to pin each vortex in order to prevent it from moving. Obviously, the lattice stiffness is central to the pinning problem. It is necessary to be able to describe the rigidity of the lattice.

Let us try to understand the effects of the pinning centers. For this purpose, we consider an ideal flux line lattice and a random distribution of pinning centers whose pin strength gradually increases from zero. The lattice will respond elastically for very small pinning force. Using the theory of elasticity of the flux line lattice, one can calculate the displacement of each vortex and the energy of the lattice. However, if one increases the pinning force, one will lose the positional long range order of the lattice. However, the stronger pinning will create dislocations and other defects in the lattice. In all cases of interest, it is generally assumed that the flux line lattice can be split into elastically independent, correlated regions of volume V_c. One can obtain the critical current for a Lorentz force which can set up in motion, this correlated region. The pinning force in this region is random and the net effect is obtained by the fluctuations in this region. If n_v be the pinning centers per unit volume, then this force is given by

$$[n_v V_c <f_p^2>]^{1/2} \qquad \text{...(4.41)}$$

One obtains the pinning force density by dividing the above force by V_c,

$$F_p = \left[\frac{n_v}{V_c} <f_p^2> \right]^{1/2} \qquad \text{...(4.42)}$$

From eq. (4.42), we note that only the square of the pinning force is involved. Obviously, this is an important result. This shows that not only the attractive but repulsive forces can also pin an array of vortices:

We must note that the result depends not only on calculation of the microscopic pinning force f_p but also on the volume V_c of the correlated region. The calculation of V_c is very complicated due to its dependence on the elastic constants of the flux line lattice as well as on the strength of the pinning center. We must note that the pinning force increases with decreasing volume V_c.

Fig. 4.9 illustrates flux pinning at defects. The hexagonal flux line lattice may be slightly distorted due to the interaction of vortex lines and pinning centers. In addition to normal precipitates, grain boundaries and dislocations can act as pinning centres. Because of extremely short coherence length ξ in HTSCs, even point defects are weak pinning centres in these materials. Flux pinning leads to irreversible magnetization curves. This effect is similar to diffusion in solids. Thermally activated flux creep is especially important at elevated temperatures in the HTSCs.

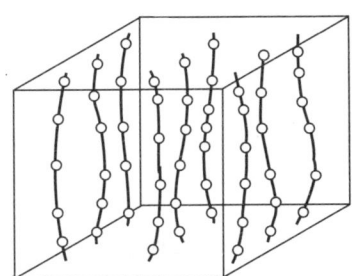

Fig. 4.9 Flux lines pinned at defects in a Type-II superconductor. As a consequence of the interaction of the flux lines and pinning centres the hexagonal lattice of the vortices is slightly distorted

4.9.1 Elasticity of Flux Line Lattice

In Voigt's notation of elastic theory, we have only three moduli C_{11}, C_{66} and C_{44} for a complete description of the elastic properties. C_{11}, the compressional modulus represents the deformation that changes only the size of the lattice parameter and not its shape. In comparison with the shear modulus C_{66} and the tilt modulus C_{44} it is large and it is therefore often neglected in the analysis of deformation of the lattice. C_{44}, the tilt modulus, describes deformation that tilts a bundle of flux lines array from the field direction while leaving its cross section in the xy-plane constant. C_{66} is the shear modulus in a plane perpendicular to the field.

One cannot describe short-scale distortion of the flux line lattice by elasticity theory, since the magnetic field does not change over distances smaller than the penetration depth. As a result, one obtains a wave number dependence of the elastic moduli.

Few important results are: (*i*) the compressional modulus and the tilt modulus, C_{44} are finite and found even increasing upto B_{c_2}. (*ii*) on the contrary the shear modulus, C_{66} vanishes at B_{c_2}. Obviously, this indicates a tendency of the lattice to develop fluid-like behaviour near B_{c_2}, instead of forming a regular lattice.

4.9.2 Correlated Volume

In the calculation of the pinning force density F_p, the first important parameter is the estimation of the volume within which the vortex lattice is almost regular. One finds that the lattice disorder is different along the z-axis, i.e., ; the field direction and in the xy-plane. Let L_c be the length along the flux line and R_c be the length in the plane, then we have

$$V_c = L_c R_c^2 \qquad \ldots(4.43)$$

In most cases, the ratio L_c/R_c a function of the ratio of the tilt and shear moduli

$$\frac{L_c}{R_c} \propto \sqrt{\frac{C_{44}}{C_{66}}} \qquad \ldots(4.44)$$

We have $L_c \gg R_c$ in conventional superconductors. Disorder is much more difficult along the field direction than perpendicular to it. In thin films with the magnetic field perpendicular to the film, L_c is often larger than the thickness of the film and the flux line lattice remains ordered along the field direction, the flux lines are straight lines.

In strongly disordered case, R_c is of the order of the lattice spacing as we are having a nearly amorphous lattice. The simplest approximation for L_c is to make it equal to the coherence length. The validity of this approximation is not known.

4.9.3 Elementary Pinning Force

The subject of the origin of the inhomogeneities in the order parameter leading to pinning is very complex. This involves point defects and such large scale defects as grain boundaries, twin boundaries, dislocations, precipitates, etc. For superconductors of technological applications, the important classes of defects are grain boundaries as in A–15 and Chevrel phase compounds or in Nb–N, and precipitates as in Nb–Ti.

One finds that the crystal defects give rise to local variations of the superconducting order parameter. These local changes couple to the flux lines lattice. One can distinguish between magnetic and core interactions in the first classification of elementary interactions.

The effects of surfaces parallel to the applied magnetic field falls in the first class. For e.g., the current distribution around a vortex core near the surface is forced to change in order that the normal component of the superfluid current vanishes at the boundary. Theoretically, one can achieve this by assuming an anti vortex image at the other side of the boundary. This results in an attractive surface vortex interaction. One finds that the net effect is a potential barrier for flux entry or exit. The second example of this magnetic interaction is the thickness variation of thin film for a field perpendicular to the surface. We must note that the vortices are pinned at points of smallest thickness, where the line energy is minimum. The penetration depth λ serves as a typical length for this class. The materials for which κ (Ginzburg–Landau parameter) is large, this kind of interaction is small and disappears with increasing magnetic field.

The coupling to the change of the order parameter is the origin of flux pinning for most defects in the *second class*. Defects are regions which deviate from the surrounding material for certain properties, e.g., elasticity, density, electron–phonon coupling, mean free path. The first three properties lead to local changes of T_c, the last to a local change of κ. These are often referred to as δT_c or $\delta \kappa$ pinning. The typical length scale is now the coherence length.

In most cases, it is found that the pinning force is proportional to Δ^2 as the condensation energy is given as a function of the square of the order parameter. In the vicinity of B_{c_2}, one obtains

$$\Delta^2 \sim 1 - \frac{B}{B_{c_2}} \qquad \ldots(4.45)$$

We must note that the pinning force always vanishes at B_{c_2}, and in most cases it vanishes linearly with the magnetic field. In that case, the pinning force is a fraction of the condensation energy divided by a typical length.

4.10 FLUX CREEP

Flux line motion is due to the Lorentz force density which acts perpendicular to the flux lines. The flux lines are pinned by defects which lead to a distortion of the flux line lattice even for no external current. At zero temperature, flux line motion is only possible if the Lorentz force density exceeds the average pinning force density. If the current density $J_{ex} < J_c(O, B)$, then J_{ex} is non dissipative.

For $J_{ex} > J_c$, one finds that flux line motion with velocity v leads to an electric field and hence to a finite voltage V. This is the flux flow regime. Now, the question arises, what happens at finite temperature? There is a finite probability that the flux lines overcome the pinning energy barriers at a finite temperature. Thus, even for $J_{ex} < J_c(T, B)$ one can have some motion and resistivity. This is termed as **thermally activated flux creep**. When $J_{ex} \gg J_c(T, B)$, we have the flux flow as mentioned.

Flux creep depends on various parameters which are not well known. However, one can make a qualitative analysis of the flux creep phenomenon in a simple way. In fact in most cases, one observes the motion of bundles of flux lines. We can say that this is a thermally activated jump from one location to another one nearby. In the simplest case the activation (or barrier) free energy $U_0(T, B)$ which is overcome by the jump of a flux line bundle is determined by the condensation energy density gained by the flux lines in a favourable region multiplied by a suitable volume.

We can see that at zero temperature, the barrier energy $U_0(O, B)$ is connected to the critical density $J_c(O, B)$. Since for $J_{ext} = J_c$ the Lorentz force density has to be equal to the pinning force density, one obtains

$$U_0(O, B) = J_c(O, B) B V_c d \qquad \ldots(4.46)$$

where d is the distance over which the flux bundle moves. d is of the order a_0.

In conventional superconductors, flux creep is a very small phenomenon, because one always has $U_0(T, B) \gg k_B T$ for all temperatures $T \leq T_c$ and $U_0/k_B T$ ranges approximately between 100 and 1000 K.

Energy of the bundle of flux lines as a function of position is shown in Fig. 4.10. There is a possibility for a bundle to jump over the barrier to an adjacent pinning site, *i.e.*, an adjacent potential will due to thermal fluctuations. Arrhenius law hints at this possibility. When one applies external field J_{ex}, the flux bundle will move with a velocity given by

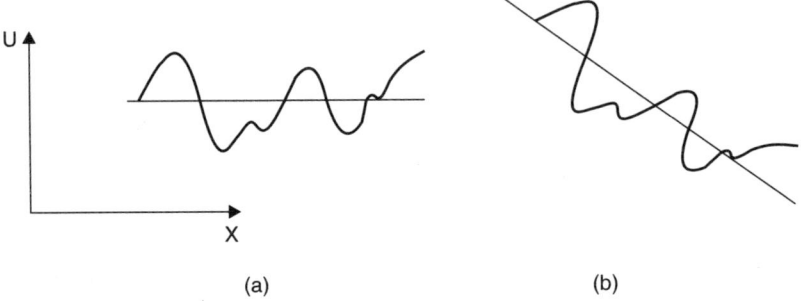

Fig. 4.10 Energy of bundle of flux lines as a function of position :
(*a*) without current and (*b*) with current

$$v = 2v_0 \exp(-U/k_B T) \sinh \frac{B J_{ex} V_c d}{k_B T}$$

$$\sim v_0 \exp(-1/k_B T)(U_0 - B J_{ex} V_c d) \qquad \ldots(4.47)$$

Here v_0 is a microscopic velocity proportional to the attempt frequency to jump over the barrier. In conventional superconductors, it may be in the range of 10 m/s. The flux creep velocity v generates an electric field $E = Bv$, hence we can observe resistivity. Let E_c, is the smallest electric field that can be measured, say 1 μV/cm, as long as the flux creep velocity v does not give rise to this field, one does not observe dissipation. One obtains the critical current for E_c by

$$E_c = Bv = BV_0 \exp(-1/k_B T)(U_0 - B J_c(T) V_c d), \qquad \ldots(4.48)$$

or
$$J_c = J_c(0)\left(1 - \frac{k_B T}{U_0} \ln \frac{B v_0}{E_c}\right) \qquad \ldots(4.49)$$

where $J_c(0)$ is given by eq. (4.46).

It is observed that in conventional superconductors, flux creep occurs only when the current is large enough to nearly overcome the barrier, *i.e.*, when the following condition holds.

$$\frac{J_c - J}{J_c} \ll 1 \qquad \ldots(4.50)$$

In contrast to the conventional superconductors, the characteristic features of the magnetic behaviour of HTSC cuprates is their large and non-exponential time-relaxation, or **"giant flux creep"** [1]. In HTSC cuprates, the ratio $U_0(B, T)/k_B T$ is considerably smaller than in conventional superconductors, because U_0 is small. Moreover the temperature of HTSC cuprates is higher than in conventional superconductors. This is why flux creep is very high in HTSC cuprates.

As stated earlier, the flux creep is a phenomenon that occurs when the driving force is almost equal to the pinning force, *i.e.*, just below the flux flow regime. However, if the pinning barrier is small and the temperature high enough to overcome the barrier, flux creep can be observed in the limit of small driving forces. In order to distinguish, the functional flux creep which occurs in accordance with the condition (4.50) from this effect, giant flux creep occurs for $J \ll J_c$, a new name has been assigned: the **thermally activated flux flow** (TAFF). We must note that TAFF is observable only when the potential barrier is low and gives rise to resistivity as shown in Fig. 4.11.

$$\rho \sim \rho_0 \exp(-U_0/k_B T) \qquad ...(4.51)$$

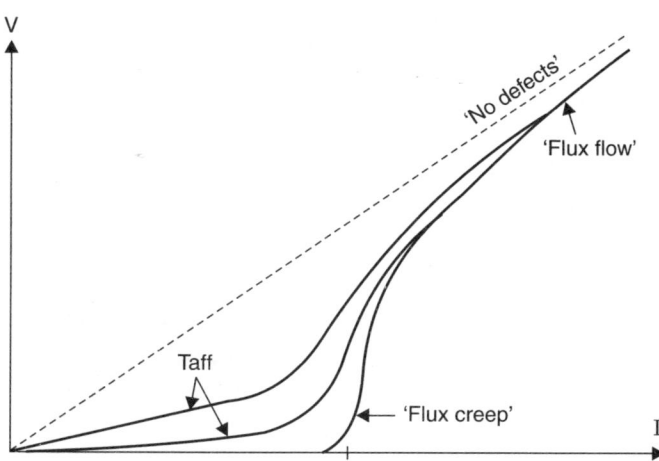

Fig. 4.11 Voltage-Current characteristic for different regimes

We have already discussed Bean model [2] which is capable of describing an irreversible M–H–loop. We have also shown that one can deduce the critical current, J_c from the magnetization curve. However, if there is a flux creep, the magnetization will relax towards the equilibrium magnetization. One can easily show that for long times, the relaxation law is logarithmic:

$$M(t) \sim -\ln \frac{t}{t_0} \quad \text{for } t > t_0 \qquad ...(4.52)$$

Obviously, one can estimate the barrier U_0 from the measurement of relaxation. If the free energy barrier is small, relaxation can occur during the time of measurements; this occurs in some HTSC cuprates. Above a given temperature, the magnetization relaxes to its equilibrium value and one obtains a reversible magnetization curve. Obviously, one would able to deduce the zero critical current by using the Bean model [2].

However, if we pass the transport current and directly measure the voltage, one can detect a signal if the current is higher than some critical value which corresponds to E_c. Thus, by carrying out the transport measurements, one would measure critical current although an analysis of magnetization measurements based on eq. (4.39) would make it appear zero.

Experimental measurements [3 – 8] of magnetic relaxation due to flux creep in HTSC cuprates, organic, Chevrel phase and heavy fermion superconductors demonstrate that their magnetic relaxation rates are large at temperatures in the millikelvin range. Some of the most striking nonthermally activated behaviour is observed in the heavy fermion superconductor UPt_3, where the normalized relaxation rate is constant from a few millikelvin to just below the T_c of 430 mK. This behaviour contrasts sharply with the predictions of the Anderson-Kim model [9, 10] of thermally activated flux creep, and can be interpreted as being due to quantum collective creep of flux vortices [11 – 13].

Remarkably, a nearly temperature dependent plateau over a wide temperature range has been observed in the normalized relaxation rate of YBCO. In one series of experiments [14], magnetic flux was trapped in melt-textured YBCO samples, and the trapped flux (magnetization) was measured as a function of time. The normalized relaxation rate S was found to be nearly constant from 10 K upto temperatures ranging from 40 K to as high as 70 K, depending on the magnetic field. Similar results were reported on single crystal.

One hypothesis which could account for possible quantum behaviour over a broad temperature range is that quantum creep of vortices (vortex tunneling) may be a dual or generalized form of Josephson tunneling which will described later on.

The surprising temperature dependence of the magnetic relaxation rates of HTSC cuprates and other type-II superconductors has forced a critical reexamination of earlier theoretical models and stimulated considerable theoretical activity [11 – 14].

The recent discovery of another striking new feature of HTSC cuprates and other type-II superconductors, known as the **irreversibility line,** has also motivated new theoretical developments. Irreversibility line, flux melting and vortex glass will be discussed latter.

REFERENCES

1. Y. Yeshurn and A.P. Malozemoff, *Phys. Rev. Lett.* 60, 2202 (1988).
2. C.P. Bean, *Phys. Rev. Lett.* 8, 250 (1962).
3. A.V. Mitin, *Sov. Phys. JETP* 66, 335 (1987).
4. A. Pollini et al., *Physica B* 165–166, 365 (1990).
5. F.L. Hamzic and I.A. Campbell, *Nature* 345, 515 (1990).
6. S. Uji et al., *Physica* C 207, 112 (1993).
7. A.C. Motta et al., *Phys. Scr.* 37, 823 (1988).
8. A.C. Motta et al., *Physica* C 185–189, 343 (1991).
9. P.W. Anderson and Y.B. - Kim, *Rev. Mod. Phys.* 36, 39 (1964).
10. P.W. Anderson, *Phys. Rev. Lett.* 9, 309 (1962).
11. G. Blatter et al., *Phys. Rev, Lett.* 66, 3297 (1991).
12. B.I. Ivlev et al., *Phys. Rev.* B44, 7023 (1991).
13. G. Blatter et al., *Rev. Mod. Phys.* 66, 1125 (1994).
14. Z. Huang, *Ph. D. Dissertation,* University of Houston (May, 1992).

Appendix 4.1—Temperature Dependencies of B_c, J_c and n_n

The critical field (B_c) and critical current density (J_c) are related through the expression

$$B_c(T) = \mu_0 \lambda(T) J_c(T) \qquad \ldots(1)$$

we may note that all three quantities, i.e., $B_c(T)$, $\lambda(T)$ and $J_c(T)$ are temperature dependent.

At absolute zero, we use the notation $B_c(0) = B_c$, etc. and eq. (1) becomes, in analogy with the equation.

$$B_0 = \mu_0 \lambda J_0 \qquad \ldots(2)$$

as
$$B_C = \mu_0 \lambda J_c \qquad \ldots(3)$$

A particular superconducting wire of radius R has a maximum current called the critical current I_c, and for a type-I superconductor, it has the value

$$I_c = 2\pi R \lambda J_c = 2\pi R B_c/\mu_0 \qquad \ldots(4)$$

The destruction of the superconducting state by exceeding the critical (transport) current I_c is called the **Silsbee effect.**

The penetration depth $\lambda(T)$ is related to the super electron density $n_s(T)$ through the expression

$$\lambda(T) = \lambda_L(T) = [m/\mu_0 n_s(T) e^2]^{1/2} \qquad \ldots(5)$$

due to London. In the two-fluid model, we have the temperature-dependent expression for the super n_s and the normal n_n, the electron densities, respectively,

$$n_s(T) + n_n(T) = n, \qquad \ldots(6)$$

where the total electron density n is independent of temperature, and at $T = 0$ we have $n_n = (0)$ and $n_s(0) = n$. The two fluids n_s and n_n interpenetrate but do not interact, and simple theory predicts the following temperature dependencies:

$$B_c(T) = B_c(0)\left[1-\left(\frac{T}{T_c}\right)^2\right] \qquad \ldots(7)$$

$$\lambda(T) = \lambda(0)\left[1-\left(\frac{T}{T_c}\right)^4\right] \qquad \ldots(8)$$

$$J_c(T) = J_c(0)\left[1-\left(\frac{T}{T_c}\right)^2\right]\left[1-\left(\frac{T}{T_c}\right)^4\right]^{1/2} \qquad \ldots(9)$$

$$n_s(T) = n\left[1-\left(\frac{T}{T_c}\right)^4\right] \qquad \ldots(10)$$

$$n_n(T) = n(T/T_c)^4 \qquad \ldots(11)$$

Figs. 1 to 4 shows the said temperature dependencies of $B_c(T)$, $\lambda(T)$, $J_c(T)$ and $n_s(T)$ respectively. We may note that some authors report exponent values or related expressions for these temperature dependences. The dashed lines in figures show the asymptotic behaviours.

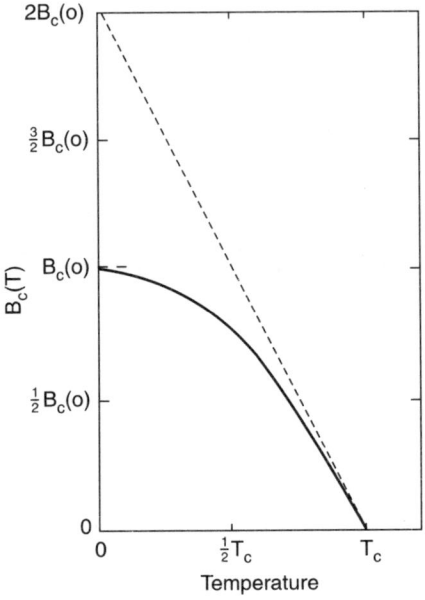

Fig. 1 Temperature dependence of the critical field $B_c(T)$. The asymptotic behaviour near $T = 0$ and $T = T_c$ are indicated by dashed lines

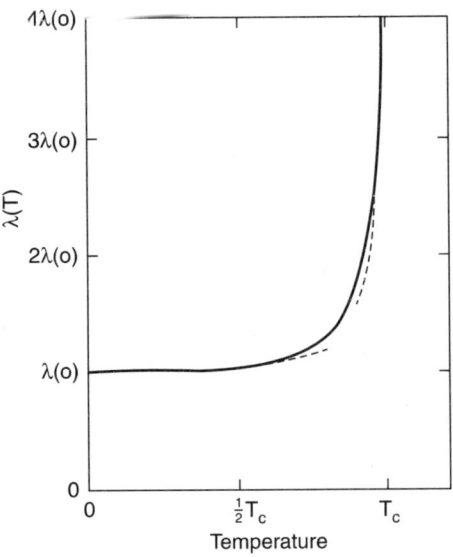

Fig. 2 Temperature dependence of penetration depth $\lambda(T)$. The asymptotic behaviour near $T = 0$ and $T = T_c$ are indicated by dashed lines

Critical Currents of Type-II Superconductors

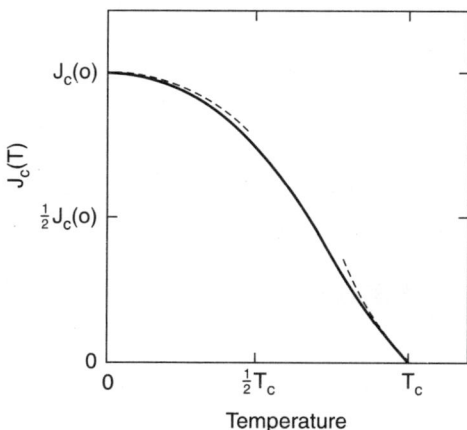

Fig. 3 Temperature dependence of critical current density $J_c(T)$. The asymptotic behaviour near $T = 0$ and $T = T_c$ are indicated by dashed lines

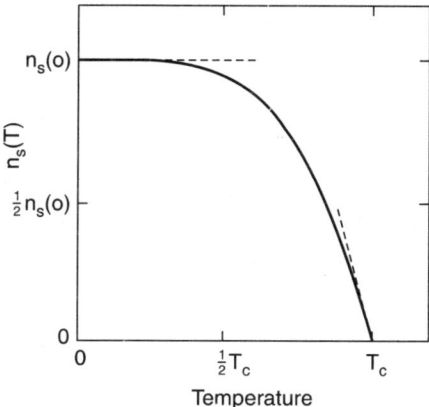

Fig. 4 Temperature dependence of density of superconducting electrons, n_s. The dashed lines indicate the slopes $dn_s/dT = 0$ at $T = 0$ and $dn_s/dT = -4n/T_c$ at $T = T_c$

Appendix 4.2—Bean Model

This model assumes that the current density can only take on the values $\pm J_c$ or 0. A sample of thickness $2a$ has a characteristic field

$$B^* = \mu_0 J_c a \qquad \ldots(1)$$

and when the applied field B_0 attains this value B^*, the fields and currents reach the centre of the sample. Let us examine the one dimensional case

$$\frac{d}{dx} B_z(x) = \mu_0 J_y(x) \qquad \ldots(2)$$

for applied fields below and above B^*.

For low applied fields, $B_0 < B^*$, the internal fields and currents only exist near the surface, with a field and current free region ($-a' < x < a'$) near the center, and the current given by

$$J(B) = J_c \qquad \ldots(3)$$

we have
$$\begin{aligned} J_y(x) &= J_c & -a \leq x \leq -a' \\ J_y(x) &= 0 & -a' \leq x \leq a' \\ J_y(x) &= -J_c & a' \leq x \leq a \end{aligned} \qquad \ldots(4)$$

Equation (2) requires that $B_z(x)$ depend linearly on x in the regions where $J_y = \pm J_c$, so one finds for the internal magnetic fields

$$\begin{aligned} B_z(x) &= B_0 \left[\frac{a'+x}{a'-a}\right] & -a \leq x \leq -a' \\ B_z(0) &= 0 & -a' \leq x \leq a' \\ B_z(x) &= B_0 \left[\frac{x-a'}{a-a'}\right] & a' \leq x \leq a \end{aligned} \qquad \ldots(5)$$

These relations match the boundary conditions $B_z(0) = B_0$ at the surface $x = \pm a$, with the magnitudes of J_c and B_0 related by the expression obtained from eq. (2)

$$J_c = \frac{B_0}{\mu_0 (a-a')} \qquad \ldots(6)$$

Fig. 1 (a) shows the plots for $B_z(x)$ and $J_y(x)$ expressions for a finite value of a' and Fig. 1 (b) shows the situation where $B_0 = B^*$ and $a' = 0$.

For high applied fields, $B_0 > B^*$, the currents and internal fields are present throughout the sample and are given by the expressions

$$\begin{aligned} J_y(x) &= J_c & -a \leq x \leq 0 \\ J_y(x) &= -J_c & 0 \leq x \leq a \end{aligned} \qquad \ldots(7)$$

$$\begin{aligned} B_z(x) &= B_0 - B^* \left[\frac{a+x}{a}\right] & -a \leq x \leq 0 \\ B_z(x) &= B_0 + B^* \left[\frac{x-a}{a}\right] & 0 \leq x \leq a \end{aligned} \qquad \ldots(8)$$

Fig. 1 shows how the internal field and current density vary with the ratio $B_0/B^*(a)$ for low fields $B_0 < B^*$, (b) for the maximum penetration field $B_c = B^*$, and (c) for high fields $B_0 > B^*$. One can see from the figure that $B_z(x)$ is symmetric about the $x = 0$ point, and $J_y(x)$ is antisymmetric about this point.

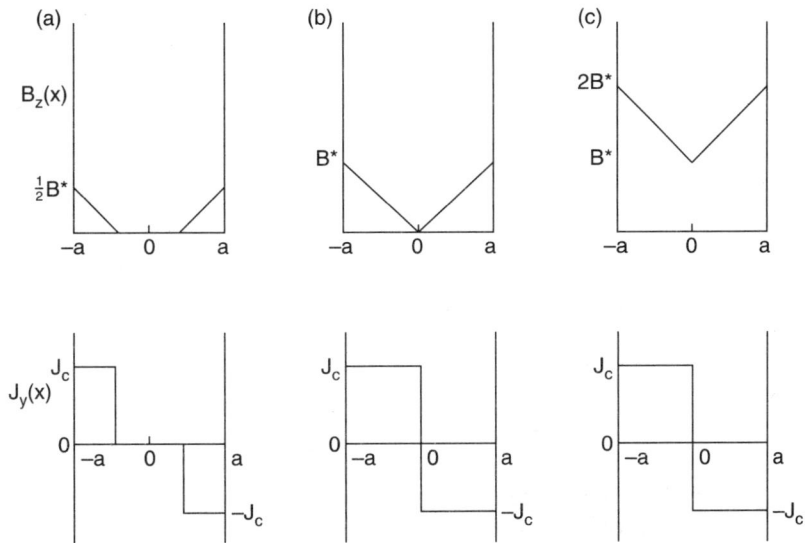

Fig. 1 Dependence of the internal magnetic field $B_z(x)$ and the current density $J_y(x)$ on the strength of the applied magnetic field B_0 for normalized applied fields given by (a) $B_0/\mu_0 J_c a = \frac{1}{2}$, and (b) $B_0/\mu_0 J_c a = 1$, and (c) $B_0/\mu_0 J_0 a = 2$ for the Bean model. There is a field free region in the centre for case (a) and case (b) is the boundary between having and not having such a region [1]

If the applied field is zero but the transport current

$$I = 2(a - a') L J_c \qquad ...(9)$$

is flowing through a wire with a rectangular cross-section of width L much greater than its thickness $2a$, then the internal magnetic field and current density will be as shown in Fig. 2 (a), which is drawn for the case $I = \frac{1}{2} I_c$, where I_c is the critical current.

$$I_c = 2a L J_c \qquad ...(10)$$

Fig. 2 (b) is drawn for the limiting case $I = I_c$, and higher applied transport currents drive the wire normal. The equations for J_y and B_z of the transport current case of Fig. (a) are the same as eqs. (4) and (5), respectively, for the applied field case of Fig. 1 (a), except for some changes in sign, as may be seen by comparing the figures. We may note that when an applied field and an (applied) transport current are present simultaneously, the situation is more complicated.

The critical states of the Bean model can be complicated when fields exceeding B^* are applied and then reversed in direction.

Bean model provides simple explanations of flux shielding whereby the average field $<B_{in}>$ inside the superconductor is lower in magnitude than the applied field B_{app}, and of flux trapping whereby $<B_{in}>$ exceeds B_{app}. These two cases are associated with current flow in opposite directions around the material.

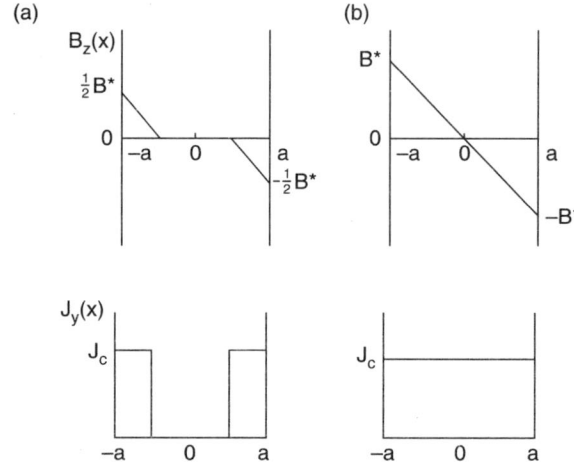

Fig. 2 Dependence of the internal magnetic field $B_z(x)$ and the current density $J_y(x)$ on the strength of the applied transport current for the Bean model. Figures are for applied current I which are (a) less than the critical current $I_c = 2a\,L\,J_c$, and (b) equal to I_c. Higher currents cause the wire to go normal [1]

Reversed Critical States and Hysteresis

Probably the most important application of the Bean model is its use to estimate the magnetization of a sample in a high applied field, $B_{app} > B^*$ and Fig. 3 shows typical hysteresis loops for low, medium, and high applied fields. Explicit expressions for the average internal field $<B>$ and the magnetization $\mu_0 M$ can be found in reference [1]. A high field hysteresis loop $(B_0 > B^*)$ furnishes the difference

$$M_+ - M_- = J_c\, a \qquad \ldots(11)$$

between the upper and lower magnetization plateaus of Fig. 3(c), where

$$\mu_0 M_+ = \frac{1}{2} B^*$$

$$\mu_0 M_- = -\frac{1}{2} B^* \qquad \ldots(12)$$

as indicated in the Figure. This gives the critical current in terms of the measured magnetization through the high-field Bean model relation

$$J_c = 2\,(M_+ - M_-)/d = 1.59 \times 10^6\, \mu_0\, \Delta M/d \quad (A/m^2) \qquad \ldots(13)$$

where current is measured in amperes, $\mu_0 \Delta M = \mu_0 (M_+ - M_-)$ is in tesla, and d is the diameter of the sample grains in meters.

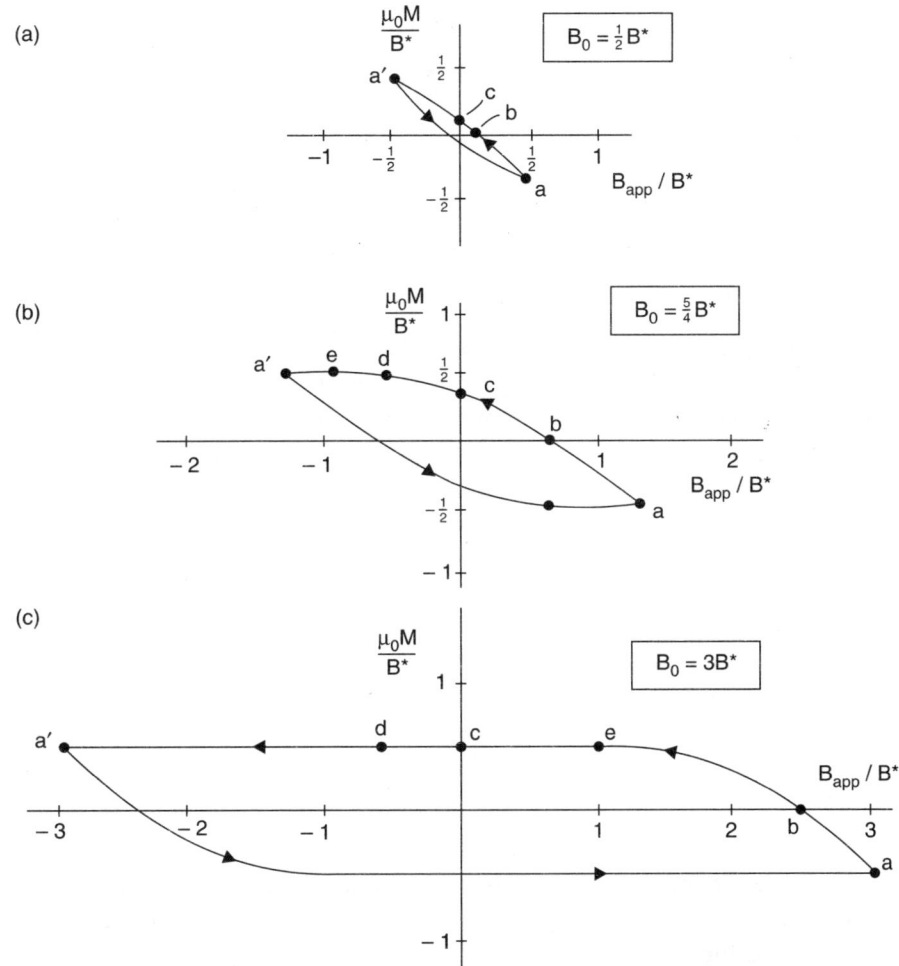

Fig. 3 Hysteresis loops of magnetization $\mu_0 M$ vs. the applied magnetic field B_{app} cycled over the range $-B_0 \leq B_{app} \leq B_0$ for the cases : (a) $B_0 = \frac{1}{2} B^*$, (b) $B_0 = (5/4) B^*$ and (c) $B_0 = 3B^*$ [1]

REFERENCE

1. P. Poole, Jr, et al., *Superconductivity*, Academic Press, New York (1995).

Microscopic Theory of Superconductivity

5.1 INTRODUCTION

So far, we have taken a purely macroscopic view of superconductivity; that is to say, we have assumed that some of the electrons behave as superconductors with the mysterious property that, unlike normal electrons, they can move through the metal without hindrance of any kind. We have then discussed what restrictions are placed on their collective behaviour by the laws of electromagnetism and thermodynamics. In this chapter, we delve a little more deeply and take a microscopic view, trying to explain from first principles how it is that this property of the superelectrons arise.

The formulation of microscopic theory of superconductivity is very difficult because of the smallness of the energy involved in the process. This energy, as calculated from the condensation energy, comes out to be ~ 10^{-8} eV/atom. In comparison to cohesive energy of metal, which is of the order of several eV/atom, this energy is very small. The difference in energy between two crystallographic structures, *fcc* and *hcp*, for instance, is of the order of 10^{-2} eV. Obviously, one has to find what interaction can lead to a state with such a small difference in energy as compared with the normal state and to such strikingly different properties.

The key to the basic interaction between electrons which gives rise to superconductivity was provided by the *isotope effect* : two different isotopes of the same metal exhibit different T_c's. The relation, valid for some simple metals, is represented by

$$T_c \sim M^{-1/2} \qquad \qquad ...(5.1)$$

where M is the atomic mass of the isotope and α is roughly 0.5. This shows that lattice vibrations (phonons) play an essential role in bringing about superconductivity in low–T_c superconductors, as Frohlich [1] and Bardeen [2] realised in the early 1950's. Another clue was provided by the early specific heat measurements [3]. These measurements has shown that, well below T_c, the electronic contribution to the specific heat was dominated by an exponential dependence of the form $C_{ex} \sim \exp[-\Delta/k_B T]$. This established the existence of an energy gap Δ in the spectrum of elementary thermal excitations above the superconducting ground state. Additional evidence for the existance of an energy gap in the superconducting state is provided by the tunnelling experiments. We must remember that under certain special circumstances a superconducting metal may not possess an energy gap. (The energy gap in the electron energy spectrum of a superconductor can be reduced by a number of agencies; for example, incorporation of magnetic impurities. It is found that magnetic impurities lower the transition temperature as well as reduce the energy gap, but that the transition temperature (T_c) can still be above zero at an impurity concentration which has reduced the energy gap to zero. In this condition, the metal is resistanceless, implying long-range coherence of the electron-pair wave, even though there is no energy gap. This is termed as **gapless superconductivity** it implies that it is the long-range coherence of the electron-pair wave, not the energy gap in the electron energy spectrum, which is the essential feature of the superconductivity. In gapless superconductors, there is at the Fermi level a minimum in the density of states, but no actual gap. In general, if a perturbation causes a superconductor to pass into the normal state through a second-order phase transition, the superconductor goes into the gapless state before it becomes normal [4].

To recapitulate, any successful microscopic theory of superconductivity must be able to explain the following general features :

(i) Superconductivity is essentially bound up with some profound change in the behaviour of the conduction electrons which is marked by the appearance of long range order and a gap in their energy spectrum of the order of 10^{-4} eV.

(ii) The crystal lattice does not show any change of properties, but must nevertheless play a very important part in establishing superconductivity because the critical temperature depends on the atomic mass (the isotope effect).

(iii) The superconducting-to-normal state transition is a phase change of the second order.

The long range order in (i) clearly means that the electrons must interact with each other. It has, of course, been appreciated for a long time that the conduction electrons in a metal interact very strongly through their coulomb repulsion, and it is surprising that the ordinary free-electron theory of metals and semiconductors, which neglects this interaction, works as well as it does. It is difficult, however, to believe that the coulomb repulsion is the interaction responsible for superconductivity because there is no known way in which a repulsive interaction can give an energy gap. Furthermore, because the energy gap is very small, the interaction responsible for it must be very weak, much weaker than the coulomb interaction. The apparent lack of any mechanism for a weak attractive interaction was for sometime the stumbling block in the way of any microscopic theory of superconductivity.

It was well known that helium-4 (^{4}He) atoms are spin zero bosons, which facilitates their condensation into a superfluid state [5] below 2.17 K. A direct Bose-like condensation of unpaired electrons, however, is not possible since electrons are spin $-\frac{1}{2}$ fermions. Schafroth et. al. [6] suggested that the superconducting state might correspond to Bose–Einstein condensation of pairs of electrons, which would behave like bosons, into localized states. One major problem with this real space pairing picture, however, is that the binding energy of a tightly bound pair would be much larger than one would realistically expect for states near the Fermi surface. In addition, the large Coulomb repulsion between the closely spaced electrons would some how need to be circumvented. Classical low temperature superconductors differ in a fundamental manner from the Schafroth's et al. [6] bound pair model, which required that the separation between mates of a pair be much less than the distance between pairs. There are, for a typical low-T_c superconductor, about one million bound pairs which have their centers of mass falling within the range of single pair function [7]. Thus, rather than weakly overlapping pairs, one has the reverse limit—very strongly overlapping pairs.

J. Bardeen, L.N. Cooper and J.R. Schrieffer (hereafter BCS) [8] took Landau theory of normal Fermi liquid [9] as their starting point. The quasiparticle (electron-like) excitations above the Fermi surface have lifetimes that are quite long for sufficiently low energies. These quasiparticles can be thought of as composite particles that incorporate the effects of screening and are weakly interacting. The interaction with the lattice (electron–phonon interaction) however, can lead to residual retarded interactions between the electrons (quasiparticles) that are attractive. In real space and time, an electron interacting with the lattice polarizes the ions in a given region, and that polarization can then attract a second electron at a slightly later time. In momentum space and the frequency domain, the effective matrix elements representing the phonon-mediated electron–electron interactions become negative for states near the Fermi surface, whose energies measured with respect to the Fermi energy, fall within a cut off $\hbar \omega_c \approx \hbar \omega_D$, where $\hbar \omega_D$ is the Debye energy [10].

Cooper [11] showed that a normal metal (with 'standard' metallic properties) could not be formed if there was a small attraction between electrons, *i.e.*, the usual ground state of an electron gas [filled Fermi sphere) is unstable with respect to an arbitrary weak attractive interaction between the electrons near the Fermi surface. In such a case, two electrons would form *'bound pair'* (often called a Cooper pair) however small the attractive interaction. The two partners of a Cooper pair have opposite momenta and spin angular momenta. A simple calculation shows that the formation of a bound electron pair lowers the energy of the system by a finite amount, so that normal state Fermi distribution no longer represents the configuration of lowest energy, *i.e.*, completely different properties for the whole ensemble of electrons would be observed. This implied that the Fermi surface would be unstable to the formation of such bound pairs if the interactions were attractive, and that the normal Fermi liquid would therefore not represent the ground state at $T = 0$. This new state containing bound pairs is identified with the superconducting ground state. The energy difference between the normal state and the superconducting state is proportional to

$$\exp(-\text{const}/|g|) \qquad \ldots(5.2)$$

where $|g|$ represents the strength of the interaction. Equation (5.2) has no series expansion in ascending powers of $|g|$ and a perturbation analysis of the transition to a superconducting state is therefore impossible. Frohlich had previously noted the difficulty with a perturbation calculation [12]. In this chapter, we first try to explain this attractive meahanism by studying the response of a normal metal to a small perturbation, *i.e.*, a test charge introduced into the metal. This leads to the concept of a dielectric 'constant'. We will also show that how a change of sign of the dielectric 'constant' changes a repulsive coulomb interaction into an attractive one and consequently leads to pairing of electrons. After this, we will briefly discuss the main results of the microscopic BCS theory.

5.2 NORMAL METAL

We know that current can easily flow in a normal metal. Obviously, this shows that electrons are nearly free to propagate in a normal metal. We are familiar with one of the simplest models of the metal due to Sommerfeld, *i.e.*, a box filled with a gas of free electrons. We can describe an electron of energy E and the momentum **p** by a wave function

$$\psi(\mathbf{r}) = \exp(i\,\mathbf{k}\cdot\mathbf{r} - \omega t) \qquad \ldots(5.3)$$

where **k** and ω are given by de Broglie relation

$$\omega = \frac{E}{\hbar} \quad \text{and} \quad |\mathbf{k}| = \frac{|\mathbf{p}|}{\hbar} = \frac{2\pi}{\lambda} \qquad \ldots(5.4)$$

The waves within the box have to be stationary, which means that boundary conditions are to be added. A well known convenient way is to introduce the Born–Von Karman boundary conditions.

If L_x is the length of the box along the x-axis, then possible values of k_x are represented by

$$k_x = \frac{2\pi}{L_x} n_x$$

where n_x is an integer. Obviously, we can define such a state by three integers : n_x, n_y and n_z. We know that electrons are fermions and hence for a given wave vector, one can have only two electrons in a given state : one with spin up and the other with spin down. In order to accomodate all the electrons in the box, one will have to fill all the states with wave vector **K** :

$$|\mathbf{K}| < K_F,$$

where k_F is the Fermi wave vector which is related to the density of electrons, n, by

$$n = \frac{K_F^2}{3\pi^2} \qquad \ldots(5.5)$$

We have the dispersion relation between the energy and the wave vector as

$$E = \frac{p^2}{2m} = \frac{\hbar^2 K^2}{2m} \qquad \ldots(5.6)$$

Thus all the states below Fermi energy E_F are filled. The density of states at this energy is given by

$$N(E_F) = \frac{m\,K_F}{2\pi^2 \hbar^2} \qquad \ldots(5.7)$$

The actual density of electrons in a metal is 10^{23} per cm^3. This means that the Fermi energy is of the order of a few eV. Some typical values estimated within free electron model are given in Table 5.1

Table 5.1 Some typical values of the characteristic quantities of a few normal and a HTSC cuprate, $Y\,Ba_2\,Cu_3\,O_{6.9}$

	$v_F \times 10^8$ cm/s	$K_F \times 10^8$/cm	$n \times 10^{22}$/cm^3	E_F (eV)	Q_D (K)
Al	2.03	1.75	18.10	11.70	390
Nb	1.37	1.18	5.56	5.32	320
Cu	1.57	1.36	8.47	7.00	340
$Y\,Ba_2Cu_3O_{6.9}$ (YBCO)	0.1	~ 0.5*	0.7	0.1	~ 400

* Estimated within a free electron model.

The electrons in a metal behave very much like free electrons found confirmation in measurements of susceptibility or specific heat. The susceptibility is found constant at low temperatures and is given by

$$\chi = 2\mu_B^2 N(E_F) \qquad \ldots(5.8)$$

The electronic specific heat as stated earlier is linear in temperature :

$$C_c = \gamma T$$

where
$$\gamma = \frac{2\pi^2}{3} N(E_F) k_B^2 \qquad \ldots(5.9)$$

Such behaviour is rather observed in most metals.

Bloch theorem* asserts that qualitatively the lattice does not change very much. Thus the energy becomes a periodic function of K. For example, if the ions form a simple cubic lattice with an interatomic distance a the E–k or the dispersion relation becomes

$$E = zt\,(\cos K_x a + \cos K_y a + \cos K_z a) \qquad \ldots(5.10)$$

which for small K has the same K^2 dependence as in eq. (5.6). We can improve our model by replacing the electron mass m with an effective mass m^* in eq. (5.6).

We must remember that the average kinetic energy of electrons is large as compared with the repulsive potential energy. The average kinetic energy of the gas is of the order of E_F, i.e., few eV and it is proportional to $n^{2/3}$. The potential energy is of the order of e^2/r_e, where r_e is the average distance between two electrons and thus it is proportional to $n^{1/3}$. Obviously, for large values of n, the repulsive coulomb energy term is negligible compared with the kinetic energy term. We can see that this situation prevails in most metals, but this term becomes important and even predominant in the low density limit. However, in most conventional superconductors, this term is not predominant and one can describe them with our simple approximation.

5.3 NORMAL STATE INSTABILITY

The gas of electrons (section 5.2) if stable in the presence of repulsion between electrons, is completely unstable in the case of attraction between them [11]. L.N. Cooper [11] in his analysis adds two electrons of opposite wave vectors just above the Fermi surface and looks for a wavefunction for those two electrons of the form

$$\psi(\mathbf{r}_1 - \mathbf{r}_2) = \sum_K g(\mathbf{K}) \exp[i\,\mathbf{K}\cdot(\mathbf{r}_1 - \mathbf{r}_2)] \qquad \ldots(5.11)$$

---

*The **Bloch theorem** states that the wave function of an electron moving in a periodic potential of the lattice has the form of a plane wave, $\exp(i\,k\,r)$, modulated by a function $u(\mathbf{K}, \mathbf{r})$ that has the same periodicity as the periodic potential

$$\psi(\mathbf{K}, \mathbf{r}) = e^{i\,\mathbf{k}\cdot\mathbf{r}}\,u(\mathbf{K}, \mathbf{r}).$$

where $g(\mathbf{K})$ represent the probability amplitude for finding one electron with momentum $\hbar \mathbf{K}$ and the corresponding electron with momentum $-\hbar \mathbf{K}$. Thus

$$g(\mathbf{K}) = 0 \quad \text{for } |\mathbf{K}| < k_F \qquad ...(5.12)$$

as all the states below K_F are already filled with electrons.

Cooper [11] assumed an attraction between these two electrons and he solved the Schrödinger equation. He was interested in calculating the wavefunction and the energy of the pair which can be measured from the Fermi energy

$$E = 2E_F + \epsilon \qquad ...(5.13)$$

Cooper [11] found that the **pair always binds**, *i.e.*, the energy ϵ is negative and given by the relation

$$\epsilon = -2\hbar w_D \exp\left[-\frac{2}{N(E_F)|g|}\right] \qquad ...(5.14)$$

where g is the strength of the attractive potential and $\hbar w_D$ is the characteristic energy of this attractive potential. The pair of these two electrons now known as **'Cooper pair'**. We must note that here two electrons bind even if g is very small. The binding energy is large enough only if $N(E_F)|g|$ is large. Obviously, one needs either large g or large density of states, $N(E_F)$ at the Fermi level. If we calculate the extension ξ of the pair, *i.e.*, the average distance between two electrons, we obtain

$$\xi \sim \frac{\hbar V_F}{|\epsilon|} \qquad ...(5.15)$$

As ϵ is small, ξ will be large, *i.e.*, hundreds of angstroms.

Now the question arises, what will happen when all electrons near the Fermi energy attracting each other? Obviously, they will tend to form pairs in order to decrease their energy. This means that the model of filling the k-states up to the Fermi level with electrons breakdown completely. In other words, in the presence of a weak attraction between electrons the ground state which we used for describing the normal metal is no longer the ground state of the system. Obviously, one will have to think of some other ground states.

5.4 ORIGIN OF ATTRACTION BETWEEN ELECTRONS CONCEPT OF DIELECTRIC 'CONSTANT'

To understand the origin of this unusual attraction between electrons, let us first describe the reaction of the charged medium to a test charge in terms of the dielectric 'constant'. In a simple metal electrons screen a test charge over a length called *Thomas-Fermi length*.

Let us consider that we introduce a test charge at point **r** in an electron gas. The electrons in a gas will rearrange themselves in order to screen the field of this added charge. One can describe the response of the system by Maxwell's equations with the introduction of the dielectric constant. The displacement vector **D** is related to the external applied charged density, our test charge ρ_t, by

$$\nabla \cdot \mathbf{D} = \rho_t \qquad \ldots(5.16)$$

The relation between the electric field vector, $\mathbf{E}$ and total charge density, ρ_{tot} is

$$\nabla \cdot \mathbf{E} = \frac{\rho_{tot}}{\epsilon_0} = \frac{\rho_t + \rho}{\epsilon_0} \qquad \ldots(5.17)$$

where ρ is the induced charge in the system due to the test charge ρ_t. One defines the dielectric constant usually as

$$\mathbf{D} = \epsilon_0 \epsilon \, \mathbf{E} \qquad \ldots(5.18)$$

where ϵ_0 is the permittivity of the free space. Thus one obtains for the dielectric constant

$$\epsilon = \frac{\rho_t}{\rho_t + \rho} \qquad \ldots(5.19)$$

If the test charge is varying in space and time, one defines the dielectric constant $\epsilon(\mathbf{q}, \omega)$ which depends on the wavelength and frequency. One can easily show from Poisson's equation that

$$\epsilon(\mathbf{q}, \omega) = \frac{V_{ext}(\mathbf{q}, \omega)}{V(\mathbf{q}, \omega)} \qquad \ldots(5.20)$$

where $V_{ext}(\mathbf{q}, \omega)$ is the Fourier component of the elctrostatic potential of the test charge in vacuum, and $V(\mathbf{q}, \omega)$ is the Fourier component of the potential of the test charge in the medium.

5.5 DIELECTRIC CONSTANT OF A GAS OF ELECTRONS

One expects that the electrical potential of the test charge in an electron gas is screened exponentially over the screening length l_s :

$$V(r) = \frac{q}{4\pi \epsilon_0 \epsilon_e \, r} \exp(-r/l_s) \qquad \ldots(5.21)$$

This means that the dielectric constant of the electron gas is given by

$$\epsilon_e(q) = 1 + \frac{K_{TF}^2}{q^2} \qquad \ldots(5.22)$$

One obtains from Thomas-Fermi theory,

$$k_{TF}^2 = l_s^{-2} = \frac{3}{2} \frac{ne^2}{E_F} \cdot \frac{1}{\epsilon_0} = \frac{4}{\pi} \frac{K_F}{a_0} \qquad \ldots(5.23)$$

where a_0 is the Bohr radius, ~ 0.5 Å. The magnitude of the screening length l_s is of the order of the Bohr radius.

We can see that $\epsilon_e(\mathbf{q})$ is *always positive* and thus the screened potential is always of the same sign as the bare potential $V_{ext}(\mathbf{q}, \omega)$ of the test charge. We must note that this dielectric constant connot turn a repulsive potential into an attractive one. Now, we will analyse the more realistic case where we add the positive ions of the lattice.

5.6 MOTION OF IONS IN METAL

Let us consider a lattice of positive ions embedded in a gas of electrons. The ions are not rigid but are vibrating around their equilibrium positions. In other words, we can say that there are elastic waves propagating in the solid at all temperature. This motion is responsible for the conventional superconductivity. Let us study these waves.

We now consider ions of charge Ze. Let n be the electron density in the solid, then the average ion density is n/Z. One can write equation of motion of an ion of mass M as

$$M \frac{d\mathbf{v}_i}{dt} = Ze\, \mathbf{E} \qquad ...(5.24)$$

In analogy with the electron current one can introduce,

$$\mathbf{J}_i = \frac{n}{Z} Ze\, \mathbf{v}_i = ne\, \mathbf{v}_i \qquad ...(5.25)$$

Thus we obtain

$$\frac{d\mathbf{J}_i}{dt} = n Z e^2 \mathbf{E} \qquad ...(5.26)$$

Let ρ_i be the density of ions. The equation of conservation for the number of ions reads

$$\frac{\delta \rho_i}{\delta t} + \nabla \cdot \mathbf{J}_i = 0 \qquad ...(5.27)$$

One obtains from last two equations

$$\frac{\delta^2 \rho_i}{\delta t^2} = -\frac{nZ e^2}{M} \nabla \cdot \mathbf{E} \qquad ...(5.28)$$

Thus Maxwell's equations give

$$\nabla \cdot \mathbf{E} = \frac{\rho_i + \rho_e + \rho_t}{\epsilon_0} \qquad ...(5.29)$$

where ρ_e is the density of electrons. One finally obtains

$$\frac{\delta^2 \rho_i}{\delta t^2} = -\Omega_i^2 (\rho_i + \rho_e + \rho_t) \qquad ...(5.30)$$

where

$$\Omega_i^2 = \frac{nZ e^2}{M \epsilon_0} \qquad ...(5.31)$$

is the ionic **plasma** frequency.

The dielectric constant of the electrons is now obtained as

$$\epsilon_e = \frac{\rho_i + \rho_t}{\rho_e + \rho_i + \rho_t} \qquad ...(5.32)$$

$\rho_i + \rho_t$ is now the test charge for electrons. Substituting eq. (5.32) in eq. (5.30), one obtains

$$\frac{\delta^2 \rho_i}{\delta t^2} = -\frac{\Omega_i^2}{\epsilon_e(q)}(\rho_i + \rho_t) \qquad ...(5.33)$$

Equation (5.33) reveals that without a test charge the ions are moving like waves. These waves, called phonons, have a wavelength dependent frequency. In this respect eq. (5.33) is an important equation. The ions move with a frequency ω_q given by

$$\omega_q^2 = \frac{\Omega_i^2}{\epsilon_e(q)} \qquad ...(5.34)$$

In the limit of small q, one obtains

$$\omega_q = \frac{\Omega_i}{\sqrt{\epsilon_e(q)}} \sim \frac{\Omega}{k_{TF}} q \qquad ...(5.35)$$

Substituting for Ω_i and k_{TF}, one obtains the velocity of sound in a metal as

$$v_s = v_F \sqrt{\frac{Z}{3} \frac{m}{M}} \qquad ...(5.36)$$

Obviously, the velocity of sound in a metal is of the order of the Fermi velocity divided by a factor ~ 100.

5.6.1 Debye Frequency (Typical Phonon Frequency)

To estimate the order of magnitude of typical phonon frequency, we start with the Debye approximation, *i.e.*, we assume that the sound velocity is frequency independent,

$$\omega_q = v_s q \qquad ...(5.37)$$

One can calculate the maximum wave vector q_D and the maximum frequency ω_D (the Debye wave vector and Debye frequency). We know that the total number of modes for phonons is $3N$ if N is the number of atoms, *i.e.*, there are N modes for each given polarization. One can obtain the total number of modes whose wave vector is smaller than q_D by dividing the volume of a sphere of radius q_D by $(2\pi)^3/V$;

$$N = \frac{V}{(2\pi)^3} \frac{4\pi}{3} q_D^3 \qquad ...(5.38)$$

Now, setting $V_0 = V/N$ for the atomic volume, the Debye wave vector is obtained as

$$q_D = \left(\frac{1}{V_0} 6\pi^2\right)^{1/3} \qquad ...(5.39)$$

and the corresponding Debye frequency is obtained as

$$\omega_D = \left(\frac{1}{V_0} 6\pi^2 v_s^2\right)^{1/3} \qquad ...(5.40)$$

Eqs. (5.39) and (5.40) provide the order of magnitude of the frequency and also frequency. One finds ω_D is of the order of 10^{12} to 10^{13} s^{-1}. One can estimate the Debye temperature (θ_D) by the relation.

$$\hbar \omega_D \approx k_B \Theta_D \qquad \ldots(5.41)$$

Θ_D is found to be of order of 10^2 K.

5.7 ORIGIN OF THE ATTRACTIVE INTERACTION

We shall now see how a test charge is screened when embedded in a medium consisting of electrons and ions. The dielectric constant is given by

$$\epsilon = \frac{\rho_t}{\rho_i + \rho_e + \rho_t} \qquad \ldots(5.42)$$

Using eq. (5.32), we obtain

$$\epsilon = \epsilon_e - \frac{\rho_t}{\rho_e + \rho_i + \rho_t} \qquad \ldots(5.43)$$

From eq. (5.30), one obtains

$$\frac{\rho_i}{\rho_e + \rho_i + \rho_t} = \frac{\Omega_i^2}{\omega^2} \qquad \ldots(5.44)$$

i.e.,

$$\epsilon = \epsilon_e - \frac{\Omega_i^2}{\omega^2} = \epsilon_e \left(1 - \frac{\omega_q^2}{\omega^2}\right) \qquad \ldots(5.45)$$

The Coulomb interaction in the medium is screened, so one finds

$$V(q, \omega) = \frac{e^2}{\epsilon_0 q^2 \epsilon(q, \omega)}$$

$$= \frac{e^2}{\epsilon_0 (q^2 + K_{TF}^2)} \left[1 - \frac{\omega_q^2}{\omega_q^2 - \omega^2}\right] \qquad \ldots(5.46)$$

Since ϵ can change sign and hence there is a range of ω for which this interaction can become attractive rather than repulsive. We further note that frequency appears in this interaction, i.e., the interaction is retarded.

Let us calculate the order of the attractive part of the potential V, the second term in eq. (5.46). We have noted that a dimensionless parameter, λ_{ep}, which appears in the BCS theory, is the product of the density of states at the Fermi level $N(E_F)$, and this attractive potential, i.e.,

$$\lambda_{ep} = N(E_F) \, |g| \qquad \ldots(5.47)$$

We have seen that a typical phonon vector is the Debye wave vector. Now, substituting all the quantities into eq. (5.46)

$$[V_{at}(q_D, 0) = -V], \text{ one obtains}$$

$$\lambda_{ep} = \frac{1}{2 + 4.7 a_0 \, (V_0 \, Z)^{-1/3}} \qquad \ldots(5.48)$$

For $Z = 1$ and $V_0^{1/3} = 3 \times 10^{-10}$ m, one obtains $\lambda_{ep} \sim 0.33$. For conventional superconductors one finds the experimental value of λ_{ep} is surprisingly close to this order of magnitude estimation [Al : 0.175 (expt.), 0.23 (calculated) ; Nb : 0.32 (expt.), 0.35 (calculated)].

So far, we have restricted ourself to show how an effective attractive interaction can occur due to the motion of ions. Now, we consider the direct interaction between an electron and the elastic wave propagating into solids, *i.e.*, phonons. We know that an electron propagating in the solid with a wave vector **k** can be scattered by a phonon of wave vector **q**. The phonon disappears and, due to momentum conservation, the electron wave vector becomes **K + q**. This is the basic process and the intensity of such a process is measured by the electron–phonon coupling constant (λ_{ep} or simply λ). In general, this coupling of electron-phonon is small and in the *BCS* theory of superconductivity, it is the product of this coupling and $N(E_F)$ enters the equations. We must note that λ is a dimensionless coupling parameter.

Now the question arises, whether there are any physical properties which depend on this electron–phonon interaction? We find one example is of electron transport. If there is a current in the metal and if electrons are scattered by phonons, the current will decrease. When temperature of the metal increases, the number of phonons also increases. This increases the scattering of electrons and consequently the resistivity of the metal. Obviously, the change of resistivity with temperature is directly related to the electron–phonon coupling. In order to have superconductivity state this coupling must be large and hence the room temperature resistivity connot be small. This helps to understand why the best metals are not superconducting. Pure metals like copper, gold and silver have very small resistivity at room temperature but are not found to exhibit superconductivity down to the lowest temperature measured.

5.8 MICROSCOPIC (BCS) THEORY

The original BCS theory assumes that the superconducting system can be described as an assembly of spin one-half fermions interacting through attractive potential; this represents a generalization of Cooper's work. The main assumptions are as follows:

(*i*) The superconducting ground state can be expressed in terms of Cooper pairs so that the states (**K**, − **K**) are simultaneously occupied or empty.

(*ii*) Various interactions in the normal and superconducting states are identical and we have to consider only the effective screened interaction.

(*iii*) The effective interaction is zero, except when two electrons of wave vectors **k** and **k**′ have energies close to the Fermi energy. Then the attractive interaction is taken as a constant, − g. If ξ_k is the energy measured from the Fermi energy, or more precisely, from the chemical potential μ, then we have

$$\xi_K = \epsilon_K - \mu \qquad \qquad ...(5.49)$$

One can impose the condition that, in order for the electrons to attract each other, the energies of both electrons have to satisfy the following criterion

$$|\xi_K| \text{ and } |\xi'_K| < k_B \Theta_D \qquad \qquad ...(5.50)$$

where Θ_D is the Debye temperature. Obviously, in *BCS* theory no detailed attempt is made to compute the actual interparticle potential; instead, the theory studies an artificial model of constant attraction of strength V acting between particles of opposite spin and momentum lying very near the Fermi surface. We must note that this model can account for most experimental results.

5.8.1 BCS Model

The next milestones were to develop the simplest model Hamiltonian containing the essential physics and to determine the superconducting ground state (ψ_0). The fact that $\Delta K \sim 1/\xi \sim 10^{-4} K_F$, suggests that ψ_0 is made up of states of electrons being excited above the normal ground state by a wave number of order ΔK. Because of the Pauli exclusion principle, electrons can only be excited into unoccupied states, and therefore it is apparent that only electronic states within a wave number range $10^{-4} K_F$ of the Fermi surface are involved in the superconducting phase transition, *i.e.*, about 10^{-4} of the electrons condense. The average spacing between these condensed electrons is roughly 10^{-6} cm. Consequently, within the volume occupied by a Cooper pair, the centres of approximately $(10^{-4}/10^{-6})^3 = 10^6$ other pairs will be found on the average. This means that the condensed pairs must overlap considerably.

This line of reasoning led to a reduction of the problem of determining the ground state of many Cooper pairs to the model (BCS) Hamiltonian:

$$H_{BCS} = \sum_{K,\sigma} \epsilon_k n_{k\sigma} - \sum_{K,K'} g_{K',K} b_K^+ b_{K'} \qquad ...(5.51)$$

The first term in eq. (5.51) gives the unperturbed energy of the electrons forming the pairs; the second is the pairing interaction in which a pair of electrons in $(K\uparrow, -K\downarrow)$ scatters to $(K'\uparrow, -K'\downarrow)$; and

$$b_K^+ = C_{K\uparrow}^+ C_{-K\downarrow}^+, \quad b_K = C_{-K\downarrow} C_{K\uparrow} \qquad ...(5.52)$$

are, respectively, *creation* and *annihilation* operators for a pair of electrons in $(K\uparrow, -K\downarrow)$. These operators obey the commutation relations of the so called imperfect Bose gas:

$$[b_K, b_{K'}^+] = (1 - n_{K\uparrow} - n_{-K\downarrow}) \delta_{KK'}$$

$$[b_K, b_{K'}] = 0 = [b_K^+, b_{K'}^+]_- \qquad ...(5.53)$$

The anticommutator of b_k and $b_{k'}$ is

$$\{b_K, b_{K'}\}_+ = 2 b_K b_{K'} (1 - \delta_{KK'})$$

from which it follows that $(b_K)^2 = 0$. According to Schrieffer [13], this point is essential to the theory and leads to the energy gap being present not only for dissociating a pair but also for making a pair move with a total momentum different from the common momentum of the rest of the pairs. It is this feature which enforces long-range order in the superfluid over macroscopic distances.

The determination of the ground state of the model Hamiltonian can now be accomplished by variational method. Since in the superconducting ground state, we are interested only in completely paired states, one may retain only the part of the first term in eq. (5.51) that connects pairs with zero net momentum

$$H_{BCS} = \sum_K 2\epsilon_K b_K^+ b_K - \sum_{KK'} g_{K'K} b_{K'}^+ b_K \qquad ...(5.54)$$

The superconducting ground state proposed by *BCS* [10] has the form

$$|\psi_0>_{BCS} = \Pi_K (u_K + v_K e^{i\phi} b_K^+) |0> \qquad ...(5.55)$$

where $|0>$ represents the vacuum state and u_k and v_k are variational parameters that may be assumed to real and, because of overall normalization of ψ_0,

$$u_K^2 + v_K^2 = 1 \qquad ...(5.56)$$

ϕ represents the phase of the order parameter. The probability v_K^2 of a pair state being occupied approaches **unity** for states far below the Fermi energy and **zero** for states far outside the Fermi sea.

The projection of the BCS state onto the *N*-particle space would lead to a function, in the coordinate representation, of the form [7]

$$<r_1, s_1; r_2, s_2, ... r_N, s_N | BCS>$$
$$= A \, \phi(\mathbf{r}_1 - \mathbf{r}_2) \chi_{12} \, \phi(\mathbf{r}_3 - \mathbf{r}_4) \chi_{34} \, ... \, \phi(\mathbf{r}_{N-1} - \mathbf{r}_N) \chi_{N-1, N} \qquad ...(5.57)$$

where A is the operator that antisymmetrizes the entire function. The function $\phi(r)$ is the relative coordinate function (not to be confused with the phase) of a pair in real space ... the same for all pairs ... and χ_{ij} is the corresponding spin function $\uparrow(i) \downarrow(j)$. Thus, within the pairing approximation, all pairs are in the **same** state in the ground state wave function. The function $\phi(r)$ in the original *BCS* theory was essentially spherically symmetric (*s*-wave symmetry). In the HTSC cuprates there is mounting evidence that this function has *d*-wave symmetry, as will be discussed in chapter 8.

The variational calculation, which is pretty standard [7], yields the following results:

$$v_K^2 = \frac{1}{2}\left[1 - \frac{(\epsilon_K - \mu)}{E_K}\right] \qquad ...(5.58)$$

where
$$E_K = \sqrt{(\epsilon_K - \mu)^2 + \Delta_K^2} = \sqrt{\xi_K^2 + \Delta_K^2} \qquad ...(5.59)$$

and μ (a Lagrange multiplier in the variational calculation) has the physical significance of the *chemical potential* (Fermi-energy); while Δ_k, called the energy gap, satisfies the so called gap equation:

$$\Delta_K = -\sum_{K'} \frac{g_{KK'} \Delta_{K'}}{2 E_{K'}} \qquad ...(5.60)$$

where $g_{KK'}$ are the matrix elements representing the effective electron-electron interactions. Excitations above the superconducting ground state are superpositions of electrons and holes that require an energy (measured with respect to the Fermi energy) given by eq. (5.59), where $\xi_K (= \epsilon_K - \mu)$ is the energy of a quasiparticle excitation above the Fermi surface for a normal Fermi liquid. The minimum energy required to create a hole-like excitation and an electron-like excitation of wave vector (momentum) **K** is the BCS energy gap, $2\Delta_K$, for states such that $K \sim K_F$ and $\xi_K \sim 0$ (Fig. 5.1). A gap thus opens up at the Fermi surface in the superconducting ground state. Δ is thus a fundamental quantity introduced by BCS and is called the half width of the energy gap or the superconducting order parameter.

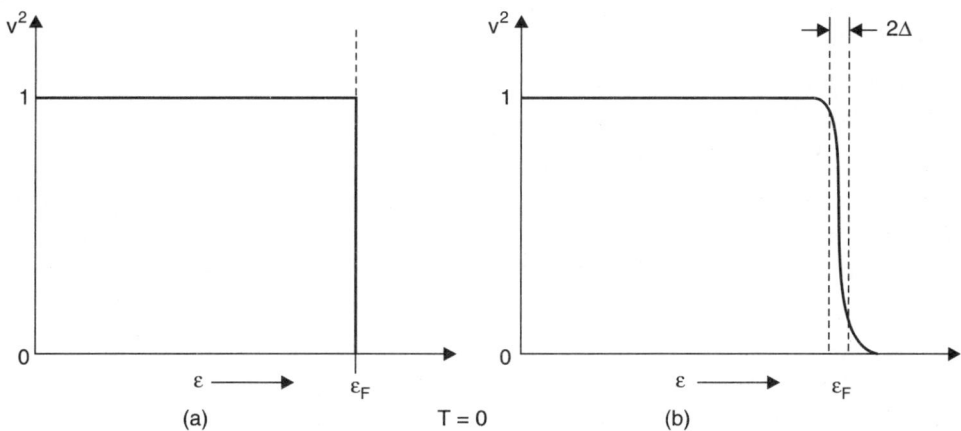

Fig. 5.1 Filling of the states **K** in the normal state (*a*) and in the superconducting state (*b*)

Fig. 5.2 shows the density of states for single electrons in the superconducting state around the Fermi energy. The energy states present in the normal state within the energy gap are shifted to edges. Therefore the density of states is strongly enhanced at the edges of the energy gap. The BCS expression for the single electron density of states $N_s(E - E_F)$ in a superconductor in the vicinity of the Fermi energy is

$$N_s(E - E_F) = N_n(E_F) \frac{|E - E_F|}{\sqrt{(E - E_F)^2 - \Delta^2}} \qquad \ldots(5.61)$$

where $|E - E_F| \geq \Delta$. The variation of the density of states $N_n(E)$ in the normal state can be neglected because $E_F \gg 2\Delta$.

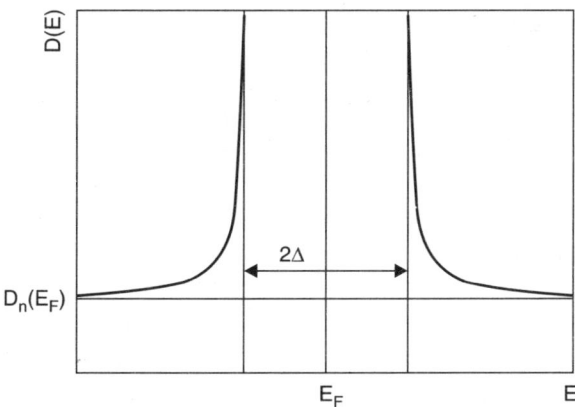

Fig. 5.2 Energy gap and density of states for single electrons in the conducting state around the Fermi energy E_F

Finally, one can point out that the density of excitations in the superconductor is given by

$$N(E) = \begin{cases} 0 & \text{for } E < \Delta \\ N(E_F) \dfrac{E}{\sqrt{E^2 - \Delta^2}} & \text{for } E > \Delta \end{cases} \qquad ...(5.61(a))$$

This means that the missing excitations for $E < \Delta$ in the superconducting state are packed up at an energy $E \geq \Delta$. This gives the divergence of the density of states at Δ.

The higher state of a superconducting system can be described as the creation of one or more quasiparticles characterized by a temperature-dependent relation

$$E(p) = \{[\Delta(T)]^2 + [(p^2 - p_F^2)/2m]^2\}^{1/2} \qquad ...(5.61(b))$$

where $p_F = m v_F$ is the Fermi momentum. Equation 5.61 (b) shows that $E(p)$ is never less than $\Delta(T)$, so that the excited states are separated from the ground state by a finite energy $\Delta(T)$, often called the energy gap. Thus a superconductor cannot absorb an arbitrarily small amount of energy.

The original BCS theory considered an isotropic system with a spherical Fermi surface and a constant negative electron-electron interaction matrix element $g_{KK'} = -g$ (within the cutoff energy $\omega_c \sim \omega_D$). In this case, the gap parameter $\Delta_k = \Delta$ is a constant independent of direction in momentum space (**K**–space), as are the parameters u_K and v_K. One obtains in the simple model for which

$$g_{KK'} = \begin{cases} |g| & (|\epsilon_K - \mu| < \hbar \omega_D \text{ and } (\epsilon_{K'} - \mu) < \hbar \omega_D) \\ 0 & \text{otherwise} \end{cases} \qquad ...(5.61(c))$$

Such isotropic pairing is known as *s*-wave pairing. However, it is evident that other non-trivial solutions to eq. (5.60) are possible if the Fermi surface is nonspherical and if the effective electron-electron interaction matrix elements strongly depend on momentum. This appears to be the case for the HTSC cuprates, as will be discussed in chapter 8.

Equation (5.60) in the light of approximation of eq. (5.61) yield,

$$\Delta_K = \Delta \text{ (for } |\epsilon_K - \mu| < \hbar \omega_D) \text{ and zero otherwise elseswhere,}$$

$$\Delta = \hbar \omega \exp[-1/N(0)|g|] \qquad ...(5.62)$$

$N(0)$ is the density of states at the Fermi surface. Δ depends on the temperature and it is fairly straightforward to generalize the gap equation to finite temperatures:

$$\Delta_K(T) = -\frac{1}{2} \sum_{k'} g_{KK'} \frac{\Delta_{K'}(T)}{E_{K'}(T)} \tan h \left(\frac{E_{K'}(T)}{k_B T} \right) \qquad ...(5.63)$$

If $g_{KK'}$ is approximated by eq. (5.61), then Δ_K is again of the form Δ (independent of K) and $\Delta(\beta)$ ($\beta = 1/k_B T$) satisfies

$$\frac{1}{N(0)|g|} = \int_0^{\hbar \omega_D} \frac{d\xi}{(\xi^2 + \Delta^2)^{1/2}} \tan h \left[\frac{1}{2} \beta (\xi^2 + \Delta^2)^{1/2} \right] \qquad ...(5.64)$$

This may be solved numerically. As T increases from zero, Δ decreases as shown in Fig. 5.3, vanishing at transition temperature T_c.

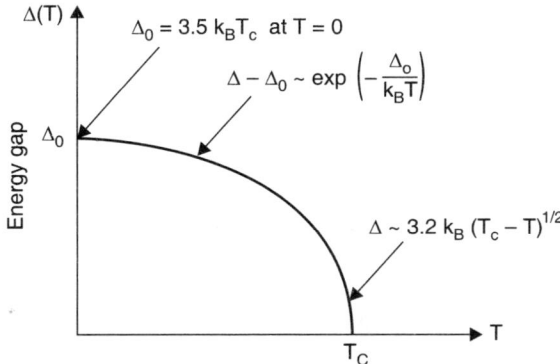

Fig. 5.3 A plot of temperature dependence of the energy gap $\Delta(T)$. Note that Δ vanishes with infinite slope as $T \to T_c$ leading to second order phase transition

Near the critical temperature the half width of the energy gap is approximately

$$\Delta(T) = 3.2\, k_B\, T_c\, (1 - T/T_c)^{1/2} \qquad \ldots(5.64\,(a))$$

The transition temperature is given by

$$\frac{1}{N(0)|g|} = \int_0^{\hbar\omega_D} \frac{d\xi}{\xi} \tan h \frac{\xi}{2k_B T_c} \qquad \ldots(5.65)$$

So in the weak coupling limit ($N(0)\,|\,g\,| \ll 1$)

$$k_B T_c = 1.14\, \hbar\, \omega_D \exp(-1/N(0)\,|\,g\,|) \qquad \ldots(5.66)$$

Because of the presence of the exponential factor in eq. (5.66), the critical temperature T_c is much smaller than the Debye temperature. This is the principal theoretical reason why T_c is so low in the superconductors discovered in the period 1911–1986. The exponential dependence of T_c on the electronic parameter $N(0)\,|\,g\,|$ also explains why the parameters pertaining to the electronic structure, which are of the order 1 eV or more, respond to phase transitions on an energy scale that is three orders of magnitude smaller ($k_B\, T_c \sim 1$ meV). Effects with such a nonanalytic dependence of transition temperature on the coupling constant cannot be obtained in any order of perturbation theory starting with the normal state as an initial state. One obtains a similar type of effect in the studies of the Kondo effect.*

ω_D in eq. (5.66) is the Debye frequency which can be extracted from the contribution of the phonons to the specific heat. ω_D varies from one metal to another but only over a small

---

*An abnormal temperature dependent effect displayed in the thermal, electrical and magnetic properties of nonmagnetic metals containing very small quantities of magnetic impurities. A striking example of *Kondo effect* is the anomalous logarithmic increase in the electrical resitivity ρ with decreasing temperature. Other properties such as specific heat, magnetic susceptibility, and thermoelectric power also display anomalous behaviour because of this effect. Theories show that Kondo temperature T_K is given by

$$T_K = \epsilon_F\, J^{1/2}\, (-1/nJ)$$

where $J \to$ exchange interaction, $n \to$ number of quantum states per unit energy often called density of states and $\epsilon_F \to$ kinetic energy of the free conduction electrons (more explicitly, the Fermi energy). A wide variety of dilute magnetic alloys exhibit Kondo effect. For a given alloy, the values of J, n and ϵ_F are unique and cannot be adjusted.

range of values. Instead of ω_D, one can use the Debye temperature ($\hbar \omega_D = k_B \Theta_D$); Θ_D ranges from 100 K to 500 K. Such a range of Θ_D (and $N(0) |g| = \lambda \sim 0.3$) implies a maximum 'BCS' value of $T_c \sim 25$ K.

Also at $T = 0°$ K,

$$\Delta_0 = \frac{\hbar \omega_D}{\sinh [1/N(0)|g|]} \qquad \ldots(5.67)$$

which in the weak coupling limit, gives

$$\Delta(0) \text{ or } \Delta_0 \simeq 2 \hbar \omega_D \exp(-1/N(0) |g|) \qquad \ldots(5.68)$$

Δ_0 is called the zero temperature gap. The critical temperature T_c is related to Δ_0 by

$$\frac{2\Delta_0}{k_B T_c} = 3.52 \qquad \ldots(5.69)$$

Moreover $\quad \dfrac{\Delta_0}{k_B T_c} = \pi e^{-\gamma} \approx 1.76 \qquad \ldots(5.70)$

Where $\gamma \approx 0.5772$ is *Euler's constant* in the weak coupling limit. $2\Delta_0/k_B T_c$ is a universal constant independent of particular material. This is frequently used as a test for the applicability of the BCS model. This ratio is in reasonably good agreement with experiment for weak coupling superconductors like Al, but tends to be larger in strongly coupled and highly anisotropic superconductors. Experimentally determined values of this ratio are typically between 3.2 and 4.6 for superconducting elements. Fig. 5.4 shows the width of the energy gap $2\Delta(0)$ versus $k_B T_c$ for various superconducting elements. The solid line corresponds to the ratio $2\Delta(0)/k_B T_c = 3.5$ predicted from the BCS theory.

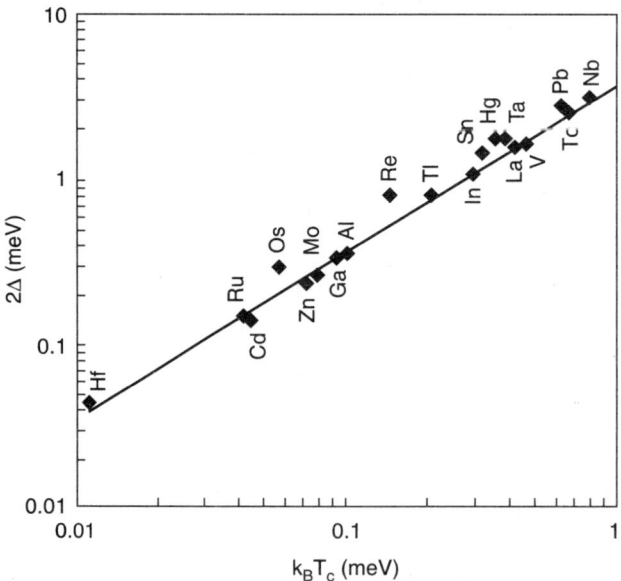

Fig. 5.4 Width of the energy gap $2\Delta(0)$ versus $k_B T_c$ for various superconducting elements. The solid line corresponds to the ratio of $2\Delta(0)/k_B T_c = 3.5$ predicted by the BCS theory

Microscopic Theory of Superconductivity

The absorption of electromagnetic waves can be used to determine the energy gap of a superconductor. As long as the phonon energy $h\nu$ is smaller than the energy gap 2Δ far infrared radiation is reflected from the surface of a superconductor. An energy of at least 2Δ is required to break up a Cooper pair in two single electrons. In addition, empty energy states are required for the single electrons resulting from pair breaking because they obey Pauli's exclusion principle. In normal conductors, empty energy states are always available above the Fermi energy E_F and therefore there exists no threshold energy for the absorption of radiation. As a consequence different far infrared reflectivities result for the normal and the superconducting state. Fig. 5.5 shows the absorption edges of selected superconductors in the far infrared reflectivity at temperature of ≈ 1.3 K. To determine the reflectivity in the normal state at this temperature a magnetic field sufficiently large to destroy superconductivity was applied.

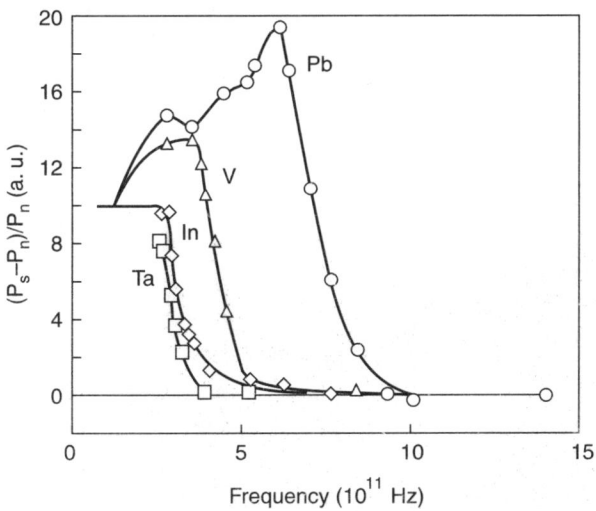

Fig. 5.5 Absorption edges of selected superconductors at T $\approx$ 1.3 K

Perhaps the most direct method of probing the BCS energy gap is to measure the tunnelling conductance of a normal metal-insulator-superconductor (NIS) tunnel junction. This technique was pioneered by Giaever [14], who used it to confirm the density of states and temperature dependence of the energy gap predicted by BCS. A plot of differential conductance dI/dV as a function of bias voltage V directly reflects the energy dependence of the density of states. For bias voltages less than Δ/e, the BCS gap suppresses the conductance at temperature below T_c. A peak in the conductance is observed when the bias voltage slightly exceeds Δ/e, due to the corresponding peak in the density of states. This method has been used extensively to characterize both classical [15] and HTSC cuprates [16, 17]. Many such measurements suggest the existence of states within the gap in HTSC cuprates, which may be due to *line nodes* where the gap Δ_k goes to zero along certain directions in K-space, reflecting their unusual pairing state symmetry.

5.8.2 Thermodynamic Functions

In the uniform system, the difference of the free energies in the superconducting state (F_S) and normal state F_N is given by

$$F_S - F_N = V \int_0^\Delta d\Delta \, (\Delta^2) \frac{\left(\frac{1}{g|N(0)|}\right)}{d\Delta} \qquad \ldots(5.71)$$

where V is volume. Using eq. (5.64), one obtains,

$$(F_S - F_N)/V = N(0) \int_0^{\hbar\omega_D} d\xi \int_0^\Delta d\Delta \, (\Delta^2) \cdot \frac{\partial}{\partial \Delta} \frac{\tanh\left(\frac{1}{2}\beta E\right)}{E} \qquad \ldots(5.72)$$

where $E = (\xi^2 + \Delta^2)^{1/2}$ and $\beta = 1/k_B T$

$$= N(0) \int_0^{\hbar\omega_D} d\xi \left[\frac{\Delta^2}{E} \tanh\left(\frac{1}{2}\beta E\right) - 2 \int_0^\Delta d\Delta \frac{\Delta}{E} \tanh\left(\frac{1}{2}\beta E\right) \right] \qquad \ldots(5.73)$$

where the second line is obtained with partial integration. The first term is just the right side of the gap equation (5.64), and the second term can be simplified by changing variables from Δ to $E' = [(\Delta)^2 + \xi^2]^{1/2}$. In this way, one finds

$$\frac{F_S - F_N}{V} = N(0) \Delta^2 \int_0^{\hbar\omega_D} \frac{d\xi}{E} \tanh\left(\frac{1}{2}\beta E\right) - 4 \frac{N(0)}{\beta} \int_0^{\hbar\omega_D} d\xi \ln \frac{\cosh \frac{1}{2}(\beta E)}{\cosh \frac{1}{2}(\beta \xi)}$$

$$= \frac{\Delta^2}{g} - \frac{4N(0)}{\beta} \int_0^{\hbar\omega_D} d\xi \left[\ln(1 + e^{-\beta E}) \right.$$

$$\left. + \frac{1}{2}\beta(E - \xi) \right] + \frac{4N(0)}{\beta} \int_0^{\hbar\omega_D} d\xi \ln(1 + e^{-\beta \xi}) \qquad \ldots(5.74)$$

Since $\hbar\omega_D \gg k_B T$, the last integral on the right may be extended to infinity. An easy calculation then shows that it equals the first temperature-dependent correction to the thermodynamic potential in the normal state

$$\frac{4N(0)V}{\beta} \int_0^\infty d\xi \ln(1 + e^{-\beta \xi}) = \frac{1}{3} N(0) V \pi^2 (k_B T)^2 \approx [F_N(T) - F_n(0)] \qquad \ldots(5.75)$$

In addition, it can readily verified that

$$-2N(0) \int_0^{\hbar\omega_D} d\xi \, (E - \xi) \approx -N(0)\left(\frac{1}{2}\Delta^2 + \Delta^2 \ln \frac{2\hbar\omega_D}{\Delta}\right)$$

$$= -\frac{1}{2} N(0) \Delta^2 - \frac{\Delta^2}{|g|} - N(0) \Delta^2 \ln \frac{\Delta_0}{\Delta} \qquad \ldots(5.76)$$

where we have made use of

$$1 = \frac{|g|N(0)}{2} \int_{-\hbar\omega_D}^{\hbar\omega_D} \frac{d\xi}{(\Delta^2 + \xi^2)^{1/2}} = |g| N(0) \int_0^{\hbar\omega_D} \frac{d\xi}{(\Delta^2 + \xi^2)^{1/2}}$$

$$\approx |g| N(0) \ln \frac{2\hbar\omega_D}{\Delta} \qquad \ldots(5.77)$$

Microscopic Theory of Superconductivity

A combination of eqs. (5.74) to (5.76) yields

$$\frac{F_S - F_N}{V} = -\frac{1}{2} N(0) \Delta^2 - N(0) \Delta^2 \ln\left(\frac{\Delta_0}{\Delta}\right)$$

$$- 4 N(0) k_B T \int_0^{\hbar\omega_D} d\xi \ln(1 + e^{-\beta E}) + \frac{1}{3} \pi^2 N(0) (k_B T)^2 \quad ...(5.78)$$

where the equalities $T < T_c \ll \theta \ll T_F$ have been assumed.

The expression (5.78) reduces at $T = 0$ to

$$F_S - F_N = -\frac{1}{2} V N(0) \Delta^2 \qquad ...(5.78(a))$$

Expression (5.78) also allows us to evaluate the leading low-temperature correction, which arises solely from the normal state contribution because $\Delta - \Delta_0$ vanishes exponentially for $T \to 0$. One can easily show that for all $T \le T_c$, $N_S \approx N_n$, where N_S and N_n are the mean number of the particles in the superconducting and normal states. This helps us to reinterpret eq. (5.78) as

$$\frac{F_S - F_N}{V} \approx -\frac{1}{2} N(0) \Delta_0^2 + \frac{1}{3} \pi^2 N(0) (k_B T)^2 \quad T \to 0 \qquad ...(5.79)$$

Since $F_S(T, 0) = F_N(T, 0) - H_c^2/8\pi$, we obtain from eq. (5.79)

$$H_c(T) = [4\pi N(0) \Delta_0^2]^{1/2} \left[1 - \frac{e^{2\gamma}}{3}\left(\frac{T}{T_c}\right)^2\right] \approx H_c(0)\left[1 - 1.06\left(\frac{T}{T_c}\right)^2\right]_{T \to 0} \quad ...(5.80)$$

where $\quad H_c(0) = [4\pi N(0) \Delta_0^2]^{1/2} \qquad ...(5.81)$

is the critical field at $T = 0$ and γ is Euler's constant whose value is ≈ 0.5772. Thus $e^\gamma \approx 1.781$.

Since $N(0)$ determines the normal-state specific heat, i.e.,

$$\frac{C_n}{V} = \frac{2\pi^2}{3} N(0) k_B^2 T \qquad ...(5.82)$$

Equations (5.70) and (5.82) together predict a **second** universal constant

$$\frac{T_c C_n(T_c)}{H_c^2(0) V} = \frac{e^{2\gamma}}{6\pi} \approx 0.168 \qquad ...(5.83)$$

which is independent of the material.

We may note that each of these parameters is measurable, and experimental confirmation is satisfactory.

It is also interesting to evaluate F_S itself, which may be obtained from eqs. (5.75) and (5.78)

$$\frac{F_S}{V} = \frac{F_N(T=0)}{V} - \frac{1}{2} N(0) \Delta^2 \left(1 + 2\ln\frac{\Delta_0}{\Delta}\right)$$

$$- 4 N(0) k_B T \int_0^{\hbar\omega_D} d\xi \ln(1 + e^{-\beta E}) \qquad ...(5.84)$$

In the present low-temperature limit, it is permissible to evaluate the integral by setting $\hbar\omega_D \to \infty$ and $\Delta \approx \Delta_0$; an approximate calculation yields,

$$\int_0^\infty d\xi \log\{1 + \exp[-\beta(\xi^2 + \Delta_0^2)^{1/2}]\} \approx \frac{1}{2} e^{-\beta\Delta_0} (2\pi\Delta_0 \beta^{-1})^{1/2}$$

Combining the above with

$$\Delta(T) \approx \Delta_0 - 2\pi \Delta_0 (k_B T)^{1/2} e^{-\Delta_0/k_B T} \quad T \ll T_C$$

and eq. (5.84), one obtains

$$\frac{F_S}{V} \approx \frac{F_N (T=0)}{V} - \frac{1}{2} N(0) \Delta_0^2 - 2 N(0) \left(\frac{2\pi\Delta_0}{\beta^3}\right)^{1/2} e^{-\beta\Delta_0} \quad \ldots(5.85)$$

The electronic specific heat in the superconducting state can then be obtained by differentiating (5.85), we have

$$\frac{C_S}{V} \approx 2N(0) \Delta_0 k_B (2\pi)^{1/2} \left(\frac{\Delta_0}{k_B T}\right)^{1/2} e^{-\Delta_0/k_B T} \quad T \to 0 \quad \ldots(5.86)$$

Expression (5.86) clearly exhibits the energy gap.

In the preceding calculations, we have considered only the low-temperature behaviour, where $\Delta - \Delta_0$ is exponentially small. Although a general evaluation of eq. (5.78) for all $T < T_c$ requires numerical analysis, it is possible to derive explicit expressions near T_c, where $\beta\Delta \ll 1$ provides a small parameter. We start from the gap equation.

$$\Delta = \frac{|g|}{\beta\hbar} \sum_n \int \frac{d^3k}{(2\pi)^3} \frac{\hbar\Delta}{(\hbar\omega_n)^2 + E_k^2} \quad \ldots(5.87)$$

which may be expanded in powers of Δ.

$$\frac{1}{|g|} = \frac{2N(0)}{\beta} \int_0^{\hbar\omega_D} d\xi \sum_n \frac{1}{(\hbar\omega_n)^2 + \xi^2 + \Delta^2}$$

$$\approx \frac{2N(0)}{\beta} \int_0^{\hbar\omega_D} d\xi \sum_n \left\{\frac{1}{(\hbar\omega_n)^2 + \xi^2} - \frac{\Delta^2}{[(\hbar\omega_n)^2 + \xi^2]^2} + \ldots\right\}$$

The derivative with respect to Δ can be easily evaluated with eq. (5.71) which gives

$$\frac{F_S - F_N}{V} = -\frac{N(0)\Delta^4}{\beta} \int_0^{\hbar\omega_D} d\xi \sum_n \frac{1}{[(\hbar\omega_n)^2 + \xi^2]^2} \quad \ldots(5.88)$$

Since $\hbar\omega_D \gg k_B T_c$, the integral may be extended to infinity.

$$\frac{F_S - F_N}{V} = -\frac{N(0)\Delta^4}{\beta} \sum_n \frac{\pi}{4} \frac{1}{|\hbar\omega_n|^3}$$

$$= -\frac{1}{2} N(0) \Delta^4 \left(\frac{\beta}{\pi}\right)^2 \sum_{n=0}^\infty \frac{1}{(2n+1)^3} = -\frac{7\zeta(3)}{8} \frac{N(0)\Delta^4 \beta^2}{2\pi^2}$$

$$= -\frac{8}{7\zeta(3)} N(0) (\pi k_B T_c)^2 \frac{1}{2}\left(1 - \frac{T}{T_c}\right)^2 \qquad ...(5.89)$$

where the symmetry of the summand has been used in the second line and the explicit form of $\Delta(T)$

$$\Delta(T) \approx k_B T_c \pi \left[\frac{8}{7\zeta(3)}\right]^{1/2} \left(1 - \frac{T}{T_c}\right)^{1/2} \approx 3.06 \, k_B T_c \left(1 - \frac{T}{T_c}\right)^{1/2} \qquad ...(5.90)$$

$$T_c - T \ll T_c$$

has been used in the last line.

We obtain the expression for critical field as

$$B_c(T) = B_c(0) \, e^\gamma \left[\frac{8}{7\zeta(3)}\right]^{1/2} \left(1 - \frac{T}{T_c}\right)$$

$$\approx 1.74 \, B_c(0) \left(1 - \frac{T}{T_c}\right) \quad T \to T_c \qquad ...(5.91)$$

where $B_c(0)$ has been taken from eq. (5.81). We may note that eqs. (5.80) and (5.91) are very similar to the phenomenological relation

$$B_c(T) = B_c(0) \left[1 - \left(\frac{T}{T_c}\right)^2\right]$$

Nevertheless there are small but distinct differences (Fig. 5.5b), and the BCS predictions provide a better fit for most simple superconductors. Since $B_c^2/8\pi$ is the actual difference in free-energy density between the normal and superconducting states, the excellent agreement between theory and experiment justifies the assumption that the Hartree-Fock energy is same in both states. This assumption is extremely difficult to justify **a priori** because the condensation energy ($\approx 10^{-7}$ eV/particle) is so much smaller than the Hartree–Fock energy (≈ 1 to 10 eV/particle). Using (5.91), one can evaluate the jump in the specific heat at T_c,

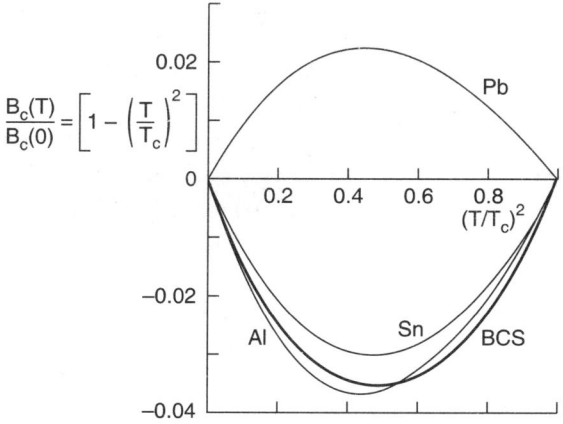

Fig. 5.5 (b) Difference between actual critical field $B_c(T)/B_c(0)$ and the empirical curve $\{1 - (T/T_c)^2\}$

$$\frac{1}{V}(C_s - C_n)\big|_{T_C} = \frac{8}{7\zeta(3)} N(0) \pi^2 k_B^2 T_c \qquad ...(5.92)$$

and a combination with eq. (5.82) yields

$$\left(\frac{C_S - C_n}{C_n}\right)_{T_c} = \frac{12}{7\zeta(3)} \approx 1.43 \qquad ...(5.93)$$

in reasonable agreement with experiment for low temperature superconductors.

5.9 PROPERTIES

(i) **Entropy.** By regarding E_k [eq. (5.59)] as representing electron excitations across the gap, one can write the expression for the entropy of a superconductor as

$$S = -2 k_B \sum_K [f_K \ln f_K + (1 - f_K) \ln (1 - f_K)] \qquad ...(5.94)$$

where $f_K = f(E_K) = [1 + \exp(\beta E_K)]^{-1}$
is the Fermi Dirac distribution function. Hence the **free energy** of the superconducting state is

$$F_S = 2 \sum_K E_K f_K - TS \qquad ...(5.95)$$

We must note that the thermodynamic properties are determined fully only if the chemical potential $\mu = \mu(T)$ and the temperature dependence of the superconducting gap $\Delta_K = \Delta_K(T)$ are explicitly determined. Since, only then is the spectrum of single particle excitations (characterized by the energy $\{E_K\}$) uniquely determined. The quantity $\Delta(T)$ is determined from eq. (5.64). The chemical potential μ is determined from the conservation of number N of particles, *i.e.*, from the condition

$$\sum_K f_K = N \qquad ...(5.96)$$

The temperature dependence of the gap in the isotropic case is schematically shown in Fig. 5.3.

(ii) **Critical Field.** By calculating the difference of the free energies $F_S - F_N$ in superconducting (F_S) and normal phases (F_N) and equating the difference with the magnetic free energy $B_c^2 V/8\pi$ (V is the volume of the system), one can obtain an approximate relation of the form

$$B_c(T) \approx B_c(0) \left[1 - 1.06 \left(\frac{T}{T_c}\right)^2\right] \qquad ...(5.97)$$

Also

$$B_c(T) \approx 1.74\, B_c(0) \left(1 - \frac{T}{T_c}\right) \quad T \to T_c \qquad ...(5.97(a))$$

These two relations are very similar to the phenomenological equation

$$B_c(T) = B_c(0)\left[1 - \left(\frac{T}{T_c}\right)^2\right] \qquad ...(5.97(b))$$

Nevertheless there are small but distinct differences, and BCS predictions provide a better fit for most simple type-I superconductors.

where $B_c(0) = [4\pi N(0)\Delta_0^2]^{1/2}$ is the critical field at $T = 0$. For applied field $B > B_c$, superconducting is destroyed because in the thermodynamic critical field B_c the spin singlet bound state is destroyed by the thermal fluctuations. The pair binding energy is then effectively overcome by the magnetic energy so that the pairs break up into single particles. Truly speaking, this type of behaviour characterizes the so called type-I superconductors.

(*iii*) **Specific heat.** One can calculate the specific heat from the standard thermodynamic analysis

$$C_{es} = -T\left(\frac{\partial^2 F_S}{\partial T^2}\right)_V \qquad ...(5.98)$$

The behaviour of specific heat in the normal metal at low temperatures is obtained as

$$C_{en} = \frac{2\pi^2}{3} N(E_F) k_B T \qquad ...(5.99)$$

one obtains at $T = T_c$ a discontinuity of the form

$$\frac{C_{es} - C_{en}}{C_{en}} = 1.43 \qquad ...(5.100)$$

i.e., the discontinuity itself is given by

$$\Delta C (= C_{es} - C_{en}) = 9.4\, N(E_F)\, k_B^2\, T_c \qquad ...(5.101)$$

Here C_{en} is the specific heat at T_c for the material in its normal state. At low temperatures, the specific heat decreases exponentially with temperature

$$C_{es} \sim \exp\left[-\frac{\Delta(0)}{k_B T}\right] \qquad ...(5.102)$$

for the special case of an isotropic gap.

However, if the gap is anisotropic $[\Delta = \Delta_k(T)]$ and has lines of zeros (along which $\Delta_k = 0$), then the low temperature dependence of C_{es} does not follow eq. (5.102) but rather a power law T^n, with n depending on the details of the gap anisotropy.

The specific heat grows with T because the number of thermally broken pairs increases with rising temperature; eventually, at $T = T_c$, $(k_B T_c \sim \Delta_0)$, all bound pairs dissociate thermally at which point C_{es} reaches maximum. If the temperature is raised further (above T_c), the excess specific heat drops rapidly to zero since no pairs are left to absorb the energy. This type of behaviour is observed in superconductors with an isotropic gap, *e.g.* Hg, Sn. We must note that this interpretation of thermal properties is based on the single particle excitation spectrum [eq. (5.59)]. We have disregarded any fluctuation phenomen near T_c, as well as collective excitations of the condensed system. One can easily show that the large coherence lengths $\xi \sim 10^3 - 10^4$ Å encountered in classical superconductors [18, 19] is related to the absence of critical behaviour near T_c. This is not the case in HTSC cuprates. Obviously, HTSC cuprates open up the possibility of studies of critical phenomena in superconducting systems.

Obviously, a measurement of the low temperature specific heat allows us to estimate the gap in the electronic spectrum. It is the existence of the gap which causes this exponential decrease. We must note that gap also exist in a semiconductor. This type of measurement gave the first confirmation of the existence of an energy gap in a superconductor.

(iv) **Penetration Depth** (λ). The spin part of the static magnetic susceptibility vanishes at $T \to 0$. This is a direct consequence of the binding of electrons in the condensed state into singlet pairs. Therefore, the Meissner effect is present at $T = 0$ because of the orbital part of the susceptibility (an electron-pair analog of the Landau diamagnetism of single electron in a normal electron gas). The expulsion of the magnetic flux from the bulk is measured in terms of the so-called London penetration depth $\lambda = \lambda(T)$, which characterizes the decay of the magnetic induction inside the sample. It decays according to

$$B(z) = B_a \exp(-z/\lambda) \qquad \ldots(5.103)$$

where the z-direction is perpendicular to the sample surface and the applied magnetic field H_a is parallel to it. The temperature dependence of the penetration depth is given by

$$\frac{\lambda(T)}{\lambda(0)} = \left[\frac{\Delta(T)\tanh\left(\frac{\Delta}{2k_B T}\right)}{\Delta_0}\right] \approx \left[1 - \left(\frac{T}{T_c}\right)^4\right]^{-1/2} \qquad \ldots(5.104)$$

This result has been derived under the assumption that the coherence length† $\xi \approx \hbar V_F/\Delta$ is much larger than λ.

We must note that for a bulk sample of dimension $d \gg \lambda$ the induction $B \equiv 0$ almost everywhere. This condition helps to determine the magnetic susceptibility χ of a superconductor regarded as an ideal diamagnet; in cgs units, $\chi \equiv M/H = -1/(4\pi)$.

(v) **Acoustic attenuation.** The sound absorption coefficient α_s in the superconducting phase is related to that in the normal phase α_n by

$$\frac{\alpha_s}{\alpha_n} = \frac{2}{1 + \exp\left(\frac{\Delta}{k_B T}\right)} \qquad \ldots(5.105)$$

This is a very simple result. This is why experimental results for (α_s/α_n) are used to determine the temperature dependence of the gap Δ.

A complete discussion of states within the BCS theory can be found in references [4, 5, 7, 19 – 28].

---

†One can estimate the coherence length in a superconductor by using uncertainty relation $\Delta p\, \xi_0 = \hbar$, where Δp is a change of electron momentum (at $\epsilon = \epsilon_F$) due to the attractive interaction, which can be estimated from the corresponding change of particle kinetic energy $\Delta E = v_F \Delta p$. Taking $\Delta E = \Delta_0$, one obtains the desired estimate of ξ.

5.10 GL THEORY AND BCS THEORY

One can derive GL theory from the microscopic BCS theory. Our basic interest is to know how the phenomenological coefficients introduced by *GL* are related to the quantities introduced in the BCS theory. Instead of giving the values of these coefficients, we will give the values of the quantities derived from these coefficients which have important physical meaning. We first take coherence length. BCS theory introduces the BCS coherence length ξ_0 which is given by

$$\xi_0 = 0.18 \frac{\hbar v_F}{k_B T_C} \qquad ...(5.106)$$

where v_F is the Fermi velocity of the order of 10^6 m/s in most metals. One obtains for $T_c \sim 10$ K, $\xi_0 \sim 1800$ Å. This length ξ_0 is temperature independent. We have seen that the relation to the temperature dependent coherence length $\xi(T)$ of the *GL* theory, depends on the mean free path l_e of the electrons. Here we cite only two limiting cases. For a clean material where the mean free path is much greater than the BCS coherence length, one finds

$$\xi(T) = 0.74 \, \xi_0 \left(\frac{T_c}{T_c - T}\right)^{1/2}, \quad l_e \gg \xi_0 \qquad ...(5.107)$$

In the dirty limit, *i.e.*, opposite limit, one finds

$$\xi(T) = 0.85 \, \sqrt{\xi_0 l_e} \left(\frac{T}{T_c - T}\right)^{1/2}, \quad l_e \ll \xi_0 \qquad ...(5.108)$$

This has an important physical consequence, *i.e.*, when one alloys superconductors with non-magnetic impurities, one in general does not change T_c and v_F very much. This means ξ_0 is rather insensitive to alloying. We must note that by decreasing the mean free path by alloying, we decrease the *GL* coherence length $\xi(T)$.

Now, we consider the penetration depth. Let us first introduce the London penetration depth, $\lambda_L = (m/\mu_0 \, n e^2)^{1/2}$.

One finds the penetration depth in the *GL* domain in the two limiting cases as

$$\lambda(T) = \frac{1}{\sqrt{2}} \lambda_L \left(\frac{T_c}{T_c - T}\right)^{1/2}, \quad l_e \gg \xi_0 \qquad ...(5.109)$$

$$\lambda(T) = 0.64 \, \lambda_L \sqrt{\frac{\xi_0}{l_e}} \left(\frac{T_c}{T_c - T}\right)^{1/2}, \quad l_e \ll \xi_0 \qquad ...(5.110)$$

It is interesting to note that contrary to the coherence length, which decreases as the mean free path becomes shorter by alloying, the penetration depth increases. One obtains the *GL* parameter $\kappa = \lambda/\xi$ in the two limits as

$$\kappa = 0.96 \frac{\lambda_L}{\xi_0}, \quad l_e \gg \xi_0 \qquad ...(5.111)$$

$$\kappa = 0.715 \frac{\lambda_L}{l_e}, \quad l_e \ll \xi_0 \qquad ...(5.112)$$

We note that κ increases with decreasing mean free path. One can produce a type-II superconductor by alloying a type-I superconducting material. One can also express these results in another way. One can use directly measurable quantities of the normal state instead of l_e and λ_L. One can estimate the mean free path from the electrical resistivity,

$$1/\rho = \frac{2}{3} N(E_F) e^2 v_F l_e \qquad ...(5.113)$$

where one can express λ_L in the form

$$\lambda_L = \left(\frac{3}{2N(E_F) v_F^2 e^2}\right)^{1/2} \qquad ...(5.114)$$

One can estimate $N(E_F)$ appearing in both eqs. (5.113) and (5.114) from the low temperature specific heat.

5.11 STRONG COUPLING EFFECTS: ELIASHBERG APPROACH

We have seen that the BCS microscopic theory provides a complete though approximate theory of both thermal as well as dynamic properties of superconductors in the weak coupling limit $N(0) |g| \ll 1$. The electron–electron interactions deriving from the electron–lattice interaction are treated in the lowest order and the electron-electron correlations are decoupled in the mean-field type approximation. Generalization of the BCS treatment concentrate on two main problems : (*i*) inclusion of the repulsive coulomb interaction between the electrons [29] and (*ii*) extension of the BCS theory to the situation with arbitrarily large electron–phonon coupling [29 – 31], by generalizing the teatment of normal metals, with electron–lattice interaction incorporated in a systematic way [32]. Both of these factors have been included in the Eliashberg approach to superconductivity [29 – 31]. The mathematical structure of Eliashberg strong coupling theory is much more complicated and we shall restrict ourself to the discussion of only the final result.

The coulomb repulsive interaction reduces the effective attractive interaction between the electrons, so that instead of eq. (5.66), one obtains in the BCS approximation

$$T_c = 1.14 \, k_B \, \Theta_D \exp\left(-\frac{1}{\lambda - \mu^*}\right) \qquad ...(5.115)$$

where μ^* is so called *pseudo-Coulomb potential* [33, 34] and $\lambda = N(0) |g|$ is the effective electron-phonon coupling.

The Eliashberg correction to the BCS theory must be evaluated numerically. The numerical solution of the *Eliashberg equation* representing higher order corrections to the BCS theory obtained by MacMillan [35] as

$$k_B T_c = 1.14 \, \hbar \, \omega_D \exp\left[\frac{1.04 \, (1+\lambda)}{\lambda - \mu^*(1+0.62\lambda)}\right] \qquad ...(5.116)$$

Fig. 5.6 illustrates the difference between values of T_c obtained by the BCS and Eliashberg theories [36].

The screened Coulomb pseudo potential μ^* is given by

$$\mu^* = \frac{\mu}{1 + \mu \ln(E_F/k_B\Theta)} \qquad ...(5.117)$$

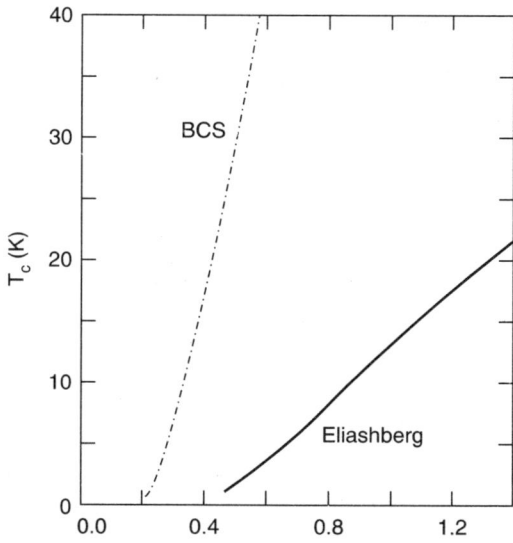

Fig. 5.6 Numerical solution of T_c versus the electron–phonon coupling constant λ for the pseudopotential $\mu^* = 0.1$. The other parameters are taken as for the superconducting element niobium. Obviously, Eliashberg theory gives a much slower increase of T_c than does the BCS theory

where μ is the screened electron-electron interaction which is reduced in order to take into account the different time scales over which the electrons and phonons interact. MacMillan has μ^* to be 0.13. The prefactor ω_D in eq. (5.117) is not a good enough measure of the effective phonon frequency which determines T_c. In 1972 Dynes [37] showed that the use of $1.21 <\omega>$ in place of ω_D improve the agreement with experiment, where $<\omega>$ is the first moment of the normalized weight function $g(\omega) = (2/\lambda\omega)\,\alpha^2 F$, where α is the electron-phonon coupling. Allen and Dynes [38] generalized the relation (5.116) as

$$T_c = \frac{\Theta_{\log} f_1 f_2}{1.2} \exp\left[-\frac{1.04\,(1+\lambda)}{\lambda - \mu^*(1+0.62\,\lambda)}\right]$$

with

$$f_1 = \left[1 + \left(\frac{\lambda}{(2.46 + 9.35\,\mu)^*}\right)^{3/2}\right]^{1/3}$$

$$f_2 = 1 + \frac{(\overline{\omega}_2/\omega_{\log} - 1)\lambda^2}{\lambda^2 + (1.82 + 11.5\,\mu^*)^2\,\overline{\omega}_2^{\,2}/\omega_{\log}^2} \qquad \ldots(5.118)$$

$$\omega_{\log} = \exp\left[\frac{2}{\lambda} \int_0^\infty \frac{\ln \omega}{\omega}\,\alpha^2 F(\omega)\,d\omega\right] \qquad \ldots(5.119)$$

and

$$\overline{\omega}_2^{\,2} = \frac{\int_0^\infty \omega\alpha^2 F(\omega)\,d\omega}{\int_0^\infty \alpha^2 \frac{F(\omega)}{\omega}\,d\omega} \qquad \ldots(5.120)$$

Recently, Kresin [39] has found an approximate solution of the Eliashberg equations which should be valid for any value of λ :

$$k_B T_c = 0.25 \, [\omega^2/\{\exp(2/\lambda_{eff}) - 1\}]^{1/2} \qquad ...(5.121)$$

where λ_{eff} is an effective coupling constant containing λ and μ^*.

Although the *Eliashberg theory* has achieved success in the analysis of the physical properties of the most of the phonon-mediated conventional superconductors, it has been known also that it is not necessarily powerful enough in analyzing the physical properties of unusual superconductors, for *e.g.*, the A = 15 compounds, HTSC cuprates, alkali doped fullerenes, heavy fermions and so on.

We see that the repulsive Coulomb interaction and the higher order electron–phonon effects combine to reduce drastically the superconducting transition temperature. This and other results [40] have led to the conclusion that the value of T_c determined within the phonon-mediated mechanism has an upper limit of the order of 30 K.

It is worthwhile to mention a very important feature of the phonon-mediated pairing. Namely the transition temperature is proportional to the Debye temperature Θ_D. Hence, T_c given by eq. (5.115) depends on the mass M of the atoms composing of lattice. In the simplest situation one expect that $T_c \sim M^{-1/2}$. A dependence of T_c on the mass M was demonstrated experimentally [41, 42] by studying the isotope influence on T_c. These observations provided a crucial argument in favour of the lattice involvement in the formation of the superconducting state. If we take into account the Coulomb repulsion between electrons, then the relation is $T_c \sim M^{-\alpha}$ with [12]

$$\alpha = \frac{1}{2} \left\{ 1 - \frac{(1+\lambda)(1+0.62\lambda)}{[\lambda - \mu^*(1+0.62\lambda)]^2} \right\} \qquad ...(5.122)$$

In the strong coupling limit ($\lambda > 1$) the exponent α is largely reduced from its initial value of 1/2. Obviously, if the value of α is small, one may interpret this fact as either the evidence for strong electron–phonon coupling or that a new nonphonon mechanism is needed to explain superconductivity.

The BCS theory is generic theory and by itself does not directly illuminate the mechanism of superconductivity. The theory may have been motivated by electron–phonon interaction, but it does not explicitly depend on it. Any attractive interaction will do. However, after the unambiguous confirmation of the electron–phonon mechanism of superconductivity in conventional metallic superconductors, some scientists turned their attention to novel non-phonon mechanisms. It was expected that these processes may occur in some exotic superconductors where the underlying physics might be different. Thus solid state physicists invented new mechanisms of pairing interactions, both under equilibrium and non-equilibrium conditions involving electronic excitations such as *e.g.*, excitons, plasmons, local pairs, bipolarons, demons etc. [31]. The expression for the critical temperature can be written as

$$k_B T_c = \Delta E \exp\left[-\frac{1+\lambda}{\lambda}\right] \qquad ...(5.123)$$

where λ describes the strength of the interaction of the electrons (Fermions) with the quasi-particles mediating the effective attractive electron–electron interaction.

5.12 OPTICAL PROPERTIES OF SUPERCONDUCTORS

One can express optical properties, of a medium in terms of the conductivity and dielectric constant. We can see that sometimes it is easier to introduce complex conductivity and discuss these properties in terms of the real and the imaginary parts, σ_1 and σ_2, of the complex conductivity σ :

$$\sigma(\omega) = \sigma_1(\omega) + i\sigma_2(\omega) \qquad \ldots(5.124)$$

For studying optical properties of a medium, we need to know $\sigma_1(\omega)$ and $\sigma_2(\omega)$ for the range of optical frequencies. We will restrict to the results obtained within the BCS theory.

At zero temperature, we find that the real part of the conductivity, $\sigma_1(\omega)$, which is related to dissipation, has a delta-function peak at zero frequency. We can see that this corresponds to an infinite conductivity. Subsequently $\sigma_1(\omega)$ is zero for frequencies smaller than $\omega_g = 2\pi/\hbar$. We can see that there is no dissipation process in that range of energy as the wave cannot break a pair and there is no single electron to be excited. Above this gap or threshold frequency ω_g, one finds that $\sigma_1(\omega)$ increases and for large frequencies becomes the same as in the normal state. This is due to the fact that superconductivity changes only the states of electrons in a range of energy Δ from the Fermi level. We know that for energies large compared with Δ, there is no difference between the normal and superconducting states of a material (Fig. 5.7).

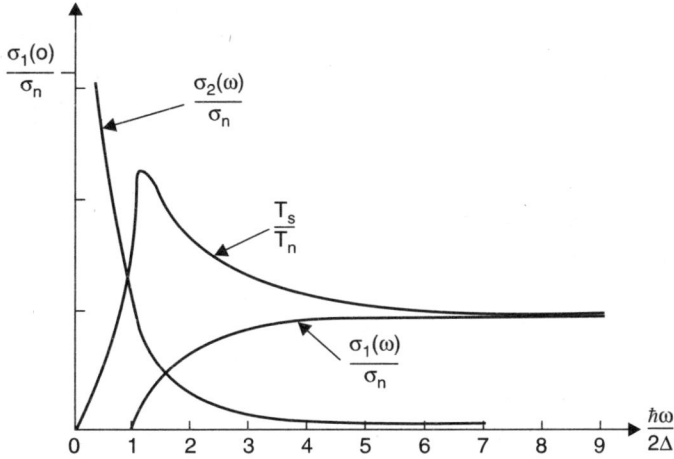

Fig. 5.7 Optical conductivity $\sigma(\omega)$ (Eq. 5.123) of a superconductor as function of frequency

We find that the behaviour of imaginary part $\sigma_2(\omega)$ is entirely determined by $\sigma_1(\omega)$ and one can analyse by using **Kramers–Kronig relations** [43]. One finds a simple formula for the temperature variation of the low-frequency limit of $\sigma_2(\omega)$.

$$\sigma_2(\omega) = \sigma_n \left[\frac{\pi \Delta}{\hbar \omega} \tanh \frac{\Delta}{2k_B T} \right] \quad \text{if } \hbar \omega \ll 2\Delta \qquad \ldots(5.125)$$

where σ_n is the conductivity of the material in the normal state.

For $T \ll T_c$, one finds

$$\sigma_2 = \sigma_n \frac{\pi \Delta}{\hbar \omega} \qquad \ldots(5.125(a))$$

and for $T \sim T_c$

$$\sigma_2 = \sigma_n \frac{\pi}{2} \frac{\Delta^2}{k_B T \hbar \omega} \qquad \ldots(5.126)$$

It is interesting to note that superconductors exhibit different behaviour from normal metals in the range of frequencies near the gap. We find that this range is in the far-infrared where $\lambda \sim 1$ mm. This is why that this range is of interest for applications of superconductors.

We find in the literature that most optical studies are done on the reflectivity of single crystals or on the transmissivity of very thin films. One finds the expression for transmissivity as

$$\frac{T_n}{T_s} = \left\{ \left[T_n^{1/2} + (1 - T_n)^{1/2} \frac{\sigma_1}{\sigma_n} \right]^2 + \left[(1 - T_n)^{1/2} \frac{\sigma_2}{\sigma_n} \right]^2 \right\}^{-1} \qquad \ldots(5.127)$$

5.13 MICROWAVE PROPERTIES

We have remarked that the existence of an energy gap means that photons of energy less than the gap energy are not absorbed. The absorption of photon (radiation) of frequency ω at temperature T depends on the ratio $\hbar \omega / 2\Delta(T)$. We find that for $\hbar \omega < 2\Delta(T)$, energy can be absorbed by excited particles which are destroyed pairs. One can show that this contributes a small resistive component to the surface impedance. For $\hbar \omega > 2\Delta(T)$, i.e. photons of higher energy see a resistance that approaches the normal state because such photons cause transitions to unoccupied normal energy levels above the gap. This mechanism leads to a strong increase in absorption as ω becomes greater than $2\Delta(T)/\hbar$.

We can describe the general behaviour as follows: at frequencies above $2\Delta(T = 0)/\hbar$ there is absorption of photons (radiations) even at $T = 0$. At frequencies below this threshold, there is no absorption of radiations (photons) at $T = 0$, i.e., the resistivity of superconductor vanishes at absolute zero. As the temperature is increased not only does the gap decrease in energy but the resistivity for photons with energy below the gap energy no longer vanishes, i.e., the absorption slowly increases initially with the increase in temperature while the absorption is due only to excited particles, and then it increases more rapidly as one surpasses the threshold energy $\hbar \omega = 2\Delta(T)$. The behaviour is depicted in Fig. 5.8.

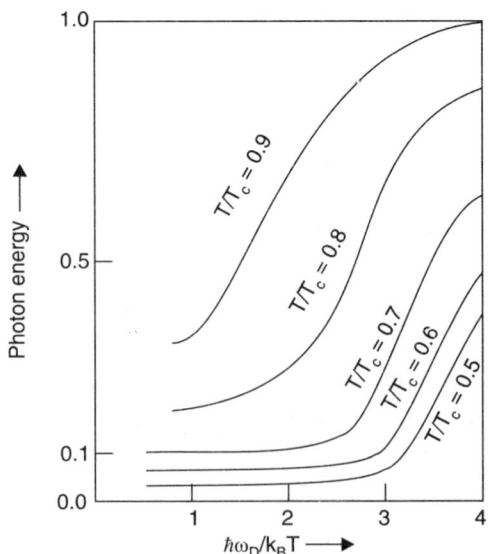

Fig. 5.8 Photon energy versus surface resistance ratio curves for aluminium at various temperatures

REFERENCES

1. H. Frohlich, *Phys. Rev.* 79, 845 (1950).
2. J. Bardeen, *Rev. Mod. Phys.* 23, 261 (1951).
3. W.S. Corak et al., *Phys. Rev.* 96, 1442 (1954) ; 102, 656 (1956).
4. K. Maki, *'Gapeless Superconductivity' in Superconductivity*, (ed.) R.D. Parks, Marcel Dekker Inc., New York (1969).
5. D.R. Tilley and J. Tilley, *Superfluidity and Superconductivity*, 3rd edition (IOP Publishing Ltd., 1990).
6. M.R. Schafroth et al., *Helv. Phys. Acta,* 30, 93 (1957) ; *Phys. Rev.* 111, 72 (1958).
7. J.R. Schrieffer, *Theory of Superconductivity,* (Benjamin/Cummings, Reading, Mass, 1964, 3rd printing revised 1983).
8. J. Bardeen, L.N. Cooper and J.R. Schrieffer, *Phys. Rev.* 106, 162 (1957) ; 108, 1175 (1957).
9. L.D. Landau, J. Exptl., *Theoret., Phys.* (USSR) 30, 1058 (1956).
10. J. Bardeen and D. Pines, *Phys. Rev.* 99, 1140 (1955).
11. L.N. Cooper, *Phys., Rev.* 104, 1189 (1956).
12. W. Frohlich, *Proc. Roy. Soc.* (London) A 223, 296 (1954).
13. J.R. Schrieffer, *Phys. Today,* p 23, (July 1973).
14. I. Giaever, *Phys. Rev Lett.* 5, 147 (1960).
15. E.L. Wolf, *Principles of 'Electron Tunneling Spectroscopy'* (Oxford Science Publications, New York, 1985).
16. M. Gurvitch et al., *Phys. Rev. Lett.* 63, 1008 (1989).
17. J.F. Zasadzinski et al., *J. Phys. Chem. Solids* 53, 1635 (1992).
18. L.D. Landau and E.M. Lifshiz, "*Statistical Physics*" Part 2, Ch. 5, 2nd ed. (1980), Pergamon Oxford.
19. A.A. Abrikosov et al., "*Methods of Quantum Field Theory in Statistical Physics*", Ch. 7, Dover, New York (1963).
20. P.G. De Gennes, "*Superconductivity of Metals and Alloys*", (Ed.) W.A. Benjamin, Reading, Massachusetts (1966).
21. M. Tinkham, "*Superconductivity*", Gordan and Breach, New York (1965).
22. J.M. Blatt, "*Theory of Superconductivity*", Academic Press, New York (1964).
23. N.N. Bogoliubov, "*A New Method in the Theory of Superconductivity*", Akad. Nauk, USSR, Moscow. Translated and Published in "*The Theory of Superconductivity*" (N.N. Bogoliubov, ed.) Gordan and Breach, New York (1968).
24. R.D. Parks, (ed.) "*Superconductivity*" (in two volumes), Dekker, New York (1969).
25. C.G. Kuper, "*An introduction to the Theory of Superconductivity*" Oxford Univ. Press (Clarendon), Oxford and New York.
26. A.C. Rose-Innes and E.H. Rhoderick, "*Introduction to Superconductivity*", Pergamon, Oxford (1969).
27. M. Cyrot and D. Pavuna, "*Introduction to Superconductivity and High–T_c Materials*", World Scientific, Singapore (1992).

28. V.L. Ginzburg and D.A. Kirzhnitz (eds.) *"High Temperature Superconductivity"*, Consultants Bureau, Plenum, New York (1982).
29. P.W. Allen and B. Mitrovic, *"In State Physics"*, H. Ehrenreich et al. (eds.) pp. 2 – 92, Academic Press, New York (1966).
30. G.M. Eliashberg, Zh. Eksperim, i. Thor, Fiz. 38, 966 (1960). [*Sov. Phys. JETP* 11, 696 (1960)].
31. K.P. Sinha and S.L. Kakani, *"High Temperature Superconductivity : Current Results and Novel Mechanisms"*, Nova Science New York (1995).
32. A.B. Midgal, *Zh. Eksp. Teor. Fiz.* 34, 1438 (1958) [*Sov. Phys.–JETP* 7, 996 (1958)].
33. P. Morel and P.W. Anderson, *Phys. Rev.* 125, 1263 (1962).
34. D.J. Scalapino, In *"Superconductivity"* Ch. 10, R.D. Parks, (ed.), Dekker, New York (1969).
35. W.L. McMillan, *Phys. Rev.* 167, 331 (1968).
36. F.S. Khan and P.B. Allen, *Solid State Commun.* 36, 481 (1980).
37. R.C. Dynes, *"Solid State Commun"*. 10, 615 (1972).
38. P.B. Allen and R.C. Dynes, *Phys. Rev.* B Condensed Matter 12, 905 (1975).
39. V.Z. Kresin, *Phys. Rev.* B 35, 8716 (1987).
40. M.L. Cohen and P.W. Anderson, *AIP Conf. Proc.* 4, 17 (1972).
41. E. Maxwell, *Phys. Rev.* 78, 477 (1950).
42. C.A. Reynolds et al., *Phys. Rev.* 78, 487 (1950).
43. C. Kittel, *"Introduction to Solid State Physics,"* John Wiley and Sons, New York, pp. 310 (1995).

Appendix 5.1—Superconductor In a Magnetic Field

Superconductor in a Magnetic Field

We know that superconductivity can be destroyed by the magnetic field. We now develop a general theory of a superconductor in a magnetic field using the microscopic theory.

Paramagnetic Effects in Superconductors

(a) **The superconducting state in the presence of magnetic field:**

The critical field $H_c(T)$ called the thermodynamical field for the transition between the normal to superconducting state is given by

$$F_s(T) - F_n(T) = \frac{H_c^2(T)}{8\pi} \qquad \ldots(1)$$

where $F_n(T)$ and $F_s(T)$ are the free energies in the superconducting and normal states respectively.

For the normal phase, in the presence of the magnetic field H, the energy is

$$F_n(T, H) = F_n(T) - \frac{1}{2}\chi_n(T)H^2 \qquad \ldots(2)$$

and for the superconducting phase, relation (2) becomes

$$F_s(T, H) = F_s(T) - \frac{1}{2}\chi_s(T)H^2 \qquad \ldots(3)$$

where χ_n and χ_s are the magnetic susceptibilities of the normal and superconducting states, respectively. From eqs. (1) to (3), one obtains the critical paramagnetic field $H_p(T)$ at which the transition from the superconducting into normal state takes place due to the paramagnetic effects. This field is given by

$$H_p(T) = \frac{H_c(T)}{\sqrt{4\pi(\chi_n - \chi_s)}} \qquad ...(4)$$

At $T = 0$, this magnitude may be expressed in terms of the microscopic properties of the metal. Indeed, relation (1) becomes

$$\frac{H_c^2(0)}{8\pi} = N(0)\frac{\Delta_0^2(0)}{2}, \quad \chi_n = 2\mu_0^2 N(0) \qquad ...(5)$$

where μ_0 is the Bohr magneton. If one takes $\chi_s = 0$ at $T = 0$, one obtains

$$\mu_0 H_p(0) = \frac{\Delta_0(0)}{2} \qquad ...(6)$$

where $\Delta_0(0)$ is the gap of the superconductor in the absence of magnetic field at $T = 0$.

This value determines the field limit above which the superconductivity cannot exist, and the field is called **Clogston–Chandrasekhar limit** [1, 2].

The existence of such a limit is connected with the fact that $\chi_s(T)$ decreases with the temperature and goes to zero at $T = 0$ for the electrons with opposite spins which form the Cooper pairs. In order to polarize the superconductor, it is necessary to break the pairs, *i.e.*, the magnetic field which should be applied is such that the Zeeman energy $\mu_0 H$ is of the order of the gap.

We consider the case of **spatial homogeneous superconductor**. The Hamiltonian can be written as

$$H_{BCS} = \sum_{P,\alpha} \varepsilon_\alpha(p) C^+_{p\alpha} C_{p\alpha} + \sum_P (\Delta(p) C^+_{p\uparrow} C^+_{-p\downarrow} + h.c.) \qquad ...(7)$$

In the external field H, we have

$$H_{BCS} = \sum_{P,\alpha} (\varepsilon_\alpha(p) + \mu_0 H) C^+_{p\alpha} C_{p\alpha} + \sum_P (\Delta(p) C^+_{+p\uparrow} C^+_{-p\downarrow} + h.c.) \qquad ...(7(a))$$

The Gorkov equations are obtained as

$$[i\omega - \varepsilon(p) + \mu_0 H] G_{\uparrow\uparrow}(p, \omega) - \Delta F^+_{\downarrow\uparrow}(p, \omega) = 1$$
$$[i\omega + \varepsilon(p) + \mu_0 H] F^+_{\downarrow\uparrow}(p, \omega) + \Delta^+ G_{\uparrow\uparrow}(p, \omega) = 1 \qquad ..(8)$$

From eqs. (8), we obtain the self consistent equation for the gap as

$$1 = g\pi T \sum_\omega \int \frac{d^3p}{(2\pi)^3} \frac{1}{\varepsilon^2(p) + \Delta^2 - (i\omega + \mu_0 H)^2} \qquad ...(9)$$

For $T = 0$, eq. (9) can be rewritten as

$$1 = g N(0) \int_{-\infty}^{\infty} \frac{d\omega}{2\pi i} \int_{-\hbar\omega_0}^{\hbar\omega_D} \frac{d\varepsilon}{\varepsilon^2 + \Delta^2 - (\omega - \mu_0 H + i\delta)^2} \qquad ...(10)$$

and after the usual algebra, one obtains

$$\ln \frac{\Delta(T)}{\Delta_0(0)} = \int_{-\infty}^{\infty} \frac{d\omega}{2\pi i} \int_{-\hbar\omega_D}^{\hbar\omega_D} d\varepsilon \left[\frac{1}{\varepsilon^2 + \Delta^2 - (\omega - \mu_0 H + i\delta)^2} - \frac{1}{\varepsilon^2 + \Delta^2 - (\omega - \mu_0 H + i\delta)^2} \right] \quad ...(11)$$

Equation (11) can be written as

$$\ln \frac{\Delta(T)}{\Delta_0(0)} = \int_0^{\hbar\omega_D} \frac{d\varepsilon}{\sqrt{\varepsilon^2 + \Delta^2}} [\theta(\sqrt{\varepsilon^2 + \Delta^2} - \mu_0 H) - 1] \quad ...(12)$$

where $\theta(z) = \begin{cases} 1 \; ; \; z > 0 \\ 0 \; ; \; z < 0 \end{cases}$

If $\mu_0 H < \Delta$, the solution of eq. (12) has the form

$$\Delta = \Delta_0(0) = 2\hbar\omega_D \exp[-1/N(0)|g|] \quad ...(13)$$

and for $\mu_0 H > \Delta$, Sarma [3] obtained the solution

$$\Delta^2 = 2\mu_0 H \mu_0(0) - \Delta_0^2(0) \quad ...(14)$$

which is allowed in the interval

$$\Delta_0(0)/2 < \mu_0 H < \Delta_0 \quad ...(15)$$

In order to obtain information about stability of the superconducting phases corresponding to the solutions (14) and (15), let us calculate the free energy of these phases using the general relations

$$F_S - F_N = \int_0^\Delta \Delta^2 \, d\left(\frac{1}{|g|}\right) \quad ...(16)$$

For the BCS solution (10), one obtains

$$F_S - F_N = -\frac{N(0)}{2} \Delta_0^2(0) \quad ...(17)$$

For the second solution (13), one obtains

$$F_S - F_N = N(0) [\Delta_0^2(0)/2 - 2\mu_0 H \Delta_0^2(0) + 2(\mu_0 H^2)] \quad ...(18)$$

and in the interval $\Delta_0(0)/2 < \mu_0 H < \Delta_0(0)$, this energy is lower than the BCS energy, and obviously, the phase is stable.

We may note that all these calculations are in the mean field approximation and the real stability seems to be questionable in the mean field approximation.

Let us consider eq. (9) at $T \neq 0$. One can perform the summation over frequencies in the usual way. One obtains

$$1 = |g| N(0) \int_0^{\hbar\omega_D} \frac{d\varepsilon}{\sqrt{\varepsilon^2 + \Delta^2}} \left[\tanh \frac{\sqrt{\varepsilon^2 + \Delta^2} + \mu_0 H}{2T} - \tanh \frac{\sqrt{\varepsilon^2 + \Delta^2} - \mu_0 H}{2T} \right] \quad ...(19)$$

Equation (13) can be written as

$$\ln \frac{\Delta(T, H)}{\Delta_0(0)} = -\int_0^\infty \frac{d\varepsilon}{\sqrt{\varepsilon^2 + \Delta^2}} [f(E + \mu_0 H) + f(E - \mu_0 H)] \quad ...(20)$$

where $E^2 = \varepsilon^2 + \Delta^2$ and $f(x)$ is a Fermi function.

In the low temperatures domain, $T \ll T_c$ for the weak fields $\mu_0 H \ll \Delta$, from (20), one obtains the solution

$$\ln \frac{\Delta(T,H)}{\Delta_0(0)} = 2 \sum_{n=0}^{\infty} (-1)^n \cosh \frac{n\mu_0 H}{2T} K_0\left(\frac{n\Delta}{T}\right) \qquad \ldots(21)$$

where $K_0(x)$ is the **MacDonald function.** One can approximate the expression (21) as

$$\ln \frac{\Delta(T,H)}{\Delta_0(0)} = -\sqrt{\frac{2\pi T}{\Delta}}\left(1 - \frac{T}{8\Delta}\right) \exp\left[-\left(\frac{\Delta - \mu_0 H}{T}\right)\right], \quad \mu_0 H \ll \Delta \qquad \ldots(22)$$

and

$$\ln \frac{\Delta(T,H)}{\Delta_0(0)} = -\sqrt{\frac{2\pi T}{\Delta}} \sum_{n=1}^{\infty} \frac{(-1)^{n+1}}{\sqrt{n}} \exp\left[-\frac{n(\Delta - \mu_0 H)}{T}\right], \quad \mu_0 H \gg \Delta \qquad \ldots(23)$$

Near the critical temperature T_c, one expands $[\varepsilon^2 + \Delta^2 - (i\omega + \mu_0 H)^2]^{-1}$ in power series in terms of Δ which is assumed to be small in this temperature range, and performing the integral over ε (in the limit $\hbar\omega_D \to \infty$), one obtains

$$\ln \frac{T}{T_C} = \psi\left(\frac{1}{2}\right) - \operatorname{Re} \psi\left(\frac{1}{2} + i\rho\right) + \frac{1}{2} f_1(\rho) \left(\frac{\Delta}{2\pi T}\right)^2 + \frac{3}{8} f_2(\rho) \left(\frac{\Delta}{2\pi T}\right)^4 + \ldots \qquad \ldots(24)$$

where
$$\rho = \frac{\mu_0 H}{2\pi T}$$

$$\psi(z) = \sum_{n=0}^{\infty} \left[\frac{1}{n+1} - \frac{1}{n+z}\right]$$

$$f_1(z) = \operatorname{Re} \sum_{n=0}^{\infty} \frac{1}{n + \frac{1}{2} + z} \quad \text{and} \quad f_2(z) = \operatorname{Re} \sum_{n=0}^{\infty} \frac{1}{\left(n + \frac{1}{2} + z\right)^5}$$

In the approximation of high fields ($\rho \gg 1$) from eq. (24), the critical temperature can be approximated as

$$T_c^2 = \frac{6}{\pi^2} (\mu_0 H)^2 \ln\left[\frac{2\mu_0 H}{\Delta_0(0)}\right]$$

If $\rho \ll 1$, eq. (24) gives

$$\frac{T_c}{T_{co}} = 1 - 7\xi(3) (\mu_0 H / 2\pi T_{co})^2$$

For the case of small field, the free energy (18) takes the form

$$F_S - F_N = N(0) \left[\frac{(\pi T)^2}{3} + (\mu_0 H)^2 - \Delta^2/2\right]$$

Making use of eqs. (18) and (24), the free energy is

$$F_S - F_N = N(0) \left[f_1(\rho) \frac{\Delta^4}{(4\pi T)^2} - f_2(\rho) \frac{\Delta^6}{4(2\pi T)^4} + \ldots \ldots \right] \qquad \ldots(25)$$

In the external field, the phase transition changes its order in the point given by

$$f_1(\rho) = 0 \qquad \ldots(26)$$

which gives $\rho_0 = 0.308$ and $T_c/T_{co} = 0.566$. These simple calculations show that

(i) if $\rho < \rho_0$ and $f_1(\rho_0) > 0$, the transition is of the *second order*.

(ii) If $\rho < \rho_0$ and $f_1(\rho_0) < 0$, the transition is of the *first order*.

The critical temperature $T_c/\Delta_0(0)$ as a function of the external field is shown in Fig. 1.

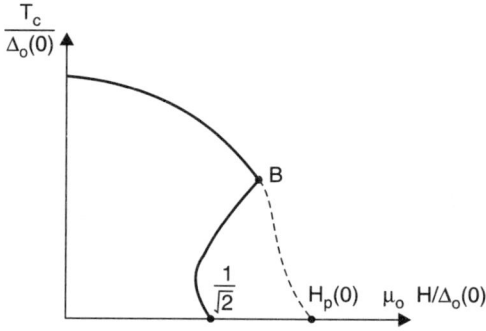

Fig. 1 The critical temperature T_c as function of the external magnetic field H

These results have been obtained by Sarma [3] for a BCS superconductor.

Fulde–Ferrell State in Superconductors

In the preceding section, we have obtained the superconducting state with an order parameter which oscillates in the space due to the action of an external magnetic field. For the case of a superconductor with $q \neq 0$, this state has been studied by Fulde and Ferrell [4] and by Takada and Izuyama [5].

In the presence of magnetic field, the BCS Hamiltonian read as

$$H_{eff} = \sum_{p,\alpha} \epsilon_\alpha(p) C^+_{p\alpha} C_{p\alpha} + \sum_p \left[\Delta_q C^+_{p+\frac{q}{2}\uparrow} C^+_{-p+\frac{q}{2}\downarrow} + h.c. \right] \qquad \ldots(27)$$

where $\epsilon_\alpha(p) = \epsilon.(p) = \mu_0 h \alpha$

$$\alpha = \begin{cases} 1 & \text{for the spin up} \\ -1 & \text{for the spin down} \end{cases}$$

h is the external magnetic field, and Δ_q is defined as

$$\Delta_q = -g \sum_p \left\langle C_{p+\frac{q}{2}\uparrow} C_{-p-\frac{q}{2}\downarrow} \right\rangle \qquad \ldots(28)$$

The Green's functions which can be obtained from the Gor'kov equations are

$$G_\alpha(p, \omega) = \frac{\omega + \epsilon_{-\alpha}\left(-\alpha p + \frac{q}{2}\right)}{[\omega + E_\alpha(p)][\omega + E_{-\alpha}(p)]}$$

$$F_\alpha^+(p, \omega) = \frac{\Delta^+}{[\omega + E_\alpha(p)][\omega + E_\alpha(p)]} \qquad ...(29)$$

where $E_\alpha(p) = \alpha\left[\frac{v_0 q}{2}x - h\right] + \sqrt{\epsilon^2(p) + \Delta_q^2}$ ...(30)

From eq. (30), we note that

$$E_\uparrow(p) + E_\downarrow(p) = 2\sqrt{(\epsilon^2 + p^2)} \qquad ...(31)$$

and these two types of solutions which describe this superconducting state, if

$$\frac{v_0 q}{2} + h < \Delta_q$$

both $E_\uparrow$ and $E_\downarrow$ are always positive in the whole k-space and the solution is a BCS one.

In the opposite case

$$\frac{v_0 q}{2} + h > \Delta_q \qquad ...(32)$$

$E_\alpha(p)$ can be negative for some values of k as one sees from eq. (30). The Cooper pairs with the negative energy must be destroyed, while these pairs remain for the states with $E_\uparrow(p)$ and $E_\downarrow(p) < 0$.

In this case, the **p**-space near the *Fermi surface* is divided into two regions: the pairing region and the depairing one. Concretely for $E_\uparrow(p) > 0$ and $E_\downarrow(p) < 0$, the electrons $\left(\mathbf{p} + \frac{\mathbf{p}}{2}; \uparrow\right)$ and $\left(-\mathbf{p} + \frac{q}{2}; \downarrow\right)$ are paired. In the region $E_\alpha(p) < 0$, the electron states $\left(\alpha p + \frac{q}{2}, \alpha\right)$ are totally blocked by the electrons with their unpaired spins, while the states $(-\alpha\mathbf{p} + \mathbf{q}/2, -\alpha)$ are unoccupied because $E_{-\alpha}(\mathbf{p})$ must be positive in this case.

This state is called the **Fulde–Ferrell (FF) state** and now we analyse the conditions for the occurrence of such a state, its stability and the properties at the finite temperatures of a superconductor in this state. If one considers the Fermi momenta $p_{0\uparrow}$ and $p_{0\downarrow}$ of the up-spin and down-spin electrons respectively and defines q as $q = \bar{q}\,(\alpha p_{0\uparrow} - p_{0\downarrow})$, the energy can be written as

$$E_\alpha(p) = \alpha\,(\bar{q}x - 1)h + \sqrt{\epsilon^2(p) + \Delta_q^2} \qquad ...(33)$$

where $p_{0\uparrow} - p_{0\downarrow} = (h/\mu_0)p_0$ ...(34)

We now examine the Fulde–Ferrell state using (33) for two cases, $0 < \bar{q} < 1$ and $\bar{q} > 1$.

In the first case $E_\downarrow(p)$ is always positive and the blocking appears only for the up spin electrons if the condition $h > \Delta_q(\bar{q}+1)$ is satisfied. The blocked region is given by the values of x and $\uparrow$ satisfying

$$-1 \leq x \leq \min(1, \phi^-)$$
$$-(\bar{q}+1)h x_1 \leq \epsilon(p) \leq (\bar{q}+1)hx_1 \qquad \ldots(35)$$

where
$$\phi^\pm(\epsilon) = \frac{h \pm \sqrt{\epsilon^2(p) + \Delta^2}}{\bar{q}\, h}, \quad x = \sqrt{1 - \frac{\Delta_q^2}{h^2(\bar{q}+1)^2}} \qquad \ldots(36)$$

In the second case ($\bar{q} > 1$), the blocking of the electrons with down spins is also possible and if $h > \Delta_q/(\bar{q}+1)$, the blocked region is

$$-1 \leq x \leq \phi^-$$
$$-(\bar{q}+1) hx_1 \leq \epsilon(p) \leq (q+1) hx_1 \qquad \ldots(36(a))$$

The blocking of the down spin electrons occurs in the case $h > \Delta_q/(\bar{q}-1)$ and its region is

$$\phi^+ \leq x \leq 1$$
$$-(\bar{q}-1) hx_2 \leq \epsilon(p) \leq (\bar{q}-1) hx_2 \qquad \ldots(37)$$

where
$$x_2 = \sqrt{1 - \Delta_q^2/h^2(\bar{q}-1)^2}$$

$$-\left\{0; 0 \leq x_1 \leq \frac{2\sqrt{\bar{q}}}{\bar{q}+1}\sqrt{1-(\bar{q}+1)^2/(\bar{q}-1)^2}\,;\, \frac{2\sqrt{\bar{q}}}{\bar{q}+1} \leq x_1 \leq 1\right. \qquad \ldots(38)$$

The parameters x_1 and x_2 defined by eqs. (36) and (38) are the measure of the blocking and take values between zero and unity.

In the limit $x_1 = x_2 = 0$, one obtains the BCS solution and for $x_1 - x_2 = 1$, the solution corresponds to normal state. The blocking induces the momentum in the Fulde–Ferrell ground state, but the direction of this momentum is opposite to that of q (the momentum of Cooper pair) and these two momenta cancel each other so that the Bloch theorem of nonexistence of mass current in the ground state is satisfied.

One of the most important results obtained in Fulde–Ferrell theory is that Fulde–Ferrell state can be a ground state if x_1 is close to unity. The other important problem is to obtain the lowest energy of the Fulde–Ferrell state. For a value of h, there is only one q which gives the lowest energy. There is also another important result is that the Fulde–Ferrell state is a ground state if its moment is zero. We may note that these results have been obtained in the Hartree–Fock approximation but seem to be quite relevant for this problem.

We now discuss the thermodynamics of the Fulde–Ferrell state. From eq. (29), the gap can be written as

$$1 = -g \sum_{p,\omega} \frac{1}{[\omega - E_\alpha(p)][\omega - E_{-\alpha}(p)]} \qquad \ldots(39)$$

Microscopic Theory of Superconductivity

Proceeding the usual way, one obtains

$$\ln \frac{T}{T_{c_o}} = \frac{1}{2}\int_{-1}^{1} dx \left[\psi\left(\frac{1}{2}\right) - \psi\left(\frac{1}{2} + i\rho(x)\right)\right] + \frac{1}{8}\left(\frac{\Delta_q(T)}{2\pi T}\right)^2 \int_{-1}^{1} dx \, \text{Re}\,\psi''\left(\frac{1}{2} + i\rho(x)\right) \quad ...(40)$$

where $\psi(z)$ is the di-gamma function and

$$\rho(x) = \frac{4(\bar{q}x - 1)}{2\pi T} \quad ...(41)$$

One can transform eq. (35) as

$$\ln \frac{T}{T_{c_o}} = g_0(\bar{q}, h, T) + \left(\frac{\Delta_g}{2\pi T}\right)^2 g_1(\bar{q}, h, T) \quad ...(42)$$

where

$$g_0 = \sum_{n=0}^{\infty} \frac{\pi T}{h\bar{q}} \left\{\tan^{-1}\frac{h(\bar{q}-1)}{2\pi T\left(n+\frac{1}{2}\right)} + \tan^{-1}\frac{h(\bar{q}+1)}{2\pi T\left(n+\frac{1}{2}\right)}\right\} - \frac{1}{n+\frac{1}{2}},$$

$$g_1 = \frac{1}{8}\frac{2\pi T}{h\bar{q}} \text{Im}\left[\psi'\left(\frac{1}{2} + i\rho(1)\right) - \psi'\left(\frac{1}{2} + i\rho(-1)\right)\right] \quad ...(43)$$

From this result, one may obtain the phase diagram as shown in Fig. 2. The transition from the normal phase in the BCS and FF phases is of the second order, but the transition from the BCS in the FF phase is of the first order. The **linear response theory** showed the existence of the Meissner effect in the Fulde–Ferrell state, which is in fact a **gapless** superconducting state.

We may note that this model is similar to that given by Sarma [3] but, only the states with $q = 0$ are considered. The energies of such states are always higher than those of the normal and BCS states.

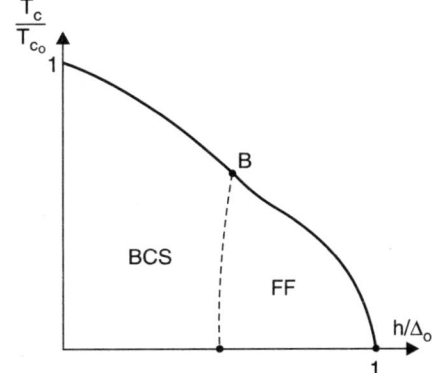

Fig. 2 The phase diagram of a superconductor with unpaired phase. The BCS is the usual states and FF state contains the blocked states

Critical Fields of a Superconductor

When the applied external field is greater than a critical field (H_c), which is temperature dependent, the superconducting state is destroyed. The important critical fields are the lower critical field (H_{c_1}) and the upper critical field (H_{c_2}).

In the $H-T$ plane below the $H_{C_1}(T)$–line, the superconductor exhibit a complete Meissner effect, and above the $H_{c_2}(T)$–line, the superconductivity is completely destroyed. Between these two lines, the superconductor is in mixed state and has a vortex structure.

We can have a qualitative discussion using the free energy as function of the magnetic field H (Fig. 3). In Fig. 3, the horizontal lines AC and DD' represent the energy of the normal state, and superconducting state respectively in the absence of magnetic field H. The energy of a type-I superconductor with a complete Meissner effect is represented by the parabolic line DEB where B denotes the thermodynamic critical field H_c. A type-II superconductor with field penetration in the form of vortices has the energy shown by the line DEC. This curve is tangent to the AC line in C which is the upper critical field H_{c_2}. The transition from the normal to the type-II superconducting state is of the second order. The penetration of the magnetic field and the occurrence of the mixed state starts with a field H_c, which is known as the lower critical field.

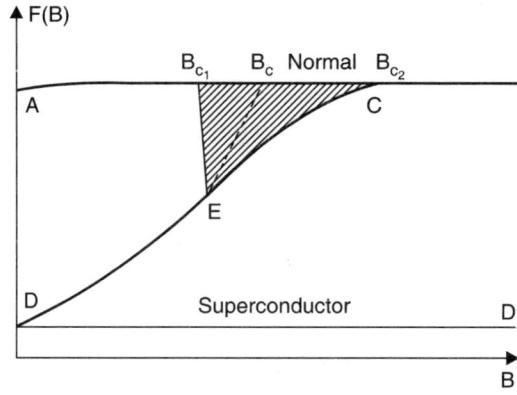

Fig. 3 The energy dependence of a superconductor as function of the magnetic field

Both of these fields, *i.e.*, B_{c_1} and B_{c_2} are temperature dependent and become zero at the critical temperature T_c.

The transition from the vortex-state in the complete Meissner state is a first order phase transition.

One can show that

$$H_{c_1} = H_c \frac{\log \chi(T)}{\chi(T)\sqrt{2}} \quad \text{or} \quad B_{c_1} = B_c \frac{\log \chi(T)}{\chi(T)\sqrt{2}} \qquad ...(44)$$

where

$$\chi(T) = \frac{\lambda(T)}{\xi(T)} = 2\sqrt{2}\ e\ \lambda^2\ H_c(T) = 2\sqrt{2}\ \frac{e\lambda^2}{\mu_0} B_c(T) \qquad ...(45)$$

REFERENCES

1. A.M. Clogston, *Phys. Rev. Lett.* 9, 266 (1962).
2. B.S. Chandrasekhar, *Appl. Phys. Lett.* 1, 7 (1962).
3. G. Sarma, *J. Phys. Chem. Solids* 24, 1029 (1962).
4. P. Fulde and R.A. Ferrell, *Phys. Rev.* 135, 550 (1964).
5. S. Takada and T. Yzugama, *Prog. Teor. Phys.* 41, 635 (1969); 43, 27 (1970).

High Temperature Superconducting Cuprates: General Survey 6

6.1 INTRODUCTION

One of the most exciting developments in science in the past three decades is the discovery of high temperature superconductivity (HTSC) in cuprates. The phenomenology of high-T_c superconductors is in many respects similar to that of conventional superconductors. On the other hand HTSC cuprates are characterised by a very short coherence length, anisotropic physical properties and considerable flux creep effects. It has been observed that the superconducting state in these systems is rather "normal" compared to the "anomalous" normal state. This discovery has led to a deluge of experimental and theoretical researches all along the world. These cuprates are close to metal-insulator transition and the stability of the insulating and the metallic phase depends on the degree of doping. Measurements of physical properties of these systems have revealed many anomalous results both in the superconducting and normal states, e.g., d-wave superconducting gap, the presence of pseudo gap in the normal state, the static or dynamic striped structure of CuO_2 planes, etc. These have posed serious theoretical challenges towards formulating the mechanisms of pairing and explanation of anomalous behaviour. Several theoretical proposals have been advanced and only a few are likely to survive in the teeth of some reliable experimental data. Before discussing these in more detail, the different families of cuprate superconductors, their crystal structures, electronic properties and their maximum T_c-values will be considered.

6.2 CUPRATES

The discovery of zero resistivity below T_c = 35 K in La_2CuO_4 in which ions of Ba^{2+} or Sr^{2+} had been introduced to substitute some of La^{3+} ions by J.G. Bednorz and K.A. Muller [1] in 1986, opened a new realm in the field of superconductivity and of many body phenomena in strongly correlated electron systems (SCES). Before the major breakthrough, most of the superconducting compounds studied were metals and alloys and the A 15 compound Nb_3Ge with T_c = 23 K held the record for the highest value of T_c [2]. The maximum value of T_c has increased steadily since 1911 to its present value of ~ 135 K for mercury based cuprate $HgBa_2Ca_2Cu_3O_{8+\delta}$ [3, 4]. When this compound is subjected to high pressure ~ 30 GPa, the onset of T_c increases to ~ 164 K [5, 6]. The dramatic evolution of critical temperatures that have been observed since 1911 are illustrated in Fig. 6.1. We have also recorded the characteristic temperatures of a variety of cryogenic liquids, as well as the lowest recorded ground temperature on earth (– 89.2°C) [7].

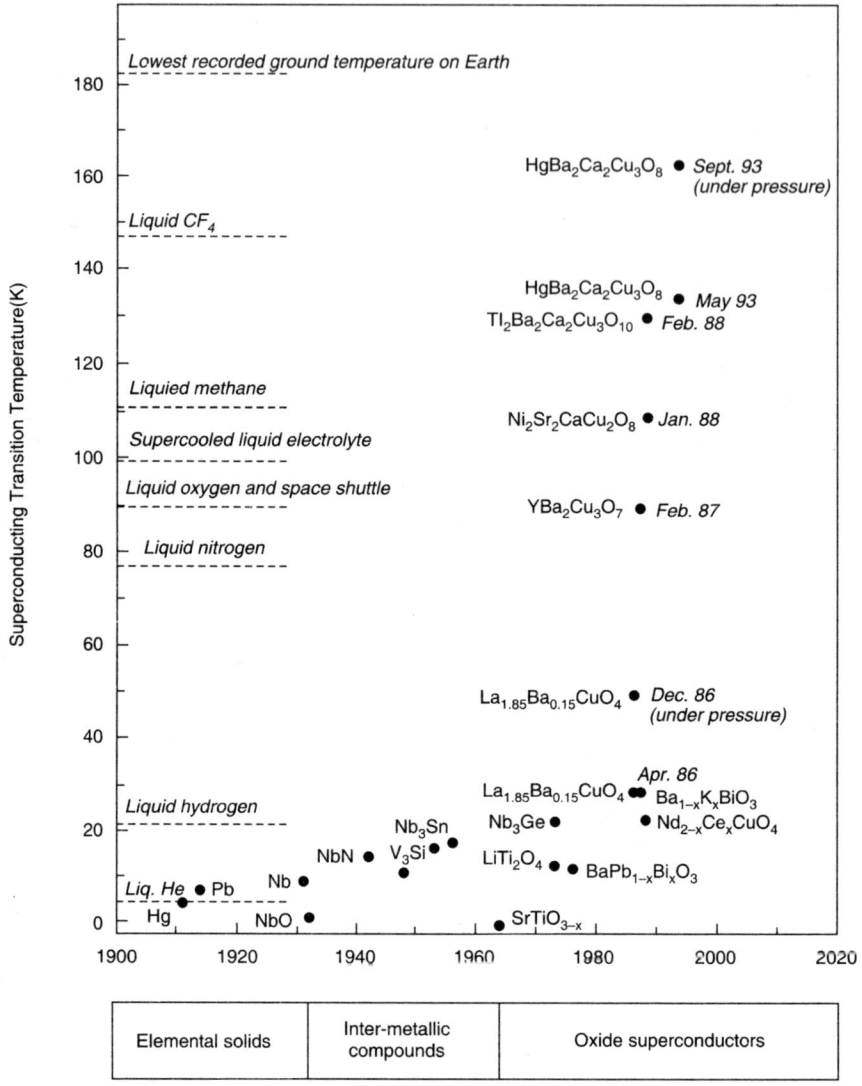

Fig. 6.1 Towards a century of superconductivity. The evolution of superconducting critical temperature since the discovery with date [7]

Since the discovery of high-T_c superconductivity in 1986 a steadily increasing number of cuprate superconductors has been discovered. Table 6.1 lists the major families of cuprate superconductors.

High Temperature Superconducting Cuprates: General Survey

Table 6.1 Major families of superconducting cuprates

Family	Symbol	Maximum T_c (K)	
$(La_{1-x} M_x)_2 CuO_4$*	214	39	[8]
$Y Ba_2 Cu_3 O_7$**	123	92	[9]
$Y Ba_2 Cu_4 O_8$	124	80	[10]
$Bi_2 Sr_2 Ca_{n-1} Cu_n O_{2n+4}$	Bi – 22 $(n-1)n$	122	[11]
$Tl_2 Ba_2 Ca_{n-1} Cu_n O_{2n+4}$	Tl – 22 $(n-1)n$	128	[12]
$Tl M_2 Ca_{n-1} Cu_n O_{2n+3}$	Tl – 12 $(n-1)n$	122	[8, 11]
$Hg Ba_2 Ca_{n-1} Cu_n O_{2n+2}$	Hg – 12 $(n-1)n$	135	[4]

*M = Sr or Ba

**Y can be replaced by rare earth elements.

6.3 LAYERING SCHEMES FOR CUPRATES

Layering schemes for La_2CuO_4 and $YBa_2Cu_3O_7$ are shown in Fig. 6.2. Single CuO_2 layers separated by double layers of LaO are characteristic of the lanthanum compounds. In $YBa_2Cu_3O_7$ two CuO_2 layers with embedded yttrium ions form CuO_2 blocks, between two of which BaO/CuO/BaO units are inserted. Fig. 6.3 shows the layering scheme for the $Tl_2Ba_2Ca_{n-1}Cu_nO_{2n+4}$ superconductors.

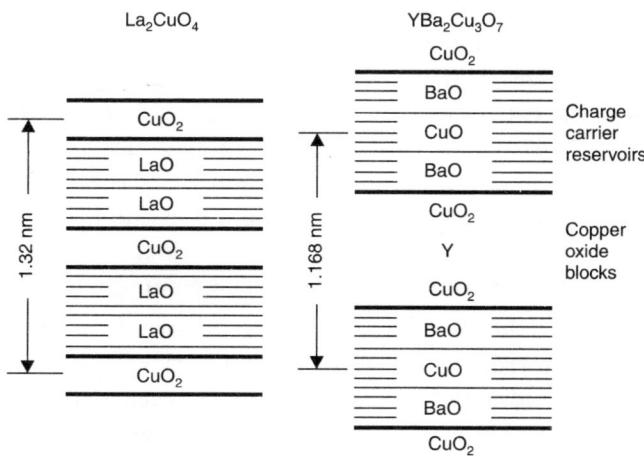

Fig. 6.2 Layering schemes for La_2CuO_4 and $YBa_2Cu_3O_7$ cuprates

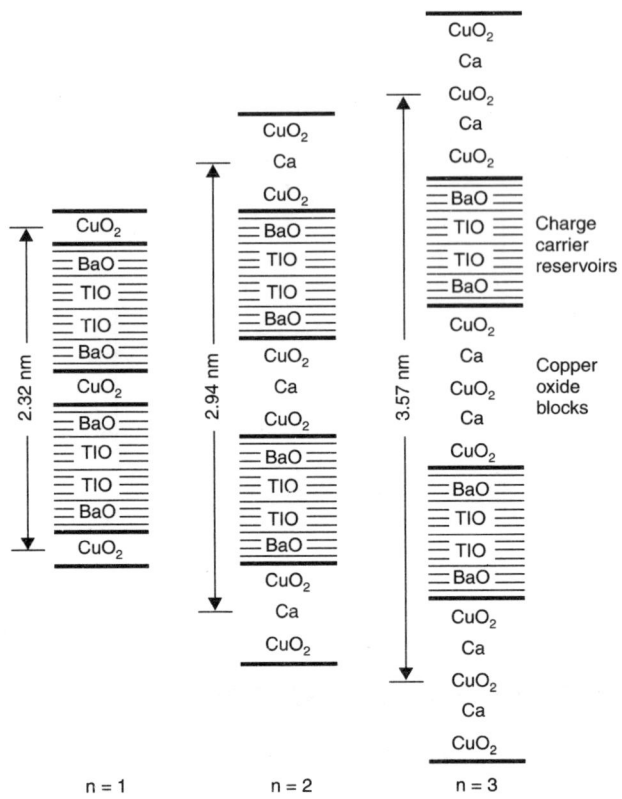

Fig. 6.3 Layering scheme for $Tl_2Ba_2Ca_{n-1}Cu_nO_{2n+4}$ cuprate superconductors. In the $Bi_2Sr_2Ca_{n-1}C_nO_{2n+4}$ cuprate superconductor counterparts Tl and Ba are replaced by Bi and Sr respectively

The layering scheme of the $Bi_2Sr_2Ca_{n-1}Cu_nO_{2n+4}$ counterparts is obtained when Tl is replaced by Bi and Ba by Sr. In both these families of cuprate superconductors the n CuO_2 layers forming the CuO_2 blocks are embedded between TlO or BiO double layers. The layering scheme for $HgBa_2Ca_{n-1}Cu_nO_{2n+2}$ cuprate superconductors is shown in Fig. 6.4. An analogous layering scheme results for the $TlM_2Ca_{n-1}Cu_nO_{2n+3}$ compounds. In contrast to the Tl – 22 $(n-1)n$ superconductors, the copper oxide blocks of the Tl – 12 $(n-1)n$ cuprates are separated only by single TlO layers. We may note that in the Bi – 22 $(n-1)n$, Tl – 22 $(n-1)n$ and Hg – 12 $(n-1)n$ cuprate families the maximum critical temperature is reached for the compound with $3CuO_2$ layers ($n = 3$) in a copper oxide block. On the other hand, in the $TlBa_2Ca_{n-1}Cu_nO_{2n+1}$ family the maximum value of T_c has been found for $n = 4$. So far, the highest T_c value of 135 K at ambient pressure has been reported for the Hg–1223 compound.

High – T_c cuprates have the Perovskite-type crystal structure and the parent compounds are Mott insulators with antiferromagnetic (AF) order. Carrier doping into this Mott insulating state induces high-T_c superconductivity, which essentially occurs in the two-dimensional (2D) CuO_2 plane. We may note that both hole and electron dopings induce superconductivity [13]. From the early stage of the research, high-T_c cuprates have been recognized to be one of SCES. Now the cuprate is confirmed to be the most established unconventional superconductor. Namely, it is predominantly believed that Cooper pairs with $d_{x^2-y^2}$ wave symmetry originates from electronic mechanism. One can understand several normal-state anomalous properties such as T-linear electric resistivity [14] and strongly enhanced AF spin correlation, considering the effect of strong electron correlation.

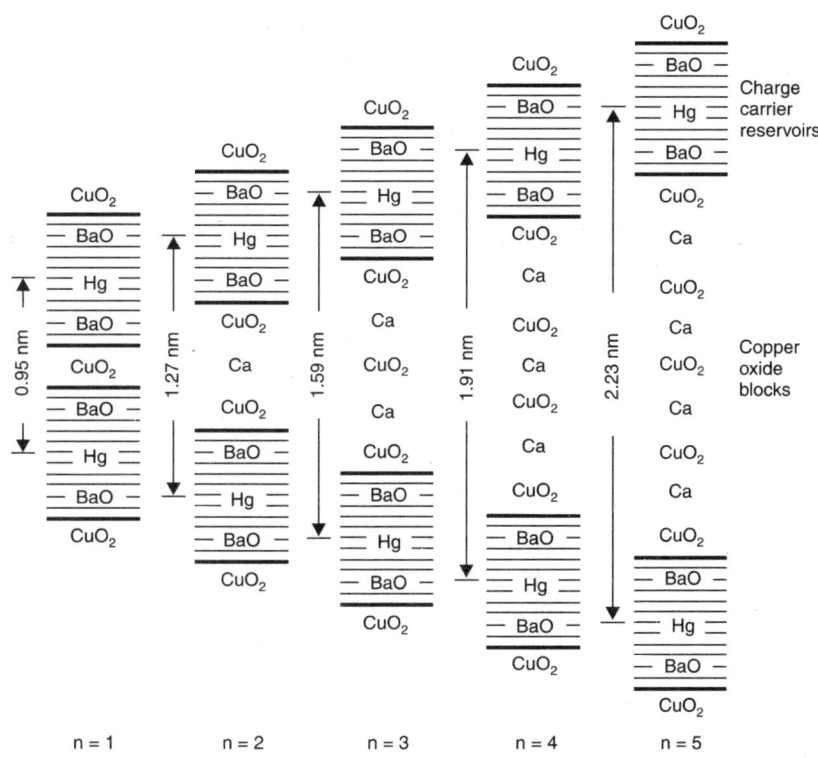

Fig. 6.4 Layering scheme for $Hg\ Ba_2\ Ca_{n-1}O_{2n+2}$ cuprates

The physical properties of the cuprate superconductors depend strongly on their chemical composition and the resulting carrier concentration. As in semiconductors the carrier concentration in the cuprate superconductors can be changed by doping. For example, let us consider the superconductor $La_{2-x}Sr_xCuO_4$. Possible oxidation state of copper are $+1, +2$ and $+3$. In the insulating parent compound La_2CuO_4 the oxidation state of copper is $+2$. The valence of La is $+3$ and that of Sr $+2$. Substitution of Sr^{2+} for La^{3+} leads to the formation, in the CuO_2 planes, of Cu^{3+} or O defects (hole), which are expected to be mobile. Clearly, an increasing number of holes is created in the CuO_2 planes with increasing Sr concentration x.

6.4 CHARGE CARRIERS IN CUPRATES

The Hall effect provides the possibility to determine the type (holes, electrons) and the concentration of the charge carriers. Fig. 6.5 shows the experimental set-up used to determine the Hall voltage, which is perpendicular to the current direction. A magnetic field **B** is applied in the z-direction perpendicular to the current direction. The Lorentz force

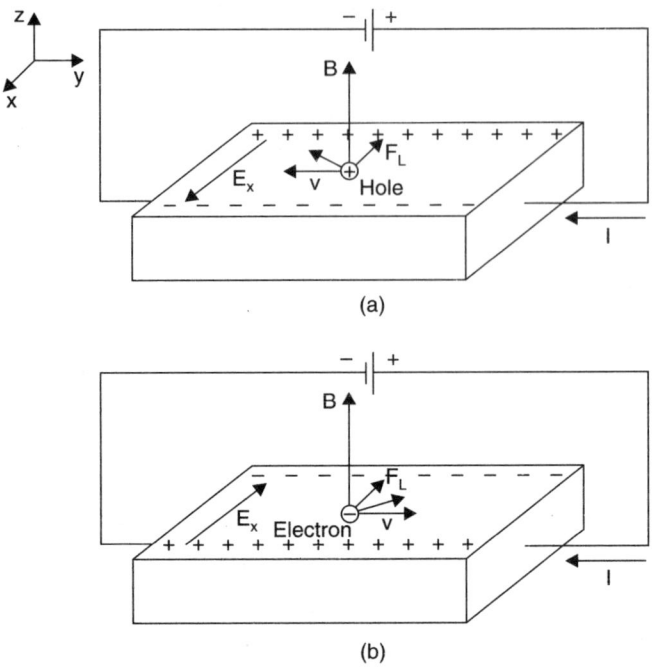

Fig. 6.5 Hall effect for holes (*a*) and electrons (*b*). The Lorentz force caused by the magnetic field applied in the *z*-direction leads to a charge separation and builds up an electric field **E** transverse to the current direction

$$\mathbf{F}_L = q(\mathbf{v} \times \mathbf{B}) \qquad \ldots(6.1)$$

acting on the charge carriers with the charge q and the average velocity $\vec{v}$ leads to a charge separation. Depending on the type of the charge carriers an electric field $\overleftarrow{E}_x$ in the + (holes) or − *x* direction (electron) is built up. At equilibrium the electric force $q\,E_x$ balances the Lorentz force $\mathbf{F}_L$

$$q\,\mathbf{E}_x = q(\mathbf{v} \times \mathbf{B}) \qquad \ldots(6.2)$$

The average velocity **v** of the charge carriers and the current density **J** are connected by the expression $\mathbf{J} = nq\mathbf{v}$, where n is the number density of electrons or holes. The resulting electric field E_x is

$$\mathbf{E}_x = \frac{1}{nq}\mathbf{J} \times \mathbf{B} \qquad \ldots(6.3)$$

where $q = +|e|$ for holes and $q = -|e|$ for electrons. The quantity $R_H = \pm 1/(n|e|)$ is termed as **Hall coefficient.** Positive and negative sign correspond to holes and electrons respectively.

Hall effect measurements clearly indicate that in the major families of high-T_c cuprate superconductors, the first family of which is $Pr_{2-x}Ce_xCu_{4-\delta}$ (where Pr can be replaced by Nd or Sm) discovered by Tokura et al. in 1989 [13]. The valence of Pr, Nd and Sm is + 3, whereas that of Ce is + 4. Substitution of Ce^{4+} for Pr^{3+}, Nd^{3+} or Sm^{3+} leads therefore to electron doping of the CuO_2 plane. Table 6.2 gives the T_c-values of these electron doped superconductors.

Table 6.2 Selected electron-doped cuprate superconductors

Superconductor	T_c (K)
$Nd_{2-x} Ce_x CuO_4$	22.8
$Pr_{2-x} Ce_x CuO_4$	22.4
$Sm_{2-x} Ce_x CuO_4$	17.6

It seems that at least two of the electron doped systems (Ln = Nd and Sm) are behaving like magnetic superconductors, *i.e.*, antiferromagnetic order of the rare earth ions coexists with superconductivity at low temperatures. These electron doped systems exhibit a variety of other magnetic phenomena including magnetic anisotropy and copper antiferromagnetism with evidence of ferromagnetic correlations.

Electron doped cuprates $Ln_{2-x}Ce_xCuO_{4-\delta}$ (Ln = Pr, Nd and Sm) are considered to be composed of two dimensional sheets of Cu–O squares with no apical oxygen. These cuprates exhibit an anomalous dependence of T_c on the concentration of doped electrons.

We may note that the evaluation of Hall effect measurements is more complicated than described above when more than one band contribute to the current transport. In this case eq. (6.3) is not valid. This more complex situation is expected to occur in the cuprates. In all cuprate superconductors the oxygen content is not well defined. In $(La, Sr)_2 CuO_{4-\delta}$ and $Y Ba_2 Cu_3 O_{7-\delta}$ the oxidation states of La and Y are + 3 and those of Sr and Ba are + 2, whereas the valence of copper depends on the concentration of excess oxygen. The formal valence $2 + p$ of copper can be determined by iodometric titration [15 – 20]. Usually, a double titration method is used to reduce systematic errors. The investigated cuprate sample has to be divided into two portions. In a first titration experiment one portion of the cuprate superconductor in question is dissolved under inert atmosphere in an acidic solution with an excess of KI. The following reactions occur in this process:

$$Cu^+ + I^- \rightarrow CuI \downarrow \quad \quad ...(6.4)$$

$$Cu^{2+} + 2I^- \rightarrow CuI \downarrow + \frac{1}{2} I_2 \quad \quad ...(6.5)$$

The resulting reaction for $[CuO]^{p+}$ is

$$[CuO]^{p+} + (2 + p) I^- \rightarrow CuI \downarrow + \frac{p+1}{2} I_2 \quad \quad ...(6.6)$$

Any Cu^+ present in the sample precipitates out as CuI. Only copper oxidised beyond Cu^+ contribute to the liberation of neutral iodine I_2. The amount of neutral iodine generated in this reaction can be determined by titration with sodium thiosulfate $Na_2 S_2 O_3$

$$I_2 + 2Na_2S_2O_3 \rightarrow 2NaI + Na_2S_4O_6 \quad \quad ...(6.7)$$

One can use the starch to determine the end point precisely because it reacts with iodine and in the presence of iodine form an intensely blue complex. When all iodine has reacted with $Na_2 S_2 O_3$, *i.e.*, at the end point of titration, the colour changes from intensively blue to colourless.

Assuming that the sample in question is single phase, and that the ratio of the metallic cation is accurately known, then one finds

$$p = M_{SC} C_{ST} V_1/W_1 - 1 \qquad ...(6.8)$$

where M_{SC} is the molar weight of the cuprate in g/mol and W_1 the weight of the sample in g. Furthermore, C_{ST} (mol/l) and V_1 (l) are the concentration and the volume of the $Na_2S_2O_3$ solution used in the titration. One can achieve reduced systematic errors by a double titration method. The second part of the cuprate sample is dissolved in HCl in the presence of air. We note that in the following reactions all copper ions are converted to the divalent state.

$$2Cu^+ + 2H^+ + \frac{1}{2} O_2 \rightarrow 2Cu^{2+} + H_2O \qquad ...(6.9)$$

$$[CuO]^{p+} + \frac{p}{2} H_2O \rightarrow CuO + \frac{p}{4} O_2 + pH^+ \qquad ...(6.10)$$

To this solution KI is added and the liberated neutral I_2 is again titrated with sodium thiosulfate $Na_2S_2O_3$ to a starch end point. Since all copper is in divalent state and hence the value of p is zero. Thus, we have

$$M_{SC} C_{ST} V_2/W_2 = 1 \qquad ...(6.11)$$

where W_2 is the weight of the second portion of the cuprate superconductor and V_2 the volume of $Na_2S_2O_3$ solution required for the titration. Combining eqs. (6.8) and (6.11), one obtains

$$p = \frac{V_1/W_1}{V_2/W_2} - 1 \qquad ...(6.12)$$

we note that eq. (6.12) contains neither the concentration C_{ST} of sodium thiosulfate solution nor the molecular weight M_{SC} of the cuprate, which depends on the average value $2 + p$ of the copper. Making use of theorem of **charge neutrality** and taking into account the known oxidation states of Y (+ 3) and Ba (+ 2) for the oxygen content of $Y Ba_2Cu_3O_8$, one obtains

$$2x = 3(2 + p) + 7 \qquad ...(6.13)$$

One finds that the situation is even more complicated for the Bi-based cuprate superconductors, because their oxygen content depends on both the Cu^{2+}/Cu^{3+} and the Bi^{3+}/Bi^{5+} ratios. One can use the reduction in hydrogen atmosphere [21] to determine the oxygen content of Bi-based high-temperature superconductors. One can measure the weight loss connected to the following reaction using a thermogravimetric system

$$Bi_2Sr_2CaCu_2O_{8+\delta} + (5 + \delta) H_2 \rightarrow 2Bi + 2Cu + CaO + 2SrO + (5 + \delta) H_2O \qquad ...(6.14)$$

We may note that the results are correct only for single-phase samples. The presence of $Bi_2 Sr_2 CuO_x$ leads to systematic errors. For the determination of the average valences of Bi and Cu a combination of iodometry and spectrometric permanganometry has also been used. One can study the effects of charge carrier density on the physical properties of the cuprate superconductors using these methods for the determination of the hole concentration.

A generic phase diagram for cuprate superconductors is shown in Fig. 6.6. These materials are near to a metal insulator transition. As the system is doped with holes the Neel temperature (T_N) for the onset of antiferromagnetism, is rapidly suppressed as the hole concentration, which is proportional to x, increases and the system eventually becomes metallic and superconducting.

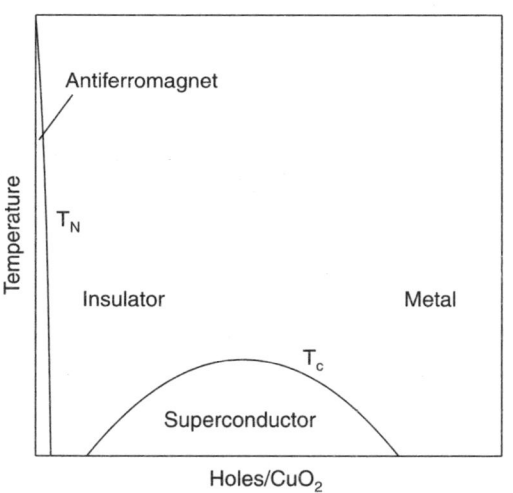

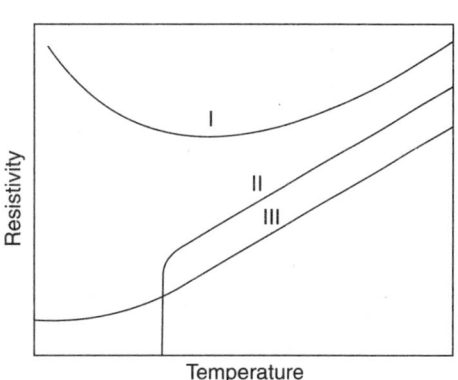

Fig. 6.6 Generic phase diagram for cuprate superconductors indicating the variation of physical properties with increasing hole concentration

Fig. 6.7 Resistance versus temperature for semiconducting (*i*) superconducting, (*ii*) and metallic phases

The conductivity increases as the copper ions become mixvalent Cu^{2+}/Cu^{3+} and the long range AF ordering becomes frustrated by the presence of doped holes in the CuO_2 planes. For $x \sim 0.15$, T_c reaches its optimum. For this optimal doping concentration one observes the strange normal state properties. For example, the resistivity of optimally doped material (*i.e.* that with highest T_c) shows a linear temperature dependence. This extends on some systems for 10 to 1000 K and extrapolating to zero resistance at zero degrees. This singular behaviour is in constrast with the observed behaviour of conventional metals where the resistivity is linear above some fraction of the Debye temperature ($\theta_D \sim 200$ K for cuprates), cuts the temperature axis at a finite temperature ($\sim \theta_D$) and saturates at high temperatures. Fig. 6.7 shows that at the lowest doping levels semiconducting behaviour with increasing resistivity at low temperatures is observed. In the superconducting region the normal state resistivity increases with increasing temperature.

There are also other anomalous normal state transport properties of these cuprates, *e.g.* the temperature dependence of the Hall effect or the logarithmic divergence of the conductivity in the normal state at low temperatures (achieved by high magnetic fields) in under doped systems. These normal state transport properties are closely related to the low energy electronic structure and the latter occupies a key position in the understanding of cuprates. For all these cuprate superconductors there is no generally accepted theory [22 – 24].

The AF insulating phase in the phase diagram consists of inplane AF ordering of the valence electrons in the CuO_2 layers. As the carrier (*e.g.*, hole) concentration is increased by doping the AF parent cuprate compound, the long range AF ordering is destroyed, although short range AF spin fluctuations persists even in the superconducting state. One can extend the phase diagram for the "**pseudogap**" as observed in optical, angle-resolved photoemission spectroscopy (ARPES), and other measurements (Fig. 6.8) [25]. Phase diagram reveal that unlike T_c, the magnitude of pseudogap, actually goes up with decreasing carrier concentration in the underdoped region.

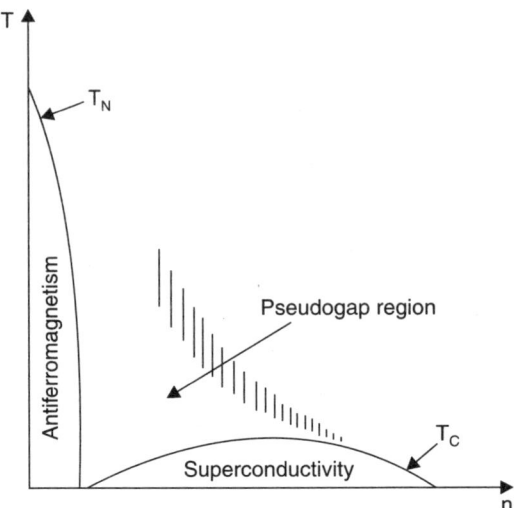

Fig. 6.8 The phase diagram of cuprate superconductors, exhibiting T_N, T_c, etc. as function of carrier concentration and pseudogap region

Fig. 6.9 shows the antiferromagnetic ordering of the copper $3d$ moments in La_2CuO_4. The Neel temperature (T_N) declines rapidly with increasing hole concentration (Fig. 6.6). A transition from an insulating to a superconducting phase is observed for further enhanced doping levels. The critical temperature increases with increasing number of holes and reaches a maximum for approximately 0.2 holes per CuO_2 unit. For further increasing carrier concentrations the critical temperature decreases again. The dependence of the critical temperature on the hole concentration will be discussed later in this chapter. At the highest doping levels the material behaves as a non-superconducting metal.

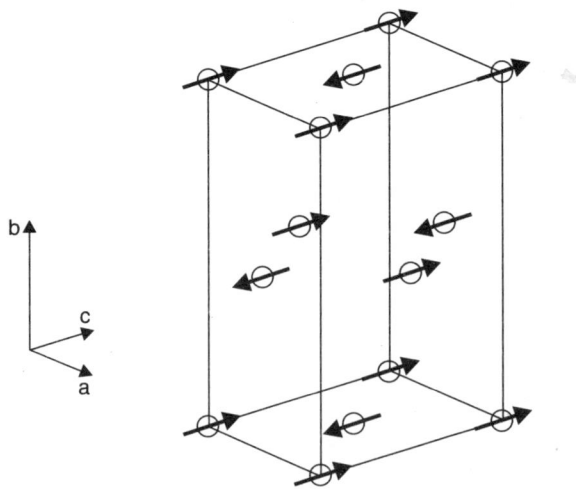

Fig. 6.9 Antiferromagnetic ordering of the copper $3d$ moments in La_2CuO_4

High Temperature Superconducting Cuprates: General Survey

We may note that in addition to the major families of cuprate superconductors (Table 6.3), there exists a large number of high-temperature superconductors which can be synthesised only under high pressure. Some of these materials and their T_c values are listed in Table 6.4. The search for new superconducting phases is stimulated by the hope of reaching even higher transition temperatures at ambient pressure than the record value of 135 K achieved for Hg-1223. Furthermore, the highest transition temperatures reached so far have been in Tl- and Hg-containing cuprates. Thallium and mercury are both volatile and poisonous elements, whose substitution by less toxic elements would be desirable. The large number of elements present in high-T_c superconductors prepared under high pressure suggests that such substitutions may be possible in future.

Table 6.3 (*i*) Some important classes of HTSC cuprates along with the maximum value of T_c observed in each class.

(*ii*) Examples of the abbreviated names (nicknames) used to denote superconducting cuprate materials.

(*i*)

Material	Maximum T_c (K)
$La_{2-x} MCuO_4$; M = Ba, Sr, Ca, Na	~ 40
$Ln_{2-x} M_x CuO_4$; Ln = Pr, Nd, Sm, Eu ; M = Ce, Th	~ 25
$Y Ba_2 Cu_3 O_{7-\delta}$	92
$LnBa_2 Cu_3 O_{7-\delta}$	~ 95
$RBa_2 Cu_4 O_8$	~ 80
$Bi_2 Sr_2 Ca_{n-1} Cu_n O_{2n+4}$ ($n = 1, 2, 3, 4$)	($n = 3$) 110
$TlBa_2 Ca_{n-1} Cu_n O_{2n+3}$ ($n = 1, 2, 3, 4$)	($n = 4$) 122
$Tl_2 Ba_2 Ca_{n-1} Cu_n O_{2n+4}$ ($n = 1, 2, 3, 4$)	($n = 3$) 122
$Hg Ba_2 Ca_{n-1} Cu_n O_{2n+2}$ ($n = 1, 2, 3, 4$)	($n = 3$) 133

(*ii*)

Material	Nickname	T_c (K)
$YBa_2 Cu_3 O_{7-\delta}$	YBCO ; YBCO-123 ; Y-123	92
$Bi_2 Sr_2 Ca_2 Cu_3 O_{10}$	BSCCO ; BSCCO-2223 ; Bi-2223	110
$Tl_2 Ba_2 Ca_2 Cu_3 O_{10}$	TBCCO ; TBCCO-2223 ; Tl-2223	122
$HgBa_2 Ca_2 Cu_3 O_8$	HBCCO ; HBCCO-1223 ; Hg-1223	133

Table 6.4 High-T_c cuprate superconductors synthesised under pressure

Family	Phase	T_c (K)
$AlSr_2Ca_{n-1}Cu_nO_{2n+3}$	Al-1223	78 [26]
	Al-1234	110 [27]
	Al-1245	83 [27]
$BSr_2Ca_{n-1}Cu_nO_{2n+3}$	B-1223	75 [28, 29]
	B-1234	110 [28, 29]
	B-1245	85 [28, 29]
$GaSr_2Ca_{n-1}Cu_nO_{2n+3}$	Ga-1223	70 [30 – 32]
	Ga-1234	107 [30 – 32]
$PbSr_2Ca_{n-1}Cu_nO_x$	Pb-1212	82 [33]
	Pb-1223	122 [34 – 36]
$CuBa_2Ca_{n-1}Cu_nO_{2n+2+\delta}$	Cu-1234	117 [37 – 39]
$(Cu_{0.5}S_{0.5})Sr_2Ca_{n-1}Cu_nO_x$	(Cu, S)-1223	$\approx$ 100 [40]
	(Cu, S)-1267	$\approx$ 60 [40]
$(Cu_{0.5}C_{0.5})Ba_2Ca_{n-1}Cu_nO_{2n+3}$	(Cu, C)-1223	67 [41]
	(Cu, C)-1223	117 [42]
$Sr_2Ca_{n-1}Cu_nO_x$	Sr-212	77 [43]
	Sr-223	109 [43]
	Sr-234	83 [43]
$Sr_2Ca_{n-1}Cu_nO_{2n+\delta}F_{2\pm x}$	Sr(F)-212	99 [44]
	Sr(F)-223	111 [44]

6.5 CRYSTAL STRUCTURES

Fig. 6.10 shows the tetragonal unit cell of the $La_{2-x}Sr_xCuO_4$ crystal structure. The lattice constants are $a = 0.3779$ nm and $c = 1.3200$ nm for $x = 0.15$ [45]. The positions of the atoms in the unit cell are given in Table 5. In the centre of the unit cell of $La_{1.85}Sr_{0.15}CuO_4$ there exists a reflection plane. For each atom position listed in Table 6.5 there exists an equivalent site at the position $z' = 1 - z$. Further atom positions result from an additional symmetry operation called body centring. The operation exchanges edge and centred atoms. Face atoms move to another face site.

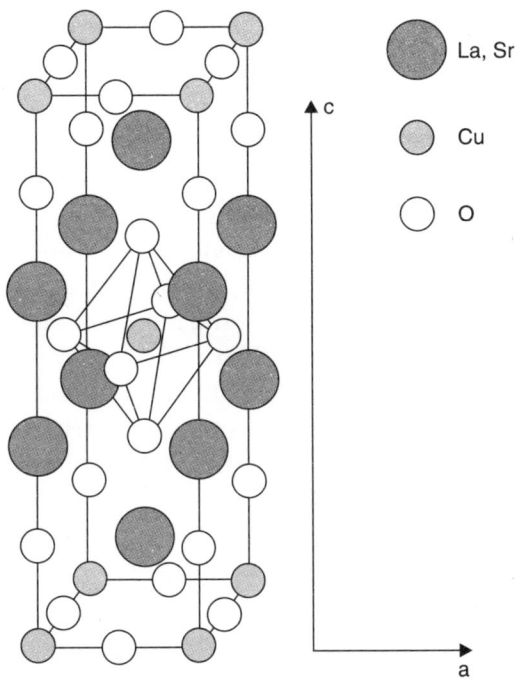

Fig. 6.10 Lattice structure of $La_{2-x}Sr_xCuO_4$ [45]

Table 6.5 Normalised atom positions in the tetragonal unit cell of $La_{1.85}Sr_{0.15}CuO_4$ with the lattice constant $a = 0.3779$ nm and $c = 1.3200$ nm [45].

Atoms	x	y	z
La or Sr	0	0	0.361
Cu	0	0	0
O1	0	0.5	0
O2	0	0	0.184

A characteristic feature of the $(La, Sr)_2 CuO_4$ structure are single CuO_2 planes at $z = 0$, 0.5 and 1. These CuO_2 layers are separated by rock-salt like two LaO layers (See Figs. 6.2 and 6.10). The Cu atoms are in the centre of an extremely elongated oxygen octahedron. Band structure calculations, however, suggest that the CuO_2 interactions along the c-axis are very small [46]. The square coordinated CuO_2 layers are a characteristic structural feature of all cuprate superconductors. Because metallic conductivity has been found for these layers, they are also called conduction planes. Along the c-axis perpendicular to these planes, insulating or semi-conducting behaviour is frequently observed.

The doped La (Sr)O planes form the block layers which capture electrons form the CuO_2 planes. From ionic picture point of view, the positive La ions are divalent and in order to conserve charge neutrality of the unit cell, the Cu ions must be in a Cu^{2+} state, this is achieved by its losing one $4s$ electron and one $3d$ electron. This create a hole in the $3d$ shell and thus Cu^{2+} has a net spin of 1/2 in the crystal. These spins interact with each other via other ions and the form the antiferromagnetic order in the undoped system.

Recent neutron scattering studies on $La_{1.6-x}Nd_{0.4}Sr_xCuO_4$ with $x \sim 1/8$ by Tranquada et al. [47] exhibit that the compound undergo a succession of transitions: first to the low temperature tetragonal (LTT) structure, then to a charged-ordered state (charge stripes) and finally, at a slightly lower temperature to a period-doubling magnetically ordered state. Emery and Kivelson [48] interpret the stripe ordering of hole-rich and hole-poor regions as being driven by frustrated kinetic phase separation $La_{1.6-x}Nd_{0.4}Sr_xCuO_4$ with $x \sim 1/8$ does not exhibit bulk superconductivity. However, Cheong et al. [49], Mason et al. [50] and Thurston et al. [51] have separately reported for low energy stripe fluctuations in optimally doped superconducting $La_{2-x}Sr_xCuO_4$. Surprisingly these experiments [49 – 51] exhibit peaks at energies as low as few MeV and in essentially the same position in k-space as in stripe ordered $La_{1.6-x}Nd_{0.4}Sr_xCuO_4$. These observations suggest that dynamical stripe correlation may play an essential role in the mechanism of pairing in these systems.

The next family of HTSC cuprates, and first to exhibit superconductivity above the boiling point of liquid nitrogen, is the so-called "123" (R – 123) family of compounds $R\,Ba_2\,Cu_3\,O_{7-\delta}$, where R = Y or any of the rare earth elements except Ce and Tb. Fig. 6.11 shows Cu Kα X-ray powder diffraction pattern for this cuprate [52]. An orthorhombic structure with the lattice parameters a = 0.38 nm, b = 0.39 nm and c = 1.17 nm has been deduced from X-ray powder diffraction data. Fig. 6.12 shows the orthorhombic unit cell (quasi-tetragonal) of $Y\,Ba_2Cu_3O_{7-\delta}$. The atom positions in the unit cell are given in Table 6.6. A tetragonal unit cell with nearly the same cell dimensions has been found for the closely related non-superconducting phase $Y\,Ba_2Cu_3O_6$. The space group for this crystal structure is $P4/mmm$. The cell volume is ~ 173 Å^3.

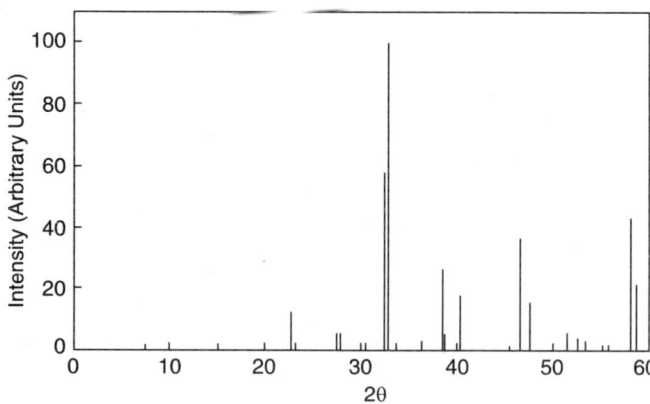

Fig. 6.11 Schematic illustration of the Cu Kα X-ray powder diffraction pattern for $Y\,Ba_2\,Cu_3\,O_{7-\delta}$ [52]

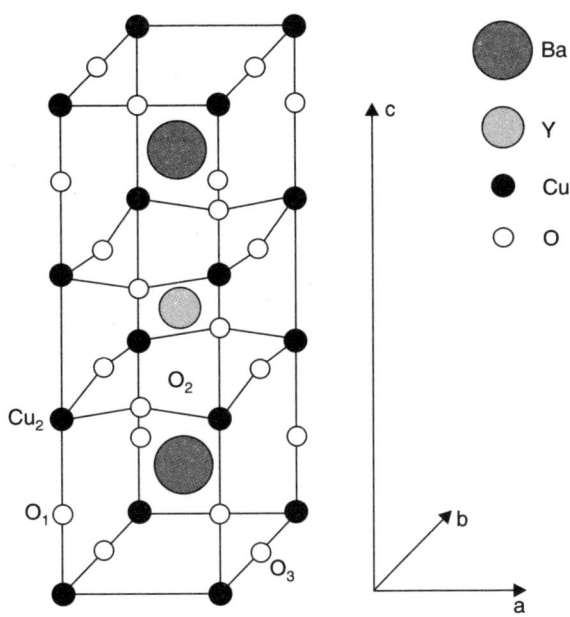

Fig. 6.12 Orthorhombic unit cell of $YBa_2Cu_3O_{7-\delta}$. We may note a special feature of the Y-123 structure, *i.e.* CuO chains in the *b*-direction

Table 6.6 Normalised atom positions in the orthorhombic unit cell of $YBa_2Cu_3O_{7-\delta}$ (Y – 123) with the lattice constants $a = 0.38$ nm, $b = 0.39$ nm and $c = 1.17$ nm. The cell volume ~ 173 Å^3 [53]

Atoms	x	y	z
Y	0.5	0.5	0.5
Ba	0.5	0.5	± 0.183
Cu_1	0	0	0
Cu_2	0	0	± 0.355
O1	0	0	± 0.159
O2	0.5	0	± 0.378
O3	0	0.5	± 0.378
O4	0	0.5	0

In contrast to $(La, Sr)_2 CuO_4$ structure the $YBa_2Cu_3O_{7-\delta}$ compound contains double CuO_2 layers with embedded Y-ions, *i.e.* the structure possesses two CuO_2 layers and one CuO chain layer per unit cell. The oxygen atoms occupy perovskite-like anion positions halfway between the copper atoms. The oxygen stoichiometry is easily varied between O_6 and O_7 due to the loosely bound oxygen atoms in the CuO chains. We may note that a special feature of the $YBa_2Cu_3O_{7-\delta}$ structure are CuO chains in the *b*-direction (Fig. 6.2 and Fig. 6.12). The oxygen stoichiometry is easily varied between O_6 and O_7 due to the loosely bound oxygen atoms in the

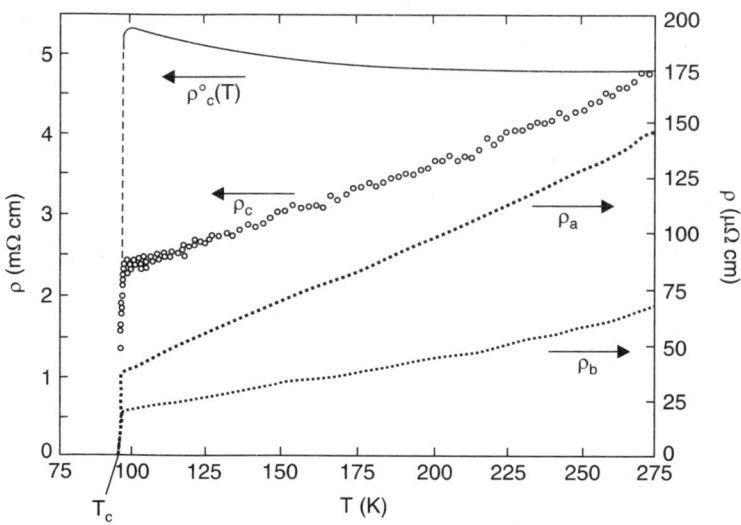

Fig. 6.13 Anisotropy of resistivity of $YBa_2Cu_3O_{7-\delta}$ [54, 55]

CuO chains. The charge neutral-formula for $YBa_2Cu_3O_7$ can be expressed as $YBa_2(Cu^{2+})_2$ $(Cu^{3+})(O^{2-})_7$ or as $YBa_2(Cu^{2+})_3 (O^{2-})_6$ (O). Oxygen stoichiometry is less than ~ 6.4, the CuO chain lose their long range order, and the compound becomes tetragonal and non-superconducting. Optimum hole doping (for maximum T_c) is achieved when the compound is fully oxygenated (i.e. $\delta \sim 0$), i.e. by increasing the oxygen concentration one gradually dopes the ab-plane with charge carriers (holes) and the compound $YBa_2Cu_3O_{6+\delta}$ eventually reaches the $YBa_2Cu_3O_7$ composition in which there are no oxygen vacancies. Detailed studies indicate that the maximum in T_c is reached for $\delta \sim 0.93$ ($T_c = 92$ K) and that for $\delta = 1.0$ the T_c is somewhat lower, $T_c = 90$ K. These studies further confirm that the best conduction channel in the normal state is along chains in the b-direction (Fig. 6.13) [54, 55]. One can create the insulating phase by reducing the hole concentration by, for e.g. annealing in a vacuum to remove oxygen or substituting Pr for $YPrBa_2Cu_3O_7$ becomes superconducting only when Pr is partially replaced with Ca and, thus for, only in thin film form [7]. This compound effectively behaves as a tunnel barrier in the temperature range where other compounds like YBCO exhibit resistivity.

The other families of HTSC cuprates are : $Bi_2Sr_2Ca_{n-1}Cu_nO_{2n+4+\delta}$ (Bi-22 $(n-1)n$); $Tl_mBa_2Ca_{n-1}Cu_nO_{2+m+2n}$ (with $m = 1, 2$); and $HgBa_2Ca_{n-1}Cu_nO_{2n+2+\delta}$ (Hg – 12 $(n-1)n$) wherein all cases, $n = (1, 2, 3,......)$ is the number of CuO_2 planes per unit formula. In these cuprate systems, several superconducting phases can coexist in the same sample. Therefore, it is of importance to study the phase composition by X-ray powder diffraction. Fig. 6.14 shows the X-ray powder diffraction patterns for $Bi_2Sr_2CaCu_2O_{8+\delta}$ (Bi-2212) and $(Bi, Pb)_2Sr_2Ca_2Cu_3O_{10+\delta}$. The synthesis of Bi-2223 has been found to be very difficult without the addition of lead. The partial substitution of Bi by Pb accelerates the formation of Bi-2223 phase.

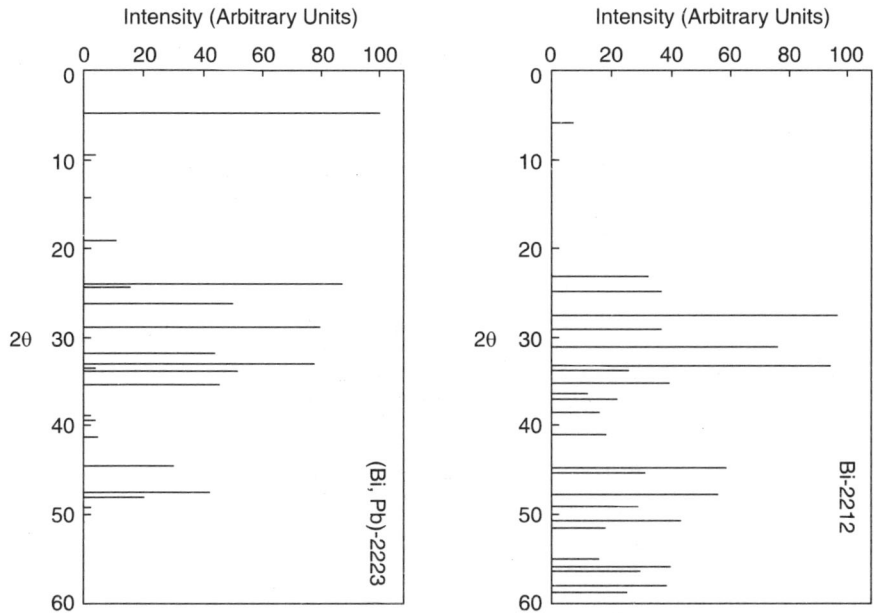

Fig. 6.14 Schematic illustration of the Cu Kα X-ray powder diffraction patterns for the (Bi, Pb)-2223 phases. (data from [56, 57])

Fig. 6.15 shows the pseudo-tetragonal unit cells of Bi-2212 and Bi-2223. The crystal structures of the Bi-based cuprates are very complex because of oxygen non stoichiometry, cation disorder and layer stacking faults. Moreover, an incommensurate super lattice structure has been observed along the b-axis of Bi-2212 based on an orthorhombic unit cell with the lattice parameters $a = 0.541$ nm, $b = 0.542$ nm and $c = 3.09$ nm.

The positions of atoms in the Bi-2212 and Bi-2223 structure are given in Tables 6.7 and 6.8 respectively. Double CuO_2 layers with embedded Ca ions are sandwiched between $SrO/Bi_2O_2/SrO$ blocks in the Bi-2212 crystal insertion of an additional Ca/CuO_2 unit leads to the Bi-2223 phase.

The crystal structure of Tl-22 $(n-1)n$ superconductors are strikingly similar to those of their Bi-22 $(n-1)n$ counterparts. However, the structural modulations generally found in Bi-based superconductors are missing. We may note that the structures of the Tl-22 $(n-1)n$ phases are close to the ideal tetragonal I4/mmm symmetry. The unit cell dimensions of these cuprates are approximately $a \cong b \cong 0.39$ nm and $c \cong (1.7 + 0.62\,n)$ nm.

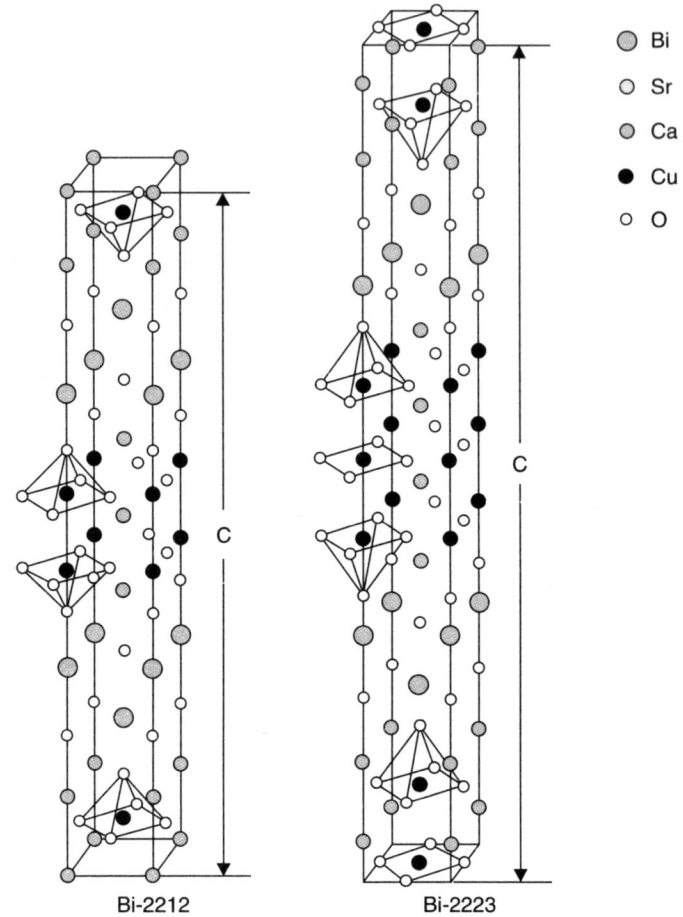

Fig. 6.15 Pseudo-tetragonal unit cells for the Bi-2212 and Bi-2223 structures. The $z = 0$ plane according to Tables 7 and 8 is in the centre of the unit cells

Table 6.7 Normalised atom positions for the Bi-2212 structure based on a pseudo-tetragonal unit cell (14/mmm)* with cell dimensions $a \cong b \cong 0.38$ nm** and $c = 3.09$ nm [46]

Atoms	x	y	z
Bi	0	0	0.199
Sr	0.5	0.5	0.109
Ca	0.5	0.5	0
Cu	0	0	0.054
O1	0	0.5	0.051
O2	0.5	0.5	0.198
O3	0	0	0.120

*The true symmetry of Bi-2212 is pseudo-orthorhombic F mmm or lower [46]

**α (pseudo-tetragonal) = α (orthorhombic)/$\sqrt{2}$

High Temperature Superconducting Cuprates: General Survey

Table 6.8 Normalised atom positions for the Bi-2223 structure based on a pseudo-tetragonal unit cell (14/mmm)* with cell dimensions $a \cong b \cong 0.38$ nm and $c = 3.82$ nm [58]

Atoms	x	y	z
Bi	0	0	0.211
Sr	0.5	0.5	0.135
Ca	0.5	0.5	0.046
Cu1	0	0	0
Cu2	0	0	0.091
O1	0	0.5	0
O2	0	0.5	0.091
O3	0	0	0.161
O4	0.5	0.5	0.213

*The atom positions are given for an orthorhombic unit cell with lattice [58]

In $Tl_2Ba_2Ca_{n-1}Cu_nO_{2n+4}$ the corresponding Bi compounds, T_c increases with the number of layers of CuO_2. It has been suggested that T_c could increase further for higher n but the compound $n = 4$ seems to have almost the same T_c as $n = 3$. The crystallographic structure shows stacking of planes; in the '2212' : TlO, TlO, BaO, CuO_2, Ca, CuO_2, BaO.

Fig. 6.16 shows X-ray powder diffraction patterns for $Tl_2Ba_2CaCu_2O_{8+\delta}$ (Tl-2212), $Tl_2Ba_2Ca_2Cu_3O_{10+\delta}$ (Tl-2223) and $Tl_2Ba_2Ca_3Cu_4O_{12+\delta}$ (Tl-2234). Because of common CuO_2 planes, the a and b lattice parameters of different cuprate superconductors are very similar. Fig. 6.16 indicates that the 110 reflexes are as expected nearly at the same position for these Tl-22 $(n-1)n$ superconductors, whereas the 002 reflexes shift to smaller 2θ-values with increasing n. This is a consequence of the insertion of additional Ca/CuO_2 units (See Fig. 6.3) and the resulting increase of the c-axis length. The positions of atoms for

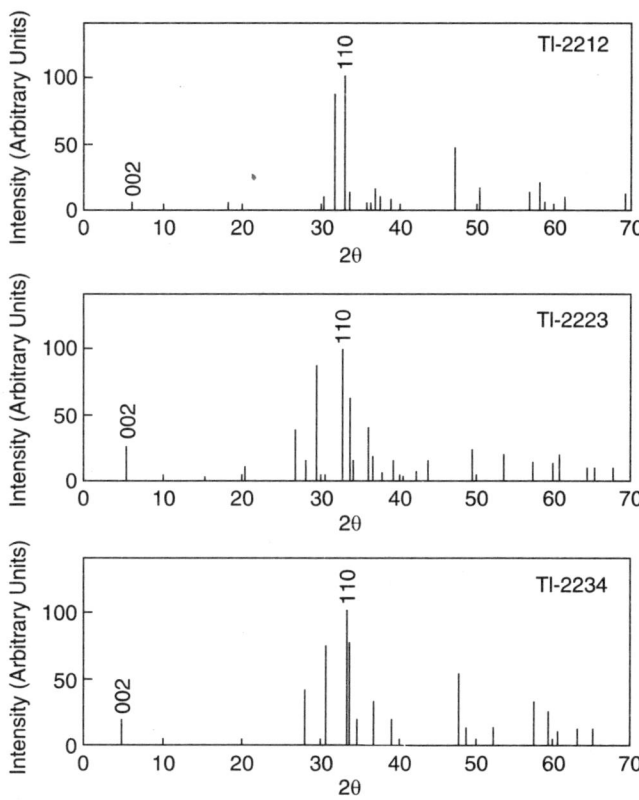

Fig. 6.16 Schematic illustration of Cu K α X-ray powder diffraction patterns for Tl-2212, Tl-2223 and Tl-2234. [59, 60]

$Tl_2Ba_2CuO_6$, $Tl_2Ba_2CaCu_2O_8$, $Tl_2Ba_2Ca_2Cu_3O_{10}$ and $Tl_2Ba_2Ca_3Cu_4O_{12}$ based on tetragonal unit cells (14/mmm) are listed in Tables 6.9, 6.10, 6.11 and 6.12 respectively.

Table 6.9 Normalised atom positions for $Tl_2Ba_2CuO_6$ ($a \cong 0.39$ nm and $c = 2.32$ nm) [46, 61]

Atoms	x	y	z
Tl	0.5	0.5	0.203
Ba	0	0	0.083
Cu	0.5	0.5	0
O1	0	0.5	0
O2	0.5	0.5	0.117
O3	0.5	0.5	0.289

Table 6.10 Normalised positions for atoms of $Tl_2Ba_2CaCu_2O_8$ ($a \cong 0.39$ nm and $c = 2.94$ nm) [46]

Atoms	x	y	z
Tl	0.5	0.5	0.21
Ba	0	0	0.12
Ca	0	0	0
Cu	0.5	0.5	0.05
O1	0	0.5	0.05
O2	0.5	0.5	0.15
O3	0.5	0.5	0.28

Table 6.11 Normalised positions of atoms for $Tl_2Ba_2Ca_2Cu_3O_{10}$ ($a \cong 0.39$ nm and $c = 3.57$ nm) [62]

Atoms	x	y	z
Tl	0	0	0.220
Ba	0.5	0.5	0.144
Ca	0.5	0.5	0.046
Cu 1	0	0	0
Cu 2	0	0	0.089
O1	0	0.5	0
O2	0	0.5	0.088
O3	0	0	0.166
O4	0.5	0.5	0.220

Table 6.12 Normalised positions of atoms for $Tl_2Ba_2Ca_3Cu_4O_{12}$ ($a \cong 0.39$ nm and $c = 4.20$ nm) [63]

Atoms	x	y	z
Tl	0.5	0.5	0.226
Ba	0	0	0.160
Ca 1	0	0	0
Ca 2	0	0	0.077
Cu 1	0.5	0.5	0.038
Cu 2	0.5	0.5	0.113
O1	0.5	0	0.038
O2	0.5	0	0.113
O3	0.5	0.5	0.178
O4	0.5	0.5	0.268

Shortly after the discovery of superconductivity above 120 K in the double-layer $Tl_2Ba_2Ca_{n-1}Cu_nO_{2n+4+\delta}$ compounds [64], a second homologous series of Tl-based cuprate superconductors was reported by Parkin et al. [65]. The composition of these compounds is represented by the formula $Tl\,Ba_2Ca_{n-1}Cu_nO_{2n+3+\delta}$ (Tl – 12(n – 1)n). The tetragonal unit cell for $TlBa_2Ca_2Cu_3O_9$ is shown in Fig. 6.17. The positions of atoms for this system are given in Table 6.13.

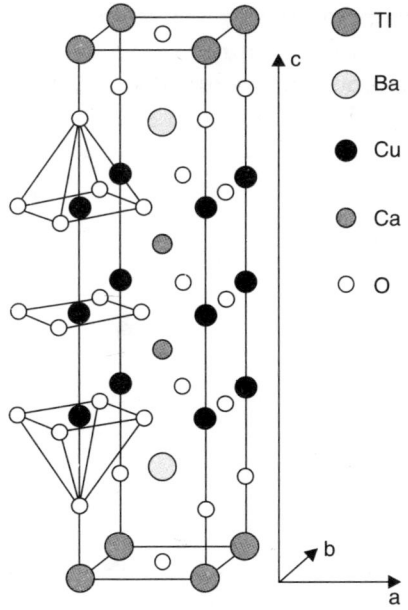

Fig. 6.17 Primitive tetragonal unit cell of Tl-1223 [65]

Table 6.13 Normalised atom positions for the $TlBa_2Ca_2Cu_3O_9$ structure based on a primitive tetragonal unit cell (P4/mmm) with lattice parameters $a = 0.38$ nm and $c \cong 1.59$ nm [65]

Atoms	x	y	z
Tl	0	0	0
Ba	0.5	0.5	0.176
Ca	0.5	0.5	0.397
Cu 1	0	0	0.5
Cu 2	0	0	0.302
O1	0	0.5	0.5
O2	0	0.5	0.304
O3	0	0	0.132
O4	0.5	0.5	0

In the Tl-12 $(n-1)n$ cuprate compounds only single TlO layers are inserted between neighbouring CuO_2 blocks. As a consequence the c lattice parameters are smaller than those of the corresponding thallium double-layer cuprates. The resulting smaller distance of adjacement CuO_2 blocks leads to an enhanced coupling of the CuO_2 conduction planes. The lattice structure of the Tl-12 $(n-1)n$ systems is primitive tetragonal (Fig. 6.17). The corresponding space group is $P4/mmm$. The lattice parameters are $a \cong 0.38$ nm and $c \cong (0.63 + 0.32\,n)$ nm.

The composition of homologous series of mercury based cuprate superconductors is represented by the formula $Hg\,Ba_2\,Ca_{n-1}\,Cu_n\,O_{2n+2+\delta}$. These monolayer mercury compounds are structurally very similar to the Tl-12 $(n-1)n$ superconductors. The synthesis of mercury based compounds frequently leads to mixtures of several superconducting phases. The X-ray diffraction pattern of a powder sample containing both the Hg-1223 and the Hg-1234 phases is shown in Fig. 6.18. In addition, minor phases are also present. Cu Kα X-ray powder diffraction patterns of single-phase Hg-1201, Hg-1212 and Hg-1223 powders are shown in Fig. 6.19.

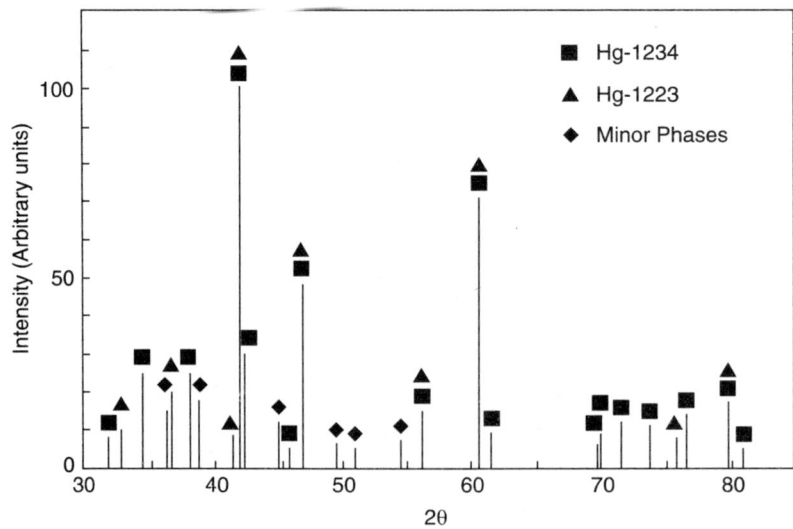

Fig. 6.18 X-ray pattern of a mixture of the Hg-1223 and Hg-1234 phases [66]

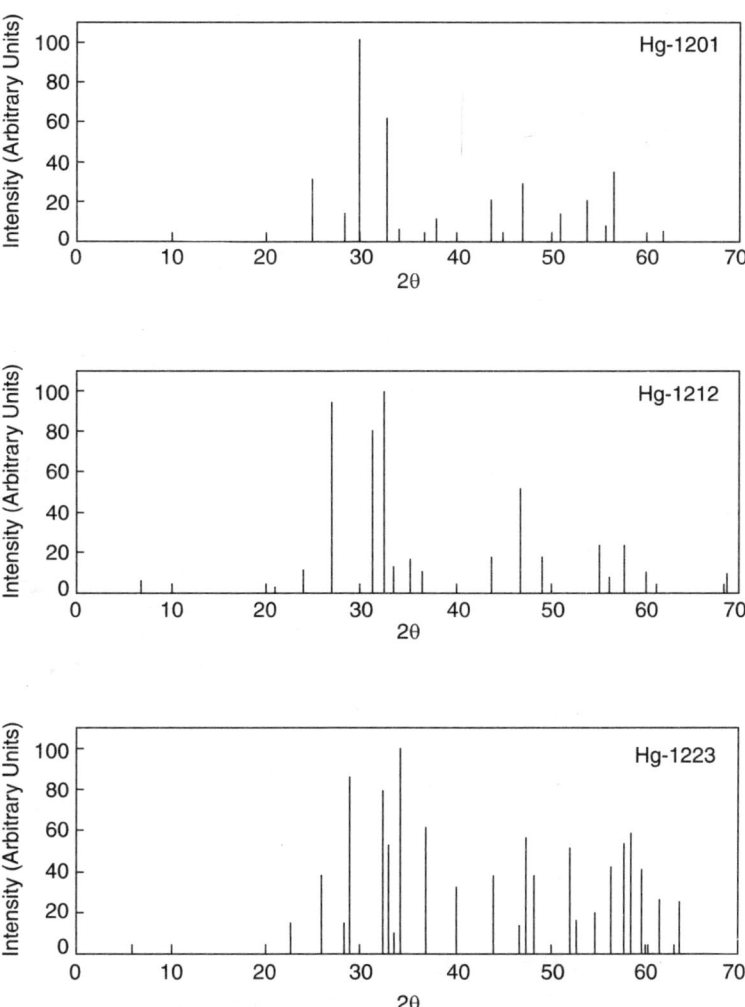

Fig. 6.19 X-ray powder diffraction patterns of single-phase Hg-1201, Hg-1212 and Hg-1223 phases. [67 – 77]

Hg-12 $(n-1)n$ compounds crystallise with the symmetry of the space group $P4/mmm$. The lattice parameters for the tetragonal unit cell are given approximately by $a \cong 0.39$ nm and $c \cong (0.95 + 0.32(n-1))$ nm [66]. The tetragonal unit cells of $Tl_2 Ba_2 CuO_{6+\delta}$ and $Hg Ba_2 CuO_{4+\delta}$ are shown in Fig. 6.20.

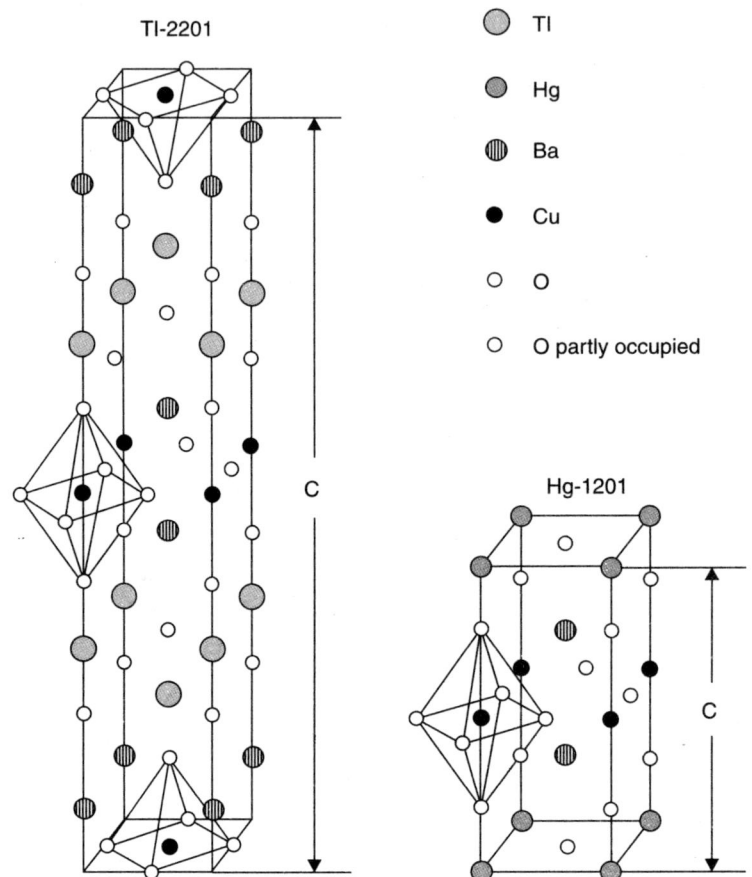

Fig. 6.20 Tetragonal unit cells of $Tl_2Ba_2CuO_{6+\delta}$ and $HgBa_2CuO_{4+\delta}$. We note that the distance between neighbouring CuO_2 planes is much smaller in the Hg-1201 than in the Tl-2201 compound

The distance d_{CuO_2} of neighbouring CuO_2 planes is much larger in the Tl double-layer compound than in Hg-1201. The reported values of d_{CuO_2} for the Tl-2201 and Hg-1201 are 1.16 and 0.95 nm respectively (See also Fig. 6.21).

In the mercury layer of $Hg\,Ba_2CuO_{4+\delta}$ only 6% of the oxygen sites are occupied. Table 6.14 provides the atom positions for $HgBa_2CuO_4$ the simplest mercury-based superconductor. Fig. 6.21 shows the distance between adjacent CuO_2 blocks for various cuprate superconductors. Large d_{CuO_2} values are characteristics of the Bi and Tl double-layer compounds. We may note that the distance of adjacent double CuO_2 planes is remarkably small for the Y-123 compound.

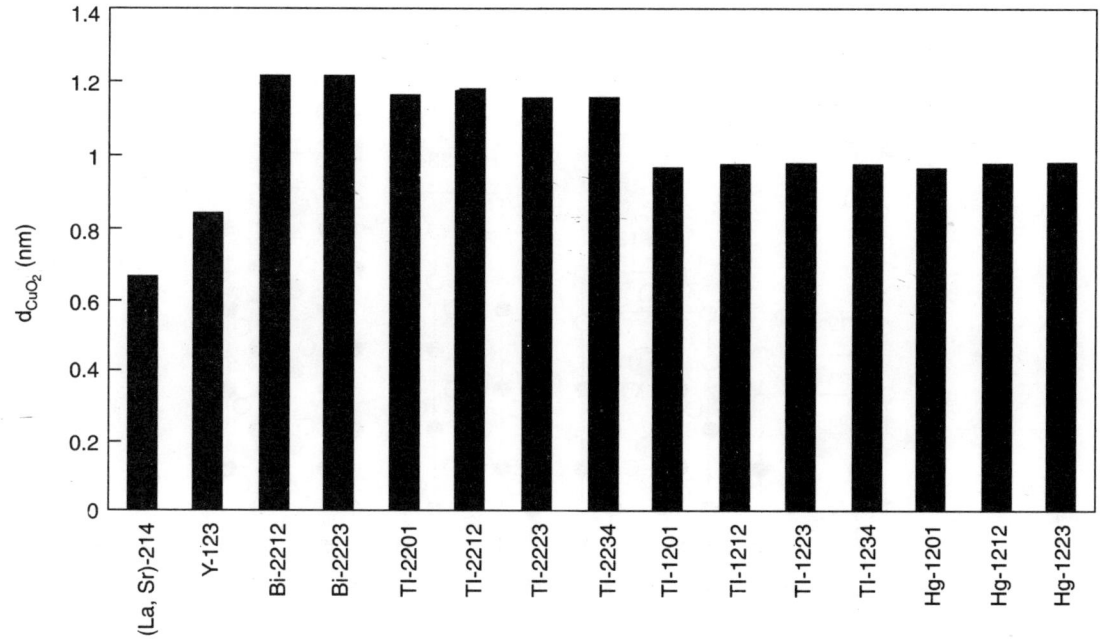

Fig. 6.21 Distances between neighbouring CuO_2 blocks for various cuprate superconductors

Table 6.14 Normalised atom positions for the simplest mercury-based superconductor $HgBa_2CuO_4$ structure based on a primitive tetragonal unit cell ($P4/mmm$) with lattice parameters $a \cong 0.39$ nm and $c \cong 0.95$ nm. The occupancy of the O3 site is only 0.06 [78]

Atoms	x	y	z
Hg	0	0	0
Ba	0.5	0.5	0.299
Cu	0	0	0.5
O1	0	0.5	0.5
O2	0	0	0.207
O3	0.5	0.5	0

The primitive tetragonal unit cells for Hg-1212, Hg-1223, Hg-1234 are shown in Fig. 6.22.

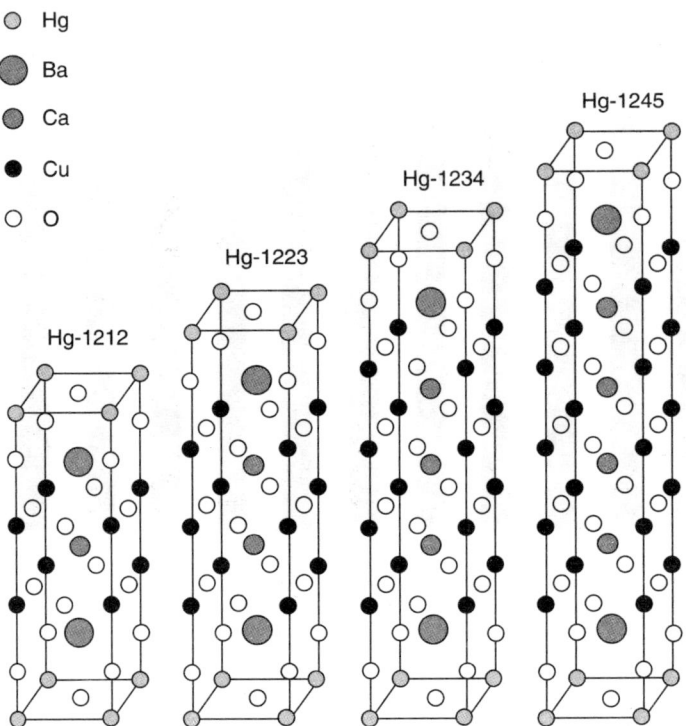

Fig. 6.22 Primitive tetragonal unit cells of Hg-1212, Hg-1223 Hg-1234 and Hg-1245 cuprates. Insertion of additional Ca/CuO$_2$ units in the Hg-12 $(n-1)n$ structures leads to enhanced c lattice parameters with increasing n[78]

The c-lattice parameter of the Hg-12 $(n-1)n$ compounds increases with the number n of CuO$_2$ planes, whereas the distance between adjacent CuO$_2$ blocks is nearly independent of n. The atom positions for Hg-1212, Hg-1223 and Hg-1234 based on primitive tetragonal unit cells ($P4/mmm$) are given in Tables 6.15, 6.16 and 6.17. We may note that for all these Hg-based cuprates the lattice parameter a is 0.385 nm.

Table 6.15 Normalised atom positions for HgBa$_2$CaCu$_2$O$_6$ (c = 1.275 nm). The occupation of the O3 site is only 0.06. [78]

Atoms	x	y	z
Hg	0	0	0
Ba	0.5	0.5	0.22
Ca	0.5	0.5	0.5
Cu	0	0	0.375
O1	0	0.5	0.38
O2	0	0	0.154
O3	0.5	0.5	0

Table 6.16 Normalised atom positions for $HgBa_2Ca_2Cu_3O_8$ ($c = 1.595$ nm). The occupation of the O3 site is only 0.06 [78]

Atoms	x	y	z
Hg	0	0	0
Ba	0.5	0.5	0.18
Ca	0.5	0.5	0.40
Cu1	0	0	0.5
Cu 2	0	0	0.302
O1	0	0.5	0.5
O2	0	0.5	0.304
O3	0.5	0.5	0
O4	0	0	0.124

Table 6.17 Normalised atom positions for $HgBa_2Ca_3Cu_4O_{10}$ ($c = 1.893$ nm). The occupations of the Hg and O4 sites are 0.79 and 0.6 respectively. [79]

Atoms	x	y	z
Hg	0	0	0
Ba	0.5	0.5	0.144
Ca 1	0.5	0.5	0.329
Ca 2	0.5	0.5	0.5
Cu 1	0	0	0.248
Cu 2	0	0	0.415
O1	0.5	0	0.251
O2	0.5	0	0.414
O3	0	0	0.091
O4	0.5	0.5	0

We may note that the low occupation of the mercury site in the Hg-1234 compound may be caused by some carbon being incorporated into the lattice structure.

Because of the complexity of the high-T_c cuprate superconductors, electron diffraction is a useful tool for the determination of the lattice parameters. Moreover, electron diffraction patterns provide information on structural defects such as incommensurate modulations and intergrowth of different cuprates. Electron diffraction patterns of Hg-1223 along the [001] and the [110] directions are schematically illustrated in Figs. 6.23 and 6.24. These diffraction patterns clearly indicate that the unit cell of Hg-1223 is tetragonal. Fig. 6.25 shows an schematic illustration of Hg-1212 along the [010] direction. The distances between adjacent direction spots in a^* and c^* directions are considerably different. This behaviour of mercury cuprate clearly shows that the length of the c-axis is much larger than the lattice parameter a. Lattice parameters of various cuprate superconductors are given in Table 6.18. We may note that the lattice parameters depend on the heat treatment conditions as well as the cation ratios.

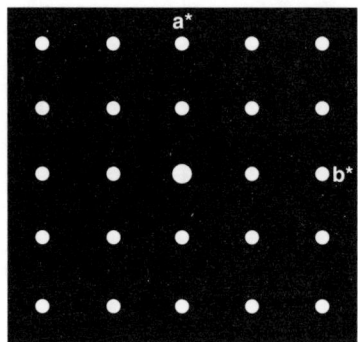

Fig. 6.23 Electron diffraction pattern for Hg-1223 along the [001] direction [66]

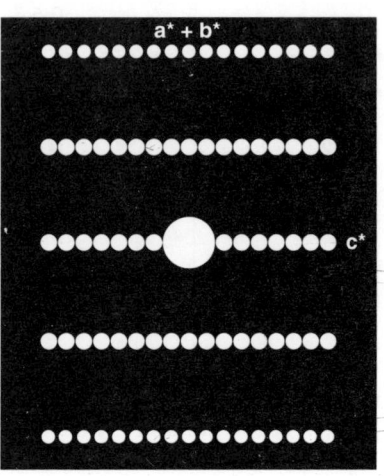

Fig. 6.24 Electron diffraction pattern for Hg-1223 along the [110] direction [66]

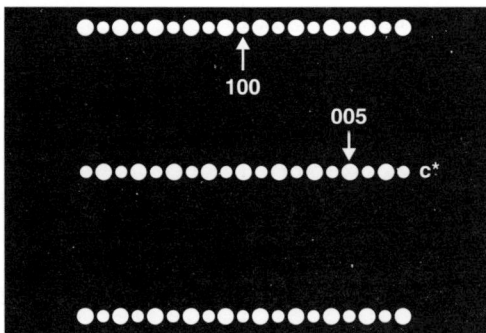

Fig. 6.25 Electron diffraction pattern for Hg-1212 along the [010] direction [70]

6.6 CRITICAL TEMPERATURES

We have seen that the physical properties of the cuprates are closely connected to the hole concentration p in the CuO_2 planes (Fig. 6.6), which is determined by the formal valence $2 + p$ of the copper atoms. Generally, the maximum critical temperature (T_c) of HTSC cuprates in question is reached for a certain value of the hole concentration.

Table 6.18 Lattice parameters of various high-T_c cuprate superconductors

Compound	Symmetry	a (nm)	b (nm)	c (nm)	T_c (K)
$La_{1.85}Sr_{0.15}CuO_4$	tetragonal	0.3779	0.3779	1.3200	36 [45]
$Nd_{1.85}Ce_{0.15}CuO_{3.93}$*	tetragonal	0.395	0.395	1.207	24 [13]
$YBa_2Cu_3O_{6.9}$	orthorhombic	0.3822	0.3891	1.1677	91 [80]
$YB_2Cu_4O_8$	orthorhombic	0.3839	0.3869	2.7243	80 [81]
$Bi_{2.2}Sr_2Ca_{0.8}Cu_2O_{8+\delta}$	orthorhombic	0.5414	0.5418	3.089	84 [56]
$(Bi,Pb)_2Sr_{1.72}Ca_2Cu_3O_{10+\delta}$	orthorhombic	0.5392	0.5395	3.6985	111 [82]
$Tl_2Ba_2CuO_{6+\delta}$	orthorhombic	0.5473	0.5483	2.3277	93 [61]
$Tl_{1.7}Ba_2Ca_{1.06}Cu_{2.32}O_{8+\delta}$	tetragonal	0.3857	0.3857	2.939	108 [83]
$Tl_{1.64}Ba_2Ca_{1.87}Cu_{3.11}O_{10+\delta}$	tetragonal	0.3822	0.3822	3.626	125 [83]
$Tl_2Ba_2Ca_3Cu_4O_{10+\delta}$	tetragonal	0.3850	0.3850	4.1984	114 [62]
$Tl_{1.1}Ba_2Ca_{0.9}Cu_{2.1}O_{7.1}$	tetragonal	0.3851	0.3851	1.2728	80 [84]
$Tl_{1.1}Ba_2Ca_{1.8}Cu_{3.0}O_{9.7}$	tetragonal	0.3843	0.3843	1.5871	110 [84]
$TlBa_2Ca_3Cu_4O_{12+\delta}$	tetragonal	0.3848	0.3848	1.9001	114 [85]
$HgBa_2CuO_{4+\delta}$	tetragonal	0.380	0.380	0.9509	94 [86]
$HgBa_2CaCu_2O_{6+\delta}$	tetragonal	0.3859	0.3859	1.2657	123 [87]
$HgBa_2Ca_2Cu_3O_{8+\delta}$	tetragonal	0.3853	0.3853	1.5818	133 [66]
$HgBa_2Ca_3Cu_4O_{10+\delta}$	tetragonal	0.3854	0.3854	1.9006	126 [66]
$HgBa_2Ca_4Cu_5O_{12+\delta}$	tetragonal	0.3852	0.3852	2.2141	110 [88]
$HgBa_2Ca_5Cu_6O_{14+\delta}$	tetragonal	0.3852	0.3852	2.526	107 [88]
$Hg_{0.5}Pb_{0.5}Ba_2Ca_4Cu_5O_{12+\delta}$	tetragonal	0.3853	0.3853	2.2172	115 [89]
$Tl_{0.5}Pb_{0.5}Sr_2Ca_2Cu_3O_9$	tetragonal	0.3815	0.3815	1.5280	118 [90]
$PbSr_2CaCu_2O_x$**	orthorhombic	0.381	0.383	1.21	70 [34]
$PbSr_2Ca_2Cu_3O_y$***	tetragonal	0.382	0.382	1.53	115 [34]
$(Cu,C)Ba_2Ca_2Cu_3O_{9+\delta}$	tetragonal	0.3859	0.3859	1.4766	67 [41]
$(Cu,C)Ba_2Ca_3Cu_4O_{11+\delta}$	tetragonal	0.3855	0.3855	1.7930	117 [41]
$BSr_2Ca_2Cu_3O_{9+\delta}$	tetragonal	0.3821	0.3821	1.3854	75 [29]
$BSr_2Ca_3Cu_4O_{11+\delta}$	tetragonal	0.3836	0.3836	1.7082	110 [28]
$BSr_2Ca_4Cu_5O_{13+\delta}$	tetragonal	0.3837	0.3837	2.022	85 [29]
$AlSr_2Ca_{2.5}Y_{0.5}Cu_3O_{9+\delta}$	tetragonal	0.3836	0.3836	1.4405	78 [26]
$AlSr_2Ca_3Cu_4O_{11+\delta}$	tetragonal	0.3839	0.3839	1.772	110 [27]
$AlSr_2Ca_4Cu_5O_{13+\delta}$	tetragonal	0.3845	0.3845	2.087	83 [27]

*Electron-doped

**$(Pb, Cu, Sr) Sr_2 (Ca, Sr) Cu_2O_x$

***$(Pb, Cu, Sr) Sr_2 (Ca, Sr)_2 Cu_3O_y$

Let us first consider the $La_{2-x}Sr_xCuO_{4-\delta}$ compound. Fig. 6.26 shows its critical temperature (T_c) versus the number of holes p per CuO_2 unit. The hole concentration p in this system depends on both the Sr content and the number of oxygen vacancies. This is why the formal valence of copper has been determined by iodometric titration [91]. The highest T_c-values have been reported in a narrow window of hole concentrations ranging from 0.15 to 0.2 holes per CuO_2. The compound is non-superconducting above a hole concentration $p \simeq 0.32$. The resistivity shows metallic behaviour above a hole concentration of $\simeq 0.15$. As stated earlier, a metal-insulator transition has been observed for low hole concentrations.

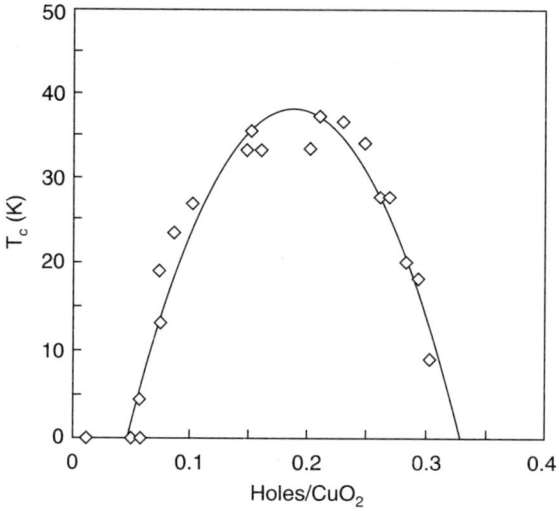

Fig. 6.26 Critical temperature (T_c) of $(La, Sr)_2 CuO_{4-\delta}$ versus the number of holes p per CuO_2 unit [91]. The solid line is just drawn to guide the eye [91]

An inverse parabolic dependence of T_c on the hole concentration similar to that shown in Fig. 6.26 is typically reported in the literature for HTSC cuprates. The main doping mechanism for $(La, Sr)_2 CuO_{4-\delta}$ is the substitution of Sr^{2+} for La^{3+}. We may note that at high doping levels the number of holes is reduced by the formation of oxygen vacancies.

The hole concentration in $YBa_2Cu_3O_{7-\delta}$ is mainly determined by the amount of excess oxygen. The insulating parent compound of this superconductor is $YBa_2Cu_3O_6$, with a tetragonal crystal structure. Due to charge neutrality, $YBa_2Cu_3O_6$ contains Cu^+ ions at the Cu 1 site (See Table 6.6).

The critical temperature T_c of $Bi_2Sr_2CaCu_2O_{8+\delta}$ as a function of the excess oxygen content δ is shown in Fig. 6.27. The maximum T_c-values of 95 K are reached for δ-values of $\simeq 0.2$. A bell-shaped dependence of the critical T_c on the amount of excess oxygen δ-results for the Bi-2212 compound. Generally, complex self doping effects occur in Bi-based HTSC, as a consequence of cation disorder. Typically some Sr and Ca sites are occupied by Bi ions with possible oxidation states + 3 and + 5 [92].

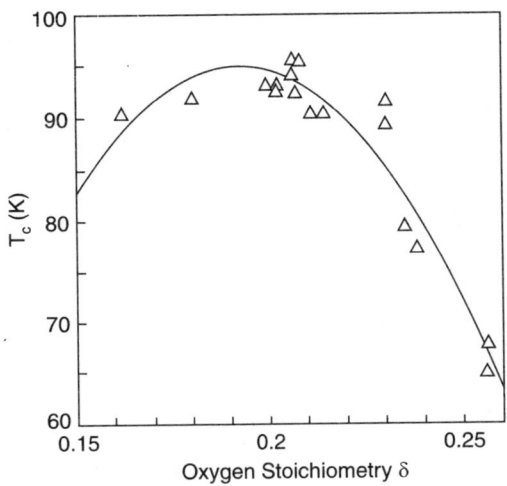

Fig. 6.27 Critical temperature (T_c) versus of $Bi_2Sr_2CaCu_2O_{8+\delta}$ versus the amount of excess oxygen δ. The solid line is drawn to help the eye [92]

Furthermore, Ca can replace Sr and vice versa. Because of these self-doping effects, some holes may be present in the Bi-based superconductors even without excess oxygen. Interstitial oxygen site O4 between the TlO double layers of $Tl_2Ba_2CuO_{6+\delta}$ is shown in Fig. 6.28.

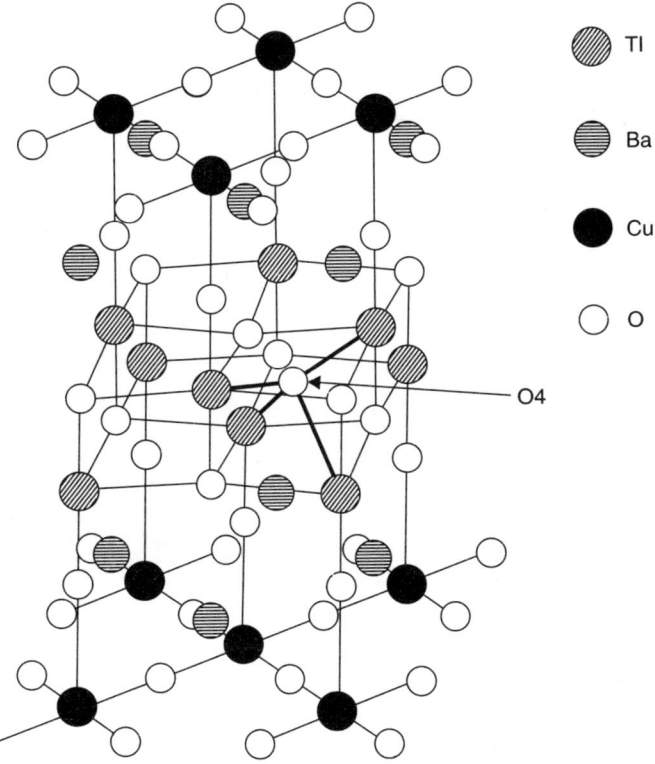

Fig. 6.28 Partly occupied interstitial oxygen site O4 between the TlO double layers of $Tl_2Ba_2CuO_{6+\delta}$ [61]

This lattice site is partly occupied by excess oxygen. The occupation of the O4 site has been determined using neutron powder diffraction data. The critical temperature (T_c) has been found to decrease with increasing amounts of excess oxygen. The resulting dependence of T_c on the occupation of this interstitial oxygen site is shown in Fig. 6.29. The excess oxygen leads to an overdoping of Tl-2201 compound; as a consequence the T_c-values are reduced.

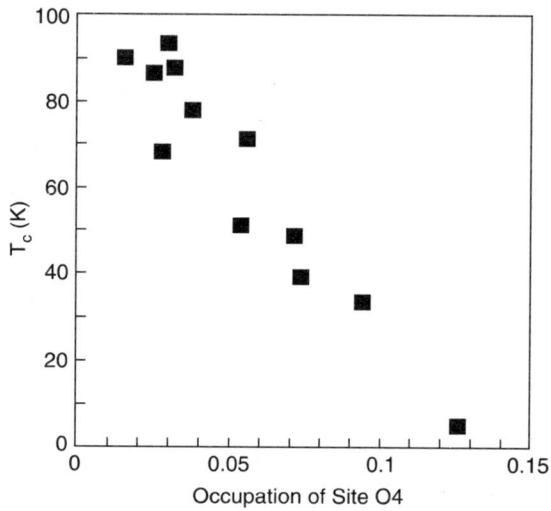

Fig. 6.29 Critical temperature (T_c) of $Tl_2Ba_2CuO_{6+\delta}$ versus the occupation of interstitial oxygen [91]

Let us now consider the critical temperature (T_c) of HTSC mercury cuprates. The critical temperature T_c of $HgBa_2CuO_{4+\delta}$ as a function of the number of holes per CuO_2 unit is shown in Fig. 6.30. The formal valence of copper has been determined by iodometric titration. Similar to the behaviour of other HTSC cuprates, a bell shaped dependence of T_c on hole concentration has been reported for Hg-1201. The critical temperature (T_c) versus the amount of excess oxygen for the $HgBa_2Ca_2Cu_3O_{8+\delta}$ compound is shown in Fig. 6.31. The maximum value of T_c (~ 135 K) is reached for a δ-value of $\simeq$ 0.2 [93].

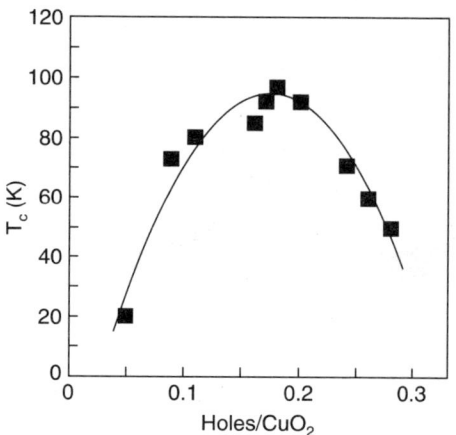

Fig. 6.30 Critical temperature (T_c) of $HgBa_2CuO_{4+\delta}$ versus hole concentration [20]. The solid line is for the guidance of eye

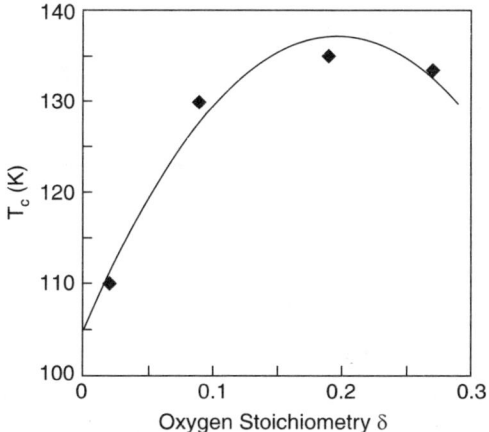

Fig. 6.31 Critical temperature (T_c) of Hg-1223 superconductor versus oxygen stoichiometry. The solid line is for the guidance of the eye [93]

A relation between the critical temperature (T_c) of optimally doped HTSC cuprates and the number n of CuO_2 layers forming a perovskite-like CuO_2 block has been obtained. Fig. 6.32 shows the maximum T_c-values of optimally doped HTSC Bi-22 $(n-1)n$ superconductors and the Y-123 compound. The T_c-values for the HTSC Bi-2201 compound are typically well below 20 K. The critical temperature T_c increases with the number of CuO_2 layers in the perovskite blocks. The maximum of $T_c = 122$ K is reached for the Bi-2223 ($n = 3$) compound. The critical temperature T_c of Bi-2234 is considerably lower than that of the Bi-2223 compound. Interestingly, the critical temperature T_c of Y-123 with 2 CuO_2 layers in the perovskite blocks is close to the optimum T_c-value for the Bi-2212 phase.

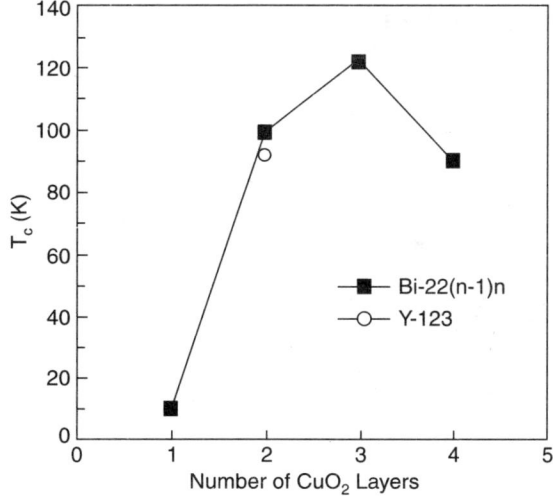

Fig. 6.32 Critical temperature (T_c) of optimally doped HTSC Bi-22 $(n-1)n$ superconductors and the Y-123 compound

A similar dependence of the critical temperature T_c on the number of CuO_2 layers has been reported for the Tl-12 $(n-1)n$ and Tl-22 $(n-1)n$ compounds [11, 94, 95]. The critical temperature T_c versus the number n of CuO_2 planes for both these families of Tl-based HTSC cuprates is shown in Fig. 6.33. The maximum T_c-value of 128 K has been reported for Tl-2223 compound [94, 95]. For the homologous series Tl-12 $(n-1)n$ a maximum value of T_c has been reported for both the $n = 3$ and 4 phases.

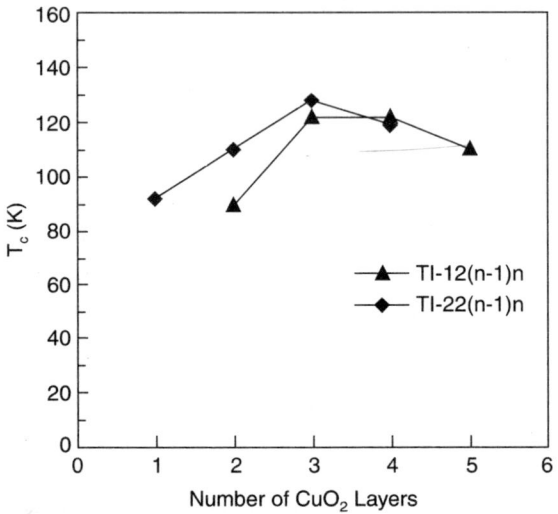

Fig. 6.33 T_c of HTSC Tl-based cuprates versus the number n of CuO_2 planes

Very recently, HTSC cuprates Hg-12 $(n-1)n$ with n upto 7 have been synthesised [88], enabling the study of the effect of the number of CuO_2 layers on T_c. The maximum values of T_c for these Hg-based HTSC-cuprates is shown in Fig. 6.34. The maximum value of 135 K is again obtained for the $n = 3$ compound. Larger numbers of CuO_2 planes in the perovskite blocks result in steadily decreasing T_c-values.

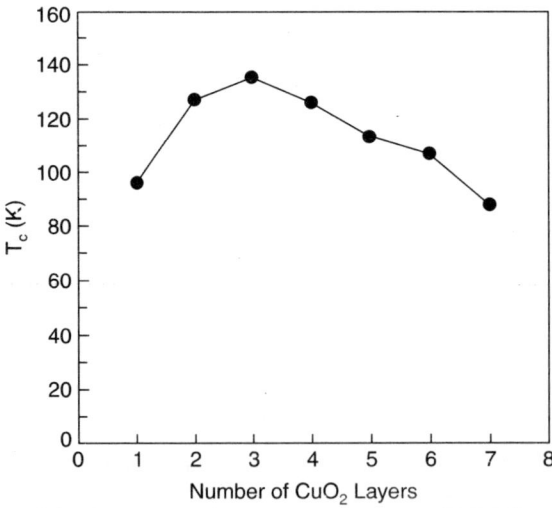

Fig. 6.34 T_c of HTSC Hg-12 $(n-1)n$ cuprates versus the number of CuO_2 planes in the perovskite blocks [88]

In compounds with $n \geq 3$ the hole concentrations of the central and outer CuO_2 planes may be considerably different. It may therefore be impossible to reach the optimum hole concentrations in all the CuO_2 layers for compounds with $n \geq 4$. This effect seems to be responsible for the observed reduction of the T_c values for Hg-12 $(n-1)n$ HTSC cuprates with more than 3 CuO_2 planes in the perovskite blocks.

The critical temperature of HTSC cuprates depends not only on the number of CuO_2 planes and the hole concentration, but also on the lattice parameters. This is why the T_c depends on the applied pressure. In addition, there exists a close connection between the lattice parameters, the hole concentration and the chemical composition. The dependence of the T_c of $HgBa_2Ca_3Cu_4O_{10+\delta}$ on the lattice parameters a and c are shown in Figs. 6.35 and 6.36 respectively. The T_c of this compound increases with decreasing a and c lattice parameters. We may note that the different T_c-values as well as the changes in the lattice parameters are a consequence of varying heat treatment conditions [28].

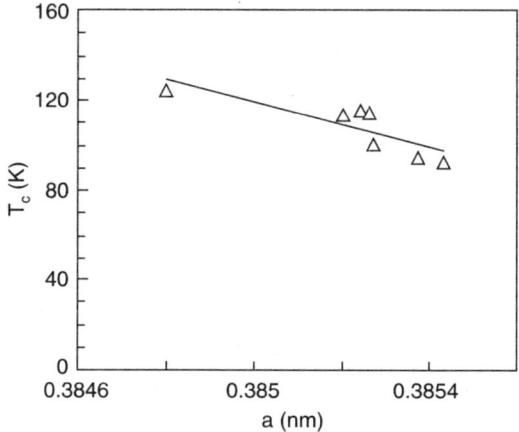

Fig. 6.35 T_c of Hg-1234 versus the lattice parameter 'a' [79]

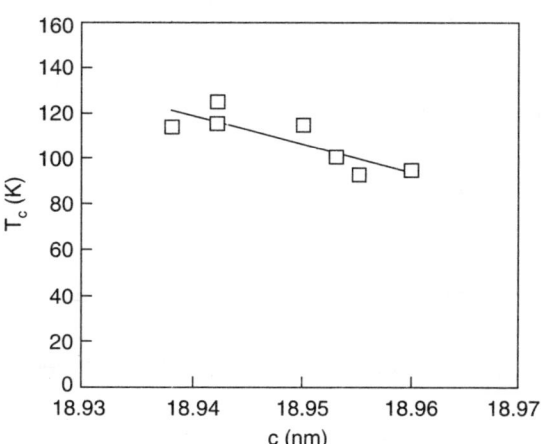

Fig. 6.36 T_c of Hg-1234 versus the lattice parameter 'b' [79]

Now, we discuss the effect of *ionic radius* of the rare earth ion R, representing the elements Yb, Er, Dy, Gd, Eu, Nd and Y, on the T_c of $RBa_2Cu_3O_{7-\delta}$. In order to separate the effects of the ionic radius and the hole concentration on the critical temperature, optimally doped R Ba_2 $Cu_3O_{7-\delta}$ samples have been used for these studies [96]. T_c-values of optimally doped R Ba_2 $Cu_3O_{7-\delta}$ samples versus the ionic radius of the rare earth ion R is shown in Fig. 6.37. The results indicate that higher T_c values result for rare earth ions with a larger ionic radius. In addition, it is also found that the lattice parameters a and b increases with increasing ionic radius of the R-ion. Based on the bond-valence sum model, it is proposed that due to the changes in the bond lengths caused by an increasing ionic radius of the rare earth ion, the number of holes (0) on plane oxygen sites increases relative to the number of Cu^{3+} on the plane Cu sites. [96].

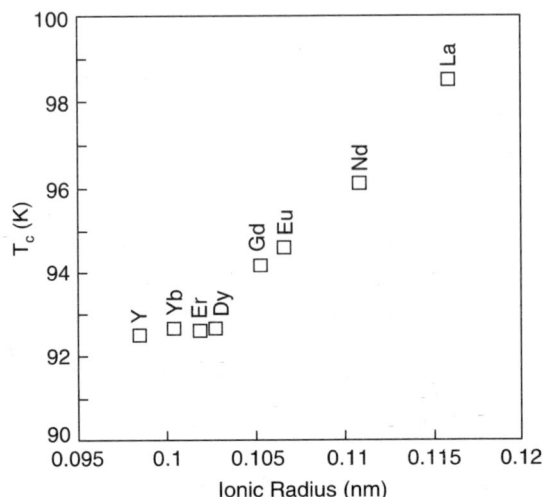

Fig. 6.37 Critical temperature (T_c) of optimally doped R Ba$_2$ Cu$_3$O$_{7-\delta}$ samples as a function of the ionic radius of the rare earth ion R [96]

We have mentioned earlier that the T_c-values of (La, Ba)$_2$ CuO$_{4-\delta}$ increase with applied pressure. A critical temperature as high as 164 K has been observed for Hg-1223 under an applied pressure of 31 GPa [100]. We note that this is about 30 K higher than the critical temperature at ambient pressure. Possible mechanism for the influence of the applied pressure on the critical temperature are changes in the hole concentration, oxygen ordering effects and intrinsic effects directly connected to the variation of the lattice parameters [97–99]. Furthermore, applied pressure may lead to a more homogeneous hole distribution on the inequivalent CuO$_2$ layers in HTSC cuprates with 3 or more such layers per perovskite block [101, 102]. Generally, the T_c of hole-doped HTSC cuprates is enhanced by applied pressure, whereas decreasing T_c-values have been observed for electron-doped cuprates [103, 104]. The initial values of the pressure derivative (dT_c/dp) for various HTSC cuprates are listed in Table 6.19.

Table 6.19 Initial pressure derivatives dT_c/dp for various HTSC cuprates

Compound	dT_c/dp (K/GPa)	Remarks	Ref.
Y Ba$_2$Cu$_3$O$_{6+\delta}$	0.4	$\delta \approx 1$	[99]
	4.3	$0.4 \leq \delta \leq 0.8$	[99]
	8.0	$\delta = 0.35$	[99]
	11.5	$\delta = 0.71$	[105]
GdBa$_{1.5}$Sr$_{0.5}$CuO$_{6+\delta}$	13.7	$\delta = 0.63$	[105]
Y Ba$_2$Cu$_4$O$_8$	5.5		[106]
GdBa$_2$Cu$_4$O$_8$	4.5		[106]

$Bi_2Sr_2CaCu_2O_{8+\delta}$	1.6	$\delta = 0.24$	[107]
	1.5	$\delta = 0.15$	[107]
	1.4	$\delta = 0.11$	[107]
		single crystal	
	-4.5	$p \parallel c$	[108]
	$+1.5$	$p \perp c$	[108]
	1.9	single crystal	[109]
$Bi_2Sr_2Ca_2Cu_3O_{10+\delta}$	1.6		[110]
$(Bi, Pb)_2Sr_2Ca_2Cu_3O_{10+\delta}$	1.7		[109]
$Tl_2Ba_2CaCu_2O_{8+\delta}$	2.0		[110]
	2.0		[111]
	2.4		[112]
$Tl_2Ba_2Ca_2Cu_3O_{10+\delta}$	5.0	$T_c(0) = 99.5$ K	[109]
	1.75	$T_c(0) \cong 129$ K	[101]
$(Tl, Pb, Bi)(Sr, Ba)_2 Ca_2Cu_3O_9{}^*$	0.13		[97]
$HgBa_2CuO_{4+\delta}$	1.72		[113]
$HgBa_2CaCu_2O_{6+\delta}$	1.70		[114]
	1.80		[114]
$HgBa_2Ca_2Cu_3O_{8+\delta}$	1.0		[101]
	1.71		[114]
	2.0		[115]
	2.3		[114]

*$(Tl_{0.7}Pb_{0.2}Bi_{0.2})(Sr_{1.8}Ba_{0.2})Ca_{1.9}Cu_3O_{9+\delta}$

We may note that the results of uniaxial and hydrostatic pressure experiments are considerably different. An uniaxial pressure applied along the c-direction of Bi-2212 single crystal leads to a reduction of the critical temperature, whereas enhanced T_c-values have been observed for hydrostatic pressure [108, 109]. In quasihydrostatic pressure experiments a solid pressure medium is used. As a consequence the specimen experiences considerable shear stresses. The very high T_c-value of 164 K achieved for Hg-1223 HTSC under quasihydrostatic pressure suggests that shear stresses contribute to the T_c enhancement.

Fig. 6.38 shows the critical temperature of $HgBa_2Ca_2Cu_3O_{8+\delta}$ versus applied pressure. We note that first, the critical temperature rises fast, at a rate of 1.0 K/GPa. Above 10 GPa the increase of T_c with increasing pressure is less pronounced.

Fig. 6.39 shows T_c versus pressure for Tl-2223 and Tl-1223. We note that first, the critical temperature of Tl-2223 increases with applied pressure and reaches a maximum value of 133 K at 4.2 GPa. The maximum enhancement of T_c due to the applied pressure is only 4 K. Further increased pressure leads to decreasing T_c values. In the investigated range $p \leq 16$ GPa, steadily increasing T_c-values have been observed for the Tl-1223 phase. However, the pressure derivative is only 0.13 K/GPa. Obviously, the pressure effects for two Tl-based HTSC are much less pronounced than those observed for the Hg-1223 compound. Large enhancements of T_c have also been reported for Hg-1201 and Hg-1212 phases [100].

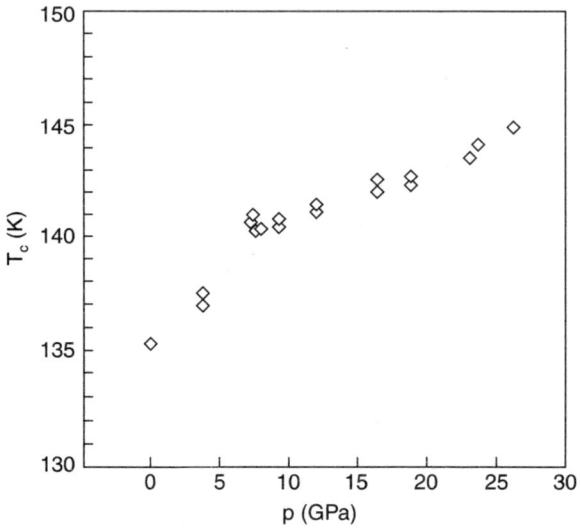

Fig. 6.38 T_c of $Hg\,Ba_2\,Ca_2\,Cu_3O_{8+\delta}$ versus applied pressure [101]

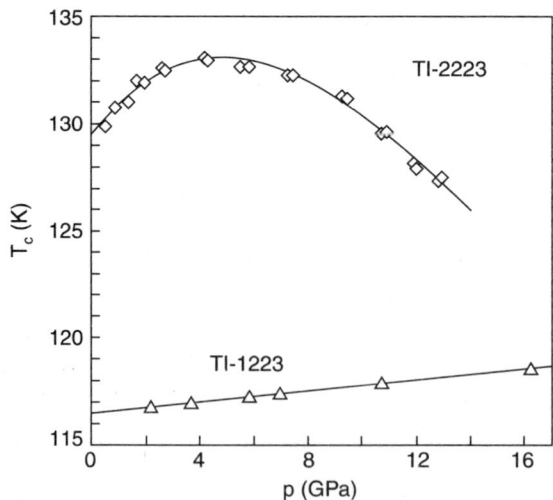

Fig. 6.39 T_c of Tl-1223 and Tl-2223 HTSC versus applied pressure [103]

Fig. 6.40 shows the T_c of the electron-doped cuprates $Sm_{1.85}Ce_{0.15}CuO_{4-\delta}$ and $Nd_{1.85}Ce_{0.15}CuO_{4-\delta}$ as a function of pressure. We note that in contrast to hole-doped HTSCs, negative pressure derivatives dT_c/dp are found [103].

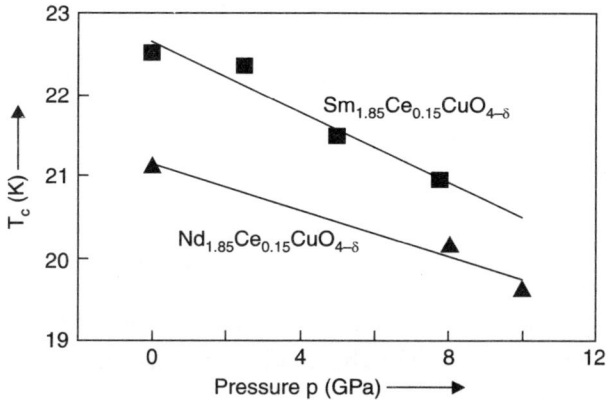

Fig. 6.40 T_c versus Pressure for electron-doped HTSC cuprates [103]

6.7 PHYSICAL PROPERTIES OF HTSC CUPRATES

A profound knowledge of the physical properties of HTSC cuprates is of fundamental importance for the development of a microscopic theory able to explain the occurrence of superconductivity well above 77 K. Moreover, some of these material data are required for the design of superconducting magnets, cables transformers, current limiters, and other power applications.

The physical properties of HTSC cuprates are generally highly anisotropic because of their layered structure. Remarkable differences from conventional superconductors result from the extremely short coherence length ξ along the crystallographic c direction perpendicular to the CuO_2 planes. HTSC cuprates are therefore extreme Type-II superconductors.

Because of the extremely short coherence length, grain boundaries can act as weak links. The consequences of the resulting granularity will be discussed later. Moreover, anisotropy and weak pinning lead to giant flux creep effects at elevated temperatures, and an irreversibility line positioned between the lower and upper critical fields is observed. Above the irreversibility line a transport current in a HTSC cuprate causes dissipation.

6.7.1 Superconducting Properties of HTSC Cuprates

One can use the anisotropic Ginzburg-Landau (GL) theory to describe the magnetic properties of HTSC cuprates. As a consequence of the layered structures of the HTSC cuprate materials the penetration depth depends on the direction of flow of the screening currents. It is assumed that the screening currents in the a and b directions are equal. Obviously, two different penetration depths λ_{ab} and λ_c result where the indices ab and c give the directions of flow of the screening currents. Due to this anisotropy, one can define two different GL parameters κ_{ab} and κ_c as

$$\kappa_{ab} = \left(\frac{\lambda_{ab} \lambda_c}{\xi_{ab} \xi_c} \right)^{1/2}$$

$$\kappa_c = \frac{\lambda_{ab}}{\xi_{ab}} \qquad \qquad ...(6.15)$$

where ξ_{ab} and ξ_c are the coherence lengths in the ab plane and along c-direction respectively. The indices of GL parameters give the direction of the applied magnetic field. One can express the dimensionless anisotropy parameter γ_0 by the ratios of the following material properties:

$$\gamma_0 = \left(\frac{m_c}{m_{ab}}\right)^{1/2} = \frac{\lambda_c}{\lambda_{ab}} = \frac{\xi_{ab}}{\xi_c} = \frac{B_{c_2,ab}}{B_{c_2,c}} = \frac{B_{c_1,c}}{B_{c_1,ab}} \quad ...(6.16)$$

where m_{ab} and m_c are the effective masses of the charge carriers for the currents in the ab plane and along the c-direction respectively. We may note that B_{c_1} and B_{c_2} are the lower and upper critical fields where the indices ab and c correspond to field directions in the ab plane and along the c direction respectively. The thermodynamic critical field B_c is given by [116, 117]

$$B_c = \frac{\phi_0}{2\sqrt{2}\,\pi \lambda_{ab} \xi_{ab}} \quad ...(6.17)$$

where ϕ_0 is the flux quantum. One finds the resulting expressions for the upper critical fields as [116, 117]

$$B_{c_2,c} = \frac{\phi_0}{2\pi \xi_{ab}^2} \quad ...(6.18)$$

$$B_{c_2,ab} = \frac{\phi_0}{2\pi \xi_c \xi_{ab}} \quad ...(6.19)$$

One can derive the coherence length and the Ginzburg-Landau parameter from the penetration depth and critical field data.

In principle the lower critical field B_{c_1} can be deduced from hysteresis loop measurements. In these experiments the applied field (magnetic) is first increased from zero to a certain maximum value and decreased. The field direction is then reversed and the field strength is

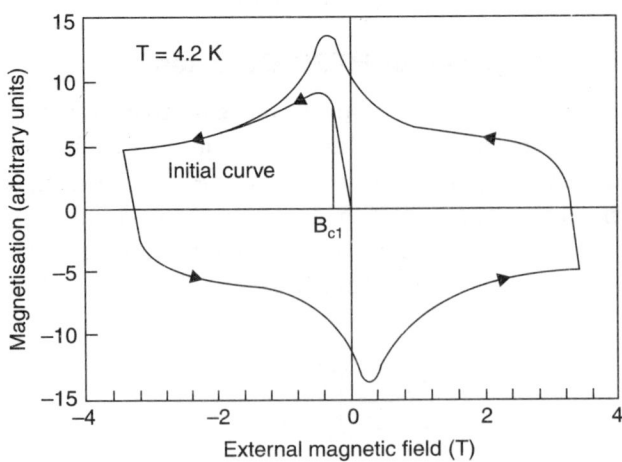

Fig. 6.41 High-field hysteresis loops for HTSC cuprate $YBa_2Cu_3O_{7-\delta}$ [118]

again increased to a maximum value and finally reduced to zero. This procedure is repeated in the following field cycles. Fig. 6.41 shows a typical high field hysteresis loop for $YBa_2Cu_3O_{7-\delta}$. The first deviation of the initial curve from linearity indicates the lower critical field B_{c_1}.

At the moment we are not considering the effects of *granularity*. Lower critical fields data for various HTSC cuprates are shown in Fig. 6.42. Typical values of B_{c_1} are between 10 and 100 mT for magnetic fields applied along the c-direction. The lower critical fields for metallic superconductors are for comparison in the range 10 – 250 mT [8].

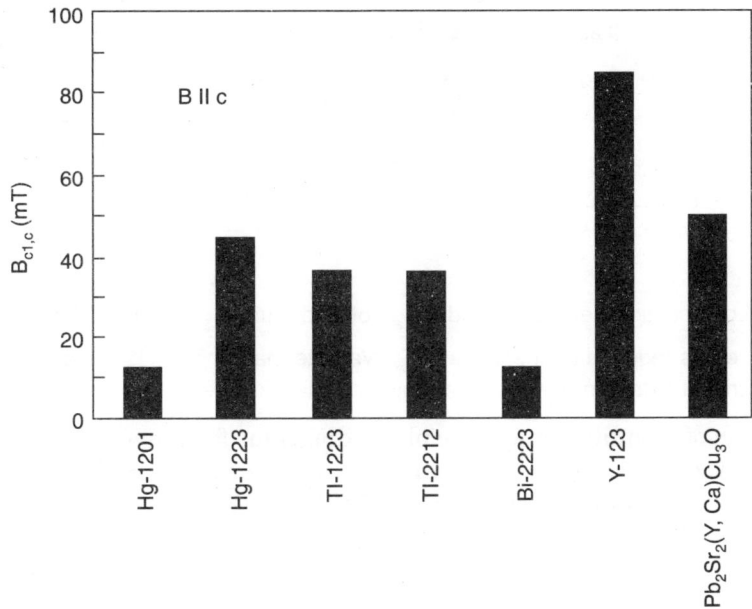

Fig. 6.42 Lower critical fields $B_{c_1,c}$ of various HTSC cuprates [8, 119 – 123]

The lower critical fields of Y-123 and Tl-2212 as a function of temperature is shown in Fig. 6.43. The solid lines in the figure are fits of the function

$$B_{c_1} = B_{c_1}(0)\,[1 - (T/T_c)^\alpha]$$

to the data. The values of α reported for Y-123 and Tl-2212 are 2 and 2.5 respectively. Generally, $\alpha = 2$ is found for conventional superconductors. Even for the Y-123 system a reasonably good fit results above $T/T_c = 0.5$ for $\alpha = 2$. The lower critical fields $B_{c_1,c}$ and $B_{c_1,ab}$ are considerably different for the Tl-2212 compound, whereas only a relatively small anisotropy has been found for Y-123. A large anisotropy of the physical properties for Bi and Tl double-layer compounds is intimately connected to the relatively large distance of neighbouring CuO_2 blocks in these systems (See Fig. 6.21).

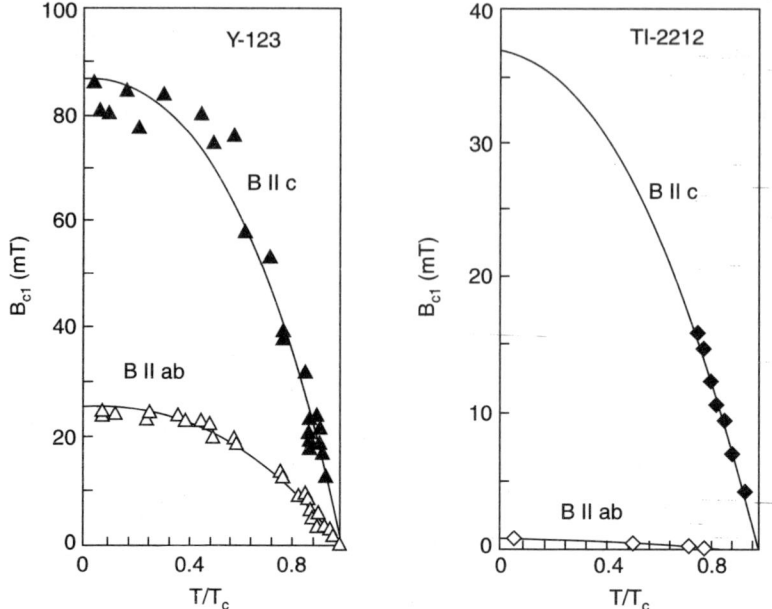

Fig. 6.43 Lower critical fields $B_{c_1,c}$ and $B_{c_1,ab}$ of $YBa_2Cu_3O_{7-\delta}$ and $Tl_2Ba_2Cu_2O_{8+\delta}$ HTSC cuprates versus reduced temperature T/T_c. We note that for Tl-2212 the anisotropy of the lower critical field is much more pronounced than for the Y-123 compound [122, 123]

A comparison of simplified magnetic phase diagrams for conventional Type-II and HTSC cuprates is presented in Fig. 6.44. We can see that the magnetic phase diagram of HTSC cuprates is much more complex than that of conventional Type-II superconductors. An outstanding feature of HTSC cuprates is the existence of an irreversibility line well below the upper critical field B_{c_2}. Above this line the vortices are movable and the critical current density is therefore zero.

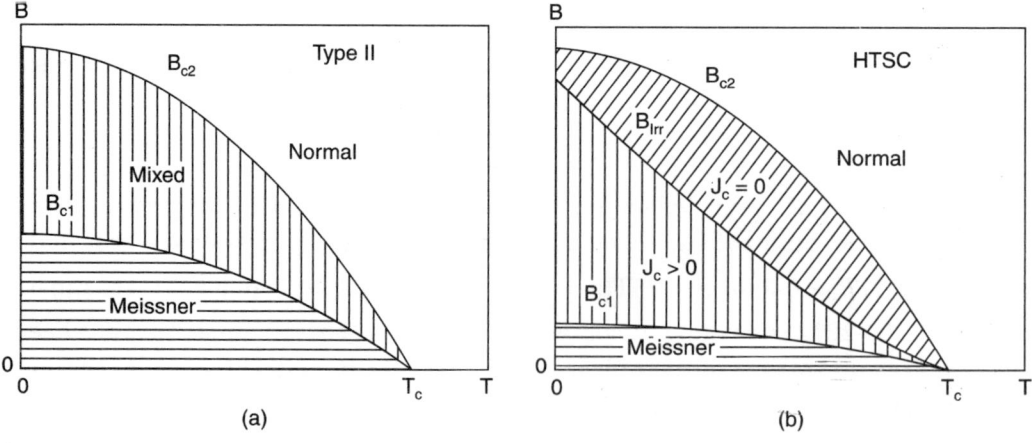

Fig. 6.44 Simplified magnetic phase diagrams for conventional Type-II (*a*) and HTSC cuprates (*b*). Above the irreversibility field B_{irr} the vortices are movable and thus the critical field density J_c is zero

In the presence of an applied magnetic field the resistance versus temperature curves of HTSC cuprates are considerably broadened. Typically extended resistance tails reaching temperatures well below T_c have been reported for HTSC cuprates [123 – 128]. The temperature dependence of the electrical resistivity of $Bi_{2.2}Sr_2Ca_{0.8}Cu_2O_{8+\delta}$ single crystal for selected magnetic fields applied along the crystallographic c-axis is shown in Fig. 6.45. The end point of resistive transition corresponds to the irreversibility temperature (T_{irr}) for the magnetic field applied. On the other hand, 90% of the normal state resistivity may be used as a criterion for the determination of B_{c_2} for the HTSC cuprate in question.

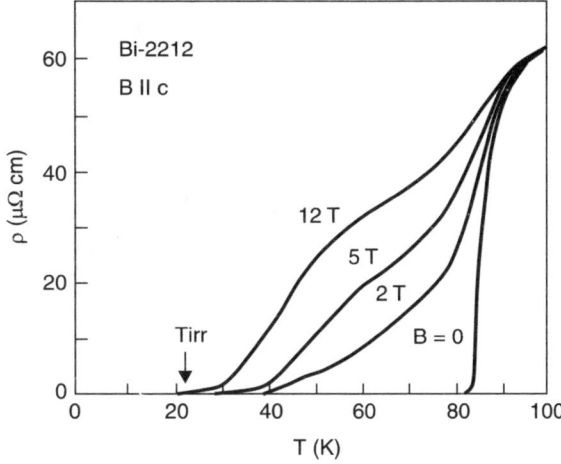

Fig. 6.45 Electrical resistivity of HTSC Bi-2212 single crystal versus temperature for magnetic fields applied along the crystallographic c-axis [128]

The upper critical field derived from the resistive transition depends strongly on the criterion used due to broad transition. We may note that studies of the reversible DC magnetisation for temperatures and fields in the vicinity of T_c and B_{c_2} provide a further possibility to measure the upper critical fields [129].

Table 6.20 Anisotropy of the upper critical fields (B_{c_2}) of selected HTSC cuprates

Compound	$B_{c_2,ab}/B_{c_2,c}$	Remarks	Reference
Y-123	5.5	single crystal	[130]
Bi-2212	15	thin film	[131]
Bi-2212	7.5	single crystal	[126]
Bi-2212	60	single crystal	[132]
Bi-2223	31	whiskers	[133]
Tl-2212	70	thin film	[127]
Tl-2223	20	single crystal	[134]
Tl-1223	8	single crystal	[134]

Table 6.20 provides the values for the anisotropy of the upper critical fields. As a consequence of imperfect texture, measurements performed on thin films may lead to reduced values of anisotropy. Moreover, the measured anisotropy depends on the accuracy of the alignment of the crystallographic axes with respect to the field direction. Because of these effects considerably different values for the anisotropy can be found in the literature for the same superconducting material.

Because of the very high critical fields of HTSC cuprates, $B_{c_2}(0)$ cannot be measured directly. From equation

$$B_c(T) = B_c(0)[1 - (T/T_c)^2]$$

One can easily show that the slope of the upper critical field at $T = T_c$ is

$$\left(\frac{dB_{c_2}}{dT}\right)_{T_c} = -\frac{2B_{c_2}(0)}{T_c} \qquad ...(6.20)$$

One uses eq. (6.20) frequently to estimate the upper critical field at zero temperature. For various HTSC superconductors, the values of B_{c_2} are shown in Fig. 6.46. The values given for HTSC cuprates are for the magnetic field applied along the c-direction. The upper critical field $B_{c_2,c}$ are typically around 100 T for HTSC cuprates. We may note that these extraordinary high upper critical fields allow the construction of very high field ($B > 20$ T) superconducting magnets. Even higher B_{c_2} values result for magnetic fields applied parallel to the ab planes of the HTSC cuprates.

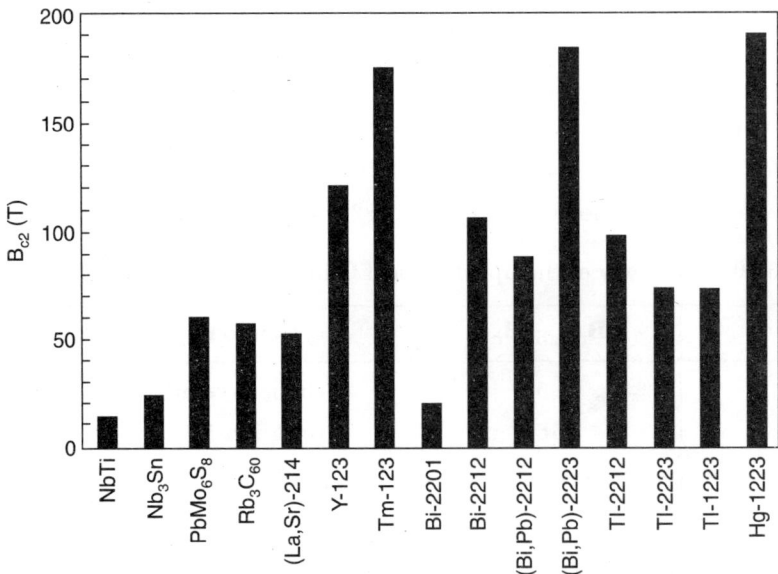

Fig. 6.46 B_{c_2} at $T = 0$ for various HTSC cuprates and Type-II superconductors [8, 116, 130, 135, 136 – 141]

Now, we consider the coherence length in HTSC cuprates. One can derive the penetration depth from reversible low-field magnetisation data or by *muon spin rotation (μSR) experiments*. Spin polarised positive muons (μ^+) are obtained from the decay of positive pions (π^+), which have a mean life of 26 ns and are produced in proton collisions

$$p + p = p + n + \pi^+ \qquad ...(6.21)$$

where n is a neutron

$$\pi^+ \rightarrow \mu^+ + \nu_\mu \qquad ...(6.22)$$

where ν_μ is a muon neutrino (Fig. 6.47). The positive muons penetrate into the superconductor under investigation. The muon decays with a mean lifetime of 2.2 μs into a positron e^+, an electron neutrino and a muon antineutrino

$$\mu^+ \rightarrow e^+ + \nu_e + \overline{\nu}_\mu \qquad ...(6.23)$$

The positron is preferentially emitted in the direction of the muon spin. In the presence of a magnetic field the muon processes at the Larmer frequency $\nu_{\mu L}$ = 135.5 MHz/T. As a consequence the resulting angular distribution $W(\phi, t)$ of the emitted positrons is modulated with the Larmer frequency $\nu_{\mu L}$ [142]

$$W(\phi, t) = 1 + A \cos(\phi - 2\pi \nu_{\mu L} t) \qquad ...(6.24)$$

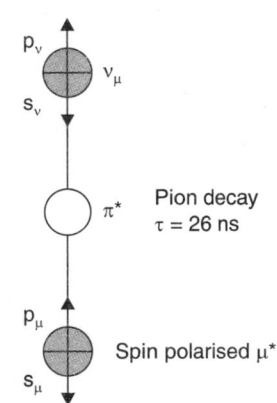

Fig. 6.47 Production of spin polarised positive muons via the pion decay. Spin S_μ and momentum p_μ of the muon are antiparallel

where A is a constant and ϕ the angle between the initial polarisation and the direction of the detector. From the time-dependent intensity of the emitted positrons information on the microscopic magnetic field at the muon site can be obtained. Slightly different microscopic magnetic fields lead to a time-relaxation of the modulation connected with the precession of the muon spin. A Gaussian distribution of the microscopic magnetic field around an average value B_{av} leads to a time relaxation function exp. ($-\sigma^2 t^2$). The resulting relation between the experimentally accessible relaxation time $1/\sigma$ and the field distribution $<\Delta B^2>$ is [142]

$$<\Delta B^2> = \frac{\sigma^2}{\gamma_\mu} \qquad ...(6.25)$$

where $\gamma_\mu = 2\pi \nu_{\mu L}$ is the **gyromagnetic ratio** of the muon, and the brackets stand for the mean value. Typically, a triangular lattice of vortices is observed in Type-II superconductors in the mixed state. The characteristic length scale for the variation of the magnetic field B in the vicinity of vortex is the penetration depth λ_L. The field distribution $<\Delta B^2>$ and the penetration depth λ_L are related by the expression [143]

$$<\Delta B^2> = \frac{0.00371}{\lambda_L^4} \phi_0^2 \qquad ...(6.26)$$

where ϕ_0 is the flux quantum. We may note that eq. (6.26) is valid for a perfect triangular lattice.

Experimentally determined values of the penetration depth λ_{ab} connected to screening currents flowing in the CuO_2 planes for various HTSC cuprates are shown in Fig. 6.48. Considerably different λ_{ab} values have been reported for $YBa_2Cu_3O_{7-\delta}$. The scatter of the reported λ_{ab} data for HTSC cuprates Y-123, Tl-2223 and Hg-1201 is indicated in light shade. A pronounced dependence of λ_{ab} on the oxygen deficiency δ has been found for $YBa_2Cu_3O_{7-\delta}$. Thus λ_{ab} increases by a factor of ≈ 10 for δ-values between 0.0 and 0.61. Obviously, the scatter in the measured λ_{ab}-values may therefore be partly due to slightly different chemical compositions. The penetration depth λ_{ab} is typically between 100 and 300 nm in HTSC cuprates. For comparison λ_L-values for several conventional superconductors are shown in Fig. 6.49. We may note that for most of these superconductors λ_L is about or less than 100 nm. The scatter of the reported data for the Cheveral phase superconductor $PbMo_6S_8$ is shown in black.

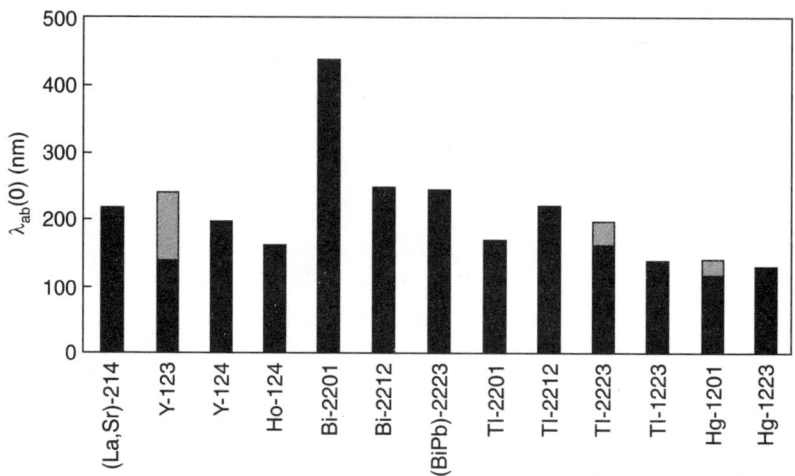

Fig. 6.48 Penetration depths $\lambda_{ab}(0)$ for various HTSC cuprates. The scatter in the data is given in light shading [8, 120, 121, 136, 141, 143 – 148]

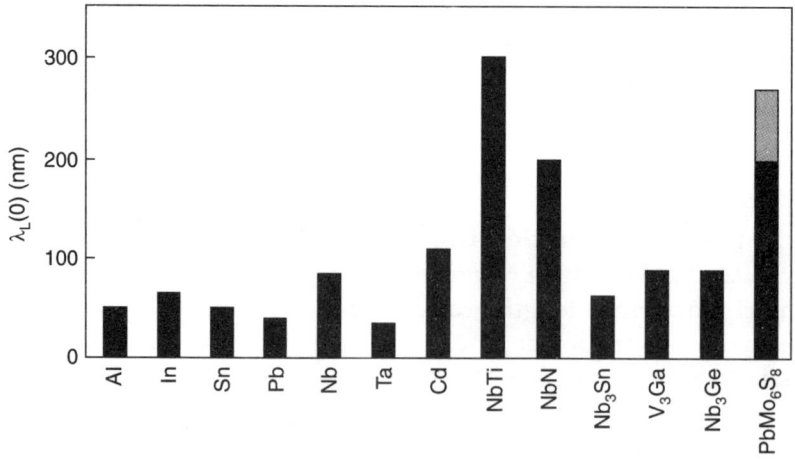

Fig. 6.49 $\lambda_L(0)$, i.e. Penetration depth of several conventional superconductors. The blank space indicates the scatter in the data for $PbMo_6S_8$ [8, 116, 143, 150]

We may note that the temperature dependence of λ_L in conventional superconductors can be well described by the empirical relation

$$\lambda_L(T) = \lambda_L(0) \left[1 - \left(\frac{T}{T_c}\right)^4\right]^{-1/2} \qquad ...(6.27)$$

The London theory predicts that the penetration depth is proportional to $n_c^{-1/2}$. The Cooper pair density n_c increases with decreasing temperature, while the density of single electron is reduced (two fluid model). Interestingly, this effect leads to a small penetration depth at low temperatures.

Penetration depth λ_{ab} for Y-123, (Hg-Cu)-1201 and Tl-2223 is shown in Fig. 6.50. The λ_{ab} data for Y-123 and (Hg-Cu)-1201 can be well represented by eq. (6.27), whereas the values for the Tl-2223 HTSC cuprate compound are close to $\lambda_{ab}(T) = \lambda_{ab}(0) \left(1 - \dfrac{T}{T_c}\right)^{0.3}$ with $\lambda_{ab}(0) = 163$ nm. The solid lines in Fig. 6.50 are fits of these two functions to the λ_{ab} data.

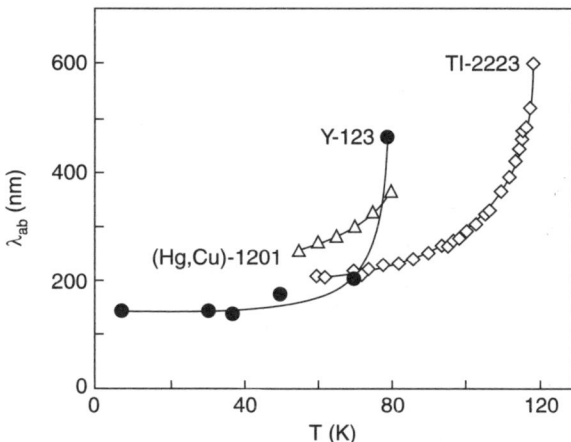

Fig. 6.50 Penetration depth λ_{ab} versus temperature Y-123, (Hg-Cu)-1201 and Tl-2223 [143, 147, 153]. The solid lines are fits of the function $\lambda_{ab}(T) = \lambda_{ab}(0)(1 - T/T_c)^{-0.3}$ with $\lambda_{ab}(0) = 163$ nm (Tl-2223) and of eq. $B(x) = B_0 \exp(-x/\lambda_L)$ (Y-123, (Hg-Cu)-1201) to the data

We may note that frequently a linear or quadratic dependence of $\lambda_{ab}(T)$ has been found in HTSC cuprates at low temperatures [136, 144, 146, 150—152]. However, such behaviour is not expected for conventional superconductors and may be caused by an unconventional pairing mechanism.

The in-plane coherence length ξ_{ab} of various HTSC cuprates is shown in Fig. 6.51 and it is typically between 1 and 4 nm. Considerably smaller values (< 0.5 nm) have been found for the out-of-plane coherence length ξ_c. Generally, ξ_c is comparable to interatomic distances in the crystal structures of HTSC cuprates. We may note that these small ξ-values are responsible for the insulating character of the charge carrier reservoirs (Figs. 6.2, 6.3 and 6.4) separating adjacent CuO_2 blocks. This means that HTSC cuprates are nearly two dimensional superconductors with intrinsic Josephson junctions along the crystallographic c-direction. As a consequence of the very short out-of-plane coherence length ξ_c, the Cooper pair density n_c can be strongly reduced even within the width of a grain boundary. Consequently the grain boundaries can act as weak links.

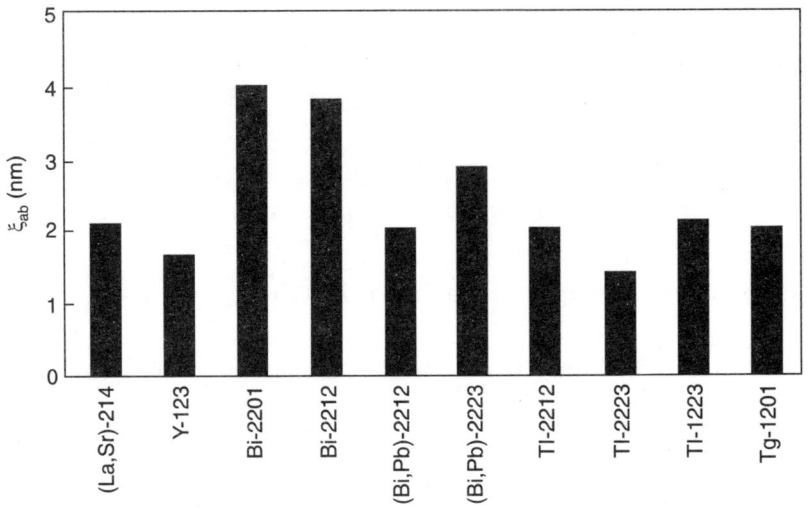

Fig. 6.51 The in-plane coherence length ξ_{ab} of various HTSC cuprates. Generally, the in-plane coherence length ξ_{ab} is considerably larger than ξ_c perpendicular to the CuO$_2$ planes

The following relations hold generally for the characteristic length scales in HTSC cuprates:

$$\xi_c < \xi_{ab} \ll \lambda_{ab} < \lambda_c \qquad ...(6.28)$$

The values of the penetration depths λ_{ab} and λ_c, the coherence lengths ξ_{ab} and ξ_c, and Ginzburg–Landau parameter κ_c are listed in Table 6.21. Generally, the κ_c-values connected to a magnetic field applied along the c-direction are around 100. Obviously, HTSC cuprates are extreme Type-II superconductors.

Table 6.21 Characteristic lengths for selected HTSC cuprates

Compound	λ_{ab} (nm)	λ_c (nm)	ξ_{ab} (nm)	ξ_c (nm)	κ_c	Ref.
(La, Sr)-214	250-410		2.1		120-200	137
Y-123	210					144
Y-123	141.5	700				143
Y-123	160-240	1360-1480				146
Y-123			1.6	0.3		130
Y-123	130		1.3	0.2	100	154
Bi-2201	438		4.0		110	136
Bi-2212			3.8	0.16		132
Bi-2212			2.7	0.18		131

Bi-2223			2.9	0.093		133
(Bi, Pb)-2212	178		2.0		89	139
(Bi, Pb)-2223	245		2.9		84	121
(Bi, Pb)-2223	88		1.35		65	140
Tl-2201			5.2	0.3		155
Tl-2212			3.1	0.68		155
Tl-2212			2.0	0.03		127
Tl-2223	163		1.36		120	147
Tl-2223	117		1.1		106	156
Tl-2234			4.5	1		155
Tl-1223	137		2.1		65	156
Hg-1201	117		2.1		56	148
Hg-1201	140		2		70	153
(Hg, Cu)-1201	247					

Another important property of HTSC cuprates is the energy gap 2Δ. In addition to tunnelling [157 – 159] and infrared reflectivity measurements [160 – 162], nuclear magnetic resonance (NMR) [163], Raman scattering [164] and high resolution angle-resolved photo-emission [165] have been used to study the gap properties of HTSC cuprates. Fig. 6.52 shows values of 2Δ versus $k_B T_c$ for selected HTSC cuprates. The ratio $2\Delta/k_B T_c$ is typically between 6 and 8, which is considerably larger than the value of 3.5 predicted by the BCS theory. We may

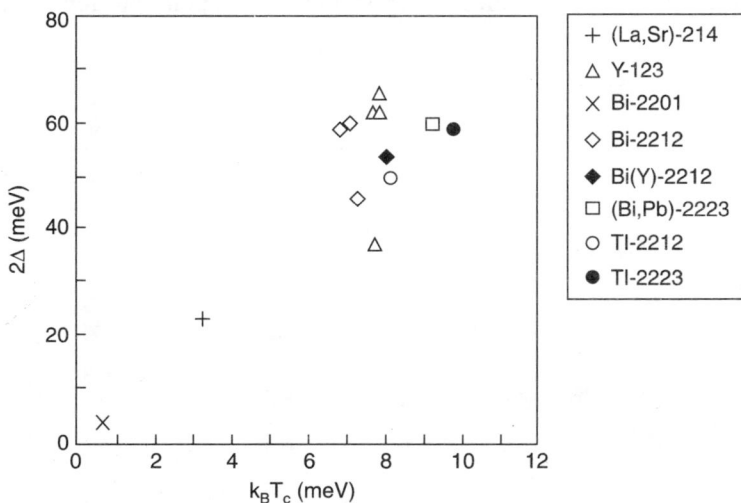

Fig. 6.52 Energy gap 2Δ versus $k_B T_c$ for selected HTSC cuprates [158 – 163, 164, 166, 172 – 173]. The ratio of $2\Delta/k_B T_c$ is typically between 6 and 8 which is well above the value of 3.5 predicted by the BCS theory

note that the energy gap 2Δ for the conventional superconductors is between 3 and 5 $k_B T_c$. Moreover the energy gap of HTSC cuprates is anisotropic [162, 166 – 167]. For Y Ba$_2$Cu$_3$O$_7$ single crystals, energy gaps of $\approx 8\, k_B T_c$ and $\approx 3\, k_B T_c$ have been obtained from infrared reflectivity measurements for the electric field parallel and perpendicular to the CuO$_2$ planes respectively [162]. The large scatter in the reported 2Δ-values may be a consequence of this anisotropy.

In conventional low temperature superconductors the electron–phonon interaction leads to the formation of Cooper pairs. Because of the dependence of the phonon frequencies on the isotopic mass M the critical temperature of these superconductors is expected to be proportional to M^β. For many simple metals β-values close to 0.5 have been found, as expected for phonon pairing mechanism. To get deeper insights into the mechanism responsible for HTSC cuprates several investigators have studied the oxygen isotope effect in these systems [168 – 171]. In these experiments ^{16}O is replaced by ^{18}O. Values of the exponent β for the Chevrel phase Pb Mo$_6$ S$_8$, the fullerene Rb$_3$ C$_{60}$ and several HTSC cuprates are shown in Fig. 6.53. A relatively strong isotope effect has been observed for PbMo$_6$S$_8$, Rb$_3$C$_{60}$ and copper free oxide superconductors Ba (Pb, Bi) O$_3$ and (Ba, K) BiO$_3$. The critical temperatures for these superconductors are 12, $\approx$ 30, 11 and 32 K respectively. These results clearly indicate that for all these superconductors the electron–phonon interaction is of importance for the pairing. On the other hand, especially for the optimally doped HTSC cuprate superconductors [174]. Only a very small isotope effect has been found. Fig. 6.54 shows T_c and β-values for Y$_{1-x}$Pr$_x$Ba$_2$Cu$_3$O$_7$ versus the Pr concentration x. Interestingly, the isotope effect is more pronounced for the compounds with larger Pr content and lower critical temperatures. The same trend is visible for (La$_{1.9}$ Sr$_{0.1}$) – 214 and (La$_{1.85}$ Sr$_{0.15}$) – 214 in Fig. 6.53. The β-values of 0.4 and 0.14 correspond to T_c-values of 33 and 39 K respectively. Like T_c, the isotope effect depends on the carrier concentration in the CuO$_2$ planes and reaches a minimum for optimally doped HTSC cuprates. An unconventional pairing mechanism may therefore be responsible for HTSC cuprates.

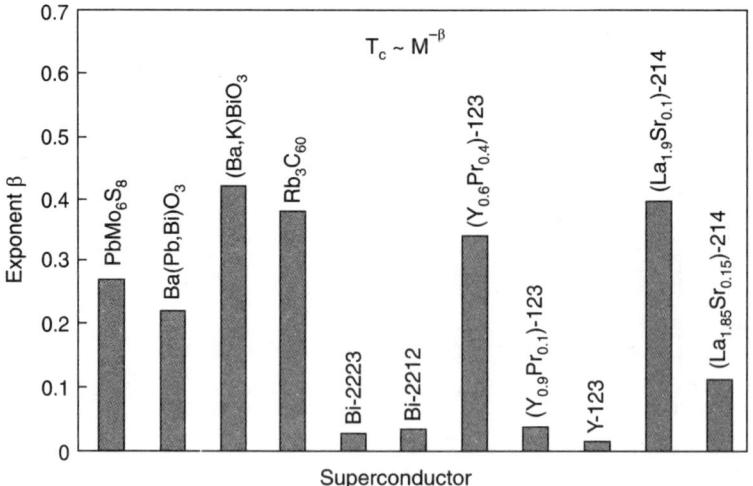

Fig. 6.53 Exponent β for selected HTSC cuprates, the fullerene Rb$_3$C$_{60}$ and the Chevrel phases PbMo$_6$S$_8$ [143]

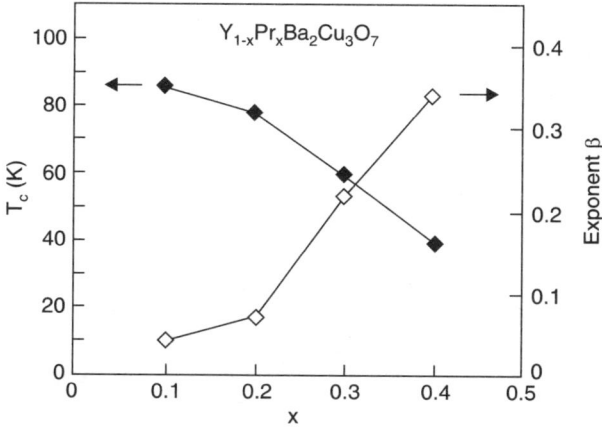

Fig. 6.54 Exponent β and T_c of (Y, Pr) – 123 versus Pr content x [143]

In magnet and power applications the critical current density is one of the most important physical properties of the HTSC cuprates. We may note that, unlike T_c, B_{c_2}, J_c is not an intrinsic material property. The achievable J_c-values are determined mainly by the microstructure of the superconductor in question, *e.g.* grain boundaries can act as weak links in HTSC cuprates, leading to small J_c-values in polycrystalline material. In the absence of grain boundary weak links, the transport critical current density is determined mainly by the pinning of the flux lines at defects. As soon as the flux lines becomes mobile, the J_c-values drop to zero and dissipation sets in. We will discuss the properties of the flux line lattice and possible pinning mechanisms later in this chapter.

Let us first discuss the anisotropy of the critical current density and epitaxial films. A simplified structure of HTSC cuprates is presented in Fig. 6.55. The currents can easily flow in

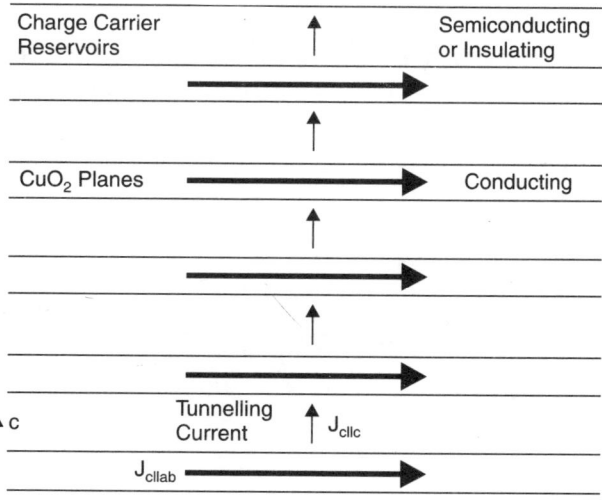

Fig. 6.55 Simplified structure of layered HTSC cuprates. Currents can easily flow within the CuO_2 planes. Because of the insulating or semiconducting nature of the charge carrier reservoirs, only tunnelling currents can flow along the *c*-direction

the CuO_2 conduction planes, whereas the charge carrier reservoirs act as insulating or semiconducting barriers. This means only tunnelling currents can flow in the crystallographic c-direction. In the CuO_2 planes of Hg-1212 films J_c-values as high as 10^8 A/cm² have been found, whereas the J_c perpendicular to the CuO_2 planes is only 5000 A/cm² [175]. Moreover, a giant anisotropy of the effective masses m_c and m_{ab} has been reported for Hg-1223 films with critical current densities of the order of $\approx 10^7$ A/cm² at 10 K. The ratio m_c/m_{ab} reaches the extraordinary high value of 32000 [176].

We may note that the further anisotropy of the J_c-values is connected to the direction of the applied magnetic field [177]. The critical field densities in epitaxial Bi-2212 films as a function of temperature and applied field is shown in Fig. 56. A relatively small reduction of the J_c values results when the magnetic field is applied along the CuO_2 (ab planes) (Fig. 6.56 a). On the other hand, the critical current densities decline rapidly when the magnetic field is parallel to the crystallographic c direction (Fig. 6.56 b). For selected values of the applied magnetic field, the resulting anisotropy $J_{c,ab}/J_{c,c}$ versus temperature is shown in Fig. 6.57.

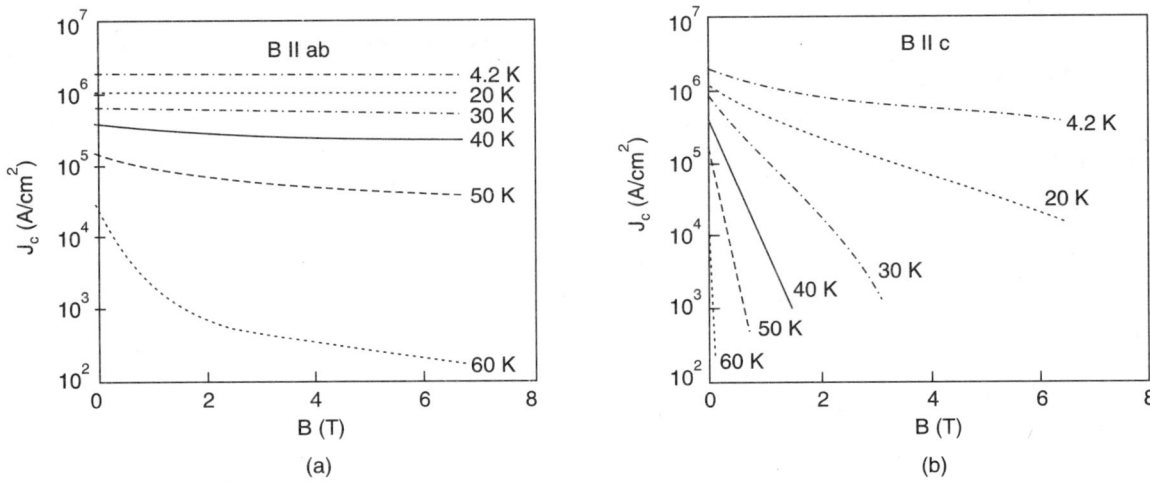

Fig. 6.56 Critical current densities in epitaxial Bi-2212 as a function of temperature and applied magnetic field B along the ab planes (a) and parallel to the crystallographic c-direction [177]

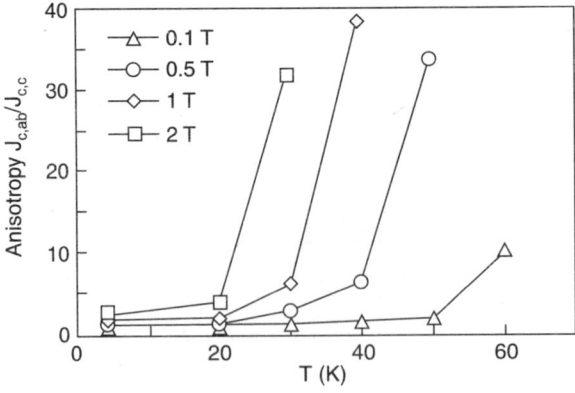

Fig. 6.57 Anisotropy of the J_c-values with respect to the direction of the applied magnetic field. We note that $J_{c,ab}/J_{c,c}$ increases with increasing temperature and magnetic field [177]

Similar high J_c-values have been reached in Bi-2223 thin films [178]. The temperature-dependence of the critical current density in three different Bi-2223 films at zero applied field is shown in Fig. 6.58. One can well represent the J_c-data by the function $J_c(T) = J_c(0)(1 - T/T_c)^\alpha$. The solid lines in Fig. 6.58 are fits based on this expression. We may note that for the two lower curves the exponent α is close to 1.4. In addition, the dependence of the J_c-values on the angle θ between the CuO_2 planes and the magnetic field direction has been also studied. The arrangement of the Bi-2223 film, the applied magnetic field, and the current direction are shown in Fig. 6.59. The results for an applied field of 4 T and temperatures of 40 and 60 K are shown in Fig. 6.60.

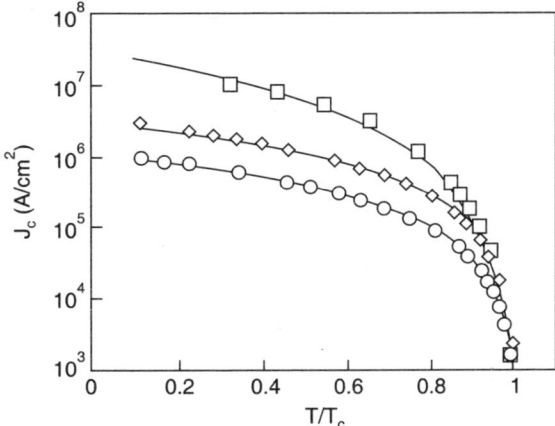

Fig. 6.58 Critical current density versus reduced temperature for three Bi-2223 films. The solid lines are fits to the function $J_c(T) = J_c(0)(1 - T/T_c)^\alpha$ [178]

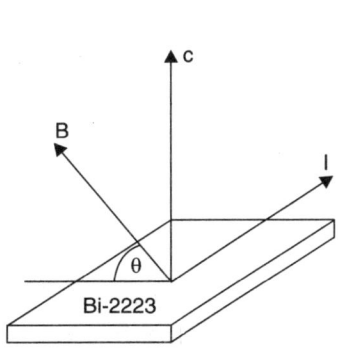

Fig. 6.59 Orientation of the applied magnetic field with respect to the current and to the crystallographic c-direction for Bi-2223 film

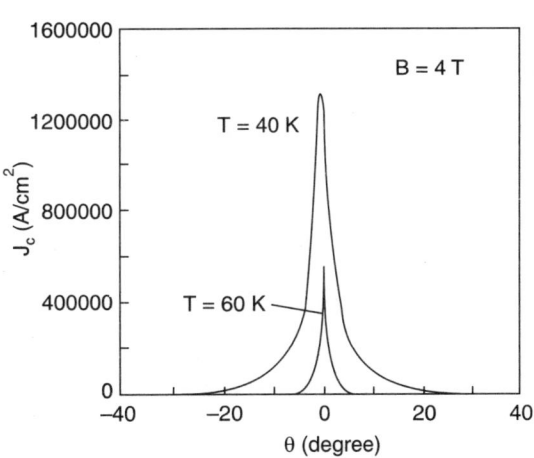

Fig. 6.60 Critical current density versus angle θ [178]

At 40 K an angle θ of 30° leads to a reduction of the critical current density by a factor of ≈ 1000 from the peak value. The same reduction results from a θ-value of less than 8° at 60 K. A detailed analysis of the data reveals [178] that

$$J_c(B, \theta) = J_{c,c}(B \sin \theta) \qquad ...(6.29)$$

This means, the reduction of the J_c-values by an applied field is determined by the field component parallel to the crystallographic c-direction.

One obvious reason for this behaviour is the anisotropy of the upper critical fields B_{c_2} (See Table 6.20). Furthermore, intrinsic pairing may be provided by the layered structure of HTSC cuprates for magnetic fields applied along the ab planes. Tachiki and Takahashi [179] proposed such a pinning mechanism. Fig. 6.61 shows that for B parallel to ab the Lorentz force is perpendicular to the different layers, whereas for B parallel to c this force is parallel to the ab planes. One can consider HTSC cuprates as made up of alternating strongly and weakly superconducting layers. For B parallel to ab the energy of the vortices in the weakly charge carrier reservoir is reduced. The layers can therefore act as intrinsic pinning sites.

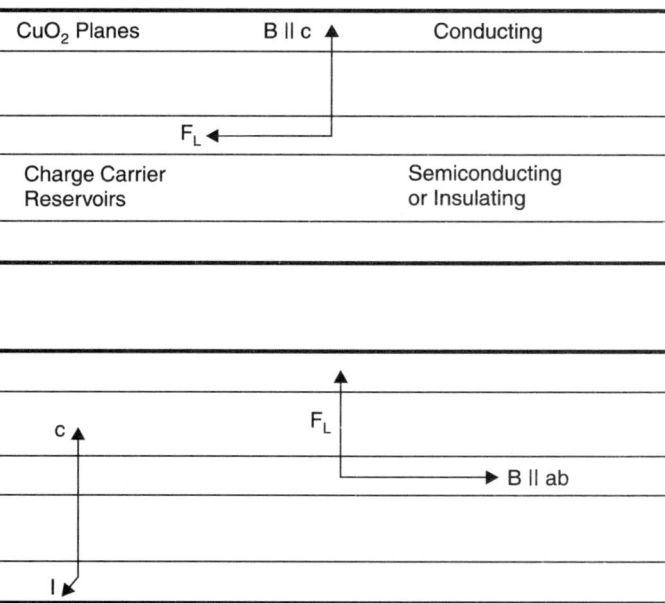

Fig. 6.61 In the layered structures of the HTSC cuprates, the weakly superconducting charge carrier reservoirs may provide intrinsic pinning for magnetic fields applied along the ab-planes

On the other hand, the results suggest that the pinning is only weak with respect to the flux motion along the ab planes. Due to the weak coupling of neighbouring CuO_2 blocks, the flux lines along the c direction may be subdivided into small segments. These nearly two dimensional vortices can easily be moved within the ab-planes, and are called pancake vortices.

Intragrain critical current densities of $YBa_2Cu_3O_7$ and $(Tl_{0.5}Pb_{0.5})Sr_2Ca_2Cu_3O_9$ as a function of applied field are shown in Figs. 6.62 and 6.63 for temperatures of 4.2 K and 77 K respectively. The intragranular critical current densities have been estimated from magnetic hysteresis measurements. The high J_c-values in Y Ba$_2$ Cu$_3$O$_7$ in the presence of an applied magnetic field are intimately connected to the small distance between neighbouring CuO$_2$ blocks is considerably smaller than in Bi and Tl double-layer compounds. Therefore the Tl-single-layer compounds, like the Y-123, show a more three dimensional superconductivity. This behaviour is reflected in the relatively high J_c-values in the (Tl-Pb)-1223 compounds, even at 77 K in the presence of an applied magnetic field.

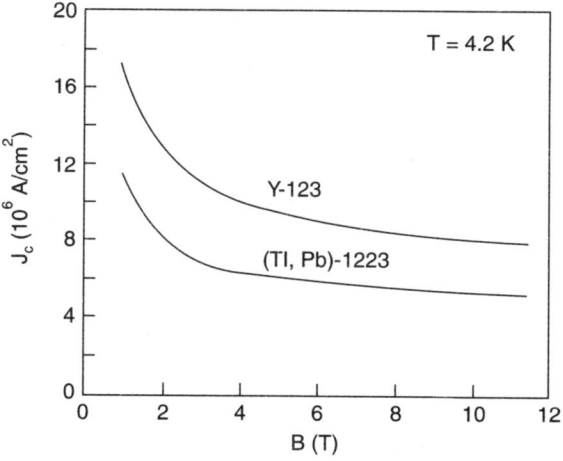 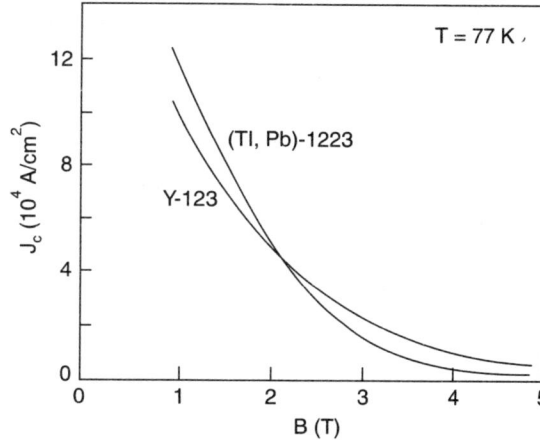

Fig. 6.62 Critical current densities of Y-123 and (Tl, Pb)-1223 versus applied field at 4.2 K [180]

Fig. 6.63 Critical current densities of Y-123 and (Tl, Pb)-1223 versus applied field at 77 K [180]

J_c-values for selected HTSC cuprates are listed in Table 6.22. In high-quality thin films, the critical current density can exceed 10^7 A/cm^2 even at 77 K. Now, we discuss the weak link behaviour of large angle grain boundaries on the transport critical densities.

Table 6.22. Critical current densities in selected HTSC cuprates

Compound	J_c (A/cm^2)	T (K)	B (T)	Ref.
Y-123 (Polycrystalline Intragranular J_c)	1.72×10^7	4.2	1	[180]
	$> 10^5$	77	1	
Y-123 (Thin Film)	5×10^6	77	0	[181]
Y-123 (Thin Film)	1.4×10^7	77	0	[182]
Y-123 (Large area thin film)	2×10^7	4.2	0	[183]
Bi-2212 (Thin Film)	10^7	4.2	0	[177]

(Contd...)

Bi-2223 (Thin Film)	10^7	30	0	
	1.3×10^6	70	0	[178]
Tl-2212 (Thin Film)	10^6	77	0	[184]
Tl-2223 (Thin Film)	7×10^5	77	0	[184]
Hg-1212 (Thin Film)	10^7	5	0	[175]
Hg-1223 (Thin Film)	$\approx 10^7$	10	0	[176]
Hg-1223 (Thin Film)	2.3×10^7	5	0	
	2.2×10^6	5	5	
	2×10^6	77	0	
	5×10^5	100	0	[185]
Hg-1223	4.4×10^5	77	0	[186]

6.7.2 Grain Boundary Weak Links

Soon after the discovery of superconductivity above 77 K in $YBa_2Cu_3O_{7-\delta}$, it was reported that the transport critical current densities $J_{c,x}$ of polycrystalline HTSC cuprates were disappointingly low. The J_c-values of polycrystalline $YBa_2Cu_3O_{7-\delta}$ are typically well below 1000 A/cm^2 at 77 K and zero applied field. In addition a magnetic field as low as a few mT is sufficient to reduce the transport critical current density by an order of magnitude. This behaviour for a silver-clad $YBa_2Cu_3O_{7-\delta}$ wire is shown in Fig. 6.64. On the other hand, $J_{c,x}$ values well above 10^7 A/cm^2 have been reached in epitaxial $YBa_2Cu_3O_{7-\delta}$ films at 77 K and $B = 0$. On the basis of these experiments, one finds that the grain boundaries present in polycrystalline

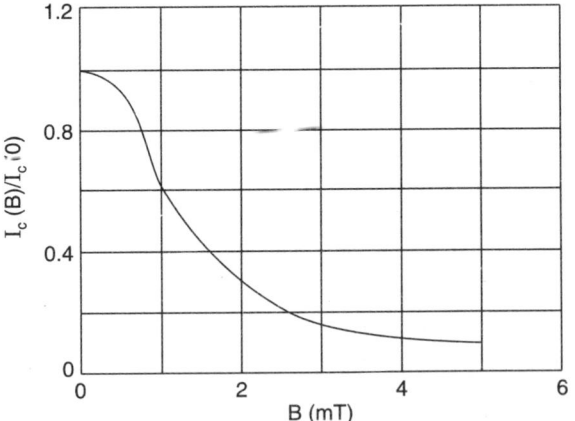

Fig. 6.64 Normalised transport critical currents at 77 K versus applied field for silver clad Y-123 wires [187]

material act as barriers for the transport current. We may note that the weak-link behaviour of the grain boundaries is closely connected to the extremely short coherence length ξ_c along the crystallographic c-direction. Typically ξ_c-values for HTSC cuprates are smaller than 1 nm. This is comparable to interatomic distances in the crystal structures of the HTSC cuprates. As a consequence, the Cooper pair density can considerably vary within the width of a grain boundary. Two grains separated by a grain boundary can therefore be considered as a Josephson weak link. This idea is supported by the large drop of the transport critical current densities caused by magnetic fields as low as a few mT. Similar pronounced effects of an applied magnetic field are the well-known Josephson–Fraunhofer diffraction patterns characteristic of superconductor-insulator superconductor (S-I-S) tunnel function. These weak links at the grain boundaries are responsible for the reduced transport critical current densities in polycrystalline HTSC cuprates [241].

The behaviour of a granular superconductor is schematically illustrated in Fig. 6.65. As a consequence of granularity, two different critical current densities, $J_{c,t}$ across the grain boundaries and $J_{c,g}$ within individual grains, coexist in the HTSC cuprates. In agreement with the thin film data, critical current densities in individual grains can exceed 10^6 A/cm^2 even at 77 K.

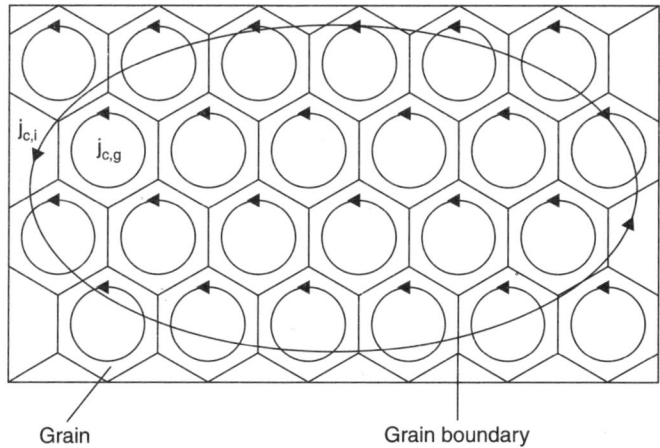

Fig. 6.65 Schematic illustration of a granular superconductor. As a consequence of weak link behaviour of the grain boundaries the intergranular screening current density $J_{c,t}$ is considerably smaller than the critical current $J_{c,g}$ within individual grains

For different field regions, magnetisation cycles of $YBa_2Cu_3O_{7-\delta}$ are shown in Fig. 6.66. Below an ill-defined magnetic field $B_{c_1}{}^*$ the magnetisation curves of polycrystalline Y Ba$_2$ Cu$_3$ O$_{7-\delta}$ are reversible, as shown in Fig. 6.66(a). The small asymmetry of the initial branch is caused by the earth's magnetic field. Irreversibility of magnetisation has been reported for applied magnetic fields as low as 3 mT. Magnetic flux starts to enter the polycrystalline Y-123 superconductor along the grain boundaries at the critical field $B_{c_1}{}^*$ (Fig. 6.66 (a)). Clearly, the field $B_{c_1}{}^*$ can be considered as the lower critical field of the weak links. The low-field hysteresis loop of Fig. 6.66(a) is therefore closely connected to the intergranular screening current density $J_{c,t}$ across the grain boundaries (See Fig. 6.65). These intergranular currents screen the applied field (magnetic) from the interior of the whole Y-123 specimen.

Above a magnetic field $B_{c_2}{}^*$ (Fig. 6.66(b)) reversible behaviour is restored. In most of the weak links this field is sufficient to destroy superconductivity and therefore this represent the upper critical field of the weak link network. On the other hand, the magnetic field is still excluded from the interior of the individual grains. As soon as the applied field exceeds B_{c_1} (Fig. 6.66(c)) magnetic flux penetrates into the individual grains. Thus B_{c_1} is the lower critical field of the individual grains, and in contrast to $B_{c_1}^*$, is an intrinsic material property of the Y-123 superconductor. Above the lower critical field B_{c_1} a large hysteresis of the magnetisation is observed, which can be attributed to the large intragranular critical current densities $J_{c,g}$ within the individual grains.

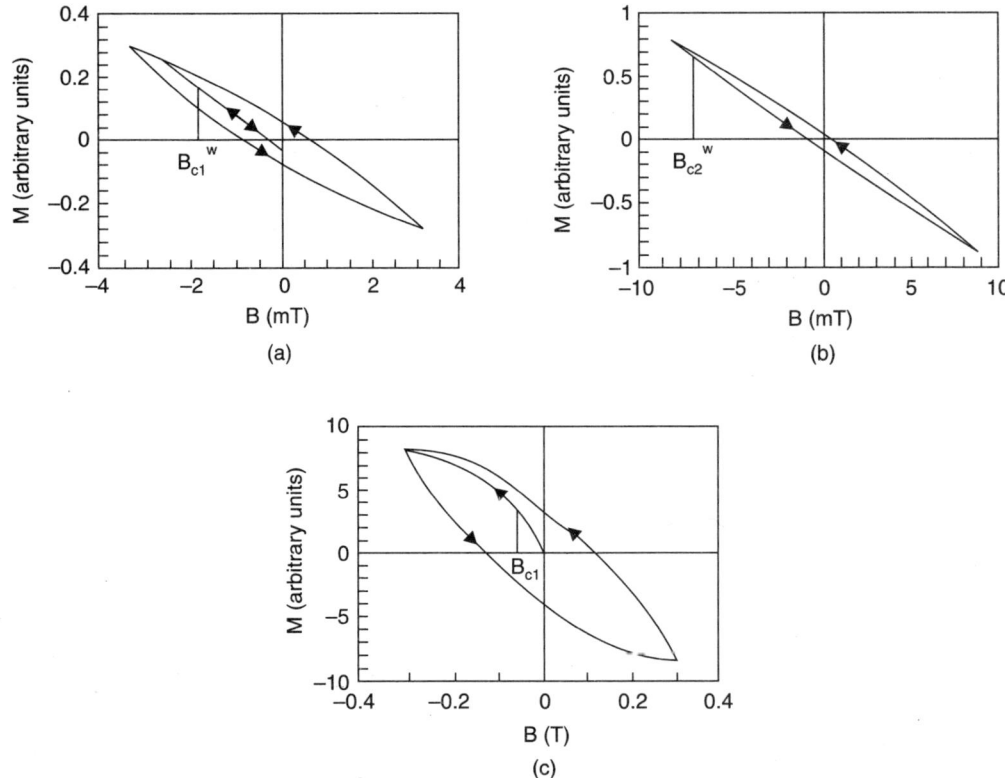

Fig. 6.66 For three different field regions, magnetisation loops of Y-123 [118]

High-field magnetisation loops of Y-123 at 4.2 (top) and 80 K (bottom) are shown in Fig. 6.67(a). At an applied magnetic field B_m, a peak in the magnetisation has been found. The magnetisation loop collapses to a straight line for sufficiently large magnetic fields as indicated in Fig. 6.67(b). This field, called the irreversibility field B_{irr}, is well below the upper critical field B_{c_2}. This reversible magnetic behaviour indicates zero critical current density $J_{c,g}$ within the individual grains. Possible reasons for the vanishing critical current densities $J_{c,g}$ could be thermally activated depinning, or flux line lattice melting.

Now, we consider the possible ways to determine the inter- and intragranular critical current densities. Four probe method can be used to determine the transport critical current.

The macroscopic critical current is limited by the grain boundaries acting as weak links. Using the Bean critical state model, the much larger microscopic critical current density within the grains can be estimated from magnetisation loops, we may note that in any comparison of critical current densities the criterion used to define J_c has to be taken into consideration.

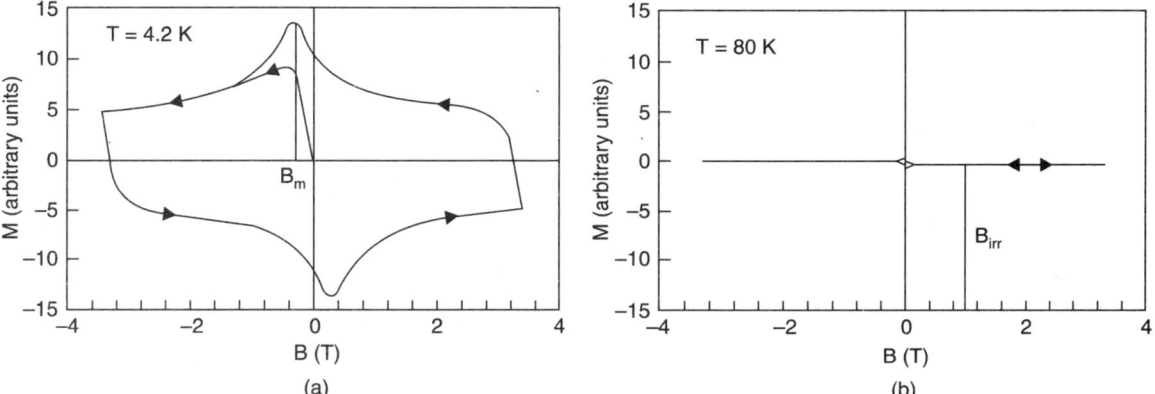

Fig. 6.67 High field magnetisation loops for Y-123 at 4.2 and 80 K (Fig. 6.67 (a)). Above the magnetic field B_{irr} the critical current density is zero, and as a consequence the magnetisation is reversible [118]

First, some criteria used to define the resistively measured transport critical current. Frequently, an electric field criterion of 1 or 0.1 μv/cm is used to determine the critical current density. Figure 6.68 indicates the influence of the selected field criterion on the critical current. Empirically it has been found that for both low and HTSC cuprate superconductors the electric field E_e and the current I in the region of interest are connected by the following expression

$$E_e = E_{eo} \left(\frac{l}{l_0}\right)^n \quad \ldots(6.29)$$

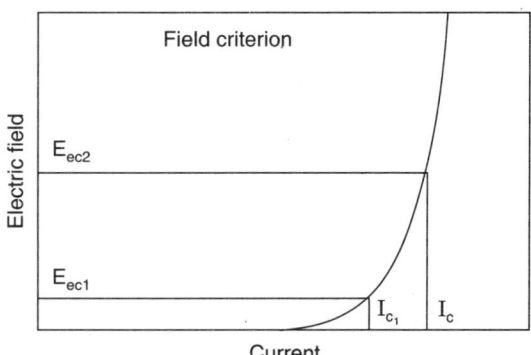

Fig. 6.68 Effect of different electric field criteria E_{ec_1} and E_{ec_2} on the critical current

where E_0 is the electric field at the current I_0.
eq. (6.29) defines the index of resistive transition n. In low-temperature superconductors the n-values at 4.2 K are typically well above 30, implying that the critical currents depend only weakly on the field criterion used. Considerably smaller values of the index of resistive measures have been found for HTSC cuprates at 77 K in the presence of an applied magnetic field. We may note that for small n-values the critical current is ill-defined.

The effect of different n-values on the critical current can be taken into consideration using an offset criterion [189 – 190]. As illustrated in Fig. 6.69, the offset critical current I_c^{offset} offset is defined by taking the tangent to the E-I curve at a selected field value E_c, and extrapolating to zero electric field. The critical currents defined by the offset and the field criterion are related by the following expression.

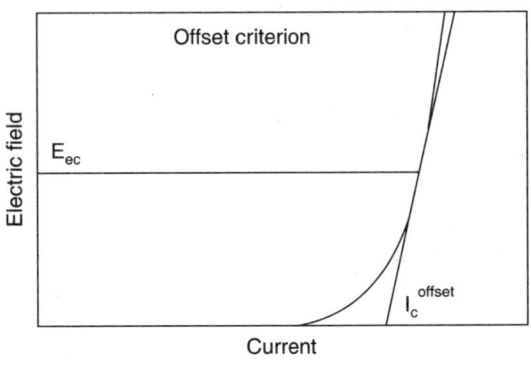

Fig. 6.69 Illustration of offset criterion to define the critical current

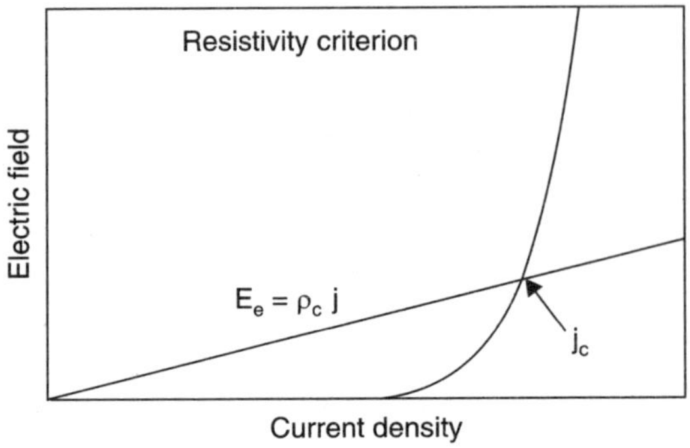

Fig. 6.70 Determination of the critical current density with a resistivity criterion

Moreover, a resistivity criterion can be used to define the critical current, as shown in Fig. 6.70. The electric field E, the current density J and the resistivity ρ of a normal conductor are connected by Ohm's law

$$E_e = \rho J \qquad \ldots(6.30)$$

Typically a resistivity of $10^{-13}\ \Omega-m$ is used to define the critical current. For comparison the resistivity of silver at room temperature is $1.6 \times 18^{-8}\ \Omega-m$.

We may note that only in single crystals can the intragranular critical current density $J_{c,g}$ be measured directly by the standard four probe method. The $J_{c,g}$-values of polycrystalline samples can be estimated from magnetisation measurements using the bean critical state model, which describes the magnetic behaviour of Type-II superconductors in the intermediate state.

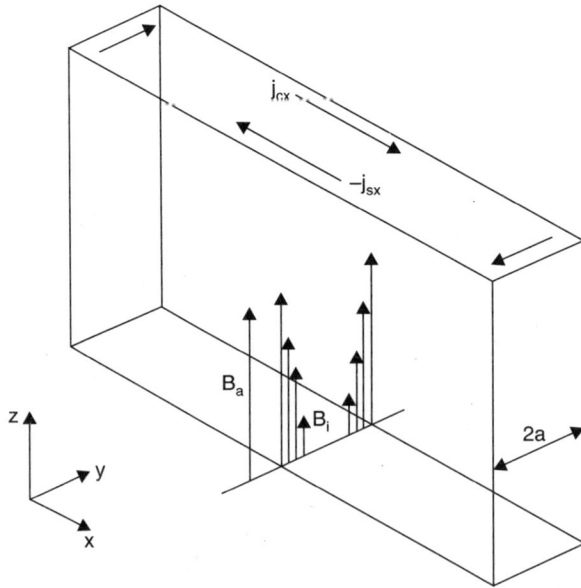

Fig. 6.71 Screening currents and internal magnetic field B_i in a thin superconducting slab of width 2a with an applied field B_a parallel to the surface

A slab of width $2a$ with a magnetic field B_a applied along the z-direction, and the resulting screening currents are shown in Fig. 6.71. As soon as the applied magnetic field B_a exceeds the lower critical field B_{c_1}, vortices containing single flux quanta enter from the surface into the interior of the superconducting slab. The demagnetization factor N for an infinite slab with the applied field parallel to the surface is zero. For not too high applied fields, the screening currents as well as the internal field B_i are zero in the central part of the slab. The screening current density J_s is equal to J_c where $B_i \neq 0$, and zero otherwise. The field dependence of the critical current density J_c is neglected in the Beam critical state model. The internal magnetic field B_i and the critical current density J_c are connected by the Maxwell equation

$$\nabla \times \mathbf{B}_i = \mu_0 \mathbf{J}_c \qquad \qquad ...(6.31)$$

where $\mu_0 (= 4\pi \times 10^{-7}$ V-s/A) is the permeability of the free space. This situation is called a critical state. The field pattern (Fig. 6.72 (a)) and the screening current distributions (Fig. 6.72(b)) in a thin slab is shown in Fig. 6.72. This situation is called a critical state. In a thin slab the field patterns (Fig. 6.32(a)) and the screening current distribution (Fig. 6.72(b)) are shown in Fig. 6.72. The initial applied field is zero. First, the magnetic field is increased to $B_0 = B_p/2$.

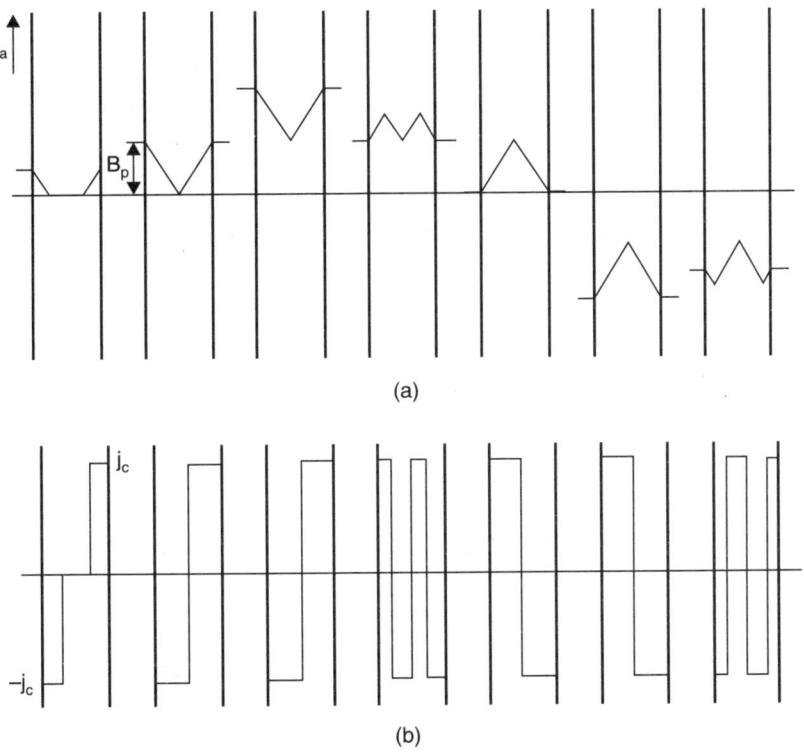

Fig. 6.72 For a cycle starting at $B_a = 0$, field pattern and screening current distribution in a thin slab. The magnetic flux reaches the centre of the slab at the penetration field $B_p = \mu_0 J_c a$. The values of the applied magnetic field B_a are from left to right $B_p/2$, B_p, $2B_p$, 0, $-2B_p$ and $-3B_p 2$

The internal magnetic field at the surface of the slab is equal to B_a. As long as the applied magnetic field is smaller than the penetration field $B_p = \mu_0 J_c a$ the internal field in the centre of the slab is zero. The screening currents are zero in the central part of the slab where $B_i = 0$ in accordance with eq. (6.32). At $B_a = B_p$ the whole superconductor is filled with screening current and magnetic flux reaches the centre of the slab. When the applied field reaches $2B_p$, the internal field in the centre of the slab becomes B_p. Thereafter the applied magnetic field is reduced to $B_a = B_p$, leading to screening currents in the reverse direction in the surface layer of the slab. The internal magnetic field exceeds the applied magnetic field because of the trapped flux. At $B_a = 0$ the trapped flux reaches its maximum value, resulting in $B_i = B_p$ in the centre of the slab. At $B_a = -2B_p$ the directions of the screening currents and the field pattern are opposite to the situation $B_a = -2B_p$. When the magnetic field is increased again, trapped flux contributes to the internal magnetic field.

The magnetic induction B_i and the magnetisation M are related as

$$B_i = \mu_0 (H_i + M) \qquad \ldots(6.32)$$

where
$$H_i = B_0/\mu_0 \qquad \ldots(6.33)$$

for the slab geometry with the applied field along the broad face. The resulting magnetisation for the slab is

$$\mu_0 M = B_i - B_a \qquad \ldots(6.34)$$

we may note that experimentally determined values of the magnetisation are typically averages over the sample volume. From Fig. 6.72, one finds that for the extreme values $\pm 2B_p$ of the applied magnetic field the average values of B_i are $\pm 3B_p/2$. Thus the resulting difference for the magnetisation in increasing and decreasing applied fields is

$$\mu_0 \Delta M = B_p = \mu_0 J_c a \qquad \ldots(6.35)$$

However, the expression (6.35) is generally valid for sufficiently large field changes $\Delta B_a \gg B_p$. From eq. (6.35) the critical current density becomes

$$J_c = \frac{2 \Delta M}{d} \text{ with } d = 2a \qquad \ldots(6.36)$$

Equation (6.36) has been frequently used to estimate intragranular critical current densities in polycrystalline HTSC cuprates. Considerable uncertainties result from ill-defined grain sizes and demagnetisation factors. Generally, it is found that the intragranular critical current densities in polycrystalline material are comparable to the large J_c-values typically measured in high-quality epitaxial thin films.

A further consequence of granularity is the dependence of the transport critical current on the magnetic history of the sample [187, 193 – 201]. The transport critical current of a silver-clad $Y Ba_2 Cu_3 O_{7-\delta}$ wire for first increasing and then decreasing magnetic field is shown in Fig. 6.73. First, the critical current steeply declines with increasing magnetic field. Above 5 mT a further enhanced applied magnetic field leads only to a small reduction of I_c. The critical current for decreasing magnetic field is considerably higher than that measured in increasing magnetic field and reaches a maximum at $B_a = 1.5$ mT. Finally, at $B_a = 0$ the critical current is only about 50% of the initial value observed in increasing magnetic fields. A possible explanation of this hysteresis of the transport critical current is the contribution of the magnetisation of the individual grains to the microscopic magnetic field at the grain boundaries. Fig. 6.74 illustrates this model [187]. At small magnetic fields flux can only penetrate into the sample along the grain boundaries, which act as weak links (Fig. 6.74 (a)). Field exclusion from the interior of the grains leads to compression of the field lines in the weak link region. For further enhanced fields magnetic flux can enter into the individual grains (Fig. 6.74 (a) top centre). The intergranular screening currents reduce the magnetic field B_i within the grains, while they enhance the field B_{gb} at the grain boundaries. These effects reduce the transport critical current across the grain boundaries for increasing applied field. On the other hand, trapped flux is present in the grains when the magnetic field is reduced from the maximum value. Due to this trapped flux the magnetic field B_i in the interior of grains exceeds the applied field. In contrast the magnetic field B_m produced by this flux is opposite to B_o outside the grains. Thus, the transport critical current across the grain boundaries exceeds the values found for increasing field (Fig. 6.74 (b)). At very low applied magnetic field B_a and B_m may nearly cancel each other, resulting in a maximum of the transport critical current. Finally, there is still a small magnetic field present at the grain boundaries for $B_a = 0$, caused by the trapped flux in the grains; as a consequence the I_c-value is considerably smaller than the zero field value of the initial branch (Fig. 6.74 (b) bottom left). The hysteresis effects for the transport critical current at 4.2 K in developmental Ag/Bi-2212, Ag/Bi-2223 and Ag Au/Tl-1223 tapes are compared in Fig. 6.75. The critical current densities of these tape conductors are much larger than the J_c-values of typically 1000 A/cm² characteristic of polycrystalline Y-123. The higher J_c-values in these tape conductors are expected as a consequence of a textured microstructure that results in better coupling of the grains. However, even in these superconductors a considerable fraction of the grains is weakly coupled as suggested by the observed hysteresis effect.

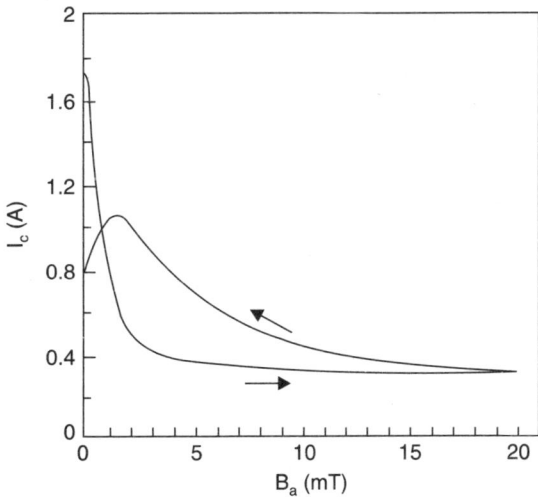

Fig. 6.73 Hysteresis of the transport critical current in a silver clad $Y Ba_2 Cu_3 O_{7-\delta}$ wire at 77 K

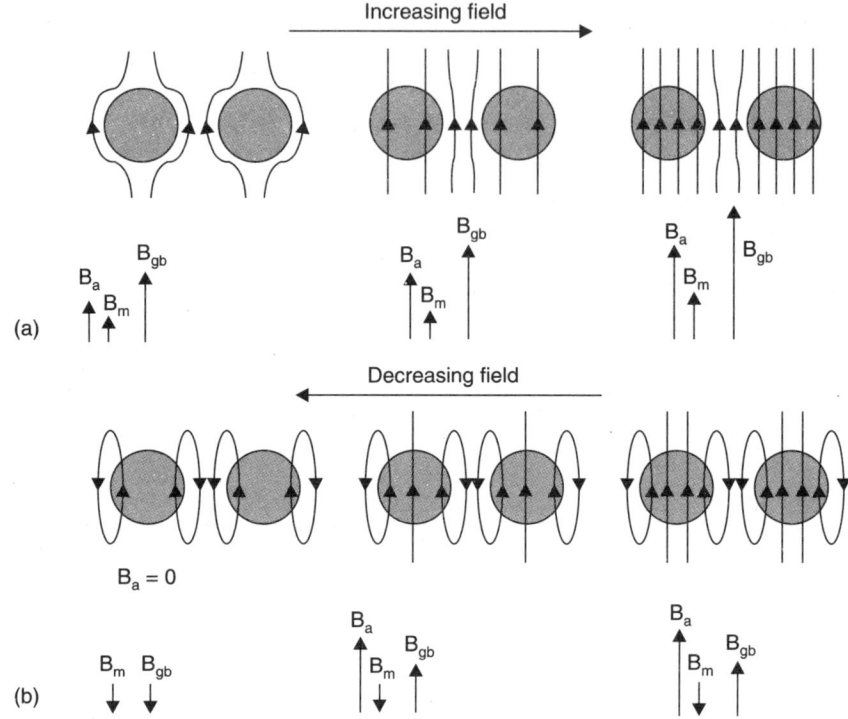

Fig. 6.74 Hysteresis of the transport critical current density. This is a consequence of the superposition of the applied magnetic field B_a and the field B_m caused by the magnetisation of the individual grains at the weakly coupled grain boundaries [187]

Fig. 6.75 indicates that the tendency of weak link formation increases slightly from Ag/Bi-2212 to Ag/Bi-2223. For the Ag Au/Tl-1223 tape, weak coupling of the grains is much more pronounced than for the two Bi-based HTSC cuprates. The strong reduction of the critical current density due to magnetic fields well below $1T$ is a further indication of grain boundary weak links.

For superconducting magnets and power applications generally, long superconducting wires are required. It is therefore of importance that high critical current densities can be achieved in polycrystalline HTSC cuprates. The observed granularity of the HTSC cuprates indicates that grain boundaries can act as weak links. On the other hand, the large J_c-values which have been achieved in highly textured films suggest that low-angle grain boundaries are strongly coupled. To clarify this aspect, the transport properties of artificial grain boundaries have been studied [202, 203]. $Y Ba_2Cu_3O_{7-\delta}$ thin films can be grown epitaxially on $Sr Ti O_3$ single-crystal substrates. This means that the crystallography prescribed in the substrate is transmitted to the Y-123 film. Using $Sr Ti O_3$ bicrystals, two Y-123 grains connected by an artificial grain boundary with a well defined misorientation angle θ can be prepared.

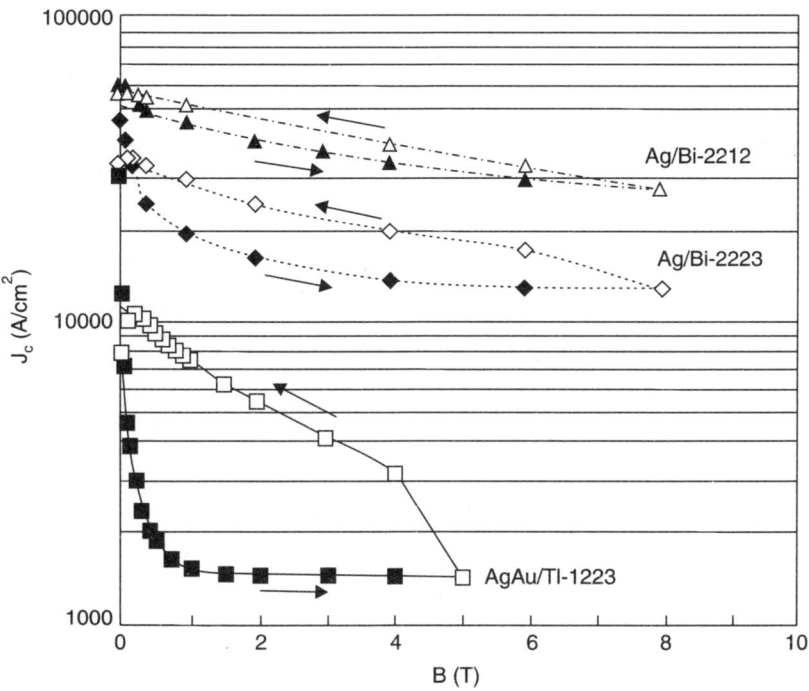

Fig. 6.75 Comparison of the hysteresis effect for the transport critical current density at 4.2 K in developmental Ag/Bi-2212, Ag/Bi-2223 and Ag Au/Tl-1223 tapes [196, 197]

The bicrystal substrates were prepared by not pressing two Sr Ti O$_3$ single crystals at a temperature of 1450°C. The Y-123 films were deposited by electron-beam evaporation of the three metal species or by laser ablation from a sintered Y-123 pellet. Slow cooling in oxygen atmosphere leads to well-oxidised Y-123 films. Fig. 6.76 illustrates the crystallography of the

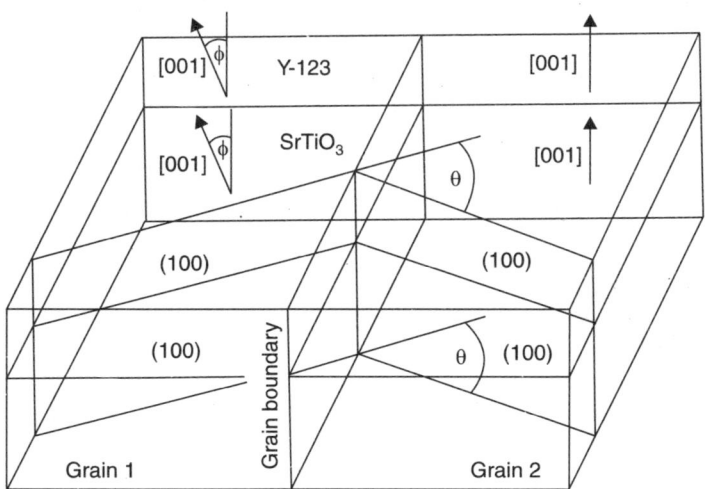

Fig. 6.76 Crystallography of the Sr Ti O$_3$ bicrystals and epitaxially grown Y-123 thin films. Grain 2 is rotated by an angle θ around the common c-axis with respect to grain 1 ([001] tilt boundary) [202]

Sr Ti O$_3$ substrate and the epitaxially grown Y-123 film. Grain 2 is rotated by an angle θ around the common c-axis with respect to grain 1. The small angle φ represents the misalignment of the c-axis of the two Y-123 films. The misorientation angle θ and φ were determined by laue diffraction method with an accuracy of 0.5°. The narrow lines of 10 μm width which were patterned by means of an excimer laser into the two grains and the grain boundary region are shown in Fig. 6.77. Aluminium wires were directly attached to the films. A four-point technique is used to measure the critical current in these three microbridges.

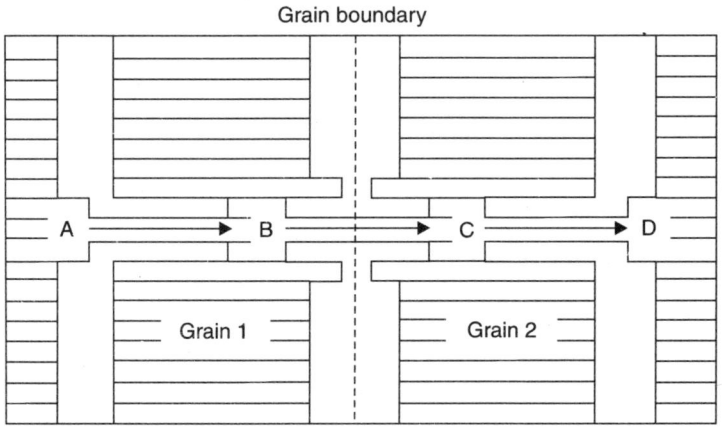

Fig. 6.77 Schematic illustration of the micro-bridges of 10 μm width used for the four-point measurements [202]

The ratio of the grain boundary critical current density $J_{c,gb}$ to the average intergrain critical current density $J_{c,g}$ in the two adjacent grains for T = 5 K as a function of the misalignment angle θ is shown in Fig. 6.78. The reduction of the critical current density across

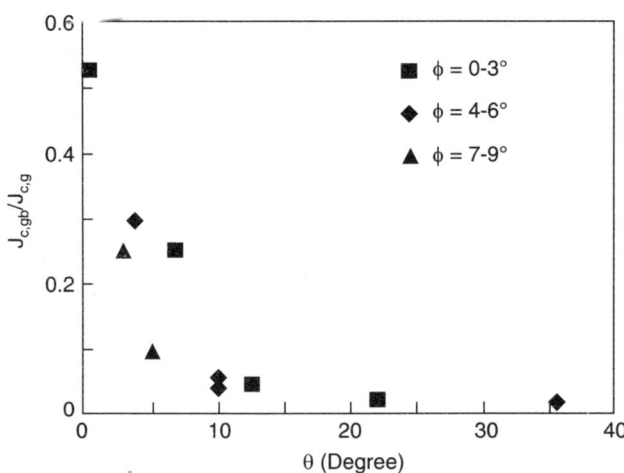

Fig. 6.78 Grain boundary critical density at 77 K versus the misalignment angle of the neighbouring grains. Some of the points are average values from the original data [202]

the grain boundary increases with increasing θ. Transport critical current densities across the grain boundary of less than 10% of $J_{c,g}$ result for misalignment angles as small as 10°. Similar detrimental effects on the transport critical current density have also been found for [100] tilt and [100] twist boundaries [203]. The $J_{c,gb}$ values decrease with increasing total misalignment angle of the adjacent grains.

The grain boundary critical current density $J_{c,gb}$ at 4.2 K normalised to the average of the $J_{c,g}$ values in the two grains as a function of the misalignment angles θ and ϕ is shown in Fig. 6.79. As expected small $J_{c,gb}$ values result for both large θ and ϕ-values. A similarly strong

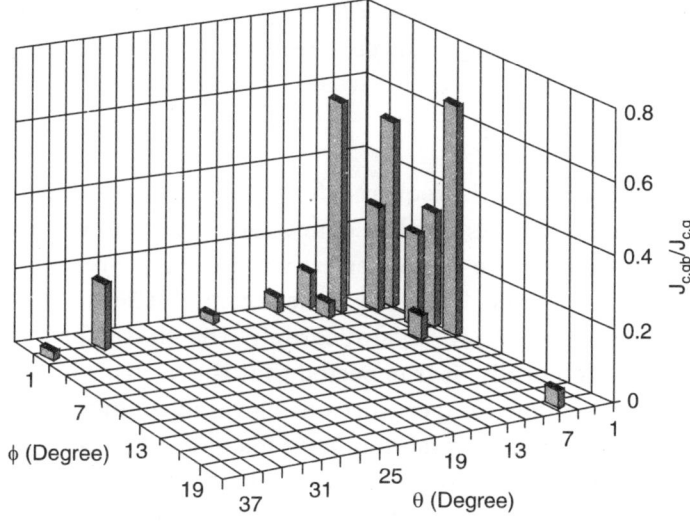

Fig. 6.79 Normalised grain boundary critical current densities $J_{c,gb}$ at 4.2 K as a function of the two misalignment angles θ and ϕ [203]

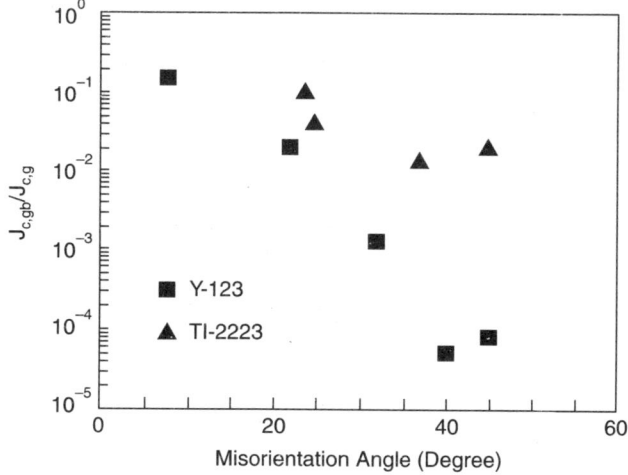

Fig. 6.80 Grain boundary critical current density at 77 K versus the misalignment angle of the neighbouring grains. Some of the points are average values from the original data [204 – 205]

reduction of the grain boundary critical current density due to large angle grain boundaries has been found for Tl-2223 films epitaxially grown on Sr Ti O$_3$ [204]. An even more pronounced effect on the $J_{c,gb}$- values has been observed for Y-123 films on yttria stabilised zirconia [205]. Fig. 6.80 presents the normalised $J_{c,gb}$ data for these two film systems.

Very recently Todt et al. [206] and Li et al. [207] have studied the properties of bulk bicrystal grain boundaries of Y-123 and Bi-2212. As in the case of the artificially thin film grain boundaries, the critical current density decreases with increasing misalignment angle of the adjacent grains. Li et al. [207] have measured a considerably lowered critical temperature in a Bi-2212 bicrystal grain boundary in addition to reduced critical current density. This result clearly suggest that a local variation of the oxygen content may be responsible for the weak-link behaviour of clean grain boundaries. However, the segregation of impurities at the grain boundaries can be a further reason for weakly coupled grains.

6.7.3 Flux Pinning

Effective pinning of the flux lines is required for loss-less current transport in a Type-II superconductor. First, some concepts used to characterise the behaviour of magnetic flux in the interior of HTSC cuprates above the lower critical field are presented. However, we will not touch different pinning theories. Typically, a hexagonal lattice of the vortices is observed in conventional superconductors. As in crystallography the characteristic property of such a lattice is long range order. Generally, it is expected that the vortices in HTSC cuprates form a flux line lattice at sufficiently low temperatures at elevated temperatures there may exist a vortex glass state. As in an amorphous solid, in a vortex glass there is short-range but no long range order. Both states are called vortex solids as long as the flux lines are pinned. The screening currents circulating around each flux line are responsible for repulsive vortex–vortex interactions, which prevent flux motion as soon as a sufficiently large fraction of the vortices is pinned. The effectiveness of this collective pinning depends on the stiffness of the flux line lattice.

The pinning energy required to create a flux line in a superconductor can be estimated from the lower critical field B_{c_1}. The gain of magnetic explosion energy and the loss of condensation energy become equal at B_{c_1}, and flux lines start to enter into the superconductor. Thus, the formation energy for each flux line is

$$E_{fl} = \int_0^{B_{c_1}} m \, dB \qquad \ldots(6.37)$$

where m is the magnetic moment connected to a single flux quantum in a flux line of length l_{fl}, and

$$m = \frac{\phi_0}{\mu_0} l_{fl} \qquad \ldots(6.38)$$

The required energy per unit length is

$$\frac{E_{fl}}{l_{fl}} = \frac{\phi_0}{\mu_0} B_{c_1} \qquad \ldots(6.39)$$

Using the Ginzburg–Landau parameter $\kappa = \lambda_L/\xi$, one finds

$$\frac{E_{fl}}{l_{fl}} = \frac{1}{4\pi\mu_0}\left(\frac{\phi_0}{\lambda_L}\right)^2 \ln\left(\frac{\lambda_L}{\xi}\right) \qquad ...(6.40)$$

where ξ is the coherence length and λ_L the penetration depth. When the flux line is in a normal region, no energy is required for the formation of the normal vortex core and the resulting pinning energy is given by eq. (6.40). Generally regions with reduced Cooper pair density can act as pinning centres.

Of further importance for the properties of the flux line lattice is the highly anisotropic nature of the HTSC cuprates. Highly conductive CuO_2 planes are separated by the insulating charge carrier reservoirs. Flux lines along the crystallographic c-direction experience strong pinning only within the CuO_2 blocks. As a consequence of the layered structure of HTSC cuprates, the flux lines may be cut into vortex segments or even pancake vortices which exist only within the strongly coupled CuO_2 layers of each perovskite block. A two-dimensional behaviour of the flux line lattice is expected for short vortex segments pancake vortices. Fig. 6.81 shows the schematically illustration of flux lines, vortex segments and pancake vortices.

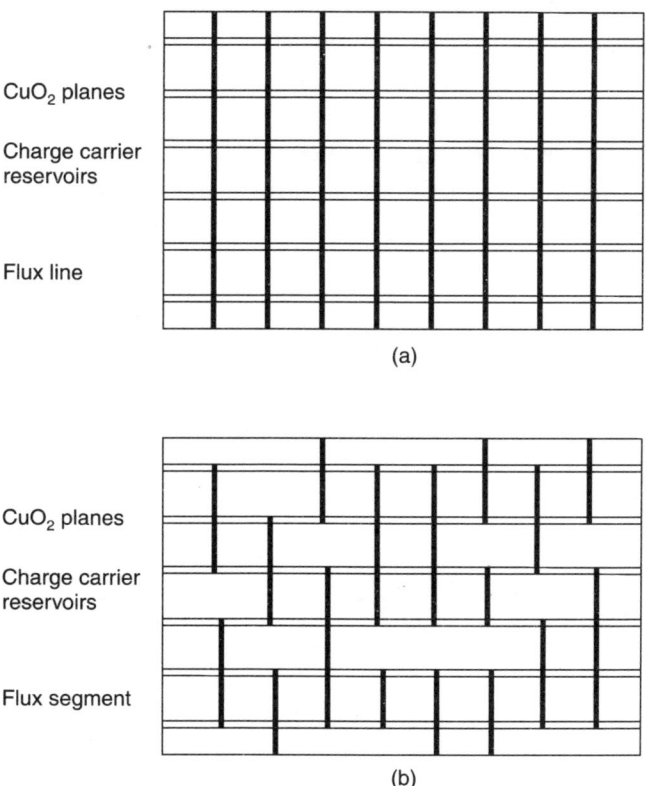

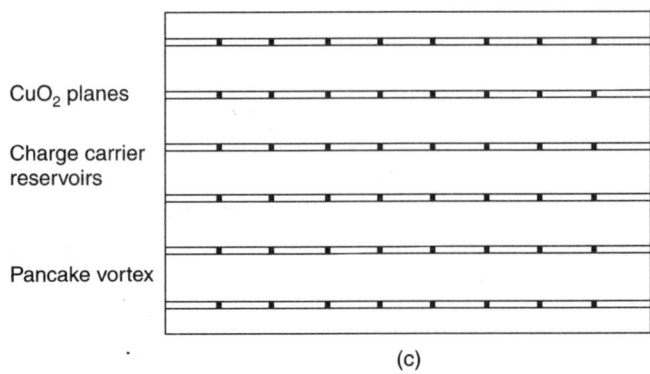

Fig. 6.81 Schematic illustration of (a) flux lines (b) vortex segments and (c) pancake vortices in HTSC cuprates

Soon after the discovery of HTSC cuprates the existence of a hexagonally correlated flux line lattice in $YBa_2Cu_3O_7$ at temperature 4.2 K was revealed by a high resolution Bitter pattern technique [208 – 209]. The experimental set-up used for the magnetic decoration of the superconducting surface is shown in Fig. 6.82. In these experiments the $Y Ba_2Cu_3O_7$ is field-cooled to 4.2 K. Thereafter the surface of the superconductor is decorated with magnetic particles. These are produced by evaporation of Nickel in the presence of ≈ 0.2 m bar helium gas. The nickel particles are preferentially deposited on those parts of the surface where a magnetic field is present, corresponding to the vortex cores of the flux lines. After warming up to room temperature the image of the flux line lattice so obtained can be examined by scanning electron microscopy. An important result of the decoration experiments is that as in conventional superconductors each vortex in $Y Ba_2Cu_3O_7$ contains a single flux quantum $\phi_0 = h/2e$ [208].

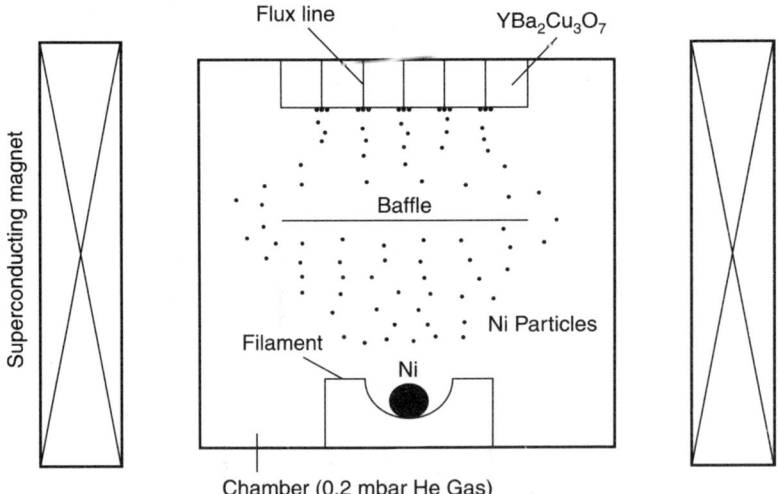

Fig. 6.82 Experimental set-up for magnetic decoration of superconductors. The chamber is immersed in liquid helium. The magnetic Ni particles are preferentially deposited at the normal vortex cores [208]

The existence of an irreversibility line and magnetic relaxation effects were first reported by Muller et al. [32]. Possible reasons for this phenomenon are a vortex glass transition, thermally activated depinning or flux line lattice melting. Irrespective of the microscopic mechanism the critical current density is zero above the irreversibility line. The irreversibility field B_{irr} has been found to be proportional $\left(1-\dfrac{T_{irr}}{T_c}\right)^\alpha$, where T_{irr} is the *irreversibility temperature* and the exponent α is close to 3/2. One can deduce the irreversibility temperature from susceptibility versus temperature data. The magnetic susceptibility is defined by the relation

$$\chi = \frac{\mu_0 M}{B} \qquad \ldots(6.41)$$

where M is the magnetisation and B the macroscopic magnetic field. Fig. 6.83 shows schematically the temperature dependence of zero-field-cooled and field-cooled susceptibility for $(La, Ba)_2 CuO_4$. Because of trapped flux the susceptibility measured for field-cooled is considerably smaller than that resulting for zero-field cooling. Above the T_{irr} temperature, both curves coincide, indicating reversible magnetic behaviour.

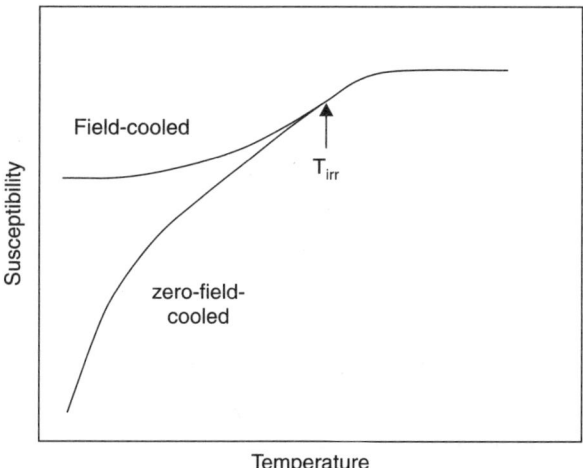

Fig. 6.83 Field-cooled and zero-field-cooled susceptibility of $(La, Ba)_2 CuO_4$ versus temperature. Above the temperature T_{irr} both curves coincide indicating reversible magnetic behaviour [210]

To study magnetic relaxation effects the superconductor under consideration can be zero-field-cooled to the desired temperature T below the critical temperature, a magnetic field is applied, and the time-dependence of the magnetisation is measured. Generally, it is found for the HTSC cuprates that the magnitude of the magnetisation is considerably reduced over long periods after the application of the magnetic field. These relaxation effects are much more pronounced than in low temperature superconductors. Like other physical properties, this relaxation is anisotropic and much faster for magnetic fields applied perpendicular to the CuO_2 planes. Relaxation effects have been observed for zero-field-cooled as well as field-cooled magnetisation of Y $Ba_2Cu_3O_7$ single crystals [211]. This result supports the idea that thermally activated flux creep is responsible for the observed magnetic relaxation in HTSC cuprates, the absence of which would indicate a glass-like behaviour [210]. We may note that the large flux creep effects in the HTSC cuprates are a consequence of both elevated temperatures and relatively small pinning energies.

Fig. 6.84 shows that thermally activated flux creep is a process similar to diffusion in a solid. Even for thermal energies k_BT well below the pinning energy V_p, there exists a finite probability for flux line hopping. This process is assisted by the Lorentz, which reduces the effective pinning potential.

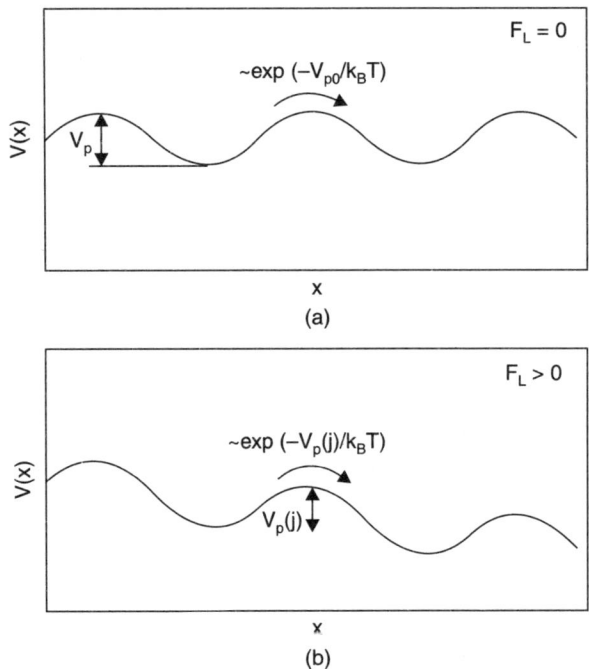

Fig. 6.84 Thermally activated flux line hopping. The Lorentz force (Fig. 6.84(*b*)) reduces the effective pinning potential V_p

We may note that in conventional flux creep theory it is assumed that the pinning potential V_p is a linear function of the current density:

$$V_p(J) = V_{po}\left(1 - \frac{J}{J_c}\right) \qquad ...(6.42)$$

where V_{po} is the *zero-current barrier height* and J_c the critical current density in the absence of thermal activation. Moreover, the hopping rate ν is given by

$$\nu = \nu_0 \exp\left[-\frac{V_p(J)}{k_BT}\right] \qquad ...(6.43)$$

where ν_0 is an attempt frequency of $10^{10} - 10^{12}$ Hz.

For $k_B T \ll V_{po}$ a logarithmic relaxation of the magnetization results from equations (6.42) and (6.43):

$$M(t) = M_0 \left[1 - \frac{k_B T}{V_{po}} \log \frac{t}{\tau_0} \right] \quad \ldots(6.44)$$

where M_0 is the initial magnetization and τ_0 a time constant connected with the attempt frequency ν_0.

Typically a logarithmic relaxation has been found for the HTSC cuprates at sufficiently low temperatures, whereas non-logarithmic behaviour has been observed at elevated temperatures [211, 214, 220]. Experimental studies of the resistive transition in high magnetic fields have revealed a logarithmic $V_p(J)$ dependence [221, 222]

$$V_p(J) = V_{po} \ln \left(\frac{J_0}{J} \right) \quad \ldots(6.45)$$

where J_0 is the critical current density corresponding to zero barrier height. This pinning potential leads to a **power law dependence** for the relaxation of the magnetisation

$$M(t) = M_0 (t/\tau_0)^{-\alpha} \quad \ldots(6.46)$$

where $\alpha = k_B T/V_{po}$ and τ_0 is again a characteristic time constant. A more complex expression for the magnetic relaxation results from collective pinning theories [224 – 226]

$$M(t) = M_0 \left[1 + \mu \frac{k_B T}{V_{po}} \ln \left(\frac{t}{\tau_0} \right) \right]^{-1/\mu} \quad \ldots(6.47)$$

where the effective pinning potential is

$$V_p(J) = \frac{V_{po}}{\mu} \left[\left(\frac{J_0}{J} \right)^\mu - 1 \right] \quad \ldots(6.48)$$

Theoretical values of μ depend on the flux bundle size. In the experimental study μ is used as a fitting parameter. For $Y Ba_2 Cu_3 O_7$ single crystals μ values varies between 0.2 and 0.6 in magnetic fields below $> T$, whereas μ increases rapidly above this field and exceeds 2 at 8 T [216].

The relaxation of the remanent magnetisation of polycrystalline (Bi, Pb)-2223 at various temperatures is shown in Fig. 6.85. After zero-field-cooling a magnetic field of 1 T was applied and immediately switched off. The magnetisation was measured with a vibrating sample magnetometer [216]. A logarithmic decay of the remanent magnetisation was observed within the time window of 1000s. The relaxation rate

$S = 1/M_0 \times dM/d \log (t_b/\tau_0)$ can be used to derive an effective activation energy V_{po}. From the single barrier expression (6.44) we have

$$S = \frac{1}{M_0} \frac{dM}{d \ln \left(\frac{t_b}{\tau_0} \right)} = -\frac{k_B T}{V_{po}} \quad \ldots(6.48(a))$$

where t_b is the time at which the first data point is measured. Because the effective pinning potential V_{po} is zero at the critical temperature, the relaxation rate S should diverge. In fact, experimentally determined values of $dM/[d\ln(t/\tau_0)]$ reach a maximum at intermediate temperatures. To avoid the divergence of the relaxation rate at T_c, both forward and backward hopping must be considered, resulting in a generalized Arrhenius law $v = v_0 \exp(-V_{po}/k_B T) \sin h(V_{po}J/k_B T J_c)$. To obtain the maximum of the relaxation rate at intermediate temperatures a distribution of activation energies have been proposed [227].

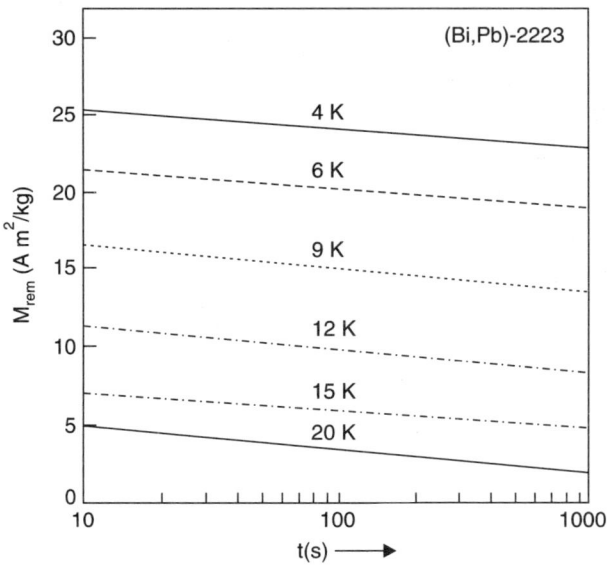

Fig. 6.85 Logarithmic relaxation of the remanent magnetisation of polycrystalline (Bi, Pb) – 2223 [216]

From the experimental data in Fig. 6.85, pinning energies of 20 – 50 meV are derived at 4K. Based on the Hagen-Griessen model, a peak in the distribution of the barrier heights has been found for $V_{po} = 20 - 30$ meV. For Y Ba$_2$Cu$_3$O$_7$ single crystals, pinning energies of ≈ 20 and ≈ 200 M eV have been reported for magnetic fields parallel and perpendicular to the crystallographic c-direction respectively [129]. The comparison of the magnetic relaxation for Y-123 and Bi-2212 single crystals is shown in Fig. 6.86. The decay of the magnetisation is nearly logarithmic. The relaxation is much more pronounced for the Bi-2212 than for the Y-123 single crystal. The enhanced flux creep in Bi-2212 seems to be a consequence of the larger anisotropy characteristic of Bi and Tl double-layer HTSC cuprates.

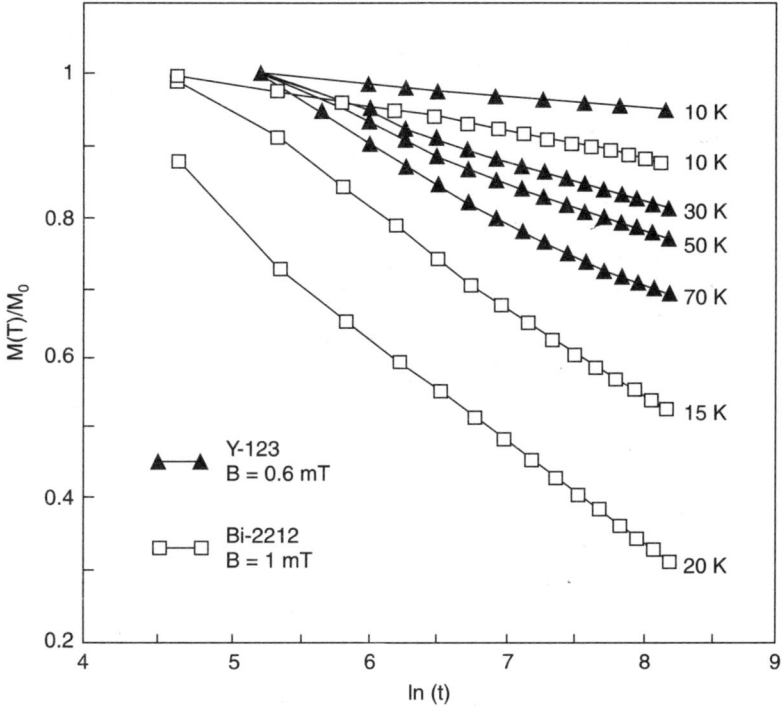

Fig. 6.86 Relaxation of the zero-field-magnetisation of Y-123 and Bi-2212 single crystals. The initial magnetisation (Bi-2212) and the M-value for $t = 200$s (Y-123) are defined as M_0 [214]

Further experimental evidence for thermally activated flux creep has been found in the broadened resistive transitions caused by magnetic fields. Fig. 6.87(a) shows the resistivity of

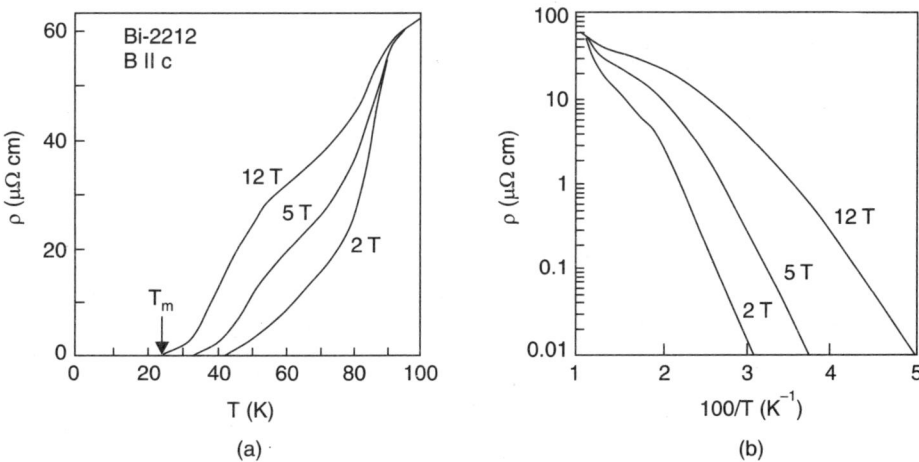

Fig. 6.87 Resistivity versus temperature (a) and corresponding Arrhenius plots (b) for a Bi-2212 single crystal. Below $\rho = 1$ $\mu\Omega$ thermally activated behaviour is reported [128]

a Bi-2212 single crystal as a function of temperature for magnetic fields of 2, 5 and 12 T applied along the crystallographic c-direction. The Arrhenius plots of these data (Fig. 6.87(b)) reveal thermally activated behaviour below a resistivity of ≈ 1 µΩ cm [128]. The flux creep resistance is given by

$$\rho = \rho_{ff} \exp\left(-\frac{V_{po}(B)}{k_B T}\right) \qquad \ldots(6.49)$$

where V_{po} is the activation energy and ρ_{ff} the resistivity for $V_{po} = 0$. The slopes of the curves in the Arrhenius plots correspond to the activation energy V_{po}. The resulting V_{po} values are ≈ 50, ≈ 40 and ≈ 30 M eV for magnetic fields of 2, 5 and 12 T applied along the c-direction respectively. Generally, the **activation energy decreases with increasing magnetic field.**

A pronounced anisotropy of the activation energy has been reported for epitaxial Hg-1212 films [228]. Arrhenius plots of the resistivity for various magnetic fields parallel and perpendicular to the CuO_2 planes of the Hg-1212 film are shown in Fig. 6.88. Below a resistivity of 2 µΩ cm the data indicate thermally activated flux creep. Obviously, the slopes of the curves increase with decreasing magnetic field. Furthermore, the activation energies are much smaller for $B \parallel c$ than for $B \parallel ab$. The activation energies for both field directions as a function of applied field are shown in Fig. 6.89.

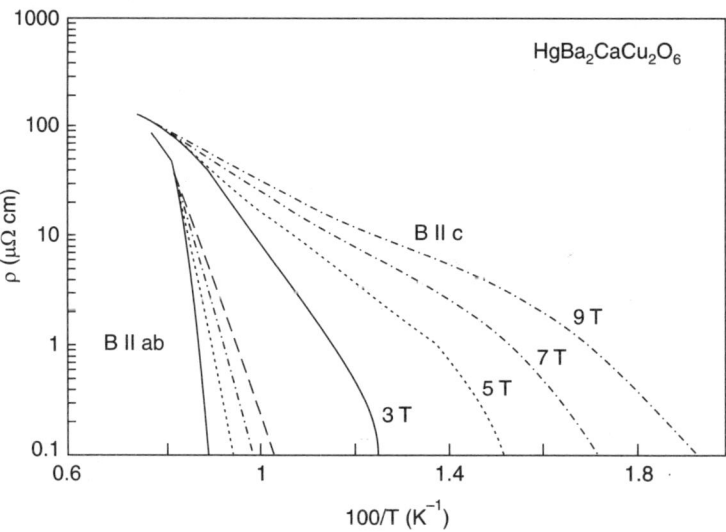

Fig. 6.88 Arrhenius plots of the resistivity of epitaxial Hg-1212 films for various magnetic fields applied along the c-direction and within the ab-plane [228]

Finally we may note that the investigation of the current voltage characteristics establishes the dependence of the activation energy V_p on the current density. The electric field E_e caused by thermally activated flux creep can be expressed as

$$E_e = {}^J\rho_{ff} \exp\left[-V_p\left(\frac{J}{J_0}, B, T\right)/k_B T\right] \qquad \ldots(6.50)$$

where ρ_{ff} is flux flow resistivity for zero activation energy. Recent studies [229] of the current-voltage characteristics for $Y\,Ba_2Cu_3O_{7-\delta}$ thin films suggest that $V_p(f) \propto (J_0/J)^\mu$ as expected for a vortex glass state (See also eqs. 6.47 and 6.48).

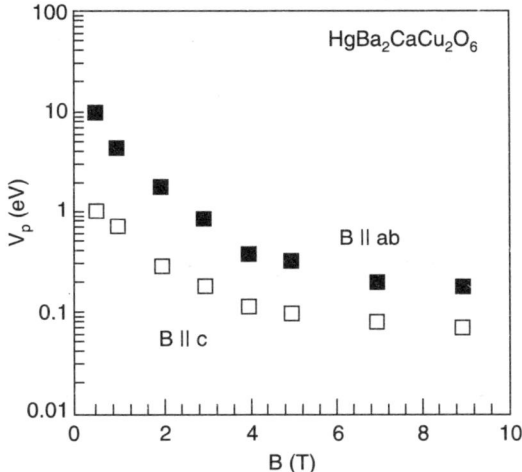

Fig. 6.89 Activation energy V_p versus applied magnetic field. The V_p values for $B \parallel ab$ are considerably higher than for $B \parallel c$ [228]

Large magnetic relaxation effects and considerably broadened resistive transitions in the presence of an applied magnetic field, characteristics of the HTSC cuprates, are consequences of relatively small pinning energies and elevated temperatures. The critical current densities are intrinsically limited by thermally activated flux creep. Generally the creep effects are more pronounced in highly anisotropic HTSC cuprates.

Because the field and temperature ranges for potential applications of HTSC cuprates are limited by irreversibility line, it is important to understand the factors which determine the position of this line in the magnetic phase diagram. In principle the irreversibility line can be deduced from resistivity and hysteresis loop measurements. The considerable scatter of the results published for the same HTSC cuprates is partly caused by the different criteria used to define the irreversibility field B_{irr}. Experimentally determined irreversibility lines of HTSC cuprates: Bi, Tl and Hg based, can be found in references [230 – 242]. Results based on the closing point of the hysteresis loop depend on the sensitivity of the experimental set-up. In addition, the precise chemical composition and the defect structures in the sample under investigation influence the position of the irreversibility line. Usually, the irreversibility temperature T_{irr} and B_{irr} are connected by the expression

$$B_{irr}(T) = B_0 \left(1 - \frac{T_{irr}}{T_c}\right)^\alpha \qquad \ldots(6.51)$$

where B_0 is a fitting parameter and the exponent α is characteristic for the HTSC cuprates in question. For instance the value of α is close to 3/2 for Y-123.

Figure 6.90 shows for several thin-film and single-crystal HTSC cuprates the irreversibility fields as a function of reduced temperature T/T_c. The magnetic field was applied along the crystallographic c-direction. The highest irreversibility fields have been found for the Y-123 compound, indicating that pinning is more efficient in HTSC cuprates with small anisotropy. The favourable physical properties of Y-123 are closely connected to the small distance between neighbouring CuO_2 blocks (See Fig. 6.21) and to the short circuiting of the insulating charge carrier reservoirs by the CuO chains. Intermediate irreversibility fields result for the Hg- and Tl-based single layer HTSC cuprates, whereas the smallest values of B_{irr} have been found for Bi-and Tl-based double layer cuprates. For example, Fig. 6.90 shows the irreversibility line of a Bi-2212 single crystal. The enhanced values of the irreversibility fields for Bi- and Tl-single layer compounds again seem to be a consequence of the reduced distances between adjacent CuO_2 blocks, as compared to the Bi- and Tl- double-layer HTSC cuprates.

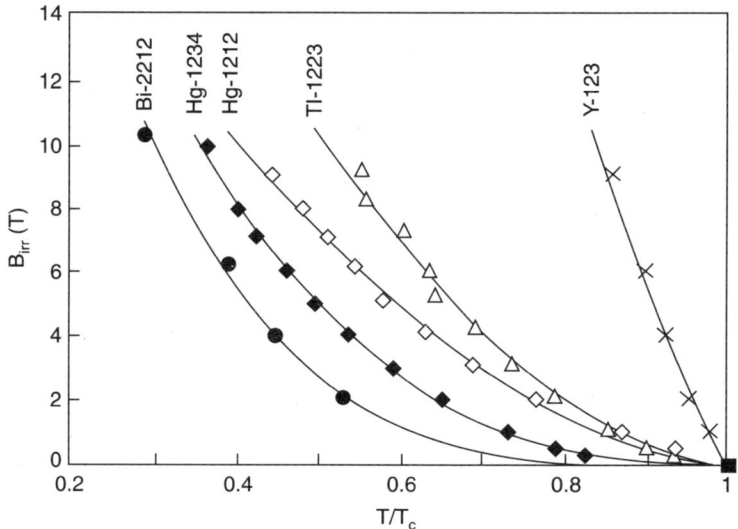

Fig. 6.90 Irreversibility field B_{irr} versus reduced temperature T/T_c of various HTSC cuprates for $B \parallel c$. The lines are fits of eq. (6.51) [228, 241]

Usually, it is expected that the CuO_2 layers within a perovskite block are strongly coupled, whereas neighbouring CuO_2 blocks are only weakly Josephson coupled. The strength of their coupling decreases rapidly with increasing distance of the CuO_2 blocks. As a consequence flux line cutting can occur, leading to ineffective pinning in highly HTSC cuprates.

Fig. 6.91 shows the irreversibility lines of several crystalline Tl-based superconductors for $B \parallel c$. The B_{irr}- values of the single-layer compounds Tl-1212 and Tl-1223 compounds are higher than those of the corresponding double-layer HTSC cuprates Tl-2212 and Tl-2223. We may note that the irreversibility fields of the Tl-1223 film indicated in Fig. 6.90 are considerably higher than the single-crystal data presented in Fig. 6.91. The results for the single-crystal Tl-HTSC cuprates suggest that the irreversibility line shifts with increasing number of CuO_2 layers perovskite unit to slightly higher fields.

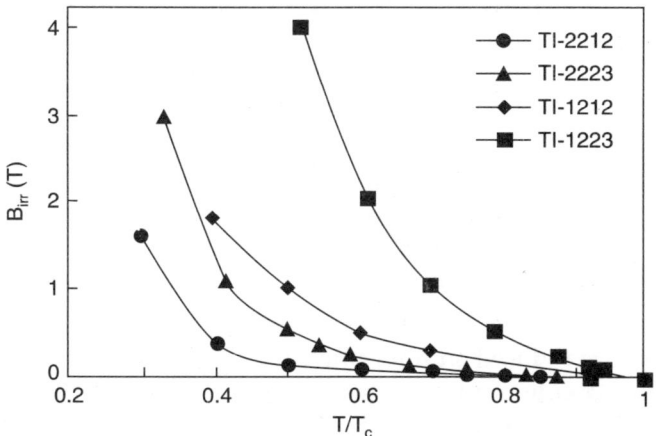

Fig. 6.91 Irreversibility lines of single crystals Tl-HTSC cuprates for $B \parallel c$ as functions of reduced temperature T/T_c [233]

Fig. 6.92 shows the irreversibility fields of polycrystalline Hg-12 $(n-1)n$ superconductors ($n = 1 - 3$) as a function of reduced temperature. The same criterion was used for the determination of the irreversibility field for all three Hg compounds. Irrespective of the different number n of CuO_2 layers per perovskite unit, the data almost collapse to a single curve. This behaviour seems to be a consequence of the equal distances between adjacent CuO_2 blocks in these three Hg-based HTSC cuprates. This means that the number n of CuO_2 layers is of minor importance for the position of the irreversibility line w.r.t. the reduced temperature. We may note the maximum of the critical temperature is reached for $n = 3$ to 4. As compared to Hg-1201 and Hg-1212 the irreversibility line of Hg-1223 is shifted to higher absolute temperatures because of the enhanced T_c-value.

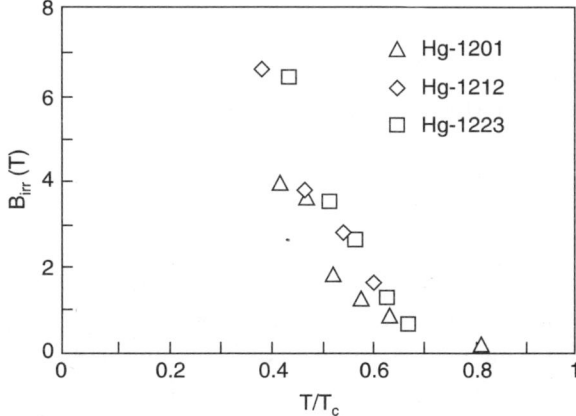

Fig. 6.92 Irreversibility lines of polycrystalline Hg-12 $(n-1)n$ cuprate superconductors as function of reduced temperature [242]

Now, we consider the possible pinning centres in HTSC cuprates. We have mentioned earlier that the strong anisotropy of the critical current density w.r.t. the direction of the applied magnetic field suggests that an intrinsic pinning mechanism exists for $B \parallel ab$. This intrinsic pinning is provided by the insulating charge carrier reservoirs inserted between the highly conductive CuO_2 blocks. Furthermore, twin [243] and colony boundaries can act as pinning centres. A colony of plate-like grains rotated around the common c-axis is shown in Fig. 6.93. Magneto-optical studies have confirmed that flux can be pinned at low-angle grain boundaries in Bi-2212 [244 – 245].

The effectiveness of intrinsic pinning in Bi-2212 can be influenced by annealing under reduced oxygen pressure [246]. As a consequence of the resulting oxygen loss from the BiO double-layers, the length of crystallographic c-axis increases, leading to weakened intrinsic pinning. In spite of reduced pinning efficiency the critical temperature is enhanced with increasing c lattice parameter. On the other hand, in the single crystals of $Bi_2Sr_2Ca_{1-x}Cu_2O_{8+\delta}$ both better pinning and the highest T_c-values have been achieved for the optimum Y-concentration of $x \cong 0.3$ [247].

In addition, such point defects as oxygen vacancies, cation substitutions or Fe, Ni, Co or Zn impurities on Cu sites can act as pinning centres [248 – 252]. Another possible way to improve flux pinning is the addition of impurity particles [253 – 256] or the precipitation of secondary phases [257 – 261]. Y_2BaCuO_5 inclusions in melt processed $Y Ba_2Cu_3O_7$

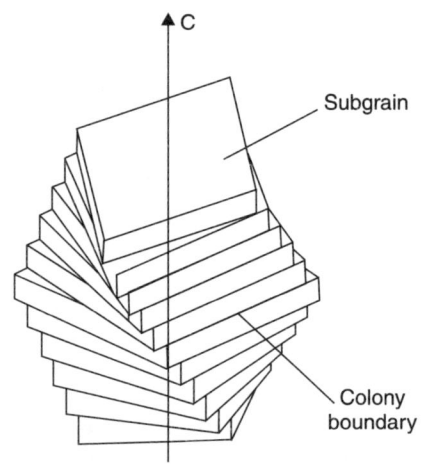

Fig. 6.93 Colony of plate-like subgrains rotated around the common crystallographic c-axis. The colony boundaries can act as pinning centres for $B \parallel ab$

have been found to provide very effective pinning [258]. It is difficult to explain this effect by pinning at the precipitates because they are much larger than the coherence length. The enhanced pinning may be connected to the strain fields present in the vicinity of the $Y_2Ba CuO_5$ inclusions. We may note that the transport critical current densities, due to segregation of secondary phases at the grain boundaries leading to weak-link behaviour.

To study flux pinning in HTSC cuprates irradiation with high energy heavy ions is a powerful tool. The energies of the incident heavy ions are typically between 100 MeV and several GeV. Each of the high energy ions produced in the irradiated HTSC cuprates an amorphous track of a few nm diameter. The most efficient pinning is expected to occur when for each vortex a columnar defect is available. The number of vortices per unit area is $N_v = B/\phi_0$, where ϕ_0 is the flux quantum. Generally, reduced T_c-values have been observed after heavy ion irradiation, whereas critical current densities are considerably enchanced due to pinning by the columnar defects. Furthermore, remarkable shifts of the irreversibility line for $B \parallel c$ to higher magnetic fields have been achieved by heavy ion irradiation. Studies of irradiation effects have been reported for Y-123, Bi-and Tl-based HTSC cuprates [262 – 277].

The irreversibility lines of Y-123, Bi-2212 and Tl-2212 before and after heavy ion irradiation are shown in Fig. 6.94. The textured Y-123 thick film was irradiated with 180 MeV Cu^{11+} ions parallel to the average c-direction at a temperature of 100 K. After a dose of 10^{12} Cu ions/cm^2 the steepness of the irreversibility line of Y-123 is even more pronounced than before irradiation [262]. The Tl-and Bi-2212 single crystals where irradiated with 6 GeV Pb ions at

300 K. The ion flux was again parallel to the crystallographic c-direction. For both 2212 cuprate compounds the irreversibility line is shifted to considerably higher reduced temperatures after irradiation with a dose of 2×10^{11} Pb ions/cm^2. The created amorphous ion tracks provide efficient pinning in these highly anisotropic superconductors below temperatures of 0.6 T_c [226]. Fig. 6.95 shows similar results for Ag/Bi-2223 tapes. Prior to irradiation the silver sheath was mechanically removed from one side of the tape because of the limited penetration depth for heavy ions in silver. The textured Ag/Bi-2223 tape was irradiated with 1 GeV Au^{23+} ions along the average c-direction. The most remarkable enhancement of the irreversibility field is observed for temperatures between 50 and 60 K [273].

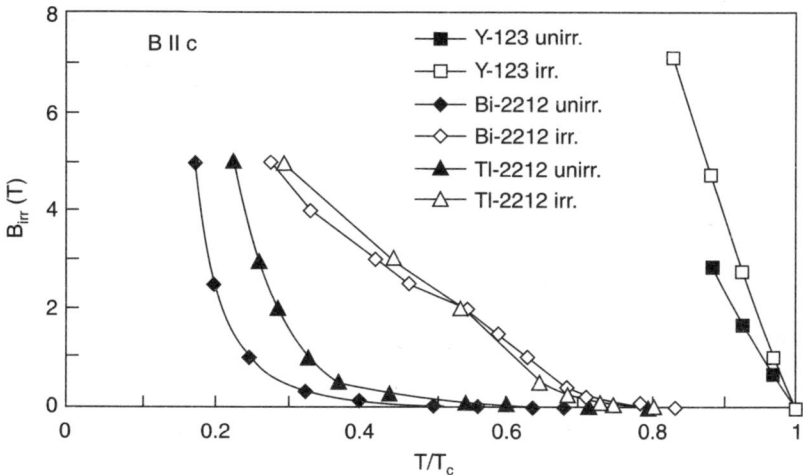

Fig. 6.94 Irreversibility fields of YBa$_2$Cu$_3$O$_7$, Bi$_2$Sr$_2$CaCu$_2$O$_8$ and Tl$_2$Ba$_2$CaCu$_2$O$_8$ HTSC cuprates prior and after heavy ion irradiation [262, 266]

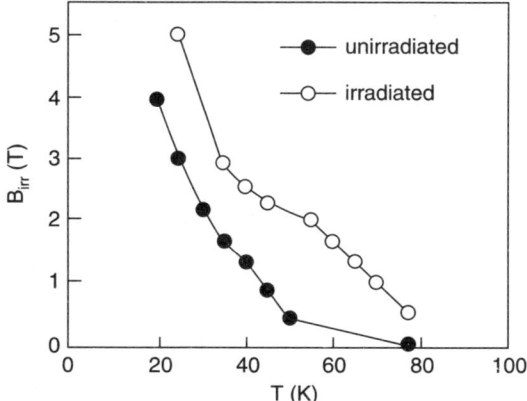

Fig. 6.95 Irreversibility field of Bi-2223 tape prior and after irradiation with 1 GeV Au^{23+} ions for $B \parallel c$ [273]

The above results clearly indicate that the amorphous tracks created by heavy ion irradiation can act as efficient pinning centres in the HTSC curpates for $B \parallel c$. As a consequence the critical current densities are considerably enhanced, especially at intermediate temperatures. An important result of these studies is that the position of the irreversibility line can be shifted to higher temperatures and fields by effective pinning centres. One can easily understand this behaviour when the irreversibility line is considered as a depinning line. In the picture of flux line melting, it is necessary to assume that the melting temperature can be increased the columnar defects.

Furthermore, irradiation at a certain angle θ_1 with respect to the *ab* planes can yield information on the nature of the flux line lattice in highly anisotropic superconductors. In such experiments the critical current density or the resistivity is measured as a function of the direction of the magnetic field with respect to the CuO_2 planes. The required angles are defined in Fig. 6.96. The pinning of flux lines of sufficient length should be most efficient when the applied magnetic field is parallel to the ion tracks, whereas such directional effects should be missing for pancake vortices which can be easily moved relative to each other. As a consequence of this behaviour inclined columnar defects ($\theta_1 \neq 90°$) are less efficient pinning centres in two dimensional superconductors.

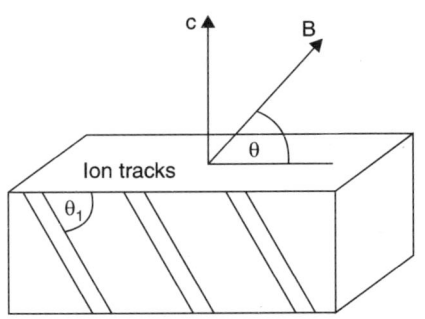

Fig. 6.96 The angles θ and θ_1 describes the direction of the applied magnetic field and the ion tracks with respect to the *ab* planes

The angular dependence of the transport critical current density in a textured Bi-2223 tape at 60 K and 1T prior and after irradiation with 10^{11} ions/cm^2 is shown in Fig. 6.97. The irradiation was performed with 2.65 GeV Au ions along the *c*-direction ($\theta_1 = 90°$). A large anisotropy and a maximum of J_c at $\theta = 0°$ are characteristic of the unirradiated Bi-2223 tapes. The small J_c-values for $B \parallel c$ are a consequence of weak pinning for this field direction.

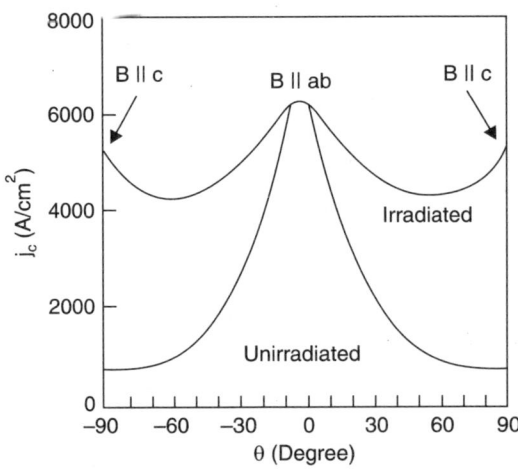

Fig. 6.97 Angle dependence of the critical current density in textured Bi-2223 tapes at 60 K and 1T before and after irradiation with 2.65 GeV Au ions. The Bi-2223 tape was irradiated with a dose of 10^{11} Au ions/cm^2 along the *c*-direction [274]

After irradiation the anisotropy is considerably reduced. Furthermore, additional maxima of J_c occur for $\theta = \pm 90°$ parallel to the amorphous tracks. The enhanced J_c-values are closely connected to the pinning provided by the columnar defects. The maximum value of J_c at $\theta = 0°$ is due to the intrinsic pinning at the layered structure. The angle-dependence of the critical current density at 60 K and 1 T after irradiation with a dose of 2.6×10^{11} Au ions/cm² at $\theta_1 = -60°$ is shown in Fig. 6.98. In addition to the maximum at $\theta = 0°$ a second peak of J_c is visible at $\theta = -60°$, indicating that the pinning at the columnar defects is most efficient when the magnetic field is parallel to the direction of the amorphous ion tracks. This behaviour suggests that at 60 K and 1 T real flux lines and not only pancake vortices exist in the Bi-2223 tape. At a temperature of 40 K the maximum of J_c for $\theta = -60°$ has vanished. This clearly reveals that at lower temperatures only two dimensional pinning is observed.

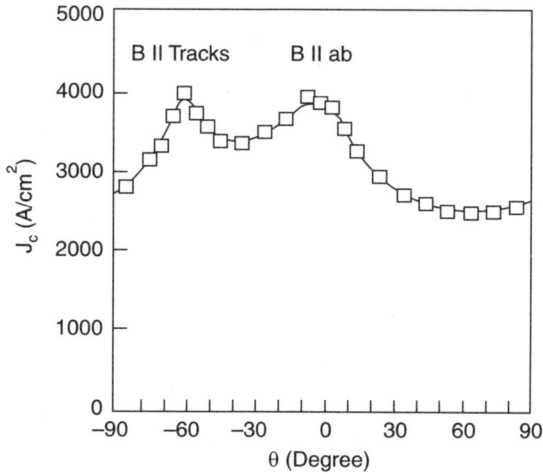

Fig. 6.98 Angle dependence of the critical current density in Bi-2223 tapes at 60 K and 1T after irradiation with 2.65 GeV Au ions at $\theta_1 = -60°$. The dose was 2.6×10^{11} Au ions/cm². A second maximum of J_c is observed for B parallel to the ion tracks [274]

A similar directional effect has been found by angle-resolved resistance measurements in a Bi-2212 single crystal. The inclined columnar effects were created by 5.8 GeV Pb ions irradiation at $\theta = 45°$. The resistance at 70 K as a function of θ is shown in Fig. 6.99. A current of 10 mA was used to determine the resistance by a standard four probe method.

Obviously the resistance is strongly reduced for $B = 0.3$ T when the applied magnetic field is parallel to the columnar defects. At a higher magnetic field of 2 T this characteristic feature has vanished. The reduced dissipation reflected by the $R(\theta)$ curve suggests that in the aligned configuration ($\theta = 45°$) highly correlated, mobile flux lines can exist in the vortex liquid above the irreversibility line [272].

Very recently it has been proposed by Hwa et al. [278], that angular dispersed **(splayed) columnar defects** may be more efficient pinning centres than amorphous ion tracks parallel to the c-direction. In the presence of splayed columnar defects the mobility of the vortices below the irreversibility line can be strongly reduced by vortex entanglement. Splayed columnar defects can be created by heavy ion irradiation at different angles [279]. In Bi-based HTSC cuprates such defects can be induced by nuclear reactions of energetic protons and Bi nuclei. The fission fragments of the Bi nuclei with energies of several tens MeV able to create amorphous tracks in the superconductor. Because GeV protons can penetrate nearly half a meter into matter, it may be possible to irradiate whole magnets with energetic protons. A promising enhancement of the irreversibility line has been observed in Bi-2212 tapes with splayed columnar defects created by nuclear reactions of Bi nuclei with 0.8 GeV protons [280].

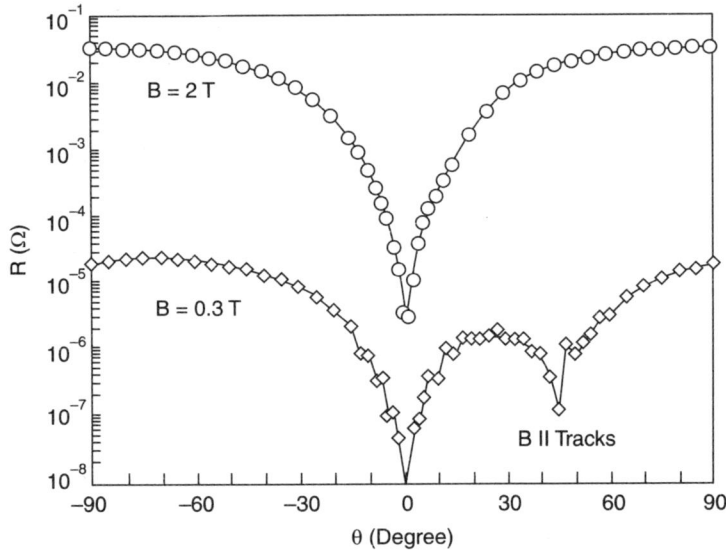

Fig. 6.99 Angle dependence of the resistance of a Bi-2212 single crystal with inclined columnar defects ($\theta = 45°$) at 70 K [272]

Now, we consider the melting transition from a vortex glass or solid to a vortex liquid. According to Lindemann criterion, an empirically found condition for the melting of a solid, at the melting temperature the root mean square displacements of the atoms from their equilibrium positions reach $\approx 10\%$ of the atomic distances of neighbouring atoms [281 – 283].

Soon after the discovery of HTSC cuprates a *glass-like behaviour* was reported in Y-123 [285] and Bi-2212 [286] materials. Theoretical calculations [284] reveal that the vortex glass phase is a true superconductor with $\rho_{lin} = (E_0/J)_{J\to 0}$, whereas a finite resistivity is characteristic of the vortex liquid phase. This vortex glass theory predicts that at the melting transition, the curvature of the electric field versus current density relation changes in a double logarithmic plot. Above the glass temperature T_g an upward curvature results, whereas a downward curvature is characteristic for the vortex glass state below T_g. Such behaviour has been confirmed experimentally for Y-123 [285] and Bi-2212 epitaxial films [286]. Similar results have also been reported for Ag/Bi-2223 tapes [287]. In this case the current shared with the silver sheath has to be carefully taken into consideration. Electric field versus current density curves for an epitaxial Bi-2223 film near the vortex glass transition in a magnetic field of 4 T applied along the c-direction is shown in Fig. 6.100. As expected from the vortex glass theory, the curvature of these curves changes at the glass temperature T_g of 18 K. Fisher's theory assumes that this melting is a second order phase transition without latent heat but with a jump in entropy. A simplified magnetic phase diagram for the vortex glass scenario is shown in Fig. 6.101. Within the solid phase there may coexist a vortex solid characterised by long range order and a vortex glass phase. In addition, regions with entangled vortices and transitions from two-to three dimensional vortex lattices may occur [288].

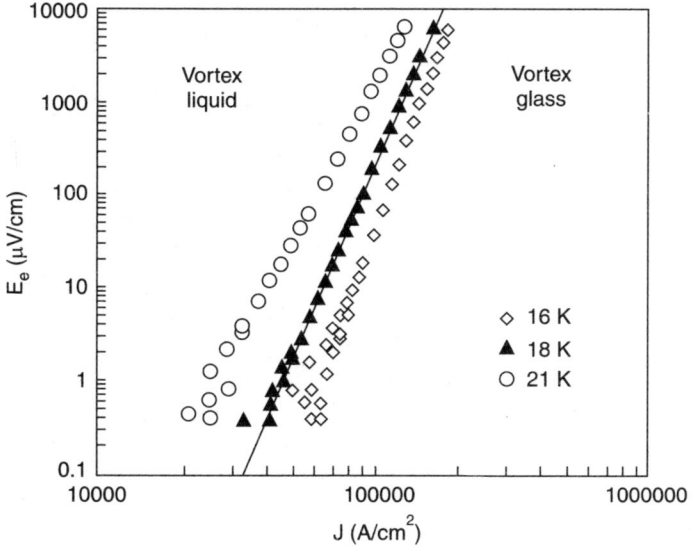

Fig. 6.100 Electric field versus current density for an epitaxial Bi-2223 film with a magnetic field of $4T$ applied along the c-direction. Below the glass temperature T_g of 18 K the E_c versus J curves show a downward curvature, whereas above T_g an upward curvature has been found. Independent of the applied magnetic field results at the glass temperature E_e/J 6.5 [286]

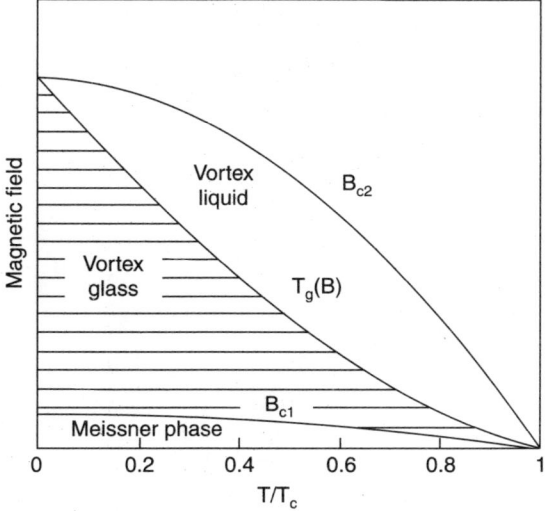

Fig. 6.101 Simplified magnetic phase diagram for HTSC cuprates with a vortex glass to liquid transition. Further transitions for instance from the vortex solid to the vortex glass phase may occur

In HTSC cuprates with a smaller degree of disorder the melting transition from a vortex solid to a vortex liquid should be of first order [289]. Recently, experimental evidence for a latent heat of $0.6\,k_B T_m$/vortex/layer has been reported in the specific heat of $Y\,Ba_2Cu_3O_7$ single crystal [290]. A similar value of $0.45\,k_B T_m$/vortex/layer has been obtained for untwinned $Y\,Ba_2Cu_3O_{7-\delta}$ single crystals from calorimetric measurements of the latent heat by means of differential thermal analysis. These results suggests that in less disordered HTSC cuprates a first order melting transition of the flux line lattice can be observed. On the other hand, a second order phase transition without latent heat may result for HTSC cuprates with a large amount of quenched disorder (vortex glass).

6.7.4 Transport Properties

We now consider the normal state resistivity, pseudogap and the thermal conductivity of HTSC cuprates. A microscopic theory of HTSC cuprates must be able to explain both the superconducting and the normal state properties of HTSC cuprates. We may note that electrical and thermal conductivity data are required for the design of bulk HTSCs for **power applications.**

The physical properties of the cubic metals are isotropic and hence the electrical and thermal conductivities can be described by single values. In materials with a more complex crystal structures these properties depend on the direction with respect to the crystallographic axes. Most of the HTSC cuprates have tetragonal or orthorhombic crystal structures. The electrical and thermal conductivity tensors of orthorhombic materials contain three independent components. The ratio of the electrical resistivities along the crystallographic a and b directions reaches values of typically two for orthorhombic HTSC cuprates. On the other hand, the electrical resistivity along the c-direction is considerably larger than the in-plane resistivity ρ_{ab}. This large anisotropy of in- and out-of-plane resistivities is a characteristic feature of the HTSC cuprates. The electrical resistivity of the HTSC cuprates shows a linear, metallic temperature dependence within the ab plane, whereas parallel to the crystallographic c-axis semiconducting behaviour has frequently been observed.

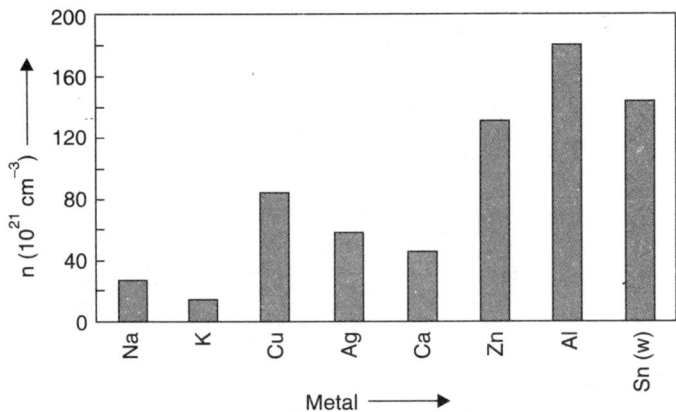

Fig. 6.102 Carrier concentration (n) in various metals [292]

One finds a general difference from simple metals is the much smaller carrier concentration in the HTSC cuprates. The carrier concentrations in various metals and HTSC cuprates are shown in Figs. 6.102 and 6.103 respectively. Generally, the carrier concentrations in HTSC cuprates are an order of magnitude smaller than in copper or silver.

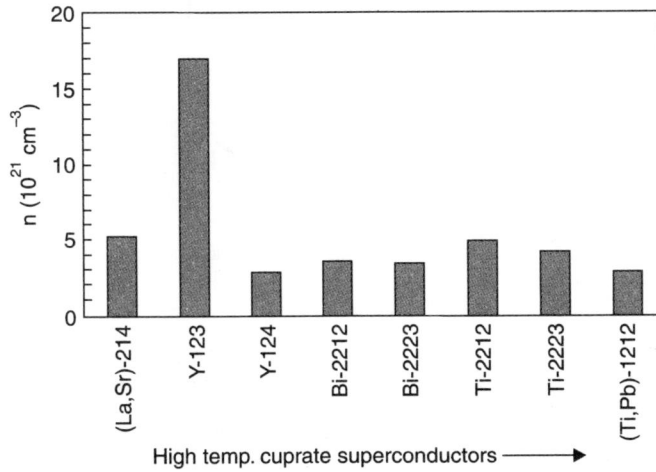

Fig. 6.103 Carrier concentration (n) in various high temperature cuprate superconductors [141]

The resistivity of polycrystalline samples is an averaged value over the different orientations of the single grains. As a consequence there is considerable scatter in the resistivity data of polycrystalline samples. However, even for single crystals, very different electrical resistivities have been reported. The scatter of the in-ρ_{ab} and out-of-plane (ρ_c) resistivity values of various $YBa_2Cu_3O_{7-\delta}$ single crystals is shown in Fig. 6.104, and Fig. 6.105 shows them for a single crystal $YBa_2Cu_3O_7$ as a function of temperature.

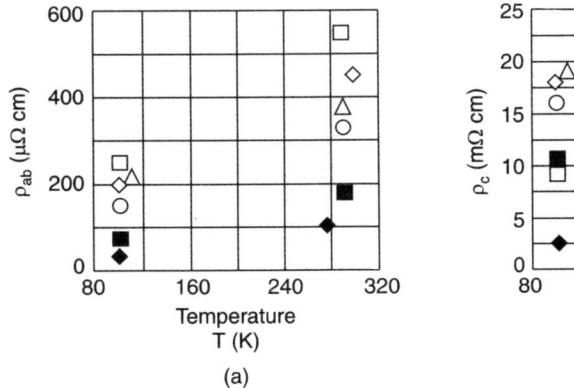
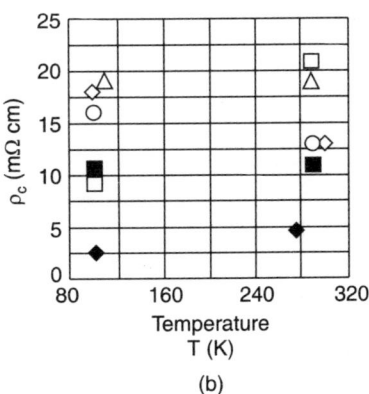

Fig. 6.104 In-(a) and out-of-plane resistivities of several $YBa_2Cu_3O_{7-\delta}$ single crystals [292]

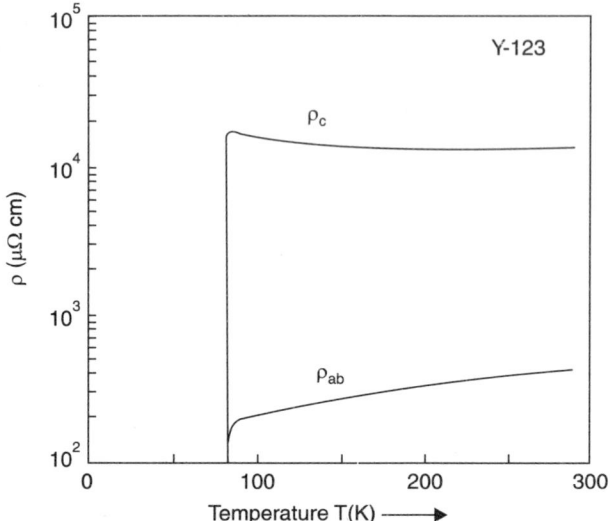

Fig. 6.105 Temperature dependence of the in- and out-of-plane resistivities for single Y Ba$_2$Cu$_3$O$_7$ single crystal [293]

We may note that the resistivity ρ_c along the c-direction shows semiconducting behaviour corresponding to enhanced ρ_c-values at reduced temperatures. As a consequence of this behaviour the resistivity anisotropy ρ_c/ρ_{ab} increases from ≈ 30 to ≈ 75 at temperatures of 290 and 100 K respectively.

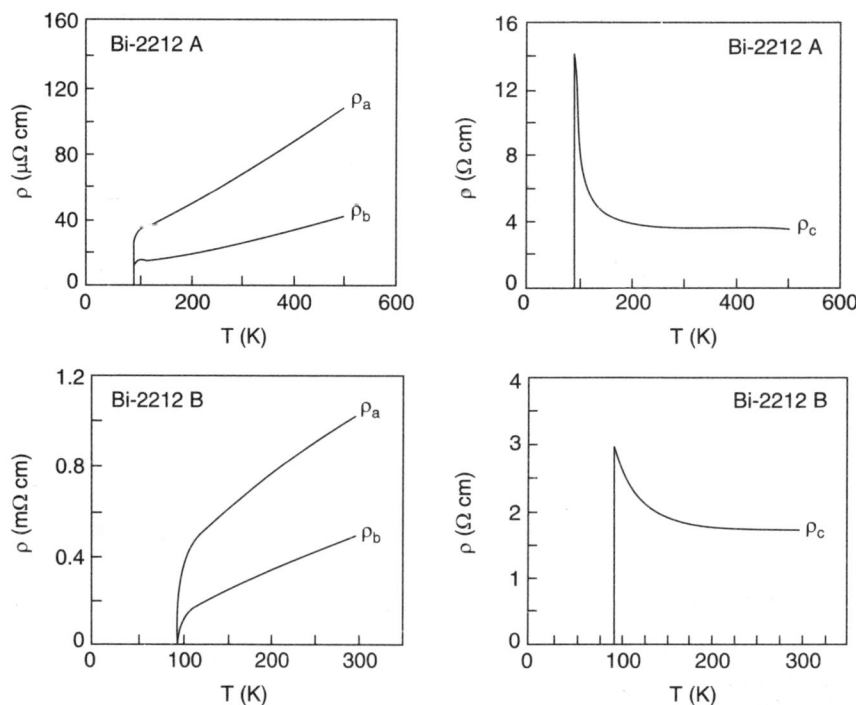

Fig. 6.106 Resistivity ρ_a, ρ_b and ρ_c versus temperature for two Bi-2212 single crystals A(top) and B(bottom) [294, 295]

The resistivities ρ_a, ρ_b and ρ_c versus temperature for two Bi-2212 single crystals A and B are shown in Fig. 6.106. The chemical composition of crystal B was determined by inductively coupled plasma analysis. Kotaka et al. [295] found the measured cation ratio as Bi:Sr:Ca:Cu = 2.14:1.77:0.92:2.00. The resistivities ρ_a and ρ_b along the CuO_2 planes show metallic behaviour for both single crystals. For crystals A and B, the in-plane resistivity anisotropies ρ_a/ρ_b at room temperature are $\approx$ 2.6 and $\approx$ 2 respectively. Moreover, the out-of-plane resistivities of both single crystals increase with decreasing temperature. We may note that the resistivity anisotropies ρ_c/ρ_a slightly above the transition temperature are considerably different, being $\approx$ 500,000 and $\approx$ 10,000 for crystals A and B respectively. However, the normal state resistivities like the critical temperatures are very sensitive to the oxygen concentration and the cation ratio that determines the carrier concentration in HTSC cuprates.

Now, we consider the effect of chemical composition on the electrical resistivity of HTSC cuprates. Both the in- and out-of-plane resistivities of $La_{2-x}Ba_xCuO_4$ single crystals at 50 and 280 K as functions of the Ba content x is shown in Fig. 6.107. The in- and out-of-plane resistivities decrease with increasing Ba concentration x. However, the out-of-plane resistivities decrease with increasing Ba concentration x. The out-of-plane resistivity ρ_c shows semiconducting behaviour and increases with decreasing temperature. The critical temperatures for the considered x-values of 0.06, 0.08 and 0.09 are 16.5, 28 and 32.5 K respectively. The lowest normal state resistivity and the highest critical temperature of 32.5 K are both found in single crystal $La_{1.91}Ba_{0.9}CuO_4$. Further, the smallest anisotropy ρ_c/ρ_{ab} of 1200 at 50 K has also been reported for this single crystal. Fig. 6.108 shows the in- and out-of-plane resistivities of $Bi_{2.1}Sr_{1.9}Ca(Cu_{1-x}Zn_x)_2O_y$ single crystals at 95 and 300 K as functions of Zn content. The in-plane resistivity shows metallic behaviour whereas ρ_c increases with decreasing temperature [297]. The in-plane resistivity ρ_{ab} increases with enhanced Zn doping. The Zn-ions, which are expected to occupy Cu sites in the Bi-2212 structure, may act as additional scattering centres for the charge carriers flowing in the ab plane. However, the out-of-plane resistivity decreases considerably with Zn doping. As a result the anisotropy ρ_c/ρ_{ab} of the Zn Bi-2212 single crystals is reduced as compared with undoped material. The effect of Sr/Ca cross substitution on the resistivity of $Bi_{1.75}Pb_{0.25}Sr_{2-x}Ca_{2+x}Cu_3O_y$ at 250 K is shown in Fig. 6.109. Single step transitions with T_c > 100 K result for x-values of 0 and 0.25. The normal state resistivity increases systematically with enhanced Ca content except for x between − 0.25 and 0 [298].

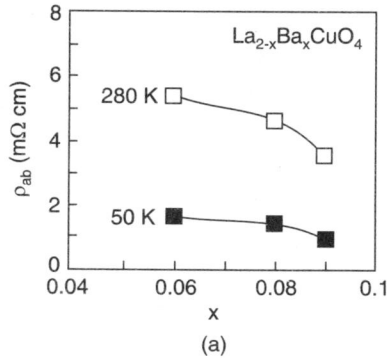
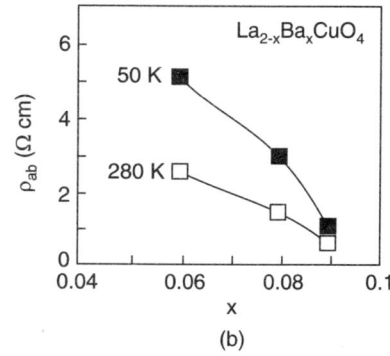

Fig. 6.107 In-(a) and out-of-plane resistivity (b) of $La_{2-x}Ba_xCuO_4$ single crystals versus Ba concentration x [296]

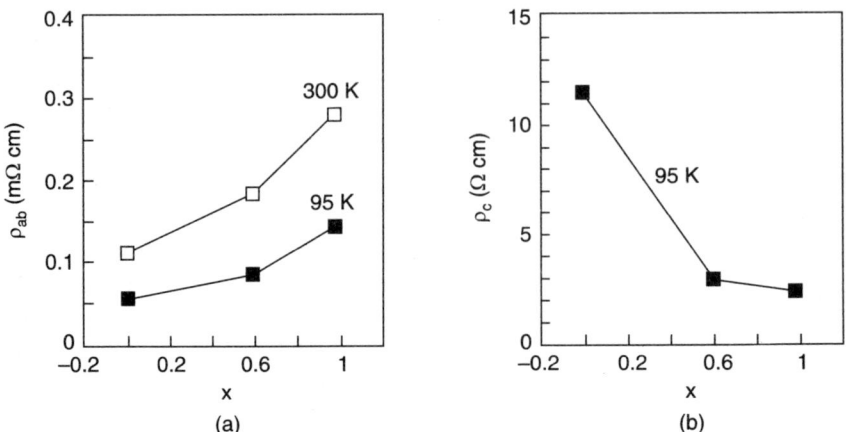

Fig. 6.108 In-(Fig. a) and out-of-plane resistivity (Fig. b) of $Bi_{2.1}Sr_{1.9}Ca(Cu_{1-x}Zn_x)_2O_y$ single crystals as function of Zn content [297]

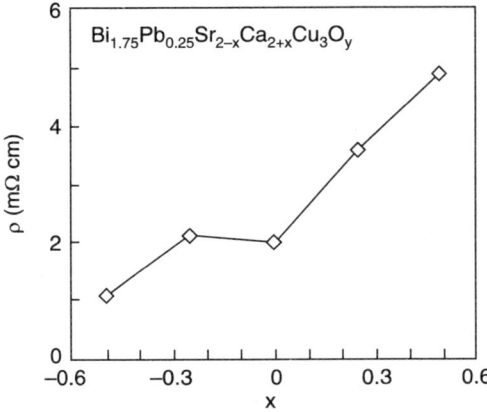

Fig. 6.109 The effect of Sr/Ca cross substitution on the normal-state resistivity of $Bi_{1.75}Pb_{0.25}Sr_{2-x}Cu_3O_y$ [298]

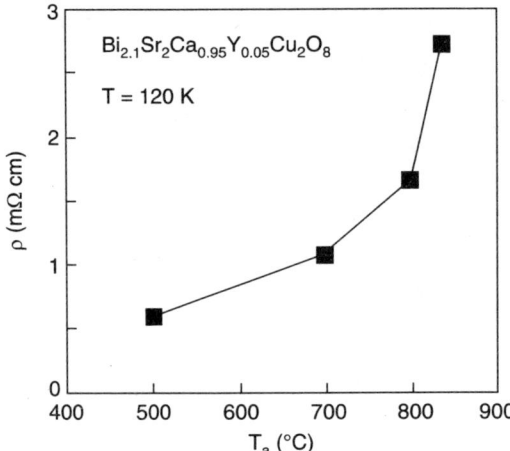

Fig. 6.110 Resistivity (ρ) of $Bi_{2.1}Sr_2Ca_{0.95}Y_{0.05}Cu_2O_8$ vs annealing temperature (T_a) [298]

We may note that the normal state resistivity depends not depend on the cation ratio but also on the heat treatment conditions and the oxygen partial pressure. The resistivity of $Bi_{2.1}Sr_2Ca_{0.95}Y_{0.05}Cu_2O_8$ at 120 K as a function of the annealing temperature is shown in Fig. 6.110. The heat treatment was performed in air. The resistivity increases considerably with increasing temperature [299]. The resistivity of Bi-2212 single crystals as a function of oxygen partial pressure is shown in Fig. 6.111 [300]. The single crystals were annealed at a temperature of 745°C. The resistivity decreases with increasing oxygen partial pressure. Forro and Cooper [300] reported that semiconducting behaviour results for $p(O_2) = 0.04$ m bar. Fig. 6.112 shows the resistivity vs. temperature data of polycrystalline as-grown and oxygen annealed Hg-1223 [73]. It is reported that annealing in flowing oxygen at 300°C not only enhances the T_c but also reduces the normal state resistivity of the Hg-1223.

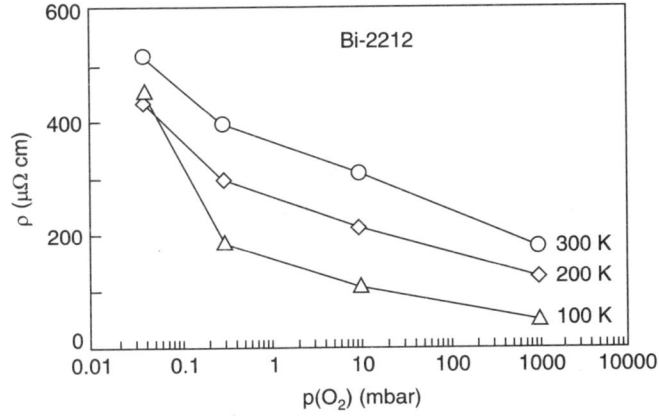

Fig. 6.111 In-plane resistivity (ρ_{ab}) of Bi-2212 single crystal vs oxygen partial pressure [300]

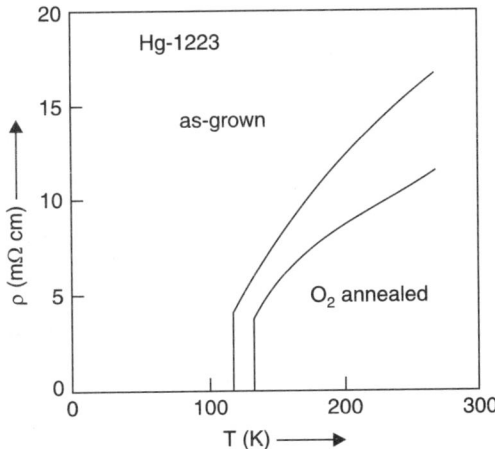

Fig. 6.112 Resistivity (ρ) of polycrystalline Hg-1223 as a function of temperature (T). Oxygen annealing leads to an enhanced T_c-value as well as a reduced normal state resistivity [73]

Table 6.23 shows the resistivity data for various HTSC cuprates. The smallest resistivity values have been measured in the ab-planes of single crystals and epitaxial films. Considerable scatter in the resistivity data has been observed especially for polycrystalline samples.

Table 6.23. Electrical resistivity (ρ) of various HTSC cuprates at $T = 250$ K

Compound	ρ(mΩ cm)	$d\rho/dT$ ($\mu\Omega$ cm/K)	Comments	Ref.
$La_{1.925}Sr_{0.075}CuO_4$	0.86	3.3	polycrystalline	[301]
$La_{1.9}Sr_{0.1}CuO_4$	0.59	2.3		
$La_{1.875}Sr_{0.125}CuO_4$	0.39	1.7		
$La_{1.85}Sr_{0.15}CuO_4$	4.3	16	polycrystalline	[45]
$La_{1.8}Sr_{0.2}CuO_4$	5	12.4	polycrystalline	[302]
$La_{1.94}Ba_{0.06}CuO_4$	7.9 $\parallel ab$	36	single crystals	[296]
	$\approx 2700 \parallel c$	≈ -4700		
$La_{1.92}Ba_{0.08}CuO_4$	6.9 $\parallel ab$	29		
	$\approx 1600 \parallel c$	≈ -2800		
$La_{1.91}Ba_{0.09}CuO_4$	5.3 $\parallel ab$	24		
	$\approx 700 \parallel c$	≈ -1900		
$YBa_2Cu_3O_7$	0.14-0.29	0.56-1.11	epitaxial films	[14]
$YBa_2Cu_3O_{7-\delta}$	0.13	0.49	thin film	[303]
$YBa_2Cu_3O_{7-\delta}$	1.35 $\parallel ab$	5.5	magnetically	[304]
	9.2 $\parallel c$	25	melt textured	
$YBa_2Cu_3O_{7-\delta}$	0.47 $\parallel ab$	2.5	melt textured	[305]
	28 $\parallel c$	110		
	5.8 $\parallel ab$	21		
	34 $\parallel c$	40		
$YBa_2Cu_3O_{7-\delta}$	0.56	1.9	polycrystalline	[306]
$YBa_2Cu_3O_{7-\delta}$	2.3	6.9	polycrystalline	[307]
$YBa_2Cu_3O_{7-\delta}$	6.8	15.5	polycrystalline	[308]
$YBa_2Cu_3O_{7-\delta}$	0.7	2.6	polycrystalline	[309]
$YBa_2Cu_{2.99}Ni_{0.01}O_y$	2.5	4.8	polycrystalline	[251]
$YBa_2Cu_{2.98}Ni_{0.02}O_y$	1.0	2.1		
$YBa_2Cu_{2.97}Ni_{0.03}O_y$	4.6	7.5		
$LaBa_2Cu_3O_{7-\delta}$	1.9	5.9	polycrystalline	[310]

...(cont.)

Compound	Value 1	Value 2	Notes	Ref.
$YBa_2Cu_4O_8$	10.0	22	polycrystalline	[311]
$YBa_2Cu_4O_8$	0.73	2.6	polycrystalline	[312]
$Y_{0.95}Ca_{0.05}Ba_2Cu_4O_8$	0.81	2.6		
$Y_{0.9}Ca_{0.1}Ba_2Cu_4O_8$	0.76	2.2		
$Bi_2Sr_2CuO_6$	≈ 25		polycrystalline	[313]
$Bi_2Sr_2CuO_6$	≈ 0.5 ∥ ab	1.6	single crystal	[314]
$Bi_{2.11}Sr_{1.89}CuO_{6+\delta}$	0.42	1.24	O_2 annealed	[315]
	1.7	4.6	Ar annealed	
$Bi_{2.15}Sr_{1.85}CuO_{6+\delta}$	0.58	1.47	O_2 annealed	
$Bi_{1.95}(Sr,La)_{2.05}CuO_y$	0.46	1.25	single crystal	[316]
$Bi_{2.2}Sr_2Ca_{0.8}Cu_2O_{8+\delta}$	0.11 ∥ ab	0.45	single crystal	[56]
$Bi_2Sr_2CaCu_2O_8$	0.059 ∥ a	0.18	single crystal	[294]
	0.023 ∥ b	0.07		
	3650 ∥ c	semiconducting		
$Bi_2Sr_2CaCu_2O_y$	2.5 ∥ ab	9.7	single crystal	[317]
$Bi_2Sr_2CaCu_2O_8$	∥ ab		single crystals	[318]
	0.47	semiconducting	0.04 mbar O_2	
	0.34	1.0	0.3 mbar O_2	
	0.26	0.9	10 mbar O_2	
	0.15	0.55	1000 mbar O_2	
$Bi_2Sr_2CaCu_2O_8$[1]	0.9 ∥ a	2.65	single crystal	[319]
	0.41 ∥ b	1.6		
	1750 ∥ c	semiconducting		
$Bi_2Sr_2CaCu_2O_8$	0.26 ∥ a	1.0	single crystal	[320]
	0.29 ∥ b	1.0		
$Bi_2Sr_2Ca(Cu,Li)_2O_8$	∥ ab		single crystals	[321]
	0.36	0.87	1.4% Li	
	0.18	0.46	3.0% Li	
	0.25	0.92	13% Li	

...(cont.)

[1] $Bi_{2.14}Sr_{1.77}Ca_{0.92}Cu_2O_{8-\delta}$

$Bi_2Sr_2CaCu_2O_8$	3.7 ∥ ab	10.9	film	[322]
(Bi,Pb)-2223[2)]	1.0	4.2	fiber	[323]
(Bi;Pb)-2223[3)]	0.96	3.6	polycrystalline	[82]
(Bi;Pb)-2223[4)]	4.9	12.4	polycrystalline	[324]
(Bi;Pb)-2223[5)]	5.6	11.6	polycrystalline	[325]
(Bi;Pb)-2223[6)]	6.4	22.8	polycrystalline	[326]
(Bi;Pb)-2223[7)]	2.8	6.0	polycrystalline	[327]
(Bi;Pb)-2223[8)]	0.97	3.2	polycrystalline	[328]
$TlBa_2CaCu_2O_y$	5.2	10.8	polycrystalline	[329]
$Tl_{1.6}Ba_2Ca_{2.4}Cu_3O_y$	2.3	10.3	polycrystalline	[94]
$Tl_{1.7}Ba_2Ca_{2.3}Cu_3O_y$	≈ 3	≈ 12	polycrystalline	[330]
$Tl_2Ba_2Ca_2Cu_3O_y$	0.96	2.6	polycrystalline	[331]
$Tl_2Ba_2Ca_2Cu_3O_y$[9)]	2.45	10.3	polycrystalline	[332]
$Tl_2Ba_2Ca_3Cu_4O_y$	31	150	polycrystalline	[333]
(Tl,Pb)-1212[10)]			polycrystalline	[334]
$x = 0.6$	19.2	≈ 46		
$x = 0.75$	14.1	≈ 19		
$x = 0.9$	5.47	≈ 18		
$x = 1$	2.64	≈ 12		
$HgBa_2CuO_{4+\delta}$	≈ 38	≈ 150	polycrystalline	[335]
$HgBa_2CaCu_2O_{6+\delta}$	20	44	polycrystalline	[75]
$HgBa_2CaCu_2O_{6+\delta}$	25	≈ 50	polycrystalline	[336]

...(cont.)

[2)]$Bi_{1.8}Pb_{0.35}Sr_{1.87}Ca_2Cu_3O_y$
[3)]$Bi_{1.8}Pb_{0.4}Sr_2Ca_2Cu_3O_y$
[4)]$Bi_{1.4}Pb_{0.6}Sr_2Ca_2Cu_3O_y$
[5)]$Bi_{1.9}Pb_{0.3}Sr_{1.6}Ca_{1.9}Cu_{3.4}O_y$
[6)]$Bi_{1.7}Pb_{0.3}Sr_2Ca_2Cu_3O_y$
[7)]$Bi_{1.84}Pb_{0.34}Sr_{1.91}Ca_{2.03}Cu_{3.06}O_{10+\delta}$
[8)]$Bi_{1.8}Pb_{0.26}Sr_2Ca_2CuO_{10+\delta}$
[9)]$Tl_{1.6}Hg_{0.4}Ba_2Ca_2Cu_3O_{10-\delta}$
[10)]$Tl_{0.5}Pb_{0.5}Sr_2Y_{1-x}Ca_xCu_2O_7$

$HgBa_2Ca_2Cu_3O_{8+\delta}$	57	220	polycrystalline	[74]
$HgBa_2Ca_2Cu_3O_{8+\delta}$	23	100	polycrystalline	[76]
$HgBa_2Ca_2Cu_3O_{8+\delta}$	23	100	polycrystalline	[75]
$HgBa_2Ca_2Cu_3O_{8+\delta}$ [11]	0.09 ∥ ab	0.6	epitaxial film	[228]
$HgBa_2Ca_3Cu_4O_{10+\delta}$	0.32 ∥ ab	1.3	single crystal	[241]
$(Cu_{0.5}C_{0.5})Ba_2Ca_2$ [12] $Cu_3O_{9+\delta}$	0.29	2.4	polycrystalline	[29]
$(Cu_{0.5}C_{0.5})Ba_2Ca_3$ [12] $Cu_4O_{11+\delta}$	0.32	3.5	polycrystalline	[29]
$BSr_2Ca_3Cu_4O_{11}$ [12]	0.42	2.9	polycrystalline	[28]
$Ca_{0.3}Sr_{0.7}Cu_2C_{0.75}O_y$	2.0	6	polycrystalline	[338]
$Sr_2CaCu_2O_{4+\delta}F_{2\pm y}$	2.2 [12]	—	polycrystalline	[44]
$Sr_2Ca_2Cu_3O_{6+\delta}F_{2\pm y}$	5.4 [12]	—	polycrystalline	[44]
$Sr_2Cu(CO_3,BO_3)O_y$	540	700	polycrystalline	[339]
$Sr_2CaCu_2(CO_3,BO_3)O_y$	260	520	polycrystalline	[339]
$Sr_2Ca_2Cu_3(CO_3,BO_3)O_y$	5.6	20	polycrystalline	[339]

[11] $T = 200$ K
[12] $T = 150$ K

6.7.5 Thermal Conductivity of HTSC Cuprates

Among the important properties of HTSC cuprates is their ability to conduct heat. There is not only an obvious technological interest in how efficiently and by what means the heat flows in these solids, but also a deep theoretical desire to understand the electronic and vibrational properties of these materials. In HTSC cuprates, such information is especially valuable due to the fact that traditional galvanomagnetic probes such as resistivity, Hall effect, and thermopower are inoperative in the (now) wide temperature range below T_c. One can find the detailed account of heat transport in Jezowski and Klamut [340] and Uher [340, 341].

The steady state thermal current density J_Q can be used to define the coefficient of thermal conductivity K as

$$J_Q = -K \nabla T \qquad ...(6.52)$$

The negative sign implies that the heat flows down the thermal gradient, *i.e.*, from the warmer to the cooler end of the sample.

Heat in a solid is carried by two distinct entities, free charge carriers (electrons or holes in cuprates), K_{el}, and quantized lattice vibrations called phonons, K_{ph}. The total thermal conductivity, K, is then

$$K = K_{el} + K_{ph} \qquad \ldots(6.53)$$

where K_{el} and K_{ph} are the electronic and phononic contributions respectively. Phonons carry the bulk of the heat in the normal state of HTSC cuprates, and it is reasonable to assume that phonons also play a prominent role at temperatures below T_c. The phonon contribution is given by

$$K_{ph} = \frac{1}{3} v_{ph} l_{ph} C_{ph} \qquad \ldots(6.54)$$

where v_{ph} is the velocity, l_{ph} is the mean free path and C_{ph} is the specific heat of phonons per unit volume. Scattering processes with lattice defects, other phonons and electrons or holes limit the mean free path l_{ph} of the phonons. We know that the phonon specific heat is proportional to T^3 at low temperatures and reaches a constant value of $3 Nk_B$ ($N \to$ number of atoms) well above the Debye temperature. From eq. (6.54) we may note that not only the phonon specific heat but also the mean free path l_{ph} depends on temperature. The number of activated photons at high temperatures is proportional to the temperature, leading to $1/T$ dependence of the mean free path. Analogously to eq. (6.54), one can write the electronic contribution to the thermal conductivity as

$$K_{el} = \frac{1}{3} v_F l_{el} C_{el} \qquad \ldots(6.55)$$

where v_F is the Fermi velocity, l_{el} the mean free path and C_{el} the electronic contribution to the specific heat per unit volume. Using

$$C_{el} = \frac{1}{3} \pi^2 D(E_F) k_B^2 T$$

and the density of states for a free electron gas $D(E_F) = 3 N_e/2E_F$ where N_e is the number of electrons, one obtains

$$K_{el} = \frac{1}{3} v_F l_{el} \frac{\pi^2 n}{m v_F^2} k_B^2 T = \frac{\pi^2 n k_B^2 T \tau}{3m} \qquad \ldots(6.56)$$

where n is the concentration and m the mass of the carriers. In eq. (6.56), we have replaced the mean free path by $v_F \tau$, where τ is the relaxation time. When the relaxation times for electrical and thermal processes are identical, the ratio of thermal and electrical conductivity becomes:

$$\frac{K_{el}}{\sigma} = \frac{\pi^2 n k_B^2 T \tau / 3m}{ne^2 \tau m} = \frac{\pi^2}{3} \left(\frac{k_B}{e}\right)^2 T \qquad \ldots(6.57)$$

Eq. (6.57) is the well known **Wiedemann-Franz law.** The behaviour of pure metals at not too low a temperature is typically well described by it. Using the expression $C_{el} = \gamma T$, one can write eq. (6.55) as

$$K_{el} = \frac{1}{3} \gamma v_F^2 \tau T \qquad \ldots(6.58)$$

The relaxation time determining the electrical conductivity is known to be proportional to T^3 for $T \ll \theta_D$ and to T^1 for $T \gg \theta_D$. According to the Wiedemann-Franz law the relaxation times for thermal and electrical processes are equal and the resulting electronic contribution to the thermal conductivity is constant for $T \gg \theta_D$, whereas it is proportional to T^2 at low temperatures.

In pure metals even at room temperature, heat is mainly transported by the electrons. On the other hand, in disordered alloys comparable electron and phonon contributions can exist. In conventional superconductors, thermal conductivity decreases rapidly below T_c. We may note that the lowering of temperature below T_c and the ensuing formation of the Cooper condensate lead to a sharp change in the electromagnetic and kinetic response of a material, and a consequent drastic modification of its heat flow pattern. The following two properties of condensate provide the overriding influence:

(i) Cooper pairs carry no entropy.

(ii) Cooper pairs do not scatter phonons.

The first condition implies that the electronic thermal conductivity vanishes rapidly below T_c. In fact, the decrease is approximately exponential and has been justified by Bardeen et al. [343].

The effect of the second condition is more subtle. Provided that the mean free path of phonons at $T > T_c$ is limited by scattering on charge carriers, on passing into the superconducting state the phonon thermal conductivity will rise because the number of normal carriers (more precisely, the density of quasiparticle excitations) rapidly decreases. The competition between the rapidly diminishing K_e on the one hand and the increasing K_p on the other will determine the overall dependence of the total thermal conductivity of any given superconductor. Because of the dominant contribution of charge carriers, a vast majority of conventional superconductors show a decrease in the ratio of the thermal conductivities in the superconducting and normal states, $K_s(T)/K_n(T)$, for $T < T_c$. In the case of elemental superconductors, the thermal conductivity may be reduced by 2-3 orders of magnitude in comparison to its normal state value. Of course at very low temperature, all conventional superconductors will behave as ordinary dielectric solids, because phonons are the only entity that can carry heat.

Phonons carry the bulk of heat in the normal state of HTSC, and it is reasonable to assume that phonons also play a prominent role at temperatures below T_c. The behaviour of K_p being limited by carrier scattering was originally treated in Bardeen et al. theory [343] and supplemented by Tewordt and Wolkhausen [344] to account for the relevant scattering processes and the anisotropy of the phonon-carrier interaction as appropriate to HTSC cuprates. The in-plane phonon thermal conductivity becomes [345]

$$K_{p,ab}(T) = (k_B/2\pi^2 v)(k_B/\hbar)^3 T^3 \int_0^{\theta_D/T} dx\, x^4\, e^x/(e^x - 1)^2 \int_0^1 d\xi\, 3/2\, (1-\xi^2)\, F(T, x, \xi) \quad ...(6.59)$$

where the overall scattering rate is given by

$$F^{-1}(T, x, \xi) = B + D_e t^4 x^4 + D_{sf} t^2 x^2 + E\, t\, x\, g(x, y)(1-\xi^2)^{3/2} + U T^4 x^2 \quad ...(6.60)$$

Coefficients B, D_p, D_{sf}, E and U in eq. (6.60), in turn, refer to phonon scattering by boundaries, point defects, sheet like faults, charge carriers, and other phonons. The function $g(x, y)$ is the ratio of the phonon-carrier scattering times in the normal and superconducting states, and its exact form depends on the superconducting energy gap. Theory can accommodate the strong coupling limit for both the s-wave and d-wave pairing mechanisms, and provides explanation for the characteristic rise and the peak in the thermal conductivity of HTSC below T_c (Fig. 6.113). The peak has been observed in samples of all major families except for Ba K BiO and Nd Ce CuO structures. While excellent fits with perfectly reasonable parameters exist for both sintered and single crystal data [347], phonons may not be the sole entity responsible for the rising thermal conductivity and the peak below T_c.

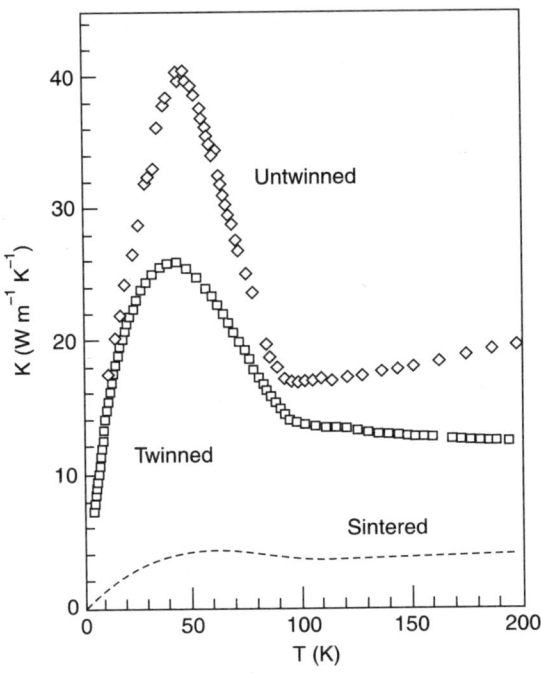

Fig. 6.113 Thermal conductivity of sintered twinned and untwinned (a-direction) samples of Y Ba$_2$ Cu$_3$ O$_{7-\delta}$ [346]

Microwave surface resistance studies [348] and ultrafast laser pump-probe studies of carrier relaxation [349] have clearly shown that the relaxation time of quasiparticles in HTSC is drastically enhanced below T_c. Such an usually long quasiparticle lifetime provides an alternative explanation for the peak in the thermal conductivity [350]. In this case the fit to the experimental data is made with the aid of the relation for the carrier contribution in the superconducting state derived by Kadanoff and Martin [351] and independently by Tewordt [352],

$$K_e^s = 1/(2k_B T^2 m) \int d^3p \, (p_z^2 \varepsilon_p^2/T) \, \text{sech}^2 (E_p/2k_B T) \simeq \phi/T \qquad \ldots(6.61)$$

Here $E_p = (\varepsilon_p^2 + \Delta_p^2)^{1/2}$, where ε_p is the normal state dispersion and Δ_p the superconducting gap. The scattering rate T is taken as $T' \sim (T/T_c)^n + w_i$, implying a power law dependence augmented by a constant term w_i that stands for the residual scattering rate due to impurities. Yu et al. [353] find that a d-wave pairing state fits best, and the relaxation rate varies as the fourth power of temperature.

The difficulty in making an unambiguous choice between the two competing interpretations rests in the fact that the charge carriers and phonons contribute roughly equally (in single crystals) to the heat transport in the normal state, and the physical processes that give rise to enhancements in either $K_p(T)$ or $K_e(T)$ below T_c have rather strong temperature dependences that lead to peaks in the thermal conductivity at virtually the same temperatures.

At very low temperatures, $T/T_c \ll 10^{-2}$, any anomalous behaviour should cease and the only mode of heat transport should be via phonons, with grains and boundaries being the dominant scatters, i.e., $K \sim T^3$. Although this is the case for insulating cuprates such as $YBa_2Cu_3O_6$ [354], the superconducting Y BCO [355] and LaSrCuO [356] show a T-linear limiting dependence, (Fig. 6.114), which is well approximated by

$$K(T) = aT + bT^3 \qquad \ldots(6.62)$$

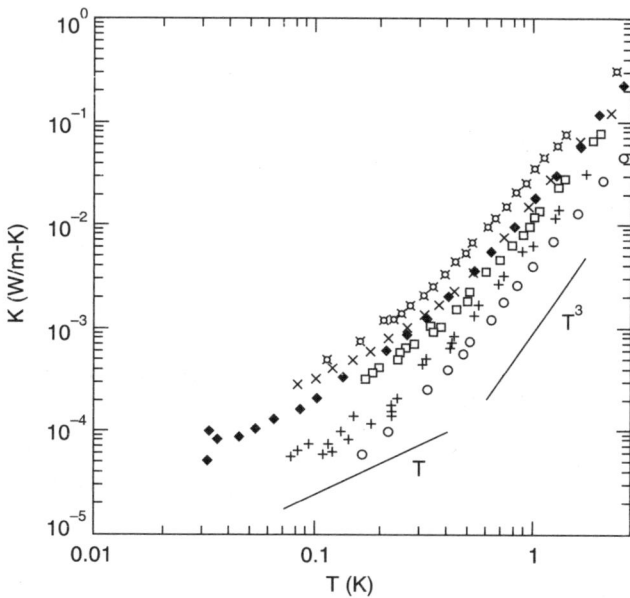

Fig. 6.114 Thermal conductivities of several samples of sintered $YBa_2Cu_3O_{7-\delta}$ at very low temperatures. The slopes for linear (T) and cubic (T^3) temperature dependences are marked in the Fig. [342]

Although various mechanisms were proposed to account for the linear term in eq. (6.62), the consensus converges on the presence of a small number (10 – 15%) of uncondensed carriers, and this provides support for the d-wave (nodes) symmetry of the superconducting state. BiSrCaCuO, on the other hand, displays a T^2-dependence below 2 K regardless of its structural form [357 – 359]. A correlation seems to exist between the T-linear limiting behaviour and the magnitude of the γ-term in the specific heat.

HTSC cuprates possess considerable structural anisotropy, which is reflected in the behaviour of the thermal conductivity. Fig. 6.115 shows the temperature dependence of the anisotropy ratio between the in-plane and the c-axis thermal conductivities, K_{ab}/K_c, for several cuprates.

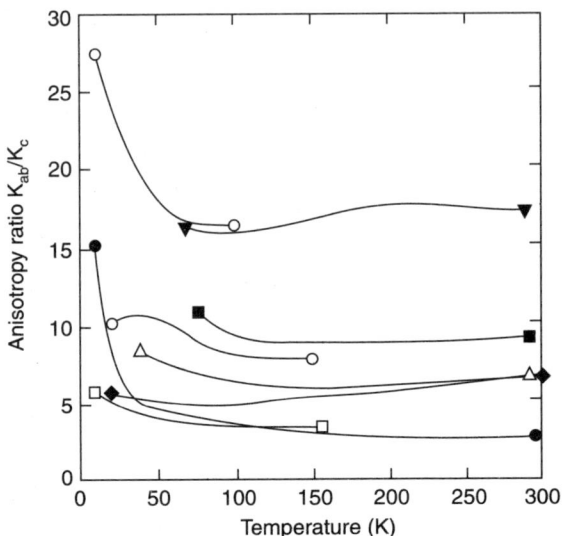

Fig. 6.115 Anisotropy ratio K_{ab}/K_c temperature dependence for crystals of HTSC cuprates: O, Y $Ba_2Cu_3O_{6.7}$ [360]; ◆Y $Ba_2Cu_3O_{7-\delta}$ [361]; ▼Y $Ba_2Cu_3O_7$ [362]; □Y $Ba_2Cu_3O_{7-\delta}$ [363]; ○$Bi_2Sr_2CaCuO_{8-x}$ [363]; △$Bi_2Sr_2CaCuO_8$[364]; ■$Tl_2Ba_2CaCu_2O_8$ [362]; ●$La_{1.96}Sr_{0.04}CuO_4$[365]

6.7.6 Effect of Magnetic Field

An external magnetic field $B > B_{c_1}$ drives a superconductor into the mixed state characterized by the presence of Abrikosov vortices of core radius ξ that contains bound excitations not too different from normal electrons. Close to critical temperature, T_c the heat is carried by unbound excitations (uncondensed electrons), and they scatter on vortices, resulting in the thermal resistance

$$W_e(B) = W_e(0)\left[1 + l_e a/\phi_0 B\right] \qquad \ldots(6.63)$$

where l_e denotes mean free path (*mfp*) of quasiparticles and a is the effective vortex cross-section. Presently, there is no general theory to cover the entire range of magnetic fields, but beyond a certain field strength the scattering weakens as the vortices start to overlap. Moreover, at the same time, tunneling between vortices drives the thermal conductivity toward its normal-state value critically on how large is l_e in relation to the coherence length ξ [366, 367]. Unbound quasiparticles may also undergo the Andreev [368] process, where an incoming electron like quasiparticle scatters on the modulation of the order parameter $\Delta(\mathbf{r})$ and transforms into an outgoing hole-like quasiparticle. This process may drastically alter the heat flow because of the near reversal of the group velocity. We may note that magnetic field, on its own, cannot distinguish between the effects due to quasiparticles and those due to phonons.

A magnetic field has a strong influence on the heat transport in HTSC single crystals and causes upto 30% reduction in the thermal conductivity [369, 370]. Fig. 6.115 (*a*) shows the thermal conductivity of *c*-axis aligned (Bi, Pb)-2223 as a function of temperature and magnetic field applied along the *c*-direction. Clearly, the maximum of the thermal conductivity below T_c is reduced with increasing magnetic field. In addition, the position of the maximum is shifted to higher temperatures. The results show interesting effects associated with the flux creep phenomenon, and this provides a way to study vortex dynamics via thermal transport measurements [371]. However, because of a similar functional dependence of the quasiparticle and phonon contributions on the magnetic field strength, one cannot assess the contributions of quasiparticles and phonons without some *a priori* information concerning the dominant carrier.

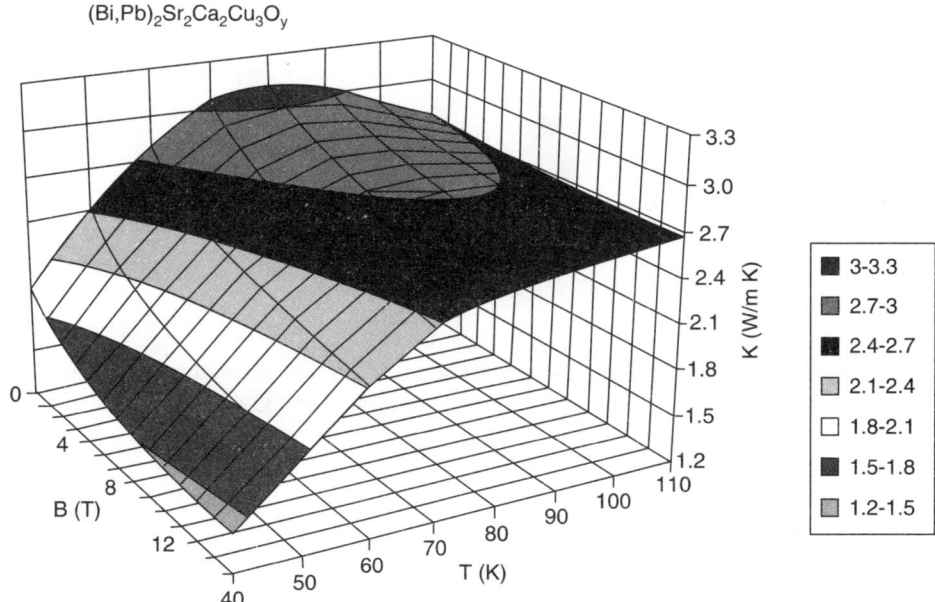

Fig. 115 (a) Thermal conductivity of c-axis aligned (Bi, Pb)-2223 as a function of temperature and magnetic field applied along the c-direction [372]

One can measure the thermal conductivity utilising a steady-state heat flow method. The apparatus is schematically illustrated in Fig. 6.116. One can adjust the temperature of the copper block at the upper end of the sample to temperatures between 4.2 and 300 K by controlled cooling with liquid helium and a resistance heater. Temperature measurements are performed with calibrated silicon diode thermometer sensors with an accuracy of ± 50 mK. The relation between the heat flux Q and the thermal conductivity K is

$$Q = KA \frac{\Delta T}{\Delta L} \qquad \ldots(6.64)$$

where A is the cross sectional area of the sample, ΔL the distance of the temperature sensors and ΔT the measured temperature difference. The heat flux P_Q at the cold end can be determined from the latent heat of evaporation of the liquid helium and the measured heater power.

The thermal conductivity of various HTSC cuprates as a function of temperature is shown in Fig. 6.117. We can see that typical values of the thermal conductivity well above T_c are between 1 and 10 W/mK. Using the Wiedemann-Franz law, one can estimate the electrical carrier contribution to the thermal conductivity from electrical resistivity [308, 309, 324, 372, 373]. From the relatively small number density of charge carriers in the HTSC cuprates this relatively small contribution to the thermal conductivity seems to be plausible.

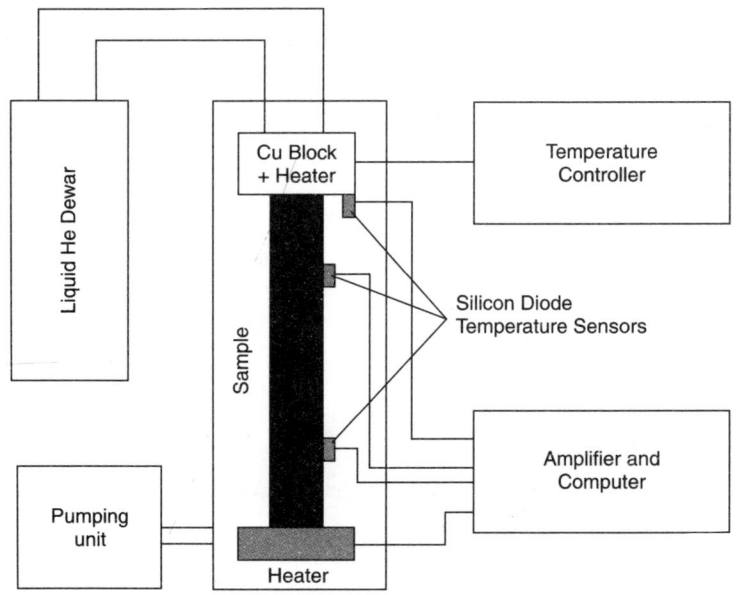

Fig. 6.116 Experimental arrangement used to determine thermal conductivity by means of steady-state heat flow method [339 (a)]

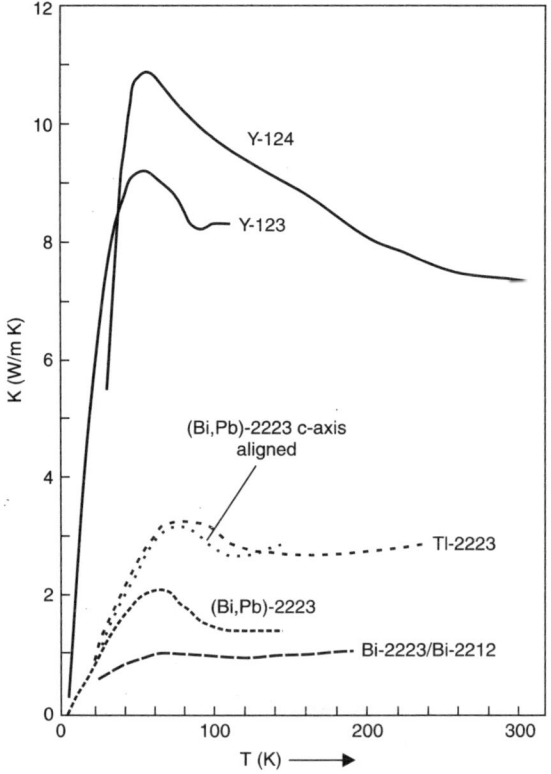

Fig. 6.117 Thermal conductivity of various polycrystalline HTSC cuprates [309, 324, 372, 376 – 378]

A remarkable feature of the thermal conductivity data is a broad maximum below the T_c. Tewordt and Wolkhausen [374] have suggested that this behaviour is caused by reduced electron–phonon scattering due to the condensation of the carriers into Cooper pairs below the T_c. Matsukawa et al. [372] have suggested an alternative to this model, *i.e.* reduced electron–electron scattering in the superconducting state. Recent experimental data for oriented Bi-2223 tapes are in agreement with a phonon contribution to the maximum in the thermal conductivity below T_c [375]. We may note that these data do not exclude a possible additional electron contribution to this maximum.

The thermal conductivity of 3-different $YBa_2Cu_3O_{7-\delta}$ crystals is shown in Fig. 6.118. We may note that no maximum is observed in the thermal conductivity of the insulating Y-123.

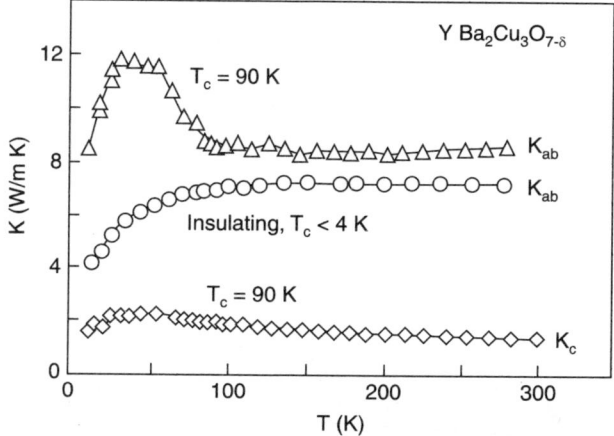

Fig. 6.118 Thermal conductivity of three different crystals of $YBa_2Cu_3O_{7-\delta}$ as a function of temperature. The results indicate that the out-of-plane thermal conductivity K_c is considerably smaller than K_{ab} [379]

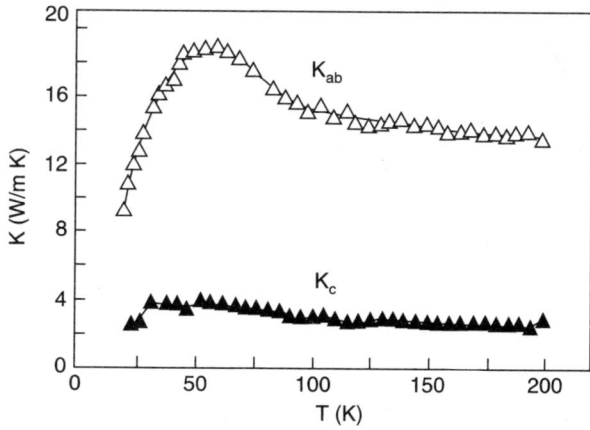

Fig. 6.119 In- and out-of-plane thermal conductivities of melt-textured $YBa_2Cu_3O_{7-\delta}$ with Y_2BaCuO_5 inclusions versus temperature [305]

Moreover, the results suggest that the thermal conductivity along the CuO_2 planes is considerably higher than that parallel to the crystallographic c-direction. We may also see that a similar anisotropy is presented in Fig. 6.119 for melt-textured Y-123 with Y_2BaCuO_5 inclusions. The anisotropy of the thermal conductivity for the melt-textured Y-123 is approximately 4.5 at 200 K.

6.7.7 Specific Heat

Specific heat has been extensively studied for many superconductors (normal as well as HTSC cuprates). We have already read that in metals, specific heat C is mainly the sum of an electronic term and a phonon contribution C_{ph} due to lattice vibrations

$$C = C_e + C_{ph} \qquad ...(6.65)$$

The electronic term is only appreciable at low temperatures and changes dramatically at the superconducting transition temperature, T_c, where the phonon contribution dominates at room temperature and is largely undisturbed by the transition at T_c. In HTSC cuprate superconductors the electronic specific heat is only a very small contribution to the total [380]. Debye temperatures θ_D for various HTSC cuprates are given in Fig. 6.120 [380]. The data are average values. The data for Y-123 are for orthorhombic crystal structure.

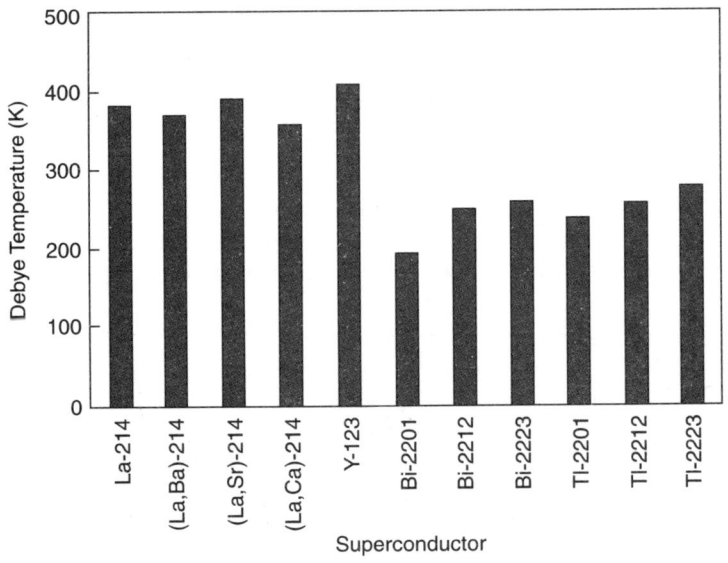

Fig. 6.120 Debye temperatures (θ_D) of various HTSC cuprates [380]

Specific heat data of $YBa_2Cu_3O_{6.92}$ single crystal in the vicinity of the superconducting transition is shown in Fig. 6.121. At T_c, in zero field a sharp jump of the specific heat is observed. This jump is considerably reduced and broadened in the presence of an applied field. A large effect results for the magnetic fields applied along the crystallographic c-direction. The specific heat of a Bi-2212 single crystal close to T_c is shown in Fig. 6.122. For the highly anisotropic Bi-2212 superconductor a more reduced peak is reported at the T_c even for zero applied field.

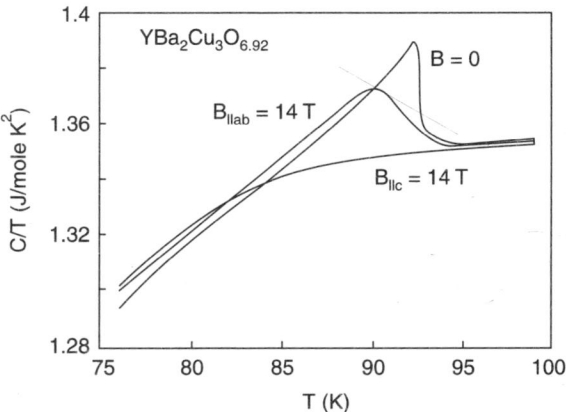

Fig. 6.121 Specific heat jump for a YBa$_2$Cu$_3$O$_{6.92}$ single crystal. The effect of an applied magnetic field is highly anisotropic [381]

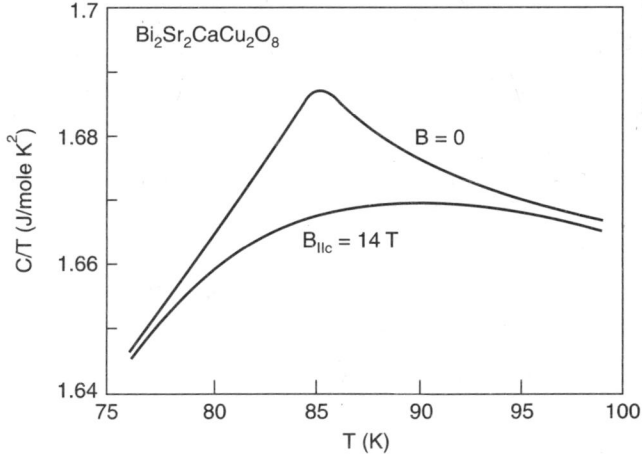

Fig. 6.122 Specific heat jump for a Bi-2212 single crystal. An applied magnetic field leads to a broadened but considerably lower peak [382]

The specific heat data for three different Y Ba$_2$Cu$_3$O$_{7-\delta}$ ceramic samples are shown in Fig. 6.123. We note that a weak anomaly is visible at the T_c. The specific heat of Y Ba$_2$Cu$_4$O$_8$ as a function of temperature is shown in Fig. 6.124. We can see that in this representation of the data the specific heat anomaly at T_c is hardly visible.

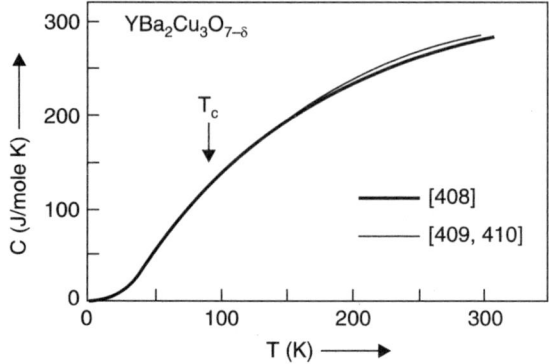

Fig. 6.123 Specific heat versus temperature (T) for three different Y-123 ceramic samples [383 – 385]

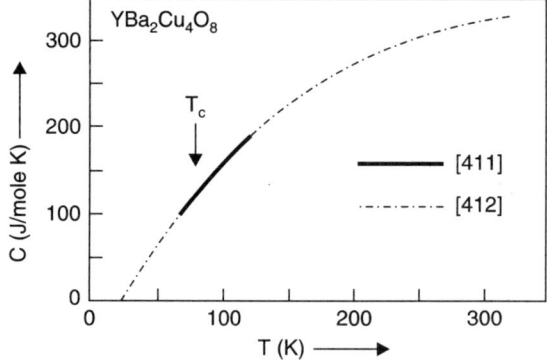

Fig. 6.124 Specific heat versus temperature for two ceramic samples of $YBa_2Cu_4O_8$ [386, 387]

Specific heat data for Bi-2212 single crystals are shown in Fig. 6.125. Fig. 6.126 shows specific heat data for three different (Bi, Pb)-2223 samples. We can see that specific heat data of these three ceramic samples are slightly different, perhaps because of slightly different chemical compositions.

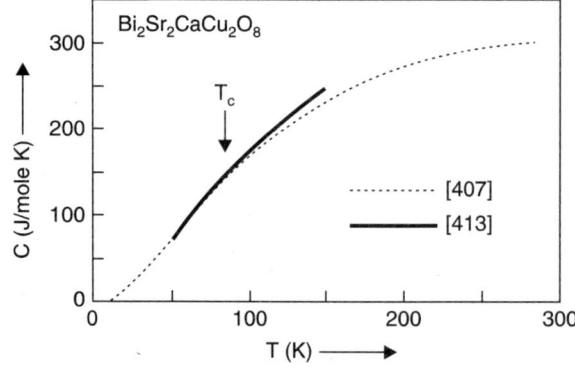

Fig. 6.125 Specific heat of two different Bi-2212 single crystals as a function of temperature [382, 388]

High Temperature Superconducting Cuprates : General Survey

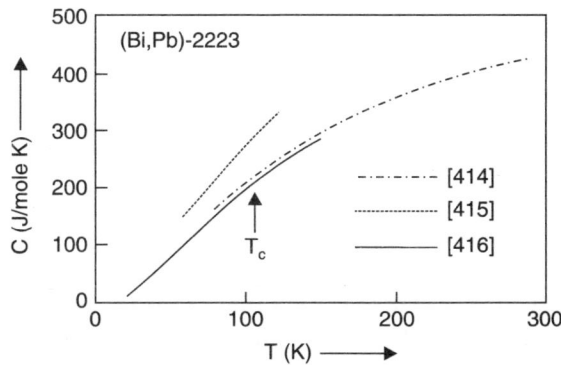

Fig. 6.126 Specific heat data of three different (Bi, Pb)-2223 ceramic samples. Due to slightly different compositions, there is small differences [389 – 391]

Specific heat of Tl-2223 and Tl-2201 is shown in Fig. 6.127. We can see that the specific heat of Tl-2223 is larger than that of Tl-2201 because of the larger number of atoms per formula unit. Fig. 6.128 shows specific heat data for two mercury based HTSC cuprates.

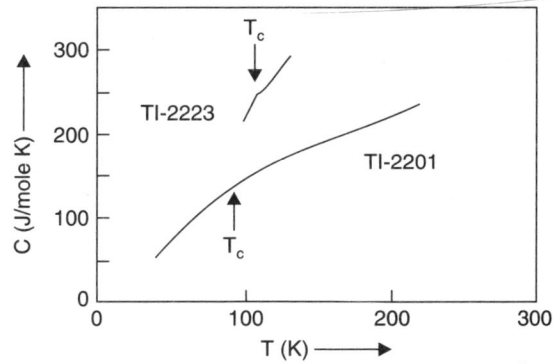

Fig. 6.127 Specific heat of Tl-2201 and Tl-2223 [392, 393]

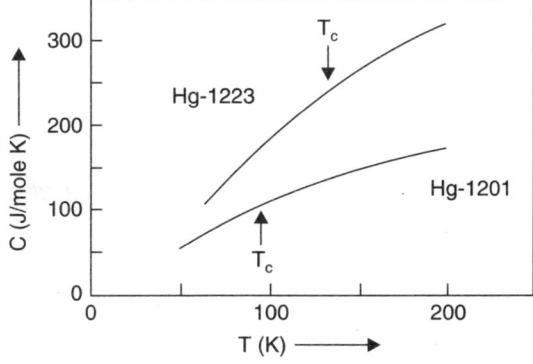

Fig. 6.128 Specific heat of Hg-1201 and Hg-1223 [394, 395]

6.7.8 Thermoelectric and Thermomagnetic Effects

Thermoelectric and thermomagnetic phenomena are sensitive probes of the nature of electronic states and interaction processes within a conductor. Obviously, they are useful tools to study the band structure and carrier dynamics of conducting solids. In superconductors, apart from providing valuable insight into the carrier transport above the T_c, these two effects are important in assessing the dynamics of vortices and quasiparticles.

When we discuss thermoelectric effects at temperature below T_c, one should keep in mind that the physics refers to the mixed state of a superconductor characterized by the presence of the flux lines or vortices. It is the dissipative motion of these vortices that gives rise to the transverse thermomagnetic effects. The vortex itself, to a first approximation, is viewed as a rigid tube of normal phase of radius ξ (coherence length) threading the superconducting phase. The core of vortex is screened from the superconducting surrounding by the encircling supercurrent J_{SC} of magnitude

$$J_{SC} = (\phi_0/2\pi \mu_0 \lambda^3) K_1 (r/\lambda) \qquad \ldots(6.66)$$

where $K_1 (r/\lambda)$ is the first order modified Bessel function, ϕ_0 is the flux quantum, λ is penetration depth, and r is the distance from the centre of the vortex. The functional form of J_{SC} together with its asymptotic behaviours for $r \ll \lambda$ and $r \gg \lambda$ are shown in Fig. 6.129. Although this description is quite adequate for low temperature conventional superconductors, the vortices in the HTSC cuprates are much less rigid and may attain a partly two dimensional character, especially in the more anisotropic cuprates. We may note that the **Lorentz force** and the **thermal force** are the two driving forces responsible for the motion of vortices.

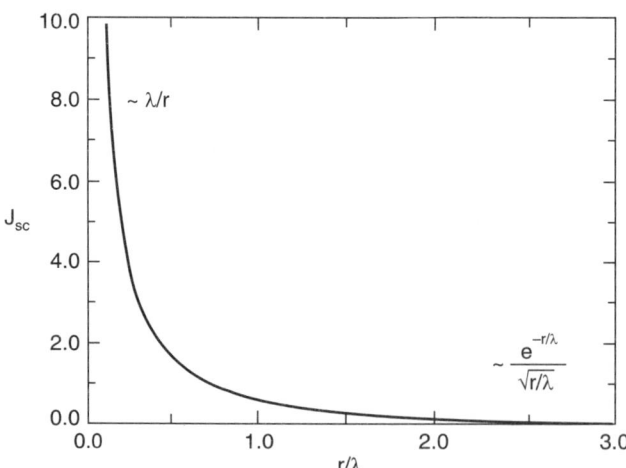

Fig. 6.129 Radial dependence of screening current J_{SC}. Asymptotic behaviour for $r \ll \lambda$ and for $r \gg \lambda$ is indicated. The actual numerical value of J_{SC} is obtained by multiplying the vertical axis by $\phi_0/(2\pi \mu_0 \lambda^3) = 2.62 \times 10^{-10}/\lambda^3$ A – m, where λ is penetration depth [396]

In addition to their large structural anisotropy, HTSC cuprates possess lower carrier density than typical metals and are very sensitive to doping and structural defects. Hence, one would expect the normal state thermopower to be somewhat larger than the typical metallic value and perhaps more complex. In reality, the thermopower in HTSC cuprates present a surprisingly clean pattern, and this allows one to generalize the behaviour across the spectrum of cuprates rather than to deal with idiosyncrasies of each perovskite family separately [397].

Perhaps the most striking feature of the thermopower in HTSC cuprates is its strong dependence on doping, $i.e.$, on the hole concentration in CuO_2 planes, and its close tie with the doping trend in the T_c. The superconducting domain of HTSC cuprates is delineated by the minimum p_{min}, and maximum p_{max}, hole concentrations per planar Cu atom. Outside of these limits there is no superconductivity. The **overdoped** samples ($p \to p_{max}$) have reduced T_c, a metallic character, and a substantially linear and negative thermopower. As the hole density decreases (by means of nonisovalent substitutions or by lowering the amount of oxygen), one obtains a certain **optimal** level of doping, which yields the highest T_c and for which the thermopower is small, with a room temperature value of practically zero. With the further decrease in the hole density ($p \to p_{min}$) in the so-called underdoped domain, the T_c decreases rapidly and the thermopower attains larger positive values, but with a temperature dependence that is essentially unchanged. Clearly, as the material is brought from the overdoped to the underdoped regime, the thermopower undergoes a steady shift upward and toward more positive values. The slope of the thermopower, -3×10^8 V/K^2, is approximately the same for all HTSC cuprates, except the yttrium compound YBCO ($i.e.$, $YBa_2Cu_3O_{7-\delta}$), which is only weakly dependent on the doping, and insensitive to the spacer layers. Furthermore, high structural and, specially, orders of magnitude higher in-plane electrical conductivity over the c-axis conductivity ensure that meaningful measurements that reveal the behaviour of in-plane thermopower can be made on sintered samples where stoichiometry is easier to control than in single crystals. Fig. 6.130 shows an canonical example of the trend in the thermopower. Obertelli et al. [399] have shown that room temperature thermopower for a variety of HTSC cuprates display the $[1 - T_c/T_{c(max)}]^{1/2}$ correlation.

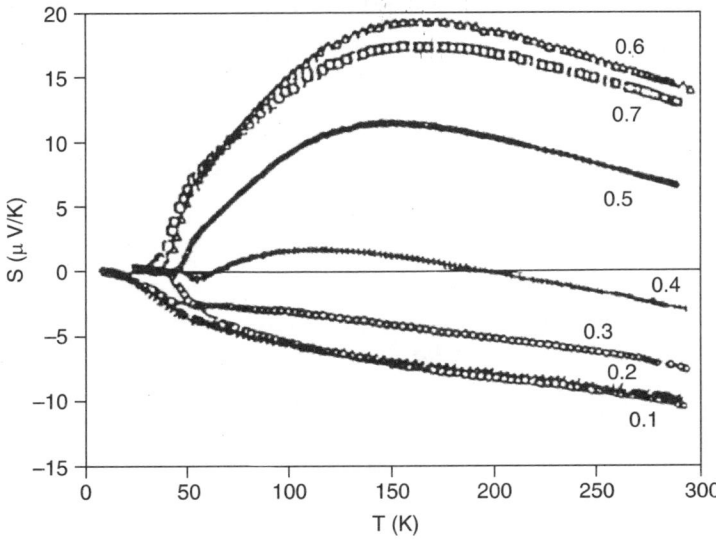

Fig. 6.130 Temperature dependence of thermopower of $Tl_{0.5}Pb_{0.5}Sr_{2-x}La_x$ CuO_5 as a function of the indicated lanthanum doping $0.1 < x < 0.7$ [398]

We have already pointed out that YBCO departs from the general trend followed by all other HTSC cuprates. The reason is the presence of CuO chains running along the b-axis of the structure that provide, in addition to the CuO_2 planes, a substantial contribution to the overall charge carrier transport and are responsible for the in-plane transport anisotropy.

Thermopower is zero in the superconducting state, the circulating current pattern comprising the flow of the normal component J_n and the counterflow of the supercurrent J_s results in a quasiparticle imbalance near the ends of the sample and may give rise to very weak thermoelectric effects [399].

The expression for the diffusion thermopower is

$$S = (k_B/e)\,(\varepsilon_c/k_B T) \qquad (\mu V/K) \qquad ...(6.67)$$

where ε_c is the conduction band energy ($\varepsilon_c \gg k_B T$). Thermopower is large in the normal state and varies as T^{-1}. From eq. (6.67) the Seeback effect is frequently interpreted as the entropy transport per unit electric charge.

In the mixed state of a superconductor there are two distinct entities that can transport entropy across the sample: vortices that transport magnetic flux and unbound quasiparticles outside of the vortex cores that transport electric charge. Vortex transport is essential for the occurrence of the transverse thermomagnetic effects, whereas quasiparticle transport gives rise to the longitudinal thermomagnetic effects (Seeback and Peltier). The Seeback effect cannot arise as a consequence of thermally driven flux of vortices and, indeed, has never been observed in the conventional low temperature superconductors. In HTSC cuprates due to a fortuitous confluence of a large region of reversibility and a power law-law rather than exponential dependence of the quasiparticle density that ensures the presence of quasiparticles at temperatures well below T_c, the Seeback effect is robust, can easily be measured, and rivals the transverse thermomagnetic effects. The **Seeback effect** in this case stems from the quasiparticle transport or, more precisely, from quasiparticles interacting with vortices.

The Seeback effect for the mixed state of a superconductor is given by [400 – 402]

$$S = S_n\,(\rho_f/\rho_n)\,[1 + \tan\theta_v \tan\theta_{qp}] + (S_\phi\,\rho_f/\phi_0)\tan\theta_v \qquad ...(6.68)$$

where θ_v and θ_{qp} are Hall angles of both vortices and quasiparticles respectively and S_n and ρ_n are the normal-state Seeback coefficient and resistivity respectively and ρ_f is the flux flow resistivity. The second term of eq. (6.68) is related to the Nernst coefficient multiplied by the tangent of the Hall angle.

If the sample contains any extended structural defects that could serve as "guide rails" and guide the motion of vortices under angle ϕ away from the direction of $-\nabla_x T$, Ghamlouch and Aubin [403] have shown that eq. (6.68) becomes

$$S = S_n\,(\rho_f/\rho_n)\,\{[1 + \tan\theta_v \tan\theta_{qp}\cos\phi] + (S_\phi\,\rho_f/\phi_0)[\tan\theta_v + \tan\phi]\}\cos\phi \qquad ...(6.69)$$

Since both Hall angles are typically very small, the dominant term in eqs. (6.68) and (6.69) is the first term and to a good approximation

$$S = S_n\,(\rho_f/\rho_n) \qquad ...(6.70)$$

Obviously, the Seeback coefficient (thermopower) in the mixed state is thus closely related to the flux-flow resistivity through the relation

$$\rho_f = \rho_n\,B_z/B_{c_2} \propto \rho_n\,\xi^2/a_0^2 \qquad ...(6.71)$$

where $\xi = (\phi_0/2\pi B_{c_2})^{1/2}$, $a_0 = (\phi_0/B)^{1/2}$ and ϕ_0 is flux quantum. Just as the flux-flow resistivity reflects the anisotropy of the structure (broader transition range for more anisotropic materials), so does the Seeback coefficient. More two dimensional cuprates have a more extend temperature range where the seeback effect is finite.

The **Peltier effect** arises as a consequence of the heat current density being carried by the electric current density along an applied electric field in zero temperature gradient. So far, only one direct measurement of the Peltier coefficient in the mixed state of HTSC has been reported [404], and results confirm the expected behaviour. Moreover, this paper also provides an alternative physical picture of the Peltier effect in the mixed state that does not rely on the use of the Kelvin relation [404].

The **Ettingshausen effect** is the transverse equivalent of the Peltier effect except that the coefficient itself is defined in terms of the transverse temperature gradient rather than the transverse heat flow. With the electric current in the x-direction and the magnetic field along the z-axis, the Ettingshausen coefficient is defined as

$$\varepsilon = \nabla_y T/(J_x B_z), \text{ with } J_y = q_y = \nabla_x T = 0 \qquad \ldots(6.72)$$

Just as the Peltier effect is the basis of the thermoelectric refrigerator, the Ettingshausen serves as the means for thermomagnetic cooling [405].

Very few measurements of the Ettingshausen effect exist in either the conventional or HTSC cuprates.

The **Righi-Leduc effect** is the thermal analogue of the **Hall effect.** A heat current q_x arising because of the applied thermal gradient $-\nabla_x T$ in the presence of magnetic field in the z-direction gives rise to a transverse temperature gradient $\nabla_y T$ given by

$$\nabla_y T = R_L B_z q_x \qquad \ldots(6.73)$$

where R_L is the Righi-Leduc coefficient. Provided the carriers scatter elastically, i.e., within the regime of validity of the Wiedemann-Franz law the Righi-Leduc coefficient relates directly to the Hall coefficient R_H as

$$R_H = L_0 T R_L \qquad \ldots(6.74)$$

where L_0 is the free-electron Lorentz number $L_0 = 2.44 \times 10^{-8} V^2 K^{-2}$. In analogy to the Hall effect, one defines the Righi-Leduc angle θ as

$$\tan \theta = \Delta_y T/\nabla_x T \qquad \ldots(6.75)$$

In a normal metal the transverse thermal gradient $\nabla_y T$ results from the asymmetric scattering of the hot and cold carriers and the sign of the Righi-Leduc effect, i.e., whether the upper edge of the sample (along the y-axis) is hotter or colder than the lower edge, is determined by the sign of the carriers.

The formalism of the Righi-Leduc effect can be applied to the mixed state of a superconductor. However, although it is necessary for the vortices to move in order to observe the Hall effect, the Righi-Leduc effect does not require any motion of the vortices. The behaviour in the mixed state is rather dramatic but not well understood.

Only two reports mention attempts to measure the Righi-Leduc effect in HTSC cuprates, both noting that the effect is very small [406, 407]. A successful use of the Righi-Leduc sample geometry to study quasiparticle scattering on a pinned vortex structure and its relevance to the issue of heat transport in HTSC cuprates below T_c [408].

The **Nernst effect** refers to generation of the transverse electric field that results from a longitudinal thermal gradient in a perpendicular magnetic field and with no electric current in any direction. The Nernst coefficient (Q) is defined as

$$Q = E_y/(\nabla_x T B_z), \text{ with } J = 0 \qquad ...(6.76)$$

However, the Nernst effect is small in normal conductors. In contrast, the Nernst signal becomes quite significant and is easily detected in the mixed state of superconductors, and this is why the Nernst effect is the most frequently studied thermomagnetic coefficient. The Nernst effect is significant for its unequivocal tie with the motion of vortices. It marks the field and temperature range where pinning is absent or very weak and vortices move freely.

The Nernst effect has been measured frequently in both the conventional as well as in many HTSC cuprates. Nernst effect behaviour in cuprates is presented in Fig. 6.131, which shows the data of Ri et al. [409] for c-axis oriented films of $YBa_2Cu_3O_{7-\delta}$ and $Bi_2Sr_2CaCu_2O_{8+x}$ together with the respective flux-flow resistivities.

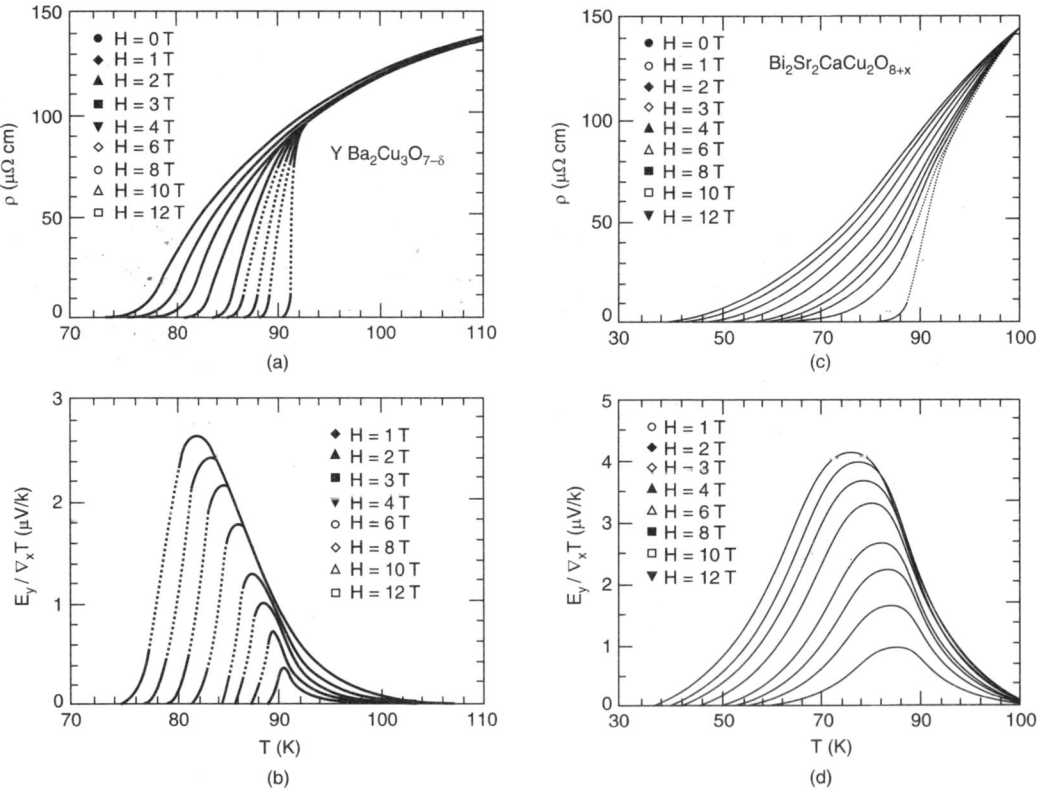

Fig. 6.131 The Nernst effect data and the associated flux-flow resistivities for c-axis oriented films of $YBa_2Cu_2O_{7-\delta}$ and $Bi_2Sr_2CaCu_2O_8$ [409]

Two important points have to be kept in mind when measuring the Nernst effect in highly anisotropic superconductors : (i) The concept of rigid, tubelike vortices need to be replaced by that of two-dimensional pancake-like vortices centered on individual superconducting planes and connected by Josephson vortices. Since Josephson vortices lack the normal core, the thermal force on such a vortex is zero and the Josephson vortices do not contribute to the Nernst effect.

(*ii*) Even if realized, the vortex antivortex unbinding above the Kosterlitz-Thouless temperature is not going to contribute to the Nernst effect in spite of the fact that it is an important resistive mechanism. Very simply, under the action of the Lorentz force the vortex and the antivortex move in opposite directions, and having opposite vorticities, they will generate resistive voltage. In contrast, under the influence of the thermal force the vortex and the antivortex both move down the thermal gradient, and the electric fields they generate cancel each other.

6.8 PSEUDOGAP

Numerous experiments have well established the fact that the underdoped HTSC cuprates exhibit a 'pseudogap' behaviour in both spin and charge degrees of freedom below a characteristic temperature T_c which can be well above the superconducting transition temperature T_c [410, 8, 411 – 419]. The presence of pseudogap in the spectrum of the elementary excitations of undoped and optimally doped HTSC cuprates implies that the electron subsystem in the normal phase is not the Fermi liquid and so the theoretical explanation of the pseudogap is recognised as the key point of understanding of HTSC cuprates. The pseudogap phenomena means the suppression of the low frequency spectral weight without any long range order. It appears below a characteristic temperature T_c^* in the underdoped regime and T_c^* approaches the superconducting transition temperature T_c near optimal doping where T_c is maximal. The pseudogap evolves smoothly into the superconducting gap. There are a lot of studies for the pseudogap from both experimental and theoretical point of view. However, the complete understanding of the phenomena remains to be obtained.

Experimental Evidence of Pseudogap

However, the existence of a Fermi surfaces in the metallic regime of high-T_c materials has been now firmly established experimentally [420 – 421]. Extensive high resolution angle-resolved photoemisssion, scanning tunnelling microscopy and inelastic neutron scattering studies have been carried out on $YBa_2Cu_3O_7$ (YBCO, $Bi_2Sr_2CaCu_2O_{8+\delta}$) (Bi-2212) and the $La_{2-x}Sr_xCuO_4$ (LSCO) systems [420 – 421]. These studies reveal the appearance of a pseudogap in the density of states before the critical temperature T_c is approached from above. This pseudogap also opens up below a characteristic temperature T_c^* for each system. Accordingly, the opening of this kind of pseudogap is considered to be a precusor effect of the superconducting phase fluctuations [422]. Furthermore, the fact that the pseudogap is accompanied by a predominantly incoherent quasiparticle spectra in some regions of the Brillouin zone is a very significant experimental discovery and may hold the crucial clue to the mechanism of HTSC cuprates.

The appearance of a sharp dispersionless peak in photoemission spectra in $Bi_2 Sr_2 Ca Cu_2O_{8+\delta}$ (T_C = 87 K) along *T-M-Z*, i.e. $(0, 0) - (\pi, 0) - (2\pi, 0)$ in the normal (105 K) and superconducting state has some special features [423]. The peak at *M* which is visible at about the energy 40 MeV in the normal state sharpens in the superconducting state. This result suggests strong local pair correlations in the normal state which amplify as one goes to the superconducting state [422].

Tunnelling spectra and neutron scattering results on YBCO also show such peaks around 40 MeV [424 – 426]. The pseudogap has also been observed by the measurement of nuclear spin lattice relaxation rate in bilayer 123 system and single layer 214 systems [427]. The temperature at which the pseudogap opens (T_{pg}) is determined at the peak of $1/T_1T$ versus T, where $1/T_1$ is the spin lattice relaxation rate. This peak temperature is comparable but larger than the superconducting T_c of systems such as $YBa_2Cu_4O_8$, $YBa_2Cu_3O_7$, $Tl_2Ba_2CuO_6$ and $La_{2-x}Sr_xCuO_4$ [Fig. 6.132] [427].

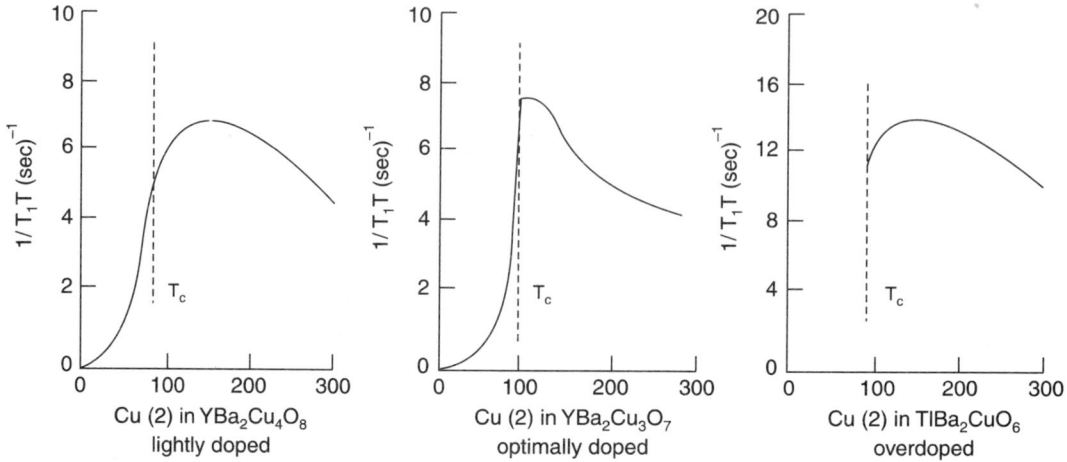

Fig. 6.132 Temperature dependence of $1/T_1T$ for Cu(2) sites in CuO_2 planes by NMR [428]

Ding et al. [421] have made a careful spectroscopic study (ARPES) of the pseudogap in the normal state of underdoped HTSC (mainly $Bi_2Sr_2CaCu_2O_{8+\delta}$) as a function of doping. The results are best summarised in the form of a schematic phase diagram shown in Fig. 6.133 [421].

The important point emerging out of these experimental studies is that T_c^* is larger for the underdoped samples than for higher doped samples. For the $T_c = 10$ K sample the pseudogap is quite large in the normal state with $T_c^* > 301$ K. The gapless Fermi surface is not recovered even at 301 K. Ding et al. [421] infer the shape and the size of Fermi surface from the minimum gap locus. It is found that the minimum gap locus for the $T_c = 10$ K sample is a large barrel as the optimal and lightly under doped samples. Its volume is smaller for the 10 K sample than the $T_c = 83$ K sample. This is an expected result.

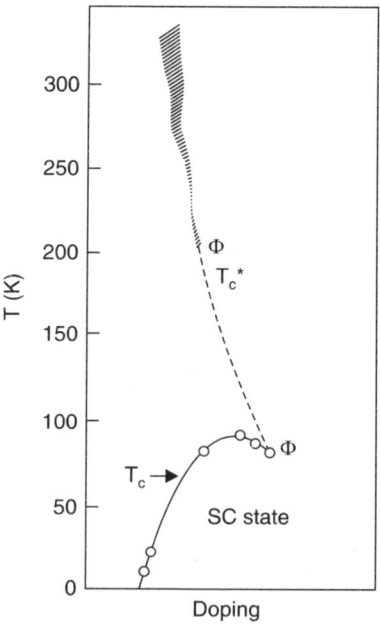

Fig. 6.133 Schematic phase diagram for Bi-2212 as a function of doping measured T_c is denoted by open circles and ϕ and T_c^* at which pseudogap closes. For $T_c = 10$ K, the symbol at 301 K is a lower bound at T_c'. The portion between T_c^* and T_c is the unusual "normal state" having a pseudogap in the excitation spectrum [429]

Norman et al. [423] found unusual dispersion and line space of the photoemission spectra in the superconducting state of $Bi_2Sr_2CaCu_2O_{8+\delta}$ (Bi-2212). Below T_c two features near the $(\pi, 0)$ point of the Brillouin zone are seen [Fig. 6.134]. There is a sharp peak at low energy and a pump at a higher binding energy. It is found that the sharp peak persists at low energy on moving towards the point (0, 0). On the other hand, the broad bump displays appreciable dispersion. This correlates significantly with the normal state dispersion.

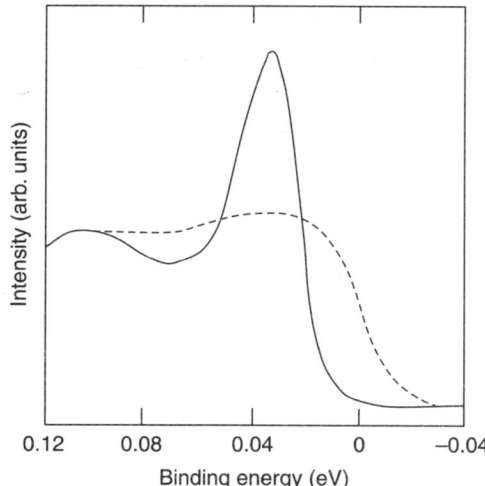

Fig. 6.134 Result at $(\pi, 0)$ point in the normal state (105 K, dashed line) and in the superconducting state (13 K – solid line) for Bi-2212 sample (T_c = 87 K) [423]

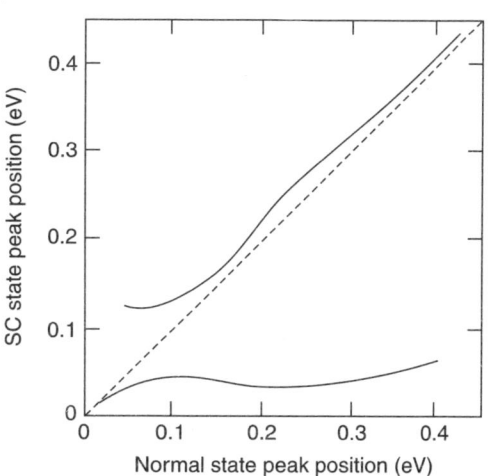

Fig. 6.135 Solid like is based on the experimental data of Norman et al. [423]. The dotted line denotes the normal state dispersion

The most important aspect of the result of Norman et al. [423] is the dispersion-less nature of the sharp peak observed by them. They further plotted the position of the low energy in the normal state. This shows a strong resemblance to the predicted feature of electrons interacting with a sharp mode in the superconducting state. An adapation of their plot is shown in Fig. 6.135 [423].

Norman et al. [423] conclude with the remark that they have found the presence of a persistent low energy phase in the photoemission spectra in Bi-2212 in the superconducting (SC) state. This exists over a large momentum range near the $\overline{M}$ $(\pi, 0)$ point. It is inferred from the dispersion and the higher binding energy hump as a function of momentum that the electrons in the SC state are interacting with a mode of resonant character with energy near 1.3 Δ_M where Δ_M is the SC energy gap at $\overline{M}$ point. This is close to 40 MeV. They further emphasise that the electron self-energy is dominated by electron-electron interaction. This is consistent with an electron-electron origin to the pairing.

Now, we present a brief account of the resonance peak observed by other techniques. Inelastic neutron scattering experiments first demonstrated the existence of a sharp collective mode ("resonance peak") in the superconducting state of optically doped (YBCO) *i.e.* YBa_2

Cu_3O_7 [426, 430]. Recent experiments show that this feature exists in the underdoped $YBa_2Cu_3O_{6+x}$ also [431, 432]. However, in the underdoped compounds the resonance peak appears even in the normal state but is considerably broadened [432]. This is smaller to the ARPES result on Bi-2212 obtained by Norman et al. [423].

Tunnelling experiments on Bi-2212 by Renner et al. [424] also finds the resonance peak but is not consistent with the picture in which w_{res} decreased with decreasing doping [424].

These experiments have directly shown the suppression of the low frequency density of states. In this connection, the gap-like structure which is similar to the normal state pseudogap is observed in vortex cores in the superconducting state under the high magnetic fields [424].

The impurity effects on the reduction of T_c reported by Tallon et al. [433] indicate the suppression of the low frequency density of states.

Loram et al. [434] have reported that the electronic specific heat is reduced well above T_c and T_c and the step height at T_c is extraordinarily small in under-doped cuprates. This indicates that the entropy has been already lost at rather high temperature.

Very recently Takahasi et al. [435] from ultrahigh resolution ($\Delta E = 7$ MeV) angle resolved (ARPES) and angle integrated (AIPES) photoemission spectroscopy performed on optimally doped $Bi_2Sr_2CaCu_2O_8$ (Bi-2212) and $La_{1.85}Sr_{0.15}CuO_4$ (LSCO) reported two different pseudogaps at E_F in HTSC cuprates. One is a 'small pseudogap' that is smoothly connected from the superconducting gap across T_c and other is a 'large pseudogap' which is seen as a depletion of the density of states near E_F and seems to be directly connected to the superconducting gap. Takahashi et al. [435] interpreted the small pseudogap originates in the superconducting pairing while the large pseudogap is closely related to the development of the antiferromagnetic correlation.

We have seen that the physical properties of HTSC cuprates are highly anisotropic as a consequence of their layered crystal structures. In addition to high T_c, HTSC cuprates are characterised by extraordinary high B_{c_2} which can even exceed 100 T at zero temperature. However, the useful magnetic field range of HTSC cuprates is limited by the irreversibility line well below the B_{c_2}. Above the irreversibility line losses are caused by flux motion which may be connected to activated depinning, a glass-like behaviour or flux line lattice melting [436]. The magnetisation of HTSC cuprates shows pronounced relaxation effects. The decay of the magnetisation may be again a consequence of glass-like behaviour or of large flux creep effects. The weak pinning in these materials is closely connected to the extremely short coherence length of less than 1 nm along the c-direction. Heavy ion irradiation can introduce additional pinning centres and as a result the irreversibility line is shifted to higher fields and temperatures. A further important consequence of the extremely short coherence length is the weak link behaviour of grain boundaries. Detailed investigations of artificial grain boundaries have indicated that low-angle grain boundaries are strongly coupled, whereas high-angle grain boundaries act as weak links. The critical current densities within single grains or in high quality thin films can exceed 10^6 A/cm^2 at 77 K and zero applied field. Generally, the critical current densities are very sensitive to magnetic fields applied along the crystallographic c-direction. An important anisotropy has also been found for the normal state resistivity of the HTSC cuprates. The in-plane resistivity typically shows a linear temperature dependence like a metal, whereas even semiconducting behaviour can result for the out-of-plane resistivity. Due to the small hole concentrations in the cuprates, the main carriers for heat transport are phonons. The thermal conductivity of HTSC cuprates is therefore orders of magnitude smaller than that of pure copper. The thermal conductivity of single crystals shows a considerable anisotropy with the smaller values parallel to the c-direction. The various experiments suggesting the relevance and continuity to superconductivity and one finds that the pseudogap phenomena are the precursor of the strong coupling superconductivity.

6.9 REPRESENTATIVE PHASE DIAGRAMS

Superconducting research is a highly active field and many superconducting systems have been investigated actively in various laboratories of the world. In this section, a group of selected phase diagrams that represent either the major industrial activities of HTSC cuprates today or HTSC cuprates that have attracted considerable attention [437 – 438].

(i) Ba-Y-Cu-O Systems

Since 1987, extensive research efforts have led to the accumulation of good amount of information concerning the crystal chemistry and phase equilibria of BYC system; this is particularly true for the subsolids relationships, as they are essential for the preparation of the HTSC compound $Ba_2 Y Cu_3 O_{6+x}$ in single phase, crystalline form. Phase diagrams of this HTSC system vary significantly depending on the annealing atmosphere (and the presence of CO_2).

Prepared in Air

Fig. 6.136 shows the complete phase diagram of the BaO $(BaCO_3) - \frac{1}{2} Y_2O_3 - CuO$ system at 950°C, first determined by Roth et al. [439]. $Y_2 Cu_2 O_5$ is the only binary oxide reported in the $\frac{1}{2} (Y_2O_3) - CuO$ system. Four phases were observed in $BaO(Ba CO_3) - \frac{1}{2} (Y_2O_3)$ subsystem. Barium rich $Ba_4Y_2O_7$ and $Ba_2Y_2O_5$ have been reported to oxycarbonates, with formulas of $Ba_4 Y_2 O_7 \cdot CO_2$ and $Ba_2 Y_2 O_5 \cdot 2CO_2$. We may note that when pure BaO is used instead of $BaCO_3$, the $Ba_4 Y_2 O_7$ and $Ba_2 Y_2 O_5$ compounds cannot be prepared. There are a total of three ternary oxides, including the HTSC cuprate $Ba_2 Y Cu_3 O_{6+x}$ (2:1:3) and ubiquitous impurity phase $Ba Y_2 CuO_5$ (1:2:1) that is known as the "green phase". Another barium-rich oxycarbonate solid solution region is known as "the other perovskite phase" and is bounded by 4:1:2, 5:1:3 and 3:1:2 compositions. The $Ba_2 Y Cu_3 O_{6+x}$ phase is known to exhibit an orthorhomic tetragonal phase transition depending on the oxygen content [440 – 441].

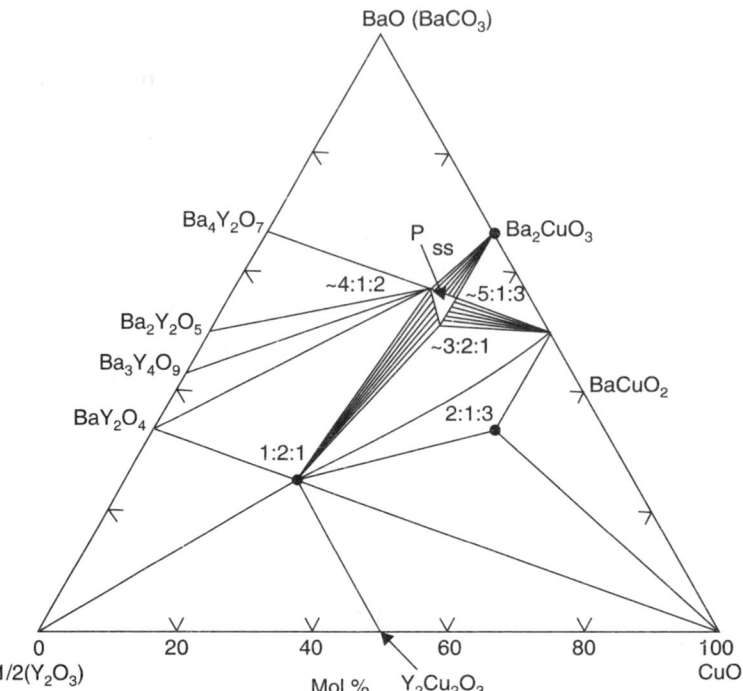

Fig. 6.136 Phase diagram of the ternary system BaO (BaCO$_3$) – 1/2 Y$_2$O$_3$ – CuO at 950°C [439]. Figure shows the positions of the Ba$_2$YCu$_3$O$_{6+x}$ (2:1:3) superconductor and the green phase BaY$_2$CuO$_5$ (1:2:1)

The high temperature reaction of CO_2 with $Ba_2YCu_3O_{6+x}$ is known to lead, upon completion, to $BaCO_3$ and copper and yttrium oxides [442]; however, oxycarbonates may form prior to complete carbonization [443]. Phase diagrams of the $Ba(O/CO_3)$-$Y(O/CO_3)$-$Cu(O/CO_3)$ pseudoternary system, as determined by powder X-ray diffraction after repeated f rings of samples at (a) 800°C and (b) 900°C in 1 atm oxygen containing ≈ 40 ppm CO_2, are shown in Figs. 6.137 and 6.138. We can easily see that in these two diagrams, the envelopes of oxycarbonate stability are shown bounded by dotted curves. The carbonate stability regions are near the BaO (BaCO$_3$) region and are bounded by the broken curves. The Ba-rich oxides in the BaO-Y(O)-Cu(O) system have a high affinity for CO_2, leading to the formation of oxycarbonates. Three oxycarbonates were identified: (1) The first phase is near the other "perovskite phase" composition (8:1:4) [439], with a homogeneity region extending Y, and a formula of $Ba_8 Y_{1+x} Cu_{4+z} (CO_3)_4 O_{11+w}$ with $u ≈ z$, $ox < 0.03$, $o < z < 0.04$. There is a pressure temperature equilibrium between oxygen gas and vacancies in the solid; for $0.05 < w < 0.08$ $x = z = 0$. (2) The tetragonal $Ba_3 Y_2 (CO_3)_u O_{6-u}$, $u ≈ 1$, phase is stable upto 960°C in purified oxygen. (3) The third carbonized phase is the important 2:1:3 phase. It appears that accommodation of the rather small "size" of cation at the Ba site is promoted by the carbonization. An extended solid solution $(Ba_{1-y} Y_y)_z Y Cu_3 (CO_3)_{0.2} O_{6.7+y}$, $o < y < 0.2$, was obtained. For $u ≈ 0.2$, $v ≈ 0.1$. the $Ba_2 Y Cu_3 (CO_3)_u O_{7-u-v}$ phase is tetragonal, with unit cell $a = 3.877 (2)$ A° and $c = 11.573 (3)$ A°. The structure of $Ba_2 Y Cu_3 (CO_3)_u O_{7-u-v}$ is basically derived from $Ba_2 Y Cu_3 O_7$ by replacing some of the oxygens by carbonate groups [444].

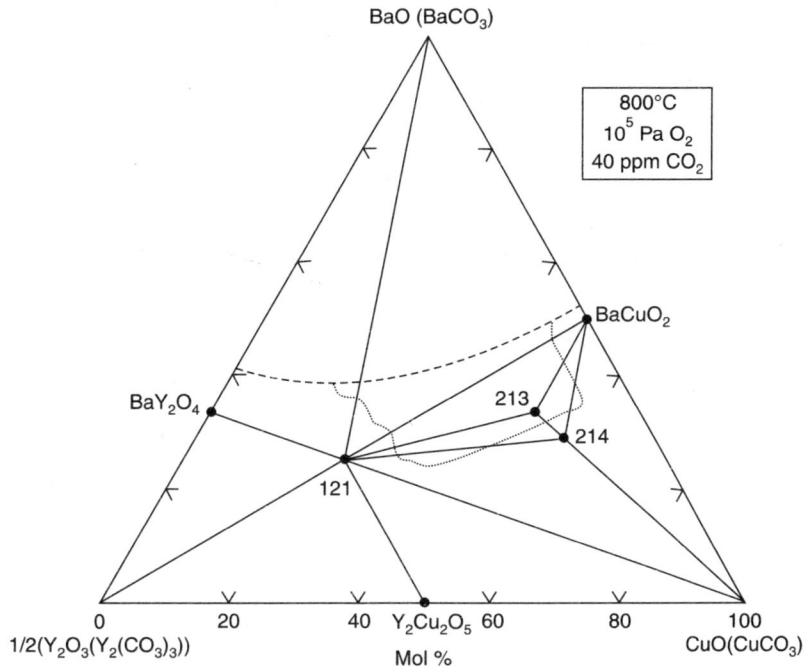

Fig. 6.137 Phase diagram of the system BaO (BaCO$_3$) – 1/2 Y$_2$O$_3$ (Y$_2$(CO$_3$)$_3$) – CuO at 800°C, 10^5 PaO$_2$, and 40 ppm CO$_2$ [443]

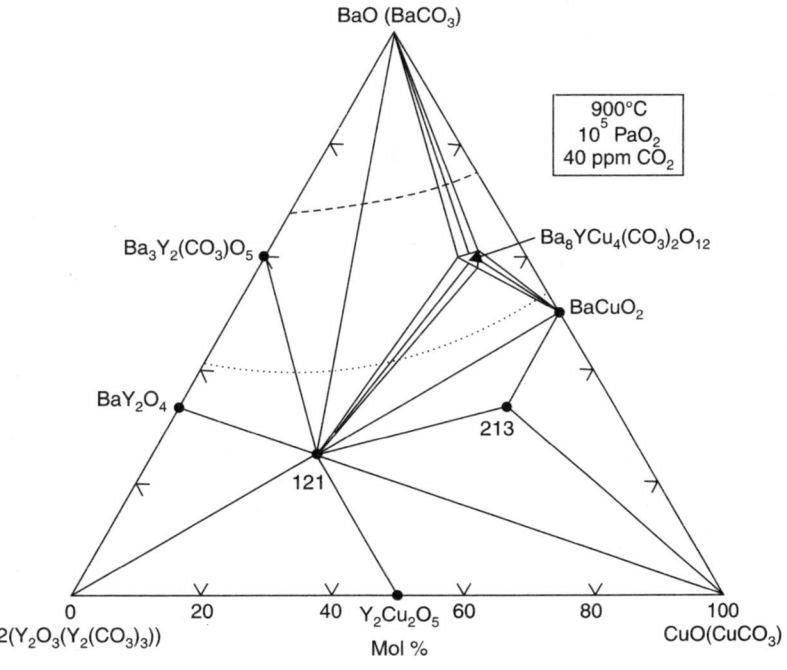

Fig. 6.138 Phase diagram of the system BaO (BaCO$_3$) – 1/2 Y$_2$O$_3$ (Y$_2$(CO$_3$)$_3$) – CuO at 900°C, 10^5 PaO$_2$ and 40 ppm CO$_2$ [443]

Prepared in Air in the Absence of CO_2

The diagram of the CO_2 free system prepared at $\approx 950°C$ is shown in Fig. 6.139. BaO_2 or $Ba(NO_3)_2$ was used as one of starting reagents. Instead of three ternary oxide phases as found in Fig. 6.137, five were found. The $Ba_2 Y Cu_3 O_{6+x}$ (2:1:3) and $Ba Y_2 CuO_5$ (1:2:1) phases are the same as those prepared in pure air, but the perovskite solid solution mentioned earlier [446] has become two point compounds with composition $Ba_4 Y Cu_3 O_{8.5}$ (4:1:3) and (5:1:2) was also found. In this diagram the $Ba_4 Y_2 O_7 CO_2$ and $Ba_2 Y_2 O_5 \cdot 2CO_2$ compounds are absent. A comparison of this phase diagram with phase diagram of Fig. 6.137 (prepared with $BaCO_3$) reveals that the tie-lines connecting the 2:1:3, 1:2:1, $BaCuO_2$ and perovskite phase are different. We may note that there are still uncertainties involved in these tie lines due to the hampered kinetics at 950°C.

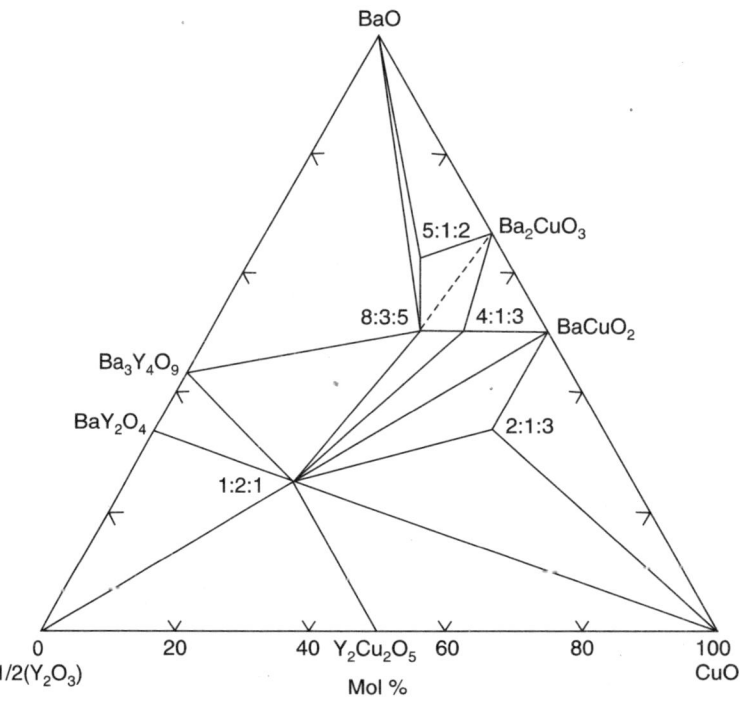

Fig. 6.139 Phase diagram of the system $BaO - 1/2 Y_2 O_3 - CuO$ at $\approx 950°C$ [445]

Prepared in Oxygen

The phase relations of the BYC system at lower temperatures of 800°C ($p(O_2) = 1$, 10^5 Pa $\leq P_{total} \leq 5 \times 10^5$ Pa, $p(CO_2) \approx 0.000$), and at 700°C, 10^5 Pa O_2 are shown schematically in Figs. 6.140 and 6.141 [447]. In these two diagrams, the 2:1:4 and 4:2:7 phases were indicated as stable at 10^5 Pa O_2 at 800°C. The 2:1:4 phase decomposes at $\geq 850°C$. However, at lower temperature, such as $\approx 700°C$, neither the 2:1:3 nor the 4:2:7 phase can formed, and a tie line exists between $Y_2 O_3 - Ba_2 Cu_3 O_{5+x}$. The $Ba_2 Cu_3 O_{5+x}$ phase only forms in the absence of CO_2. This phase is only stable at $Po_2 \approx 0.2$ if the CO_2 content is well below that of the ambient temperature. It decomposes after formation if heated in air.

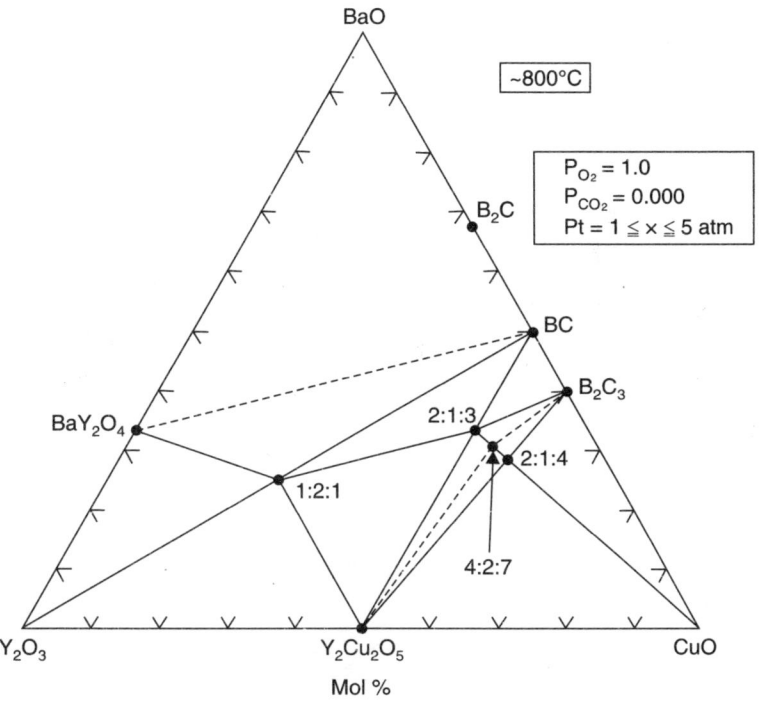

Fig. 6.140 Phase diagram of the system BaO – 1/2 Y_2O_3 – CuO. Schematic phase relations in O_2 at 800°C (P_{O_2} = 1.0 and 1 atm ≤ P_{total} ≤ 5 atom, P_{CO_2} ≤ 0.000) [447]

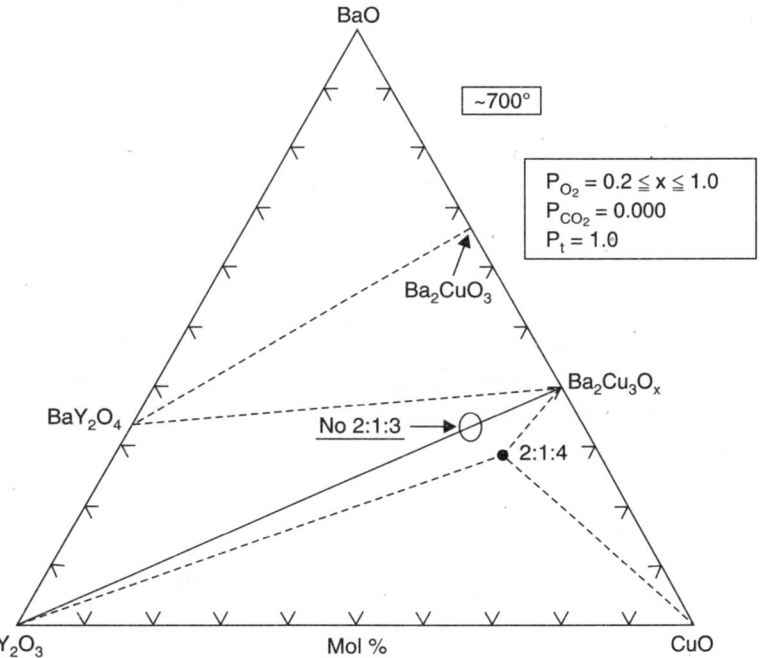

Fig. 6.141 Phase diagram of the system BaO – 1/2 Y_2O_3 – CuO. Schematic phase relations in O_2 at ≈ 700°C and 1 atm O_2 [447]

Effect of Oxygen Partial Pressure

Fig. 6.142 shows for the 213, 214, and 427 phase regions of the BYC system vs. the oxygen partial pressure plotted vs. inverse temperature over the range between 500 and 1000°C and 10^{-2} Pa $< p(O_2) < 10^7$ Pa [448]. The approximate boundaries between the phases are indicated by the dotted curves.

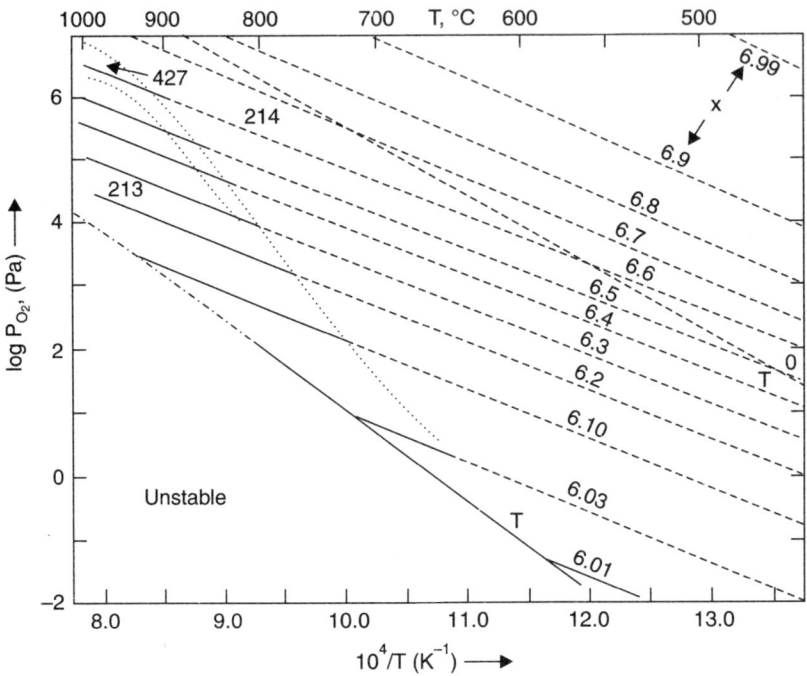

Fig. 6.142 Oxygen pressure P_{O_2} vs $1/T$ for oxygen contents x in the range 6.01 to 6.99 near 2:1:3 (–) and 2:1:4 (– – –) compositions in the BaO – 1/2 Y_2O_3–CuO system [448]. The decomposition (–) and orthorhombic-to-tetragonal transition (– – –) lines are shown as well as the phase transition boundaries (......). The orthorhombic phase exists above the (– – –) line. Lines of constant oxygen content are shown by solid line in the 2:1:3 stability region and dashed lines in the 2:1:4 region where 1:2:3 is metastable

The approximate boundaries between the phases are indicated by the dotted curves. Included in this plot for the 213 system, the decomposition line (thick solid line [449]), and the orthorhombic tetragonal transition line (both dashed line) [450, 451], lines of constant oxygen content (full lines) in the 213 stability region, and dashed lines where 123 is metastable. The, $2\,\text{CuO} \rightarrow \text{Cu}_2\text{O} + \frac{1}{2}\,\text{O}_2$ phase boundary above 900°C (alternating dots and dashes) lies close to the 123 decomposition boundary. The stability boundary between 213 and 214 phases continues to decrease to lower oxygen partial pressure as temperature decreases. This 213/214 phase boundary occurs at considerably higher temperatures than that of the 213 tetragonal/orthorhombic transition, indicating that the fully oxidized orthorhombic 213 superconductor is not thermodynamically stable at any temperature or pressure. The slope of the 213/214 phase boundary is so large that at lower temperature (< 600°C) the 213/214 boundary will intersect the 213 decomposition line. The oxygen-depleted tetragonal 213 phase is unstable at all oxygen pressures below 600°C.

Liquidus Diagrams

Melt processing investigations of this HTSC material for viable commercial applications constitute a major activity within the HTSC research community. The liquids information for the Ba-Y-Cu-O system is critical for crystal growth and melt processing. The primary phase field for $Ba_2 Y Cu_3O_{6+x}$, and univariant reactions in the phase diagram near the CuO-rich corner have been reported [452 – 455].

(ii) Ba-R-Cu-O Systems (R → Lanthanide)

The discovery that the substitution of the most of lanthanide (3+) ions, R, for Y also produced a high temperature superconductor with $T_c \approx 90$ K has provided numerous alternative materials for investigations of possible desirable properties [456]. Proceeding from the La system, which has the largest ionic size of R, toward the Er system with a smaller ionic size, a general trend of phase formation, solid solution formulation and phase relationship was found to be correlated with the ionic size of R. The ternary phase compatibility diagrams of the systems BaO (Ba CO_3) – $\frac{1}{2} Y_2O_3$–CuO and BaO ($BaCO_3$) $\frac{1}{2} R_2O_3$–CuO in the vicinity of the CuO corners (most relevant to the processing of the high-T_c materials), where R = La, Nd, Sm, Eu, Gd, Er, are shown schematically in Fig. 6.143 (a) to (f) [457]. Features of the progressive changes in the

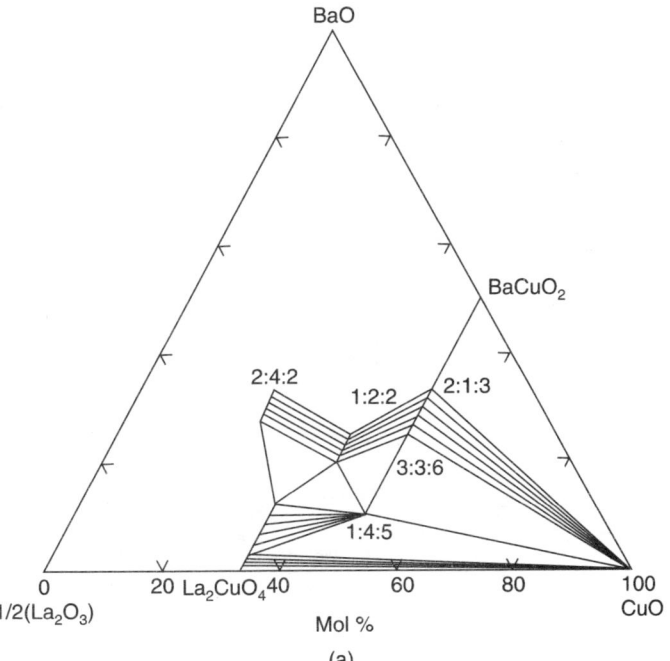

(a)

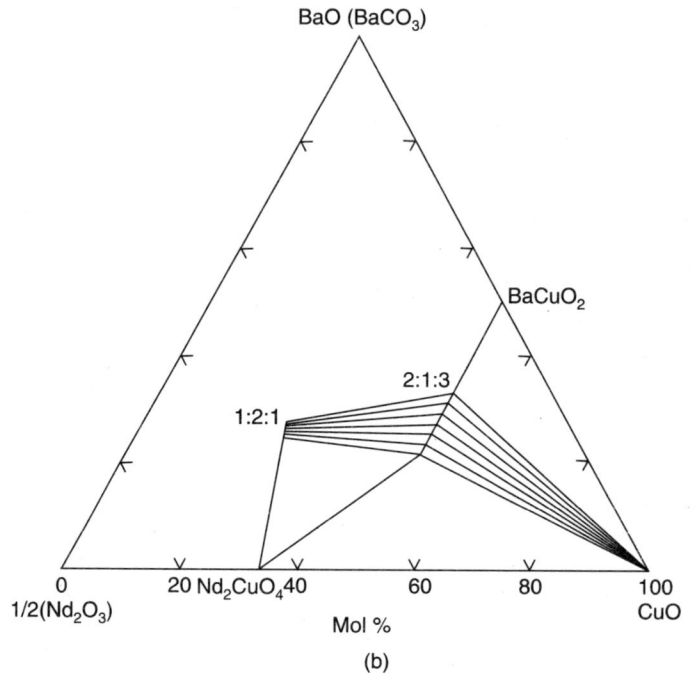

(b)

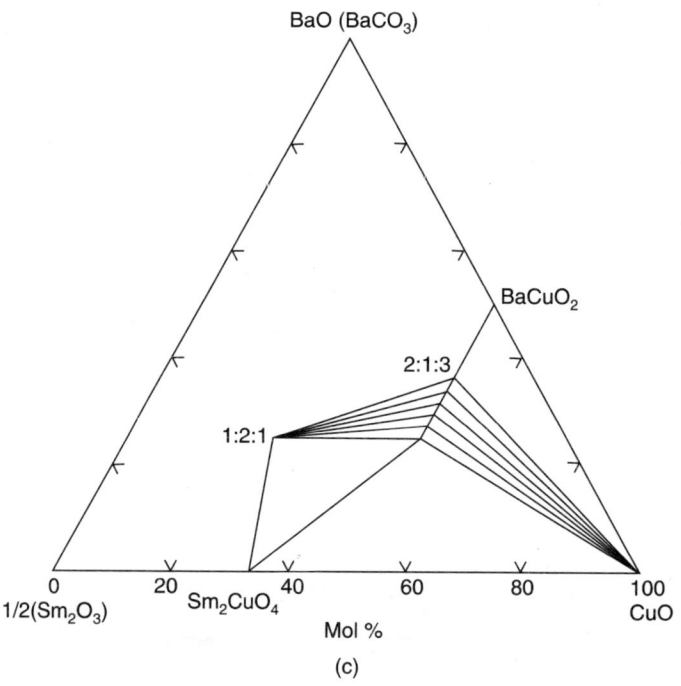

(c)

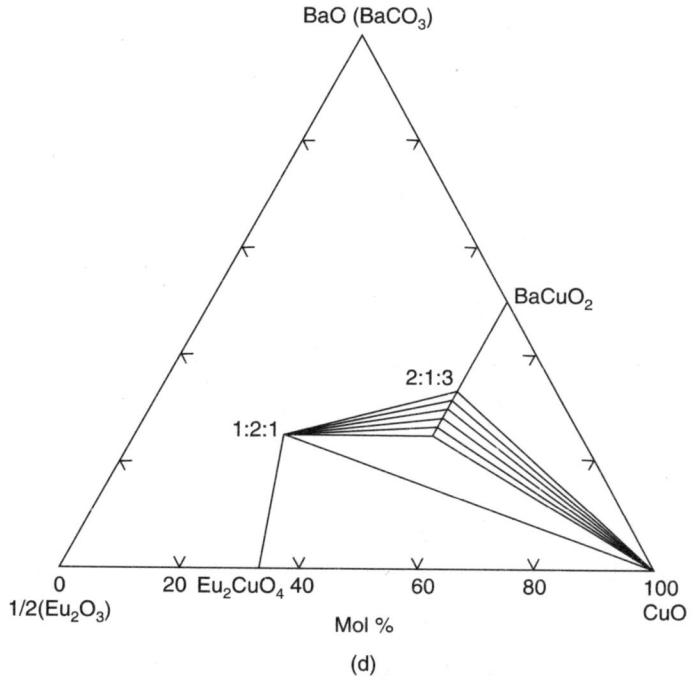

(d)

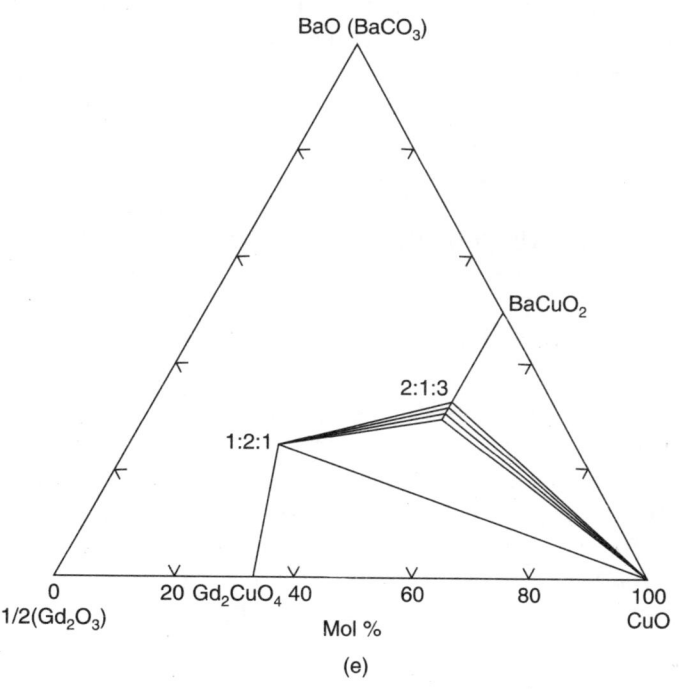

(e)

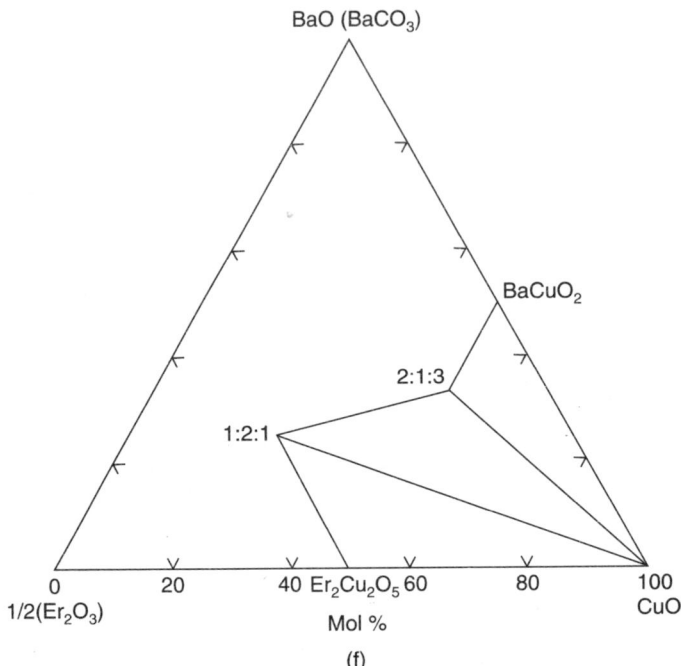

Fig. 6.143 Sub-solidus phase equilibria of the BaO – 1/2 R_2O_3 – CuO system near the CuO corner : (a) R = La (b) Nd (c) R = Sm (d) R = Eu (e) R = Gd and (f) R = Er [457]

appearance of these ternary diagrams near the CuO corner include the following: (a) The La system has the largest number of ternary compounds and solid-solution series; this number decreases as the size of R decreases (b) The superconductor phase, $Ba_2RCu_3O_{6+x}$, for the first half of the lanthanide family, i.e. R = La, Nd, Sm, Eu and Gd which are relatively larger in size, exhibit a solid solution of $Ba_{2-z}R_{1+z}Cu_3O_{6+x}$ with a range of formation that decreases as the size of R decreases; this solid solution region terminates at Dy and beyond, where the superconductor phase assumes a point stoichiometry. The size compatibility between Ba^{2+} and R^{3+} is a predominant factor governing the formation of this solid solution. As the mismatch between R^{3+} and Ba^{2+} increases, the range of substitution decreases. The approximate upper limit of the solid solution range of z of $Ba_{2-z}R_{1+z}Cu_3O_{6+x}$ are La : 07, Nd : 0.7, Sm = 0.7, Eu; 0.5 and Gd: 0.2 (c) A trend is observed regarding the tie-line connections between BaR_2CuO_5, CuO the superconductor phases $Ba_{2-z}R_{1-z}Cu_3O_{6+x}$, and the binary phase R_2CuO_4, or $R_2Cu_2O_5$. We may note that the binary phase R_2CuO_4 is replaced by the binary phase $R_2Cu_2O_5$ after the tie-line connection changes.

More complete diagrams of the systems with R = La, Nd are shown in Figs. 6.144 and 6.145, respectively. It is within the Ba-La-Cu-O system that the first 30 K HTSC phase in polycrystalline form, $Ba_xLa_{5-x}Cu_5O_{5(3-y)}$, was discovered by Bednorz and Muller [1]. The isothermal section of the Ba-La-Cu-O system [458] shows a total of five solid solutions : $Ba_{2+x}La_{4-2x}Cu_{2-x}O_{10-2x}$ [242], $BaLa_4Cu_5O_{13+x}$ (145), $Ba_xLa_{2-x}CuO_{4-(x/2)+\delta}$ (021) and $Ba_{1+x}La_{2-x}Cu_2O_{6-(x/2)}$, (122) and a solid solution $Ba_{3+x}La_{3-x}Cu_6O_{14\pm x}$ that spans from the 213 composition to the 336 composition. We may note that the limits of most of these solid solutions have not been quantified. The solubility limits for $Ba_{2+x}La_{4-2x}Cu_{2-z}O_{10-2x}$ were reported to be $0.15 \leq x \leq 0.25$. The tie-line connectivity of the figure is schematic.

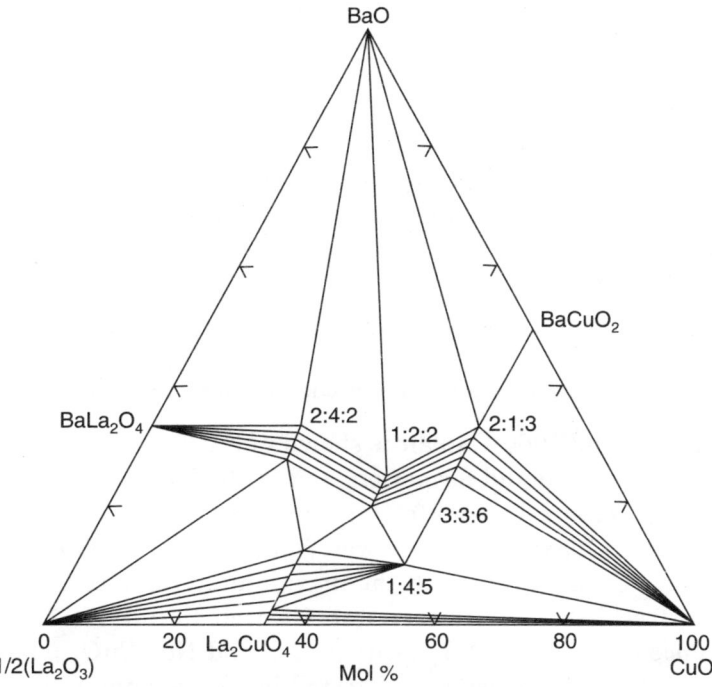

Fig. 6.144 Phase diagram for the system BaO – 1/2 La$_2$O$_3$ – CuO at 950°C in air [458]

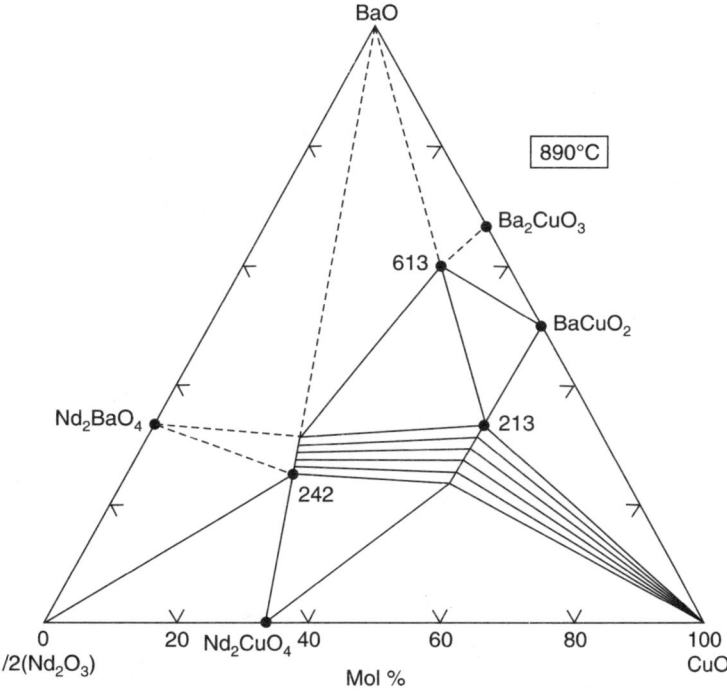

Fig. 6.145 The subsolidus phase diagram of the BaO – 1/2 Nd$_2$O$_3$ – CuO system around the Ba$_{2-x}$Nd$_{1+x}$Cu$_3$O$_z$ compound in air at 890°C [459]

Fig. 6.145 shows the ternary diagram of the Ba-Nd-Cu-O system at 890°C in air. In the barium rich region, samples were annealed in air with $CO_2 < 3$ ppm. A total of three phases were found in this system. In addition to the solid solution of the superconductor (213), $Ba_{2-x} Nd_{1-x} Cu_3 O_{7-\delta}$ ($0.04 \leq x \leq 0.6$) and $Ba_{2+x} Nd_{4-2x} Cu_{2-x} O_{10-2x}$ (242) (x is negligible), a 6:1:3 phase (orthorhombic: $a = 3.886$ (2), $b = 3.984$ (2) and $c = 13.001$ (5) A°) is also found. The existence of the $Ba_{2-x} Nd_{1+x} Cu_3 O_z - Ba_{2+x} Nd_{4-2x} Cu_{2-x} O_{10-2x}$ two phase field enables one to select a starting composition that leads to composite superconductors of these two phases two phases that are completely devoid of the minor second phases that segregate at $Ba_{2-x} Nd_{1+x} Cu_3 O_z$ grain boundaries after a solid state sintering.

Goodilin et al. [460] conducted a melting study of the $Ba_{2-x} Nd_{1+x} Cu_3 O_z$ solid solution. A sequence of schematic quasiternary sections of the Cu-rich corner of the $BaO - \frac{1}{2} Nd_2 O_3 -$ CuO system between 970 and 1060°C in air is shown in Fig. 6.146 (a), (g). Fig. 6.146 (a) shows the phase relationships below liquid formation, and at the bottom Fig. (b) shows the onset of liquid formation in the system. Fig. (c) to (f) shows that the stability region of $Ba_{2-x} Nd_{1+x} Cu_3 O_z$ is extensive and its maximum range was found in air at 995°C ($0 \leq x \leq 1$). At 995 – 1045°C the $Ba_{2-x} Nd_{1+x} Cu_3 O_z$ phase with the maximum x value coexists in the copper rich corner with the Ba-free $Nd_2 CuO_4$ phase and Cu-rich liquid. At 1060°C (Fig. (g)); the simultaneous presence of the $Ba_{2+x} Nd_{4-2x} Cu_{2-x} O_{10-2x}$ and $Nd_2 CuO_4$ phases was found, and only the $Ba_{2-x} Nd_{4-x} Cu_{2-x} O_{10-2x}$ phase was detected in equilibrium with Nd-poor $Ba_{2-x} Nd_{1+x} Cu_3 O_z$ solid solution.

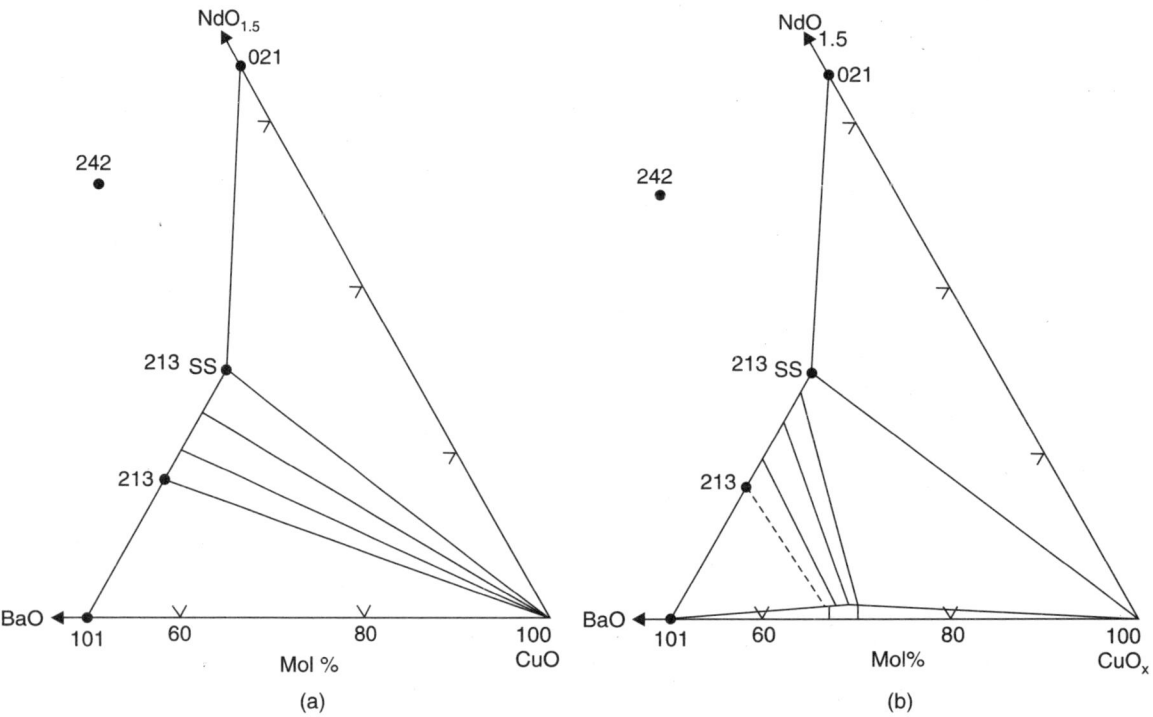

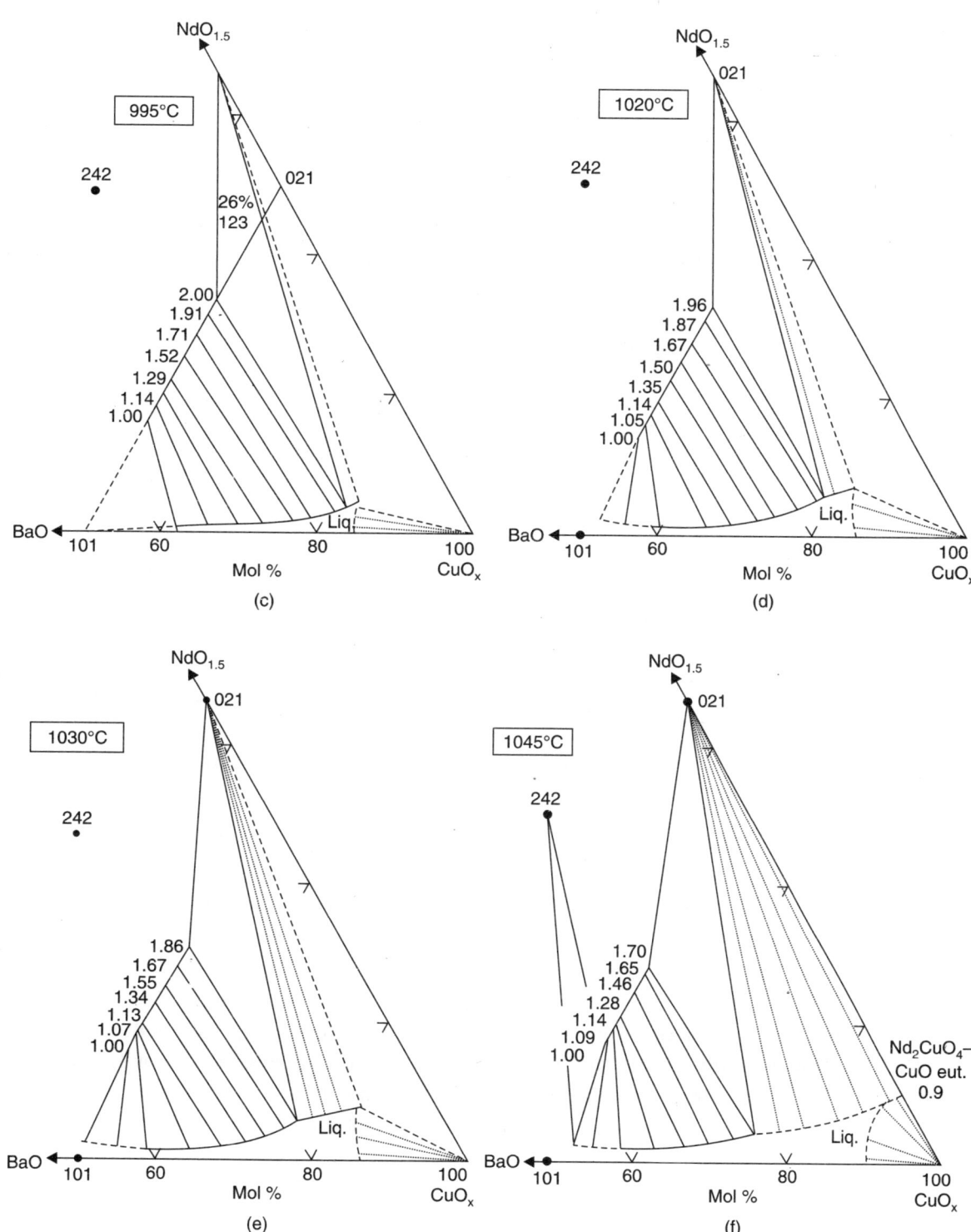

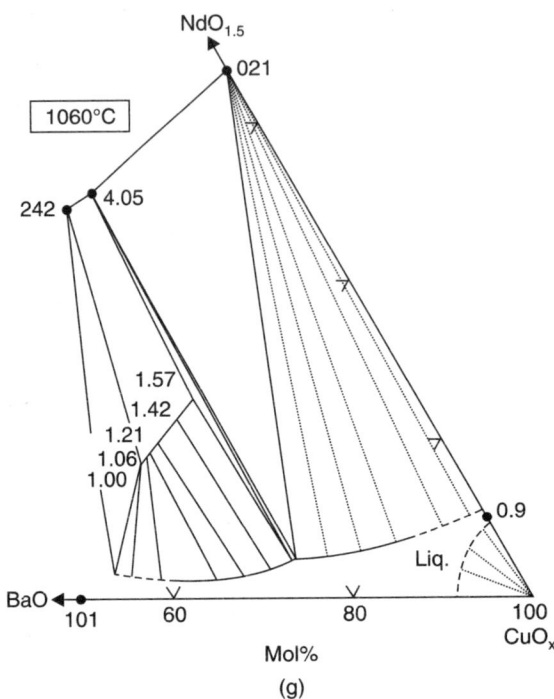

Fig. 6.146 Schematic quasiternary sections of Cu-rich corner of the BaO – 1/2 Nd$_2$O$_3$ – CuO system phase diagram in air at different temperatures (a) phase relations below melting (b) beginning of liquid formation in the system (c) 995°C (d) 1020°C (e) 1030°C (f) 1045°C (g) 1060°C [460]. The numbers on (e), (f) and (g) correspond to the x-value in the solid solution Ba$_{2-x}$Nd$_{1-x}$Cu$_3$O$_z$.

(iii) (La, Sr)-Cu-O Systems

In this system, the composition, La$_{2-x}$Sr$_x$CuO$_{4-y}$ in polycrystalline form was found to exhibit onset T_c temperatures in the 20 K range. The dependence of the tetragonal to the orthorhombic phase transition temperature on the hole concentration of oxygen-deficient La$_{2-x}$Sr$_x$CuO$_{4-y}$ is shown in Fig. 6.147. Fig. 6.148 shows a schematic phase diagram of the SrO-La$_2$O$_3$-CuO system at 980°C in air [462]. The precise solid solution limit has not been determined. Extensive solid solutions were observed for quite a few phases. The superconducting solid solution series La$_{2-x}$Sr$_x$CuO$_4$ that crystallized with the K$_2$NiF$_4$ structure was confirmed to exist with complete substitution of $0 \leq x \leq 1$. This solid solution is in equilibrium with all the three end members of this ternary oxide system, the 1:4:5 phase and another solid solution series, La$_{2-x}$Sr$_{1+x}$Cu$_2$O$_{6-x/2+\delta}$ (1:2:2). The 1:2:2 phase is an end member of the La$_{2-x}$Sr$_{1+x}$Cu$_2$O$_{6-x/2+\delta}$ series, which has the Sr$_3$Ti$_2$O$_7$ structure, and has a smaller solid solution range of $0 \leq x \leq 0.14$.

High Temperature Superconducting Cuprates: General Survey

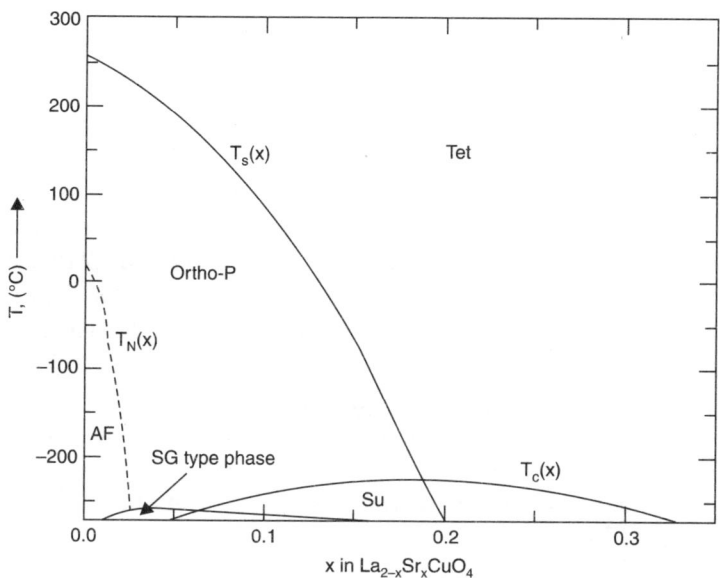

Fig. 6.147 Temperature diagram of the system $SrO-La_2O_3-CuO$, section of $La_{2-x}Sr_xCuO_4$ [461]. T_c refers to the superconducting transition temperatures of the orthorhombic paramagnetic phase and the tetragonal phase, and $T_N \rightarrow$ the Neel temperature between the orthorhombic antiferromagnetic phase and the paramagnetic phase; ortho $-P$ = orthorhombic paramagnetic phase; AF = orthorhombic antiferromagnetic phase; Tet = tetragonal phase; SG = spin glass type phase; Su = superconductor phase; T_S = temperature of the orthorhombic $-P$ and the tetragonal phases

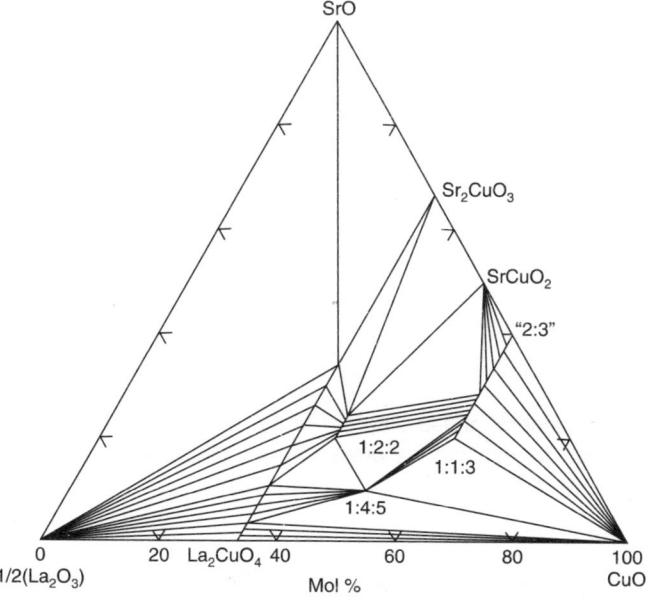

Fig. 6.148 Phase diagram of the system $SrO-La_2O_3-CuO$ at 980°C in [462]

(iv) (Bi, Pb)-Sr-Ca-Cu-O Systems

The ability of bismuth to form lamellar oxides was recognized in 1987, and the single layer Bi-Sr-Cu-O compound having a critical superconducting temperature ranging from 9 to 22 K was discovered by Michel et al. in 1987 [463]. Superconducting bismuth cuprates were later reported to form a family of layered structure phases with ideal formulas $Bi_2 Sr_2 Ca_{n-1} Cu_n O_{4+2n}$ with $n = 1, 2$ and 3, depending on the number of $(CuO_2)_n$ layers. These phases exhibit variations of cation ratio. Three well known superconductor phases in the BSCCO system are usually referred to as the single-layered 2201 (Bi:Sr:Ca:Cu) phase, the two layered 2212 (Bi:Sr:Ca:Cu) phase, and three-layered 2223 (Bi:Sr:Ca:Cu) phase. Among these three phases, the two most widely investigated are the 80 K Pb-free two layered 2212 phase [464 – 466] and the 110 K Pb-doped 3-layered 2223 ((Bi, Pb):Sr:Ca:Cu) phase [467 – 469].

The interpretations of phase relations in a quaternary system is the same as those in the binary and ternary systems. However, in the quaternary system it is more difficult to present the relations in a simple form due to the presence of many variables. Fig. 6.149 shows the tetrahedral system used to illustrate the position of the three superconductors in the Pb-free BSCCO system [470].

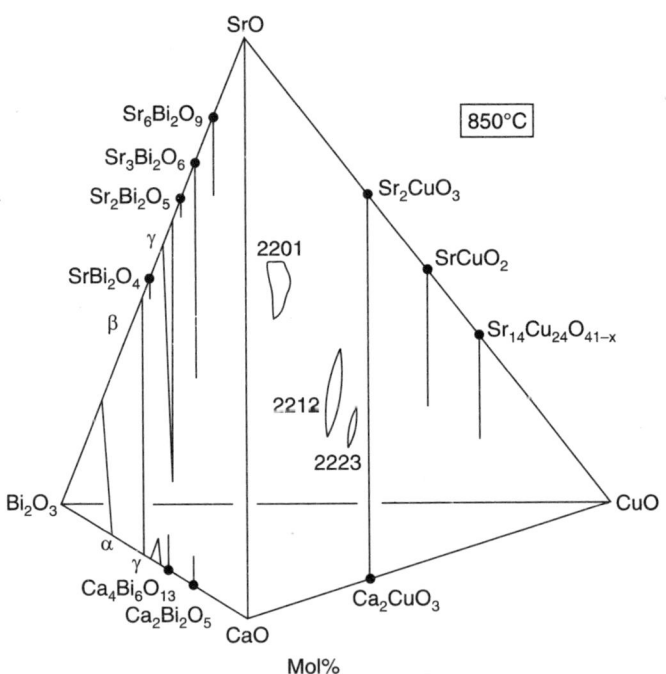

Fig. 6.149 Tetrahedral phase diagram of the Bi_2O_3-SrO-CaO-CuO system of 850°C exhibiting the presence of various phases and locations of 2201, 2212 and 2223 phases [470]

Phase Diagrams for One CuO_2 Layer

The phase equilibria of the Bi-Sr-Cu-O system at 875 – 925°C is shown in Fig. 6.150 [471]. Four ternary oxide compounds were found. We may note the coexistence of the solid solution $Bi_{2.2-x}Sr_{1.8+x}CuO_z$ (usually referred as the Raveau 11905 phase) and the $Bi_2Sr_2CuO_6$ [2201]

phase. It is this 11905 that is the single layered 9–20 K superconductor, whereas the 2201 phase does not superconduct. The Raveau phase and 2201 phase are in equilibrium with each other. The Raveau solid solution was found for the approximate range $0 \le x \le 0.15$ for $Sr_{1.8-x}Bi_{2.2+x}Cu_{1\pm x/2}O_z$. This phase is structurally similar to the $n = 1$ member of $Sr_2Bi_2Ca_{n-1}Cu_nO_{2n+4}$. The 2201 phase was found to be monoclinic and CuO deficient (< 1 mol %) and only has a small homogeneity region. The observed X-ray diffraction of this phase does not match that of the Raveau phase. In the literature, the 2201 symbol is commonly used in place of Raveau phase and may be interpreted as a part of the extended single phase region of the Raveau phase (Fig. 6.149). The Raveau phase was found to melt at 870°C [472]. The other two ternary compounds viz. $Bi_4Sr_8Cu_5O_x$ and $Bi_2Sr_3Cu_2O_x$ are not superconductors.

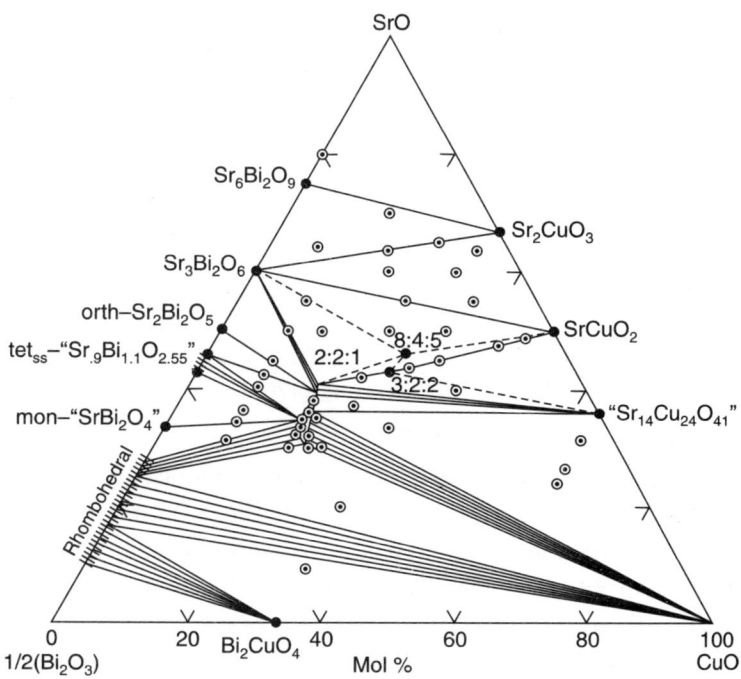

Fig. 6.150 Phase diagram of the system Bi_2O_3-SrO-CuO at 875-925°C in air [471]

Phase Diagrams for Two CuO_2 Layers

Solid solution region. Various investigations of the solid solution boundaries of the 2212 phase have appeared in the literature. Fig. 6.151 summarizes some of the published data and indicates the range in size and shape reported for the solid solution field upto 1993 [473 – 478]. Disagreements among the reported solid solution regions indicate the complicated nature of the equilibria in this system. We may note that many factors influence the experimental results, including the sensitivity of the phase assemblages to processing conditions, the sluggish kinetics of phase formation, and the very closely spaced phase stability fields of the high-T_c phases.

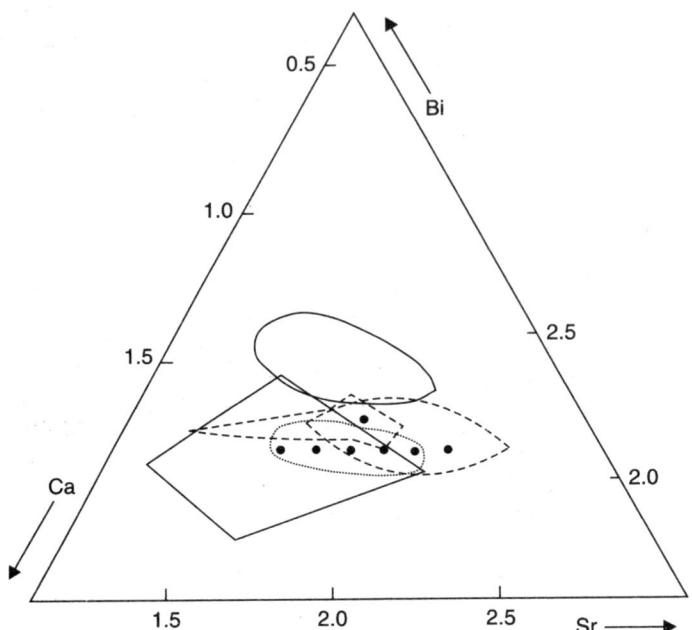

Fig. 6.151 Summary of various determinations of the single-phase solid solution region of the 2212 phase in the Bi_2O_3-SrO-CaO-CuO system [473]. The cross-section shown corresponds to the formula $Bi_{2+y} Sr_{3-y-x} Ca_x Cu_2 O_{8+x}$ [473, (• • •)] 850°C, air [474, (. . .) 865°C, oxygen]; [475, (- - -) 850°C, air], [476, (- - -) 860°C, air], [477, (–) varying temperature], [478, (–) 830°C, air]

Fig. 6.152 shows the temperature stoichiometry dependence of the $Bi_{2.8} Sr_{3-y} Ca_y Cu_2 O_{8+x}$ composition [479]. We can see that in this projection, the 2212 phase exhibits an extended single phase region with variable Sr, Ca, Bi and oxygen content. This single-phase region approximates a half-moon shape, with the greatest width of y at about 800°C. With the increasing temperature the extension of the single-phase region shrinks and is shifted to Sr-richer compositions. We may note that the four-phase regions outside the single phase area vary depending on the y value of the formula. This high-temperature annealing of the Ca-rich 2212 phase leads to precipitation of Ca_2CuO_3 and a liquid, whereas annealing of the Sr-rich 2212 phase leads to the formation of $Bi_2Sr_3Cu_2O_{88}$ cuprates, and liquid. At temperatures above 870°C the Sr-rich 2212-phase decomposes. The ratio Sr:Ca of the critical composition of the 2212 phase was determined to be about 2:1 ($Bi_{2.18} Sr_2 CaCu_2 O_{8+\delta}$). At the maximum temperature, the 2212 phase melts to 2201 + cuprates $\rightarrow L$.

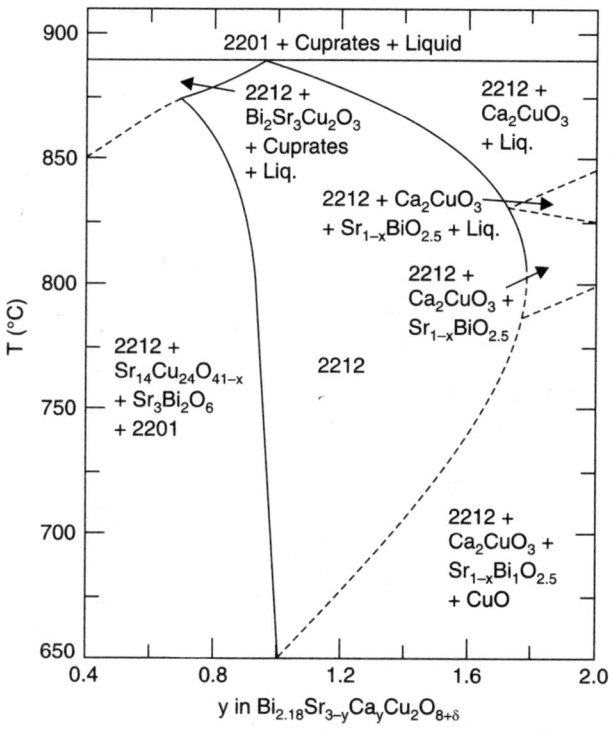

Fig. 6.152 Temperature (Sr, Ca) stoichiometry phase relations for the 2212 phase with composition $Bi_{2.18} Sr_{3-y} Ca_y Cu_2O_{8+\delta}$. 2201 is used to represent the Raveau superconducting phase [479]. Initial melts of four phase equilibrium volumes containing the 2212 phase are given in Table 6.24

Table 6.24. Initial melts of four-phase equilibrium volumes containing the 2212 phase of BSCCO (See Fig. 6.152)

Sample ID	Four-phase equilibria	DTA $T(°C)$	Melt Composition			
			Bi	Sr	Ca	Cu
1	2212–2110–119x5–CuO	25	41.7	16.1	22.3	19.9
2	2212–0x21–2110–CuO	830	30.4	7.4	26.6	35.6
3	2212–0x21–CaO–2110	838	35.2	20.8	24.6	19.4
4	2212–014x24–CuO–119x5	856	27.6	25.1	8.8	38.5
5	2212–2110–119x5–CaO	856	37.8	18.0	19.6	24.6
6	2212–014x24–0x11–CuO	861	26.2	24.2	22.3	27.3
7	2212–014x24–0x11–0x21	863	25.3	23.6	18.9	32.2
8	2212–119x5–2310–CaO	873	29.8	23.2	21.9	25.1

...(contd.)

9	2212–4805–0x21–CuO	875	25.9	21.8	14.2	38.1
10	2212–2201–2310–014x24	877	23.9	32.8	11.7	31.6
11	2212–4805–0x11–CuO	877	26.4	26.3	12.7	34.6
12	2212–119x5–2201–2310	877	32.8	30.7	4.5	32.0
13	2212–119x5–014x24–2201	878	27.7	26.9	2.5	42.9
14	2212–0x21–014x24–2310	885	28.7	21.4	20.8	29.1
15	2212–4805–0x11–0x21	887	29.9	24.7	14.9	30.5
16	2212–2310–0x21–CaO	889	26.9	24.7	17.6	30.8

Primary Crystallization Field of the 2212 Phase

The initial melt compositions for the 16 four-phase volumes involving 2212 phase are shown in Table 6.24. The Bi concentration was reported to cover a range from 24 to 42%, Sr from 7 to 33%, Ca from 2 to 27%, and Cu from 19 to 43%. Figs. 6.153 and 6.154 shows the two views of three

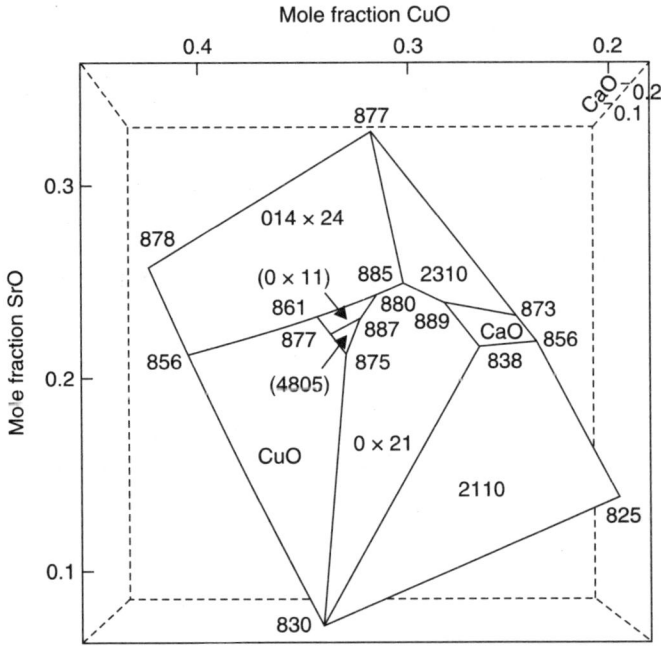

Fig. 6.153 Primary crystallization field of the Bi-Sr-Ca-Cu-O 2212 phase using orthogonal coordinate [480]

dimensional pictorialization of the 2212 crystallization volume. In these figures approximate temperatures of the four-phase initial melting equilibria are also indicated. We can see that the volumes in these diagrams are expressed in cartesian coordinates. The polygonal areas on the surface of the crystallization volume have been labelled according to the presence of the second solid in equilibrium with 2212. The maximum melting temperature in Fig. 6.153 is

889°C, corresponding to the initial melting equilibrium for the four-phase volume 2212-2310-O_x21-CaO. The overall wedge-like shape of the volume is apparent in Fig. 6.154. In the lowest melting region (low SrO), the volume is very thin and terminates along a sharp edge. The higher SrO end of the volume is much wider and is terminated by the 2201 crystallization surface. Inside the volume, the 2212 phase is in equilibrium with liquid. At the corners, it is in equilibrium with three phases and with L. At the edges, 2212 is in equilibrium with two phases plus L, and on each face 2212 is in equilibrium with L and the labelled phase.

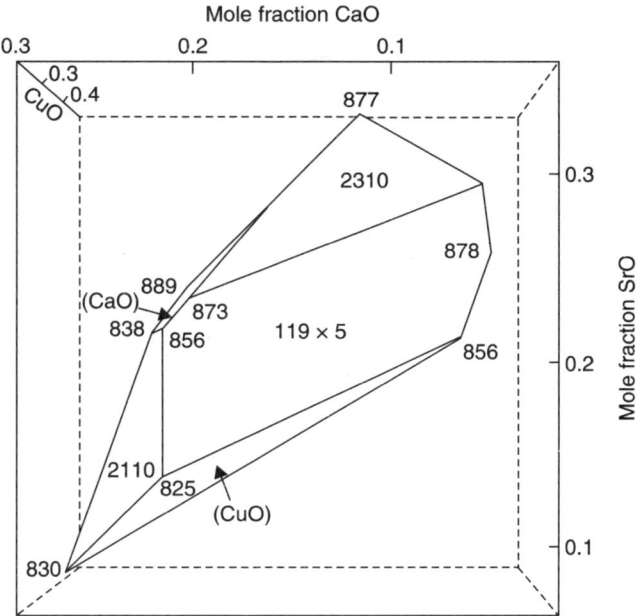

Fig. 6.154 View of the primary crystallization field of the Bi-Sr-Ca-Cu-O 2212 phase approximately orthogonal to the view shown in Fig. 152 [480]

Effect of silver (Ag) addition. The use of silver (Ag) as an additive is widespread in BSCCO tape and wire processing. Except at very high temperature, the presence of silver does not affect the stability of the 2212 phase [481]. A schematic temperature composition phase relations of the $Bi_2Sr_2CaCu_2O_x$ composition and silver is shown in Fig. 6.155. Liquid immiscibility between oxide and Ag liquids in 8 – 98 at % range at all temperature is found [482]. The solubility of Ag in the specific 2212 composition was less than detection limits, but that silver depressed the melting temperature of mixtures with the $Bi_2Sr_2CaCu_2O_x$ composition through the formation of a entectic with ≈ 4% Ag.

With the addition of excess silver (≈ 30 Wt%), the initial melting temperatures of the 2212 four-phase equilibrium volume were all lowered, by amounts ranging from 4 to 22°C [480] silver entered the melt, with a mole fraction of 6 to 8%. The outline of Ag-free and Ag-containing melts are shown as projections (normalized in terms of the CaO, CuO and B_2O_3 contents) in Fig. 6.156. Ag addition is manifested in a reduced content of the melts.

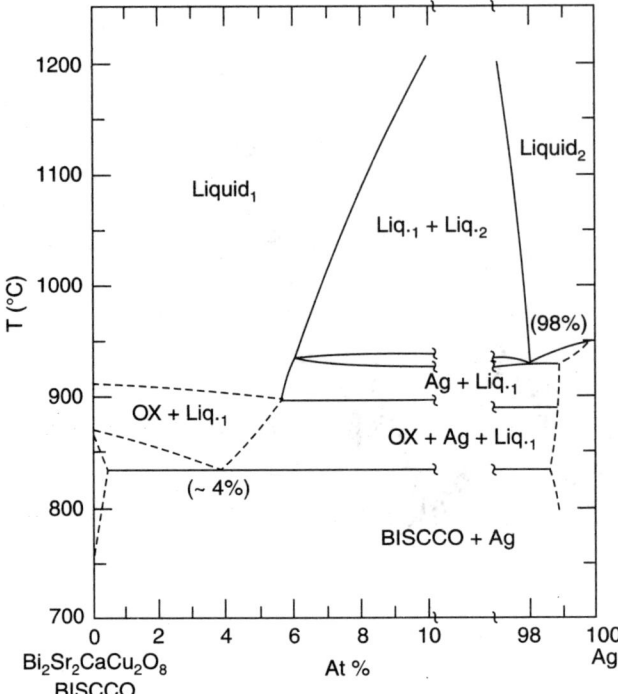

Fig. 6.155 Temperature composition phase relation in the system of $Bi_2Sr_2CaCu_2O_8$ Ag [482]. OX stands for oxygen

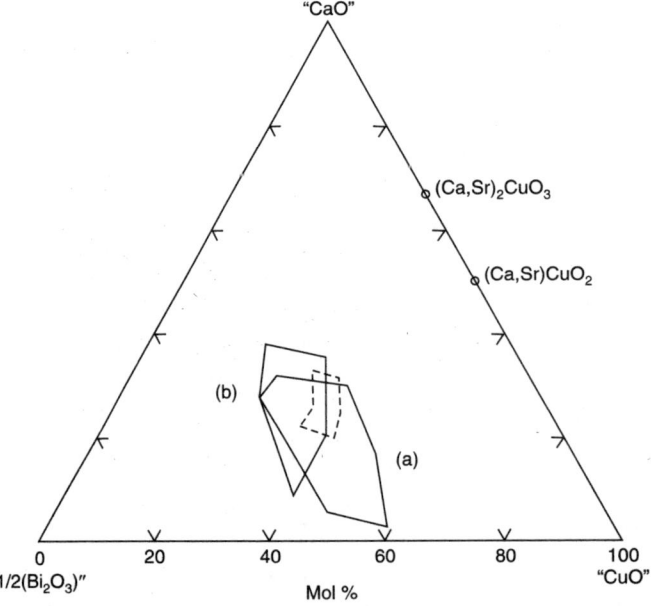

Fig. 6.156 Primary crystallization field of the Bi-Sr-Ca-Cu-O 2212 phase projected onto the ternary plane and shown in broken lines, (a) without the addition of Ag and (b) with Ag added [480]. The projection of the 2212 solid solution is also shown in broken lines

Phase Diagrams for Three CuO_2 Layers

Phase equilibrium investigations of the lead free three layered HTSCs are much less numerous than those of the Pb-containing system. Fig. 6.157 represents one such study [483], showing the section of $Bi_2O_3 - \frac{1}{2}(SrO + CaO)$ at 850°C through the quaternary system Bi_2O_3-SrO-CaO-CuO. We may note that the 2223 phase is also surrounded by very flat two, three, and four phase equilibria. Obviously, a small deficiency of CuO and/or Bi_2O_3 during the preparation of the 2223 phase sample results in a significant decrease of the volume content of the 2223 phase. On the other hand, an excess of Bi_2O_3 and CuO results in the **formation** of the 2223 phase in addition to CuO and liquid. At 850°C, the 2212 and the 2223 Pb-free superconductors are in equilibrium with a liquid of composition close to a Ca-rich 2201 phase. However, the phase only exists above 840°C.

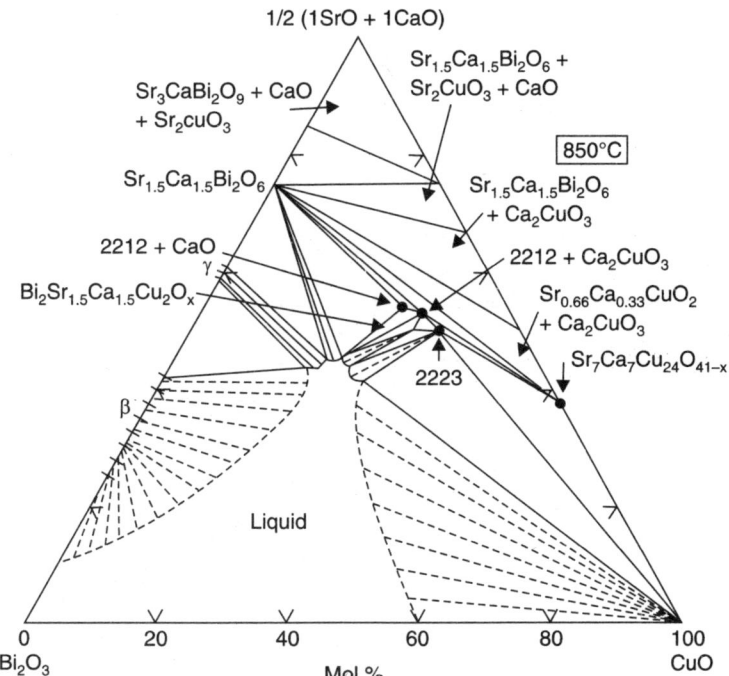

Fig. 6.157 Section Bi_2O_3 – 1/2 (SrO + CaO) – CuO through the quaternary phase diagram of the system Bi_2O_3-SrO-CaO-CuO at 850°C in air, showing the equilibrium of 2212 and 2223 with liquid [483]

The schematic projection diagram of the solid solution region of the Pb-2223 phase of composition $Bi_{2.27-x} Pb_x Sr_2 Ca_2 Cu_3 O_{10+d}$ as a function of temperature is shown in Fig. 6.158 [484]. A single phase region has been identified between $x = 0.18$ and $x = 0.36$. We may note that this schematic diagram is intended to illustrate phase existence regions only. There is a relatively narrow temperature range within which a single phase can form (about 838 – 860°C). Below and above this temperature range with x values of $0.1 > x > 0.36$, a multiple phase region is reported. For $x > 0.36$, 2223 was reported in equilibrium with numerous phases, including $Pb_4 (Sr, Ca)_5 CuO_x$, indicating that the maximum Pb solubility was exceeded. For $x < 0.18$, only small amount of 2223 were detected with the 2212 phase predominant over the entire temperature range. The solid solution region of a composition $Bi_y Pb_x Sr_2 Ca_2 Cu_3 O_{10+d}$ at

850°C and the surrounding phase fields as a function of the x and y values is illustrated in Fig. 6.159 [484]. We can see that the solid solution has a triangular shape, and Pb substitution has widened the homogeneity range of the 2223 phase.

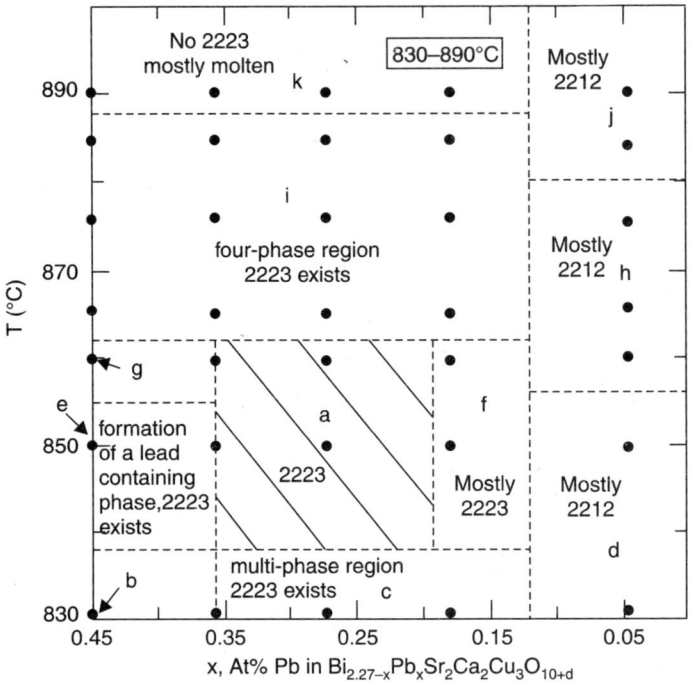

Fig. 6.158 Temperature vs. lead content x plot exhibiting phase regions for $Bi_{2.27-x}Pb_xSr_2Cu_3O_{10+d}$ [484]

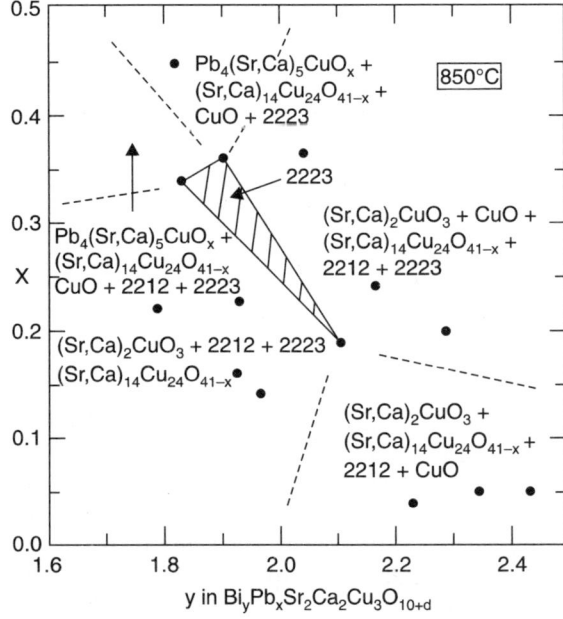

Fig. 6.159 Phase field surrounding the 2223 phase ($Bi_yPb_xSr_2Ca_2Cu_3O_{10+d}$) as a function of Pb content vs. Bi content y at 850°C [484]

In another subsolidus equilibrium study by Wong-Ng et al. [485], the Pb-2223 phase was found to be in equilibrium with 11 phases: 1×20 [$(Ca, Sr)_2 Pb O_4$], O_x 21, CaO, CuO, 3221 [$(Pb, Bi)_3 (Sr, Ca)_5 O_x$]), O_x 11 [Ca-rich], O_x 11' [Ca-poor], 2310, 119 $x5'$, and 2212. Equilibria involving the complete 29 subsolidus five phase volumes and their initial melting temperatures of the Pb-2223 phase in the (Bi, Pb)-Sr-Ca-Cu-O are given in Table 6.25. Due to large number of phases and closely spaced phase compositions, the Pb-2223 phase compatibilities, similar to those of the 2212 and Pb-free 2223 phases, include a number of relatively "flat" or shallow, five-phase equilibrium volumes. Among the 29 five-phase volumes, 16 involve the 2223-2212 phases that are mutually stable in a topologically constant manner. Because the 2212 and 2223 phases have similar structures (being members of the same homogeneous series), and the 2212 phase is a precursor for the formation of the 2223 phase, their mutual solid state compatibilities are extensive.

Table 6.25. Set of 29 Pb-2223 five-phase volumes (7.5 vol. % O_2/92.5 Ar) of the (Bi, Pb)-Sr-Ca-Cu-O system (See Fig. 6.159)

2223–2212–1x20–119x5–2310	2223–2212–014x24–1x20–119x5	2223–2212–0x11–2310–119x5
2223–2212–1x20–2310–CaO	2223–2212–2310–0x11–3221	2223–2212–119x5–0x11–CuO
2223–2212–0x21–1x20–CuO	2223–2212–014x24–119x5–CuO	2223–2212–0x21–014x24–CuO
2223–2212–1x20–0x11–3221	2223–2212–014x24–0x21–1x20	2223–2212–0x11–0x11'–CuO
2223–2212–1x20–CuO–0x11	2223–2212–3221–CaO–1x20	2223–2212–0x11–0x11'–2310
2223–2212–3221–CaO–2310	2223–0x11'–2310–0x11–119x5	2223–0x11'–CaO–2310–0x11
2223–1x20–0x11'–CuO–0x11	2223–0x11'–014x24–119x5–CuO	2223–0x11'–CaO–0x11–3221
2223–1x20–CuO–0x21–014x24	2223–1x20–CuO–0x11'–01424	2223–2310–3221–CaO–0x11
2223–1x20–CaO–0x11'–2310	2223–1x20–2310–014x24–0x11'	2223–1x20–0x11'–3221–0x11
2223–1x20–CaO–0x11'–3221	2223–0x11'–2310–119x5–014x24	

(v) Tl-Ba-Ca-Cu-O Systems

The studies of phase equilibrium in the Tl-Ba-Ca-Cu-O system are not extensive as those in the BYC, BRC or the BSCCO systems, partly because of the additional processing parameters of vapour pressure and also because of the toxicity of Tl-containing compounds. The four-component composition tetrahedron for the TlO_x-BaO-CaO-CuO system [486] is shown in Fig. 6.160. The nominal compositions of the presently known and confirmed superconducting phases were found to lie in the plane determined by TlO_x, 2BaO.CuO, and CaO.CuO (Fig. 6.161). Moreover, the nominal composition of each quaternary superconductor phase lies at the intersection of one of the two lines originating at the component CaO-CuO and terminating at Tl-1201 and Tl-2201 compositions. A second set of lines originates at TlO_x and terminates at the Tl-free compositions 0212, 0223 and 0234. Therefore, travelling along and any one of the second family of lines corresponds to adding or deleting thallium from the previous compound.

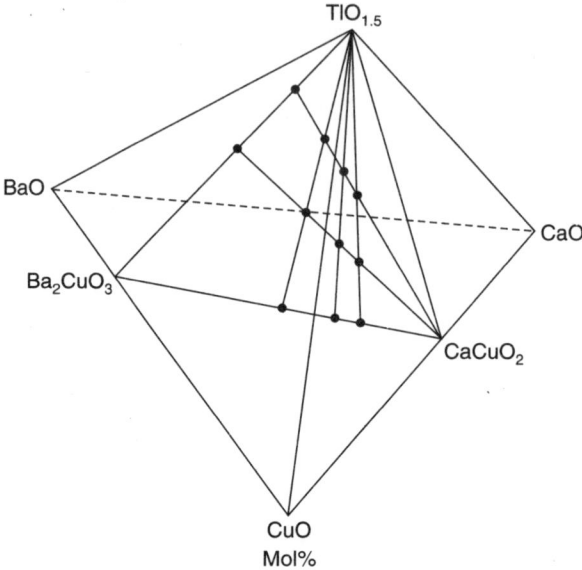

Fig. 6.160 Four-component composition tetrahedron for the TlO_x-BaO-CaO-CuO system. The nominal composition of the superconducting phases in the system all lie in the plane determined by TlO_x, 2BaO.CuO and CaO-CuO (84) [486]

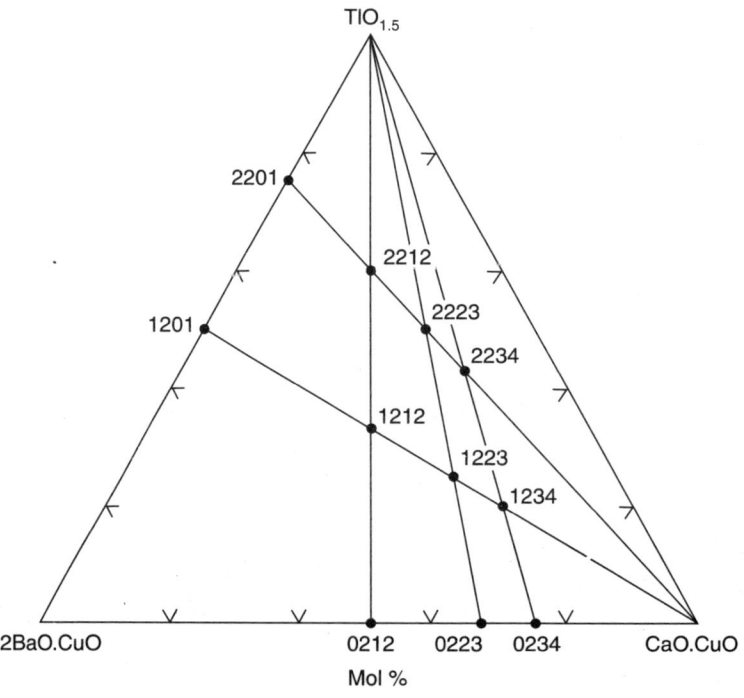

Fig. 6.161 Pseudoternary composition diagram for triangular plane of Fig. 6.160 of the TlO_x-BaO-CaO-CuO system, exhibiting the location of superconducting phases [486]

Fig. 6.162 shows the phase diagram of the system containing the 20 K superconductor $Tl_2Ba_2CuO_6$ (2201) [487]. The $Tl_2Ba_2CuO_6$ phase exists in both tetragonal and orthorhombic forms, and is in equilibrium with Tl_2BaO_4, $Tl_6Ba_4O_{13}$, $Tl_2Ba_2O_5$, $BaCuO_2$ and CuO. The pseudobinary cut of the system $Tl_2Ba_2O_5CuO$ is shown in Fig. 6.163. The $Tl_2Ba_2CuO_6$ phase melts incongruently into $Tl_2Ba_2O_5$ and L.

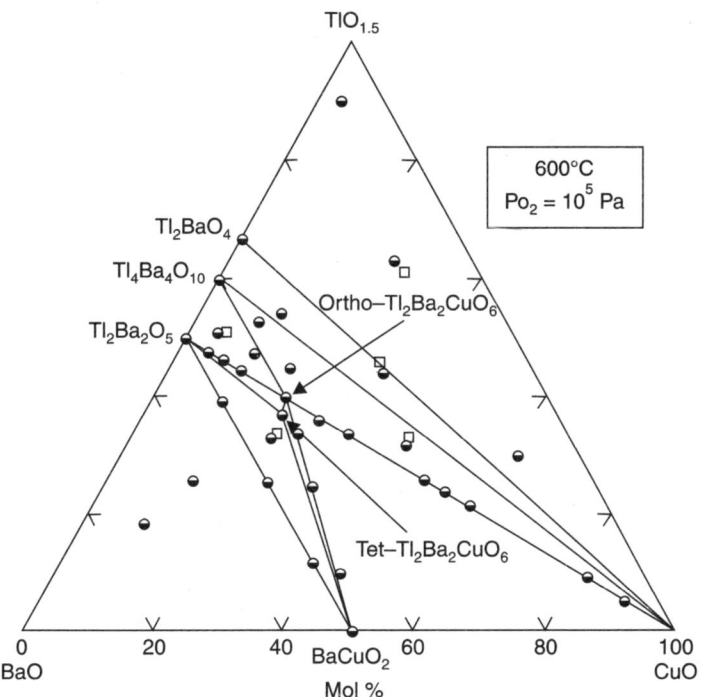

Fig. 6.162 Phase diagram showing the subsolidus relationships at 600°C and 10^5 Pa O_2 of the system of BaO-CuO-Tl_2O_3.
Ortho = orthorhombic; Tet = tetragonal; [487].

The solid solution extent of Tl-2212 phase is illustrated in the system $Ba_2CaCu_2O_x$-Tl_2O_3 system as a function of temperature (Fig. 6.163) [488]. A single phase region around the 2212 composition at higher T_c occurred with less than Tl_2O_3 than the 2212 stoichiometry.

(vi) Hg-Ba-Ca-Cu-O Systems

The mercury containing HTSC cuprates $Hg\,Ba_2\,Ca_{n-1}O_{2n+2+\delta}$ where $n = 1, 2, 3......$ have fascinated scientists due to their high-T_c. Not only does this family of cuprates exhibit the highest T_c (= 134 K) among the high-T_c systems known to date [489] but also, under high pressure, these materials show many unusual properties, including pressure-induced T_c enhancements [490]. In general a parabolic variation of T_c with δ and with pressure has been observed over a wide range of superconductivity values [491 – 493]. Pressure induced T_c enhancement is still under active investigation.

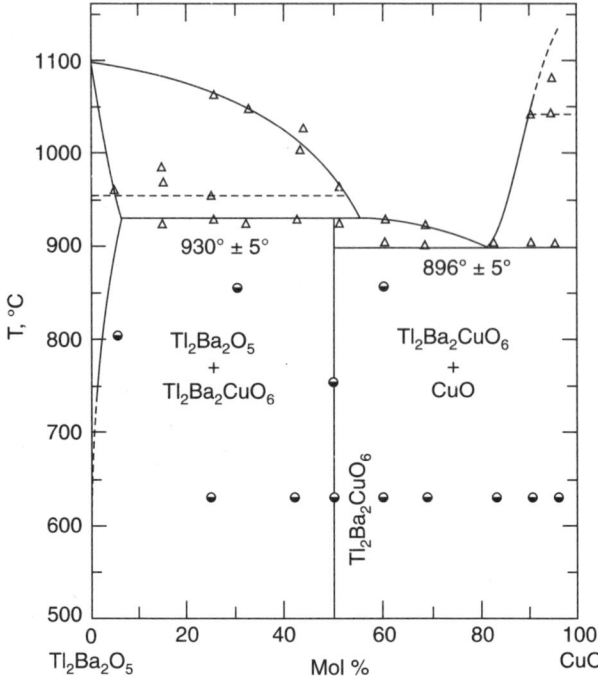

Fig. 6.163 Temperature vs composition plot for the binary system of $Tl_2Ba_2O_5$-CuO [487]

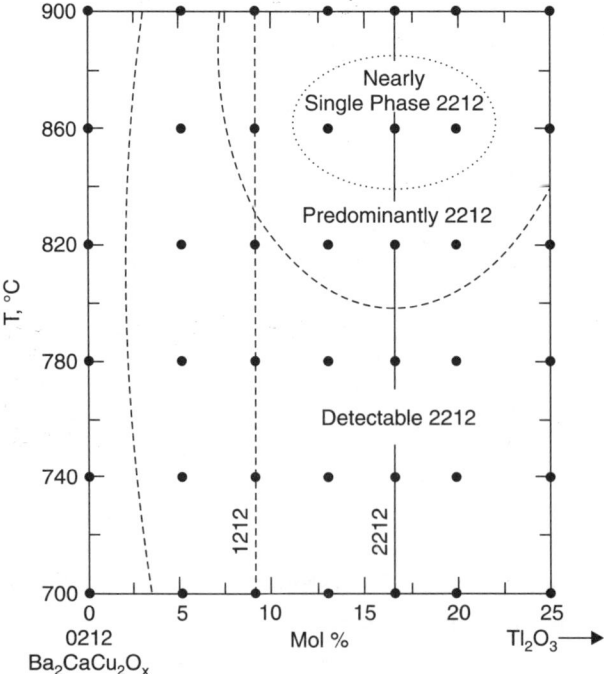

Fig. 6.164 Temperature vs composition plot showing the solid solution extent of the Tl-2212 phase as a function of temperature [488]

Due to the experimental difficulties involved in the determination of Hg vapour pressure, and the toxicity of Hg, very little activity has occurred on phase equilibrium studies. The 1201, 1212, 1223, and 1234 phases in this system have been characterized rather extensively. The highest pressures reached the onset value of 164 K at around 30 GPa.

We have seen that the physical properties of HTSC cuprates are highly anisotropic as a consequence of layered crystal structures of cuprates. In addition to high transition temperatures, the HTSC cuprates are characterised by extraordinary high critical fields B_{c_2} which can even exceed 100 T at zero temperature. However, the useful magnetic field range of HTSC cuprates is limited by the irreversibility line well below the upper critical field. Above the irreversibility line losses are caused by flux motion, which may be connected to thermally activated depinning, a glass-like behaviour or flux line lattice melting [494]. The magnetization of HTSC cuprates shows pronounced relaxation effects. The decay of the magnetisation may again be a consequence of glass like behaviour or large flux creep effects. The weak pinning in these materials is closely connected to the extremely short coherence length of less than 1 nm along the c-direction. Heavy ion irradiation can introduce additional pinning centres and as a result the irreversibility-line is shifted to higher fields and temperatures. A further important consequence of the extremely short coherence length is the weak-link behaviour of grain boundaries. Detailed investigations of artificial grain boundaries have indicated that low-angle grain boundaries are strongly coupled, whereas high-angle grain boundaries act as weak links. The critical current densities within single grains or in high quality thin films can exceed 10^6 A/cm^2 at 77 K and zero applied field. Generally, the critical current densities are very sensitive to magnetic fields applied along the crystallographic c-direction. An important anisotropy has also been found for the normal state resistivity of HTSC cuprates. The in-plane resistivity typically shows a linear temperature dependence like a metal, whereas even semiconducting behaviour can result for the out-of-plane resistivity. The pseudo-gap phenomena is most challenging and puzzling phenomenon. The thermal conductivity of HTSC cuprates is orders of magnitude smaller than that of pure copper, since due to the small hole concentrations in these systems the main carriers for heat transport are phonons. The thermal conductivity of single crystals shows a considerable anisotropy with the smaller values parallel to the c-direction. BYC, BRC and BSCCO systems have been studied most extensively because of their viable industrial applications in the coated-conductor and wire/tape development area. For the coated-conductor research, materials compatibility is an important issue. Therefore, information on interactions of HTSC cuprates films with substrates and other buffer layers is essential. It will be important to have phase equilibria data that include the high T_c phases and the buffer materials. In the next chapter we will discuss theory of high T_c cuprates. Now, we study the electronic states of HTSC cuprates.

6.10 ELECTRONIC STATES OF THE HTSC CUPRATES

The HTSC cuprates, such as $La_{2-x}Sr_xCuO_4$ (the so called 214 compounds) and $YBa_2Cu_3O_{7-\delta}$ (the so called 123 compounds), have one common structural unit: the quasi-two-dimensional structure that is approximated by CuO_2 planes, one of which is drawn schematically in Fig. 6.165. We discuss mainly the role of these planes since it is widely accepted that the electronic properties of those subsystems are the main factor determining the observed superconductivity, antiferromagnetism, and localization effects in these systems. In stoichiometric La_2CuO_4 or $YBa_2Cu_3O_7$, the formal valence of copper (Cu) is 2+, *i.e.*, it corresponds to one hole ($3d^9$) electron configuration. In a strictly cubic structure, with Cu^{2+} surrounded by O^{2-} ions in an octahedral arrangement, the highest band is doubly degenerate and of e_g symmetry, *i.e.*, composed of $d_{x^2-y^2}$ and $d_{3z^2-r^2}$ orbitals. However, in HTSC cuprates, the octahedra are largely elongated in the direction perpendicular to the CuO_2 planes, so that the bands are further split; it is commonly

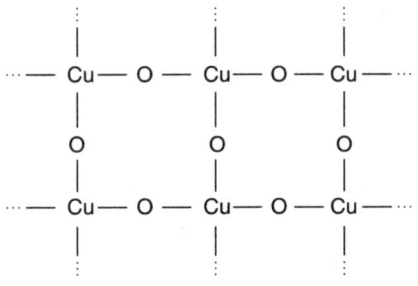

Fig. 6.165 Schematic representation of the CuO_2 planes in HTSC cuprates in the tetragonal phase. The Cu-Cu distance $a \approx 1.9$ Å for La_2CuO_4

assumed that the antibonding orbital $d_{x^2-y^2}$ is higher in energy and hence half filled. These d states hybridize with the oxygen $2p_x$ and $2p_y$ orbitals of σ type (Fig. 6.166). In Fig. 6.166 both the bonding and antibonding configurations are shown. The latter corresponds to the signs of the two *p*-orbitals shown in brackets.

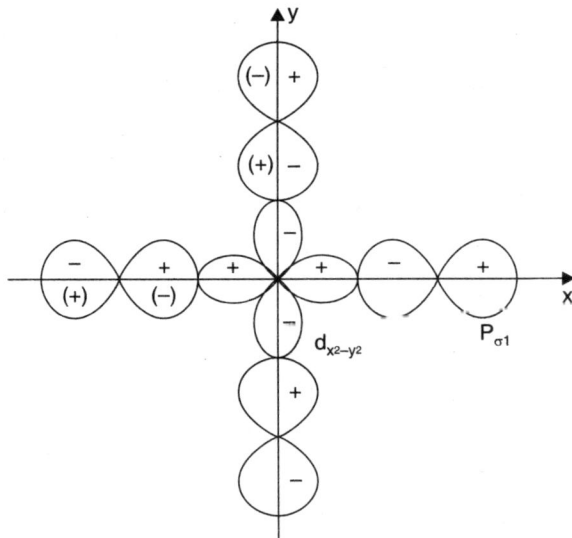

Fig. 6.166 The configuration of the $3d_{x^2-y^2}$ and p_σ orbitals for bonding configurations. The reverse sign for two *p*-orbitals, *i.e.* those in brackets, represent the hybridized configuration for the antibonding state

One can obtain a simple description of the electronic states for the planar CuO_2 system by introducing a single band representing Cu d electrons in the tight binding approximation. For the square configuration of the Cu atoms (which reflects the tetragonal structure of La_2CuO_4), such a description of band energies has the form

$$\varepsilon_K = 2t\,(\cos K_x a + \cos K_y a) \qquad \ldots(6.77)$$

where t is the so called Ropping or Bloch integral $<i\mid v\mid j>$ between the nearest neighbouring ions i and j and a is the Cu-Cu distance. For La_2CuO_4 and $YBa_2Cu_3O_{6.5}$, this band is half filled, with the Fermi surface determined from the condition $\varepsilon_K = \mu = 0$. As shown in Fig. 6.167, this leads to a square in reciprocal space connecting the points $(\pi/a)\,(\pm 1, 0)$ with the points $(\pi/a)\,(0, \pm 1)$. The oxygen electrons in the $2p$ states are regarded as playing only a passive role of a transmitter of the individual d electrons from one $d_{x^2-y^2}$ state to its neighbour (we may note that the O^{2-} valence state has completely filled p-shells). If the number of electrons in that band is decreased (for $e.g.$, by substituting Sr for La in 214 compounds) then the Fermi surface shrinks and gradually transforms into a circle. Within this modes, La_2CuO_4 should be metallic. However, at $T < T_N \approx 240$ K, this compound orders antiferromagnetically, and the ground state is then insulating. The fact that this system

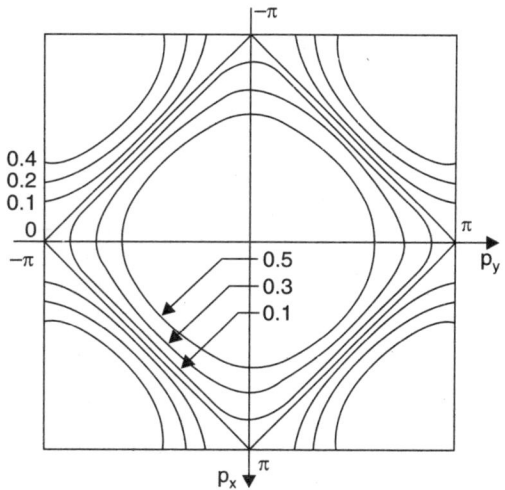

Fig. 6.167 The shape of the two dimensional Fermi surface for the band energy of the form of eq. (1). The values specified represent μ/t as a parameter. The square shape corresponds to $\mu = 0$ or equivalently to $n = 1$ [495]

remains insulating above the Neel temperature (T_N) means that the stoichiometric La_2CuO_4 and $YBa_2Cu_3O_{6.5}$ are Mott insulators, not a Slater-split band antiferromagnet; for the latter, the split-band structure for $T < T_N$ should coalesce into one band as $T \to T_N$. The presence of paramagnetic insulating state for both La_2CuO_4 and $YBa_2CuO_{6-\delta}$ supports the view that those oxides should be regarded as narrow band systems characterized by strong electron–electron interactions $(U < U_c)$, as originally proposed by Anderson [496]. An antiferromagnetic ground state is then expected since the kinetic exchange interaction between the strongly correlated electrons takes place [497].

One faces a principal problem, when holes occur in the Mott insulator, $i.e.$, when one considers $La_{2-x}Sr_xCuO_4$ or $YBa_2Cu_3O_{6.5-x}$. However, for small x the kinetic energy of the holes and the exchange energy of electrons may become comparable or the latter may become even larger than the former. In such situations, the motion of the holes will be influenced by the setting in of almost instantaneous spin-spin correlations. Obviously, if such metallic states (if formed) cannot be regarded as Fermi liquid with slowly evolving spin fluctuations; instead the resonance between various spin fluctuations, instead, the resonance between various spin configuration must be built into the electron wave function characterizing its itinerant state. The decomposition of the resonating spin configurations into spin pair-singlet configurations into spin pair-singlet configurations constitute an important characteristic of the RVB theory [496, 498]. Some experimental evidence for the **quantum spin-liquid state** above the Neel temperature (T_N) has been provided by neutron quasi-elastic scattering [499], these results were subsequently interpreted [500].

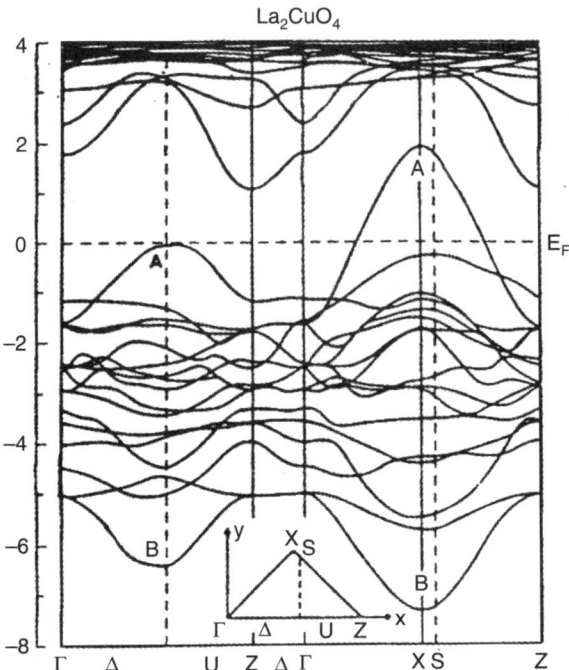

Fig. 6.168 Energy bands for La_2CuO_4 calculated within a local density approximation for the assumed crystal structure are body-centered tetragonal. A portion of the x-y plane in the extended Brillouin zone scheme is also shown in the inset. The portions *B* and *A* correspond to the bonding and antibonding parts of the hybridized band [495]

The interpretation of the metallic state in terms of a single, narrow band requires the presence of both $3d^9$ (Cu^{2+}) as well as $3d^8$ (Cu^{3+}) states. Most of the X-ray spectroscopical studies [501–503] conclude that the satellite peak corresponding to $3d^8$ configuration is actually absent. Therefore, in order to explain both the insulating properties of $La_2Sr_xCuO_4$ for $x \simeq 0.04 - 0.05$, one introduces hybridized $2p - 3d$ states for the holes introduced by doping. Band structure calculations by Mattheiss [504] for $La_{2-x}Sr_xCuO_4$ shown in Fig. 6.168, justify a reasonable description within a simple two-dimensional tight binding model with only the Cu $3d_{x^2-y^2}$ and p_σ orbitals on oxygens taken into account. Namely, the structures denoted by A and B in Fig. 6.168 correspond respectively, to antibonding and bonding hybridized bands, with respective band energies

$$E_{K\pm} = \frac{\varepsilon_p + \varepsilon_d}{2} \pm \left[\left(\frac{\varepsilon_p - \varepsilon_d}{2}\right)^2 + 4V^2\left(\sin^2\frac{K_x a}{2} + \sin^2\frac{K_y a}{2}\right)\right]^{1/2} \quad \ldots(6.78)$$

where ε_p and ε_d are atomic level positions for the $3d$ and $2p$ states respectively. Detailed calculations [48] lead to a nonzero bandwidth of the $2p$ band because of the p-p overlap; then $\varepsilon_p \to \varepsilon_p + \varepsilon_K$, with $\varepsilon_K = 2t_p [\cos(K_x a/\sqrt{2}) + \cos(K_y a/\sqrt{2})]$.

The band structure calculations should be regarded as providing input parameters for the parameterized models, which include electron correlations more accurately. On the basis of various estimates [505 – 506] of these parameters, one can assume that they fall in the range

$$|\varepsilon_p - \varepsilon_d| = 0 - 2 \text{ eV}$$
$$|V| \approx 1 - 1.5 \text{ eV}$$
$$|t| \leq 0.5 \text{ eV}$$
$$|t_p| \leq 0.5 \text{ eV}$$

and
$$U \leq 8 - 10 \text{ eV}$$

From the above estimates of the parameters, one finds that $|V| \sim |\varepsilon_p - \varepsilon_d|$. Clearly, one may not be able to use the perturbation expansion in $V/(\varepsilon_p - \varepsilon_d)$ of the Anderson lattice Hamiltonian. Such a perturbation expansion was used [507] when transforming the hybridized model represented by the Hamiltonian given by (See chapter 8)

$$H = \varepsilon_f \sum_{i\sigma} N_{i\sigma} + \sum_{K\sigma} \varepsilon_K n_{K\sigma} + U \sum_i N_i\uparrow N_i\downarrow + \frac{1}{\sqrt{N}} \sum_{K\sigma} (V_K e^{i\mathbf{K}\cdot\mathbf{R}} a_{i\sigma}^+ C_{k\sigma} + H.C.)$$

into an effective narrow-band model represented by (See chapter 8).

$$E(R) = -\frac{W}{2} + \pi^2 |t| \left(\frac{a}{R}\right)^2 + \frac{4\pi}{3}\left(\frac{R}{a}\right)^3 \frac{zt^2}{U}$$

On the basis of the facts that the present day band calculations do not provide paramagnetic insulating states for the stoichiometric materials, such as La_2CuO_4, and that the antiferromagnetic ground state is difficult to achieve within the local density approximation. One will have to look for other models.

We note that the band structure of La_2CuO_4 is quasi two dimensional. This is evident from the fact that (i) there is very little dispersion in the energy bands along the c-axis, and (ii) the charge distribution is negligibly small between the layers. By contrast, the energy bands in the basal plane centre of the peak in $\chi(q)$ is shifted away from the boundary of the BZ.

6.10.1 Fermi Surface Nesting in $La_{2-x}(Ba, Sr)_x CuO_4$

Xu et al. [508] have studied in detail the Fermi surface of $La_{2-x}(Ba, Sr)_x CuO_4$ based on the rigid band model. As is evident from Fig. 6.169 (a), the Fermi surface for $x = 0$ consists of hole-like cylinders along the z-axis in the BZ centered about the x-points. At the concentration $x = 0.17$, at which E_F touches the van Hove singularity, the Fermi surface undergoes a dramatic change from hole-like cylinders centered at the corner X points to electron cylinders centered at T (See Fig. 6.169 (b)). We can see that there is significant Fermi surface nesting in the [110] direction where a large portion of the Fermi surface is spanned by common wave vector $\mathbf{q}$. Xu et al. [508] have shown that the resulting generalized susceptibility $\chi(\mathbf{q})$ (calculated from the constant matrix element approximation) is substantially enhanced near the zone boundary points X and N. For $x > 0$, the magnitude of the spanning vector $|\mathbf{q}|$ is reduced so that the centre of the peak in $\chi(\mathbf{q})$ is shifted away from the boundary of the BZ.

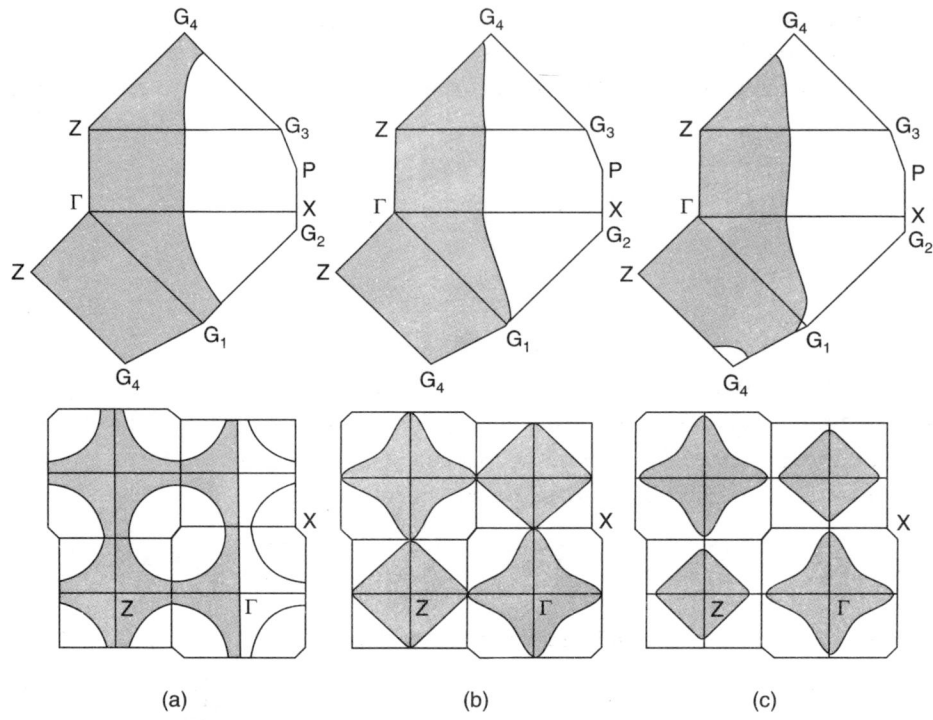

Fig. 6.169 Fermi surface of $La_{2-x}Ba_xCuO_4$ in the extended zone scheme: (a) $x = 0$ (b) $x = 0.17$ at the van Hove singularity and (c) $x = 0.2$, [508]

Fermi surface nesting is likely to drive a phonon branch soft, thereby enhancing the electron–phonon interaction. The near perfect Fermi surface nesting also suggests the possibility of charge density wave or spin density wave instabilities. In the case of pure La_2CuO_4, where $\mathbf{q} = 2k_F = (1/2, 1/2, 0)$, the strong electron–phonon interaction may induce some kind of lattice distortion and open a semiconducting band gap at E_F. The distorted system is stabilized by the gain in the electronic energy near the gap. It is interesting to note that the intensity of the peak in $\chi(\mathbf{q})$ reaches its maximum value near the concentration of $x = 0.15$, where T_c is maximized. Within this *Peierl's instability* picture, the essential role of the Ba or Sr doping is to shift $\mathbf{q}$ away from the boundary of the Brillouin Zone (BZ), thereby stabilizing the tetragonal phase which is metallic and therefore superconducting. Mattheiss [504] and Weber [509] have considered a Fermi surface induced planar breathing type displacement of O atoms away from the central Cu site, which is semiconducting with orthorhombic symmetry. However, the space group is different from that of room temperature La_2CuO_4. Instead, an antiferromagnetic ordering was reported in the orthorhombic phase in which the Neel temperature (T_N) depends sensitively on the amount of oxygen vacancy. Similarly, a very small amount of Ba or Sr doping ($x \simeq 0.02$) could destroy the long range antiferromagnetic order [510].

Tetragonal-Orthorhombic Transition

Based on group theoretical analysis, Kasowski et al. [511] demonstrated that the two Cu atoms in the observed orthorhombic phase remain equivalent so that the energy bands must be degenerate on certain faces of the new BZ. This would indicate that La_2CuO_4 is metallic, contrary to the resistivity measurement. This metallic behaviour is confirmed by several first principles energy band calculations of the orthorhombic phase, which is predicted to be metallic with very little changes from that of the undistorted tetragonal phase. To reconcile this with the observed semiconducting behaviour requires an additional perturbation that breaks up the symmetry of the two Cu atoms, possibly by long range ferromagnetic order, by oxygen vacancies or by Ba and Sr doping. Another possibility proposed by Kasowski et al. [511] is that at very low temperature La_2CuO_4 becomes a semiconductor due to an orthorhombic to monoclinic structural phase transition. A third possibility proposed by Tanigawa et al. [512], is that the orthorhombic phase of La_2CuO_4 is metallic with an unusual temperature dependence of its resistivity due to strong electron correlations. This is supported by the observation of Fermi surfaces on orthorhombic La_2CuO_{4-y} in positron annihilation experiments. The measured Fermi surfaces of the non stoichiometric samples appear to be more complicated than that predicted by the band structure for stoichiometric La_2CuO_4. We may note that the topology of the Fermi surface is extremely sensitive to doping or oxygen stoichiometry.

Effects of Oxygen Vacancies

Unlike the La atom, which does not contribute much to the occupied conduction band, the oxygen atoms actively participate in the bonding and antibonding states. However, the effects of oxygen vacancies cannot be understood purely on the basis of the rigid band model. Sterne and Wang [513] have studied the effects of oxygen vacancies on the electronic properties of La_2CuO_4 using a super cell geometry. The super cell is constructed by doubling the tetragonal unit cell in the two planar directions with 28 atoms in the unit cell. One out of the 16 oxygen atoms in the super cell was removed from the two dimensional Cu-O_2 network which formally corresponds to $La_2CuO_{3.75}$.

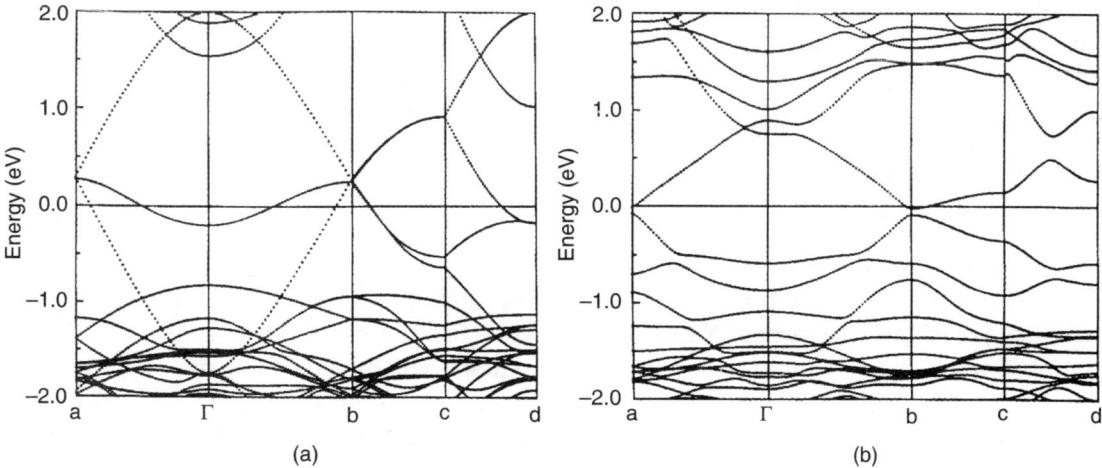

Fig. 6.170 Energy bands of (a) La_2CuO_4 folded back to the reduced zone of the 28 atom super cell, and (b) $La_2CuO_{3.75}$ with one out of 16 oxygen atoms removed from the two dimensional Cu-O plane in the super cell [513]

Fig. 6.170 indicates the effects of the oxygen vacancy, where the band structure of La$_2$CuO$_4$ in this 28 atom super cell (Fig. 6.170(a)) with the corresponding gaps at E_F in Fig. 6.170(b). Oxygen vacancies alter the band structure significantly near E_F with the four fold degeneracy at **a** and **b** completely lifted. Removing an oxygen atom reduces the coordination numbers of the neighbouring copper atoms, which leads to a narrower conduction bandwidth. We may note that the system is very close to being a semiconductor. The electron pocket around **b** is compensated by a hole pocket elsewhere in the BZ, but the overlap is clearly very small.

Another consequence of the oxygen vacancy is to produce an upward shift of E_F, which can be seen by comparing the fourfold degenerate levels at **a** and **b** that lie above E_F in Fig. 6.170(a) with the corresponding gaps at E_F in Fig. 6.170(b). One can draw a similar conclusion by comparing the band structures of Y Ba$_2$ Cu$_3$O$_7$ and that of Y Ba$_2$ Cu$_3$O$_6$. We can understand this from the ionicity of the oxygen atom is a manner similar to the case of Ba or Sr doping. If we assume that the oxygen atom is present as O^{2-}, then removing an oxygen atom will leave two extra electrons which are accomodated in part by an upward shift of E_F. It is assumed that removing a neutral oxygen atom will leave two extra electrons which are accommodated in part by upward shift of E_F. It is assumed that removing a neutral oxygen atom with four valence p-electrons will at the same time remove six valence p-states, thereby pusing E_F upward. One finds that 0.3 electrons remain in a sphere of radius 1.9 a.u. around the vacancy site, which is stabilized by the Madelung potential somewhat similar to the F-centre in ionic oxides. These results indicate the importance of self-consistency to model oxygen vacancy. Papaconstantopoulos et al. [514] have used the tight binding coherent potential approximation to understand the disorder phase vacancies and found that deoxygenation does not raise E_F, contrary to simple doping arguments. However, they have taken vacancy on site energy as infinite in their calculation, so that all charge is expelled from the vacancy sites. It remains interesting to see if the discrepancy reflects errors of non-self consistent charge distribution in the CPA calculations or effects of disorder not included in the **super cell model.**

We may note that so far, the calculations have not included the relaxations of atoms around the vacancy site. Depending on the size of the relaxation, there may be some quantitative changes to the results, should be valid. Band gaps were found to be opening up near E_F in a superlattice of ordered oxygen vacancies. The vacancy-vacancy separations in the superlattice are reasonably far apart, so that band gaps will probably be smeared into a deep valley in the density of states (DOS) in samples of randomly arranged oxygen vacancies. Accordingly, the static dielectric screening should be reduced in oxygen samples.

6.10.2 Electronic Properties of YBa$_2$Cu$_3$O$_{7-x}$

The absence of 2 + y oxygen atoms from the perfect triple perovskite YCuO$_3$(Ba CuO$_3$)$_2$ is the unique feature of the structure of YBa$_2$Cu$_3$O$_{7-x}$ [515]. In the limit $x = 0$, the vacancies are due to the absence of the oxygen atoms from (i) the Y-plane which separates the adjoining CuO$_2$ layers, and (ii) half of the oxygen sites in the Cu-O plane between the Ba-O layers which leads to the formation of the Cu-O chain. There are three weakly coupled Cu-O layers in the unit cell, consisting of two quasi-two-dimensional Cu-O$_2$ planes and a one dimensional Cu-O chain. In the unit cell, there are three weakly coupled Cu-O layers, consisting of quasi-two-dimensional Cu-O$_2$ planes and a one-dimensional Cu-O chain. Along the c-axis, the oxygen atom in the

Ba-O plane (O_{Ba}) is significantly closer to the four-fold coordinated Cu atoms in the chains (1.85 $A°$) than the five fold coordinated Cu atoms in the planes (2.30 $A°$). As x is increased from 0 to 1, the oxygen atom in the chain layer is removed and some of the oxygen atoms are moved to the empty sites, which eventually leads to the orthorhombic to tetragonal transformation. One notable change in the structure as the oxygen atoms are removed is that O_{Ba} atom moves away from the Cu-O_2 plane towards the Cu-O chain along the c-axis, so that the Cu atoms in the chain layer are more isolated and the interlayer coupling between Cu-O_2 planes are further reduced. Cava et al. [516] found a remarkable correlation between T_c and the oxygen vacancy concentration. T_c is around 90 K for $0 \leq x \leq 0.2$ before decreasing sharply to 60 K, where it remains upto $x = 0.5$. Beyond $x = 0.5$, T_c drops sharply and antiferromagnetic order has been observed near $x = 0.7, 0.85$ and 1.0 [517 – 519].

Band Structure of Y Ba_2 Cu_3O_7

Mattheiss and Hamann [520] and Massidda et al. [521 – 522] independently reported the band structure of $YBa_2Cu_3O_7$. In the absence of detailed neutron analysis, the CuO chains were chosen along the a-axis rather than the experimentally determined b-axis in the calculations of Mattheiss and Hamann. The resulting band structure differs somewhat from that of Massidda et al. and subsequent calculations, especially near E_F, due to slightly different CuO distance in the chain. Away from E_F, they found that the electronic structure of $YBa_2Cu_3O_7$ consists of occupied O_{2s} bands centered at -15 eV, Ba_{5p} bands centered at -10 eV, as well as unoccupied Ba_{5d} and Y_{4d} bands above 3.4 eV. Like La_2CuO_4 states near E_F are dominated by pd σ bands of Cu-O_2 planes.

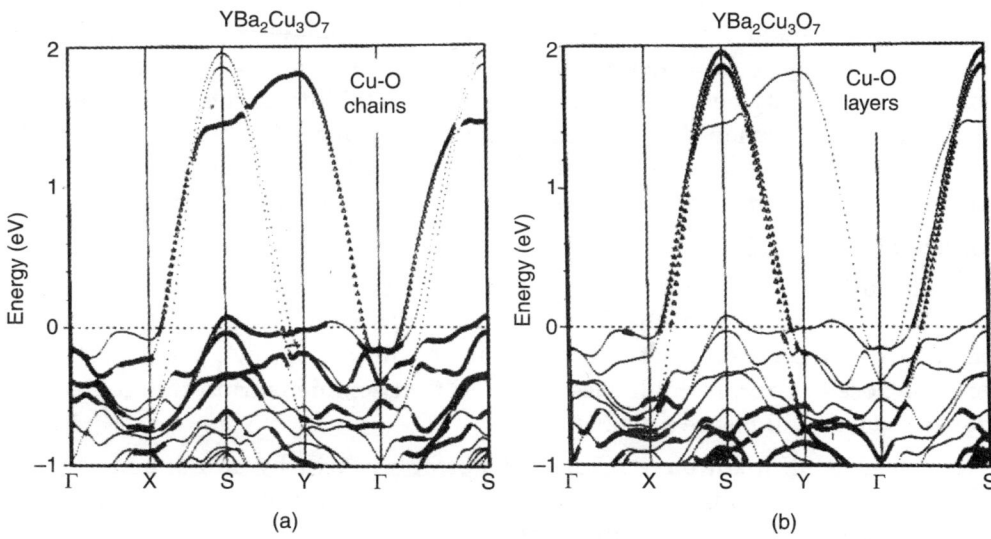

Fig. 6.171 Energy bands of Y Ba_2 Cu_3O_7 along some high symmetry directions for $k_z = 0$ of the orthorhombic BZ [523]. (a) States with more than 60% of their charge on the Cu-O chains are emphasized with the large symbols, (b) those with more than 80% charge on the two Cu-O layers are emphasized

Fig. 6.171 shows the band structure of Krakauer and Pickett [523], which is in close agreement with the calculations of Massidda et al. [521] and Ching et al. [524]. The states with more than 60% of their charge on the Cu-O chains are emphasized with the large symbols in Fig. 6.171(a), while states with more than 80% charge on the two Cu-O layers are shown in Fig. 6.171(b). Among the four bands that cross E_F, two of them have the majority of their charge on the Cu-O_2 plane corresponding to one band for each layer. These bands are similar to those of La_2CuO_4 except that they are less than half-filled. The Cu-O chains are responsible for the two other bands : (i) a steep band crossing E_F, which is strongly dispersive only in the chain direction, i.e., along the X-S and T-Y directions, and (ii) a flat band just at E_F along the Y-S direction.

Massidda et al. [521], by examining the projected DOS pointed out that the strong pd σ bond in the Cu-O_2 planes leads to a rather wide bandwidth of ≥ 8 eV. The corresponding width for the Cu-O chains is noticeably narrower. Along the c-axis, the O_{Ba} atoms lie significantly closer to the copper atoms in the chain (Cu_{chain}) than those in the plane (Cu_{plane}), which leads to some noticeable dispersion in the two chain bands along the c-axis [521]. Massidda et al. [521] have discussed the reasons that O_{Ba} atoms lie closer to Cu_{chain} than Cu_{plane}. First of all, the shorter $Cu_{chain} - O_{Ba}$ bond is stronger so that only the bonding states are filled, leaving their antibonding partners empty. By Contrast, the $Cu_{plane} - O_{Ba}$ bonding and antibonding states, which have a smaller splitting, are fully occupied. Secondly, the Cu_{plane} atoms are five fold coordinated, with the O_Y atom on the opposite site of the Cu_{plane} missing, while the fourfold coordinated Cu_{chain} atoms are balanced by the O_{Ba} atoms above and beneath them. These two effects together exert a Coulombic repulsion on the O_{Ba} atom and push them away from the Cu_{plane}.

Detailed analysis of the charge density distributions confirmed the one-dimensional nature of the Cu-O chain as well as the two-dimensional nature of the Cu-O_2 planes. The ionic Y and Ba atoms were found to act as electron donors and do not otherwise participate, even if the fully substituted material ordered magnetically at very low temperature. The lack of conduction electron density near the Y-site explains the stability of the high superconducting critical temperature (T_c) when the isolated Y atoms are replaced by strongly magnetic rare-earth such as Gd or Er [525, 526]. Alp et al. [527] confirmed the absence of exchange coupling of the Gd ion with the conduction electrons, responsible for superconductivity.

Fermi Surface of $YBa_2Cu_3O_7$

Yu et al. [522] have made a detailed study of the Fermi surface and charge distribution of $YBa_2Cu_3O_7$. With the exception of the flat band at E_F, a possible nesting feature can be seen in the Fermi surfaces corresponding to three other bands. Unlike La_2CuO_4, however, the spanning vectors are not close to a reciprocal lattice vector so that commensurate charge density waves or spin density waves will not be possible.

Semdskjaer et al. [528] have measured Fermi surfaces by two-dimensional angular correlation of the position annihilation radiation (2D – ACPAR) technique on a single crystal $YBa_2Cu_3O_{7-x}$. The momentum distribution shows structures associated with the higher zones, which indicates positron annihilation with extended rather than localized electrons. Three nearly cylindrical Fermi surface sheets have been observed. There is a square piece centered at the S-point, which corresponds to the two bands of predominantly Cu-O_2 layer character which are too close to one another to be resolved. Qualitatively chain character appears to be reasonable considering the facts that (i) they are much more sensitive to oxygen vacancies and (ii) there are some discrepancies in different theoretical predictions for one of the chain bands with very large effective mass. However, further study will be necessary to establish a more quantitative comparison with band theory.

Band Structure of $YBa_2Cu_3O_6$

The band structure of tetragonal $YBa_2Cu_3O_6$ by Yu et al. [529] is shown in Fig. 6.172. Due to the missing oxygen atoms along the chain, the 1D chain bands in $YBa_2Cu_3O_7$ that are emphasized with large symbols in Fig. 171 (a) are completely absent. In their place, there are two completely filled degenerate states (d_{yz} and d_{xz}) of Cu_{chain} of Γ and M. These orbitals form π-bonding bands with O_{Ba} atoms. The two Cu-O_2 bands remain the same as in $YBa_2Cu_3O_7$ but there is an upward shift of E_F due to the ionicity of the missing oxygen atoms. Accordingly, there are significant changes in the Fermi surface and the hole count in the Cu-O_2 bands. In particular, the band along ΓM in Fig. 6.172 suggests possible Fermi surface nesting along the (110) direction, which may explain the occurrence of antiferromagnetic order near $x \cong 1$. Yu et al. [529] found an enhancement of the Stoner factor $S = N(E_F) I$ in $YBa_2Cu_3O_{7-x}$ from 1.12 ($x = 0$) to 1.38 ($x = 1$), even though the DOS at E_F is actually reduced.

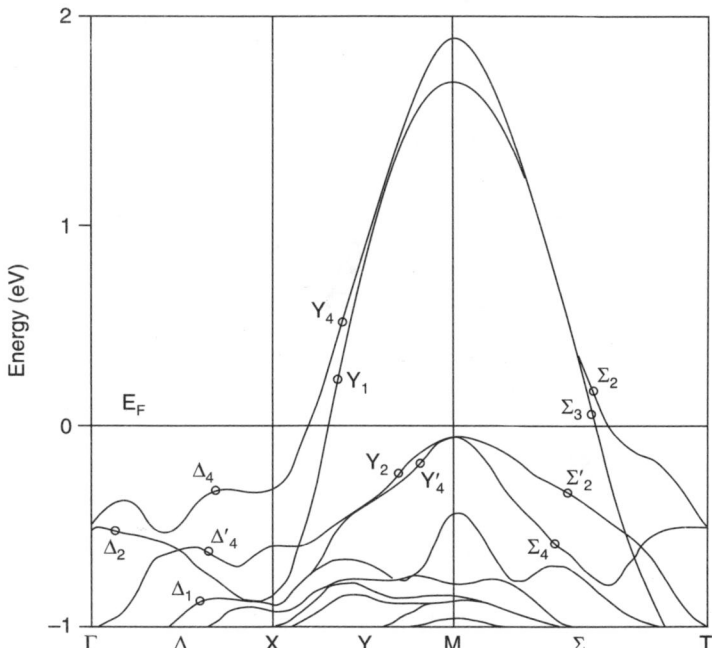

Fig. 6.172 Energy bands of $YBa_2Cu_3O_6$ along some high symmetry directions [529]

Oxygen Vacancies and the Phase Diagram of $YBa_2Cu_3O_{7-x}$

The phase diagram of the HTSC cuprates is very sensitive to the filling factor of the $pd\sigma$ antibonding bands, which is a measure of the hole concentration of the Cu-O_2 layers. In the half filled limit, the compound is an antiferromagnetic insulator due to Fermi surface nesting. Superconductivity occurs away from the half filled limit, where the spanning vector is no longer commensurate with the periodicity of the lattice, so that the long range antiferromagnetic order is suppressed. In the case of $La_{2-x}(Ba, Sr)_x CuO_{4-y}$ this can be controlled by doping and oxygen vacancy. In the case of $YBa_2Cu_3O_{7-y}$, the CuO chain layer essentially dopes the Cu-O_2

layers even in the limit $y = 0$. We will also see that the Cu-O chain layer may also play an additional role of providing a metallic intervening layer that enhances the interlayer coupling among the Cu-O$_2$ layers and therefore raises T_c significantly.

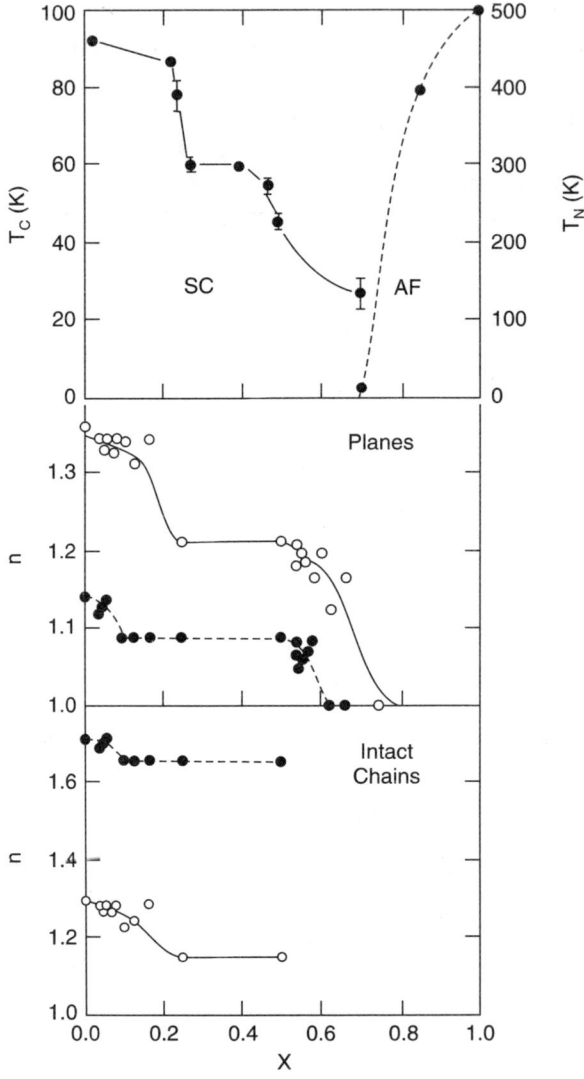

Fig. 6.173 Experimental phase diagram of YBa$_2$Cu$_3$O$_{7-x}$ (top) together with hole counts calculated for a plane (centre) and an intact chain (bottom) [530]. Two sets of parameters were considered in the theoretical calculations, with the solid circles reproducing the LDA band structure and the open circles representing adjusted parameters to line up the centre of the Cu-O$_2$ plane band and the intact chain band

The top part of Fig. 6.173 shows the experimental phase diagram of YBa$_2$Cu$_3$O$_{7-x}$ as a function of oxygen vacancy concentration. In the superconducting phase, T_c shows plateaus, which have been explained by Zaanen et al. [530] using a tight binding model for a unit cell doubled along the a-axis to accommodate two in equivalent chains. One of the chains is either intact ($0 \leq x \leq 0.5$) or empty ($0.5 \leq x \leq 1$) while the other chain has periodic oxygen vacancies arranged according to the nonstoichiometry. The electronic structure is described by removing the linking oxygen atoms, without changing the tight binding parameters. Starting from the empty chain, each additional oxygen atom introduces one antibonding level and two holes. If the level is above E_F, the antibonding level will absorb two holes and there is no doping in the Cu-O_2 layer. The hole count on the Cu-O_2 layer is modified only when the level lies below E_F, which is excluded in certain ranges of x. As is evident from the central part of Fig. 6.173, the resulting hole count on the Cu-O_2 layers shows a plateau that correlates nicely with T_c.

6.10.3 Electronic Structure of the Bismuth and Thallium HTSC

The discovery of HTSC in bismuth [531, 532] and thallium [533, 534] based cuprates has provided valuable insight into the mechanism of superconductivity. The structures of these materials have been determined [535 – 538] and T_c depends on the number of Cu-O_2 layers in a very simple way: The more the Cu-O_2 layers, the higher the T_c e.g., $Bi_2Sr_2CuO_6$, which has only one CuO_2 layer, has a comparatively low T_c of about 6 – 22 K [539], while the two-layer compound $Bi_2Sr_2CaCu_2O_8$ superconducts at 84 K [535]. Aside from the small orthorhombic distortion in the unit cell and the superlattice structure along the orthorhombic b direction, both structures are body-centered tetragonal. The structure of the one-layer $Bi_2Sr_3CuO_6$ ($Tl_2Ba_2CuO_6$) consists of a square Cu-O_2 plane above which is a layer of Sr-O (Ba-O) followed by two Bi-O (Tl-O) layers and another Sr-O (Ba-O) layer before the whole structure repeats with the copper oxygen layer shifted to the body centered tetragonal position. The two-layer structure differs in that the copper oxygen plane is replaced by two Cu-O_2 layers separated by a layer of calcium. The three-layer structure includes an additional set of Ca:Cu-O_2 layers of the two-layer structure.

Sterne and Wang [540] have calculated the electronic structure for the one and two layer bismuth compounds and compared them to see what may be responsible for the large difference in their transition temperatures. One finds essentially identical Cu-O_2 bands in both compounds, but the Bi-O planes which lie between the Cu-O_2 layers are metallic in the HTSC compounds but are almost insulating in the low T_c compound. Since the metallic nature of the chain layer bands in $YBa_2Cu_3O_{7-x}$ also decreases with decreasing T_c as oxygen vacancies are introduced. It seems that the metallic nature of the layers between the Cu-O_2 layers could enhance T_c significantly.

Band Structure of $Bi_2Sr_2CaCu_2O_8$

The band structure for the HTSC $Bi_2Sr_2CaCu_2O_8$ compound is shown in Fig. 6.174(a) [540], which is very similar to the calculations of Hybersten and Mattheiss [541], Krakauer and Pickett [542] and Freeman et al. [543]. There are two Cu-O_2 $pd\sigma$ antibonding bands extending from about 1 eV below E_F to 1.5 eV above E_F, similar to HTSC compounds discussed earlier. The new feature is the presence of two bismuth p-electron bands which dip below E_F by about 0.7 eV around the K-point. These interact with the Cu-O_2 antibonding bands producing a complicated band structure around K. One finds a shallow band of mixed Bi and Cu-O_2 character which barely crosses E_F and more dispersive band which has almost pure bismuth character crossing E_F, indicating that the bismuth oxide layers are metallic. At higher energies of about 1 to 1.5 eV above E_F, the characters of the Bi p bands mix with the p orbital of the oxygen atoms on the Bi and Sr layers, indicating that these states are derived from the Bi-O antibonding bands.

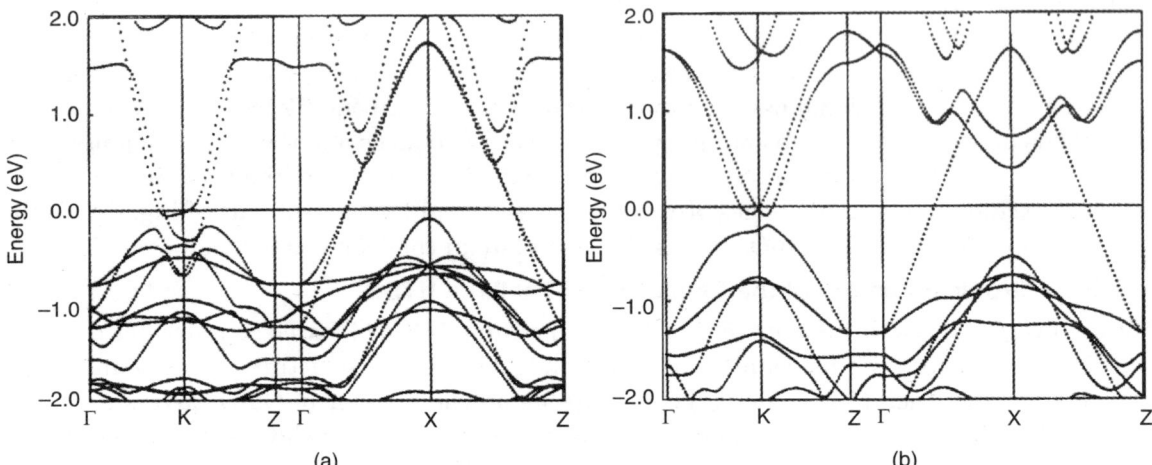

Fig. 6.174 Energy band structures around the Fermi energy for (a) HTSC $Bi_2Sr_2CaCu_2O_8$ and (b) the low T_c superconductor $Bi_2Sr_2CuO_6$ [540]. We may note the presence of the Cu-O antibonding band in both systems and the difference in the bismuth bands around the K-point

Band Structure of $Bi_2Sr_2CuO_6$

The band structure low T_c superconductor $Bi_2Sr_2CuO_6$ is shown in Fig. 6.174 (b). The copper oxygen antibonding band looks essentially identical to the corresponding bands in the $Bi_2Sr_2CaCu_2O_8$ compound. Around the K-point, the two Bi p-bands barely cross and extend less than 0.1 eV below E_F, compared with 0.7 eV in the case of the higher T_c system. Moreover, the low T_c system shows a gap between the Bi-p bands and the antibonding Cu-O bands at K which is not present in HTSC compounds so that the metallic nature of these layers will be greatly reduced.

Electronic Structure of Thallium Compounds

T_c-increases with increasing Cu-O_2 layers in the thallium compounds $Tl_2Ba_2CuO_6$, $Tl_2Ba_2CaCu_2O_8$ and $Tl_2Ba_2Ca_2Cu_3O_{10}$ have T_c's 80, 110 and 125 K respectively. The band structures of the two and three layers thallium compounds have been first reported by Yu et al. [549] and that of the one-layer compounds by Hamann and Mattheiss [550]. The antibonding Cu-O_2 band, whose filling factor is controlled by charge transfer to the Tl-O band that crosses E_F. In general, the Tl-O bands lie higher in energy relative to the valence band complex than the corresponding Bi-O bands in the Bi-compounds, so that it is the Tl_{6s}-O_{2p} state that crosses E_F. Detailed charge density wave (CDW) countours show strong interplanar coupling between the two Tl-O layers that were absent in the Bi-compounds.

Near E_F, each Ca:Cu-O_2 layer introduces one antibonding band which is half filled. However, the three antibonding CuO_2 bands are nearly degenerate in $Tl_2Ba_2Ca_2Cu_3O_{10}$, with filling factors very similar to that $Tl_2Ba_2CaCu_2O_8$, which indicates that there is charge transfer from the Cu-O_2 bands to the Tl-O bands as more Ca:Cu-O_2 layers are added. Thus, T_c correlates with the metallic nature of the intervening Tl-O bands, similar to the Bi compounds.

Kasowski et al. [551] have calculated the electronic structure of the one to four-layer Tl compounds. In addition to the antibonding Cu-O_2 bands and Tl-O bands, they found narrow Cu-O bands that lie very close to and sometimes overlaps E_F. The narrow Cu-O bands are unique feature of their method, which is also present in their early work on La_2CuO_4 [552], Y $Ba_2Cu_3O_7$ [553], and the Bi compounds [554]; but is absent from the band structures shown in previous sections. Due to the flatness and the multiplicity of these narrow bands, the hole concentration can become quite large even for a very small overlap. This is why, this method is not particularly accurate to predict the filling factors of the Cu-O_2 and the Tl-O bands. For example, they found only one antibonding Cu-O_2 and intersecting E_F in $Tl_2Ba_2CuO_6$ that is significantly less than half filled, and the Tl-O band appears to be insulating. However, there is one electron to be shared between the antibonding Cu-O_2 and the Tl-O bands. If the Tl-O layer was indeed insulating, then the Cu-O_2 band has to be exactly half filled and the system would an antiferromagnetism insulator as found in La_2CuO_4.

The intervening metallic Tl-O layers play the dual role of doping the Cu-O_2 layers and enhancing the interlayer coupling between the Cu-O_2 layers. These effects increase with increasing number of Ca:Cu-O_2 layers and therefore increasing T_c. However, the degree of charge transfer among the Tl-O and Cu-O_2 layers appears to be much weaker than that of the Bi compounds considering the magnitude of the superconducting temperatures. One can understand this apparent discrepancy from neutron diffraction measurements on many different samples of $Tl_2Ba_3CuO_6$, which have shown that T_c varies from 4 to 70 K depending on the precise structural arrangement [555]. In particular, there is an orthorhombic phase with a well ordered structure which is non superconducting, while the superconducting material is pseudo-tetragonal with disorded oxygen within the Tl-O plane.

Lee et al. [556] have reported significant frequency shifts of thallium nuclear magnetic resonance in the superconducting state of the HTSC $Tl_2Ba_3Ca_3Cu_4O_{10+x}$, which suggest that the Tl-O layers participate directly in the superconductivity.

REFERENCES

1. J.G. Bednorz and K.A. Muller, *Z. Phys. B* 64, 189 (1986).
2. J. Muller, *A 15 Type Superconductors*, *Rep. Proc. Phys.*, 43:(1980).
3. A. Schilling et al., *Nature* 363, 36 (1993).
4. J.J. Capponi et al., *Physica C* 235-240, 146 (1994).
5. C.W. Chu et al., *Nature* 363, 323 (1993).
6. E. V. Antipov et al., *Physica C* 215, 1(1993).
7. A. S. Alexandrov and P. P. Edwards, *Physica C* 331, 97(2000).
8. C.P. Poole, et al., *Superconductivity*, Academic Press, San Diego (1995).
9. M.K. Wu, et al., *Phys. Rev. Lett.* 58, 908 (1987).
10. E. Kaldis et al., *Physica C* 185-189, 190 (1991).
11. R. Hott et al., *Phys. B1*, 48, 355 (1992).
12. A. Maignan et al., *Physica C* 219, 407 (1994).
13. Y. Tokura et al., *Nature* 337, 345 (1989).

14. B. Hopfengartner et al., *Phys. Rev. B* 44, 741 (1991).
15. E.H. Appleman et al., *Inorganic chemistry* 26, 3237 (1987).
16. A.I. Nazzal et al., *Physica C* 153-155, 1367 (1988).
17. Y. Maeno, *Physica C* 185-189, 587 (1991).
18. W.M. Chen et al., *Physica C* 270, 349 (1996).
19. W.M. Chen et al., *Physica C* 270, 155 (1996).
20. A. Fukuoka et al., *Physica C* 265, 13 (1996).
21. P. Krishnaraj et al., *Physica C* 234, 318 (1994).
22. K.P. Sinha and S.L. Kakani, *High Temperature Superconductivity: Current Results and Novel Mechanisms*, Nova Science Pubs., New York (1995).
23. S.L. Kakani, *Superconductivity: Current Topics*, Bookman Associates, Jaipur (2001), Chapter 3.
24. K.P. Sinha and S.L. Kakani, *Proc. Natl. Acad. Sciences*, India, LXXII, Section A, Pt. III, 153 (2002).
25. J.H. Miller, Jr. and J.R. Claycomb, (Preprint) 97 : 091.
26. A.T. Matveev, E. Takayama-Muromachi, *Physica C* 254, 26 (1995).
27. M. Isobe et al., *Physica C* 234, 120 (1994).
28. E. Takayama-Muromachi et al., *Physica C* 241, 137 (1995).
29. T. Kawashima et al., *Physica C* 254, 131 (1995).
30. M. Isobe et al., *Physica C* 222, 310 (1994).
31. E. Takayama-Muromachi and M. Isobe, *Jpn. J. Appl. Phys.* 33, L 1399 (1994).
32. J. Ramirez-Castellanos et al., *Physica C* 251, 279 (1995).
33. X.-J. Wu et al., *Physica C* 251. 279 (1995).
34. X.-J. Wu et al., *Physica C* 266, 261 (1996).
35. T. Tamura et al., *Physica C* 249, 111 (1995).
36. T. Tamura et al., *Physica C* 277, 1 (1997).
37. C.-Q. Jin, et al., *Physica C* 223, 238 (1994).
38. X.–J. Wu et al., *Physica C* 223, 243 (1994).
39. M. A. Alario-Franco et al., *Physica C* 222, 52 (1994).
40. E. Takayama-Muromachi et al., *Physica C* 252, 221 (1995).
41. T. Kawashima et al., *Physica C* 224, 69 (1994).
42. H. Kumakura et al., *Physica C* 226, 222 (1994).
43. T. Kawashima, E. Takayama-Muromachi, *Physica C* 267, 106 (1996).
44. T. Kawashima et al., *Physica C* 257, 313 (1996).
45. M. Decroux et al., *Europhys. Lett.* 3, 1035 (1987).
46. R.M. Hazen, "Crystal Structure of High Temperature Superconductors", In *Physical Properties of High Temperature Superconductors II*, D.M. Ginsberg, ed., World Scientific, Singapore (1990).
47. J.M. Tranquada et al., *Nature* 375, 561 (1997) *Physica C* 282-287, 166 (1997).
48. V.J. Emery and S.A. Kivelson, *Nature* 374, 434 (1995).
49. S.W. Cheong et al., *Phys. Rev. Lett.* 67, 1791 (1991).
50. T.E. Mason et al., *Phys. Rev. Lett.* 68, 1414 (1992).

51. T.R. Thurston et al., *Phys. Rev. B* 46, 9128 (1992).
52. R. Beyers et al., *Appl. Phys. Lett.* 50, 1918 (1987).
53. A. Junod et al., *Europhys. Lett.* 4, 247 (1987).
54. M. Cyrot and M. Pavuna, *"Introduction to Superconductivity and High-T_c materials"*, World Scientific, London, Singapore, New Jersey (1992).
55. T.A. Friedman et al., *Phys. Rev. B* 42, 6217 (1990).
56. S.A. Sunshine et al., *Phys. Rev. B* 38, 893 (1988).
57. G. Calestani et al., *Physica C* 158, 217 (1989).
58. C.J.D. Hetherington et al., *Appl. Phys. Lett.* 53, 1016 (1988).
59. R.M. Hazen et al., *Phys. Rev. Lett.* 60, 1657 (1988).
60. M.R. Presland et al., *Physica C* 277, 170 (1997).
61. J.L. Wagner et al., *Physica C* 277, 170 (1997).
62. D.M. Ogborne and M.T. Weller, *Physica C* 223, 283 (1994).
63. T. Kajitani et al., *Physica C* 161, 483 (1989).
64. Z.Z. Sheng and A.M. Hermann, *Nature* 332, 55 (1988).
65. S.S.P. Parkin et al., *Phys. Rev. Lett.* 61, 750 (1988).
66. E.V. Antipov et al. *Physica C* 215, 1 (1993).
67. S.M. Loureiro et al., *Physica C* 243, 1 (1995).
68. H.R. Khan et al., *Physica C* 229, 165 (1994).
69. M. Itoh et al., *Physica C* 212, 271 (1993).
70. R.L. Meng et al., *Physica C* 214, 307 (1993).
71. P.G. Radaelli et al., *Physica C* 216, 29 (1993).
72. M. Paranthaman, *Physica C* 222, 7 (1994).
73. H.M. Shao et al., *Physica C* 232, 5 (1994).
74. K. Isawa et al., *Physica C* 222, 33 (1994).
75. R.L. Meng et al., *Physica C* 216, 21 (1993).
76. Z.L. Huang et al., *Physica C* 217, 1 (1993).
77. O. Chmaissem et al., *Physica C* 212, 259 (1993).
78. M. Cantoni et al., *Physica C* 215, 11 (1993).
79. S.M. Loureiro et al., *Physica C* 257, 117 (1996).
80. R.J. Cava et al., *Phys. Rev. Lett.* 58, 1676 (1987).
81. I.G. Gopalakrishnan et al., *Physica C* 182, 67 (1991).
82. W. Carrilo-Cabrera, W. Gopal, *Physica C* 161, 373 (1989).
83. S.S.P. Parkin et al., *Phys. Rev. Lett.* 60, 2539 (1988).
84. R. Beyers et al., Appl. Phys. Lett. 53, 432 (1988).
85. D.M. Ogborne and M. Weller, *Physica C* 230, 153 (1994).
86. S.N. Putilin et al., *Nature* 362, 226 (1993).
87. E.V. Antipov et al., *Physica C* 218, 348 (1993).
88. J.J. Capponi et al., *Physica C* 256, 1 (1996).

89. H. Schwer et al., *Physica C* 245, 7 (1995).
90. M.D. Marcos and J.P. Attfield, *Physica C* 270, 267 (1996).
91. J.B. Torrance et al., *Phys. Rev. Lett.* 61, 1127 (1988).
92. Th. Schweizer et al., *Physica C* 225, 143 (1994).
93. A. Bertinotti et al., *Physica C* 250, 213 (1995).
94. R.S. Liu et al., *Physica C* 182, 119 (1991).
95. A. Maignan et al., *Physica C* 170, 350 (1990).
96. G.V.M. Williams and J.L. Tallon, *Physica C* 258, 41 (1996).
97. G. Triscone et al., *Physica C* 272, 21 (1996).
98. A.K. Klehe et al., *Physica C* 257, 105 (1996).
99. C.C. Almasan et al., *Phys. Rev. Lett.* 69, 680 (1992).
100. L. Gao et al., *Phys. Rev. B* 50, 4260 (1994).
101. D. Tristan Jover et al., *Physica C* 235-240, 893 (1994).
102. D. Tristan Jover et al., *Physica C* 218, 24 (1993).
103. J. Beille et al., *Solid State Commun.* 77, 141 (1991).
104. J.J. Markert et al., *Phys. Rev. Lett.* 64, 80 (1990).
105. M. Baran and I. Fita, *Physica C* 261, 125 (1996).
106. M. Baran et al., *Physica C* 241, 383 (1995).
107. R. Sieburger et al., *Physica C* 181, 335 (1991).
108. N. Watanabe et al., *Physica C* 235-240, 1309 (1994).
109. R. Kubiak et al., *Physica C* 166, 523 (1990).
110. J.E. Schirber et al., *Physica C* 157, 237 (1989).
111. S. Han Shun-hui et al., *Physica C* 156, 113 (1988).
112. B. Morosin et al., *Physica C* 152, 223 (1988).
113. A. -K. Klehe et al., *Physica C* 223, 313 (1994).
114. Y.S. Yao et al., *Physica C* 224, 91 (1994).
115. H. Takahashi et al., *Physica C* 218, 1 (1993).
116. T.P. Orlando and K.A. Delin, *"Foundation of Applied Superconductivity,"* **Addison-Wesley Publishing Company, Reading, Massachusetts (1991).**
117. M. Tinkhan, *Physica C* 235-240, 3 (1994).
118. S. Senoussi et al., *J. Appl. Phys.* 63, 4176 (1988).
119. R. Usami et al., *Physica C* 243, 19 (1995).
120. G. Triscone et al., *Physica C* 264, 233 (1996).
121. I. Matsubara et al., *Physica C* 256, 33 (1996).
122. Dong-Ho Wu and S. Sridhar, *Phys. Rev. Lett.* 65, 2074 (1990).
123. Ch. Heinzel, *Europhysics Letters* 13, 531 (1990).
124. J.Z. Sun, et al., *Appl. Phys. Lett.* 54, 763 (1989).
125. E. Zeldov et al., *Phys. Rev. Lett.* 62, 3093 (1989).
126. J.N. Li et al., *Appl. Phys. A* 47, 209 (1988).

127. J.H. Kang et al., *Appl. Phys. Lett.* 53, 2560 (1988).
128. T.T.M. Palstra et al., *Phys. Rev. Lett.* 61, 1662 (1988).
129. A.P. Malozemoff "Macroscopic Magnetic Properties of High Temperature Superconductors," In *Physical Properties of High Temperature Superconductors. Vol. I*, D.M. Ginsberg, ed., World Scientific, Singapore (1990).
130. U. Welp et al., *Phys. Rev. Lett.* 62, 1908 (1989).
131. J.H. Kang et al., *Appl. Phys. Lett.* 52, 2080 (1988).
132. T.T.M. Palstra et al., *Phys. Rev. B* 38, 5102 (1988).
133. I. Matsubara et al., *Phys. Rev. B* 45, 7414 (1992).
134. O. Laborde et al., *Physica C* 162-164, 1619 (1989).
135. S. Foner et al., *Phys. Rev. B* 46, 14936 (1992).
136. M. Akamastu et al., *Physica C* 235-240, 1619 (1994).
137. A.J. Panson et al., *Appl. Phys. Lett.* 50, 1104 (1987).
138. J.J. Neumeier et al., *Appl. Phys. Lett.* 51, 371 (1987).
139. L. Zhang et al., *Phys. Rev. B* 45, 4978 (1992).
140. F. Shi et al., *Phys. Rev. B* 41, 6541 (1990).
141. D.R. Harshman and A.P. Mills, Jr. *Phys. Rev. B* 45, 10684 (1992).
142. G. Schatz and A. Weidinger, *"Nuclear Condensed Matter Physics"*, John Wiley and Sons, Chichester, England (1996).
143. D.R. Harshman et al., *Phys. Rev. B* 39, 851 (1989).
144. Ju Young Lee and T.R. Lemberger, *Appl. Phys. Lett.* 62, 2419 (1993).
145. L. Krusin-Elbaum et al., *Phys. Rev. Lett.* 62, 217 (1989).
146. N. Athanassopoulous, and J.R. Cooper, *Physica C* 259, 326 (1996).
147. G. Brand Statter et al., *Physica C* 235-240, 1845 (1994).
148. G. Le Bras et al., *Physica C* 271, 205 (1996).
149. W. Bucket, *Superleitung, Grundlagen and Anwendungen*, VCH, Weinheim, Germany (1994).
150. M.R. Beasley, *Physica C* 209, 43 (1993).
151. T. Shibauchi et al., *Physica C* 264, 227 (1996).
152. B.J. Feenstra et al., *Physica C* 278, 213 (1997).
153. D. Pelloquin et al., *Physica C* 273, 205 (1997).
154. L. Krusin-Elbaum et al., *Phys. Rev. B* 39, 2936 (1989).
155. H. Mukaida et al., *Phys. Rev. B* 42, 2659 (1990).
156. J.R. Thompson et al., *Phys. Rev. B* 48, 14031 (1993).
157. T. Ekino et al., *Physica C* 235-240, 1899 (1994).
158. R. Escudero et al., *Physica C* 166, 15 (1990).
159. N. Hudakova et al., *Physica C* 246, 163 (1995).
160. R.T. Collins et al., *Phys. Rev. Lett.* 59, 704 (1987).
161. Z. Schlesinger et al., *Phys. Rev. Lett.* 59, 1958 (1987).
162. R.T. Collins et al., *Phys. Rev. Lett.* 63, 422 (1989).
163. H. Seidel et al., *Europhys. Lett.* 5, 647 (1988).

164. B. Friedl et al., *Phys. Rev.* Lett. 65, 915 (1990).
165. R. Manzke et al., *Europhys. Letters* 9, 477 (1989).
166. G.D. Mahan, *Phys. Rev. B* 40, 11317 (1989).
167. L.N. Bulaevskii and M.V. Zyskin, *Phys. Rev. B* 42, 10230 (1990).
168. T.A. Faltens et al., *Phys. Rev. Lett.* 59, 915 (1987).
169. K.J. Leary et al., *Phys. Rev. Lett.* 59, 1236 (1987).
170. L.C. Bourne et al., *Phys. Rev. Lett.* 58, 2337 (1987).
171. B. Batlogg et al., *Phys. Rev. Lett.* 58, 2333 (1987).
172. W.W. Warren et al., *Phys. Rev. Lett.* 59, 1860 (1987).
173. I. Takeuchi et al., *Physica C* 158, 83 (1989).
174. R.P. Huebener et al., *Phys. B* 1, 50, 163 (1994).
175. L. Krusin-Elbaum, et al., *Nature* 373, 679 (1995).
176. Y. Tsabba and S. Reich, *Physica C* 269, 1 (1996).
177. P. Schmitt et al., Physica *C* 168, 475 (1990).
178. H. Yamasaki et al., *IEEE Trans. Appl. Supercond.* 3, 1536 (1993).
179. M. Tachiki and S. Takahashi, *Cyrogenics* 32, 923 (1993).
180. R.S. Liu et al., *Appl. Phys. Lett.* 60, 1019 (1992).
181. L. Schultz et al., *IEEE Trans. Magn.* 27, 990 (1991).
182. D. Kumar et al., *Appl. Phys. Lett.* 62, 3522 (1993).
183. R.A. Rao et al., *Appl. Phys. Lett.* 69, 3911 (1996).
184. G. Samadi et al., *Physica C* 268, 307 (1996).
185. S.H. Yun et al., *Appl. Phys. Lett.* 69, 3423 (1996).
186. H. Moriwaki et al., *Appl. Phys. Lett.* 69, 3423 (1996).
187. J.E. Evetts and B.A. Glowacki, *Cryogenics* 28, 641 (1988).
188. J. Mannhart et al., *Phys. Rev. Lett.* 61, 2476 (1988).
189. J.W. Ekin, *Appl. Phys. Lett.* 55, 907 (1989).
190. J.W. Ekin et al., *Appl. Phys. Lett.* 61, 858 (1992).
191. C.P. Bean, *Phys. Rev. Lett.* 8, 250 (1962).
192. C.P. Bean, *Rev. Mod. Phys.* 36, 31 (1964).
193. K. Kwasnitza and Ch. Widmer, *Cryogenics* 29, 1035 (1989).
194. K. Kwasnitza and Ch. Widmer, *Physica C* 171, 211 (1990).
195. K. Kwasnitza and Ch. Widmer, *Physica C* 202, 75 (1992).
196. K. Shibutani et al., *Appl. Phys. Lett.* 64, 924 (1994).
197. R.E. Gladyshevskii et al., *Physica C* 255, 113 (1995).
198. M.P. Maley et al., *IEEE Trans. Magn.* 27, 1139 (1991) 253.
199. F.A. List et al., *Physica C* 275, 220 (1997).
200. W.D. Lee et al., *J. Appl. Phys.* 77, 3942 (1995).
201. R. Wesche, *Physica C* 246, 186 (1995).
202. D. Dimos et al., *Phys. Rev. Lett.* 61, 219 (1988).

203. D. Dimos et al., *Phys. Rev. B* 41, 4038 (1990).
204. E. Sarnelli et al., *Appl. Phys. Lett.* 65, 362 (1994).
205. J. A. Alaseo et al., *Physica C* 247, 263 (1995).
206. V.R. Todt et al., *Appl. Phys. Lett.* 69, 3746 (1996).
207. Q. Li et al., *Appl. Phys. Lett.* 70, 1164 (1997).
208. P.L. Gammel et al., *Phys. Rev. Lett.* 59, 2592 (1987).
209. G.J. Dolan et al., *Phys. Rev. Lett.* 62, 827 (1989).
210. K.A. Muller et al., *Phys. Rev. Lett.* 58, 1143 (1987).
211. Y. Yeshurun A.P. Malozemoff, *Phys. Rev. Lett.* 60, 2202 (1988).
212. P.W. Anderson, *Phys. Rev. Lett.* 9, 309 (1962).
213. P.W. Anderson and Y.B. Kim, *Rev. Mod. Phys.* 36, 39 (1964).
214. Y. Yeshurun et al., *Cryogenics* 29, 258 (1989).
215. L. Civale et al., *Phys. Rev. Lett.* 65, 1164 (1990).
216. M. Mittag et al., *Physica C* 174, 101 (1991).
217. P. Svedlindth et al., *Physica C* 176, 336 (1995).
218. A. Spirgatis et al., *Cryogenics* 33, 138 (1993).
219. M. Niderost et al., *Physica C* 235-240, 2891 (1994).
220. A. Gupta et al., *Physica C* 272, 33 (1996).
221. E. Zeldov et al., *Appl. Phys. Lett.* 56, 680 (1990).
222. E. Zeldov et al., *Appl. Phys. Lett.* 56, 1700 (1990).
223. V.M. Vinokur et al., *Phys. Rev. Lett.* 67, 915 (1991).
224. Y. Ren and P.A.J. de Groot, *Cryogenics,* 33, 357 (1993).
225. M.V. Feigel'man et al., *Phys. Rev. Lett.* 63, 2303 (1989).
226. M.V. Feigel'man et al., *Phys. Rev. B* 43, 6263 (1991).
227. C.W. Hagen and R. Griessen, *Phys. Rev. Lett.* 62, 2857 (1989).
228. M. Rupp et al., *Appl. Phys. Lett.* 67, 291 (1995).
229. P. Berghuis et al., *Physica C* 256, 13 (1996).
230. T.W. Li et al., *Physica C* 274, 197 (1997).
231. N. Ihara and T. Matsushita, *Physica C* 257, 223 (1996).
232. M. Kiuchi et al., *Physica C* 260, 177 (1996).
233. A Wahl et al., *Cryogenics* 34, 941 (1994).
234. D.N. Zheng et al., *Cryogenics* 33, 46 (1993).
235. L. Zhang et al., *Physica C* 268, 287 (1996).
236. U. Welp et al., *Appl. Phys. Lett.* 63, 693 (1993).
237. Z.J. Huang et al., *Phys. Rev. B* 49, 4218 (1994).
238. O. Laborde et al., *Physica C* 235-240, 2717 (1994).
239. J.G. Wu et al., *Cryogenics* 36, 17 (1996).
240. K. Isawa et al., *Appl. Phys. Lett.* 64, 1301 (1994).
241. J. Lohle et al., *Physica C* 266, 104 (1996).

242. U. Welp et al., *Physica C* 218, 373 (1993).
243. H. Asaoka et al., *Physica C* 268, 14 (1996).
244. N. Nakamura et al., *Appl. Phys. Lett.* 61, 3044 (1992).
245. T. Egi et al., *Appl. Phys. Lett.* 66, 3680 (1995).
246. R. Funahashi et al., *Appl. Phys. Lett.* 64, 646 (1994).
247. G. Villard et al., *Appl. Phys. Lett.* 69, 1480 (1996).
248. T.W. Li et al., *Physica C* 257, 179 (1996).
249. R. Noetzel et al., *Physica C* 260, 290 (1996).
250. H. Ma et al., *Europhys. Lett.* 10, 375 (1989).
251. G.K. Bichille et al., *Cryogenics* 31, 833 (1991).
252. M. Wakata et al. *Cryogenics* 32, 1046 (1992).
253. N. Adamopoulos et al., *Physica C* 242, 68 (1995).
254. K. Fossheim et al., *Physica C* 248, 195 (1995).
255. S. Sengupta et al., *Physica C* 264, 34 (1996).
256. Y. Fuwa and K. Ogawa, *Physica C* 277, 54 (1997).
257. K. Matsumoto et al., *Cryogenics* 30, 5 (1990).
258. M. Murakami et al., *Cryogenics* 32, 930 (1992).
259. H. -L Su et al., *Physica C* 249, 241 (1995).
260. P. Majewski et al., *Physica C* 249, 234 (1995).
261. P. Majewski et al., "Precipitation and Pinning in Bi-2223 Ceramics". In *'High Temperature Superconductors : Synthesis, Processing and Applications—II*, U. Balachandran and P.J. McGinn, eds., TMS Warrendale, USA (1997).
262. H. Kumakura et al., *Physica C* 251, 231 (1995).
263. A. Ignatiev et al., *Appl. Phys. Lett.* 70, 1474 (1997).
264. H. Kumakura et al., *J. Appl. Phys.* 72, 800 (1992).
265. J.R. Thompson et al., *Appl. Phys. Lett.* 60, 2306 (1992).
266. V. Hardy et al., *Physica C* 191, 85 (1992).
267. H. -W. Neumuller et al., *Cryogenics* 33, 14 (1993).
268. X. Gao et al., *Physica C* 250, 325 (1995).
269. H. Frank et al., *Physica C* 259, 142 (1996).
270. A.K. Pradhan et al., *Physica C* 264, 109 (1996).
271. C. Goupil et al., *Physica C* 278, 23 (1997).
272. F. Warmont et al., *Physica C* 277, 61 (1997).
273. L. Civale et al., *Physica C* 208, 137 (1993).
274. P. Kummeth et al., *Appl. Phys. Lett.* 65, 1302 (1994).
275. V. Hardy et al., *Physica C* 178, 255 (1991).
276. J.E. Tkaczyk et al., *Appl. Phys. Lett.* 62, 3031 (1993).
277. D.G. Steel et al., *Physica C* 265, 159 (1996).
278. T. Hwa et al., *Physica C* 257, 16 (1996).
279. V. Hardy et al., *Physica C* 257, 16 (1996).

280. L. Krusin-Elbaum et al., *Appl. Phys. Lett.* 64, 3331 (1994).
281. G. Blatter et al., *Rev. Mod. Phys.* 66, 1125 (1994).
282. F. Lindemann, *Phys. Z* 11, 69 (1910).
283. A. Houghton et al., *Phys. Rev. B* 40, 6753 (1989).
284. M.P.A. Fisher, *Phys. Rev. Lett.* 62, 1415 (1989).
285. R.H. Koch et al., *Phys. Rev. Lett.* 63, 1511 (1989).
286. H. Yamasaki et al., *Cryogenics* 35, 263 (1995).
287. H. Mawatari et al., *Cryogenics* 35, 161 (1995).
288. D. Ertas and D.R. Nelson, *Physica C* 272, 79 (1996).
289. I.F. Herbut and Z. Tesanovic, *Physica C* 255 324 (1995).
290. A. Junod et al., *Physica C* 275, 245 (1997).
291. A. Schilling et al., *Nature* 382, 791 (1996).
292. Ch. Kittel, *'Introduction to Solid State Physics'*, John Wiley and Sons, New York (1976).
293. T. Penny et al., *Phys. Rev. B* 38, 2918 (1988).
294. S. Martin et al., *Appl. Phys. Lett.* 54, 72 (1989).
295. Y. Kotaka et al., *Physica C* 235-240, 1529 (1994).
296. M.K.R. Khan et al., *Physica C* 258, 315 (1996).
297. D.-S. Jeon et al., *Physica C* 253, 102 (1995).
298. S.M. Green et al., *J. Appl. Phys.* 66, 728 (1989).
299. J.L. Tallon et al., *Appl. Phys. Lett.* 54, 1591 (1989).
300. L. Forro and J.R. Cooper, *Europhys. Lett.* 11, 55 (1990).
301. H. Takagi et al., *Phys. Rev. B* 40, 2254 (1989).
302. R.J. Cava et al., *Phys. Rev. Lett.* 58, 408 (1987).
303. P.C. McIntyre et al., *Appl. Phys.* 68, 4183 (1990).
304. N. Wendling et al., *Physica C* 235-240, 1517 (1994).
305. M. Ikebe et al., *Cryogenics* 34, 57 (1994).
306. J. Van den Berg et al., *Europhysics Lett.* 4, 737 (1987).
307. I. Schildermans et al., *Physica C* 278, 55 (1997).
308. B.M. Terzijska et al., *Cryogenics* 32, 53 (1992).
309. U. Gottwick et al., *Europhysics Letters* 4, 1183 (1987).
310. T. Wada et al., *Appl. Phys. Lett.* 52, 1989 (1988).
311. S. Adachi et al., *Physica C* 233, 149 (1994).
312. X.G. Zheng et al., *Physica C* 271, 272 (1996).
313. C. Michel et al., *Z. Phys. B* 68, 421 (1987).
314. F. Rullier-Albenque et al., *Physica C* 254, 88 (1995).
315. J. Hejtmanek et al., *Physica C* 264, 220 (1996).
316. R. Jin et al., *Physica C* 250, 395 (1995).
317. Donglu Shi et al., *Appl. Phys. Lett.* 54, 2358 (1989).
318. L. Forro et al., *Europhys. Lett.* 10, 371 (1989).

319. Y. Kotaka et al., *Physica C* 235-240, 1529 (1994).
320. M.A. Quiijada et al., *Physica C* 235-240, 1123 (1994).
321. T. Horiuchi et al., *Physica C* 221, 143 (1994).
322. C. Stolzel et al., *Physica C* 204, 15 (1992).
323. W. Carrillo-Cabrera et al., *Appl. Phys. Lett.* 55, 1032 (1989).
324. K. Mori et al., *Physica C* 162-164, 152 (1989).
325. T. Uzumaki et al., *Appl. Phys. Lett.* 54, 2253 (1989).
326. S. Bansal et al., *Physica C* 173, 260 (1991).
327. S.K. Bandyopadhyay et al., *Physica C* 228, 109 (1994).
328. V. Plechacek et al., *Cryogenics* 30, 11 (1990).
329. S. Nakajima et al., *Physica C* 170, 443 (1990).
330. S.T. Kaneko et al., *Physica C* 178, 377 (1991).
331. B.T. Ahn et al., *Appl. Phys. Lett.* 60, 2150 (1992).
332. Y.X. Jia et al., *Physica C* 234, 24 (1994).
333. M. Kikuchi et al., *Physica C* 158, 79 (1989).
334. R. Vijayaraghawan et al., *Physica C* 179, 183 (1991).
335. I.C. Chang et al., *Physica C* 223, 207 (1994).
336. Y.T. Ren et al., *Physica C* 217, 6 (1993).
337. K. Yamaura et al., *Physica C* 229, 183 (1994).
338. M. Uehara et al., *Physica C* 229, 310 (1994).
339. S. Yang et al., *IEEE Trans. Appl. Supercond.* 5, 1471 (1995).
340. A. Jezowski and J. Klamut in *'Studies of High-Temperature-Superconductor'* (A. Narlikar, Ed.), Vol. 4, p. 263, Nova Science Publishers, New York, 1990.
341. C. Uher, *J. Superconductivity* 3, 337 (1990).
342. C. Uher, in *'Physical Properties of High Temperature Superconductors*, (D.M. Ginsberg, Ed.), Vol. 3, p. 159. World Scientific, Singapore, 1992.
343. J. Bardeen, G. Rickayzen and L. Tewordt, *Phys. Rev.* 113, 982 (1959).
344. L. Tewordt and Th. Wolkhausen, *Solid State Communi.* 70, 839 (1989).
345. S.D. Peacor, et al., *Phys. Rev. B* 43, 8721 (1991).
346. C. Uher et al., *J. Supercond.* 7, 323 (1994).
347. S.D. Peacor et al., *Phys. Rev. B* 44, 9508 (1991).
348. D.A. Bonn et al., *Phys. Rev. Lett.* 68, 2390 (1992).
349. J.M. Chwalek et al., *Appl. Phys. Lett.* 57, 1696 (1990).
350. R.C. Yu et al., *Phys. Rev. Lett.* 69, 1431 (1992).
351. L.P. Kadanoff and P.C. Martin, *Phys. Rev.* 124, 670 (1961).
352. L. Tewordt, *Phys. Rev.* 128, 12 (1962).
353. R.C. Yu et al., *Phys. Rev. Lett.* 69, 1431 (1992).
354. J.L. Cohn et al., *Phys. Rev. B* 38, 2892 (1988).
355. U. Gottwick et al., *Europhysics Lett.* 4, 1183 (1987).
356. C. Uher and J.L. Cohn, *J. Phys. C* 21, L 957 (1988).

357. S.D. Peacor and C. Uher, *Phys. Rev. B* 39, 11559 (1989).
358. D.-M. Zhu et al., *Phys. Rev. B* 40, 841 (1989).
359. G. Sparn et al., *Physica C* 162-164, 508 (1989).
360. M. Sera et al., *Solid State Commun.* 70, 839 (1989).
361. S.J. Hagen et al., *Phys. Rev. B* 40, 9389 (1989).
362. S.C. Cao et al., *Phys. Rev. B* 44, 12571 (1991).
363. V.B. Efimov and L.P. Mezhov-Deglin, *Low Temp. Phys.* 23, 204 (1997).
364. M.F. Crommie and A. Zettl, *Phys. Rev. B* 43, 408 (1991).
365. D.T. Morelli et al., *Phys. Rev. B* 41, 2520 (1990).
366. C. Caroli and M. Cyrot, *Phys. Kondens. Mater.* 4, 285 (1965).
367. K. Maki, *Phys. Rev.* 158, 397 (1967).
368. A.F. Andreev, *Sov. Phys. JETP* 19, 1228 (1964).
369. T.T.M. Palstra et al., *Phys. Rev. Lett.* 64, 3090 (1990).
370. N.V. Zavaristky et al., *Physica C* 180, 417 (1991).
371. R.A. Richardson et al., *Phys. Rev. Lett.* 67, 3856 (1991).
372. M. Matsukawa et al., *Cryogenics* 37, 255 (1997).
373. H. Fujishiro et al., *Physica C* 235-240, 1533 (1994).
374. L. Tewordt and Th. Wolkhausen, *Solid State Commun.* 70, 839 (1989).
375. S. Cartelazzi et al., *Physica C* 273, 314 (1997).
376. S.D. Peacor and C. Uher, *Phys. Rev. B* 39, 11559 (1989).
377. C. Uher et al., *Physica C* 235-240, 1377 (1994).
378. F. Yu, et al., *Physica C* 267, 308 (1996).
379. S.J. Hagen et al., *Phys. Rev. B* 40, 9349 (1989).
380. A. Junod, *"Specific heat of High Temperature Superconductors II,* D.M. Ginsberg (ed.), World Scientific, Singapore (1990).
381. E. Janod et al., *Physica C* 234, 269 (1994).
382. A. Junod et al., *Physica C* 229, 209 (1994).
383. V.G. Bessergenev et al., *Physica C* 245, 36 (1995).
384. A. Mirmlstein et al., *Physica C* 241, 301 (1995)
385. L. Ghivelder et al., *Physica C* 194, 97 (1992).
386. A. Schilling et al., *Physica C* 169, 237 (1990).
387. A. Junod et al., *Physica C* 168, 47 (1990).
388. H. Kierspel et al., *Physica C* 262, 177 (1996).
389. J.E. Gordan et al., *Physica C* 185-189, 1351 (1991).
390. M. Ausloos et al., *Physica C* 235-240, 1767 (1994).
391. N. Kobayashi et al., *Physica C* 219, 265 (1994).
392. A.K. Bandopadhyay et al., *Physica C* 165, 29 (1990).
393. A. Mirmelstein et al., *Physica C* 248, 335 (1995).
394. B.F. Woodfield et al., *Physica C* 235-240, 1741 (1994).

395. O. Jeandupeux et al., *Physica C* 216, 17 (1993).
396. Charles P. Poole, Jr. *'Handbook of Superconductivity,* Academic Press, San Diego, USA (2000).
397. A.B. Kaiser and C. Uher, in *'Studies in High Temperature Superconductors'* Vol. 7 (A.V. Narlikar, Ed.) p. 353, Nova Science, New York (1991).
398. C.K. Subramaniam et al., *Supercond. Sci. Technol.* 7, 30 (1994).
399. V.L. Ginzburg, Sov. *Phys. Usp.* 34, 101 (1991).
400. R.P. Nuebener et al., *Physica C* 181, 345 (1991).
401. H.C. Ri et al., *Phys. Rev. B* 47, 12312 (1993).
402. E.A. Meilikhov and R.M. Farzetdinova, *J. Supercond.* 7, 897 (1994).
403. H. Ghamlouch and M. Aubin M, *Physica C* 269, 163 (1996).
404. G. Yu. Logvenov et al., *Phys. Rev. B* 46, 11102 (1992).
405. H.J. Goldsmid, *Electronic Refrigeration*, Pion Limited (1986).
406. M. Galffy et al., *Phys. Rev. B* 41, 11029 (1990).
407. N.V. Zavaritsky et al., *Physica C* 180, 417 (1991).
408. K. Krishana et al., *Phys. Rev. Lett.* 75, 3529 (1995).
409. H.C. Ri, et al., *Phys. Rev. B* 50, 3312 (1994).
410. B. Batlogg et al., *'Proceedings of 10th Anniversary of HTSC workshop on Physics, Materials and Applications'*, World Scientific, Singapore (1996).
411. D. Pavuna, in the *Gap Symmetry and Fluctuations in High-T_c Superconductors,* Bok et al., (Eds.), Plenum Press, New York (1998).
412. H.R. Ott et al., *J. Low Temp. Phys.* 99, 251 (1995).
413. T. Hanaguri et al., *Physica C* 317-318, 345 (1999).
414. P. Wolfle, *Physica C* 317-318, 55 (1999).
415. T. Timusk and B. Slatt, *Rep. Prog. Phys.* 62, 61 (1999).
416. Y. Yanase and K. Yamada (Preprint).
417. A.S. Alexandrov, in *'Proceedings of International School of Physics,* Enrico Fermi Course CXXX VI, G. Iasonisi, G. Schrieffer and M.L. Chiofalo (Eds.), IOS Press, Amsterdam (1998).
418. F. J. Owens and C. P. Poole, Jr., *'The New Superconductors', Plenum*, New York (1996).
419. A. Bianconi, et al., *Phys. Rev. B* 54, 12015, *Phys. Rev. Lett.* 76, 3412 (1996).
420. D.S. Marshall, et al., *Phys. Rev. Lett.* 76, 4841 (1996).
421. H. Ding et al., *Nature* 382, 51 (1996); *Phys. Rev. B* 55, 14872 (1997).
422. K.P. Sinha, Mod. *Phys. Lett. B* 12, 805 (1998).
423. H.R. Norman et al., *Nature* 392, 157 (1998).
424. Ch. Ramner et al., *Phys. Rev. Lett.* 80, 149 (1998).
425. Y. Wilde et al., *Phys. Rev. Lett.* 80, 153 (1998).
426. H.F. Fong et al., *Phys. Rev. Lett.* 75, 321 (1996); *Phys. Rev. B* 54, 6708 (1996).
427. H. Yusouka, *Hyperfine Interactions* 105, 27 (1997).
428. M.B. Walker, *Phys. Rev. B* 53, 5835 (1986).
429. K. A. Kouznestov et al., *Phys. Rev. Lett.* 79, 3050 (1997).
430. J. Rossat-Mignod, et al., *Physica C* 185-189, 86 (1991).

431. P. Dai et al., *Phys. Rev. Lett.* 77, 5425 (1996).
432. P. Bourges, et al., *Europhys. Lett.* 38, 313 (1997).
433. J.L. Tallen et al., *Phys. Rev. Lett.* 79, 5294 (1997).
434. J.W. Loram et al., *Physica C* 235-240, 134 (1994).
435. T. Takahashi et al., *J. Phys. Chem. Solids* 62, 41 (2001).
436. M. Tinkham, *'Introduction to Superconductivity'*, McGraw Hill, New York (1996).
437. J.D. Whitler and R.S. Roth, *'Phase Diagrams for High-T_c Superconductors I'*. The American Ceramic Society (1991).
438. T.A. Vanderah et al., *'Phase Diagrams for High T_c Superconductors II'*, American Ceramic Society (1991).
439. R.S. Roth et al., in *'Ceramic Superconductors II'* (M.F. Yan, Ed.), Publ. American Ceramic Society (1988).
440. W. Wong-Ng. et al., *J. Amer. Ceram. Soc.* 81, 1829 (1998).
441. J.D. Jorgenson et al., *Phys. Rev. B* 36, 3608 (1987).
442. Y. Gao et al., *J. Mater. Res.* 5, 1363 (1990).
443. P. Karen and A. Kjekshus, *J. Solid State Chem.* 94, 298 (1991).
444. P. Karen et al., (Preprint 1999).
445. D.M. De Leeuw et al., *Physica C* 152, 39 (1988).
446. J. Ritter, *Powder diff.* 3, 30 (1988).
447. R.S. Roth et al., *J. Res. Natl. Inst. Stand. Technol. (USA)* 95, 291 (1990).
448. D.E. Morris et al., *Physica C* 168, 153 (1990).
449. R. Bormann and J. Noelting, *Appl. Phys. Lett.* 54, 2148 (1989).
450. T.B. Lindermer and A.L. Sutter, Jr., *Oak Ridge Natl. Lab Report ORNL/TM-10827 (1988)*.
451. E.D. Specht et al., *Phys. Rev. B : Condens. Matt,* 37, 7426 (1989).
452. T. Aselege and K. Keefer, *J. Mater. Res.* 3, 1279 (1988).
453. G. Krabbles et al., *J. Solid State Chem.* 103, 420 (1993).
454. G. Krabbles et al., *Trans. Mater. Res. Soc. Jpn.* 19 A, 463 (1994).
455. W. Zhang and K. Osamura, pp. 437-440 in *Adv. Supercond. III, Proc. 3rd Int. Symp. Supercond.* International Superconductivity Technology Centre, Sendai, Japan, November 6-9, 1990 (K. Kajimura and H. Hajakawa, eds.) Springer-Verlag, Tokyo (1991).
456. Y. Le-page et al., Phys. Rev. B 36, 3617 (1987).
457. W. Wong-Ng et al., *J. Solid State Chem.* 84, 117 (1990).
458. D. Klibanow et al., *J. Am. Ceram. Soc.* 71, C-267 (1988).
459. S.I. Yoo and R.W. McCallum, *Physica C* 210, 147 (1993).
460. E. Goodilin et al., *Physica C* 289, 251 (1997).
461. R.J. Birgeneau and G. Shirane, in *"Neutron Scattering Studies of Structural and Magnetic Excitations in Lamellar Copper oxides"*—A review, Prog. High Temp. Superconduct. (D.M. Ginsberg, Ed.) p.p. 151-211, World Scientific Publishing, Singapore (1989).
462. Hahn, J. et al., *Chemtronics* 2, 126 (1987).
463. C. Michel et al., *Z. Phys. B* 68, 421 (1987).
464. J.I. MacManus-Driscoll and J.C. Bravman, *J. Am. Ceram. Soc.* 77, 2305 (1994).

465. P. Majewski et al., *J. Electronic Mater.* 22, 1259 (1993).
466. Y.S. Sung and E.E. Hellstrom, *Physica C* 255, 266 (1995).
467. N. Merchant et al., *Appl. Phys. Lett.* 65 (8), 1039 (1994).
468. Q.Y. Hu et al., *Physica C* 250, 7 (1995).
469. J. Jiang and J.S. Abell, in *'High Temperature Superconductor: Synthesis, Processing and Large Scale Applications* (U. Balachandran et al., eds.) p. 13. The Minerals, Metals and Materials Society, (1996).
470. B. Hettich et al., in *'High Temp. Superconduct.*, Proc. ICMC '90 Conference, Mater. Aspects High-Temp. Supercond. Garmisch-Partenkichen, Germany (1990), Vol. I, (H.C. Freyhardt et al., Eds.), p. 399 DGM Information sges, Oberursel, Germany, 1991.
471. R.S. Roth et al., *J. Res. Natl. Inst. Stand. Technol.* USA 95, 291 (1990).
472. M. Neveriva et al., *Physica C* 199, 328 (1992).
473. K. Knizek et al., *Physica C* 216, 211 (1993).
474. T.G. Holesinger et al., *Physica C* 217, 85 (1993).
475. P. Majewski et al., *J. Electronic Mater.* 22, 1259 (1993).
476. B. Hong and T.O. Mason, *J. Am. Ceram. Soc.* 74, 1045 (1991).
477. S.J. Golden et al., *J. Am. Ceram. Soc.* 74, 123 (1991).
478. R. Muller et al., *Physica C* 203, 299 (1992).
479. P. Majewski et al., *Physica C* 249 [34], 234 (1995).
480. W. Wong-Ng et al., *J. Amer. Ceram. Soc.* 81, 1829 (1998).
481. J.L. Driscoll et al., in *Appl. Superconduct.* Ist Goettingen, Germany, October 4-9, 1992 Vol. I. (H.C. Freyhardt, Ed.) p. 325, DGM Information sges, Oberursel, Germany, (1993).
482. W. McCallum et al., in *Processing of Long Lengths Supercond.*, Proc. Symp. Pittsburgh, Pennsylvania, October 17-21, 1993 (U. Balachandran, et al., eds.), p. 195, Minerals, Metals and Materials Society, Warrendale, PA (1994).
483. P. Majewski et al., *Adv. Mater.* 3[1] 67 (1991).
484. S. Kaesche et al., *J. Electron. Mater.* 24 (12) 1829 (1995).
485. W. Wong-Ng et al., *J. Mater. Res.* 12 [11], 2855-2865 (1997).
486. M.P. Siegal et al., *J. Mater. Res.* 12 (11), 2825 (1997).
487. T.K. Jondo et al., *J. Alloys Compd.* 195 (1-2) (1993).
488. L.P. Cook et al., in *'Superconductivity and Ceramic Superconductors'* (K.M. Nair and E.A. Geiss, Eds.) Ceramics Trans. 13, American Ceramics Society, Westerville, OH, 1990.
489. E.V. Antipov et al., *Physica C* 215, 1 (1993).
490. C.W. Chu, in *'Proceedings of Quantum Theory of Real materials'*, Festschrift to Honor Marvin Cohen, (J.R. Chelikowsky and S.G. Louie, Eds.), p.415 Kluwer Academic, Norwell, MA (1996).
491. S. M. Loureiro, "The Synthesis and Structure of Superconductors of Mercury", Ph. D. thesis, CNRS, Grenobles, France (1997).
492. A. Fukuoka et al., *Physica C* 265, 13 (1996).
493. D.T. Jover et al., *Phys. Rev. B* 54, 4265 (1996).
494. W. Zhen et al., *Appl. Phys. Lett.* 70, 114 (1997).
495. P. Wrobel and L. Jacak, *Mod. Phys. Lett. B* 2, 511 (1988).

496. P.W. Anderson, *Science* 235, 1196 (1987).
497. P.W. Anderson, *Phys. Rev.* 115, 2 (1959).
498. P.W. Anderson, *"Frontiers and Borderlines in Many Particle Physics"* (1987).
499. G. Shriane et al., *Phys. Rev. Lett.* 60, 1057 (1988).
500. S. Chakravarty et al., *Phys. Rev. Lett.* 60, 1057 (1988).
501. M. Nucker et al., *Z. Phys. B* 67, 9 (1987).
502. A. Fujimori et al., *Phys. Rev. B* 35, 8814 (1987).
503. P. Steiner et al., *Z. Phys. B* 69, 449 (1988).
504. L.F. Mattheiss, *Phys. Rev. Lett.* 58, 1028 (1987).
505. P. Fulde, *Physica* 153-155, 1769 (1988).
506. T. Zannen et al., *Physica* 153-155, 1636 (1988).
507. F.C. Zhang and T.M. Rice, *Phys. Rev. B* 37, 3759 (1988).
508. J.H. Xu et al., *Phys. Lett. A* 120, 489 (1987).
509. W. Weber, *Phys. Rev. Lett.* 58, 1371 (1987).
510. T. Fujita et al., Jpn. *J. Appl. Phys.* 26, L 368 (1987).
511. R.V. Kasowski et al., *Solid State Commun.* 63, 1077 (1987).
512. S. Tanigawa et al., MRS Symposium Proc. 99, 57 (1988).
513. P.A. Sterne and C.S. Wang, *Phys. Rev. B* 37, 7472 (1988).
514. D.A. Papaconstantopoulos et al., *Phys. Rev. Lett.* 61, 211 (1988).
515. J.D. Jorgensen et al., *Phys. Rev. B* 36, 5731 (1987).
516. R.J. Cava et al., *Nature* 329, 423 (1987).
517. J.M. Tranquada et al., *Phys. Rev. Lett.* 60, 156 (1988).
518. J.H. Brewer et al., *Phys. Rev. Lett.* 60, 1073 (1988).
519. J.W. Lynn, *Phys. Rev. Lett.* 60, 2781 (1988).
520. L.F. Mattheiss and D.R. Harman, *Solid State Commun.* 63, 395 (1987).
521. S. Massidda et al., *Phys. Lett. A* 122, 198 (1987).
522. J. Yu et al., *Phys. Lett. A* 122, 203 (1987).
523. H. Krakauer and W.E. Pickett, in *'Novel Superconductivity'* edited by S.A. Wolf and V.Z. Kresin, Plenum, New York (1987) p. 501.
524. W.Y. Ching et al., *Phys. Rev. Lett.* 59, 1333 (1987).
525. D.W. Murphy et al., *Phys. Rev. Lett.* 58, 1888 (1987).
526. P.H. Hor et al., *Phys. Rev.* Lett. 58, 1891 (1987).
527. E.E. Alp et al., *Phys. Rev. B* 36, 8910 (1987).
528. L.C. Semdskjer et al. (Preprint).
529. J. Yu et al., in *'Novel Superconductivity'*, edited by S. A. Wolf and V.Z. Kresin, Plenum, New York (1987), p. 367.
530. J. Zaanen et al., *Phys. Rev. Lett.* 60, 2685 (1988).
531. H. Maeda et al., *Phys. Lett.* 27, L 209 (1988).
532. C. Michel et al., *Z. Phys. B* 68, 421 (1987).
533. Z.Z. Sheng et al., *Phys. Rev. Lett.* 60, 937 (1988).

534. S.S.P. Parkin, *Phys. Rev. Lett.* 60, 2539 (1988), ibid 61, 750 (1988).
535. S.A. Sunshine et al., *Phys. Rev. B* 38, 983 (1988).
536. R.M. Hazen et al., *Phys. Rev. Lett.* 60, 1174 (1988).
536. J.M. Tarascon et al., (Preprint).
537. M.A. Subramanian et al., *Science* 239, 1015 (1988).
538. R.M. Hazen et al., *Phys. Rev. Lett.* 60, 1657 (1988).
539. J.B. Torrance et al., *Solid State Commun.* 66, 703 (1988).
540. P.A. Sterne and C.S. Wang, *J. Phys. C* 21, L 949 (1988).
541. M.S. Hybertsen and L.F. Mattheiss, *Phys. Rev. Lett.* 60, 1661 (1988).
542. H. Krakauer and W.E. Pickett, *Phys. Rev. Lett.* 60, 1665 (1988).
543. A.J. Freeman et al., *Physica C* 153-155 1225 (1988).
544. J. Yu et al., *Physica C* 152, 273 (1988).
545. D.R. Hamann and L.F. Mattheiss, *Phys. Rev. B* 38, 5138 (1988).
546. R.V. Kasowski et al., *Phys. Rev. B* 38, 6470 (1988).
547. R.V. Kasowski et al., *Phys. Rev. B* 36, 7248 (1987).
548. F. Herman et al., *Phys. Rev. B* 36, 2309 (1987).
549. F. Herman et al., *Phys. Rev. B* 36, 5248 (1987).
550. A.W. Hewat et al., *Physica C* 153-155, 369 (1988).
551. M. Lee et al., (Preprint).

Hubbard Model, Anderson Lattice Model and Superconductivity   7

7.1 INTRODUCTION: NARROW BAND SYSTEMS

The modern theory of metals derives from the concept of a free electron gas, which obeys the Pauli exclusion principle [1]. The principal influences of the lattice periodic potential on the individual electron states are to renormalize their mass and to change the topology of the Fermi surface. Landau [2] was the first to recognize the applicability of the electron-gas concept to the realistic situation where the repulsive Coulomb interaction between particles is small compared to the kinetic energy of electrons. Landau incorporated the interaction between electrons into a further (many-body) renormalization of the effective mass and investigated the physical properties, *e.g.*, specific heat, magnetic susceptibility, sound propagation, and thermal and electric conductivities in terms of quasiparticle contributions.

An important next development was contributed by Mott [3], who pointed out that if the Coulomb interaction between the electrons is sufficiently strong, *i.e.*, comparable to the band energy of the quasiparticles, then electrons in a solid would have to localize on the atoms, with one electron per atom. This qualitative change of the nature of the single electron states from those for a gas to those for atoms is called the *metal-insulator* or the *Mott transition*. An empty (unoccupied) state in the Mott insulator (*i.e.*, that without electrons available) will act as a *mobile hole*. This means that in these circumstances, the transport of charge takes place via the correlated hopping of electrons through such hole states. In the Mott-insulator limit, these hole states play a crucial role in establishing the superconductivity of oxides, as discussed in chapter 8.

The paramagnetic or magnetically ordered states of electrons comprising the Mott insulator distinguishes this class of materials from ordinary band (Bloch-Wilson) insulators or intrinsic semiconductors, the latter are characterized at $T = 0$ by a filled valence band and empty conduction band, separated by a gap. The electrons in the filled valence band are paired; hence Bloch-Wilson insulators are diamagnetic.

Now, we consider, how we can treat Mott insulators and metals within a single microscopic description of electron states by generalizing the band theory of electron states so as to describe Mott insulators within the same microscopic model. Hubbard in 1964 proposed the first step in this direction. Hubbard showed by use of a relatively simple model that as the interaction strength (characterized by the magnitude U of the inter atomic Coulomb repulsion) increases and becomes comparable to the band energy per particle (characterized by the bare bandwidth W), the original band of single particle states splits into two halves. Obviously, the Mott insulator may be modeled by a lattice of hydrogenic-like atoms with one electron per atom, placed in the lowest $1s$ state. Fig. 7.1(a) and (b) shows the distinction between the normal metallic and Mott insulating states. From Fig. 7.1, we can see that the metal (Fig. 7.1a) is depicted as an assembly of electrons represented by the set of plane waves characterized by the wave vector **k** and spin quantum number $s = \sigma/2$, where $\sigma = \pm 1$.

The transformation to the Mott localized state may take place only if the number of electrons in the metallic phase is equal to the number of parent atoms, *i.e.*, when the starting band of free electron states is half-filled. The collection of such unpaired spin moments will lead to paramagnetic behaviour at high temperatures. As the temperature is lowered, the system undergoes a magnetic phase transition: in the case of Mott insulators, the experimentally observed transition is almost always to antiferromagnetism (Fig. 7.1b), where each electron with its spin moment up is surrounded by the electrons on nearest neighbouring sites with spins in the opposite direction (down).

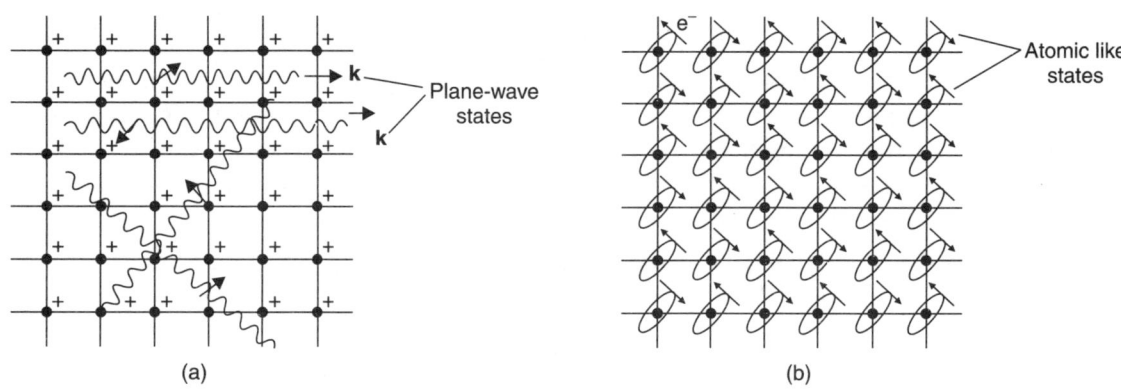

Fig. 7.1 (*a*) Schematic representation of a normal metal as a lattice of ions and the plane waves with wave vector **k** representing free electron states (*b*) model of Mott insulator as a lattice of atoms with electrons localized on them. We may note that the ground state configuration is usually antiferromagnetic, *i.e*, spins antiparallel to each other

Such spin configuration reflects a two-sublattice (Neel) antiferromagnetic state. The actual magnetic structure of Cu^{2+} ions in La_2CuO_4 is shown in Fig. 7.2.

If the number of electrons in the band of electron states is smaller than the number of available atomic sites, then electron localization cannot be complete because empty atomic sites are available for hopping electrons. However, for the half-filled band case, as the ratio U/W increases, half of the total number of single-particle states in the starting band are gradually pushed above the Fermi level E_F. An increase in the ratio U/W may be achieved by lengthening the interatomic distance, thus reducing W, which is directly proportional to the wave function overlap for the

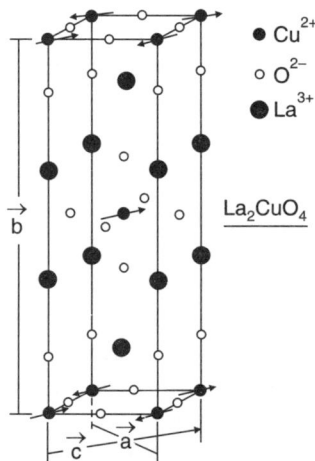

Fig. 7.2 The magnetic structure of La_2CuO_4. The neighbouring Cu^{2+} ions in the planes have their spins (each representing $3d^9$ configuration) antiparallel to each other

two states located on the nearest neighbouring sites. The splitting of the original band into two Hubbard sub-bands eliminates double (single-singlet) occupations of the same energy state ε. Effectively, one can say that this pattern reflects the situation of electrons being separated from each other as far as possible; however, the correspondence between the Hubbard split-band situation is shown in Fig. 7.3, and the electron disposition in the spin lattice in real space is by no means obvious and requires a more detailed treatment that relates the two descriptions of the Mott insulator. Now, we deal this problem.

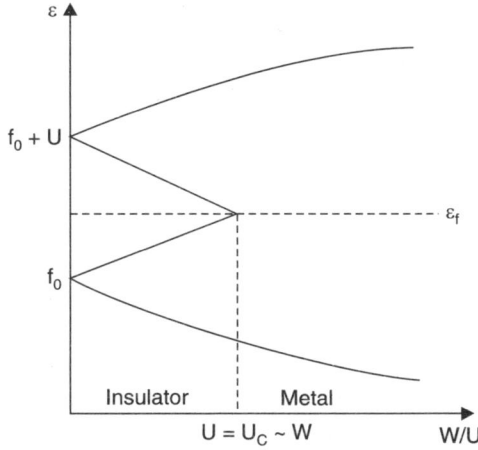

Fig. 7.3 The Hubbard splitting of the states in a single half-filled band for the strength of the intra-atomic Coulomb interaction $U > U_c$. The state with filled lower Hubbard sub-band for $U > U_c$ is identified with that of Mott insulator [4]

(i) Hubbard Model

The simplest model of describing narrow band systems and of correlated electrons is the one-state Hubbard model [4]. Hubbard model which despite its simplicity, exhibits many properties characteristic of superconductors. The Hamiltonian has the form

$$H = \sum_{K,\sigma} \varepsilon_{K\sigma} n_{K\sigma} + U \sum_{i} n_{i\uparrow} n_{i\downarrow} \qquad ...(7.1)$$

where ε_K is the single-particle (band) energy per electron with the wave vector **K**, U is the magnitude of interatomic Coulomb repulsion between the two electrons located on the same atomic site i, $n_{K\sigma}$ is the number of electrons in the single particle state $|K\sigma\rangle$, and $n_{i\sigma}$ is the corresponding quantity for the atomic state $|i\sigma\rangle$. This simple Hubbard Hamiltonian describes the localization versus delocalization aspect of electron states since the first term provides the gain in energy ($\varepsilon_K < 0$) for electrons in the band state $|K\sigma\rangle$, whereas the second term accounts for an energy loss ($U > 0$) connected with the motion of electrons throughout the system that is hindered by encounters with other electrons on the other side. $U > 0$ is defined by

$$U(R) = \int \phi^*(r-R) V_c(R) \phi(r-R) d^3r \qquad ...(7.2)$$

Relation (7.2) provides the Coulomb repulsion energy associated with orbital $\phi(r-R)$ on an atom at position R. The competitive aspects of the two terms in eq. (7.1) is expressed explicitly if the first term is transformed by the Fourier transformation to the site $\{|i\sigma\rangle\}$ representation. Then, eq. (7.1) may be rewritten as

$$H = \sum_{tj\sigma} t_{ij} a_{i\sigma}^+ a_{j\sigma} + U \sum_i n_{i\uparrow} n_{i\downarrow} \qquad ...(7.3)$$

where

$$t_{ij} = \frac{1}{N} \sum_K \varepsilon_K \exp[i\mathbf{K}.(\mathbf{R}_j - \mathbf{R}_i)] \qquad ...(7.4)$$

is the Fourier transform of the band energy ε_k and $a_{i\sigma}^+$ is the creation (annihilation of electrons in the (**Wannier**) state centred on the site $\mathbf{R}_j$. The first term in eq. (7.3) represents the motion of an electron through the system by a series of hops $j \to i$, which are described in terms of destruction of the particle at site j and its subsequent recreation on the neighbouring site i. The width of the corresponding band in this representation is given by

$$W = 2 \sum_{j(i)} |t_{ij}| = 2z|t| \qquad ...(7.5)$$

where z is the number of nearest neighbours ($n.n$) and t is the value of t_{ij} for the $n.n$ pair $\langle ij \rangle$. Thus the Hamiltonian (eq. (7.3)) is parameterized through the bandwidth W and the magnitude U. In actual calculations, it is the ratio U/W that determines the localized versus collective behaviour of the electrons in the solid. We may note that the hopping amplitude $t > 0$ is a measure of contribution from an electron hopping from one site to another. As it is clear from eq. (7.5), the hopping amplitude t may also be written in the form of an integral. The Hamiltonian (7.3) exhibits an electron-hole symmetry, which is of some importance because most HTSC cuprates are hole types with a close half-filled band.

The electronic configurations of several atoms that occur commonly in HTSC cuprates, are given in Table 7.1. The notation used in nl^N, where n is the principal quantum number, the orbital quantum number $l = 0$ for an s state, $l = 1$ for a p-state, $l = 2$ for a d-state, and N is the number of electrons in each l-state. A full l-state contains $2(2l + 1)$ electrons corresponding to 2, 6 and 10 for s, p and d states, respectively. The Cu^{2+} ion ($3d^9$) may be looked upon as a filled d-shell ($3d^{10}$) plus one $3d$ hole, and in the cuprates this hole is $d_{x^2-y^2}$ orbital in the CuO_2 plane.

Table 7.1 Electronic configurations of some selected atoms commonly used for band structure calculations [6]

Atom no.	Symbol	Core[b,c]	Atom configuration	No. of valence electrons	Ion	Ion configuration	No. of electrons
8	O	Be 4	$[2s^2]2p^4$	4	O^{1-}	$2p^5$	5
					O^{2-}	$2p^6$	6
14	Si	Ne 10	$3s^2 3p^2$	4	Si^{4+}	–	0
19	K	Ar 18	$[3p^6]4s$	1	K^+	–	0
20	Ca	Ar 18	$4s^2$	2	Ca^{2+}	–	0

(contd.)...

23	V	Ar 18	$3d^34s^14p^1$	5	V^{3+}	$3d^2$	2
29	Cu	Ar 18	$3d^{10}4s^1$	11	Cu^{1+}	$3d^{10}$	10
					Cu^{2+}	$3d^9$	9
					Cu^{3+}	$3d^8$	8
38	Sr	Kr 36	$5s^2$	2	Sr^{2+}	–	0
39	Y	Kr 36	$4d^15s^2$	3	Y^{3+}	–	0
41	Nb	Kr 36	$4d^35s^15p^1$	5	Nb^{4+}	$4d^1$	1
50	Sn	– 46	$5s^25p^2$	4	Sn^{4+}	–	0
56	Ba	Xe 54	$[5p^6]6s^2$	2	Ba^{2+}	–	0
57	La	Xe 54	$5d^16s^2$	3	La^{3+}	–	0
80	Hg	– 78	$[5d^{10}]6s^2$	2	Hg^{2+}	$[5d^{10}]$	0
81	Tl	– 78	$[5d^{10}]6s^2p^1$	3	Tl^{3+}	$[5d^{10}]$	0
82	Pb	– 78	$[5d^{10}]6s^26p^2$	4	Pb^{4+}	$[5d^{10}]$	0
83	Bi	– 78	$[5d^{10}]\,6s^26p^3$	5	Bi^{3+}	$[5d^{10}]6s^2$	2
					Bi^{4+}	$[5d^{10}]6s^1$	1
					Bi^{5+}	$[5d^{10}]$	0

[a]Core electrons listed in square brackets are sometimes included in the basis set.
[b]The core of Sn is Kr plus the fourth transition series ($4d^{10}$) closed shell.
[c]The core of Tl, Pb, and Bi is Xe plus the rare earth ($4f^{14}$) and fifth transition series ($5d^{10}$) closed shells.

The various s, p, and d wavefunctions called orbitals have the unnormalized analytical forms given in Table 7.2 and spacial electronic charge distribution of the d orbitals is sketched in Fig. 7.4, the sign on each lobe being the sign of the wavefunction. Fig. 7.5 shows the sigma (σ) bonding between oxygen p_x and p_y orbitals and copper $d_{x^2-y^2}$ orbitals in a cuprate CuO_2 plane.

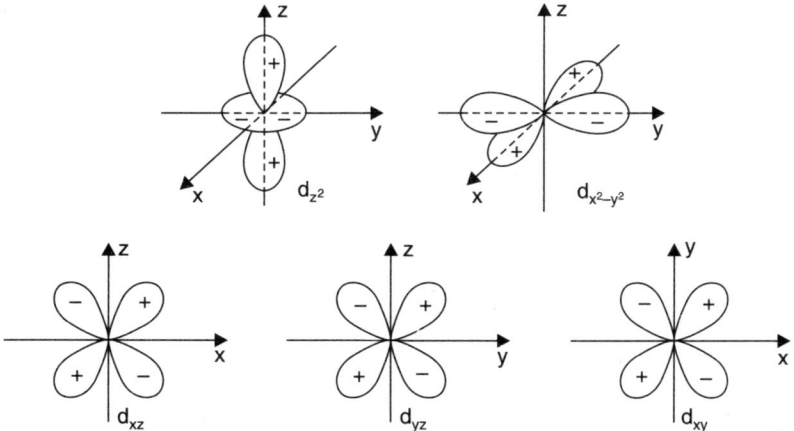

Fig. 7.4 Spatial distribution of electron density for the five d-orbitals. The signs (±) on the lobes are for the wavefunction; the sign of the electric charge is the same for each lobe of a particular orbital [7].

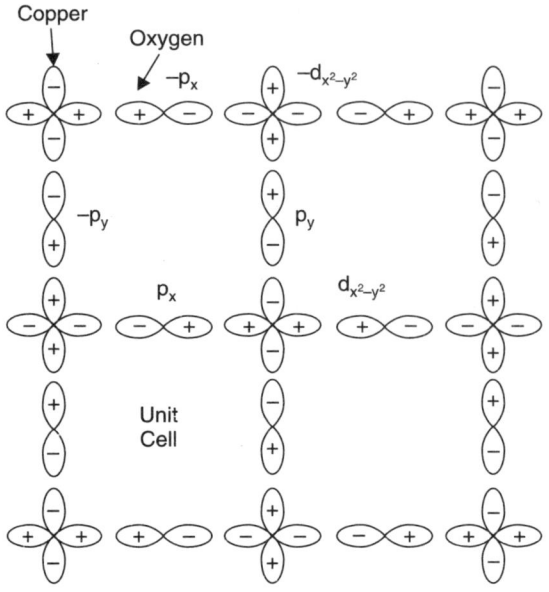

Fig. 7.5 Orbitals used for the three-state model of Cu-O planes. Each copper contributes a $d_{x^2-y^2}$ orbital and each oxygen contributes either a p_x or a p_y orbital, as shown. The unit cell contains one of each type of ion, and hence one of each type of orbital. The figure shows four unit cells [6].

Table 7.2 Unnormalized analytical expressions in cartesian and Polar coordinates for the s, p and d orbitals [6]

Orbital	Cartesian form	Polar form
s	1	1
p_x	$\dfrac{x}{r}$	$\sin\Theta\cos\phi$
p_y	$\dfrac{y}{r}$	$\sin\Theta\sin\phi$
p_z	$\dfrac{z}{r}$	$\cos\Theta$
d_{xy}	$\dfrac{xy}{r^2}$	$\sin^2\Theta\sin\phi\cos\phi$
d_{yz}	$\dfrac{yz}{r^2}$	$\sin\Theta\cos\Theta\sin\phi$
d_{zx}	$\dfrac{zx}{r^2}$	$\sin\Theta\cos\Theta\cos\phi$
$d_{x^2-y^2}$	$\dfrac{x^2-y^2}{r^2}$	$\sin^2\Theta(\cos^2\phi-\sin^2\phi)$
d_{z^2}	$\dfrac{3z^2-r^2}{r^2}$	$3\cos^2\Theta-1$

$^a l = 0, 1,$ and 2, respectively.

(ii) Hubbard Subbands and Hole States

The normal-metal case is represented in eq. (7.1) by the limit $W/U \gg 1$; the first (band) term then dominates. On the other hand, the complimentary limit $W/U \ll 1$ corresponds to the limit of well separated atoms, since the excitation energy of creating double occupancy on a given atom (with the energy penality $\varepsilon \sim \mu$) far exceeds the band energy of individual particles. The transition from the metallic to the atomic type of behaviour takes place when $W \sim U$; this is also the crossover point where the single band in Fig. 7.3 splits in two. The actual dependence of the density of states for interacting particles is shown in Fig. 7.6.

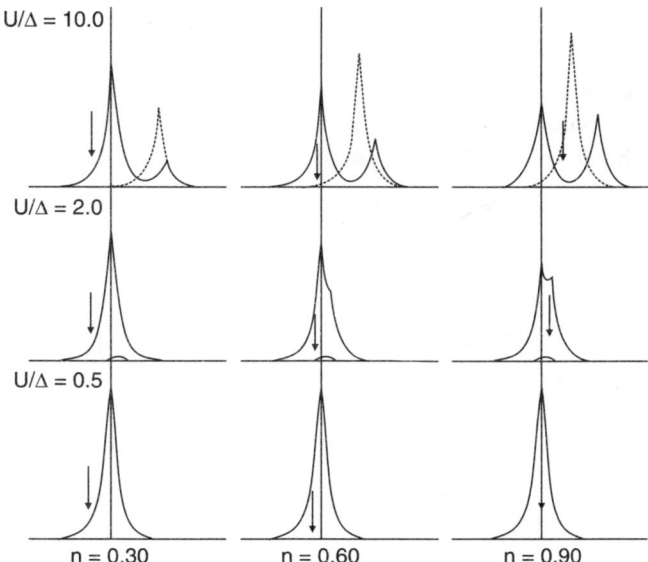

Fig. 7.6 The Hubbard splitting of the states for different band fillings $n = 0.3$, 0.6 and 0.9 and for different $U/W = 0.5$, 2 and 10 respectively. The x-axis is the axis of particle energy; the y-axis is the value of density of states. The arrow indicates the position of the Fermi energy, whereas the dotted line represents the inverse lifetime of the quasiparticle state in the pseudogap. [8]

These curves were drawn for the *Lorentzian shape of the density of states* (DOS), i.e., for a starting band with a characteristic width Δ:

$$\rho^\circ(\varepsilon) = \frac{\Delta}{\pi} \frac{1}{(\varepsilon - t_0)^2 + \Delta^2} \qquad \ldots(7.6)$$

where t_0 determines the position of the centre of the band (usually chosen as $t_0 = 0$). Spalek et. al. [9] have shown that with a growing magnitude of interaction (U/Δ) the density of states given by eq. (7.6) splits into two parts described by the density of states.

$$\rho(\varepsilon) = \frac{\Delta}{\pi}\left[\frac{1-n/2}{(\varepsilon-t_0)^2+\Delta^2} + \frac{n/2}{(\varepsilon-t_0-U)^2+\Delta^2}\right] \qquad \ldots(7.7)$$

The first term describes the original DOS (eq. 7.6), with the weighing factor $(1 - n/2)$, whereas the second represents the upper subband (on the energy scale) and with the weighing factor ($n/2$) and shifted by an amount U. These two terms and the corresponding two parts of DOS in Fig. 7.6. describe the Hubbard subbands. The dashed line in Fig. 7.6 represents the inverse lifetime of single-electron states placed in the **pseudogap**, while the arrows point to the position of Fermi energy in each case. For $n = 1$, the Fermi level falls in a pseudogap, where the lifetime of those quasiparticle states is very short. This is reminiscent of the behaviour encountered in an ordinary semiconductor, where the states in the band gap are those with a complex wave vector **K**.

To display the similarity between the **Mott insulator** as a two-band system in which the Hubbard subbands assume a role similar to the valence and conduction bands in an ordinary semiconductor is plotted in Fig. 7.7, which shows the position of Fermi level as a function of number of electrons, n, per atom in the system. As n moves past unity, a jump in E_F occurs for $U/\Delta \gg 1$. This is exactly what happens in the ordinary semiconductor when the electrons are added to the conduction band. This feature reveals once more that the states near the upper edge of the lower Hubbard subband (*i.e.*, the states near E_F for n close to but less than unity) can be regarded as hole states. One can easily see that those states are the ones with a high effective mass.

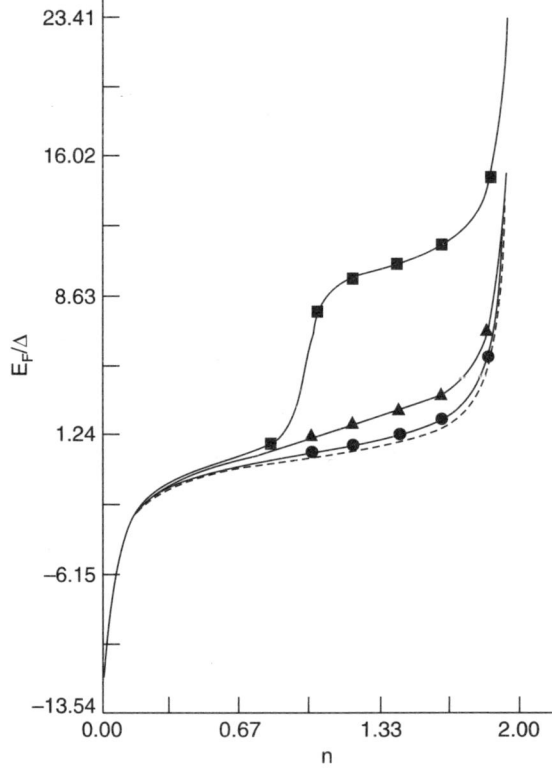

Fig. 7.7 The position of Fermi level E_F as a function of band filling n, for different values of interaction (from bottom to the top curve). $U/\Delta = 0$, 0.5, 2 and 10. For $U/\Delta = 10$, the Fermi level jumps between the subbands when $n = 1$ [8].

We must note that the Hubbard subband structure is characteristic of magnetic insulators and cannot be obtained with a standard band theoretical approach to the electron states in solids. The N states in the lower Hubbard subband are singly occupied; this is directly related to the picture of unpaired spins drawn in Fig. 7.1(b) and is one of the reasons for calling the electron states for such interacting systems correlated electron states. The other reason arises because the proper description of the electronic states near the localization threshold (the Mott transition) requires that one incorporate two-particle correlations into the quasiparticle states. The Hubbard split-band picture is only the first step in the proper description of electron states. These correlations will lead to a vary heavy mass of quasiparticles near the Mott transition: the heavy mass indicates a strong reduction of bare bandwidth W as the localization threshold is approached from the metallic side.

(iii) Localized Versus Itinerant Electrons: Metal-Insulator Transitions

The Hubbard split band picture of unpaired electronic states in a narrow band (Fig. 7.3 and 7.6) provides a rationale for the existence of a paramagnetic insulating ground state of the interacting electron system. The corresponding experimentally observed metal-insulator transitions (MITS) at finite temperature are very spectacular (Fig. 7.8), where the resistivity (on a logarithmic scale) is plotted as a function of inverse temperature for a canonical system $(V_{1-x}Cr_x)_2O_3$ (the data are from Ref. 10). The number of transitions (1, 2, or 3) depends on the Cr content. We may note the presence of an intervening metallic state between the

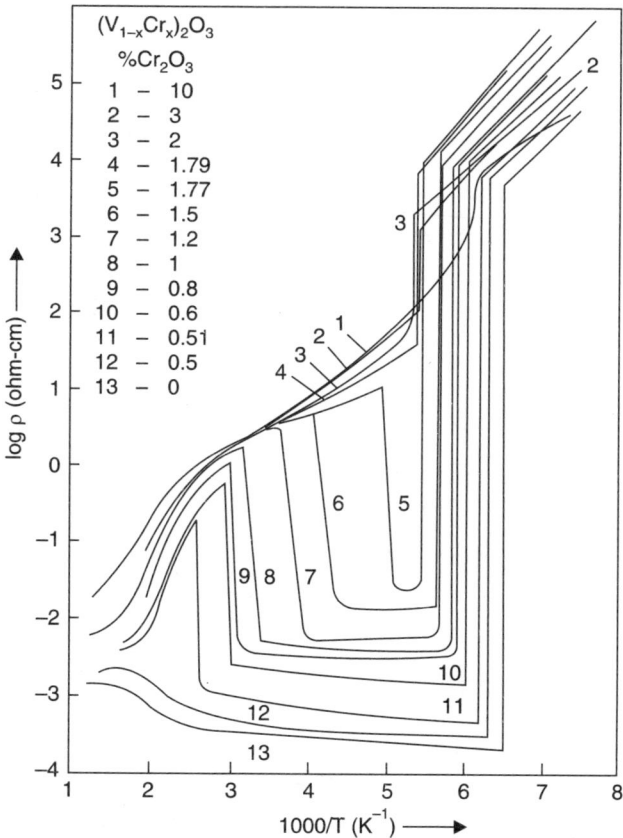

Fig. 7.8 Experimental measurements [10] pertaining to the variation of resistivity ρ in the logarithmic scale with inverse temperature $100/T$ for the $(V_{1-x}Cr_x)_2O_3$ system. The atomic content of Cr_2O_3 in V_2O_3 for each curve is specified.

antiferromagnetic insulating (AFI) and paramagnetic insulating (PM) states, as well as the reentrant metallic behaviour at high temperatures for $0.005 < x < 0.00178$. Let us discuss the physical implications of a model of interacting narrow-band electrons for $U \sim W$ starting from the Hamiltonian eq. (7.3) [10 – 12]. The main features of the ground state and thermodynamic properties are as follows.

In the absence of interactions ($U = 0$), the band energy per particle is $\bar{\varepsilon} = -(W/2) n(1 - n/2)$, where $0 \le n \le 2$ is the degree of band filling; for $n = 1$, this reduces to $\bar{\varepsilon} = -W/4$. When the interactions are present, the band is narrow; this is because of a restriction on the electron motion caused by their repulsion. One way of handling this restriction is to adjoin to the bare bandwidth a multiplying factor ϕ. This leads to renormalized density of states for quasiparticles (Fig. 7.9). The factor ϕ is a function of the particle–particle correlation function $\eta \equiv \langle n_{i\uparrow} n_{i\downarrow}\rangle$, the expectation value for the double occupancy of a representative lattice site. The quantity η is calculated for $T = 0$ self-consistently by minimizing the total energy E_G (per site), composed of the band energy $E_B = \phi \bar{\varepsilon}$ and Coulomb repulsion energy U_η, where $\phi = 8\eta(1 - 2\eta)$ [9, 11]. These two energies represent the expectation values of the two terms in eq. (7.3) for the case of a half-filled band. The optimal values of the quantities are given by

$$\eta_0 = \frac{1}{4}\left(1 - \frac{U}{U_c}\right) \qquad \text{...(7.8}(a)\text{)}$$

$$\phi_0 = 1 - \left(\frac{U}{U_c}\right)^2 \qquad \text{...(7.8}(b)\text{)}$$

and

$$E_G = \left(1 - \frac{U}{U_c}\right)^2 \bar{\varepsilon} \qquad \text{...(7.8}(c)\text{)}$$

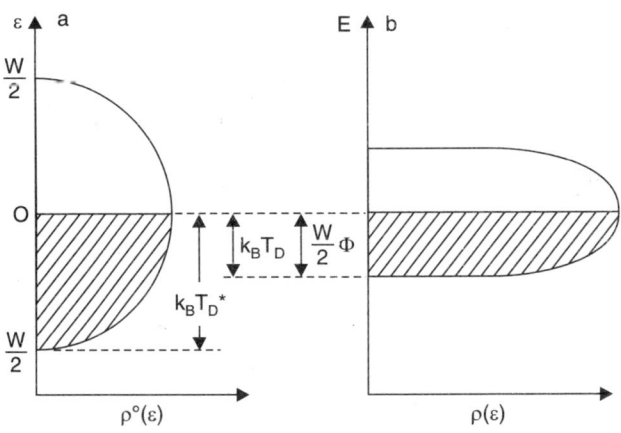

Fig. 7.9 Representation of bare (ρ_0) and quasiparticle (ρ) densities of states. The band narrowing factor ϕ for interacting electrons (b) is specified. The degeneracy temperature T_D for the interacting electrons and that corresponding to non-interacting electrons (T_D^*) are also indicated. Situation drawn corresponds to the half-filled case ($n = 1$) for which the Fermi energy can be chosen as $E_F = 0$

With $U_c = 8 |\bar{\varepsilon}| = 2W$. Thus, as U increases, η_0 decreases from 1/4 to 0. At the critical value $U = U_c$, $E_B = 0$ and there are no double occupancies for the same lattice site; this signals the crossover by the system from the itinerant (band) to the localized (atomic like) state. The point $U = U_c$ corresponds to a true phase transition at $T = 0$; the last statement can be proved by calculating the static magnetic susceptibility, which is [11]

$$\chi = \frac{\chi_0}{\phi_0 \left[1 - U\rho \dfrac{1 + I/2}{(1+I)^2} \right]} \qquad ...(7.9)$$

where $I = U/U_c$, ρ is the density of bare band states at $\varepsilon = E_F$; and χ_0 is the magnitude susceptibility of band electrons with energy ε_k at $U = 0$. As $\phi \to 0$, *i.e.*, $U \to U_c$, the susceptibility diverges. The localized electrons are represented in this picture by noninteracting magnetic moments for which the susceptibility is given by the Curie law $\chi = C/T \to \infty$ as $T \to 0$. Clearly, the metal-insulator transition is a true phase transition; η_0 may be regarded as an order parameter, and the point $U = U_c$ as a critical point. Now, we consider metallic phase, which permits a generalization of the previous results to the case $T > 0$. First, the increase of magnitude of interaction U reduces the band energy according to $E_B = - W \phi_0/4$. Eventually, E_B becomes comparable to the interaction part $U\eta$; they exactly compensate each other at $U = U_c$. The resultant electronic configuration (localized *vs.* itinerant) is then determined at $T > 0$ by the very small entropy and the exchange interaction contributions. One can estimate the entropy of the metallic phase in the low-temperature regime may be estimated by using the linear specific heat for electrons in a band narrowed by correlations, namely, $C_v = \gamma T = (\gamma_0/\phi_0) T$, where $\gamma_0 = 2\pi^2 k_B^2 \rho/3$ is the linear specific heat coefficient (per one atom) for uncorrelated electrons. *i.e.*, $U = 0$. Hence, the entropy $S = \gamma T = C_v$. Combining this relations with eq. 7.8(c), one obtains the explicit expression for the free energy of the metallic phase [12, 13]:

$$\frac{F}{N} = \left(1 - \frac{U}{U_c} \right)^2 \bar{\varepsilon} - \frac{1}{2} \frac{\gamma_0}{\phi_0} \qquad ...(7.10)$$

This is the free energy per one atomic site. On the other hand, if the exchange interaction between the localized moments is neglected then each site in the paramagnetic state is randomly occupied by an electron with its spin either up or down. The free energy F_1 for such insulating system of N moments is provided by the entropy term for randomly oriented spins, *i.e.*,

$$\frac{F_1}{N} = - k_B T \ln 2 \qquad ...(7.11)$$

Now, a system in thermodynamic equilibrium assumes the lowest F state. The condition for the transition from the metallic to local-moment phase is specified by $F = F_1$. The phase transition determined by this condition can be seen explicitly when we note that the free energy varies with T either parabolically (eq. (7.10)) or linearly (eq. (7.11)), depending on whether the system is a paramagnetic metallic (*PM*) or a paramagnetic simulating (*PI*) phase. As shown in Fig. 7.10, several of those curves intersect in one or two points depending on the value U/W. We may note that these intersection points determine the stability limits of *PM* and *PI* phases. The lowest curve for the *PM* phase lies below the straight line for the *PI* state; there is no transition, *i.e.*, the metallic Fermi liquid state with the effective mass enhancement $m^*/m_0 \lesssim 2.5$ is stable at all temperatures. As U/W increases, the parabolas fall higher on the free energy (F/WN) scale and the possibilities for the transitions open up. The two higher curves illustrate the case in which the intersections with the straight lines occur at J and K and at L and M, respectively; at low and high temperatures the parabola lies below the straight line for F_1/WN, so that the metallic phase is stable in those T regions. At intermediate temperatures, the PI phase is stable. The loci of the intersections move further apart on the k_BT/W scale as U/W is increased (Fig. 7.10(b)) where the phase boundaries; this part of the figure represents the temperature of the transitions (the intersection point on Fig. 7.10(a)) versus the relative magnitude of intersection U/W. One finds that the PM phase is stable at low temperature; thus, reentrant metallic behaviour is encountered at high T.

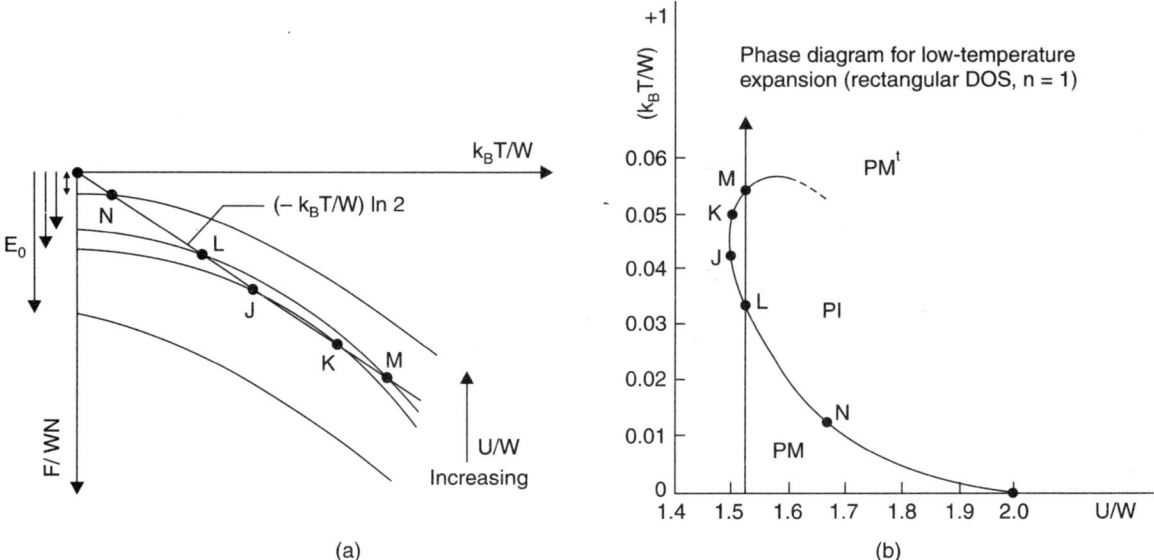

Fig. 7.10 Plots of free energies for the paramagnetic Mott insulator (the straight line starting from the origin) and the correlated metal (the parabolas). The parabolic curves points of crossing at L and J correspond to a discontinuous-metal-insulator transition, while those crossing at K and M correspond to the reverse. (b) Schematic representation of the phase diagram between paramagnetic metallic (PM and PM') and paramagnetic insulating (PI) phases. The points of crossing from (a) are also shown. The vertical line with an arrow represents a sequence of the transition seen in Fig. 7.8 for $0.005 \leq x \leq 0.018$ and in the paramagnetic phase

The explicit form of the curve drawn in Fig. 7.10(b) is obtained from the coexistence condition $F = F_1$, which leads to the following expression for the transition temperature [12, 13]:

$$\frac{k_B T_c}{W} = \frac{3}{2\pi^2}\left[1-\left(\frac{U}{U_c}\right)^2\right]\left\{\ln 2 \pm \left[(\ln 2)^2 + \frac{\pi^2}{3}\frac{1-U/U_c}{1+U/U_c}\right]^{1/2}\right\} \qquad \ldots(7.12)$$

The root T_- represents the low temperature part, *i.e.*, that for $k_BT/W \leq 0.049$; the $T_\pm$-part is the one above the point where both curves meet; this takes place at the lower critical value of $U = U_{lc}$ such that

$$\frac{U_{lc}}{U_c} = 1 - \frac{3\sqrt{2}}{2\pi} \frac{1}{\rho |\bar{\varepsilon}|^{1/2}} \approx 0.75 \qquad \ldots(7.13)$$

Below the value of $U = U_{lc}$, the correlated Fermi liquid is stable at all temperatures. Ultimately, for $1.58 \leq U/W \leq 2.0$, only one intersection (at low T) of the curve remains. Obviously, in this regime of U/W, the reentrant metallic behaviour is achieved gradually as the temperature increases.

(iv) Strongly Correlated Electrons: Kinetic Exchange Interaction and Magnetic Phases

In the limit $W \ll U$, the ground state of the interacting electron system will be metallic only if the number of electrons N_e in the system differs from the number N of the atomic sites. Similarly, only then can charge transport take place via the hole states in the lower Hubbard subband (for $N_e < N$), i.e., when the transport of electrons can be represented via hopping from site to site, avoiding the doubly occupied configurations on the same site. This restriction on the motion of individual electrons is described above in terms of the band narrowing factor ϕ, which in the normal phase, is now equal to [9, 11] $\phi = (1 - n)/(1 - n/2)$.

This shows that the effective quasiparticle bandwidth $W^* = W\phi$ is nonzero only if the number of holes $\delta = 1 - n > 0$.

For $W \ll U$, there is one class of dynamic processes that is important in determining the magnetic interactions, namely, the virtual hopping processes, with the formation of a doubly occupied site configuration in the intermediate state. Fig. 7.11 shows such processes, where one electron hops onto the site occupied by an electron with opposite spin, and then hops back to the original site. During such processes, the electrons can exchange positions (and this yields to the spin reversal of the pair with respect to the original configuration) or the same electron hops back and forth. The corresponding effective Hamiltonian, including the virtual-hopping processes in first nontrivial order, has the form

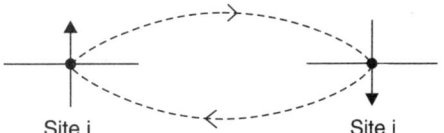

Fig. 7.11 Virtual hopping processes between singly and doubly occupied atomic sites and that lead to an antiferromagnetic exchange interaction between the neighbouring sites. This interaction is responsible for the antiferro magnetism in most of the Mott insulators

$$H = \sum_{K,\sigma} \phi_\sigma \varepsilon_K n_{K\sigma} + \sum_{ij} (2t_{ij}^2/U)\left(\mathbf{S}_i \cdot \mathbf{S}_J - \frac{1}{4} n_i n_j\right) \qquad \ldots(7.14)$$

wherein general the band narrowing factor $\phi_\sigma = (1 - n)/(1 - n_\sigma)$, $n_\sigma = \langle n_{i\sigma}\rangle$ is the average number of particles per site with the spin quantum number σ, and $n_i = n_{i\uparrow} + n_{i\downarrow}$ is the operator of the number of particles on given site i. We may note that in the paramagnetic state $n_\sigma = n_{-\sigma} = n/2$, and ϕ_σ reduces to $\phi = (1 - n)/(1 - n/2)$, the value for the normal state.

The effective Hamiltonian (eq. 7.14) represents approximately the original Hubbard Hamiltonian for $W \ll U$ [14]. When $n \to 1$, $\phi \to 0$, and eq. (7.14) reduces to the Heisenberg Hamiltonian with antiferromagnetic interaction, which is the reason why most Mott insulators order antiferromagnetically. In the limit of a half-filled band, one finds that the effective bandwidth $W^* \equiv W\phi = 0$, thus proving that electrons in that case are localized on atoms. The nature of the wave function for these quasi-atomic states has not yet been satisfactorily analyzed, though some evidence given later shows that they should be treated as **soliton** states.

For $n < 1$, the normalized band (the first term) and the exchange parts in eq. (7.14) do not commute with each other. Clearly, this means that for the narrow-band system of electrons represented by eq. (7.14) the spin dynamics influences the nature of itinerant quasiparticle states of energies $\phi\,\varepsilon_k$. Interestingly, as $n \to 1$, the two terms in eq. (7.14) may contribute equally to the total energy. The critical concentration of electrons n_c for which these two terms are comparable is

$$n_c \simeq 1 - \frac{1}{2z}\frac{W}{U} \sim 0.02 - 0.05$$

We have plotted schematically the commonly accepted phase diagram describing the possible magnetic phases in the plane n–U/W in Fig. 7.12. Close to the case of one electron per atom, the antiferromagnetic (AF) phase is stable for any arbitrary strength of intersection. As intermediate filling, the ferromagnetic phase (F) may be stable. On the low intersection side, the ferromagnetic phase terminates at points where the **Stoner criterion** is met, i.e., when $\rho°(E_F)\,U = 1$, where $\rho°(E_F)$ is the value of the bare density of states (per spin) at the Fermi level.

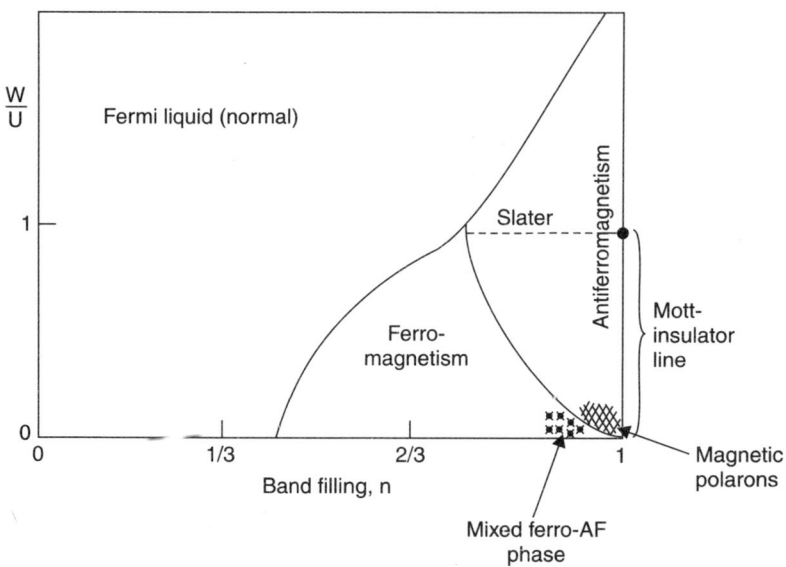

Fig. 7.12 Commonly accepted magnetic phase diagram

We may note that peculiar features appear in the corner where $n \approx 1$ and $W/U \ll 1$, i.e., where the number of holes is small, so that the exchange interaction contribution to the total system energy is either larger than or comparable to the band energy part $\phi\bar\varepsilon$. In such a situation, a mixed ferro-antiferromagnetic phase is possible [15 – 17]. When the number of holes is very small, each hole may form a **magnetic polaron** with a ferromagnetic cloud accompanying it: the hole is self trapped within the cloud of ferromagnetic polarization is created.

(v) Magnetic Polarons

Nagaoka [17] proved that in the limit $W/U \to 0$ the ground state of the Mott insulator with one hole involves ferromagnetic ordering of spins. This is because in this limit the antiferromagnetic exchange term in eq. (7.14) variables and the band energy is lowest when $\phi_{\sigma=\uparrow} = 1$ and $\phi_{\sigma=\downarrow} = 1 - n$. One can thus choose an equilibrium state with $n_\uparrow = n$, $n_\downarrow = 0$, i.e., a state with all spins pointing up.

Mott [3] has pointed out that if W/U is small but finite, a hole may create locally a ferromagnetic polarization of the spins in a sphere of radius R, surrounded by a reservoir of antiferromagnetically ordered spins. Fig. 7.13 shows the situation schematically. The energy of such a hole accompanied by a cloud of saturated polarization can be estimated roughly as [18].

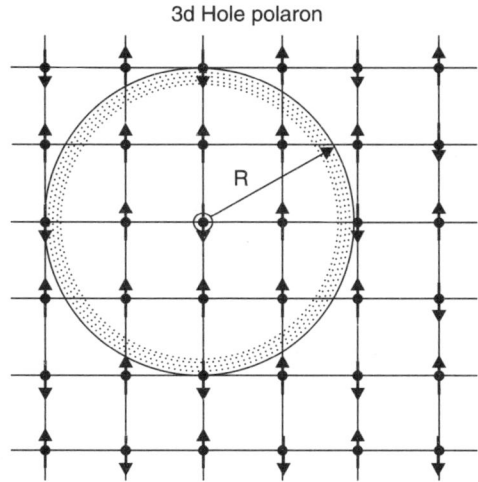

Fig. 7.13 Magnetic polaron state, i.e., one hole in the antiferromagnetic Mott insulator. This hole produces ferromagnetic polarization around itself and may become self trapped

$$E(R) = -\frac{W}{2} + \pi^2 |t| \left(\frac{a}{R}\right)^2 + \frac{4\pi}{3}\left(\frac{R}{a}\right)^3 \frac{zt^2}{U} \qquad ...(7.15)$$

where a is the lattice constant and t is the hopping integral t_{ij} between the z nearest neighbours. The first term in eq. (7.15) is the band energy of a free hole in a completely ferromagnetic medium, the second term represents the loss of kinetic energy due to the hole confinement and the third term involves the antiferromagnetic exchange energy penality paid by polarizing the spins ferromagnetically within a volume $4\pi R^3/3$. Minimizing eq. (7.15) w.r.t. R, one obtains the optimal number of spins contained in the cloud.

$$N = \frac{4\pi}{3}\left(\frac{\pi U}{W}\right)^{3/5} \qquad ...(7.16(a))$$

and the polaron energy

$$E_0 = -\frac{W}{2}\left[1 - \frac{5\pi^2}{3z}\left(\frac{W}{z}\right)^{2/5}\right] \qquad ...(7.16(b))$$

Eq. (7.15) holds for a three dimensional system: for a planar system, the factor $(4/3)\pi(R/a)^3$ in the last term should be replaced by the area $\pi(R/a)^2$. One then obtains the corresponding optimal values as

$$N = \pi \left(\frac{2\pi U}{W}\right)^{1/2} \qquad ...(7.17(a))$$

and
$$E_0 = -\frac{W}{2}\left[1 - \frac{2\pi^2}{z}\left(\frac{W}{2\pi U}\right)^{1/2}\right] \qquad ...(7.17(b))$$

These size estimates are required when one discusses the hole states at the threshold for the transition from antiferromagnetic to superconductivity in HTSC cuprates. We must note that U/W must be appreciably larger than unity in order to satisfy the requirement $N \gg 1$. In other words, the condition $R \gg a$ must be met, so that the spin subsystem (and the hole dynamics) may be treated in the continuous-medium approximation, the condition under which eq. (7.15) can be derived.

(vi) The spin liquid

The difference between the electron liquid of strongly correlated electrons (represented, for example, by the holes in the lowest Hubbard subband) and the Fermi liquid can be shown clearly in the limit of relatively high temperatures $W^* \ll k_B T \ll U$, where the quasiparticle band states with energies ($\phi_{\varepsilon k}$) are populated equally, independent of their energy. Namely, if N_e electrons are placed into N available states, the number of configuration for a phase with excluded double occupancies of each state is [19]

$$2^{N_e} \frac{N!}{N_e!(N-N_e)!} \qquad ...(7.18(a))$$

The first factor is the number of spin configurations for the singly occupied sites, while the second specifies the configuration entropy – the number of ways to distribute N_e spinless particles among N states. This leads to molar entropy in the form

$$S_L = R[n \ln 2 - n \ln n - (1-n)\ln(1-n)] \qquad ...(7.18(b))$$

where $n = N_e/N$ is the degree of the subband filling and R is the gas constant. The above reduces to $S_L = R \ln 2$ for $n = 1$, i.e., to the entropy of the N spins (1/2) on the lattice. By contrast, in a Fermi liquid that obeys the Fermi-Dirac distribution double occupancies are not excluded as shown in Fig. 7.14. Then the corresponding number of configurations is

$$\left(\frac{N}{N_e/2}\right)^2 = \frac{N!}{(N_e/2)!(N-N_e/2)!} \qquad ...(7.19(a))$$

and the corresponding molar entropy is

$$S_F = R[2 \ln 2 - n \ln n - (2-n)\ln(2-n)] \qquad ...(7.19(b))$$

Obviously, for $n = 1$, $S_F = 2S_L = 2R \ln 2$. We must emphasize that only the value for S_L reproduces correctly the entropy of N localized paramagnetic spins (the electronic part of the entropy for magnetically disordered states of Mott insulator). Clearly, in accord with intuitive reasoning, the Fermi-Dirac distribution that allows for double state occupancy cannot be applied to a strongly correlated electron liquid, which is termed as **spin liquid**. The state of such a liquid reduces to that of the spin system on the lattice if $N = N_e$ (for the Fermi liquid case, the ground state is then a metal with a half-filled band).

Hubbard Model, Anderson Lattice Model and Superconductivity

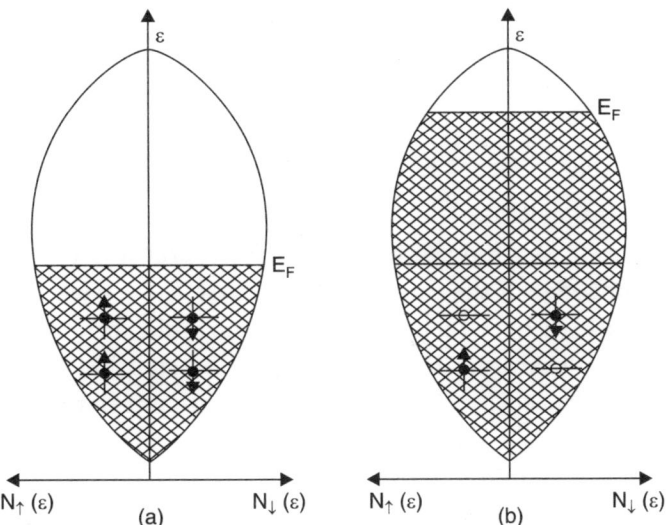

Fig. 7.14 Schematic representation of the difference in the *k*-space occupation for the ordinary fermions (left) and the strongly correlated electrons (right). Both spin subbands with $\sigma = \uparrow$ and $\downarrow$ are drawn. We may note that the holes drawn in (b) do not appear: they are only shown to indicate a single occupancy of each single particle state. The position of the Fermi level is different for the same number of electrons in the two situations [19].

To handle the regime of low temperatures and of arbitrary number of holes, we will generalize these results. One observes that in Fig. 7.6 the band states for $U \ll W$ are split for any arbitrary degree of band filling (refer eq. (7.7)). Clearly, in enumerating the distribution of particles in the lower Hubbard subband, one must exclude double occupancies of the same energy (ε) state. Since the quasiparticle energy is labeled by the wave vector **K**. One can equivalently exclude the double occupancies of given state $|\mathbf{K}\rangle$. Under this assumption Spalek and Wojcik [19] obtained the following expression for the statistical distribution

$$\bar{n}_{K\sigma} = (1 - \bar{n}_{K-\sigma}) = \frac{1}{1 + \exp[\beta(E_{K\sigma} - \mu)]} \quad \ldots(7.20(a))$$

where $\beta = (k_B T)^{-1}$, $\bar{n}_{k\sigma}$ is the average occupancy of the state $|\mathbf{K}\sigma\rangle$, and μ is the chemical potential that is determined from the conservation of the total number of particles:

$$N_e = \sum_{K\sigma} \bar{n}_{K\sigma} \quad \ldots(7.20(b))$$

The corresponding molar entropy is now obtained as

$$S_L = -\frac{R}{N} \sum_K [(1-\bar{n}_K)\ln(1-\bar{n}_K) + \bar{n}_{K\uparrow}\ln\bar{n}_{K\uparrow} + \bar{n}_{K\downarrow}\ln\bar{n}_{K\downarrow}] \quad \ldots(7.20(c))$$

with $\bar{n}_K = \bar{n}_{K\uparrow} + \bar{n}_{K\downarrow}$.

We may note that the distribution function (eq. 7.20(c)) differs from the ordinary Fermi-Dirac formula by the factor $(1 - \bar{n}_{K,-\sigma})$, which expresses the conditional probability that there should exist no second particle with spin quantum number **k**(– σ) if the state **k**σ is to be occupied by an electron as shown in Fig. 7.14(b). If $E_{K\sigma} = E_K$, i.e., particle energy does not depend on its spin direction, then eq. 7.20(a) takes the form

$$\bar{n}_K = \frac{1}{1 + (1/2) \exp[\beta(E_K - \mu)]} \qquad ...(7.20(d))$$

We can see that relation 7.20(d) is the same type of relation that applies to the occupation number of simple donors. If we drop the index **K** and ε represents the position of donor level with respect to the bottom edge of the conduction band. At $T = 0$, each state is singly occupied. This is the principal feature by which the present formula differs from the Fermi-Dirac distribution at $T = 0$, as illustrated in Fig. 7.15. The distribution (eq. 7.20(a)) leads to the doubling of the volume enclosed by the Fermi surface in spin-liquid state compared to the Fermi-liquid state. At low temperatures, the application of the distributions (eq. 7.20(a)) or (eq. 7.20(c)) yields the Fermi-liquid like properties: a linear T dependence of the specific heat (of large magnitude if $n \to 1$) and of entropy. At high temperatures, the new distribution leads to the entropy of the form of eq. 7.18(b) and local-moment behaviour in the form of Curie-Weiss law for the susceptibility. Hence the properties of spin liquid governed by the distribution (eq. 7.20(a) to 7.20(d)) interpolate between those of a metal and those of local moments. We may note that such behaviour is observed in many correlated system, for e.g., in heavy fermion systems.

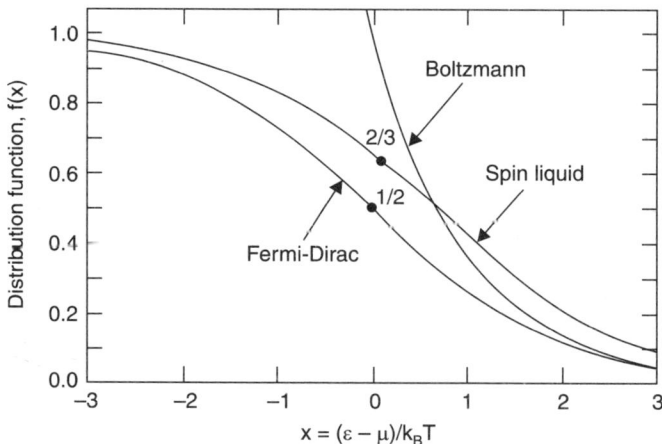

Fig. 7.15 Comparison of the Fermi-Dirac and Boltzmann distribution for $\bar{n}_{k\sigma}$ with that for the strongly correlated electrons (the spin-liquid phase): the total occupancy $n_k = n_{k\uparrow} + n_{k\downarrow}$ has been taken in the latter case.

Entropy expression (eq. 7.20(c)) can also be rewritten for the paramagnetic state in the following form :

$$S_L = -nR \ln 2 - k_B \sum_K [n_K \ln n_K + (1 - n_K) \ln(1 - n_K)] \qquad ...(7.21)$$

The first part of eq. (7.21) represents the entropy of spin moments, while the second represents the entropy of spinless fermions. An alternative decomposition has been put forward [21] in which the dynamics of correlated electrons is decomposed into that of neutral fermions called spinons and the chared bosons called holons. Within this picture, the onset of superconductivity is considered as a combined effect of Bose condensation of holons with a simultaneous formation of a coherent paired state by the fermion counterpart [22-24].

The above model of spin liquid deals only with its statistical properties in the $U \to \infty$ limit. The problem now arises as to what happens when the spin part of the form of second term in eq. (7.14) is included. The problem of the resultant ground state of holes in a Mott insulator is still a matter of intensive debate [22-24]. The state called the *resonating valence bond* (RVB) state has been invoked [24] specifically to deal with this problem. We will discuss it in detail in next chapter. Fig. 7.16 shows schematically, the formation of RVB state for the case without holes (a) and with one hole (b). The connecting lines represent bonds, across which the two electrons from spin singlet pairs. The resonating nature of bonds is connected with the idea that the RVB ground state is a coherent superposition of all such paired configurations. The dynamic nature of this spin dimerization is connected with the terms $(S_i^- S_j^- + S_i^- S_j^-)$ in the exchange part of the Hamiltonian (eq. (7.14)). There is the possibility that the RVB state (which, for obvious reasons, differs from the ordinary Neel antiferromagnet) is a ground state for the planar CuO_2 planes in HTSC oxides, such as $La_{2-x} Sr_x CuO_4$, where the long-range magnetic order is destroyed for $x \approx 0.02 - 0.03$.

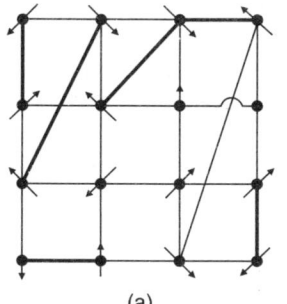 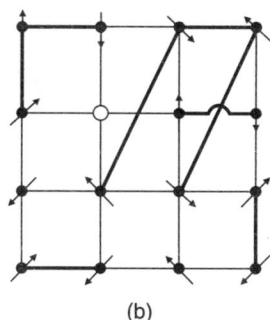

(a)   (b)

Fig. 7.16 Schematic representation of singlet-spin pairing forming the RVB state. One must take all paired configurations to calculate the actual ground state. (a) represents the RVB state for Mott insulator and (b) is that with one hole. The latter case will contain an unpaired spin as indicated.

7.2 THE HYBRIDIZED SYSTEMS

Most of the strongly correlated systems are encountered in oxides and in several classes of organic and inorganic compounds. In oxides, the $3d$ orbitals of cations such as Cu^{2+} and Ni^{2+}, hybridize with the $2p$ orbitals of oxygen, particularly if the atomic $(3d)$ states are energetically close to the $2p$ states. The properties of correlated and hybridized states can be properly studied with the help of the **Anderson lattice model Hamiltonian**, which is of the form

$$H = E_F \sum_{i\sigma} N_{i\sigma} + \sum_{K\sigma} \varepsilon_K n_{K\sigma} + U \sum_i N_{i\uparrow} N_{i\downarrow} + \frac{1}{\sqrt{N}} \sum_{K\sigma} (V_K e^{i\mathbf{K}\cdot\mathbf{R}} a_{i\sigma}^+ C_{Ka} + H.C)\ ...(7.22)$$

The first term is the energy of the atomic electrons positioned at E_F, the second term represents the energy of band electrons, the third term represents the intra atomic Coulomb repulsion between two electrons of opposite spins, while the last term describes the mixing of atomic with band electrons due to the energetic coincidence (degeneracy) of those two sets of states. We may say that in heavy fermions, the atomic states are $4f$ states, whereas they are $3d$ states of Cu^{2+} ions in HTSC cuprates, respectively. Here, $N_{i\sigma} = a_{i\sigma}^+ a_{i\sigma}$, and $n_{K\sigma} = C_{K\sigma}^+ C_{K\sigma}$ are the number of particles in the given atomic (i) or **K** states, respectively. In Hamiltonian (7.22), the following parameters appear: the atomic level position ε_f, the width W of a starting band states with energies $\{\varepsilon_K\}$, the magnitude U of the Coulomb repulsion for two electrons located in the same atomic site, and the degree of hybridization (mixing, $\mathbf{V}_K$,) characterized by its magnitude V.

One will have to distinguish two completely different situations: (i) $U > W > |\varepsilon_f| \gg |V|$, and ($ii$) $U > W > |V| \gtrsim |\varepsilon_f|$. Case ($i$) applies when the starting (bare) atomic level is placed deeply below the Fermi level and the atomic states admix weakly to the band states. In case (ii), the hybridization is large and is responsible for strong mixing of the two starting sets of states. The band structure corresponding to the hybridized band states in the absence of electron–electron interactions i.e., $U = 0$ is depicted in Fig. 7.17. One finds a small gap in the hybridized band structure; it occurs around the bare atomic level position ε_f and separates two hybridized bands. Those two bands, which have the energies

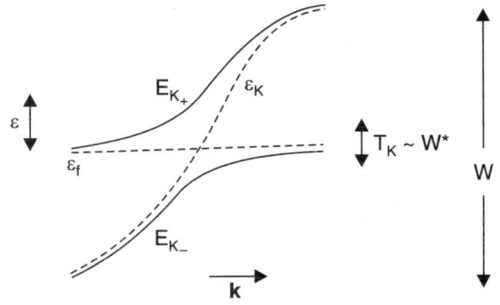

Fig. 7.17 Schematic representation of the hybridized bands with energies $E_{K\pm}$, which are formed by mixing the band states (with energy ε_K) and atomic states (located at $\varepsilon = \varepsilon_f$). The original band has the width W much wider than the packed structure of the width W^*.

$$E_{K\pm} = \frac{\varepsilon_K + \varepsilon_f}{2} \pm \left[\left(\frac{\varepsilon_K - \varepsilon_f}{2}\right) + |V_K|^2\right]$$

Corresponding to the bonding and antibonding types of states in molecular systems. Fig. 7.18 shows the structure of hybridized bands, an example of the density of states for each band. One finds that strongly peaked structures occur in the regions near the gap. If the Fermi level falls within these peaks, a strong enhancement of the effective mass should take place solely because of these peculiarities of the band structure. In some situations, only a **pseudogap** caused by the hybridization is formed (Fig. 7.19). This is so if the hybridization matrix element V depends on the wave vector **k** and if along some directions in reciprocal space $V_K = 0$.

We may note that the inclusion of the interaction in eq. (7.22) renders the treatment of the Anderson lattice Hamiltonian much more complicated. So far no rigorous solutions of this problem has been obtained. A large variety of approximate treatments have been proposed [25 – 28], in all of which the principal task was to provide a satisfactory description of heavy fermion materials [29 – 31]. In effect, the limiting case of almost localized strongly correlated electrons was studied, which, among others, provides a quasiparticle electronic structure similar to that shown in Fig. 7.18, with a very strong enhancement of the DOS near the Fermi surface.

This leads to very heavy *quasiparticles*, which, in some systems, may undergo transitions either to antiferromagnetism or to superconducting states. In this respect, we find that heavy fermion systems are analogous to the HTSC cuprates, though with much lower T_c.

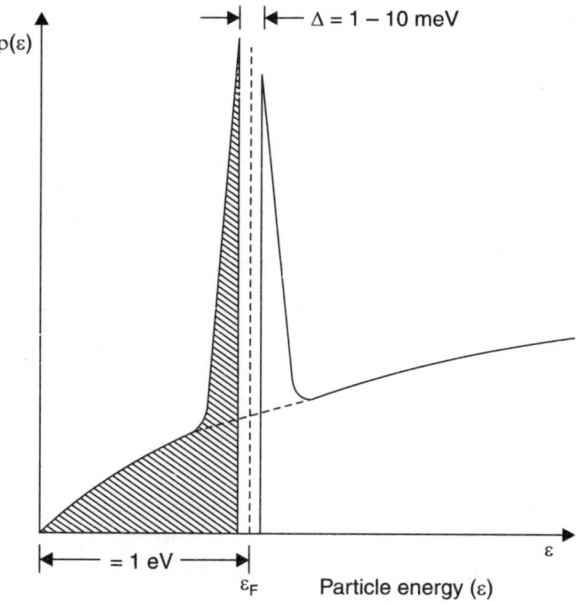

Fig. 7.18 Density $\rho(\varepsilon)$ of hybridized states versus particle energy ε. We may note that the hybridization gap Δ_h may be very small. Compared to the total width of the band states, the position of Fermi level ε_F corresponds to the filled lower band

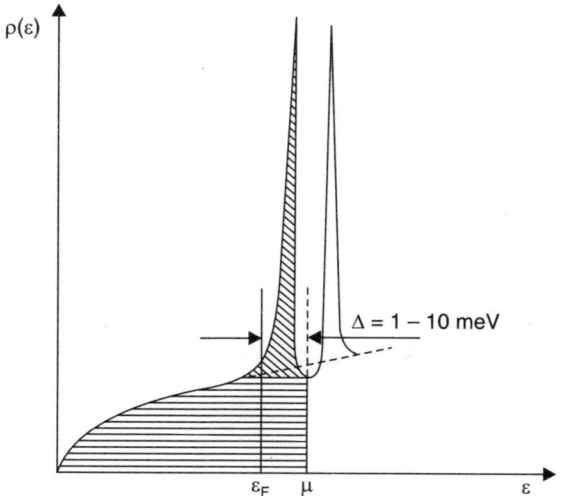

Fig. 7.19 Same as in Fig. 7.18 but with **pseudogap** among the hybridized bands

7.3 NOVEL MECHANISMS OF ELECTRON PAIRING

The binding of two fermions into a boson is a prerequisite for condensation of microscopic particles into a coherent macroscopic state. This condensation may take the form of Bose–Einstein condensation if the interaction energy between the pairs is much smaller than the binding energy of a single pair. Such a Bose condensed state of charged particles may exhibit the principal properties of the superconducting state, *e.g.* **Meissner-Ochsesnfeld effect** [32]. We have seen that in the BCS theory pair condensation occurs under a completely different condition, *i.e.*, when the states of different pairs overlap strongly so that the motion of one widely separated pair takes place in the mean field of all other pairs.

The pairing of particles in the BCS theory is described in momentum (reciprocal) space, where it is assumed that the quasiparticles states with a well defined Fermi surface are formed first: the pairing involves electrons from the opposite points on the Fermi surface ($\mathbf{k}, -\mathbf{k}$) and generates either a spin singlet state (as in the classic superconductors) or a higher angular-momentum state (as in superfluid 3_{He} [33]). Because of a small coherence length ($\xi \sim 10$ Å), the new superconductors offer an opportunity for exploring the possibility of pairing in real (coordinate) space. Moreover, since the carrier concentration determined from the Hall effect measurements [34] for HTSC cuprates is at least one order of magnitude smaller than that for ordinary metals. The pairing mechanism in cuprates is still not clear. There is rapidly accumulating evidence that pairing interaction in HTSC cuprates is strong electron–electron and makes it unlikely that electron pairing in these materials is caused by extremely strong electron–phonon interaction. Furthermore, the electron–phonon interaction does not allow for a connection (or strictly speaking, competition) between the observed superconductivity and antiferromagnetism [5]. This is one of the reasons for an intensive search for a purely electronic mechanism of pairing. We will discuss this in detail in the chapter 8.

EXCHANGE INTERACTIONS AND REAL SPACE PAIRING

7.3.1 Narrow Band Systems

We have already discussed an approximate Hamiltonian (eq. (7.14)), which includes the antiferromagnetic exchange interactions between the correlated electrons in the limit $U/W \gg 1$. The precise form of this Hamiltonian is second order in W/U is

$$H = \sum_{ij\sigma}{}' t_{ij} b_{i\sigma}^+ b_{j\sigma} + \sum_{ij}{}' \frac{2t_{ij}^2}{U} \left(S_i \cdot S_j - \frac{1}{4} v_{ij}\right) \qquad ...(7.23)$$

where the primed summation means that $i \neq j$. In this Hamiltonian doubly occupied site i configurations $|i \uparrow \downarrow\rangle$ are excluded. This exclusion is reflected by the presence of creation ($b_{i\sigma}^+$) and annihilation ($b_{i\sigma}$) operators for electrons in the state $|i\sigma\rangle$, which are defined as

$$b_{i\sigma}^+ \equiv a_{i\sigma}^+ (1 - n_{i-\sigma}) \quad \text{and} \quad b_{i\sigma} = a_{i\sigma}(1 - n_{i-\sigma}) \qquad ...(7.24)$$

So that

$$v_{i\sigma} = b_{i\sigma}^+ b_{i\sigma} \quad \text{and} \quad v_i = \sum_\sigma v_{i\sigma} \qquad ...(7.25)$$

The spin operator is defined as

$$S_i \equiv (S_i^+, S_i^-, S_i^z) \equiv [a_{i\uparrow}^+ a_{i\downarrow}, a_{i\downarrow}^+ a_{i\uparrow}, (n_{i\uparrow} - n_{i\downarrow})/2]$$

we may not that the same representation at the operator S_i can be written in terms of projected operators $b^+_{i\sigma}$ and $b_{i\sigma}$. The factor $(1 - n_{i-\sigma})$ in eq. (7.24) imposes explicitly the restriction that the creation or the annihilation of electrons in the state $|i\sigma\rangle$ can take place only if there is no second electron already on the same site. Thus, $v_i = \sum_\sigma n_{i\sigma}(1-n_{i-\sigma})$ enumerates only the singly occupied sites ($v_i = 0$ or 1). In other words, the N states corresponding to the doubly occupied site configurations have been projected out. Thus, eq. (7.23) describes the dynamics of strongly correlated electrons for $N_e \leq N$ of electrons. Also, in performing the summations in eq. (7.23), one usually considers only the pairs $\langle ij \rangle$ of nearest neighbours; in this approximation the parameters $J_{ij} = J$ and $t_{ij} = t$ can be chosen as constants.

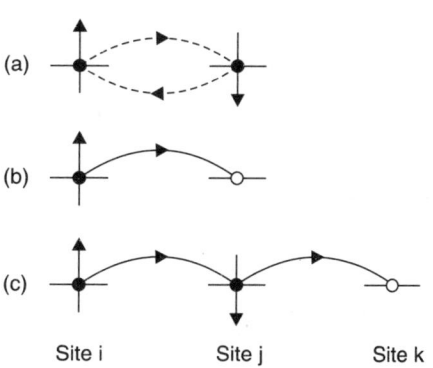

Fig. 7.20 Various hopping processes in narrow band systems in a partial band filling case: (a) virtual hopping processes leading to kinetic exchange interaction: (b) single particle hopping representing the band energy of correlated electrons; (c) contribution to the pair hopping process; this process gives the pairing contribution in eq. (7.27) with $k \neq i$

The first term in eq. (7.23) describes the single particle hopping of electrons from the singly occupied to the empty atomic sites; the second term describing the exchange interaction induced by virtual hopping between the sites i and j, while the three site part describes the motion of electrons with spin σ from the singly occupied site located at i to the next nearest neighbouring empty site k via the occupied configuration (with electron of opposite spin) located at size j. The various contributions to eq. (7.23) are represented graphically in Fig. 7.20.

If one introduces a new pair of creation and annihilation operators in coordinate space by

$$\tilde{b}^+_{ij} = \frac{1}{\sqrt{2}}(b^+_{i\uparrow}b^+_{j\downarrow} - b^+_{j\downarrow}b^+_{j\uparrow}) \qquad ...(7.26(a))$$

and

$$\tilde{b}_{ij} = \frac{1}{\sqrt{2}}(b_{i\downarrow}b_{j\uparrow} - b_{i\uparrow}b_{j\uparrow}) \qquad ...(7.26(b))$$

then the Hamiltonian (eq. 7.23) with inclusion of the three-site part can be written in the following very suggestive closed form [35, 36] :

$$H = \sum_{tj\sigma} t_{ij} b^+_{i\sigma} b_{j\sigma} - \sum_{tjk} (2t_{ij}t_{jk}/U)\tilde{b}^+_{ij}\tilde{b}_{kj} \qquad ...(7.27)$$

The first term represents, as before, the dynamics of single electrons moving between the empty sites regarded as holes; the second term combines the last two terms in eq. (7.23) and expresses the dynamics of the singlet pairs [cf. eqs. (7.26 a) and (7.26 b)]. The division in eq. (7.27) into single-particle and pair parts is in analogy to the BCS Hamiltonian; however, here, the operators are expressed in coordinate space. The term with $i = k$ in the pairing part enumerates the spin-singlet pairs of neighbouring spins; the terms with $i \neq k$ represent pair hopping of such singlet pair bonds. Thus, in the language of operators [eq. (7.26)], one adds the bonds dynamics to that of single electrons. Moreover, the forms of eqs. (7.23) and (7.27) are completely equivalent; hence the pairing effect and the antiferromagnetism should be directly linked within this formalism (they are two different expressions of the same part of H).

It is difficult to diagonalize the Hamiltonian [eq. (7.27)] to obtain the eigen values of the system. Part of the problem arises from the fact that the single particle operators $\tilde{b}_{i\sigma}$ and $b_{i\sigma}^+$ do not obey the fermion anticommutation relation, and that the pair operators b_{ij}^+ and $\tilde{b}_{ij}$ do not obey boson commutation relations. Additionally, the two terms in eq. (7.27) do not commute, so that the itinerant characteristics of the electrons and the pair-binding effects combine and produce a paired metallic phase, particularly if the two terms are of comparable magnitude. We have already seen earlier that if the number of holes $\delta \equiv 1 - n < \delta_c \sim 0.02$, then the pairing (or exchange) part dominates and antiferromagnetism sets in. Detailed calculations by Anderson [24] lead to the boundary line between the antiferromagnetic and ferromagnetic phase (Fig. 7.21). The energy of the completely saturated ferromagnetic phase (CF) indicated does not depend on the value of exchange integral $J_{ij} \equiv 2\, t_{ij}^2/U$.

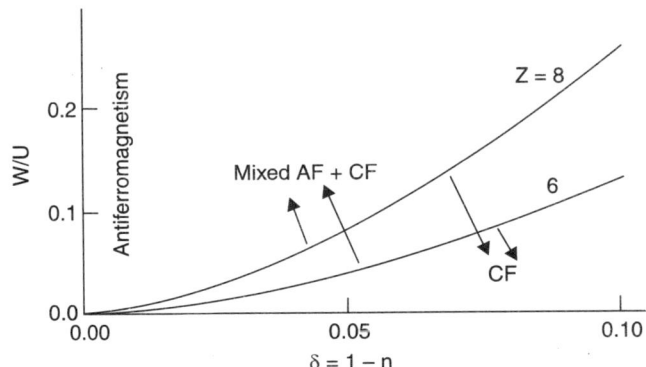

Fig. 7.21 Phase boundary between mixed ferromagnetic (CF)–antiferromagnetic (AF) phase and pure ferromagnetic phase for the simple cubic ($z = 6$) and bcc cubic ($z = 8$) structures. A similar type of phase boundary can be obtained for other structures [37].

Let us now discuss the superconducting phase for which the pairing part in eq. (7.27) plays a crucial role. To make the problem tractable at this point, one replaces the operators (eq. 7.26) by fermion operators [38, 39], *i.e.*,

$$\tilde{b}_{ij}^+ \to a_{i\sigma}^+ \qquad b_{i\sigma} \to a_{i\sigma} \qquad \ldots(7.28(a))$$

and introduces the replacement

$$\tilde{b}_{ij}^+ \to b_{ij}^+ = \frac{1}{\sqrt{2}}\, (a_{i\downarrow}^+ a_{j\uparrow}^+ - a_{i\downarrow}^+ a_{j\uparrow}^+) \qquad \ldots(7.28(b))$$

and
$$\tilde{b}_{ij} \to b_{ij} = \frac{1}{\sqrt{2}} (a_{i\downarrow} a_{j\uparrow} - a_{i\downarrow} a_{j\downarrow}) \qquad (7.28(c))$$

Simultaneously, one renormalizes the parameters t_{ij} and J_{ij} in such a manner that they contain the restrictions on particle dynamics due to the projection of doubly occupied site configurations in the expression for the ground-state energy. Within the Gutzwiller-Ansatz approximation [9], eqs. (7.26) reduce the starting Hamiltonian to the form

$$H = \sum_{ij\sigma} t_{ij} a_{i\sigma}^+ a_{j\sigma} \sum_{ijK} (2t_{ij} t_{jK}/U) b_{ij}^+ b_{Kj} \qquad ...(7.29)$$

where $\delta = 1 - n$. This Hamiltonian has been solved within the mean-field approximation equivalent to the BCS approximation [40, 41] and with neglect of the pairing terms with $K \neq i$. This leads to the following self-consistent equations for $\Delta_K \neq 0$:

$$\frac{J}{N} \sum_K \frac{\gamma_K^2}{E_K} \tanh\left(\frac{\beta E_K}{2}\right) = 1 \qquad ...(7.30)$$

with $J = 2t^2/U$, $E_K = [(\varepsilon_K - \mu)^2 + |\Delta_K|^2]^{1/2}$ and $\gamma_K = [\cos(K_x a) + \cos(K_y a)]$ for a planar configuration of the lattice. One must supplement this equation with the equation for the chemical potential in the superconducting phase of the form

$$\frac{1}{N} \sum_k \left(1 - \frac{\varepsilon_K}{E_K}\right) \tanh\left(\frac{\beta E_K}{2}\right) = n \qquad ...(7.31)$$

In solving eq. (7.30), solutions of the following type have been considered:

(i) **Extended wave** [40, 42]

$$\Delta_K^{(S)} = \Delta[\cos(K_x a) + \cos(K_y a)] \qquad ...(7.32(a))$$

(ii) **d-wave** [41, 42]

$$\Delta_K^{(d)} = \Delta[\cos|K_x a) - \cos(K_y a)] \qquad ...(7.32(b))$$

(iii) **Mixed s-d phases** [43]:

$$\Delta_K^{(sd)} = s\Delta_K^{(s)} + d\Delta_K^{(s)} \qquad ...(7.32(c))$$

The mixed phase was found to be the most stable close to the half-filled band case. For the half-filled band case, the ground-state energy for the s and d waves states are the same.

The type of solution obtained within the mean field approximation is illustrated through Fig. 7.22, where the temperature dependence of the specific heat is shown for a different number of holes δ and for $|t|U = 0.1$ and with inclusion of nearest-neighbour repulsive Coulomb interaction V. A discontinuity of $C(T)$ at $T = T_c$ takes place for each δ. For comparison, the dotted lines represent the specific heat for the normal phase.

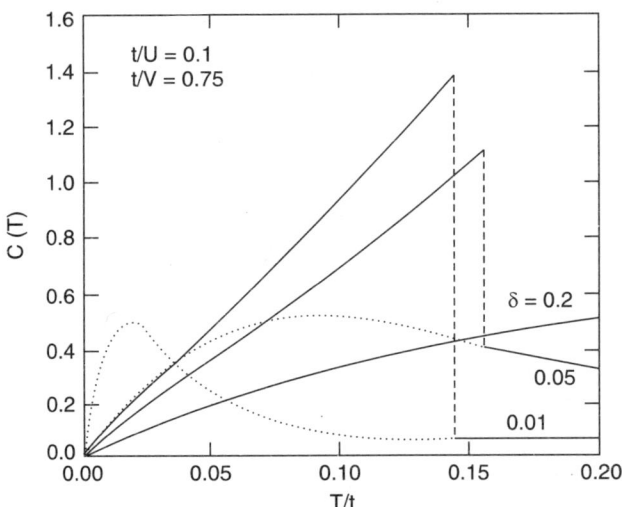

Fig. 7.22 Temperature dependence of the specific heat C(T) within the mean-field approach to the exchange mediated pairing in a narrow band. The dotted line represents C(T) for the normal phase, while the discontinuity occurs at the transition to the superconducting phase [42].

There is a major problem with the standard mean field solutions discussed in Refs. 35, 40-44, namely yields a nonzero (in fact, maximal or almost maximal) value of the superconducting transition temperature T_c for the half-filled band case, which corresponds to the Mott-insulating state. This is a spurious result; it appears because by performing the transformations, eq. (7.28) the double site occupancies reappear again for $n < 1$. To remove some of unphysical features of the meanfield, a new formalism has been proposed [42-44] in which auxiliary (*slave*) bosons are introduced. In this formalism, some of the properties of the projected operators [eqs. (7.24) and (7.26)] are already preserved in the mean-field approximation involving boson and fermion fields on the same footing. The transition temperature T_c now vanishes, as it should, in the limit $n = 1$. According to Zou and Anderson [45] the slave bosons represent holes in the Mott insulators and are regarded as charged, while the fermions are neutral. These entities are called **holons** and **spinons** respectively.

The holon-spinon language is introduced formally by noting that the projected operators [eq. 7.24] are represented as

$$b_{i\sigma}^+ = b_i f_{i\sigma}^+ \quad \text{and} \quad b_{i\sigma} = b_i^+ f_{i\sigma} \qquad ...(7.33)$$

where b_i and b_i^+ are annihilation and creation operators located at the atomic site i, while $f_{i\sigma}^+ \equiv a_{i\sigma}^+$ and $f_{i\sigma} \equiv a_{i\sigma}$ are the commonly used fermion operators. Making use of eq. (7.33), eq. (7.23) takes the form

$$H = \sum_{tj\sigma}' t_{ij} b_i b_j f_{i\sigma}^+ f_{j\sigma} - \sum_{tj\sigma}' (2 t_{ij}^2/U) b_{ij}^+ b_{ij} + \sum_i \lambda_i (b_i^+ b_i + \sum_\sigma f_{i\sigma}^+ f_{i\sigma} - 1) - \mu \sum_{i\sigma} n_{i\sigma}^+ \qquad ...(7.34)$$

(three site terms)

The first two terms represent, respectively, the coupled holon-spinon hopping and the binding of spinons into singlet pairs. The third term expresses the fact that the number of holons and spinons is equal to unity on each site; the Lagrange multiplier λ_i thus explicitly provides for the removal of double occupancies. The fourth term represents the conservation of the number of electrons. Now, in a further approximation, one decouples fermions from bosons and then solves the two part self-consistently. In the mean field treatment λ_i is taken as the same at each site ($\lambda_i \to \lambda$) and in which one introduces the replacement $<b_i\, b_i^+> = <b_i><b_i^+> = |b_i|^2 = 1 - n$.

The superconducting solution is described in terms of two correlation functions: $\Delta_B \equiv <b_i^+\, b_j^+> \approx <b^+>^2$, characterizing the Bose condensation of holons, and $\Delta_F \equiv <f_{i\uparrow}\, f_{j\downarrow}>$, characterizing the gap in the spectrum of fermion excitations (the site indices i and j denote a pair $<ij>$ of nearest neighbours). The nonzero Δ_B occurs only below a temperature T_B, which is called as Bose condensation temperature, whereas the nonzero Δ_F appears only below $T = T_{RVB}$, characterizing the mean field solution within the AVB theory. The superconducting phase is characterized by nonzero values of both Δ_B and Δ_F simultaneously. This is because in the mean field approximation $<b_{i\uparrow}\, b_{i\downarrow}> = \Delta_B \Delta_F$. Hence, the lower of the two temperatures (T_B and T_{RVB}) determines the superconducting transition temperature.

In Fig. 7.23 [37], we have plotted these two temperatures as a function $\delta = 1 - n$. We must note that in order to have $T_B \neq 0$ a small nonzero overlap $t_z = 0.1\, t$ was taken in the direction perpendicular to the square planar configuration of the atoms. One finds that $T_c \to 0$ as $\delta \to 0$, as should be the case.

We may note that the above treatment is based on several approximations that need further explanation. The most serious of them is that in the limit of $\delta = 0$ this approach does not properly reproduce the characteristics of the Mott insulator that has the properties of the spin (1/2) Heisenberg antiferromagnet. Moreover, the three site terms that provide the Bose pairing $b_i^+\, b_k^+$ have been ignored. One can solve these two problems [46] by (i) replacing the operators in eq. (7.24) in **K** space by the operators $b_{k\sigma}^+$ and $b_{k\sigma}^+$ with the statistical mechanical distribution for $<b_{k\sigma}^+\, b_{k\sigma}^+>$ given by eq. 7.20(c), i.e. by the statistical distribution for the **spin liquid** [19]; and (ii) by transforming the three-site terms in eq. (7.27) to **K** space and

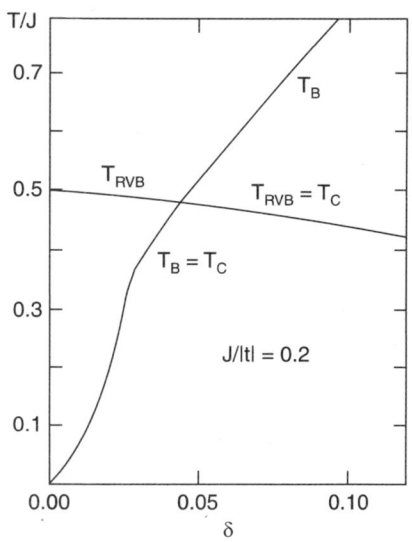

Fig. 7.23 Critical temperature (T_c) (the thick line) versus $\delta = 1 - n$. The temperatures T_{RVB} and T_B are those characterizing the onset of coherency for the spinons and the Bose condensation of holons. We may note that T_c is determined by the lower of the two temperatures [37].

introducing the *BCS*-type approximation by taking only the terms of the type $b^+_{K\uparrow} b^+_{K\downarrow} b_{-K'\downarrow} b_{-K'\uparrow}$. In effect, the $\underline{K}$ dependence of the gap $\Delta_K \equiv \langle b^+_{K\sigma} b_{K\sigma} \rangle$ can be derived explicitly: it is of extended *s*-wave form [eq. 7.32(*a*)].

Wheatley et al. [47] have proposed model involving paired boson condensation (characterized again by the boson $\langle b^+_i b^+_i \rangle$ amplitude) because of coherent tunneling of quasiparticle holon pairs between the CuO_2 planes. This approach [47] explains the increase of T_c if the number of CuO_2 neighbouring planes increases. The last result is supported by the experiment results for Bi and Tl compounds for which $T_c > 100$ K. However, some degree of interplanar coupling is needed in any theory involving the Bose condensation since it is absent in strictly two-dimensional model.

7.3.2 Hybridized Systems

The electron states near the Fermi surface in the HTSC cuprates such as La_2CuO_4 involve hybridization of electrons of atomic like $3d_{x^2-y^2}$ states of copper with $2p_\sigma$ of oxygen (Fig. 7.5). One can describe these by the Anderson lattice Hamiltonian of type of eq. (7.22) with a width of the bare *p*-band $W \approx 4$ eV, the position of the $3d^9$ level at $\varepsilon_f \equiv \varepsilon_d - \varepsilon_p \sim 1$ eV, $U \leq 10$ eV, and the hybridization magnitude $|V| \simeq 1.5$ eV [48]. The hybridization is intersite in nature, *i.e.*, it involves the $2p$ and $3d$ orbitals located on different sites. Therefore, the effective hybridization energy is $V_z \simeq 6$eV, where $z = 4$ is the number of nearest neighbouring O atoms in the plane for a given Cu atom. We see that $V_z > \varepsilon_f$; hence, the $3d$ and $2p$ states mix strongly, *i.e.*, the *d* electrons can be promoted to $2p$ hole states and vice versa. Additionally, $2p$ electrons can be promoted to form the $3d^{10}$ configurations of the excited states. If $V_z \geq \varepsilon_f$, but $|V|_z \ll \varepsilon_f + U$, the above two promotion mixing events are low and high energy processes, respectively. Fig. 7.24 shows the schematically drawn situation, where U is assumed to be by far larger than $|\varepsilon_f|$, W, or $|V|_z$. Let us first consider this limiting situation [49].

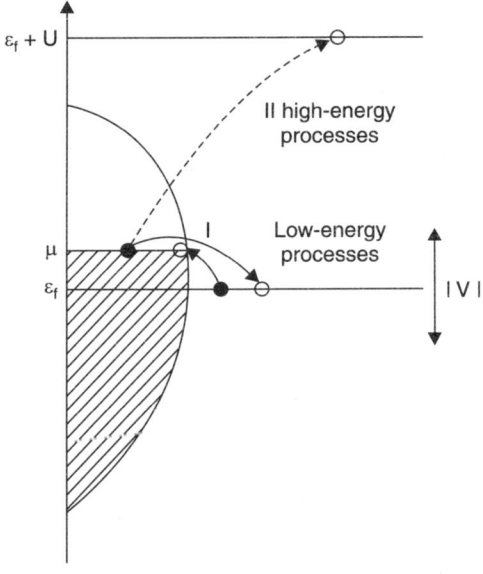

Fig. 7.24 Division of the charge-transfer (p-d) process in low- and high-energy parts. The processes II give rise to the Kondo and superexchange interactions when treated perturbationally to second- and fourth-order respectively.

The high energy processes take place only as virtual events *i.e.*, with electron hopping from *p* state to the highly excited $3d$ state and back. Such virtual *p-d-p* processes have been drawn schematically in Fig. 7.25, where site *m* labels the $2p_\sigma$ state of the oxygen anion O^{2-}

centered at $\mathbf{R}_m$ and site i labels $3d_{x^2-y^2}$ due to the Cu^{2+} ion centered at $\mathbf{R}_i$. Then, the effective Hamiltonian can be rewritten in the real space language, and for large U as [49]

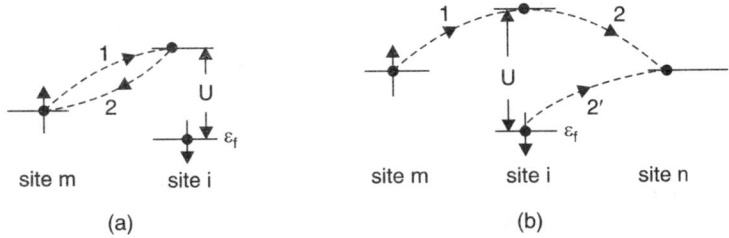

Fig. 7.25 Schematic representation of the hopping processes induced by the high-energy mixing processes. (a) virtual hopping, and (b) three site processes. The hoppings labeled 2 and 2' are alternative processes.

$$H = \sum_{K,\sigma} \varepsilon_K n_{K\sigma} + \varepsilon_f \sum_{i\sigma} b^+_{i\sigma} b_{i\sigma} + \sum_{im\sigma} (V_{im} b^+_{i\sigma} C_{m\sigma} + V^*_{im} C^+_{im\sigma} b_{i\sigma})$$

$$- \sum_{imn} \frac{2 V^*_{mi} V_{im}}{U + \varepsilon_f} \tilde{B}^+_{im} \tilde{B}_{in} \quad ...(7.35)$$

The first term describes the band energy of itinerant ($2p_\sigma$) electrons, while the third term represents the residual mixing since, as in the case of narrow-band electrons, the operators ($b^+_{i\sigma}$) and ($b_{i\sigma}$) are projected operators [eq. (7.24)] for the starting $3d$ states. The last term represents the so-called hybrid (interorbital) pairing with the pairing operators

$$\tilde{B}^+_{im} = \frac{1}{\sqrt{2}} (b^+_{i\uparrow} C^+_{m\downarrow} - b^+_{i\downarrow} C^+_{i\uparrow}) \quad ...(7.36(a))$$

and

$$\tilde{B}_{im} = \frac{1}{\sqrt{2}} (b_{i\downarrow} C_{m\downarrow} - b_{i\downarrow} C_{i\uparrow}) \quad ...(7.36(b))$$

The meaning of effective Hamiltonian [eq. (7.35)] is as follows: The first three terms provide eigenvalues representing the hybridized quasiparticle states with the structure discussed earlier. The last term provides a singlet pairing for those hybridized states. It expresses (for $m = n$) the Kondo interaction between the p and $3d$ electrons of the form

$$\sum_{im} \frac{2|V_{im}|^2}{U + \varepsilon_f} (\mathbf{S}_i \cdot \mathbf{S}_m - \frac{1}{4} V_i n_m)$$

It is antiferromagnetic in nature, with the exchange integral

$$J_{im} = \frac{2|V_{im}|^2}{U + \varepsilon_f} \sim 0.5 \text{ eV}$$

hence the pairing results in a spin singlet state. We may note that eq. (7.35) represents hybridized correlated states in the so-called fluctuating regime in which $U \gg |V_{im}| \gtrsim \varepsilon_f$. This is the reason why one cannot completely transform out the hybridization. Also, the occupancy n_f of the atomic level is a non integer because the strong hybridization induces a redistribution of the particles among starting atomic and band states.

When both U and $|\varepsilon_f|$ are much larger than $|V_{im}|$, one can transform out the hybridization completely and obtain instead of eq. (7.35) the following effective Hamiltonian.

$$H = \sum_{\vec{K},\sigma} \varepsilon_K n_{K\sigma} + \varepsilon_f \sum_{i\sigma} b_{i\sigma}^+ b_{i\sigma} + \sum_{ijm\sigma} \frac{V_{mi}^* V_{mj}}{\varepsilon_f} b_{i\sigma}^+ b_{j\sigma} (1 - n_{m\sigma}) \times \sum_{imn} \frac{2 V_{mi}^* V_{in} U}{\varepsilon_f (\varepsilon_f + U)} \tilde{B}_{im}^+ \tilde{B}_{in} \ldots (7.37)$$

We have now a two bond system: the $3d$ electrons acquire a bandwidth $W^* \sim (V^2/\varepsilon_f)(1 - <n_{m\sigma}>)$. The spin singlet pairing is again of interband type. The part with $m = n$ in last term is equivalent to the **Kondo interaction** [50]. Here, the lattice version of this Hamiltonian provides both pairing the itinerancy to the bare atomic electrons.

We may note that the hybrid pairing introduced in this section expresses both the **Kondo interaction** (the two side part) and pair hopping. It is therefore considered that this may be suitable for a discussion of superconductivity of Kondo lattice effects in heavy fermion systems. The pairing part supplements the current discussions of Anderson lattice Hamiltonian in the $U \to \infty$ limit [26, 27, 28]. We may note that the Kondo interaction mediated pairing introduced above represents the strong-coupled version of spin-fluctuation mediated pairing for almost localized systems introduced previously [51-54].

An approach using the **slave boson language** for hybridized systems has also been formulated by Newns [55] and contains a principal feature of the effective Hamiltonian (7.35) and Newns has also discussed the solution in the mean-field approximation. Fig. 7.26 shows the dependence of the superconducting transition temperature T_c versus hole concentration x_h. A comparison is shown with experimental data obtained by Torrance et al.[56]; dependence of T_c the full concentration range of holes is shown in Fig. 7.27. The superconductivity appears for HTSC $La_{2-x} Sr_x CuO_4$ only for $0.04 \lesssim x_h < 0.34$.

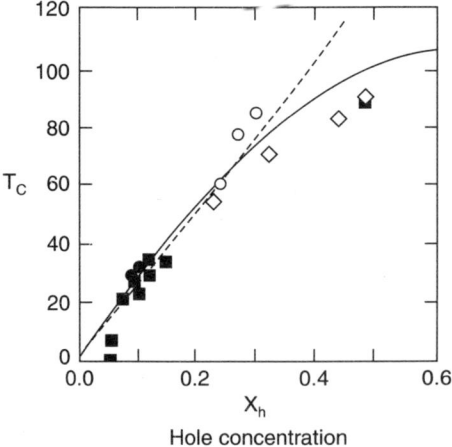

Fig. 7.26 The superconducting transition temperature T_c versus hole concentration X_h. Experimental data for $La_{2-x} Sr_x CuO_4$ (squares) and data for $Y Ba_2 Cu_3 O_{7-y}$ (circles and diamonds) [55]

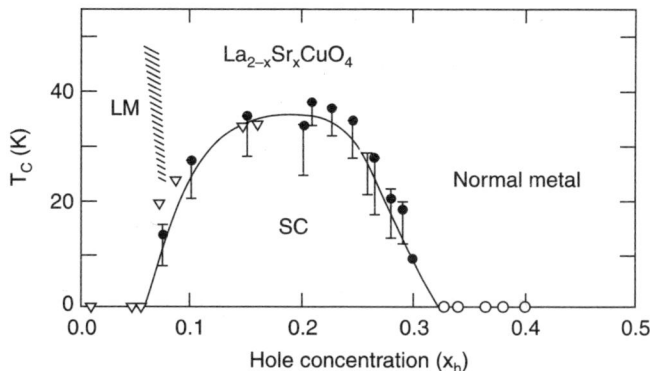

Fig. 7.27 The superconducting transition temperature (T_c) versus hole concentration (x_h) for HTSC $La_{2-x}Sr_xCuO_4$ in the full range. LM stands for regime of local moments (insulating phase) [56].

7.3.3 Properties of Correlated Electrons

So far we have discussed two alternative models and mechanisms of exchange mediated pairing: narrow band model with d-d kinetic exchange mediated pairing, and the hybridized model with d-p Kondo interaction mediated pairing. The hybridized model may be considered as a basis of narrow-band behaviour in real oxides and in heavy fermion systems since the direct d-d (or f-f) overlap of the neighbouring atomic wavefunctions is extremely small. Now, starting from the hybridized (Anderson lattice) model, we present a brief overview of the narrow-band properties of the correlated electrons.

Let us first discuss the quasiparticle states in $U \to \infty$ limit. The simplest approximation is to reintroduce ordinary fermion operators $a_{i\sigma}^+$ and $a_{i\sigma}$ in eq. (7.35) and readjust the hybridization accordingly [57]. In effect, one obtains the hybridized bands of the form of

$$E_{K\pm} = \frac{\varepsilon_p + \varepsilon_d}{2} \pm \left[\left(\frac{\varepsilon_p - \varepsilon_d}{2}\right)^2 + 4V^2\left(\sin^2\frac{K_x a}{2} + \sin^2\frac{K_y a}{2}\right)\right]^{1/2}$$

where ε_p and ε_d are atomic level positions for the $3d$ and $2p$ states, respectively. i.e.,

$$E_{k\pm} = \frac{\varepsilon_f + \varepsilon_k}{2} \pm \left[\left(\frac{\varepsilon_f - \varepsilon_k}{2}\right)^2 + 4|\tilde{V}_K|^2\right]^{1/2} \quad \ldots(7.38)$$

where $\tilde{V}_K \equiv q^{1/2} V_K$, and $q \equiv (1 - n_f)/(1 - n_{f/2})$ for $0 \leq n_f \leq 1$, while V_K is the space Fourier transform of V_{im}. For the case of CuO_2 layers [58].

$$|\tilde{V}_K|^2 = qV^2\left[\sin^2\left(\frac{K_x a}{2}\right) + \sin^2\left(\frac{K_y a}{2}\right)\right] \quad \ldots(7.39)$$

If the Fermi level falls into the lower hybridization band and $n_f = 1 - \delta$, with $\delta \ll 1$, then it can be shown that the quasiparticles describing the hybridized states are of mainly quasi-atomic character. In other words, in the effective Hamiltonian [eq. (7.29)] the pairing takes place between heavy quasiparticles. This limiting situation describes qualitatively the situations in Heavy fermions with Kondo interaction mediating the pairing. By contrast, if the Fermi level falls close to the top of the upper hybridization band (as in the case for HTSC cuprates, since the p band is almost full and $3d$ level is almost half-filled), then the pairing is mainly due to the band electrons ($2p$ holes in the case of HTSC cuprates). These results have been obtained by constructing explicitly the eigen states corresponding to the eigenvalues [eq. (7.38)] and taking the limits corresponding to heavy fermions ($n_f \to 1$) and HTSC oxides ($n = n_d + n_p \approx 3$).

7.3.4 Mott-Hubbard Insulators, Charge Transfer Insulators and Mixed Valent Systems

We now consider the problem concerning the Mott localization in systems with hybridized d-p states. The systems such as NiO, CoO or MnO regarded as classic Mott insulators are, strictly speaking hybridized $3d - 2p$ systems. However, these cases are to a good approximation ionic systems in the sense that the electronic configuration in, for $e.g.$, NiO, is $Ni^{2-}\ O^{2-}$. Then, the valence $2p$ band is completely full and plays only a passive role in effective d-d charge transfer process [59, 60]. In this approach, the kinetic exchange interaction between d-electrons (induced by virtual d-d electrons) is expressed as a fourth-order effect in the hybridization V since the virtual d-d transition involves a sequence of d-p and p-d transitions in the fourth order.

The possible macroscopic states of hybridized systems are as shown in Fig. 7.28 as a schematic classification of possible states of hybridized systems modeled by the periodic Anderson Hamiltonian eq. (7.22). The parameter W/U characterizes the degree of correlation of quasi-atomic electrons that may acquire a nonzero bandwidth mainly due to hybridization; the parameter V/ε_f characterizes the degree of mixing of the states involved. If the d (or f) atomic level lies deeply below the top of the valence band ($V/\varepsilon_f \ll 1$), then we have either Mott-Hubbard (M-H) or charge-transfer (C-T) insulators; for the former the band gap Δ is due to $d^n - d^{n+1}$ excitations ($i.e.$, $\Delta \sim U - W$), whereas for the latter it is due to $d^n p^2 \to d^{n+1} p^2$ charge transfer transitions. The

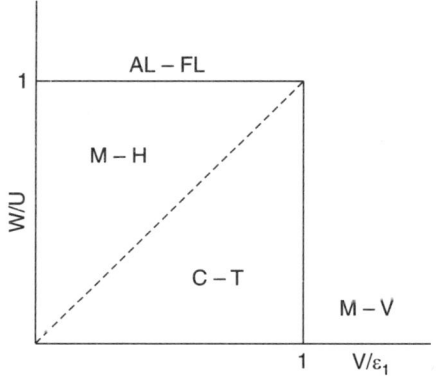

Fig. 7.28 Schematic representation of the regimes of charge-transfer (C – T) and Mott-Hubbard (M – H) insulating states, as well as of mixed-valent (M – V) and almost-localized Fermi-liquid (AL – FL) metallic states

atomic $3d$ (or $4f$) electrons are unpaired in both the C-T and M-H states. If $V/\varepsilon_f \gtrsim 1$ and $W/U \ll 1$, then we enter mixed-valent (M-V) and (close to the border with M-H) heavy fermion regimes. On the other, if $W/U \gtrsim 1$, then irrespective of the value of V/ε_f we encounter the correlated metal regime that we call an almost-localized Fermi liquid ($AL\ FL$).

7.3.5 Magnetic Interactions, Polarons and Pairings

The antiferromagnetism is stable only close to the half filling of the d band (Fig. 7.21). In the case of hybridized model, one has to calculate explictly the contributions to the d-p and d-d

interactions. In the case of the hybridized model, one has to calculate explicitly the contributions to the d-p and d-d interactions within the perturbation expansion for the Anderson lattice model but with the high-energy mixing processes (cf. Fig. 7.24) treated in this manner [49, 50]. One obtains the magnetic part of the effective Hamiltonian to fourth order as

$$H_m \simeq J_{pd} \sum_{im} \left(\mathbf{S}_i \cdot \mathbf{S}_m - \frac{1}{4} n_i n_m \right) + J_{dd} \sum_{<ij>} \left(\mathbf{S}_i \cdot \mathbf{S}_J - \frac{1}{4} N_i N_J \right)$$
$$+ J_{pp} \sum_{<mm'>} \left(\mathbf{S}_m \cdot \mathbf{S}_{m'} - \frac{1}{4} n_m n_{m'} \right) \quad ...(7.40)$$

where the first term represents the p-d Kondo-type interaction, with the exchange integral

$$J_{pd} \approx \frac{2|V|^2}{U + \varepsilon_f} \left[1 - \frac{|V|^2}{U + \varepsilon_f} (n_d + n_p + 1) \right] \quad ...(7.41)$$

The second term expresses the d-d (kinetic exchange) interaction, with $J_{dd} = |V|^4/(U + \varepsilon_f)^3$, and the last term represents the interaction between p holes, with $J_{pp} = |V|^4 n_d / (U + \varepsilon_f)^3 \approx J_{dd}$. The antiferromagnetic p-d and d-d interactions are not compatible; in the hole language, the p-hole polarizes its surroundings ferromagnetically (Fig. 7.29). We may note that the hole may be located in any O^{1-} ion,, so its position in the volume of radius R is not fixed. A simple estimate of the canting angle θ between the neighbouring $3d$ spins $\mathbf{S}_i$ and $\mathbf{S}_m$ caused by the hole polarization gives

$$\cos \frac{\theta}{2} \approx \frac{J_{pd} - 2 J_{dd}}{2 J_{dd}} \quad ...(7.42)$$

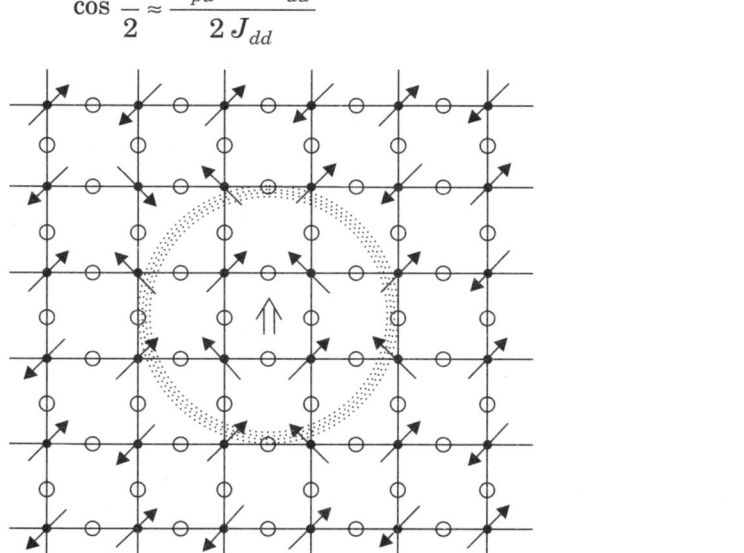

Fig. 7.29 Schematic representation of a $2p$ hole polaron in the planar CuO_2 structure. Cu^{2+} ions are indicated by the arrows, while O^{2-} ions are indicated by open circles. The hole creates a canted spin configuration with resultant ferromagnetic polarization and autolocalizes in it. This is the reason why HTSC cuprates remain insulating when the concentration of the hole does not exceed $x_c \sim 0.04 - 0.05$.

Taking $J_{pd} \approx 0.5$ eV and $J_{dd} \simeq 50\,K$, one obtains the canting angle θ through the relation $\cos(\theta/2) \approx 25\,x_p$. The energy E_c of the system with a single hole carting the surrounding spins is

$$E_c = -\frac{1}{2}\frac{(J_{pd}-2J_{dd})^2}{J_{dd}} = -J_{dd}z \qquad \ldots(7.43)$$

This energy is lower than the energy $(-J_{dd}z)$ of the antiferromagnetic (Neel) state of antialigned d-spins due to Cu^{2+} copper ions. Now, we estimate the radius R of the hole polaron with aligned spin, as depicted in Fig. 7.29. One also obtains the expression for the energy E_p of a single polaron:

$$E_p = \frac{E_0}{(R/a)^2} - \frac{1}{2}\frac{(J_{pd}-2J_{dd})^2}{J_{dd}}zx^2p \qquad \ldots(7.44)$$

where $\bar{x} = (a/R)^2$ is the probability to find a p hole on a given oxygen atomic site within the radius R. Minimizing with respect to R for the two dimensional case, one obtains

$$\frac{R}{a} = \left[\frac{J_{pd}-2J_{dd}}{J_{dd}E_0}\right]^{1/2} \approx J_{pd}\sqrt{\frac{z}{J_{dd}E_0}} \approx 4 \qquad \ldots(7.45)$$

A metal-insulator transition takes place when the neighbouring polarons overlap, i.e., when $Rx_{pc}^{-1/2} = 1$; this yields the critical hole concentration $x_c \approx 0.07$. One can also estimate this critical concentration by equating the band energy of holes, which is $-(W/2)\,x_p(1-x_p)$, with the magnetic energy gain per hole due to aligning the neighbouring d-spins $(-J_{pd}^2/2J_{dd})\,zx_p^2$. This leads again to $x_c \approx 0.068$, in rough agreement with the observed value $x_c \simeq 0.04 - 0.05$. For $x > x_c$, the ground state of the system is metallic. Within the exchange-mediated mechanism, all interactions in eq. (7.40) are antiferromagnetic. Hence, in general, one has p-d pairing characterized by the operators of eq. (7.36), the p-d pairing [62, 63] characterized by the operators

$$p^+_{mm'} = \frac{1}{\sqrt{2}}(C^{+}_{m\uparrow}C^{+}_{m'\downarrow} - C^{+}_{m\downarrow}C^{+}_{m'\uparrow}) \qquad \ldots(7.46(a))$$

and

$$p_{mm'} \equiv \frac{1}{\sqrt{2}}(C_{m\downarrow}C_{m'\uparrow} - C_{m\uparrow}C_{m'\downarrow}) \qquad \ldots(7.46(b))$$

All three types of pairing may contribute to the superconducting ground state. However, the d-p interaction is much stronger, hence the d-p hybrid type of pairing is in the limit $U > W > |V| \gtrsim \varepsilon_f$, the dominant one. As stated, this type of pairing may appear effectively as a d-d or p-p type of pairing in the hybridized basis, depending on whether the Fermi level lies close to the top of the lower or upper hybridized bands, respectively. For the sake of completeness, one can write the full effective Hamiltonian with all pairings specified, namely

$$H = \sum_{\vec{K},\sigma}\varepsilon_K n_{K\sigma} + \varepsilon_f b^+_{i\sigma}b_{i\sigma} + \sum_{im\sigma}(V_{im}b^+_{i\sigma}C_{m\sigma} + V^*_{im}C^+_{m\sigma}b_{i\sigma})$$

$$+ J_{pd}\sum_{(imn)}\tilde{B}^+_{im}\tilde{B}_{in} + J_{dd}\sum_{(ijK)}\tilde{b}^+_{ij}\tilde{b}_{Kj} + J_{pp}\sum_{(mm'm'')}p^+_{mm'}p_{m''m'} \qquad \ldots(7.47)$$

In deriving this result, one does not assume that $|V| \ll \varepsilon_f$; therefore eq. (7.47) is applicable to the situation with fluctuating valence. Next, by introducing a slave boson representation [eq. (7.33)], one obtains most general Hamiltonian for treatment of pairing in correlated systems [61]. A decisive role of hybrid p-d, d-d, and p-p pairings in HTSC cuprates is still unclear. However, eq. (7.47) can serve as a basis for the discussion of antiferromagnetism and superconductivity in heavy fermion systems. In heavy fermion systems, the role of itinerant $2p$ states in played by hybridized $5d$-$6s$ conduction bands, while the role of $3d$ electrons is played by the $4f$ electrons.

7.4 COEXISTENCE OF ANTIFERROMAGNETISM AND SUPERCONDUCTIVITY

One can visualize the possible coexistence of antiferromagnetism (AF) and superconductivity (SC) phases, by considering a narrow band model with the two dimensional (almost square) Fermi surface as drawn in Fig. 7.30. Obviously, the band energy of electrons located on the Fermi surface has the property $\varepsilon_{\vec{K}-\vec{Q}} = -\varepsilon_{\vec{K}}$, where $\mathbf{Q} \equiv (\pi a_1, \pi/a) = 2k_F$. This is so called **nesting condition**; any system with this property is unstable with respect to formation of the spin density wave (SDW) state with the wave vector $\mathbf{Q}$. The SDW state is identical with a two sublattice AF state if the magnitude of magnetic moment in the ordered state is $\mu \sim \mu_B$, where μ_B is the Bohr magneton. We may note that $\mathbf{Q}$ connects two single particle states on the opposite sites of the Fermi surface since $-\mathbf{K}_F + \mathbf{Q} = \mathbf{K}_F$. Moreover, both SDW and SC states couple electrons with the opposite spins.

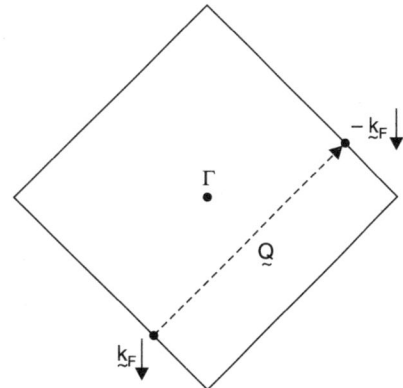

Fig. 7.30 Two dimensional Fermi-surface for a half filled band. The opposite points of the surface are related by the wave vector $\mathbf{Q} = (\pi/a)\,(1,\,1)$

This is why AF and SC states are compatible for $n \approx 1$. In HTSC cuprates, there is no clear evidence that these two phases coexist. One can easily show [35, 36] that close to the half-filled narrow band case, $T_N/T_C \sim 6-8$. The analysis of AF–SC coexistence within the Anderson lattice Hamiltonian is yet not clear.

7.5 CONCLUSIONS

In this chapter, we have mainly concentrated on the properties of correlated electrons in normal, antiferromagnetic and superconducting phases in HTSC cuprates, in which the last two are phases caused by antiferromagnetic exchange interactions. We have also discussed two theoretical models: the **Hubbard model** of correlated narrow band ($3d$) electrons and the **Anderson lattice model** of correlated and hybridized electrons, involving $2p$ and $3d$ states in the case of HTSC cuprates. Anderson lattice model is regarded as more general and applicable to both HTSC and heavy fermion systems; in some limiting situations as discussed previously hybridized bands exhibit a narrow band behaviour.

The principal novel feature of the metallic phase involving either $3d$ (in HTSC cuprates) or $4f$ (in heavy fermion systems) electrons is that for the half-filled band configuration the itinerant electrons states transform into a set of localized constituting the Mott insulator. The Mott insulator, the spin-liquid and the heavy-fermion states are the primary phases of correlated electrons different from the normal metal state. Fig. 7.31 depicts this difference, where the arrows point to both common features for normal and correlated metals, as well as to those specific to the system correlated systems.

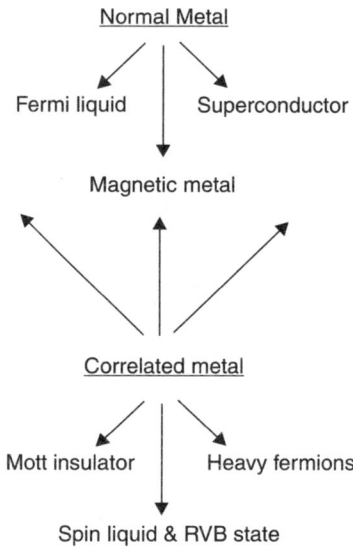

Fig. 7.31 Schematic representation of the difference between the normal metal and correlated metal. Only the latter state may lead to the Mott localization, as well as to heavy fermion and spin-liquid metallic phases.

The qualitative division between Fermi liquid and spin liquid states is sketched in Fig. 7.32, where the various thermodynamic phases have been specified for each class. The complementary regimes are those with $U/W \ll 1$ and $U/W \gg 1$. Most of the metallic systems can be located between these two limiting situations. It remains to be proven more precisely that the Mott-Hubbard boundary separating Fermi-liquid for the Mott insulator for $n = 1$ extends to the part of the diagram with $n \neq 1$. This is a fundamental problem, related in the case of strongly interacting systems to the question of validity of the **Luttinger theorem**[1] and to the problem of validity of the Bloch theorem for a correlated metal. Also the question of applicability of the Fermi-liquid concept in the limit $U/W \gg 1$ is connected with that concerning the properly defined existence of fermion quasiparticles[2], interacting only weakly among themselves. One should emphasize that the discussion of the standard meanfield treatment of

---

[1] The **Luttinger theorem** states that as long as the metallic state is stable the volume encircled by the Fermi surface remains independent of the strength of the electron–electron interaction. This theorem is not valid when the Mott transition takes place and the Fermi surface disappears. The volume also doubles when metal is described by a spin liquid.

[2] The holons and spinons are regarded as solitons.

superconductivity reduces the whole problem to the single-particle approach with a self-consistent field $\sim \Delta_k$. It is yet not completely clear what types of collective excitations (in addition to quasiparticle states) are needed to make the theory complete. The introduction of holons as bosons and spinions as fermions [20, 21] seems to be just one possibility; more natural seems to be treatment of holons as spinless fermions and of spinons as boson operators that reflect magnon like properties of local moments.

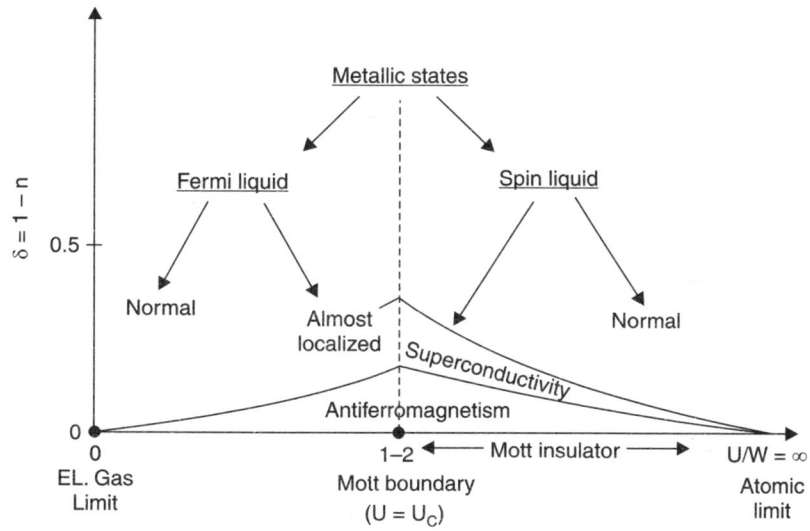

Fig. 7.32 Qualitative distinction between Fermi-liquid and spin-liquid states. The Mott boundary $U = U_C$ roughly separates the two limiting phases.

It would be proper to have some look concerning the analogy of the studies of magnetism and superconductivity. Heisenberg in 1928 introduced the exchange interaction $J_{ij} \mathbf{S}_i \cdot \mathbf{S}_j$ between the magnetic atoms with spins $\{S_i\}$. The ferromagnetic state was understood in terms of a molecular field $\mathbf{H}_i \sim \langle \mathbf{S}_i \rangle$, which was related to the direct exchange integral J_{ij}. Later, various exchange interactions have been introduced, such as super exchange, double exchange, *RKKY* interaction, the **Bloembergen-Rowland interaction, Hund's rule exchange,** and **Kinetic exchange** to explain magnetism in specific systems, such as oxides, rare-earth metals, and transition metals. However, all these new theories provided a description in terms of a single-order parameter the magnetization $\langle \mathbf{S}_i \rangle$; the particular feature of the electron states in each case (localized states, itinerant states, or a mixture of two states) is contained only in the way of defining the order parameter or the exchange integral. By analogy, the BCS theory provided a concept of a superconducting order parameter ($\Delta_\mathbf{k}$), which is universal for all theories of singlet superconductivity. New mechanisms of pairing in HTSC cuprates, heavy fermions and other exotic superconductors should provide a novel interpretation to the coupling constant $V_{\mathbf{k}_\mathbf{k}}$, as well as provide some details concerning the specific features of the system under consideration: the gap anisotropy, the role of hybridization, etc. [64-67]. We will discuss these issues in chapters 8 and 9.

REFERENCES

1. A. H. Wilson, *"The Theory of Metals"*, Cambridge Univ. Press, London and New York (1958).
2. G. Baym and C. Pethick, in *"The Physics of liquid and Solid Helium"*, K.H. Bennemann and J.B. Ketterson, eds. Ch. 3.
3. N.F. Mott, *"The Metal-Insulator Transitions"*, Taylor and Francis, London (1974).
4. J. Hubbard, *Proc. Royal. Soc.* London A 281, 401 (1964).
5. D. Vaknin et al, *Phys Rev.* B 37, 7473 (1988).
6. C.P. Poole, Jr. et al. *'Superconductivity'*, Academic Press, New York, (1995)
7. C.J. Ballhausen, *Introduction to Ligand Field Theory*, McGraw Hill, New York, (1962).
8. M.Ac quarone et al., *J. Phys. C*15, 959 (1982).
9. J. Spalek et al., *Phys. Rev. B* 28, 6802 (1983).
10. H. Kuwamoto et al., *Phys. Rev B* 22, 2626 (1980).
11. W.F. Brinkman and T.M. Rice, *Phys. Rev.* B2, 4302 (1970).
12. J.Spalek et al., *Phys. Rev. B*39, 4175 (1989).
13. J. Spalek et al., *Phys. Rev. Lett.* 59, 728 (1987).
14. P.W. Anderson, *Phys. Rev.* 115, 2(1959).
15. P.B. Vischer, *Phys. Rev B* 10, 943 (1974).
16. J. Spalek et al., *Phys. Stat. Sol. (b)* (1981)
17. Y. Nagaoka et al., *Phys. Rev.* 147, 392 (1966).
18. M. Heritier and P. Lederer, *J. Phys.* 38, L-209 (1977).
19. J-Spalek and W. Wojcik, *Phys. Rev. B* 37, 1532 (1988).
20. S.A. Kivelson et al., *Phys. Rev. B.* 35, 8865 (1987).
21. P.W. Anderson and Z. Zou, *Phys. Rev. Lett,* 60, 132 (1988).
22. P.W. Anderson, *The theory of superconductivity in the High–Tc cuprate superconductors*, princeton University Press, 1997.
23. H. Fukuyama et al., *Physica C* 153-155, 1630 (1988).
24. P.W. Anderson, et al., *Physica C* 153-155, 527 (1988).
25. P.A. Lee et al., *Comments Condens. Matter. Phys.* 12, 99 (1986).
26. T.M. Rice, in *'Proc.Int.school/phys. Enico Fermi-1987: 'Frontiers and Borderlines in Many Particle Physics'*.
27. D.M. Newns and N. Read, *Adv. Phys.* 36, 799 (1987).
28. P. Fulde et al., in *Solid State Physics'*, H. Ehrenreich and D. Turnbull, eds.), Vol. 41. Academic Press, San Diego, California.
29. G.R. Stewart, *Rev. Mod. Phys.* 56, 7, 55 (1984)
30. G.R. Stewart, *Nature* (London) 320, 124 (1986).
31. F. Steglich, in *"Theory of Heavy Fermions and Valence Fluctuations"* (T. Kasuy and T. Saso, eds.) p.23ff, Springer Verlag, Berlin and New York.
32. M.R. Schafroth, *Phys. Rev.* 100, 463 (1955).
33. A. Legget, *Rev. Mod. Phys.* 47, 331 (1975).
34. M.W. Shafer et al., *Phys. Rev B* 36, 4047 (1987).

35. J. Spalek, *Phys. Rev.* B 37, 533, (1988).
36. M. Acquarone, *Solid State Commun.* 66, 937 (1988).
37. Y. Suzumura et al., *J.Phys. Soc. Jpn.* 57, 2768 (1988).
38. P.W. Anderson, *Science* 235, 1196 (1987).
39. G. Baskaran et al., *Solid State Commun.* 63, 973 (1988).
40. G. Baskaran et al., *Solid State Commun.* 63, 973, (1987).
41. M. Cryot, Solid State Commun. 62, 821 (1987).
42. A.E. Ruckenstein et al., *Phys Rev.* B 36, 857 (1987).
43. G. Kotliar, *Phys. Rev.* B 37, 3664 (1988).
44. Y. Isawa et al. *Physica* 148 B, 391 (1987).
45. Z. Zou and P. W. Anderson, *Phys. Rev.* B 37, 627 (1988).
46. J. Spalek, *Phys. Rev. B* 40 (preprint).
47. J.M. Wheatley et al., *Phys. Rev.* B37, 5897 (1988).
48. F. Milla, *Phys. Rev. B* 38, 11358 (1988).
49. J. Spalek, *Phys. Rev. B* 38, 208, (1988); *J. Solid State Chem.* 76, 224 (1988).
50. J. R. Schrieffer and P.A. Wolff, *Phys. Rev.* 149, 491 (1966).
51. P. W. Anderson, *Phys. Rev. B* 30, 1549 (1984).
52. K. Miyake et al., *Phys. Rev. B.* 34, 6554 (1986).
53. D. J. Scalapino et al., *Phys. Rev. B* 34, 8190 (1986).
54. M. R. Norman, *Phys. Rev. Lett.* 59, 232 (1987); *Phys. Rev. B* 37, 4987 (1988).
55. D. M. Newns, *Phys. Rev. B* 36, 5595 (1988); *Phys. Scripta* T 23, 113 (1988).
56. J. B. Torrance et al., *Phys. Rev. Lett*, 61, 1127 (1988).
57. T. M. Rice and K. Ueda, *Phys Rev. Lett.* 55, 995 (1985); *Phys. Rev. B* 34, 6420 (1986).
58. K. Miyake et al., *J. Phys. Soc.* Jpn. 57, 722 (1988).
59. P. W. Anderson, *Phys. Rev.* 115, 2 (1959).
60. P. W. Anderson, in 'Solid State Physics' Vol. 14 pp 99-213 (1963), Academic Press, New York (Eds. F. Seitz and D. Turnbull).
61. J. Spalek (preprint).
62. V. J. Emery, *Phys. Rev. Lett,* 58, 2794 (1987).
63. V. J. Emery and G. Reiter, *Phys-Rev.* B 38, 4547 (1988).
64. K. P. Sinha and S. L. Kakani, *"High Temperature Superconductivity, Current Results and Novel Mechanisms,* Nova Science Publishers, New York (1995)
65. S. L. Kakani, *"Superconductivity : Current Topics"*, Arihant Publishing House, Jaipur (India), 2001.
66. J. Orenstein and A. J. Mills, *Science* 288, 468 (2000).
67. C. J. Tsues and J. R. Kirtley, *Rev.Mod. Phys.* 72, 969 (2000).

Theory of High Temperature Superconductivity in Cuprates 8

8.1 INTRODUCTION

In the foregoing chapters we explained some of the theoretical aspects of superconductivity. This chapter deals with theories of HTSC cuprates. We have seen that HTSC cuprates are so unusual, so unexpected, that there is still plenty of controversy concerning these materials. Many electronic properties of these substances are unusual in the normal state and cannot be explained in terms of the Fermi-Liquid theory. Aside from some unusual properties observed in the underdoped region associated with the so-called **pseudo gap** formation, typical examples of these properties include the following:

(i) The resistivity in the ab-plane has more or less linear dependence on T, i.e., $\rho \propto T$ for an unusually large range of temperatures (upto several hundred Kelvin). At high temperatures the resistivity does not saturate as in common metals; at low temperatures the T^2 term behaviour, common to many metals, is generally not observed.

(ii) The optical conductivity $\sigma(\omega)$ has a long tail proportional to $1/\omega$ upto the order of several tenths of an 1 eV, while the true Lorenzian part of the Drude conductivity seems to have unsually low weight.

(iii) The Raman scattering intensity has a flat background extending to several tenths of an eV.

(iv) In photoemission experiments the *quasiparticle peak* is unusually broad and accompanied by an intense incoherent background, leading to a very small spectral weight of the quasiparticle, if any exists.

(v) Nuclear magnetic resonance (NMR) of copper shows an unusually flat temperature dependence of the longitudinal relaxation time T_1 above room temperature for a certain range of doping excepting the very doping region.

The superconducting state of HTSC cuprates present the following characteristics:

(i) The gap ratio $2\Delta(0)/k_B T_c$ which is 3.53 in the BCS low temperature conventional superconductors is found much large in HTSC cuprates.

(ii) In conventional superconductors the coherence length is 10^{-4} cm. In HTSC cuprates, it is thousands of times smaller and equal to ~ 10 – 40 Å. A small value of the coherence length characterizing the spatial stretching of wave function of a Cooper pair points towards a strong coupling of quasiparticles in the pair. Such a strong coupling results in high critical temperature.

(iii) The dependence of T_c on the concentration of charge carriers has a non-monotonic character. The maximum value of T_c is attained at a relative small density of charge carriers equal to ~ 10^{21} cm^{-3}. In conventional superconductors, the temperature T_c rises monotonically with rising concentration. At small concentration of charge carriers their Fermi energy in the conduction band of HTSC cuprates has a small value about 0.4 eV.

(iv) Due to large coherence length (~ 10^{-4} cm) in conventional superconductors only a 10^{-4}th part of electrons placed near the Fermi surface participate in forming Cooper pairs.

In the HTSC cuprates, the space distribution of the wave function of Cooper pairs is small, therefore many charge carriers, involved in the conduction band structure, participate in their formation.

(v) Abrikosov and Gorkov [1] have shown that in BCS theory, the T_c monotonically fast decreases increasing concentration v of magnetic impurities. In HTSC cuprates, the T_c is weakly dependent on v is smaller than v_{cr} and suddenly vanishes at $v \geq v_r$.

(vi) A Josephson current has been detected in HTSC cuprates.

(vii) The flux quantum in HTSC cuprates is $\phi_0 = \pi/2e$.

All these experimental results clearly indicate that the simple nearly free electron model is not valid in these HTSC cuprates. It is clear that there are so many unusual features, so much conflicting data, that a comprehensive understanding of HTSC cuprates (comparable to BCS for low temperature conventional superconductors is still many years away. After this survey of exceptional features in HTSC cuprates, we now analyse the various theoretical models presented so far. On the other hand, magnetic fluctuations indicate that the Coulomb interaction in these cuprates is important as compared with the kinetic energy of electrons. Although much progress, both experimental and theoretical, has been made in understanding HTSC cuprates since their discovery in 1986, there is still no consensus about the underlying mechanism causing superconductivity in these systems. In this chapter, we have presented some theoretical models illustrating underlying physics relevant to high T_c in these systems [2-23].

We have seen that superconductivity requires the presence of Cooper pairs (consisting of either real space or non-real space particles) and long range phase coherence among the Cooper pairs. In the BCS theory for conventional metallic superconductors, the mechanism responsible for pairing and establishment of the phase coherence are identical. i.e., two electrons in Cooper pair are attracted by phonons and phase coherence among the Cooper pairs is established also by phonons. Both phenomena occur almost simultaneously at critical temperature T_c. In HTSC cuprates, there is almost a consensus that these two mechanisms occur at different temperatures at least, in the underdoped regime: some kind of pairing exists above T_c. The magnitudes of energy gaps (pseudo gap: Δ_p and coherence gap Δ_c) which correspond to the pairing process (above T_c) and establishment of long-range phase coherence (at T_c) have difference dependence on hole concentration (p) in CuO$_2$ planes of HTSC cuprates [5]. The magnitude of Δ_c, which is proportional to T_c, has the parabolic dependence on p, whereas the magnitude of the pseudo-gap (Δ_p), increases linearly with decrease of hole concentration.

Most physicists argue that the same microscopic mechanism is operating in all the HTSC cuprates. All HTSC cuprates found to date are based on compounds in which planes of CuO$_2$ are separated by a number of layers of intervening perovskite structures composed of oxygen and various counter ions such as La, Y, Ca, Bi, Th, Hg, etc., The fact that cuprate compounds in which single CuO_2 layers may be spaced by 10Å or more of intervening insulating material can still have quite high. T_c's suggest that the basic superconducting unit is a two dimensional CuO_2 layer. Anderson [10, 24, 25] argues strongly that at least two CuO_2 adjacent layers are needed for high-T_c, but this claim remains controversial. The main controversy arises with regard to the origin of the attraction of paired holes which form the condensate and on the appropriate description of the normal state.

The search for non-phonon electronic coupling mechanism in HTSC cuprates is motivated by several difficulties of the conventional BCS theory. First of all, the temperatures of the order of 100 K and higher cannot be understood as originating from the electron–phonon coupling [26]. The HTSC cuprates have also a remarkably low density of states at the Fermi level for their high-T_c. But perhaps the most spectacular is a rather low isotope effect in $La_{2-x}Sr_xCuO_4$, and its disappearance in $YBa_2Cu_3O_{6+x}$. The substitution of ^{18}O for ^{16}O gives $\alpha \approx 0.02$ in YBCO and $\alpha \approx 0.15$ in LBCO. However, a weak isotope effect is not conclusive; there are BCS phonon assisted conventional superconductors such as Ru and Zr that do not show an isotope effect for reasons that do not apply here.

We have mentioned in the previous chapter that the symmetry of the order parameters in these HTSC cuprates is d-wave, i.e. it changes sign under a 90° rotation. The sign change means that the gap may vanish at the points on the Fermi surface. ARPES has shown that $\Delta(\mathbf{K}) \sim \Delta_0 [(\cos K_x a - \cos K_y a)]$, where $\mathbf{K}$ is the wave vector. This is the $d_{x^2-y^2}$ form of the gap, and is maximal for momenta parallel to the Cu-O-Cu band and vanishing for momenta at angles of 45° to this bond. The four Fermi surface points at which the gap magnitude vanishes are the nodes. Obviously, this establishes superconductivity in cuprates to be d-wave.

Many theoretical models have been proposed to explain the mechanism and anisotropic properties of HTSC cuprates. Crudely speaking, one can place them under following the four categories [2-15, 19-26].

(*i*) Modified phonon mechanisms

(*ii*) Magnetic coupling mechanisms

(*iii*) Electronic coupling mechanisms

(*iv*) Exotic superconductivity mechanisms.

The occurrence of very small isotope effect suggests a limited involvement of phonon-induced pairing mechanism. Obviously, the phononic contribution is always there to a small measure [5]. Very exotic superconductivity has been predicted and linked with quantum Hall effect. However, this seems fairly unlikely in HTSC cuprates. Many theories were confronted with necessity to solve the problem of finding a new pairing mechanism without applying the electron–phonon interaction. There were many approaches. Some of the investigators returned to **exicton, plasmon, polaron** pairing mechanisms [2, 4]. Anderson in 1987 [24] proposed a **resonating valence band** (RVB) model and developed further by him and collaborators [10]. In many works, the **magnon pairing mechanism** was studied. Among the models incorporating the non-phonon pairing mechanisms one should note the works where the Hubbard model in systems with repulsive interaction only. It is assumed that the pairing effect is caused only by the kinematic interaction. Charge fluctuation mechanisms can also lead to spin correlations. In correlated charge transfer mechanisms the excitation of charge (electrons or holes) takes place pairwise [27, 5]. Having so many various possible mechanisms driving the high-T_c superconductivity in cuprates, any phenomenological concept may be very useful in identifying which of these possibilities are realistic and which should be excluded. Varma et al. [28] proposed a universal phenomenological approach to the properties of the normal phase of HTSC cuprates. In the present chapter, we have presented the broad outlines of these models. There are so many unusual features, so much conflicting data, that a comprehensive understanding of HTSC's (comparable to BCS for LTSCs) is still very difficult.

8.2 NORMAL STATE FERMI SURFACE

We have read that superconductivity is caused by an instability in the normal state, it is first necessary to have some model for the normal state before trying to understand the superconducting state in cuprates. For any metal, common normal-state measurements, *e.g.* resistivity, tunnelling conductance, NMR, pseudo gap, etc. provide data that are only compatible with certain theoretical models, and thus the possible choices among models are restricted. We may note that for HTSC cuprates, this limitation is very important.

8.2.1 Momentum Space

The simplest case of all is that of a metal with one single numerical value for the Fermi energy, E_F. At absolute zero, all electronic states below that energy level are occupied, and all above are unoccupied. At any finite temperature, there is a slight tail in the Fermi distribution function, which governs the occupancy of the available energy levels. The Fermi function is [29]

$$\frac{1}{\exp[(E - E_F)/k_B T] + 1}$$

At any temperature of interest, this Fermi function drops suddenly from 1 to 0 very near $E = E_F$. The difference in a superconductor is that there is an energy gap exactly at Fermi level, so that the tail is forced to extend slightly further on either side of E_F. A comparison of two situations is shown in Fig. 8.1.

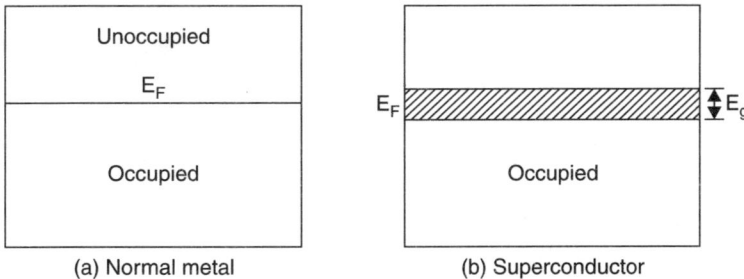

Fig. 8.1 Comparison of the nature of the Fermi level (E_F) in a normal metal (*a*) and a superconducting metal (*b*). A key feature of BCS theory is that the energy gap is centered right at the Fermi level (E_F) [30].

In the three dimensions of momentum space, this kind of uniform filling of low-energy levels amounts to filling up a sphere, because $E = K^2/2m$ (and $K^2 = K_x^2 + K_y^2 + K_z^2$). A sketch of this is presented in Fig. 8.2, where the shaded area represents the partially filled states near the Fermi level. We may note that the boundary between filled and unfilled states is termed the Fermi surface. As simple as this surface in Fig. 8.2, it is still fuzzy and diffuse at any finite temperature (although the diffuseness is exaggerated here for emphasis).

The occupied states below the Fermi level comprise the Fermi sea. The many electrons interact weakly with one another, in a disordered way. Continuing the analogy of terminology, these electrons are said to form a **Fermi liquid.** The normal state is a conventional **Fermi liquid.**

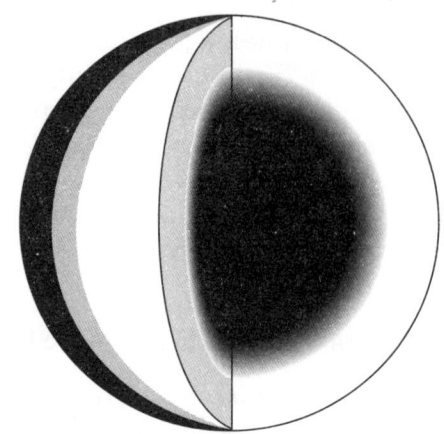

Fig. 8.2 A sketch of the Fermi surface in a conventional simple metal. The x-, y- and z- axes correspond to dimensions in momentum space : k_x, k_y and k_z. The radial distance outward (k^2) is increasing energy. The black inner region denotes fully occupied states in the metal; the light outer region denotes empty states. The transition from totally filled to total empty takes place over a narrow band of energy states centered on the Fermi level (E_F).

As soon as a material becomes anisotropic, the simplicity of the Fermi surface of the Fermi surface goes away. To begin with, the Fermi surface becomes a prolate ellipsoid if the effective mass in one direction is much larger than in the other two. Then the Fermi level of energy can be different along different axes K_x, K_y and K_z. Depending on the direction within momentum space, the sequential filling of levels may be in an entirely different energy band. We say that different bands cross, the Fermi energy at different momentum vector **K**.

We have seen in chapter 6 that the HTSC cuprates are extremely anisotropic materials and so, even in the normal state, the HTSC's have a radically altered Fermi surface. Fig. 8.3 is the result of a band-structure calculation [31-33] for YBCO, and bears no resemblence to a sphere. The Fermi surface is **electron-like** in some places, and **hole-like** in others. Some parts of it are due to the Cu-O_2 planes and some due to the Cu-O chains. We may note that a number of conditions *e.g.* magnetic fluctuations, charge fluctuations, and undulations in the CuO_2 planes have been omitted from the band structure calculations. When this kind of complexity is combined with the experimental limitations associated with imperfect samples, any theory will be tentative only.

Fig. 8.3 Three-dimensional picture of the calculated Fermi surfaces of $YBa_2Cu_3O_7$, slightly broadened and extended periodically.

The alternatives to the Fermi liquid theory of HTSC include **bipolarons**, in which the charge carriers bind into pairs to form bosonic (not fermionic) excitations [34], **resonance valence bond coupling** [24], a **marginal Fermi liquid** [28], **anyons** [35], etc.

8.3 EXPERIMENTAL RESULTS

Experimental data about the Fermi surface has helped to clarify the picture about HTSC cuprates to some extent. It is instructive to consider briefly the experimental aspects of determining the shape of the Fermi surface. There are basically three ways to measure a **Fermi surface**: (i) **Angle-Resolved Photoelectron Spectroscopy** (ARPES); **Angular Correlation Annihilation Radiation** (ACAR); and **de Hass-van Alphen resonance** (dHvA). All these three methods have certain advantages, and certain problems as well. For example, **dHvA** sees electrons orbiting the Fermi surface and determines their elapsed orbiting time through a resonance with an applied magnetic field. But when samples contain voids and impurities, scattering events interrupt the electron trajectories, degrade the data, and leave the interpretation uncertain.

Photoemission preferentially sees the surface of a sample, so if there is surface contamination, the results can be misleading. Photoemission data have been used to deduce the presence of a Fermi surface at $T < 50$ K. Since (i) this technique must be performed in a vacuum, (ii) it is very sensitive to surface effects, and (iii) $Y_1Ba_2Cu_3O_{7-x}$ loses oxygen to a vacuum at temperatures above 50 K, experiments performed above 50 K showed no Fermi surface. Subsequently, experiments performed at other laboratories provided some evidence that $Y_1Ba_2Cu_3O_{7-x}$ has a Fermi surface. The experiments used positron annihilation data together with complicated computer computations to deduce the existence of a Fermi surface at temperatures bracketing T_c.

The nearly two dimensional structure of the HTSC cuprates means that these materials are far different from low temperature conventional (LTSC) superconductors. One very important consideration is this: what kinds of electrons (or holes) engage in the pairing mechanism that causes superconductivity? In a simple LTSC, the answer is trivial: the free electrons in the conduction band. In the HTSC cuprates, where superconductivity occurs primarily in CuO_2 planes, this is not necessarily so.

It is beyond the scope of this chapter to summarize all serious theoretical approaches, instead we will explain as many aspects of the problem as we can. Alternative views can be found in several reviews [2 – 28] or by scanning several conference proceedings.

8.4 RESONATING VALENCE BOND (RVB) THEORY

Shortly after the discovery of HTSC cuprates, Anderson in 1987 [24] proposed a one band large–U Hubbard model. [**The Hubbard Model** [37] is defined by three specifications:

(i) **The Hamiltonian**

$$H = \sum_{<ij>\sigma} t_{ij} C^+_{i\sigma} C_{j\sigma} + U \sum_i n_{i\uparrow} n_{i\downarrow}$$

The first term represents the motion of an electron through the system by a series of hops $j \to i$, which are described in terms of destruction of the particle at site j and its recreation on neighbouring site i. The width of the corresponding band in this representation is given by

$$W = 2 \sum_{j(i)} |t_{ij}| \approx 2z|t|$$

where z is the number of nearest neighbours ($n.n$) and t is the value of t_{ij} for the $n.n$ pair $<ij>$. Obviously, the above Hamiltonian is parameterized through the bandwidth W and the magnitude U. In actual calculations, it is the ratio U/W that determines the localized versus collective behaviour of the electrons in the solid.

(ii) a lattice which defines the pattern of bonds $<ij>$ on which the electron can hop with matrix element t_{ij}, and

(iii) a filling factor x, which we define such that $x = 0$ is the half filled case (a standard reference point) and x counts the number of holes as in $La_{2-x}Sr_xCuO_4$. The Coulomb energy U inhibits double occupancy near half filling to explain the mechanism and properties of HTSC in CuO_2 layers of cuprates, which were suggested to be key building blocks. Since the HTSC region in cuprates is close to a region where a **Mott-like metal-insulator**[1] transition occurs, Anderson [24] realized that a reasonable model for HTSC cuprates is essentially one band large-U Hubbard model. An exactly half filling, such model reduces to the **Heisenberg Hamiltonian**, which is well known to describe an antiferromagnetism (AFM).

There are various versions of the Hubbard hypothesis, some of which claim ultimately to give "weak" superconductivity via spin fluctuations. We will focus on versions which hope to generate "strong" superconductivity, by first condensing into a new state (claimed by Anderson [1.] to be a **resonating valence bond** (RVB) state). This new state is not a conventional Fermi liquid, so the low T superconducting condensate is not conventional BCS.

One approach to understanding the strong-coupling Hubbard model is exact diagonalization of finite size systems. Let us consider the $x = 0$ case. With one site and one electron there are two states, $\uparrow$ and $\downarrow$, both having zero energy. With two sites and two electrons

---

[1]**Mott Insulator**: A Mott insulator is a material in which the conductivity vanishes as temperature tends to zero, even though band theory would predict it to be metallic. NiO, $LaTiO_3$ and V_2O_3 are few examples out of the many known examples. However, the HTSC cuprates are only Mott insulators known to become superconducting when the electron concentration is changed from one per cell. A Mott insulator is fundamentally different from a conventional (band) insulator. In the latter system, conductivity is blocked by the Pauli exclusion principle. When the highest occupied band contains two electrons per unit cell, electrons cannot move because all orbitals are filled. In a Mott insulator, charge conduction is blocked instead by electron–electron repulsion. When the highest occupied band contains one electron per unit cell, electron motion requires creation of a doubly occupied site. If the electron-electron repulsion is strong enough this motion is blocked. The amount of charge per cell becomes fixed, leaving only the electron spin on each site to fluctuate. Doping restores electrical conductivity by creating sites to which electrons can jump without incurring a cost in Coulomb repulsion energy.

Virtual charge fluctuations in a Mott insulator generate a "super exchange" interaction, which favours anti-parallel alignment of neighbouring spins. In many materials, this leads to long-range antiferromagnetic order [36].

there are six states, each described by a **Slater determinant:** $|1> = (1\uparrow, 1\downarrow)$, $|2> = (2\uparrow, 2\downarrow)$, $|3> = (1\uparrow, 2\downarrow)$, $|4> = (1\downarrow, 2\uparrow)$, $|5> = (1\uparrow, 2\uparrow)$, and $|6> = (1\downarrow, 2\downarrow)$. If we neglect t in comparison to U, H is diagonal in this basis. States 1 and 2 have energy U, and 3, 6 have energy O. After t is turned on, a triplet of states of energy O remains, ($|5>$, $|6>$, and $2^{-1/2}(|3> + |4>)$) and one state, $2^{-1/2}(|1> - |2>)$, has energy U. The singlet state plays a special role:

$$|VB> = 2^{-1/2}(1\uparrow 2\downarrow - 1\downarrow 2\uparrow) = 2^{-\frac{1}{2}}(|3> - |4>).$$

It is a **Heitler-London "valence bond" state.** Two eigenstates are mixtures $|VB>$ and ionized states $2^{-1/2}(1\uparrow 1\downarrow + 2\uparrow 2\downarrow)$, with eigenvalues $U/2 \pm \sqrt{(U/2)^2 + 4t^2}$. For $t \ll U$, the ground state is the $|VB>$ state with a weak admixture of ionized states, and the energy is $-4t^2/U$.

As the number of "atoms" N increases, the size of the space grows rapidly, as $(2N)!/(N!)^2 \sim 4^N$. Of these, exactly 2^N correspond to states with only one electron per site, giving a 2^N, fold degenerate ground state of energy 0 when $t = 0$. These states (denoted $|\Sigma>$) are characterized by the spin orientation on each site. It is this very large ground state degeneracy which makes the problem so interesting and difficult.

For even a modest nine atom (3×3) 2d array, direct diagonalization of the 48, 620 matrix is not easy. A considerable simplification occurs if t is small, and the problem is truncated to the 2^N-order space (512 for $N = 3 \times 3$) of singly occupied states. This truncated problem has spins at each side but no "charge" degrees of freedom. Hence it describes an insulator, as is reasonable for small enough t/U. The H-matrix is O in this truncated basis, and it necessary to do first order degenerate perturbation theory in the parameter t/U. This introduces matrix elements.

$$<\Sigma'|H_{eff}|\Sigma> = \sum_n \frac{<\Sigma'|H|n><n|H|\Sigma>}{E_0 - E_n} \qquad ...(8.1)$$

between the states $|\Sigma, |\Sigma'>$ of the singly occupied subspace $E(\Sigma) = E_0 = 0$. The intermediate states $|n>$ always have one site doubly occupied ($E_n = U$). The resulting matrix (1.8) turns out to be the same as the matrix of

$$H_{eff} = J \sum_{<ij>} \left(\sigma_i \cdot \sigma_j - \frac{1}{4}\right) \qquad ...(8.2)$$

where $\sigma = (\sigma_x, \sigma_y, \sigma_z)$ are the Pauli matrices. When applied to the two-site problem, eq. (2) has eigenvalues O (triply degenerate) and $-J$ (singly degenerate) leading to the identification $J = 4t^2/U$. The additive constant $-J/4$ per bond is then dropped from eq.(2), which is the antiferromagnetic Heisenberg Hamiltonian. In a three-dimensional simple cubic lattice eq. (2) has an antiferromagnetically ordered low temperature phase (the **"Neel" state**), a phase transition at $J_N \sim J$, and a high temperature paramagnetic phase with disordered local moments. It is believed that this provides a correct qualitative description for systems such as NiO and CuO which are antiferromagnetic insulators at low T and remain insulating even above T_N irrespective of whether the number of electrons per unit cell is even or odd.

We may note that in one dimension, the Heisenberg antiferromagnet has no long range spin order, even at $T = 0$. The ground state is given by the famous "Bethe ansatz" [2, 3], and is very complicated. It is easy to see why ordered spins are not favoured, by making variational estimates of ground state energies using simple trial wavefunctions. The simple antiferromagnetic state $|Neel>$ is $|1_\uparrow, 2_\downarrow, 3_\uparrow, 4_\uparrow, ... >$ and has energy per site $-J/4$, which is also the energy per bond, and comes entirely from the $S_{iz} S_{jz}$ part of relation (8.2). A "valence bond" state $|VB>$ of the form $|VB(1,2), VB(3,4), ... >$ does better. Each valence bond has energy $-3J/4$, with equal contributions from the three cartesian components $S_{i\alpha} S_{j\alpha}$. Since there are two sites per band, the energy per site is $\left(-\frac{3}{8}J\right)$, 50% lower than the Neel state.

The **"quantum fluctuations"** carried by the operator

$$\frac{1}{2}(S_i^+ S_j^- + S_i^- S_j^+) = S_{ix} S_{iy} + S_{jx} S_{jy} \qquad ...(8.3)$$

are responsible for destroying the ordered state for small spin ($s = 1/2$) and low dimensionality. In a two dimensional square lattice, the ground state of eq. (2) is not known. It may be ordered antiferromagnetically, but it is quite certain that no long-range magnetic order persists above $T = 0$.

For a $2d$ triangular lattice, the best "Neel" state has three spin sublattices oriented at $120°$. It is not likely that much long range magnetic order occurs in the ground state. The triangular geometry causes "frustration". Anderson [4, 5] argued that in this case the ground state would be a **"resonating valence bond"** (RVB) state

$$|RVB> = \sum_p c(p) |VB(1,2), VB(3,4), ... >$$

where the sum runs over all permutations p of bond arrangements. It is possible that such a state is also the ground state of the $2d$ square lattice. Even if this is not true, still when second-neighbour exchange coupling is introduced or else a finite hole concentration x, these sources of frustration could stabilize RVB state. Anderson [24] proposed that an RVB state with finite x would be a novel kind of superconductor. The scale of T_c would be related to $J \sim 1000$ K rather than $w_D \sim 200$ K.

It is hypothesized that the insulating state of the pure La_2CuO_4 is the **resonating valence bond** (RVB) state or **quantum spin liquid** proposed by Anderson [38] and Fazekas and Anderson [39]. There are pre-existing spin singlet pairs in the RVB state and they become charged superconducting pairs by strong enough doping. We many note that RVB state is not a conventional Fermi liquid, so the low temperature superconducting condensate is not conventional BCS.

One can roughly describe the RVB state as a coherent superposition of singlet pairs of electrons ("bonds") at neighbour copper sites in the $2D$ square lattice of a CuO_2 layer. The bonds among the vertices of a plaquette delimited by four neighbouring sites are then allowed to "resonate" between the two possible configurations, such as shown in Fig. 8.4, thus reducing the energy of the superposition by a quantity of the order of the superexchange interaction J.

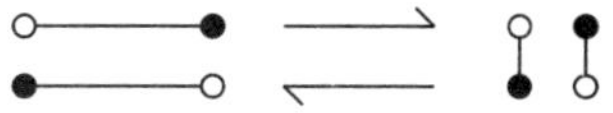

Fig. 8.4

In second quantization a singlet RVB between sites i and j is created by

$$b_{ij}^+ = \frac{1}{\sqrt{2}} (C_{i\uparrow}^+ C_{j\downarrow}^+ - C_{i\downarrow}^+ C_{j\uparrow}^+) \qquad ...(8.4)$$

In direct space notation, one can conveniently express the RVB coherent state as

$$| \text{RVB} > = P_N (\Sigma\, g(l)\, C_{i\uparrow}^+ C_{i+l\downarrow}^+)^N |0> \qquad ..(8.5)$$

where P_N is a **Gutzwiller projector** over the Hilbert subspace of N singly occupied sites, $C_{i\sigma}^+$ creates a fermion on site i, l runs over all its neighbour sites and $g(l)$ is a link dependent pair wave-function, in general permitting for different symmetries [40].

The basic assumption of RVB model is that all the essential physics is contained in the nearly half filled Hubbard Hamiltonian:

$$H = t \sum_{<i,j>,\sigma} (C_{i\sigma}^+ C_{j\sigma}^+ + h.c) + \sum_i \mu\, n_i + U \sum_i n_{i\uparrow} n_{i\downarrow} \qquad ...(8.6)$$

where t is the Cu-Cu hopping integral and U is Coulomb repulsion when two electrons exist in the same orbital and site, $(C_{i\sigma}^+, C_{i\sigma})$ are the fermion (creation, annihilation) operators with $n_i = n_{i\uparrow} + n_{i\downarrow}$ and $n_{i\sigma} = C_{i\sigma}^+ C_{i\sigma}$. The best evidence that superconductivity in Cu-O based superconductors has something to do with the Hubbard Hamiltonian (8.6) comes from diffraction experiments showing ordered antiferromagnetism in La_2CuO_4 [41] and $YBa_2Cu_3O_{6+x}$ [42]. In the "Hubbard" hypothesis:

"The fundamental physics of the oxide superconductors is contained in the Hamiltonian (8.6) on a 2-d square lattice for small number of holes $0 \leq x \leq 0.2$".

Since the properties of the Hubbard model in this limit are not yet under good theoretical control, there is no proof or disproof, and testing the hypothesis is regarded as a key issue. Many workers suspect that it is wrong but that a modified Hubbard hypothesis (with added features such as electron–phonon effects, two bands, or next neighbour interactions) is right, with superconductivity still driven by the coulomb U term. We will discuss this in detail later in this chapter.

The Hamiltonian (8.6) can be transformed to order (t^2/U) into usual kinetic exchange Hamiltonian,

$$H_T = -t \sum_{<ij>} (1-n_{i,-\sigma})\, C_{i\sigma}^+ C_{j\sigma}\, (1-n_{j,-\sigma}) + \mu \sum_i n_i - \frac{t^2}{U} \sum_{\substack{<ij> \\ \sigma\sigma'}} [n_{i\sigma} n_{j\sigma} + C_{i\sigma}^+ C_{j\sigma'}^+ C_{i\sigma'} C_{j\sigma}]$$

$$...(8.7)$$

Here U is estimated to be in the range 5-10 eV and $t \sim 1$ to 2 eV. This could give $t^2/U \sim 1000$ to 2000 K. The double occupancy inherent in the free particle picture has been eliminated by using the **Gutzwiller** projection operator

$$P_N = \pi_l (1 - n_{i\uparrow} n_{i\downarrow}) \qquad ...(8.8)$$

It is very difficult to handle analytically the Gutzwiller projection operator, P_N. There are however numerical techniques to handle this projection operator exactly namely a Monte-Carlo summation over spatial configurations which have been used by Gros [43] and by Yokoyama and Shiba [44] and several other basis.

For constructing a RVB state, one has to create $N/2$ nearest neighbouring pair bounds with the operators represented by eq. (8.4). These are then placed on the lattice in some array and added up in equal phases. We must remember that b_{ij} represents a boson operator. Then the Bose condensed state is

$$\left[\sum_{<ij>} b_{ij}^+ \right]^{N/2} \bigg| V_{ac} \qquad ...(8.9)$$

which is equivalent to BCS state of some kind.

Elementary excitations are created by breaking a valence bond, and can be identified as topological defects or solitons, as in Schrieffer's theory of conduction in polyacetylene [45]. Such elementary excitations are neutral spin $\frac{1}{2}$ fermions (also called as 'spinons') and charge $\pm e$ spinless bosons (also called 'holons' in the case of charge $+ e$) [45, 46]. In this way, the quantum numbers of the original particles are redistributed or 'fractionized' [47], among new elementary excitations, a procedure which is common to the boronization of *'Tomonaga-Luttinger liquid'* in one dimension [48]. On the other hand, the observation of true electron or hole states involves recombination of such excitations.

This is at the basis of the origin of pseudogap as observed in c-axis conductivity and photoemission experiments. The formation of triplet excitations out of a liquid of spins arranged in singlet resonating valence bonds is an energy costing process. Removing a physical (charged, spinny) electron from a plane during a c-axis conduction process in the normal state thus requires overcoming an energy gap $\sim J$ (spin gap, or pseudo gap). With the onset of phase coherence at $T = T_c$, the pseudogap evolves into the superconducting gap, while the spin and charge degrees of freedom recombine, thus giving rise to electron- and hole-like quasi particle states, available for the formation of Cooper pairs, whose nature is rather conventional.

Baskaran et al. [49] considered the said half filled Hubbard band with an optimization procedure, which projects out kinetic energy completely, leading to the Hamltonian

$$H = -J \sum_{<ij>} b_{ij}^+ b_{ij} \qquad ...(8.10)$$

where $J = 4t^2/U$. It has a *BCS* like solution (RVB of Anderson)

$$|G> = P_n \pi_K (U_K + V_K C_{K\uparrow}^+ C_{-K\downarrow}^+) |0> \qquad ...(8.11)$$

with $\qquad \sum_K \dfrac{V_K}{U_K} = \sum_K \alpha(K) = 0, \quad |\alpha(K)| = 1 \qquad ...(8.11(a))$

and $V_K = \pm U_K$ assuming on whether the wave vector **K** is inside or outside the pseudo Fermi surface, determined by the quasiparticle operators α_K and β_K.

$$\alpha_K = \left(\frac{1}{\sqrt{2}}\right)(C_{K\uparrow} + C^+_{K\downarrow}) \quad \text{if } |K| < K_F$$

$$= \left(\frac{1}{\sqrt{2}}\right)(C_{K\uparrow} - C^+_{-K\downarrow}) \quad \text{if } |K| > K_F \qquad ...(8.11(b))$$

Similar relations hold for β_K. The quasi particle energy becomes

$$E_K = \Delta J |\gamma_K| = \Delta J |\cos K_x a + \cos K_y a| \qquad ...(8.12)$$

we must note that $\alpha(K)$ does not change sign across a Fermi surface as is pointed out by Anderson.

As a consequence of signs in eq. 8.11(b), $\alpha^+_K \alpha^+_{K'}$, creates a charged excitation if **K** and **K'**, are both, on the same side of the pseudo-potential, and a neutral spin excitation if they are on the opposite side. When we project out charged excitations in the insulating state, only the latter spectrum will persist, it will resemble closely the excitation spectrum of a simple Fermi gas. There will be a gap for charged excitations, which will more or less gradually disappear with temperature. Spin excitations will have no gap but their spectrum is restricted in **K**, ω space by the condition that **K** and **K'** be on opposite sides of the pseudo Fermi surface.

A relevant feature of the RVB model is the unconventional dependence of the mean-field critical temperature T_c on the hole concentration thin x away from half filling ($x = 0$). At $x = 0$, the one band Hubbard Hamiltonian is equivalent to the antiferromagnetic Heisenberg Hamiltonian,

$$H = J \sum_{<ij>} \sigma_i \cdot \sigma_j \qquad ...(8.13)$$

with $\sigma = (\sigma_x, \sigma_y, \sigma_z)$ are the Pauli matrices, which is locally gauge invariant. Local gauge invariance cannot be broken, and indeed the system does not undergo a superconducting transition. Only at $x > 0$, where the appropriate Hubbard Hamiltonian is only globally gauge invariant, can this symmetry be broken and the system acquire off-diagonal long range order [50]. At low doping, phase fluctuations dominate, and the mean-field T_c is strongly suppressed, while T_c drops sharply at large doping, as indeed observed experimentally.

As stated earlier, the quasi-particles excitations that can be generated over the ground state are of two types: (*i*) The **spinon**, which is just an isolated unpaired spin and does not carry a charge, for *e.g.*, Cu^{2+}_σ. **Spinons** can propagate freely through the copper lattice. Spinons are spin 1/2 entities, hence fermions and their distribution in the lowest energy states, subjects to the exclusion principles defines a **pseudo-Fermi energy**. (*ii*) The second type of excitation envisaged in the *RVB* model is a single spinless charge on copper site. It is named as 'holon' and is produced by the process

$$Cu^{2+}_\sigma + \text{hole}_{-\sigma} = Cu^{3+}_{\uparrow\downarrow}.$$

We may note that Cu^{3+} is not in the spin triplet state owing to crystal field energy splitting. Holons being in spin zero state are bosons. In this model, as stated earlier, superconductivity arises from the Bose condensation of holons in doped materials ($x > 0$). Spinon excitation is gapless but there will be a gap for charged excitation which disappears with increasing temperature. Baskaran et al. [49] obtained the following expression for the true energy in relation to order parameter Δ as

$$\text{Gap} = \frac{3x\,J\,\Delta/2}{\sqrt{1+(3\Delta\,J/2W)^2}} \qquad ...(8.14)$$

where $2W$ is the bandwidth ($\sim tx$). For some choice of parameters (t, U, x), Bhaskaran et al. [49] found 2 Gap/T_c close to the BCS value.

The RVB state essentially represents a quantum fluid in two dimensional planes of CuO_2. There are some unusual normal state predictions for resistivity and other properties [10]. In brief, one can say that RVB mean field theory [49] brought out the neutral spinons and their pseudo Fermi surface in the insulating state and superconductivity in the doped case.

Assuming a link-dependent variational Ansatz for the mean field $<b_{ij}^+>$, corresponding to the inclusion of inter-size interactions, Kotliar [40] found that an RVB state is not incompatible with a d-wave symmetry order parameter, that the transition temperature is higher for the latter than for the s-wave analog, and that such symmetries are close in energy, thus allowing for a mixed symmetry paired state, depending on the microscopic details and on specific properties of the material.

Affleck and Marston [51] and Kotliar [40] independently brought out an energetically better mean field state namely the **d-RVB** or the **flux state.** Kivelson et al. [45] have discussed **short range RVB** in some detail focussing on spinon and holon excitations as stated earlier. **Slave Boson theories** and **gauge theories** by Baskaran and Anderson [50] Baskaran [52], Zou and Anderson [53], Ioffe and Larkin [54], Wiegman [55], Lee and Nagaosa [56], Fukuyama [57] followed suit and there were intense activities and speculations, including the possible parity violating superconducting states [58-61] with connection to **Laughlin's quantum Hall state.** Baskaran and Anderson [50] cite {gauge} as a **gauge theory.** Wen et al. [62] later found that this gauge field captures the physics of chiral fluctuations among the interacting spins in the low-doped regime. The d-RVB state [40, 63] can be thought of a uniform RVB state [49] in which pi fluxes are condensed at low temperatures. There are good theoretical indications that the $2d$ Heisenberg model has a long range antiferromagnetic order. Several static and quasi static phenomenon theory inspired non-linear sigma analysis [64]. However, the dominant correlations in the ground state is that of a d-RVB state as suggested by Hsu [65]. It is meaningful to think of the ordered state as a spinon density wave in a spin liquid state. The antiferromagnetic order is fragile and disappears at about $x \sim 1.5$ of doping. Recent ARPES study by Wells et al. [66], Kin et al. [67] and Laughlin et al. [68] is underdoped and insulating layer cuprates point out that the d-RVB with its massless Dirac like spinon spectrum is a good reference state to describe the Mott insulating and the underdoped state. This questions the real relevance of the non-linear sigma model in its present form, in the doped quantum melted region [69].

There are many theoretical and experimental results which are difficult to reconcile with RVB predictions. Some rigorous calculations show that the ground state of the system in question (two-dimensional network of spin half) is a Neel state. Further, single band Hubbard model can induce antiferromagnetism but not superconductivity – a difficulty also applicable to other theories based on this model [70].

The RVB theory based on the non Fermi liquid state in which the distinct excitation, 'spinon' and 'holon' exist. The spinon pairing occurs at T^*(so called 'spin gap') and holons condensate at T_c [71]. Since 'spin gap' is based on the singlet pairing, the RVB mechanism explains the various experiments for under-doped cuprates. The experimental observation of antiferromagnetic order in insulating $LaCuO_4$ is consistent with the predicted low temperature behaviour of the Hubbard model in the strong coupling limit. However, the continuities from normal states to the superconducting state and from over doped to under-doped cuprates are not necessarily obvious, and the physical origin of the spin-charge separation is also not clear.

The physics in the underdoped regime is complicated by disorder long range Coulomb interaction, charge localization and microphase separation effects. This is the origin of **stripe,** (spin and charge inhomogeneities) phase. Balents et al. [72] have discussed the issue of how an insulator to superconductor transition takes place in the ground state by their theory of '**nodal Fermi liquids**'. Their reference state is a d-wave superconductor, where quantum fluctuations induced by Coulomb correlations drive a metal (or insulator) superconductor transition. This theory captures some of the physics of t–J model and over emphasizes pair fluctuation of charges. There are also some fundamental questions whether one can have a boson metal ground state for low doping [73].

Anderson [10] carried out an analysis of interacting fermion systems and found that there are two fundamentally different fixed points, '**Fermi liquid theory**' and '**Luttinger liquid theory**'. A Fermi liquid is characterized by a single particle Green's function endowed with simple poles, whose location in the complex plane defines unambiguously the energy-momentum relation of the quasiparticle scattering yields a finite decay rate of order $(k - k_F)^2$. Thus the life time of a quasiparticle grows linearly with the distance in energy from the Fermi surface. In a Fermi liquid, one finds that the spectral function is related to the retarded Green's function.

$$A(\mathbf{K}, \omega) = -\frac{1}{\pi} Im\, G(\mathbf{K}, \omega + i\delta) = Z_k\, \delta(\omega - \xi_k) \qquad ...(8.15)$$

where $\xi_K = \varepsilon_K - \mu$ represents the dispersion relation of the noninteracting particles measured with respect to Fermi energy, and $G(\mathbf{K}, \omega + i\delta) = G_R(\mathbf{K}, \omega)$ is the single-particle retarded Green's function. Here, Z_k is a renormalization factor, generally momentum dependent and defined as minus the residue of the Green's function around its pole at $\omega = \xi_k$. One finds that for a Fermi liquid $0 < Z_k \leq 1$, the latter equality holding in the limiting case of a noninteracting gas [74]. The Luttinger liquid is a state in which charge and spin acquire distinct spectra and correlations having unusual exponents. These include most interacting 1-D systems, and some higher (especially) 2–D systems in which the band spectrum is bounded above, (for *e.g.* with Mott-Hubbard gaps and an upper Hubbard band). In dimensions $D = 1$, Fermi liquid theory breaksdown, and a pure Fermi system, without symmetry breaking, is known to belong to the universality class of **Tomonga-Luttinger Model** [75-76]. At variance with a Fermi liquid, it describes a normal (*i.e.*, metallic) interacting Fermi liquid in $D = 1$, whose low energy properties are characterized, above all, by anomalous power law decay in the correlation function, leading to a continuous (*i.e.*, non step-like) momentum distribution, with a power law singularity at the Fermi surface, the exponent η being non-universal, and by a single particle density of states vanishing as ω^η near the Fermi level, implying the absence of well-defined quasi-particles, with finite lifetime. These are replaced by dynamically independent bosonic charge and spin-density collective modes, each propagating, with a (generally) different velocity. This is so-called separation of spin and charge. Moreover, for a Luttinger liquid, one has $Z_K \to 0$ in a nontrivial way, implying an anomalous (infinite) wave function renormalization.

The two key factors that lead to the breakdown of the Fermi liquid and four parameters that are necessary for the description of the low energy physics of Tomonaga– Luttinger liquid are :

(*i*) The internal degrees of freedom of the electron, spin and charge, are decoupled and move independently with different velocities v_σ and v_ρ (σ and ρ represents spin and charge; respectively). Obviously, elementary excitations of the Tomonaga-Luttinger liquid are not quasi particles but collective modes for the spin and charge degrees of freedom, *i.e.*, the spinon with spin 1/2 and holon with charge e. This phenomenon is termed as spin-charge separation.

(*ii*) Various physical quantities obey non-universal power law, showing behaviours as functions of length, energy and temperature. The power-law scaling is described by the coupling constants k_ρ and k_σ for charge and spin fluctuations respectively. Since the spin-rotation invariance requires k_σ [77], k_σ has no significant effect on the quantities. This means, k_ρ plays a crucial role in scaling laws.

One of the methods used to investigate the spin-charge separation is to compare the spin and charge correlation functions, defined as

$$S(q, \omega) = \sum_v |<v|S_q^z|0>|^2 \delta(\omega - E_v + E_0) \qquad ...(8.16(a))$$

and
$$N(q, \omega) = \sum_v |<v|N_q|0>|^2 \delta(\omega - E_v + E_0) \qquad ...(8.16(b))$$

with
$$S_q^z = \left(\frac{1}{\sqrt{N}}\right) \sum_i \exp(iqR_i) S_i^z$$

and
$$N_q = \left(\frac{1}{\sqrt{N}}\right) \sum_i \exp(iqR_i)(ni - n) \qquad ...(8.16(c))$$

where S_i^z is the Z-component of the spin operator, $n_i = n_{i\uparrow} + n_{i\downarrow}$, and N is the number of sites. In the small q limits, these reduces to [77]

$$S(q, \omega) = \frac{2}{\pi} \frac{v_\sigma q^2}{\omega^2 - v_\sigma^2 q^2}$$

and
$$N(q, \omega) = \frac{2}{\pi} \frac{v_\rho q^2}{\omega^2 - v_\rho^2 q^2} \qquad ...(8.16(d))$$

The spectral weight appears at $\omega = V_\sigma |q|$ and $\omega = v_\rho |q|$ for the spin and charge correlation functions respectively. Clearly, this is a manifestation of the spin charge separation, and the positions of the two poles, $v_\sigma |q|$ and $v_\rho |q|$, correlated to the energy of the spinon and holon with momentum q respectively. On the other hand, each correlation function near $q = 2K_F$, K_F being the Fermi momentum, has spectral weight at both $\omega = v_\sigma |q - 2K_F|$ and $\omega = v_\rho$

$|q - 2K_F|$ [77].

Another important quantity which manifests spin-charge separation is the single-particle excitation or the spectral function $A(K, \omega)$, defined by

$$A(K, \omega) = A_-(K, \omega) + A_+(K, \omega)$$

and

$$A_\pm(K, \omega) = \sum_{\nu\sigma} |<\nu| a_{K\sigma} |0>|^2 \delta(\omega \mp E_\nu \pm E_0)$$

where A_- (A_+) is the electron-removal (electron-addition) spectral function, and $a_{K\sigma} = C_{K\sigma}^+$ and $C_{K\sigma}$ for A_+ and A_- respectively, $C_{K\sigma}^+$ ($C_{K\sigma}$) being the creation (annihilation) operator for an electron with momentum K and spin σ. $|\nu>$ is ν^{th} eigenvector with eigenvalue E_ν. $|0>$ denotes the ground state. One of the difficulties posed by a non-Fermi liquid in $D < 3$ is the possibility of defining a Fermi surface, since the momentum distribution function $n(K)$ is expected to be smooth at $K = K_F$ (equivalently, $Z_K \to 0$ in a rather special way, and quasiparticles are ill-defined objects on the basis of experiments in HTSC cuprates (especially ARPES), one can easily assume the existence of a $(D-1)$ dimensional submanifold in **K** space, starting from which one defines the excited states, characterizing the low-energy properties of the system.

Metzner and Dicastro [78] addressed the issue of dimensional crossover from a **Luttinger liquid** in $D = 1$ to a **Fermi liquid** in $D = 3$, employing a nonperturbative diagrammatic approach, permitting to recover exact ward identities in continuous dimensionality. $1 < D < 2$. They reported that charge and spin obey asymptotically valid conservation laws down to $D = 1$ [79]. Obviously, for $1 < D < 2$ the low energy properties, *e.g.*, spectral function, the momentum distribution, etc. can be described by a **tomographic Luttinger liquid model** [10], defined as an angular sum over one-dimensional Luttinger liquids, each intersecting the two-dimensional Fermi "surface" in a left and a "right" Fermi point. Such a tomographic Luttinger liquid model has been predicted to exhibit the separation of charge and spin degrees of freedom [80], typical of a one-dimensional Luttinger liquid. However, this is an highly controversial issue.

Anderson [10] has set two crucial experimental tests of RVB Luttinger-liquid interlayer tunneling theory of HTSC cuprates. He finds that the theory for the HTSC cuprates normal state permit some predictions about the effects of superconductivity on two aspects of the one particle Green's function: ARPES and tunneling for a high-T_c insulator–high-T_c junction. One of the most basic features of the RVB-Luttinger liquid charge and spin into two nearly independent excitations in the normal state, leading to the expansion of the unusual Fermi liquid quasiparticle pole in the electron or hole single particle Green's function into a cut encompassing a wide band of frequencies. This causes a characteristic shape of photoemission spectrum and also responsible for the lack of coherent metallic conduction perpendicular to the plane. A second consequence of the way in which the superconductivity enters in these systems must appear in tunneling in highT_c insulator–high T_c junctions, especially those fabricated perpendicular to the c-axis.

8.5 t-J MODEL

Zhang and Rice (ZR) [81] argued that a good starting point for modelling strong correlation effects in HTSC cuprates should be the Hubbard model or the *t-J* models close to half filling. Zaanen and Oles [82] remarked that doped holes locate in oxygen orbitals in the localized limit. They couple strongly to the Cu spins by a second order Kondo-like antiferromagnetic interaction [81, 82, 83] and argued that this coupling dominates the behaviour of the excess holes and leads to the formation of local singlets. The singlets are formed at a given site by a Wannier state of oxygen hole centered at this site and the respective hole localized at Cu. The hopping of oxygen holes appears now to be equivalent to the motion of a singlet between the respective two nearest neighbour Cu sites. The motion of the singlet and the interaction between the localized spins may be described by a *t-J* model, given by

$$H_{t-J} = -t \sum_{<ij>, \sigma} (\tilde{C}_{i\sigma}^+ \tilde{C}_{j\sigma} + h.c) + J \sum_{<i,j>} \mathbf{S}_i \cdot \mathbf{S}_j \qquad ...(8.17)$$

where $\tilde{C}_{i\sigma}^+ = C_{i\sigma}(1 - n_{i,-\sigma})$ is the annihilation operator of an electron with σ at site i with the constraint of no double occupancy, **S** is the spin operator with $S = 1/2$ at site i. The summation over $<i, j>$ means that each bond $i - j$ is included once. Since the Zhang-Rice (ZR) singlet and localized spin exchange their positions by hopping process, the t-term corresponds to the hopping of the ZR singlet. The term J represents the exchange interaction ($J > 0$) between the localized spins. This Hamiltonian is called a *t-J* model and become the simplest effective model to discuss the pairing in HTSC cuprates. The following form of the effective singlet-band *t-J* model has also been used [84].

$$H_{t-J} = -t \sum_{<ij>, \sigma} (\tilde{C}_{i\sigma}^+ \tilde{C}_{j\sigma} + h.c) + J \sum_{<i,j>} \left(\mathbf{S}_i \cdot \mathbf{S}_j - \frac{1}{4} n_i n_j \right) \qquad ...(8.18)$$

The local spin operator S is defined as

$$S_i = \frac{1}{2} \sum_{\sigma, \sigma'} \sigma_{\sigma\sigma'} \tilde{C}_{i\sigma}^+ \tilde{C}_{i\sigma'} \qquad ...(8.19)$$

The coupling constants are related through $J = 4t^2/U$.

The *t-J* model is known for many years since it was derived from the Hubbard Hamiltonian in the strong coupling limit [81]. For applying this model to the HTSC cuprates, it is assumed that electron- or hole-doping causes the spin defects with electric charge on copper sites so that the defects can be looked upon as holes on Cu-sites. It remains controversial whether the *t-J* model is sufficient to describe the holes in a CuO_2 plane since it neglects completely the charge transfer process and thus describes the situation where the pairing could result only entirely from the magnetic exchange interaction *J*. On one hand, there are arguments provided by the exact diagonalization of finite clusters that the low energy physics should be well described by the *t-J* model since (*i*) in this region the spectra of the realistic Hamiltonian map onto the ones of the effective one with somewhat renormalized parameters [85] and (*ii*), the low energy physics of Cu_2O_7 and Cu_2O_8 clusters is well described by the *t-J* model extended by a next nearest hopping, t' [86]. On the other hand Emery and Reiter [87] have shown that the low energy physics of the effective t-J model does not necessarily correspond to that of the two band model. One can clearly see the distinction between the two models when the Coulomb interactions become large [87, 88]. The t-J model gives a 2^N fold degenerate ground state (GS) for one hole per unit cell, where N is the number of lattice sites, while the three-band model still has an AF-exchange interaction and the ground state has an AF-order. Gooding and Elser [88] have shown the ground state of a 4×4 cluster with one dopes hole is ferromagnetic in the $t - J$ model and has an intermediate spin value of $S = 5/2$ and $k = (\pi/2, \pi/2)$ for the full Hamiltonian projected onto the subspace without double occupancies.

Anderson [24] first suggested the superconducting ground state of $t - J$ model. This idea follows in a natural way from the possibility of representing the exchange term, J of the effective Hamiltonian in a form of local singlets on the bonds given by eq. (8.10) with eq. (8.4). Baskaran et. al [49] have shown that the state of local singlets which resonate on the bonds is formally equivalent to the superconducting BCS state projected on the subspace of fixed particle number. The state of the resonating singlets is called a RVB state as discussed earlier. It is close energetically to the antiferomagnetic order in the states with long range order and thus a part of magnetic energy is lost, while they can move freely in the RVB states. Indeed, if we treat the kinetic energy H_t in eq. (8.18) in mean field approximation, one finds that the ground state is superconducting at finite concentration of excess holes [50]. However, one must consider this result only as a qualitative one since strong electron correlations expressed by the projection operators $(1 - n_{i,-\sigma})$ are then not treated rigorously [81]. Provided a more adequate treatment of the projection operators on the subspace without the double occupancies at Cu sites within the Gutzwiller approximation [89, 90]. This approximation results in obtaining the energy of the system from the effective renormalization Hamiltonian.

$$H'_{t-J} = g_t H_t + g_s H_J \qquad ...(8.20)$$

Here, the factors g_t and g_s renormalize, respectively the averages $<C^+_{i\sigma} C_{j\sigma}>$ and $<\mathbf{S}_i \mathbf{S}_j>$, obtained for free fermions. One obtains the result of the Gutzwiller approximation as

$$g_t = \frac{2x}{1+x}, g_s = \frac{4}{(1+x)^2} \qquad ...(8.21)$$

Here x stands for the doping of the CuO_2 plane, i.e. $n = 1 + x$. Considering a realistic ratio of $t/J = 5$, one obtains that the superconducting d-wave state is stable and that the superconducting order parameter $g_t <C^+_{iJ}>$ increases from δ [81].

The $t - J$ model has also been used to study the stability of commensurate flux phases by representing the hopping element by the one dependent of local phase, $t_{iJ} = t \exp(i\phi_{ij})$. For $J/t > 1$, the ground state of the $t - J$ model is then found to be a commensurate flux phase, with the flux $\phi = \sum \phi_{ij}$ pinned to the dopping [91, 92]. These states exhibits the Meissner effect, i.e. they are superconducting. However, their significance for the real HTSC cuprates seems to be rather limited as these states are stable only for an unrealistically large ratio $J/t > 1$.

Many studies of $t - J$ model suggest that this model is able to describe a large body of the low energy physics of HTSC cuprates [10, 93]. This seems to be true for many normal state properties of HTSC cuprates where exact numerical predictions of the t-J model are available for the comparison with the experiment. Whether the phenomenon of HTSC in cuprates itself can be explained within this model is not clear. There are several calculations yielding instabilities of the normal state with respect to d-wave superconductivity [94 – 97] and also reasonably high values for the transition temperature [98 – 101]. Moreover, these calculations, however, use often somewhat uncontrolled assumptions, making definite conclusions difficult. There is also the view [10] that the larged observed values of T_c are not directly related to large mean field T_c's in isolated CuO_2 planes described by the t-J model.

ARPES measurements on $Sr_2CuO_2Cl_2$ by Wells et al. [102] have shown that the t-J model does not explain the experimental data near $\mathbf{K} = (\pi, 0)$: the t-J model predicts that the energies of the quasiparticle at $(\pi, 0)$ and $(\pi/2, \pi/2)$, are very similar, while the experimental energy at $(\pi, 0)$ is lower than that at $(\pi/2, \pi/2)$. Recent theoretical studies by Tohyama et al. [103] and Tohyama and Maekawa [104] have shown that this discrepancy can be resolved by introducing hopping matrix elements to second and third nearest neighbours (t' and t'') into the $t - J$ model. The hopping process may exist between not only the first nearest neighbour pairs but also the second and third neighbour ones connecting sites at distance $\sqrt{2}\ a$ and $2a$ respectively. The latter is represented by

$$H_{t't''} = -t' \sum_{<ij> 2\text{nd}}^{\sigma} C_{i\sigma}^+ C_{j\sigma} - t'' \sum_{<ij> 3\text{rd}} (C_{i\sigma}^+ C_{j\sigma} + h.c.) \qquad ...(8.22)$$

$<i, j>$ 2nd and $<i, j>$ 3rd run over second and third nearest neighbour pairs, respectively. The total Hamiltonian $H = H_{tJ} + H_{t't''}$, is named the **t - t' - t'' - J model**. This model explains the systematic evolution of line shape in the normal state from undoped to overdoped cases. However, the low energy features related to superconducting and pseudogaps have not been explained so far. Instead, for the $t - J$ model without t' and t'', the gauge theories based on the spin charge separation have not been applied and some of the features are explained.

For LSCO, the ratios t'/t and t''/t from $t' - t' - t'' - J$ model are estimated to be – 0.12 and 0.08 respectively, and $J/t = 0.4$ ($t = 0.35$ eV). [105]. To obtain various excitation spectra, Tohyama et al. [105] performed the exact-diagonalization calculation on a $\sqrt{18} \times \sqrt{18}$ – site cluster with two holes ($x = 0.11$).

It is controversial whether $t - J$ model (also $t - t' - t'' - J$ model) and related models have the **stripe-type** ground state [106]. A possible requirement for the appearance of a stable stripe phase is the presence of the long range part of the Coulomb interaction [107, – 109] and/or the coupling to lattice distortions.

The t-J model and their modifications [5] have helped to clarify several optical and magnetic properties of HTSC cuprates [104] but superconducting mechanisms remains unclear. Studies on generalized t-J models suggest a magnetic origin of superconductivity but the numerical results seem to require, either a super-exchange J, or a three site term [104], which is beyond the realistic range for cuprates. Jackelmann et al. [110–111] have shown that the constraint of the double occupancy in these t-J and generalized t-J models reduces the mobility of the superconducting pairs. On the other hand, the search for signals of superconductivity in the Hubbard model have been negative so far [93, 112, 113].

8.6 SPIN FLUCTUATION MECHANISM

Moriya et al. [114–116] have considered the mechanism of superconductivity as due to the exchange of the antiferromagnetic spin fluctuations. The possibility of d-wave superconductivity in the Hubbard model has been discussed using the random phase approximation (RPA) by Shimahara and Takada [117] or the fluctuation exchange approximation by Bickers et al. [118].

Moriya et al. [114–116] consider the following effective coupling between quasiparticles in the weak coupling theory

$$H_{int} = -\frac{1}{N^2} \sum_{K,K',\vec{q}} \sum_{\alpha,\beta,\gamma,\delta} J(\mathbf{q})(\tau_{\alpha\beta} \cdot \tau_{\gamma\delta}) a^+_{\vec{K}+\vec{q},\alpha} a_{\vec{K},\beta} a_{\vec{K}'-\vec{q},\gamma} a_{\vec{K},\delta} \qquad ...(8.23)$$

where τ are the Pauli matrices. For the effective interaction, Moriya et al. [114–116] used the particle-particle vertex part defined by

$$J(\mathbf{q}) = \frac{U}{1 - 2U\,\bar{x}\,(\mathbf{K})} = U + 2U^2 \chi(\mathbf{q}) \qquad ...(8.24)$$

where U is the intraatomic exchange constant. $\chi(\mathbf{q})$ is the susceptibility and $\bar{\chi}(\mathbf{q})$ is its irreducible part.

From the interaction given by eq. (8.23) matrix elements for even and odd parity pairings are obtained as

$$H_{int} = \frac{1}{2N^2} \sum_{K,K'} \sum_{\sigma} \Big\{ V^{odd}_{K,K'} \left(a^+_{K,\sigma} a^+_{-K,\sigma} a_{-K',\sigma} a_{K',\sigma} + a^+_{K,\sigma} a^+_{-K,\bar{\sigma}} a_{-K,\bar{\sigma}} a_{K,\sigma} \right)$$

$$+ V^{even}_{K,K'} a_{K,\sigma} a_{-K,\bar{\sigma}} a^+_{-K',\bar{\sigma}} a^+_{-K,\bar{\sigma}} \Big\}$$

with
$$V^{odd}_{K,K'} = -\frac{1}{4}[J(\mathbf{K}-\mathbf{K}') - J(\mathbf{K}+\mathbf{K}')] \qquad ...(8.25)$$

and
$$V^{even}_{K,K'} = \frac{3}{4}[J(\mathbf{K}-\mathbf{K}') + J(\mathbf{K}+\mathbf{K}')] \qquad ...(8.26)$$

where $\bar{\sigma}$ denotes the opposite spin of σ.

In the weak coupling theory, the transition temperature is expressed as

$$T_c = 1.14\,\omega_c \exp(-1/\lambda) \qquad ...(8.27)$$

where ω_c is the cut-off energy. The value of λ is determined by the largest eigenvalue of the linearized gap equation

$$\rho < V_{KK'}\,\Delta(\mathbf{K}') > = -\lambda\,\Delta(\mathbf{K}) \qquad ...(8.28)$$

where ρ is density of states at the Fermi energy and $\langle......\rangle$ denotes the average on the Fermi surface. One can determine the symmetry of the pairing by the eigenfunction belonging to the largest eigenvalue (D_{4n}). However, the weak coupling theory is not sufficient, since the coupling constant λ is of the order of unity.

Moriya et al. [116] have considered the strong coupling theory for the superconductivity due to the antiferromagnetic spin fluctuations.

The antiferromagnetic spin fluctuations mediate attractive interactions for certain channels of the scattering between the quasiparticles. At the same time they contribute to the self energy of the quasi-particles, which leads to an energy shift and a quasiparticle damping. As regards superconductivity, the damping is important since it causes depairing effects for any unconventional type of Cooper pairings in contrast to the ordinary s-wave case. Therefore, one must treat both the effects on the same footing. One can achieve this by solving the following Dyson-Gor'kov equations for the normal and anomalous Green's functions simultaneously.

$$[i\omega_n - \xi_K - \sum\nolimits^{(1)}(\mathbf{K}, i\omega_n)] G(\mathbf{K}, i\omega_n) - \sum\nolimits^{(2)}(\mathbf{K}, i\omega_n) F^+(\mathbf{K}, i\omega_n) = 1 \qquad ...(8.29)$$

$$[-i\omega_n - \xi_K - \sum\nolimits^{(1)}(-\mathbf{K}, i\omega_n)] F^+(\mathbf{K}, i\omega_n) + \sum\nolimits^{(2)}(-\mathbf{K}, i\omega_n) G(\mathbf{K}, i\omega_n) = 0 \quad ...(8.30)$$

where $\xi_K = \varepsilon_K^0 - \mu$ is the single particle energy measured from the chemical potential, $\sum\nolimits^{(1)}(\mathbf{K}, i\omega_n)$ the normal self energy potential, $\sum\nolimits^{(2)}(\mathbf{K}, i\omega_n)$ the anomalous self energy.

At the level of one-Boson exchange, one finds the self energies as

$$\sum\nolimits^{(1)}(\mathbf{K}, i\omega_n) = \frac{U}{N} \sum_p T \sum_{\omega_m} [1 + U\chi(\mathbf{K}-\mathbf{p}, i\omega_n - i\omega_m)] G(\mathbf{p}, i\omega_m) \qquad ...(8.31)$$

$$\sum\nolimits^{(2)}(\mathbf{p}, i\omega_n) = \frac{U}{N} \sum_p T \sum_{\omega m} [1 + U\chi(\mathbf{K}-\mathbf{p}, i\omega_n - i\omega_m)] F(\mathbf{p}, i\omega_m)] \qquad ...(8.32)$$

where $\chi(\mathbf{K}, i\omega_n)$ is the dynamical susceptibility for the boson Matsubara frequencies. At the input of theory, one may use the spin fluctuations which is determined by the normal state properties. In this scheme, one obtains the normal Green's function by solving the full non-linear equation

$$G(\mathbf{K}, i\omega_n) = \frac{1}{i\omega_n - \xi_K - \sum\nolimits^{(1)}(\mathbf{K}, i\omega_n)} \qquad ...(8.33)$$

with the self energy expressed by relation (8.31).

The transition temperature is determined as the temperature below which the linear equation for the anomalous self energy has a non-trivial solution. By normalizing the anomalous self-energy as

$$f(\mathbf{K}, i\omega_n) = \sum\nolimits^{(2)}(\mathbf{K}, i\omega_n) |G(\mathbf{K}, i\omega_n)|,$$

the linearized equation takes the following form

$$f(\mathbf{K}, i\omega_n) = -\frac{U}{N}\sum_p T \sum_{\omega_m} |G(\mathbf{K}, i\omega_n)| \, [1 + U\chi(\mathbf{K}-\mathbf{p}, i\omega_n - i\omega_n)]$$
$$|G(\mathbf{p}, i\omega_m)| \, f(\mathbf{p}, \omega_m) \quad ...(8.34)$$

Therefore T_c is defined as the temperature where the largest eigenvalue of the Kernel becomes unity. One can determine the symmetry of the superconducting state by the symmetry of the corresponding eigen function. The largest eigen value of eq. (8.34) is obtained by using the Lanczos algorithm [119].

The spin fluctuation mechanism explains fairly well the high-T_c and some other anisotropic properties observed in HTSC cuprates. However, the phenomenon of pseudo-gap remains unexplained. There are also deep issues [10] related to the incompatibility of real space super exchange and Fermi liquid background apart from the fact that normal state anomalies and inter layer regularities in T_c are not explained satisfactorily. **Spin-bag** theory due to Schrieffer et al. [120] is essentially a spin fluctuation theory with a real space tinge to it. It also suffers from similar criticism.

8.7 SPIN-FERMION MODEL

Morr and Pines [121] have put forward the view that the resonance peak observed in inelastic neutron scattering experiments on underdoped and optimally doped samples of $YBa_2Cu_3O_{6+x}$ comes from spin-wave excitation. This model is based on spin-fermion interactions. The spin-wave mode dispersion is taken as

$$\omega_{\underline{q}}^2 = \Delta_{SW}^2 + C_{SW}(\underline{q} - Q)^2 \qquad ...(8.35)$$

where Δ_{SW} is the spin-wave gap, C_{SW} is the spin-wave velocity and $Q = (\pi, \pi)$. Usually, for systems having strong antiferromagnetic correlations, the dispersion of spin-wave modes is linear, i.e. $\omega_q \propto q$ [122]. It is further assumed by them that at high temperatures the spin wave mode is strongly doped owing to its coupling to be minimal. Thus the resonance should be observable if the conditions $\Delta_{SW} < \omega_c \approx 2\Delta_{sc}$. They claim that in underdoped systems, the spin damping present at higher temperatures is reduced. Accordingly the resonance mode remains visible. The spin wave mode in optimally doped systems is overdamped and hence invisible.

The bonding and anti-bonding tight binding quasi particle bands for a two-layer system such as $YBa_2Cu_3O_{6+x}$ are

$$\varepsilon_K = -2t_1[\cos(K_x a) + \cos(K_y a)] - 4t_2 \cos(K_x a)\cos(K_y a) \pm t_\perp - \mu \qquad ...(8.36)$$

whereas mentioned earlier, t_1 and t_2 are the hopping matrix elements in the in-plane nearest and next-nearest neighbours respectively and $t_\perp$ is the transfer integral between nearest neighbours on different planes. Using the spin-fermion model of Monthoux and Pines [123], the spin-wave propagator, χ is written as [121]

$$\chi^{-1} = \chi_0^{-1} - \text{Re}\,\Pi - i\,\text{Im}\,\Pi \qquad ...(8.37)$$

where χ_0 is the bare propagator and Π is irreducible particle-hole bubble. The form of $(\chi_0^{-1} - \mathrm{Re}\,\Pi)$ is chosen by Morr and Pines [121] in such a way that they reproduce the inelastic scattering results in the normal state of the underdoped $YBa_2Cu_3O_{6+x}$ systems. Thus

$$\chi_0^{-1} - \mathrm{Re}\,\Pi = [1 + \xi^2(\underline{q} - \underline{Q})^2 - \omega^2/\Delta_{SW}^2]/\alpha\xi^2 \qquad ...(8.38)$$

where ξ being the magnetic correlation length, $\Delta_{SW} = C_{SW}/\xi$, α is a constant.

After making several assumptions, Morr and Pines [121] calculated the imaginary part of Π, which described the damping due to the decay of spin excitation into a particle-hole pair. The calculation is confined to the odd channel which involves quasiparticle excitations between the bonding and antibonding bands. To the lowest order in the spin-fermion coupling g_{sf} (for $W > 0$) and in the superconducting state.

$$\mathrm{Im}\,\Pi_{odd} = (3\pi g_{sf}^2/8) \sum [1 - f(E_{\underline{K}+\underline{q}}^+) - f(E_{\underline{K}}^-)]$$

$$\times [1 - (\varepsilon_{\underline{K}+\underline{q}}^+ \varepsilon_{\underline{K}}^- + \Delta_{K+q}\Delta_K)/E_{\underline{K}+\underline{q}}^+ E_{\underline{K}}^-]\,\delta(\omega - E_{\underline{K}}^- - E_{\underline{K}+\underline{q}}^+)$$

$$+ [f(E_{\underline{K}}^-) - f(E_{\underline{K}-\underline{q}}^-)][1 + (\varepsilon_{\underline{K}+\underline{q}}^+ \varepsilon_{\underline{K}}^- + \Delta_{K+q}\Delta_K)/E_{\underline{K}+\underline{q}}^+ E_{\underline{K}}^-]$$

$$\times \{\delta(\omega + E_{\underline{K}}^- - E_{\underline{K}+\underline{q}}^-) - \delta(\omega - E_{\underline{K}}^- + E_{\underline{K}+\underline{q}}^+)\} \qquad ...(8.39)$$

as before $f(x)$ is Fermi function and $E_{\underline{K}}^\pm = [(\varepsilon_{\underline{K}}^\pm)^2 + |\Delta_K|^2]^{1/2}$ is the dispersion of the bonding and anti-bonding bands in the superconducting state. They assume a d-wave gap $\Delta_K = \Delta_{sc}[\cos(k_x a) - \cos(k_y a)]/2$. In Fig. 8.5, the calculations of $\mathrm{Im}\,\Pi_{odd}$ at $Q = (\pi, \pi)$ as frequency for the normal (solid line) and superconducting states are shown.

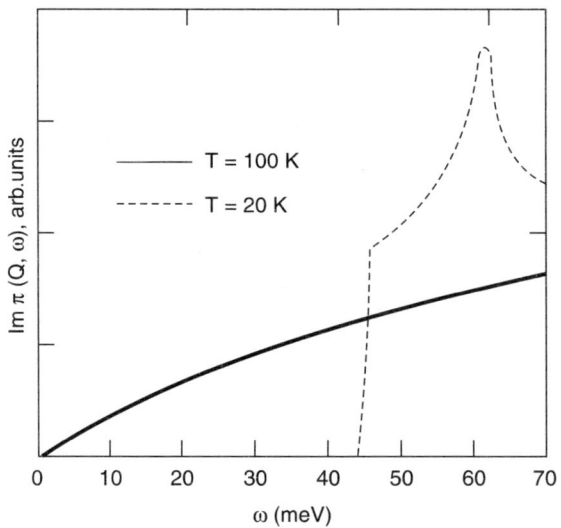

Fig. 8.5 $Im\,\Pi_{odd}$ at $Q(\pi, \pi)$ [121].

The results are for $YBa_2Cu_3O_7$ using the values $t_1 = 300$ MeV, $t_2 = 0.40\,t_1$, $t_\perp = 0.3\,t_1$ and $\mu = -1.27\,t_1$. These correspond to a 22% hole concentration in the planes. $\Delta_{sc}(T=0) \approx 25$ MeV. They have taken that $2\Delta_{SW}$ has the same value from the underdoped to overdoped regions. However, Δ_{SW} increases linearly in this region.

Having determined the behaviour of $Im\ \Pi_{odd}$ at $Q\ (\pi, \pi)$, the results for χ_{odd} at $Q = (\pi, \pi)$ as a function of frequency is obtained. The relationship

$$\chi_{odd}(Q, \omega = \Delta_{SW}) = [Im\ \Pi_{odd}(Q, \Delta_{SW})]^{-1} \qquad ...(8.40)$$

is used in the computation. Results are shown in Fig. 8.6.

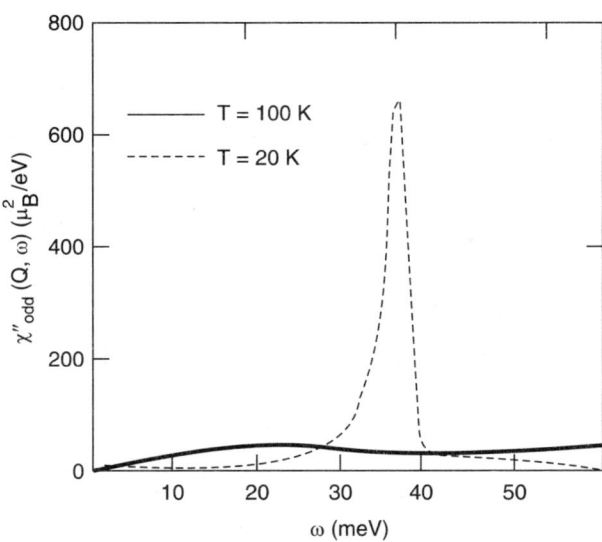

Fig. 8.6 χ'' at $Q\ (\pi, \pi)$ as a function of frequency [121].

The solid line gives the normal state behaviour and the dashed one for the superconducting state. Their argument is that in the normal state, the spin excitations are over damped. Hence χ'' shows a flat maximum at $\omega = 20$ MeV. In the superconducting state this spin damping is so strongly reduced, that the spin-wave becomes very sharp at a frequency chosen to be at 41 MeV to have agreement with experiment. However, several conditions are imposed and the parameters are adjusted to have agreement. The claim that spin-wave model provides a natural explanation is not sustainable [124-126].

Several authors [127-130] have proposed pairing mechanisms based on the antiferromagnetic or spin density wave (SDW) gap formation or their precursor. In these mechanisms 'hot spot' which is a part of the Fermi surface in the vicinity of the antiferromagnetic Brillouin zone, exists near $(0, \pi)$ and plays a special role. Since the quasi particles at 'hot spot' are strongly scattered by the antiferromagnetic spin fluctuations the gap structures appears at the 'hot-spot'. Datum and Tewordt [130] have made calculations by treating the pseudo gap phenomena as a precursor of the magnetic instability (antiferromagnetism or SDW). However, considering that the ground state is actually superconductive, one can hardly expect that the magnetic order and superconductivity are continuously connected beyond the superconducting transition. Yanase and Yamada [131] are of the view that the strong spin fluctuations near the magnetic critical point may lead to the gap-like structures. However, in that case, the gap formation first takes place at 'hot spot' because the interaction is strong in the vicinity of $Q\ (\pi, \pi)$ and $\Omega = 0$. Yanase and Yamada [131] have further remarked that 'hot spot' comes to exist in the wide region near $(0, \pi)$ as the spin fluctuations lead to the transformation of the Fermi surface. This means that the gap structure would have a similar shape to the $d_{x^2-y^2}$ wave superconducting gap. However, in HTSC cuprates it is not easy to occur. The interaction caused by the superconducting fluctuations is strong in the vicinity of q (0, 0), whereas the important K-points for the self-energy on the Fermi surface are sure to be on the Fermi surface. On the other hand, in case of the anti-ferromagnetic spin fluctuations, the interaction is strong in the vicinity of Q (π, π). Obviously, the important K-points for the quasi-particles with the momentum K exist in the vicinity of **K + Q**. Yanase and Yamada [131] have further remarked that they are on the Fermi surface only when K on the 'hot-shot'. As a result, the quasi-particles slightly apart from 'hot-spot' are not directly scattered by the strong interaction. This reveals that the pseudogap formation is not impossible but difficult to be attributed to the magnetic interaction on the numerical point of view.

8.8 INTERLAYER PAIR-TUNNELING (ILPT) MECHANISM

The coherent single particle hopping between CuO_2 planes blocked in the normal state, where as in principle there is no restriction for the coherent tunneling of pairs. One can view this as a second order effect in inter chain hopping $t_\perp$ which becomes operative as soon as pairs are formed, *i.e.* immediately below T_c, in the superconducting state. A superconducting pair is a singlet object, *i.e.* tunneling of one pair is associated to transport of charge, but not of spin. Obviously, interlayer pair tunneling is not frustrated by an in-plane ground state characterized by spin-charge separation.

One may say that ILPT mechanism is of kinetic origin, i.e. the kinetic energy associated with c-axis dynamics, inaccessible for the single particles in the normal state, becomes available in the superconducting state. Obviously, this causes the lowering of the total electronic energy. We must remember that this is a situation which has no counterpart in conventional Fermi-liquid based superconductivity, where the kinetic energy is enhanced upon going into the superconducting state, while being over compensated by a reduction in potential energy. It is interesting to note that superconductivity within the ILPT model arises via a dramatically opposite mechanism, *i.e.* instead of having the gain in potential energy to overcome the loss of kinetic energy, it is the gain in kinetic energy which is the driving mechanism. We must note that ILPT is not a pairing mechanism in itself, since it presupposes that the singlet electron pairs are already formed by some instability within the CuO_2 planes, before ILPT mechanism becomes operative. Obviously, ILPT mechanism is logically, rather than physically independent from the underlying pairing mechanism. However, the pairing symmetry, among other properties, can be formally determined by the symmetry of the in-plane potential kernel $V_{kk'}$ only.

Anderson [132], soon after his proposal of the RVB for the ground state of the two dimensional Hubbard model close to half filling suggested that the driving mechanism yielding such high-T_c in HTSC cuprates might be the Josephson like tunneling of pairs between adjacent CuO_2 planes [133]. The observed correlation between T_c and the number of layers n in HTSC cuprates have been considered to be the evidence for such mechanism [134, 135]. (In Bi-Tl- and Hg-based HTSC cuprates the T_c increases monotonically with n, apparently tending to an asymptotic constant value as n increases).

Coherent tunneling of pairs between adjacent planes may occur essentially in two ways :

(i) *By Josephson like tunneling*

$$H_J = \frac{1}{t} \sum_{<ll'>} \sum_{K} |t_\perp(\mathbf{K})|^2 \, C^{(l)+}_{K\uparrow} C^{(l)+}_{-K\downarrow} C^{(l')}_{-K\downarrow} C^{(l')}_{K\uparrow} \qquad ...(8.41)$$

or (ii) by *"Pair rearrangement"*

$$H_R = \frac{1}{t} \sum_{<ll'>} \sum_{K} |t_\perp(\mathbf{K})|^2 \, C^{(l)+}_{K\uparrow} C^{(l')+}_{K\uparrow} C^{(l)+}_{K\uparrow} C^{(l)}_{K\uparrow} \qquad ...(8.42)$$

where $C^{(l)+}_{K\sigma}$ creates an electron in the state $(K\sigma)$ on layer l, $<ll'>$ label adjacent Cu O$_2$ layers, t is main in-plane hopping amplitude, *i.e.* we can say it as nearest neighbour hopping and H_J has been restricted to singlet pairs with zero relative momentum (Cooper pairs). The single-particle (coherent) hopping amplitude $t_\perp(\mathbf{K})$ possesses a non-trivial dependence on the two dimensional wave vector $\mathbf{K}$. Anderson [135] pointed out that in addition to the absence of coherent single particle hopping, group state average of H_R vanishes which is peculiar to the ILPT mechanism.

Chakravarty et al. [136] have discussed the main consequences of an ILPT contribution to a conventional superconducting mechanism in respect to Anderson's RVB model. They have derived a BCS-like self-consistent equation for the energy gap Δ_k for the case of a bilayer complex. The assumption of a phonon-based pairing kernel

$$V_{KK'} = -V \text{ for } |\xi_K|, |\xi_{K'}| < \hbar \omega_D \qquad ...(8.43)$$

leads to an *s*-wave pairing order parameter, whose anisotropy in *k*-space is however rendered highly unconventional due to the local nature of ILPT term. One easily obtains the high values of T_c even in that case, without requiring but a fine turning of the in plane coupling strength. One finds,

$$\frac{1}{\lambda} \approx \frac{T_c}{T_c - T_J} + \log \frac{\omega_D - T_J}{T_c - T_J} \qquad ...(8.44)$$

where $\lambda = VN(0)$ and $T_J = |t_J|^2/t$. One obtains $T_c \approx T_J/(1-\lambda)$, for small λ, and interlayer pair tunneling dominates the determination of T_c whereas $T_c \approx \omega_D \exp(t - 1/\lambda)$, if $T_J \ll T_c$, where the BCS limit is recovered [10, 136].

Within the band theory, one finds that the amplitude for the interlayer hopping between the odd and even conduction bands is basically given by

$$t_\perp(K) = \frac{t_\perp}{4}[\cos(K_x a) - \cos(K_y a)]^2 \qquad ...(8.45)$$

at the energy scale involved. Here $t_\perp$ is the interplane hopping parmeter, whose value is mostly determined by vertical Cus-Cus hopping, with smaller contributions from transverse Cus-$Op_{x,y}$ hopping between adjacent layers. Hopping through insulating spacers such as (BaO) Hg (BaO) in Hg-based, multilayered cuprates is estimated to be one order of magnitude smaller, although it retains the same dependence on $\mathbf{K}$. (Eq. (8.45)).

The peculiar **K**-dependence of $t_\perp(\mathbf{K})$ represented by eq. (8.45) involves sharp maxima at $x = (\pi/a, 0)$ and symmetry related points, and a nodal line along $T - M$ ($K_x = K_y$), direction, although $t_\perp(\mathbf{K})$ possesses full s-wave symmetry. Although $t_\perp(\mathbf{K})$ vanishes in coincidence with a d-wave gap function but it is not devoid of physical consequences, e.g. enhanced exponent in the power law dependence of the penetration depth [137]. All such features are inherited by the pair-tunneling amplitude $T_J(\mathbf{K})$, which is thus endowed with a highly nontrivial, anisotropic structure in **K** space.

Within the ILPT model, Chakravarty [138] obtained the following relation relating the Josephson's plasma frequency to the tunneling energy as

$$\frac{1}{2} m \omega_{pc}^2 = \sum_k T_J(\mathbf{K}) |b_K^2| \qquad ...(8.46)$$

where $b_K = \Delta_K \chi_K$ is the pair amplitude and $m = \dfrac{\hbar^2 \varepsilon A}{16\pi e^2 d}$ is an effective mass, related to the junction area A (base of a unit cell) the interlayer distance d, and dielectric constant ε of the interlayer medium. It is claimed that ILPT mechanism may enhance several superconducting properties and account for numerous experimental features of HTSC cuprates.

Angilella [139] have considered the issue of the mixing of symmetry channels in the superconducting order parameter for a bilayer superconductor in the presence of an ILPT mechanism. Angilella [139] generated superconductivity within each individual CuO_2 layer through a Hubbard-like in plane potential, including primarily an on-site repulsion and nearest-neighbour interaction, which has been strongly enhanced through the inclusion of the interlayer tunneling amplitude t_J. The model Hamiltonian considered by Angilella [139] is

$$H = \sum_{K,\sigma,i} \xi_K^i C_{K\sigma}^{i+} C_{K\sigma}^i + \sum_{KK'ij} \overline{V}_{KK'}^{ij} C_{K\uparrow}^{i+} C_{-K\downarrow}^{i+} C_{-K\downarrow}^j C_{K'\uparrow}^j \qquad ...(8.47)$$

where $C_{K\sigma}^{i+}$ ($C_{K\sigma}^i$) creates (destroys) a fermion on the layer $i(i = 1, 2)$, with spin projection σ along a specified direction, wave vector **K** belonging to the first Brillouin zone of a 2D square lattice, and band dispersion $\xi_K^i = \varepsilon_K^i - \mu$, measured relative to the chemical potential μ. The second term in eq. (8.47) describes an effective pair interaction, already restricted to the singlet channel only with

$$\overline{V}_{KK'}^{ij} = \frac{1}{N} U_{KK'} \delta_{ij} - T_J(k) \delta_{KK'}(1 - \delta_{ij}) \qquad ...(8.48)$$

where N is the number of sites in the square lattice, $U_{KK'}$ measures the coupling interaction within each plane, and $T_J(K)$ is the tunneling matrix element adjacent layers.

A meanfield treatment of the Hamiltonian (8.47) yield the approximate expression

$$H_{MF} = \sum_{K\sigma i} \xi_K^i C_{K\sigma}^{i+} C_{K\sigma}^i + \sum_{Kj} [\Delta_K^i C_{K\uparrow}^{i+} C_{-K\downarrow}^{i+} + h.c.] \qquad ...(8.49)$$

where the auxillary complex scalar field, *i.e.* the gap function is introduced as

$$\Delta^i_K = \sum_{jK'} \overline{V}^{ij}_{KK'} b^j_{K'}$$

$$= \frac{1}{N} \sum_{jK'} U_{KK'} b^+_{K'} - T_J(\mathbf{K}) b^j_K (1-\delta_{ij}) \qquad ...(8.50)$$

with $b^i_K = <C^i_{-K\downarrow} C^i_{K\uparrow}>$.

Although predictions based on ILPT mechanism agree with several aspects of the HTSC phenomenology, *e.g.* high-T_c, its dependence on the number of layers in layered cuprates, the absence of Hebel-Slitcher coherence peak in the NMR relaxation rate etc. However, the estimates of the electronic condensation energy from the ILPT mechanism found inconsistent with several direct and indirect measurements of that same quantity in single layered cuprates, *e.g.* one layers Tl and Hg cuprates. Baskaran [140] made a suggestion that part of this could be accounted for through the particle-hole pair tunneling mechanism. However, this still does not solve the problem completely. There are also other suggestions [141]. This has become a challenge and perhaps a revision of part of Anderson's central dogma to the effect that 'a single layer of cuprate may be superconducting' may be called for. Obviously, one has to go back to the original one layer RVB mechanism and see how it can explain the large T_c of one layer materials. RVB gauge theory ideas have been pursued a lot along these directions [142, 143] including some instant on ideas [144] and an idea [145] of mechanism is also operating in addition to the inter plane kinetic mechanism.

Yin et al. [146] considered a model within the context of Interlayer tunnelling theory HTSC cuprates. They consider a reduced Hamiltonian in a subspace in which both the states ($\mathbf{K}\uparrow$) and ($-\mathbf{K}\downarrow$) are both simultaneously occupied or unoccupied.

$$H_{red} = \sum \varepsilon_K C^+_{K\sigma i} C_{K\sigma i} - \sum V_{K,K'} C^+_{K\uparrow i} C^+_{-K'\downarrow i} C_{-K\downarrow i} C_{-K'\uparrow i}$$

$$- \sum T_J [C^+_{K\uparrow i} C^+_{-K\downarrow i} C_{-K\downarrow j} C_{K\uparrow j} + h.c.] \qquad ...(8.51)$$

Where ($C^+_{K\sigma i}$, $C_{K\sigma i}$) are fermion (creation, annihilation) operators for state $|K\sigma i>$, $i = 1, 2$ is the layer index, σ spin K = in plane wave vector ε_K is the single particle energy and V is the in-plane pairing interaction and T_J is the Josephson type inter-layer tunnelling matrix element. In the mean field analysis, they find the gap equation

$$\Delta_K = (1/[1 - \chi_K T_J(K)]) \sum V_{KK'} \chi_{K'} \Delta_{K'} \qquad ...(8.52)$$

where $\chi_{K'} = (1/2E_K) \tanh(E_K/2T)$ is the pair susceptibility and $E_K = [(\varepsilon_K - \mu)^2 + \Delta^2_K]^{1/2}$, μ being the chemical potential. Further the factorization $V_{KK'} = V_{K_g K'_g}$ was resorted to by assuming that $g_K = 1/2[\cos(K_x a) - \cos(K_y a)]$ has the *d*-wave symmetry. The full gap at $T = 0$ on the Fermi surface, taking the gap to be real turns out to be

$$g_{KF} = g_{KF} \Delta_0 + (T_J(K_F)/2) S_{gn}(\Delta_{KF}) \qquad ...(8.53)$$

where Δ_0 is a positive definite integral.

For the above model, the imaginary part of the spin susceptibility $\chi(q, \omega)$ expression is

$$\chi(q, \omega) = \sum [A^+_{K,q} F^-_K / \Omega'_{K,q}(\omega)] + (A^-_{K,q}(1 - F^+_{K,q})/2x$$

$$\{(1/\Omega^{2+}_{K,q}(\omega)) - (1/\Omega^{2-}_{K,q}(\omega))\} \quad \ldots(8.54)$$

Here
$$A^\pm_{K,q} = \frac{1}{2}[1 \pm (\varepsilon_K - \mu)(\varepsilon_{K+q} - \mu) + (\Delta_K \Delta_{K+q})/E_K E_{K+q})]$$

$$\Omega'_{K,q}(\omega) = [\omega - (E_{K+q} - E_K) + i\delta]$$

$$\Omega^{2\pm}_{K,q}(\omega) = [\omega \pm (E_{K+q} + E_K) + i\delta] \quad \ldots(8.55)$$

$F^\pm_{Kq} = f(E_{K+q}) \pm f(E_K)$; $f(x) \to$ the Fermi function. For computation, they choose the in plane one electron dispersion as

$$\varepsilon_K = -2t_1[\cos(K_x a) + \cos(K_y a)] + 4t_2 \cos(K_x a) \cos(K_y a)$$

The values of parameters are $t_1 = 0.25$ eV, $t_2 = 0.45\, t_1$, and $\mu = -0.315$ eV which corresponds to an open Fermi surface with band filling of 0.86; $V = 0.2$ eV, $N(0)$ 0.92. For $T_J = 0.075$ eV the $Im\, \chi$ at $T = 0$ is shown in Fig. 8.7. They also computed $Im\, \chi$ for $T \neq 0$. The result at $T = 20$ K and $T = 0$ are almost indistinguishable. The unusual peak in the neutron scattering at 40 MeV is thought to arise from the inter-layer tunnelling model in contrast to the BCS model which shows only a step discontinuity [123]. The domination of T_J makes the density of states sharply peak at $T_J/2$. The intensity and the position of the peak normalized to zero temperature values decreased with increasing temperature. Its position at $T = 0$ is 41.2 MeV. They claim that this vindicates the interlayer tunnelling theory of HTSC cuprates.

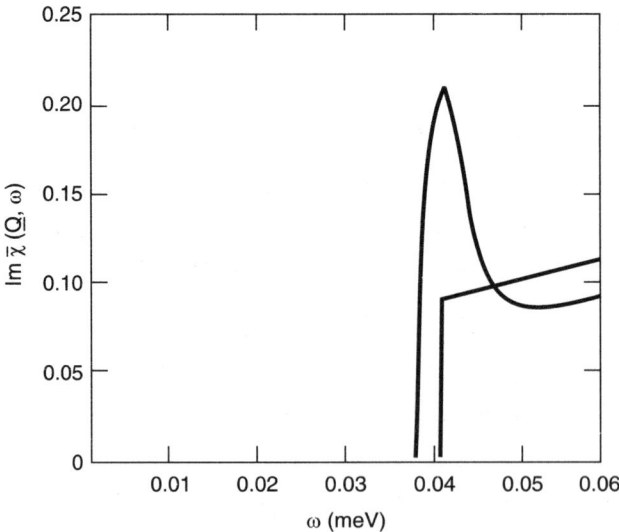

Fig. 8.7 $Im\, \chi$ at $T = 0$ for $T_J = 0.075$. The step discontinuity at the edge is for the BCS theory ($T_J = 0$) [146].

Chattopadhyay and Das [147] and Das and Chattopadhyay [148] motivated by the doping dependence of the transport properties in the HTSC cuprates, mainly of their c-axis resistivity and optical conductivity assume that the coupling between adjacent CuO_2 planes can be provided by a mixing of both single-particle and pair tunneling. The relative amount of the two effects is assumed to depend strongly on temperature and on the pseudogap, in a rather phenomenological way, so that pure ILPT dominates in the underdoped region, whereas a Fermi liquid like picture, where in single particle hopping is restored, is progressively recovered as doping (or temperature) is increased. They remarked that the extended ILPT model still accounts for the experimental absence of bilayer splitting.

8.9 COLLECTIVE EXCITATION COUPLING TO QUASI PARTICLES

Shen and Schrieffer [149] studied the anomalous momentum, temperature and doping dependence of the spectral photoemission line shape in data from $Bi_2 Sr_2 Ca Cu_2 O_{8+\delta}$ (Bi-2212). They focus on the anisotropic excitation pseudogap in the normal state of underdoped Bi-2212. The broad feature at 100 – 200 MeV that is always present near $(\pi, 0)$ is absent in spectra at the Fermi surface along $(0, 0)$ to (π, π) directions in underdoped samples. In overdoped samples, the features are absent in both these directions. They suggest that these results arise from a stronger dressing of the photo hole for $K \cong (\pi, 0)$ in the underdoped materials than in other cases. The hole couples strongly to collective modes of momentum q whose spectral function $\chi''(q, \omega)$ peaks near $Q(\pi, \pi)$ for the underdoped case. This coupling is weak because $\chi''(q, \omega)$ is weak and broad in momentum for the overdoped case. There is no discussion of the nature of the collective excitations which couple strongly to quasi-particles. These collective excitations provide the glue that leads to pairing of quasi-particles. Treating these collective excitations as some kind of bosons, they calculate **photo-hole spectral function** on the basis of a coupled fermion-boson model with the choice of band structure corresponding to the Fermi surface (Fig. 8.8) [149]. The quasiparticle peak appears when the resonance condition $\omega - \varepsilon_K - \Sigma_1(K, \omega) = 0$, is satisfied for the case when the level with $\Sigma_2(K, \omega)$ 0 is small. The self energies $\Sigma_1(K, \omega)$ and $\Sigma_2(K, \omega)$ are calculated within the ranbow approximation [149]. The main difference between phonons and the envisaged collective excitation is the following. The phonon anomalies in conventional low temperature superconductors have their signature in the energy axis. The collective excitations, on the other hand, have their characteristic features both in the momentum and the energy spread. They find that the quasiparticle excitations spectrum can not be resolved from the higher energy loss features in the normal state. However, below T_c it becomes resolution limited sharp peak with a dip separating it from the loss features. The striking fact is that in the underdoped samples the gap (pseudogap) already opens in the normal state.

Controzzi [151] have made a study of simple model of strongly correlated electrons in an anisotropic 3D lattice. The main feature of the model is the interlayer tunnelling mechanism which has been represented as pure correlated hopping. Within the BCS mean field treatment, this model shows superconductivity with strongly density-dependent transition temperature with a maximum at low doping. The opening of the gap in the excitation spectrum makes the state stable with respect to direct electron hopping between the planes.

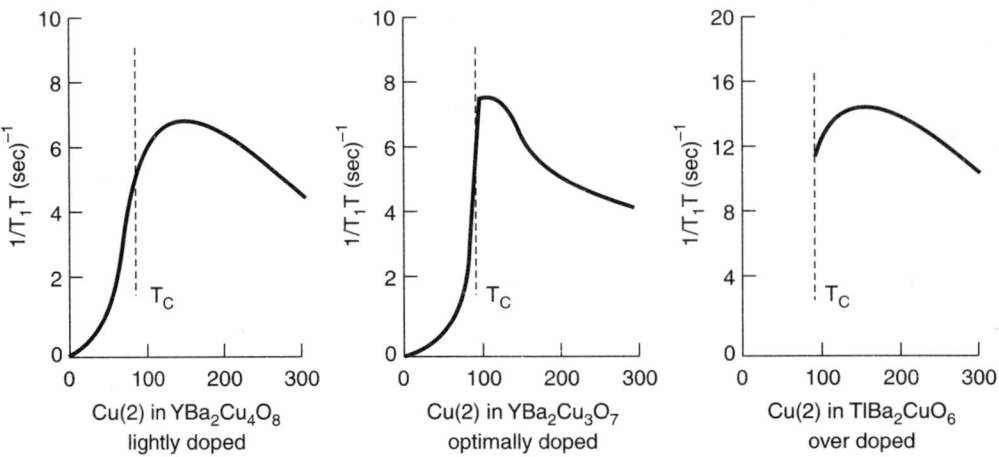

Fig. 8.8 Temperature dependence of $1/T_1T$ for Cu (2) sites in Cu O_2 planes by NMR [150].

8.10 INTER LAYER AND INTRALAYER EFFECTS IN HTSC CUPRATES

We have read that all the HTSC cuprates possess layered structure and there are one or more Cu O_2 planes separated by charge reservoir blocks. In the thallium and bismuth families, one, two or three of these Cu-O layers are sandwiched between layers of thallium or bismuth oxide and the superconducting transition temperature (T_c) increases with the number of layers in the sandwich upto three layers. Several authors [152 – 155] have studied the influence of the layered structure on the unusual properties of these HTSC cuprates and proposed that the interlayer interactions play a significant role in enhancing T_c. These calculations are mainly based on one-band two layer and two-band two-layer within the Hubbard tight binding Hamiltonian. Hofmann et al. [153], Tesanovic [152] and Bishop et al. [156] have used Green's function technique within the mean field approximation, while Raine and Mattews [155] have used ingenious transformation of operators within Hubbard-III decoupling scheme [157]. These authors consider intra and interlayer pairing and calculated the transition temperature T_c. They have shown the importance of interlayer interaction in establishing the superconducting state and long range order in HTSC cuprates. However, all these authors in their calculations have neglected the excitonic type interlayer correlations. Recently Ajay and Tripathi [158], and Kakani and coworkers [159] have studied separately the effects of inter and intra layer interactions on HTSC cuprates. The model Hamiltonian considered by these authors is

$$H = H_{\text{intra}} + H_{\text{inter}} \quad \ldots(8.56)$$

where

$$H_{\text{intra}} = \sum_{r,K,\sigma} \varepsilon_K C^+_{r,K,\sigma} C_{r,K,\sigma} + U \sum_{r,K',K,q} C^+_{r,K+q\uparrow} C^+_{r,K'-q\downarrow} C_{r,K'\downarrow} C_{r,k\uparrow} \quad \ldots(8.56\,(a))$$

and

$$H_{\text{inter}} = (-)t \sum_{r,s,K,\sigma} C^+_{r,K,\sigma} C_{s,K,\sigma} + W \sum_{r,s,K,K'} C^+_{r,K+q,\sigma} C^+_{s,K'-q,\sigma'} C_{s,K',\sigma'} C_{r,K,\sigma} \quad \ldots(8.56(b))$$

where r and s are layer indices. For two layer systems, when $r = 1(2)$, then $s = 2(1)$.i $C^+_{K,\sigma}(C_{K\sigma})$ are the creation (annihilation) operators of the charge carriers in the Cu O_2 plane with wave vector **K** and spin σ. The first term in eq. 8.56(a) represents the energy of free charge carriers within the Cu O$_2$ plane, while the second describes the attractive interlayer interaction [153, 156]. The first term in eq. 8.56(b) is the direct hopping term represents the attractive interlayer interaction. W may contain contributions from exciton, plasmons, phonon, or any other proposed mechanism mediated interaction and direct Coulomb interaction between the charge carriers of different layers.

Following Green's function technique and equation of motion method, Patra and Tripathi [158] and Kakani and Kashyap [159] independently obtained the following expression for T_c,

$$\gamma_c = \frac{1}{4N} \sum_K [\tanh \overline{\epsilon}_{2K}/2k_B T_c - \tanh \overline{\epsilon}_{1K}/2k_B T_c] \qquad ...(8.57)$$

where γ_c is the excitonic type interlayer correlation, related with interlayer interaction W through the relation $\bar{t} = t + W_r \gamma_c$, $\overline{\epsilon}_{2K} = \epsilon_K + U<n_2> + W<n_1>$ and $\overline{\epsilon}_{1K} = \epsilon_K + U(n_1) + W<n_2>$,

where
$$<n_1> = \frac{1}{N} \sum_{K_1} <C^+_{1K_1\downarrow} C_{1K_1\downarrow}>$$

and
$$<n_2> = \frac{1}{N} \sum_{K,\sigma} <C^+_{2K\sigma} C_{2K\sigma}>$$

The numerical study (within one-band two layer model) exhibit that intra- and interlayer attractive interactions can enhance superconducting transition temperature and long range order in HTSC layered cuprates having two layers per unit cell [160, 161].

Kakani and Kashyap [159] have also made study of isotope effect, specific heat, density of states and critical field. They obtained the following expressions for specific heat, density of states, free energy and critical field for intra and inter layer interactions as:

Specific Heat

$$C^e_{s_1} = \frac{N(0)}{N} \left(\frac{1}{2k_B T^2}\right) \int \epsilon_K [(\omega + \epsilon_{2K}) \operatorname{sech}^2 (\beta E_{2K}/2) + (\omega + \epsilon_{1K}) \operatorname{sech}^2 (\beta E_{1K}) d\epsilon_K \quad ...(8.58)$$

$$C^e_{s_2} = \frac{N(0)}{N} \left(\frac{1}{2k_B T^2}\right) \int \epsilon_K [(\omega + \epsilon_{1K}) \operatorname{sech}^2 (\beta E_{1K}/2) - (\omega + \epsilon_{2K}) \operatorname{sech}^2 (\beta E_{2K}/2)] d\epsilon_K$$

$$...(8.59)$$

Density of States Function $N(\omega)/N(0)$

$$N_1(\omega)/N(0) = \frac{1}{2\pi}[\omega/(\omega^2 - \Delta_2^2)^{1/2} - \omega/(\omega^2 - \Delta_1^2)^{1/2}] \qquad ...(8.60(a))$$

$$N_2(\omega)/N(0) = \frac{1}{2\pi}[\omega/(\omega^2 - \Delta_1^2)^{1/2} - \omega/(\omega^2 - \Delta_2^2)^{1/2}] \qquad ...(8.60(b))$$

Free Energy $(F_{SN} = F_S - F_N)$

$$(2F_{SN}/V)_I = -\left[N(0)\frac{\Delta_1^2}{2} - 4N(0)/\beta \exp\left(\frac{-\beta\Delta}{2}\right)(2\pi \Delta_1 \beta^{-1})^{1/2} + \frac{4N(0)}{\beta}(1/\beta)(\pi^2/12)\right]$$

$$...(8.61(a))$$

$$(2F_{SN}/V)_{II} = -\left[N(0)\frac{\Delta_2^2}{2} - \frac{4N(0)}{\beta}\exp\left(\frac{-\beta\Delta}{2}\right)(2\pi\Delta_2\beta^{-1})^{1/2} + \frac{4N(0)}{\beta^2}\left(\frac{\pi^2}{12}\right)\right] \quad ...(8.61(b))$$

Critical Field

$$(H_c)_1 = \left[8\pi\left|\{-N(0)\Delta_1^2/4 - (4N(0)/\beta)\}(e^{-\beta\Delta_1/4})(2\pi\Delta_1\beta^{-1})^{1/2}\right.\right.$$

$$\left.\left.+\frac{4N(0)}{\beta}(1/2\beta)\frac{\pi^2}{12}\right\}\right|\right]^{1/2} \quad ...(8.62(a))$$

$$(H_c)_{II} = \left[8\pi\left|\{-N(0)\Delta_2^2/4 - (4N(0)/\beta)\}e^{-\beta\Delta_2/4}(2\pi\Delta_2\beta^{-1})^{1/2}\right.\right.$$

$$\left.\left.+4N(0)\beta\left(\frac{1}{2\beta}\right)\frac{\pi^2}{12}\right\}\right|\right]^{1/2} \quad ...(8.62(b))$$

These studies reveals to some extent that interlayer interactions play some role in establishing high T_c in cuprates. Within a one band two layer model (which can be extended to a multilayer system) the excitonic type interlayer correlations (γ_c) seems to be responsible to enhance T_c to some extent in the study of HTSC layered cuprates. Raine and Matthews [155] have shown that the present mean field calculations can be extended to two band-two-layer tight binding model. Baskaran found that the interlayer and intralayer mechanisms do not help each other and also dominate in different regime of doping.

8.11 BIPOLARONIC MODEL

The basic idea of bipolaronic superconductivity is the reconsideration of the electron–phonon interaction in the narrow band crystals. Anderson [163] and Chakravati et al. [164, 165] showed the occurrence of bipolarons in the limit of strong electron–phonon coupling ($g \geq 2.5$). The formation of the bipolarons in the non-metallic transition metal compounds were shown independently by Lakkis et al. [166], Chakravarty et al. [167] and Alexandrov and Ranninger [168] and this is in fact equivalent to the occurrence of the Cooper pairs. Alexandrov and Ranninger [169] showed that in a narrow band crystal the strong electron–phonon coupling gives rise to the bipolarons, which are equivalent under certain conditions to a superfluid charged Bose system. Alexandrov et al. [170] reconsidered the model, using the observation that the strong coupling condition

$$g\,N(0) > 1 \quad ...(8.63)$$

is similar to that for small polaron formation

$$\frac{2z\,g^2\,\omega}{D} > 1 \quad ...(8.64)$$

where $2z\,g^2\,\omega$ is the effective phonon-mediated attraction and $N(0) \sim D^{-1}$ is the density of states at the Fermi level, D is the band width, ω is the characteristic phonon frequency in the system and z is the number of nearest neighbours. Following the condition (8.64), the electronic band width D is drastically reduced and the narrow polaronic bandwidth is obtained as

$$W = D\exp(-g^2) \quad ...(8.65)$$

when the attractive interaction between two polarons is stronger than the Coulomb repulsion, a condensed state appears. The properties of this state being determined by the value of the bipolaron binding energy

$$A \cong 2\epsilon_p - U \qquad ...(8.66)$$

where U is coulomb replusion and $\epsilon_p = g^2 \omega$.

When the polaron–polaron interaction is attractive but small, i.e. $A \ll W$, the small polarons form spatially overlapping Cooper pairs with superconducting properties similar to the usual BCS superconductivity. The critical temperature has not exactly the same expression because

$$\omega \gtrsim \epsilon_0 - W \qquad ...(8.67)$$

which is in fact a violation of the adiabatic limit.

Alexandrov et al. [170] considered another superconducting state called "bipolaronic superconductivity". This is supposed to appear due to the existence of local pairs of fermionic carriers which are bound together by an attractive mechanism. One can describe a narrow band superconductor in the site representation by the Hamiltonian

$$H = \sum_{i,i'} T_{i,i'} C_i^+ C_{i'} + \sum_{i,q} [g_i(q)(C_i^+ C_i b_q + h.c.)] + U \sum_{i,i'} C_i^+ C_{i'}^+ C_{i'} C_i$$

$$+ \sum_q \omega(q) b_q^+ b_q \qquad ...(8.68)$$

where $T_{i,i'}$ denotes the hopping integral, $g_i(q)$ the electron-phonon coupling, $U_{i,i'}$ the Coulomb interaction, and $\omega(q)$ the phonon energy. Making use of Lang and Firsov [171] transform S_1, one obtains

$$H_p = \hat{S}_1 H \hat{S}_1^{-1} = H_0 + H_1 \qquad ...(8.69)$$

where $\hat{S}_1 = \exp\left[\sum_{iq} (\omega^{-1}(q) C_i^+ C_i b_q g(q) + h.c.\right]$

H_0 and H_1 in eq. (8.69) are given by

$$H_0 = \sum_i (T_i - \epsilon(p)) C_i^+ C_i + \sum_{i,i'} V_{i,i'} C_i^+ C_{i'}^+ C_{i'} C_i + H_{ph} \qquad ...(8.70)$$

and

$$H_1 = \sum_{i,i'} \sigma_{i,i'} C_i^+ C_{i'} \qquad ...(8.71)$$

where H_{ph} is the last term from eq. (8.68) and

$$\epsilon_p = \sum_q \omega^{-1}(q) |g(q)|^2 \qquad ...(8.72)$$

$$V_{i,i'} = U_{i,i'} - \sum_q \frac{1}{\omega(q)} |g(q)|^2 \exp[i\mathbf{q}\cdot(\mathbf{i}-\mathbf{i}')] \qquad ...(8.73)$$

The energy ϵ_p is called as the polaronic level shift, $V_{i,i'}$ is the polaron–polaron interaction (which has to be attractive for the superconducting state) and $\sigma_{i,i'}$ is the kinetic energy of the small polaron. Instead of eqs. (8.72) and (8.73), if we use the thermal average values (with the phonon density matrix) the energy $\sigma_{i,i'}$ comes out to be

$$\sigma_{i,i'} = T_{i,i'} \exp(-g^2) \qquad \text{...(8.74)}$$

$$g^2 = \sum_q \omega^{-2}(q) \cot g \frac{\omega(q)}{2T} |g(q)|^2 [1 - \cos \mathbf{q} \cdot (\mathbf{i} - \mathbf{i}')] \qquad \text{...(8.75)}$$

For describing the superconducting state, one has to consider the interaction $V_{i,i'}$. Alexandrov and Ranninger [169] considered the case $|V_{i,i'}| \gg \sigma_{i,i'}$ in which the bipolarons, in the low density limit, have an excitation spectrum as a **quantum superfluid**.

There is also another possibility to obtain superconducting state, i.e. by considering the properties of strongly coupled polarons. Alexandrov et al. [170] considered $\sigma|v|$ as a small perturbation and eliminated the single polaronic state from eq. (8.69). Now, defining a new transformation S_2 by

$$<f| S_2 |p> = \sum_{i,i'} <f|\hat{\sigma}_{i,i'} C_i^+ C_{i'} |p> (E_f - E_p)^{-1} \qquad \text{...(8.76)}$$

Applying (8.76) to the Hamiltonian (8.69).

$$\tilde{H} = \exp(S_2) H_p \exp(-S_2) \qquad \text{...(8.77)}$$

One obtains

$$\tilde{H} = \sum_{m=m'} [\tilde{v}(m-m') b_m^+ b_m b_{m'}^+ b_{m'} - t(m-m') b_m^+ b_{m'}] \qquad \text{...(8.78)}$$

In the limit $A \gg \omega$, one obtains $t(m-m')$ and $\tilde{v}(m-m')$ as

$$t(m-m') = (2T_{m,m}^2/A) \exp(-g^2) \qquad \text{...(8.79)}$$

$$\tilde{v}(m-m') = 4v(m-m') + 2T_{m,m}^2/\Delta \qquad \text{...(8.80)}$$

and in the opposite limit $T \leq \Delta \leq \omega$, one obtains the above parameter as

$$t(m-m') = (2T_{m,m}^2/A) \exp(-2g^2) \qquad \text{...(8.81)}$$

$$\tilde{v}(m-m') = 4v(m-m') + (2T_{m,m}^2/A) \exp(-g^2) \qquad \text{...(8.82)}$$

The above results reveal that the effective bipolaronic mass and the effective interaction are strongly dependent on the ratio A/ω and both of them increase rapidly with increasing A/ω. Using the pseudospins,

$$S_m^z = \frac{1}{2} - b_m^+ b_m, \quad S_m^x = \left(\frac{1}{2} b_m + b_m^+\right)$$

$$S_m^y = \frac{1}{2}(b_m - b_m^+)$$

The transformed Hamiltonian (8.78) reads as

$$\tilde{H} = \sum_m S_m^z [\mu + \sum_{m'=m} \tilde{v}(m-m') S_{m'}^z] + \sum_{m \neq m'} t(m-m') [S_m^z S_{m'}^z + S_m^y S_{m'}^y] \qquad \text{...(8.83)}$$

where μ is the chemical potential of bipolarons. This can be determined from the condition of conservation of the number of bipolarons

$$\frac{1}{N} \sum_m <S_m^z> = \frac{1}{2} - n \qquad \text{...(8.84)}$$

Alexandrov and Ranninger [169] analysed this model within the mean field approximation and obtained the energy of excitations, a gap proportional to t, which gives an incorrect result leading to an exponential dependence of the specific heat. We must note that the true excitation spectrum consists of pseudomagnons with a gapless dispersion. Alexandrov et al. [170] have shown that by taking into consideration the quantum fluctuations, one can perform the thermodynamics of this system in the random phase approximation.

Defining the order parameter for the superconducting state by

$$S^x = <S^x_m> = \frac{1}{2}[R^2 - (2n-1)^2]^{1/2} \quad ...(8.85)$$

One obtains R by the self-consistent equation

$$\frac{1}{R} = \frac{1}{N}\sum_N \frac{1}{\omega(\mathbf{K})} \coth\frac{\omega(\mathbf{K})}{2T} \quad ...(8.86)$$

where the excitation spectrum is given by

$$\omega(\mathbf{K}) = R[t - t_K \cos^2\theta + \tilde{v}_K \sin^2\theta)(t - t_K)]^{1/2} \quad ...(8.87)$$

with
$$\cos\theta = (2n-1)R \quad ...(8.88)$$

Here t_K and $\tilde{v}_K$ being the Fourier transforms of $t(m)$ and $v(m)$. Making use of the condition $S^z = 0$, one can obtain T_c, the critical temperature from eq. (8.86) by the equation

$$\frac{1}{2n-1} = \frac{1}{N}\sum_K \coth[(2n-1)(t - t_K)/2T_c] \quad ...(8.89)$$

In the case of dilute system $|n| \ll 1$, one obtains

$$T_c = \frac{3.31(na^{-3})^{3/2}}{m^*}(1 - 0.54\,n^{2/3}) \quad ...(8.90)$$

where $m = (ta^2)^{-1}$ is the effective bipolaronic mass in a lattice with a constant 'a'. In the high density limits, $|2n-1| \ll 1$, one obtains from eq. (8.89)

$$T_c \cong \frac{t}{2}\left[c^{-1} - \frac{(2n-1)^2}{3}\right] \quad ...(8.91)$$

where c is a constant which depends on the lattice type. We note that this model is equivalent to the negative U Hubbard Hamiltonian but with a temperature-dependent narrow band W. However, the phenomenological negative U Hubbard Hamiltonian is applicable to polaronic system only in the limit $\omega \gg U$ and with D being replaced by a temperature-dependent W. We find that in the limit $\omega \ll \Delta$, the bipolaronic Hamiltonian can be parametrized by an extended negative-U Hubbard model in the strong coupling limit.

Alexandrov and Edwards [172] argued that HTSC in cuprates derives from the Bose-Einstein condensation (BEC) of small (bi) polarons in the electronically active Cu O_2 planes. Alexandrov [173] have shown that bipolaronic theory of superconductivity provides a parameter-free fit of the superconducting critical temperature and the upper critical field of cuprates. The theory describes their non-Fermi liquid normal state, including the in plane and out of plane resistivity, Hall effect, magnetic susceptibility [174]. Alexandrov [175] have further shown that two distinct energy scales, the d-wave superconducting order parameter and **charge segregation** in the form of **stripes** in cuprates are unified as a result of the formation of mobile bipolarons in the normal state and their BEC. Alexandrov [175] have also shown that both the d-wave superconducting order parameter and the striped charged distribution result from the bipolaron (centre of mass) energy band dispersion rather than from any particular interaction. The expression obtained for T_c on the basis of bipolaron condensation mechanism by Alexandrov [176] and Alexandrov and Kabanov [177] reads as

$$T_c = 1.64 \left(\frac{e R_H}{\lambda_{ab}^4 \lambda_c^2} \right)^{1/3} \qquad ...(8.92)$$

where R_H is in-plane Hall-coefficient given by

$$R_H = \frac{1}{2e \, n_B} \frac{4 m_x m_y}{(m_x + m_y)^2} \qquad ...(8.93)$$

Here m is the bare band mass, n_B is the bipolaron density and λ_{ab} and λ_c are in plane and out-of-plane penetration depths expressed as

$$\lambda_{ab} = \left[\frac{m_x m_y}{8\pi \, n_B \, e^2 (m_x + m_y)} \right]^{1/2}$$

and
$$\lambda_c = \left[\frac{m_c}{16 \pi \, n_B \, e^2} \right]^{1/2} \qquad ...(8.94)$$

Very recently, Ashkenazi [178 – 180] by considering the quasiparticles of HTSC cuprates remarked that these consist of : **polaron like 'stripons'** carrying charge and associated primarily with large-U orbitals in stripe-like inhomogeneities, 'quasi-electrons' carrying charge and spin, and associated with hybridized small-U and large-U orbitals, and 'svivons' carrying spin and lattice distortion. Ashkenazi [178] has shown that this electronic structure leads to the systematic behaviour of spectroscopic and transport properties of the HTSC cuprates and also the occurrence of a normal state pseudogap. High-T_c pairing results from transitions between pair states of stripons and quasi-electrons through the exchange of svivons.

A model of coexisting bipolarons, free electrons, and BCS pairs was proposed by Ranninger and Robaskiewicz [181] for the intermediate electron-phonon coupling. The polaronic model has been reconsidered by Nasu [182] on the basis of an attractive analysis of the electron phonon coupling. The effects of the electron–phonon coupling are in fact determined by the magnitude of the phonon energy $\omega(q)$, the electron-phonon coupling g, the transfer energy t. In the case of $g \ll t$, excitations appear in the system, called as **large polarons**. These are in fact electrons interacting with the phonon cloud, which is very thin but extended. If $g \gg t$, the excitations become **small polaron excitations** consisting of electrons with phononic cloud in the same site, and with a considerable mass enhancement. As the energy $\omega(q)$ increases, the difference between the large and small polarons disappears because of the quantum phononic effects. If $\omega \gg g$, then the phonon can follow the motion of the electron and this situation is in fact the inverse adiabatic limit in which some molecular crystals are expected to appear. A phase diagram given by Nasu [182] as a function of t, g and $\omega(q)$ shows the existence of the superconducting region and a bipolaronic insulator region. In the mean-field approximation, the superconducting gap is maximum for ω-t. In the inverse adiabatic limit, there is a region where the small polarons dominate and the superconducting state becomes more stable than the **Charge-Density State** (CDS) because the retardation effect is absent. The collective excitations of the superconducting state change their nature continuously from the BCS pair breaking excitations to the superfluid type, with the increase of g/t.

Alexandrov and Ranninger [183] obtained the angle resolved photoemission spectrum (ARPES) for a system of small polarons in the normal and in a BCS-like superconducting state. Interpretations of ARPES are based exclusively on the electron correlation approach to HTSC. These authors [183] have shown that ARPES in HTSC cuprates may be compatible with the picture of small polarons as intrinsic charge carriers in these materials.

Emin [184, 185] has attempted to demonstrate that bipolaronic superconductivity due to 'small' bipolarons can not exist. According to Emin, bipolaronic superconductivity could exist however, if the bipolarons were 'large' bipolarons. Alexandrov and Ranninger [186] have shown that Emin's conjectures are inconsistent on a theoretical ground and moreover are not borne out by the experimental findings.

8.12 LOCHON (BOSON) FERMION MODEL

The pairing interaction mediated by lochons (local charged bosons or local singlet pairs or bipolarons) has a long history. It was first suggested by Ganguly et al. [187], for the pairing of fermions in conventional superconductors to achieve the enhancement of superconducting transition temperature. The model envisages the existence of centres or complexes which can harbour lochons (local bosons, pairs or sometimes refers to as negative U centres) which can undergo double charge fluctuation in interaction with fermions of the metallic matrix of the system. The interaction mechanism involves the splitting of a lochon (local boson) into a pair of fermions (belonging to a wide band) and the inverse process in which there is a confluence of a fremion pair to give the localized boson. The discovery of HTSC cuprates led to the revival of this mechanism by several groups [2, 5, 188 – 192].

Formalism for the Combined Lochonic and Phononic Mechanism

The Hamiltonian for the combined mechanism in the Nambu Spinor representation is given by [188, 193 – 194]

$$H = H_0 + H_{cp} + H_{cl} \qquad ...(8.95)$$

where

$$H_0 = \sum \varepsilon_K \psi_K^+ \sigma_z \psi_K + \sum \omega_q a_q^+ a_q + \sum E_l b_l^+ b_l \qquad ...(8.95(a))$$

$$H_{cp} = \sum_{q, K} P_q \phi_q \psi_K^+ \sigma_z \psi_{K'} \qquad ...(8.95(b))$$

$$H_{cl} = \sum_{K, l} B_l \psi_K^+ [\sigma_+ b_l + \sigma_- b_l^+] \psi_K \qquad ...(8.95(c))$$

In the above relations, the carrier (fermion) field operators are denoted as [195, 196] :

$$\psi_K = \begin{pmatrix} C_{K\uparrow} \\ C_{K\downarrow} \end{pmatrix}, \psi_K^+ = (C_{K\uparrow}^+ \; C_{-K\downarrow}) \qquad ...(8.96)$$

$C_{K\sigma}(C_{K\sigma}^+)$ being the fermion (hole) annihilation (creation) operators in the state $|K \sigma >$, K is wave vector, σ spin index and ε_K is the corresponding single particle energy. Further $\phi_q = (a_q + a_{-q}^+)$ is the phonon field operator, $a_q^+(a_q)$ being the phonon creation (annihilation) operator corresponding to the energy ω_q; P_q is the coupling constant involving carrier-phonon interaction. The entities

$$b_l^+ = C_{l\uparrow}^+ C_{l\downarrow}^+, \; b_l = C_{l\downarrow} C_{-l\uparrow} \qquad ...(8.97)$$

are lochon creation and annihilation operators. From their structure, it is clear that they are composites of fermion's in the singlet spin state and localized in the orbital state $|l>$ at site l, E_l represents lochon energy and B_l is the coupling constant for carrier-lochon interaction. Further $\sigma_{\pm}(\sigma_x \pm i\sigma_y)/2$ and σ_x, σ_y, σ_z are Pauli matrices. In the interaction processes discussed above both charge and spin are conserved although the number of free fermions and localized fermions is not independently conserved. The total number for each spin is conserved. Thus we must have

$$n_\sigma(\text{total}) = \sum_K n_{K\sigma} + \sum_l n_{l\sigma}$$

where
$$n_{K\sigma} = C_{K\sigma}^+ C_{K\sigma} \; ; n_{l\sigma} = C_{l\sigma}^+ C_{l\sigma}$$

These considerations demand that there be a common chemical potential (μ) for the itinerant and localized charge carriers [190, 191]. While the fermions have the ususal anti-commutation relations, for lochons the relations are :

$$[b_l, b_{l'}^+] = [1 - (n_{i\uparrow} + n_{i\downarrow})] \delta_{ll'} \qquad ...(8.97(a))$$

$$[b_l, b_{l'}] = [b_l^+ b_{l'}^+] = 0 \qquad ...(8.97(b))$$

The lochon operators satisfy the additional relations $b_l^{+2} = b_l^2 = 0$, a feature of the exclusion principles.

Superconducting Transition Temperature (T_c)

The order parameter in the temperature Green Function formalism is expressed as [193, 197]

$$Z\Delta(\omega_n, K) = T/(2\pi)^d \left[\sum_{\omega'_n} dK' T(i(\omega_n - \omega_n')) (K - K') F^+(i\omega_n', K') \right] \qquad ...(8.98)$$

where $\omega_n = 2\pi n T$ for phonons and lochons and $\omega_n = (2n + 1)\pi T$ for fermions, d is the dimensionality and T is the temperature (and k_B and $\hbar$ are taken as unity). With this definition the anomalous Green's function [195]

$$F^+(i\omega_n, K) = \Delta(i\omega_n, K)/[(\omega_n^2, + \varepsilon^2 + \Delta^2(i\omega_n, K)] \qquad ...(8.99)$$

In contradiction to Kresin [197], where two dimensional plasmon modes are considered in addition to strong electron–phonon coupling, in our formulation the vertex funciton T is a sum of the phononic and lochonic parts

$$T = T_{ph} + T_{lo} \qquad \text{...(8.100)}$$

with
$$T_{ph} = \lambda_p D_{ph} [i(\omega_n - \omega_n'); K - K'] \qquad \text{...(8.101)}$$

λ_p being the dimensionless carrier-phonon coupling constant and the phonon Green's function [195].

$$D_{ph} = \omega^2 / [(\omega_n - \omega_n')^2 + \omega^2] \qquad \text{...(8.102)}$$

Similarly,
$$T_{lo} = \lambda_l D_{lo} [i(\omega_n - \omega_n')] \qquad \text{...(8.103)}$$

where λ_l is the carrier-lochon coupling constant and the lochon Green function

$$D_{lo} = 1/[i(\omega_n - \omega_n') - E_l] \qquad \text{...(8.104)}$$

The Coulomb repulsion has not been explicitly written in the above; its effects is incorporated further. Similarly, the explicit form of the renormalization function has not been written. This is treated as a renormalization parameter with the form

$$Z = 1 + \lambda_p \qquad \text{...(8.105)}$$

or simply $Z = 1$, as the phonon coupling is considered to be weak.

The solution of eq. (8.98) near $T = T_c$ after the usual contour integration and with two cutoffs corresponding to phonon and lochon modes turns out to be

$$1 = \lambda_p \ln(1.13\, \omega_p/T_c) + \lambda_l \ln(1.13\, \omega_l/T_c) \qquad \text{...(8.106)}$$

Here ω_p and ω_l are the Lorentzian cutoffs (in temperature units) for the phonon and lochon mechanism respectively. Thus in the combined mechanism, we have for $Z = 1$,

$$T_c = 1.13(\omega_p)^{r_p} (\omega_l)^{r_l} \exp[-1/(\lambda_p + \lambda_l)] \qquad \text{...(8.107)}$$

where
$$r_p = \lambda_p / [\lambda_p + \lambda_l],\ r_l = \lambda_l / [\lambda_p + \lambda_l] \qquad \text{...(8.108)}$$

$$r_p + r_l = 1,\ \lambda_l = \overline{\lambda}_l - \overline{\mu}_c \qquad \text{...(8.09)}$$

$\overline{\mu}_c$ is the dimensionless coupling constant associated with Coulomb repulsion between the free carriers. The carriers in HTSC cuprates are subjected to strong correlation in the conducting planes. Thus one has to project out the probability of two carriers starting from the same site or ending at the same site in the CuO_2 planes. Accordingly $\overline{\mu}_c$, with this restriction, will have a very small value. The coupling constant for lochon-induced interaction will depend on the concentration of carriers as well as lochons. As they are not independently conserved, this dependence (eq. 8.95(c)) will have the form $f(x_c) = x_c(1 - x_c)$ where x_c is the carrier concentration. Thus

$$\overline{\lambda}_l = \overset{\circ}{\lambda}\, x_c(1 - x_c) \qquad \text{...(8.110)}$$

$\overline{\lambda}_l$ has the maximum value at $x_c = 1/2$. It will be denoted by λ_l^m. If the formation of lochons (local pairs) is caused by interaction with some specific optical phonons or excitons, there will be a reduction factor $\exp(-g^2)$ arising from the polaronic modulation of mixing between wide band and localized state. It is temperature independent for $T < \omega_0$, ω_0 being the frequency of the boson model involved. Thus

$$\overline{\overset{\circ}{\lambda}}_l = \overline{\overset{\circ}{\lambda}}_l \exp(-g^2) \qquad \qquad ...(8.111)$$

Further for normalised dimensionless from x_c will range from 0 to 1. Owing to the dependence of λ_l on x_c as occurs in eq. (8.110), the critical temperature T_c, which follows from eq. (8.107), will vary with x_c. It will have a maximum value for certain concentration and will decrease for other ranges. This fact also explains why T_c goes down on overdoping when the system becomes a better metal. The dependence of T_c on carrier concentration and its peaking at a certain value of x_c may be further modulated if some remnants of the van Hove singularity in the densities of states $N(\varepsilon)$ remain for the quasi two dimensional layers. Also, the renormalization of lochon energy from doping may produce some asymmetry in the T_c versus x_c curve. Fig. 8.9 shows the variation of T_c against x_c for various HTSC cuprates in the symmetrical case.

Superconducting Gap at $T = 0$

At $T = 0$, the gap $\Delta(0)$ is given by

$$\Delta(0) = \lambda_p \int_0^{\omega_p} d\varepsilon \, \Delta_p(0)/[\varepsilon^2 + \Delta_p(0)]^{1/2} + \lambda_l \int d\varepsilon \, \Delta_l(0)/[\varepsilon^2 + \Delta_l^2(0)]^{1/2} \qquad ...(8.112)$$

where $\Delta_p(0)$ and $\Delta_l(0)$ represents the gap for the pure phonon and lochon parts. Then the gap for the combined mechanism turns out to be

$$\Delta(0) = \Delta_p(0) \ln[\omega_p + (\omega_p^2 + \Delta_p^2(0))^{1/2}/\Delta_p(0)]^{\lambda_p}$$

$$+ \Delta_l(0) \ln[\omega_l + (\omega_l^2 + \Delta_l^2(0))^{1/2}/\Delta_l(0)]^{\lambda_l}$$

$$...(8.113)$$

Using eqs. (8.107) and (8.113), one obtains the ratio

$$2\Delta(0)/T_c = 1.75[\{\Delta_p(0) \ln [\omega_p + (\omega_p^2 + \Delta_p^2(0))^{1/2}/\Delta_p(0)]^{\lambda_p}$$

$$+ \{\Delta_l(0) \ln [\omega_l + (\omega_l^2 + \Delta_l^2(0))^{1/2}/\Delta_l(0)\}]^{\lambda_l}$$

$$\times \{[\exp[1/(\lambda_p + \lambda_l)]/\omega_p^{r_p} \, \omega_l^\eta]\} \qquad ...(8.114)$$

It is clear from the relation (8.114), that there is no universal value of $2\Delta(0)/T_c = 3.52$ as in the BCS theory [198]. The ratio may range from 3 to 6 depending on the values of the parameters involved for each system. This is in general agreement with the results for the cuprates.

The superconducting condensation energy is found from the above as

$$W_N - W_S = \frac{1}{2} N_c(0) \Delta^2(0) \qquad \qquad ...(8.115)$$

where $\Delta(0)$ is given by (110) and $N_c(0)$ is the density of states per unit cell at the Fermi surface.

For cuprates, the estimated values are reported to be an order of magnitude larger than the BCS value [199]. There is uncertainty in extracting values from experimental results which are quoted in the range 9 to 36 K [199].

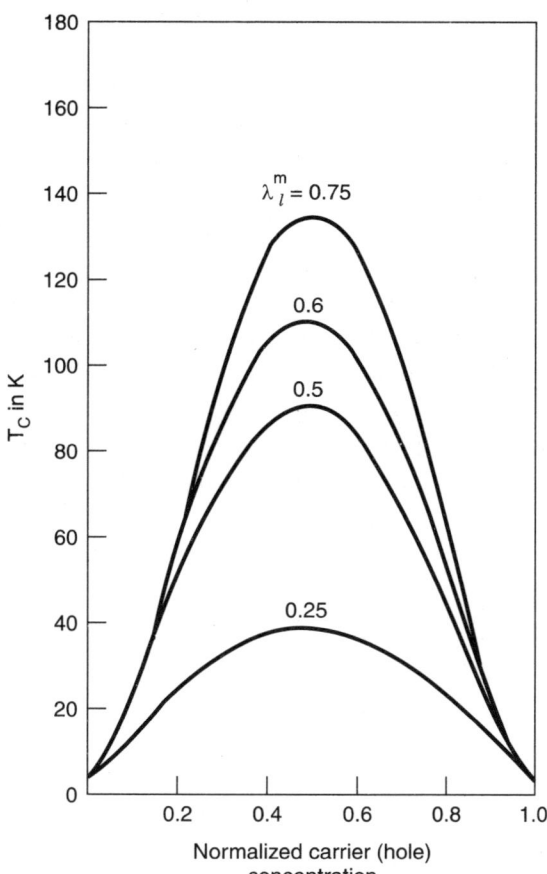

Fig. 8.9 Transition temperature, T_c versus normalized carrier concentration x_c. The value of $\lambda_p = 0.25$, $\omega_{pl} = 100$ K and $\omega_l = 500$ K for all values of the curves, λ_l^m values for each curve is shown for each bell-shaped curve at the maximum.

Variation of Isotope Effect on Doping

Initial experiments on cuprates showed that the isotope effect is very small at the maximum T_c value of each system. Measurements of oxygen isotope effect in these systems revealed that on decreasing T_c by appropriate doping the isotope exponent tends towards larger values. The carrier concentration when one moves from T_c^{max} for x_c^{max} towards lower values of x_c. We now consider the expression for T_c given by eq. (8.107). We may note that

$$\omega_p = \text{constant } M_0^{-1/2} \quad \ldots(8.116\,(a))$$
$$g^2 = \text{constant } M_0^{-1/2} \quad \ldots(8.116\,(b))$$

where M_0 is the oxygen mass in dimensionless unit. Then the oxygen exponent turns out to be [193]

$$\alpha_0 = -\delta \ln T_c / \delta \ln M_0 \quad \ldots(8.117)$$
$$= \frac{1}{2} \left[\lambda_p/(\lambda_p + \lambda_l) + g^2 \ln(\omega_l/\omega_p) \{\lambda_p \lambda_l/(\lambda_p + \lambda_l)^2\} + g^2 \{\lambda_l/(\lambda_p + \lambda_l)^2\} \right] \quad \ldots(8.117)$$

From the above, it is clear that when lochon part dominates and the T_c value is near the maximum, the isotope exponent is small. With λ_l decreasing α_0 tends towards larger values and for $\lambda_l = 0$, $\alpha_0 = 1/2$. The entire expression (8.118) is capable of positive (even $\alpha_0 > 1/2$) and negative values depending on the values of the parameter involved. The variation of α_0 with x_c (and in turn on T_c) is controlled by the dependence of λ_l given by eq. (8.110) and the increase of g^2 with decreasing x_c. Based on eq. (8.110), the variation of α_0 against (T_c/T_c^{max}) is shown in Fig. 10. The oxygen isotope exponent increases as the transition decreases (with decreasing x_c). It tends to the value 0.5 when the lochon part becomes insignificant and only phonon contribution remains. The behaviour shown in Fig. 8.10 is not universal but depends on the values of the parameter chosen[193]. The one shown corresponds to doped YBCO system, for example $YBa_{2-x}La_xCu_3O_7$ [200]. The general trend for other cuprates will be similar except for variations arising from changed values of the parameter for each cuprate system. However, if there are structural changes, bulk or local arrangements of atoms at some doping region, there may be concomitant sudden changes in the parameters.

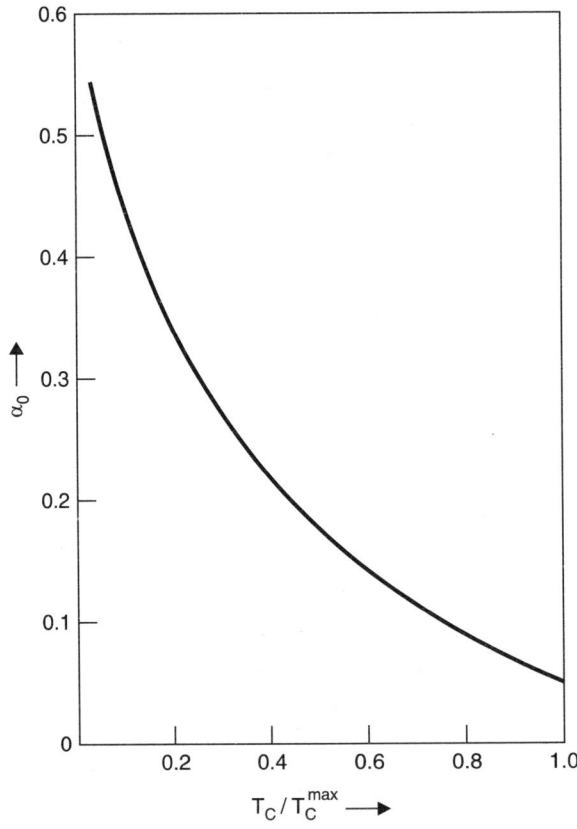

Fig. 8.10 The oxygen isotope exponent α_0 as a function of T_c/T_c^{max}, T_c in turn depends on carrier concentration. The figure corresponds to $\lambda_l^{max} = 0.5$ and $g^2(x_c) = 0.03, 0.0375, 0.05, 0.075, 0.15$ for x_c ranging from 0.5, 0.4, 0.3, 0.1. Other parameters have the same values as given in the caption of Fig. 8.9.

Pairing Symmetry in Lochon Model

Considering a system having single conducting layer per unit cell, besides the adjoining dielectric (semiconducting) layer such as $Tl_2 Ba_2 Cu O_{6+\delta}$, authors stated that both s- and d-wave pairing is possible for layered systems, e.g. cuprates [201 – 203]. Experimental measurement towards the symmetry of the gap parameter does indicate a superposition of s- and d-waves [204]. It is important to assess the relative values of the corresponding critical temperatures. There is a danger of overestimating one symmetry relative to the other if the values of the parameter involved are not reasonably chosen [205].

For a two-dimensional plane with square lattice configuration of atomic distribution in $Cu O_2$ planes, the coupling constant involving fermion-lochon interaction will have the general form [201, 203].

$$B_l(k) = \phi_k B \qquad \ldots(8.119)$$

where
$$\phi_k = m_s \phi_{sk} + m_d \phi_{dk} \qquad \ldots(8.120)$$

with

$$\phi_{sk} = [\cos(K_x a) + \cos(K_y a)]$$
$$\phi_{dk} = [\cos(K_x a) - \cos(K_y a)] \qquad \ldots(8.121)$$

where a is lattice constant of the square. For normalised ϕ_k the mixing coefficients must satisfy

$$|m_s|^2 + |m_d|^2 = 1 \qquad \ldots(8.122)$$

where ϕ_{sk} and ϕ_{dk} belong to different irreducible representations of the square symmetry.

Making use of a suitable canonical transformation the lochon induced pairing interaction takes the form K [189, 190, 206].

$$H_{lK} = \sum V_l(K, K') C^+_{K'\uparrow} C^+_{-K'\downarrow} C_{-K\downarrow} C_{K\uparrow} \qquad \ldots(8.123)$$

where
$$V_l(K, K') = V_l(\phi_k, \phi_{k'}) \qquad \ldots(8.124)$$

for ϕ_k taken to be real; $V_l = (B^2/E_l)$

The equation for the superconducting gap $\Delta(K)$ has the well known form

$$\Delta(K) = \sum_{K'} V_l(K, K') \Delta(K') \{\tan h\, (\beta E_{K'/2})/2\varepsilon_{K'}\} \theta(\omega_l - |\varepsilon_{K'}|) \qquad \ldots(8.125)$$

where $E_{K'} = [\varepsilon^2_{K'} + \Delta^2(K')]^{1/2}$; $\omega_l \rightarrow$ cutoff frequency. By writing $\Delta(K) = \phi_k \Delta$, the K dependence of the gap function is transferred to ϕ_k. Thus we get the relation for the sum of the gap equation as

$$1 = V_l \sum_{K,l} (m_s^2 \phi_{sK}^2 + m_d^2 \phi_{dK}^2) \{\tan h\, (\beta E_K/2\varepsilon_K)\} \theta(\omega_l - |\varepsilon_K|) \qquad \ldots(8.126)$$

At the critical temperature, the gap parameter vanishes and one have to evaluate

$$1 = V_l \sum_l (m_s^2 \phi_{sK}^2 + m_d^2 \phi_{dK}^2) \{\tan h\, (\beta \varepsilon_K/2\varepsilon_K)\} \theta(\omega_l - |\varepsilon_K|) \qquad \ldots(8.127)$$

For the evaluation of eq. (8.127), one requires the density of states (DOS) for a two-dimensional square lattice at the Fermi energy $N(\varepsilon_F)$.

Considering the nearest neighbour (NN) hopping of carriers only, the two-dimensional dispersion relation is given by

$$\varepsilon_K = -2t[\cos(K_x a) + \cos(K_y a)] \qquad ...(8.128)$$

where t is the NN hopping integral, and a the lattice constant of the square. The bandwidth is $8t$. The DOS for the above dispersion contains the van Hove singularity near half filling which is close to the Fermi energy. According to Mohan [207] the van Hove singularity is not responsible for the onset of superconductivity in cuprates although it may influence the coupling constant to a certain measure.

Wilson [208], in a analysis of Seeback data of cuprates has suggested that the negative U-model (the same as lochon model) will have singularity like features at the Fermi energy very similar to van Hove singularity [209, 210]. Thus one can replace van Hove singularity by "negative-U singularity". Without going into the case of this singularity like feature, the following density of states has been used in this region

$$N(\varepsilon_F) = (2/\pi D) \ln |(D/\varepsilon_F)| \qquad ...(8.129)$$

where D is the width of the singularity at the Fermi energy. The cosine functions in eq. (8.121) can be expanded in their arguments. Retaining terms upto second order only and evaluating at the Fermi energy with the constraint $\delta(\phi_{sK} - |\varepsilon_f|/2t)$, one obtains

$$(\phi_{sK}^2) = (|\varepsilon_F|/2t)^2 \qquad ...(8.130)$$

$$(\phi_{dK}^2) = [4 - 4(|\varepsilon_F|/2t) + (|\varepsilon_F|/2t)^2] \qquad ...(8.131)$$

The transition temperature T_{cs} and T_{cd} for the s-wave and d-wave components are obtained as

$$T_{cs} = 1.13\, \omega_l \exp[-1/[V_l(|\varepsilon_F|^2/4t^2)((2/\pi D)\ln D/(|\varepsilon_F|))] \qquad ...(8.132)$$

$$T_{cd} = 1.13\, \omega_l \exp[-1/[V_l(4 - 4(|\varepsilon_F|/2t) + ((|\varepsilon_F|/2t^2))((2/\pi D)\ln D/(|\varepsilon_F|))] \qquad ...(8.133)$$

The relative values of T_{cs} and T_{cd} depend on the ratio $(|\varepsilon_F|/2t)$. If

(i) $|\varepsilon_F| \ll 2t$, then $T_{cd} \gg T_{cs}$;

(ii) For $|\varepsilon_F| \gg 2t$, $T_{cs} \gg T_{cd}$;

(iii) When $|\varepsilon_F| \sim 2t$, $T_{cs} \sim T_{cd}$.

We shall now consider choice (iii) in some detail. The singularity width does not span the full bandwidth. Accordingly, we choose $D \le 4t$. The solution of the full eq. (8.126) turns to be with due cognizance of the relation (8.125),

$$T_c = 1.13\, \omega_l \exp[-1/[V_l(|\varepsilon_F|^2/4t^2)(2/\pi D)\ln|4t/(|\varepsilon_F|)]] \qquad ...(8.134)$$

Now, for $0 < x < 2$, the function $\ln x$ can be expanded as $\ln x = (x-1)$ + (higher order terms). Retaining, lower order terms, eq. (8.134) can be written as

$$T_c = 1.13\, \omega_l \exp[-\pi t/2V_l(|\varepsilon_F|/4t)(1 - |\varepsilon_F|/4t)] \qquad ...(8.135)$$

The Fermi energy for a two-dimensional system $|\varepsilon_F| \propto n_c$; where n_c is the areal carrier density on each CuO_2. After appropriate renormalization, the expression can be recast in the form

$$T_c = 1.13\, \omega_l \exp[-1/\lambda_l(x_c)] \qquad \ldots(8.136)$$

where
$$\lambda_l(x_c) = \lambda_l^\circ\, x_c\, [1 - x_c] \qquad \ldots(8.137)$$

and x_c represents the carrier concentration and $\lambda_l^\circ = 2V_1 N_r(\varepsilon_F)$, $N_r(\varepsilon_F)$ = density of states without the factor $x_c(1 - x_c)$. This is precisely the form suggested by Sinha and coworkers [189, 211, 190, 206] earlier (see eq. (8.110)).

We may note that eqs. (8.135) and (8.136) contain contributions from both s-wave and d-wave pairing. Owing to the dependence given by eq. (8.137) the T_c will change with doping showing a maximum at some value; the form will be parabolic (bell shaped) which is indeed observed in cuprates [c.f. Fig. 8.9.]. We may note that there will be phonon contribution towards the pairing although this is not discussed here. The expression for T_c for the combined mechanism involving both lochon and phonon is given in eqs. (8.106) to (8.109).

Domanski, et al. [212] have used an equivalent model but call it boson-fermion model. In the atomic limit they calculate the spectral properties of the fermions. They find that on lowering the temperature the features develop into a three pole structure in the vicinity of the Fermi level. This three pole structure arises from the local bonding, antibonding and non-bonding states between bosons (lochons) and fermions. Thus the pseudogap in the density of states of the fermions is connected with the strong local pair correlations in the normal state. However, in the normal state global phase coherence is absent. It develops only in the superconducting state below T_c. Ranninger and Romano [213] have extended this treatment of boson (lochon fermion, the pseudogap and incoherent quasi particle features in high-T_c superconductors. The model is essentially the lochon-fermion model to examine the interrelation between interaction except that local bosons are called localized bipolarons. Earlier they had treated bipolarons as quasi particles without any internal structure [213]. This was generalized by taking into account the internal polaronic structure of the bipolaronic bosons. These bosons are now composed of charge and lattice degrees of freedom existing together in the coherent quantum state.

In the site representation, the boson (lochon) fermion model is written as

$$H = (W - \mu) \sum C_{i\sigma}^+ C_{i\sigma} - t \sum C_{i\sigma}^+ C_{j\sigma} + (E_B - 2\mu) \sum b_i^+ b_i$$
$$+ v \sum [b_i^+ C_{i\downarrow} C_{i\uparrow} + C_{i\uparrow}^+ C_{i\downarrow}^+ b_i] - \hbar \omega_0 \alpha \sum b_i^+ b_i (a_i + a_i^+)$$
$$+ \hbar \omega_0 \sum (a_i^+ a_i + 1/2) \qquad \ldots(8.138)$$

The fermion and boson creation operators have been defined earlier. Here W is the base fermionic half bandwidth, the boson (lochon) energy level is denoted by E_B and the boson-fermion pair exchange coupling constant by V. The chemical potential μ is common to both fermions and bosons. Note that indices i denote effective sites of units made of adjacent clusters of the metallic fermions and dielectric bosonic subsystem. Boson and fermion operators are assumed to commute with each other owing to small overlap of the oscillator wave functions at different sites. Various eigenstates of the above Hamiltonian are considered and the pair distribution functions (PDF) are calculated within this manifold. The parameters are chosen as hole concentration ≈ 0.25, total concentration of particles n_{tot} = 2. The lochon (bipolaron) level E_B is put above the bare electronic level W. The precise position of lochon level is adjusted such that

$$E_B = 2W + \hbar \omega_0 \alpha^2 - \delta E_B \qquad \ldots(8.139)$$

with $\delta E_B = 0.025$, giving $n_F \approx 0.75$. Further in units of W, $\alpha = 2.5$ v = 0.25, $\omega_0 = 0.1$, and $\delta E_B = 0.025$, $\xi = X(M \omega_0/\hbar)^{1/2}$, dimensionless parameter, M = mass of the oscillator. In Fig. 8.11, the pair distribution function (PDF) is illustrated for various temperatures in units of W[213]. PDF shows a double peak structure and changes into a single peak structure as the temperature is lowered below T_c, the temperature below which the pseudogap opens up. The two peak positions correspond to two local lattice deformations alternatively occupied by local (local pairs). There are experimental indications of such dynamical local lattice fluctuations [213].

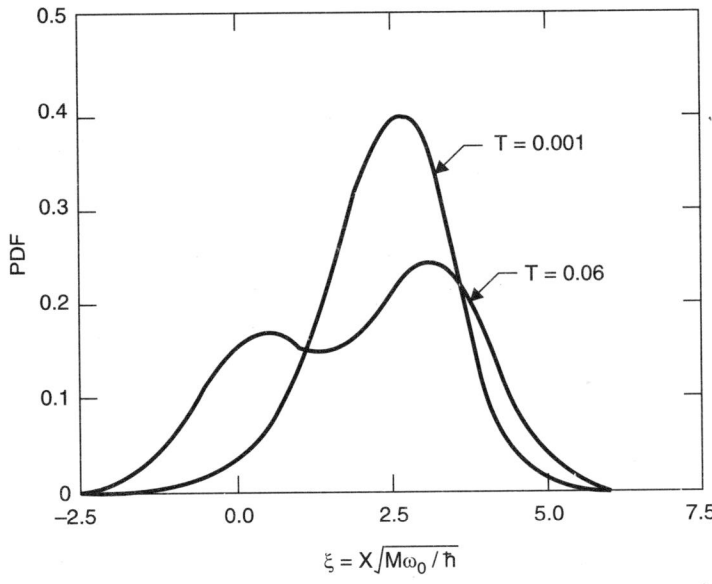

Fig. 8.11 Pair distribution function (PDF) for various temperatures in units of W.

The intensity of the photo-emission spectrum $I_{PES}(\omega)$ is evaluated from a single site boson (lochon)-fermion system. This is given by $I_{PES}(\omega) = I_E(\omega) f(\omega)$, where $f(\omega)$ is the Fermi distribution function and I_E is the emission part of total one particle fermionic spectral function. I_{PES} is calculated for different temperatures (in units of W). It is found that for high temperature ($T \approx 0.06$) a broad spectral function is observed which is close to a typical Fermi liquid. As the temperature is lowered the spectral function starts showing a pseudogap. At the same time a broad incoherent contribution appears which extends over energy region of the order of the order of half bandwidth (≈ 0.5 eV). Further, it is almost temperature independent at low temperatures. This feature is vindicated by experiment [212]. All the above aspected are depicted in Fig. 8.12 : [213].

Theory of High Temperature Superconductivity in Cuprates

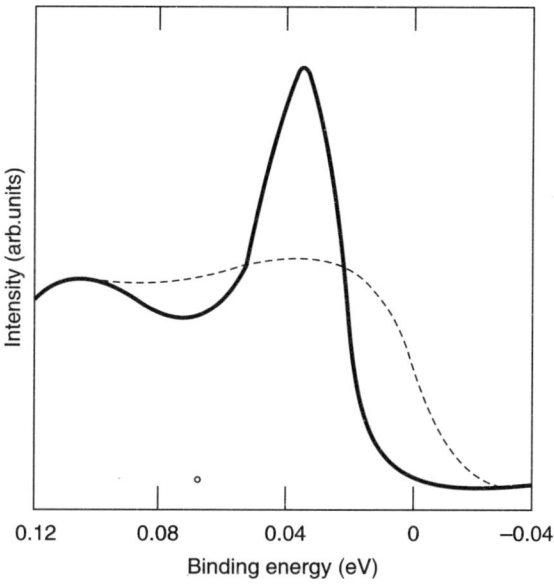

Fig. 8.12 I_{PES} versus ω. $\Delta\omega = 0.05$. As before $\alpha = 2.5$, $\nu = 0.5$, $\omega_0 = 0.1$, $\delta E_B = 0.025$ [214].

The temperature behaviour of the pseudogap is also shown in their original figure [213]. The evolution of the electronic density of states as a function of temperature is shown in Fig. 8.13.

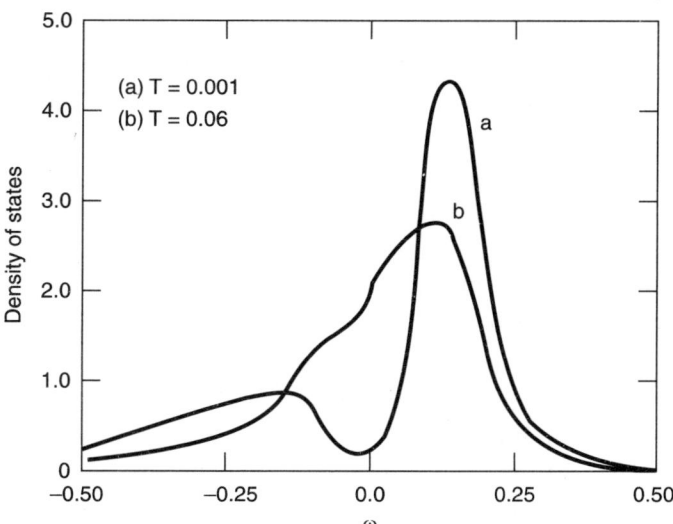

Fig. 8.13 $\Delta\omega = 0.05$; α, σ, ω_0 and δE_B are given in the same values as before [215].

However, pure fermion-lochon mechanism has its limitations.

8.13 MARGINAL FERMI LIQUID (MFL) MODEL

The unusual properties of the normal state HTSC lead to the idea that there exists a state of interacting Fermi systems that is drastically different from the Landau theory of Fermi liquid. Varma et al. [216] and Varma [217] proposed a phenomenological theory, which may be consistent with a variety of microscopic (*e.g.* magnetic or phonon) mechanisms is known as marginal Fermi liquid (MFL) model based upon the idea that a non-Fermi liquid behaviour appears in the normal state of HTSC if the quasi particle scatter from a bosonic system (charge or spin fluctuations) that have a flat energy spectrum over the frequency range $0 < T < \omega_c$, where ω_c is a cut-off frequency. In this picture, the width of the excitation spectrum is proportional to the temperature or the frequency whichever is larger. The normal state of HTSC, called by Varma et al. [216], MFL has as the main point, the assumption that the polarizability of the strongly interacting system has the form

$$Im\ \pi(\mathbf{q}, \omega) \cong N(0)\ \tanh\ (\omega/T) \qquad \ldots(8.140)$$

or
$$Im\ \pi(\mathbf{q}, \omega) \sim N(0)\ \omega/T\ \text{for}\ \omega < T$$
$$\sim N(0)\ \sin\ \omega\ \text{for}\ T < \omega < \omega_c \qquad \ldots(8.141)$$

Here $N(0)$ stands for the unrenormalized one-particle density of states. Such a form gives the one particle self-energy

$$\sum (\mathbf{K}, \omega) \sim N^2(0)\ (\omega_{\ln} x/\omega_c - i\pi/2x) \qquad \ldots(8.142)$$

where $x = \max(|\omega|, T)$, and ω_c is a cut-off frequency, which is remarkably different from that of a Fermi liquid. MFL model shows that there is no scale for low energy excitations other than temperature and this is to be constrasted with the Fermi liquid model where the scale is the Fermi energy E_F for holes or J for a hole in the Heisenberg antiferromagnet.

In a normal Fermi liquid, one finds that the Green function for a fixed momentum $\mathbf{K}$, $G(\mathbf{K}, \omega)$ consists of a quasi particle part and of an incoherent part. As usually, the quasi particle is characterized by its weight, Z_K and its lifetime, $T_K = Im\ \Sigma(\mathbf{K}, \omega)$. The lifetime for a Fermi liquid behaves as $T_K \sim -\omega^2$ when the Fermi level is approached by ω, while the present form of $\pi(\mathbf{K}, \omega)$ gives the lifetime which vanishes linearly in this limit, $T_K \sim |\omega|$. In addition, the quasiparticle weight vanishes logarithmically in the same limit, *i.e.* $Z_K \sim \ln |\omega| E_K$. Such an exotic system of fermions has been called by Varma et al. [216] a MFL.

Surprisingly, the MFL concept leads to the behaviour which agrees well with numerous experimental data : (*i*) the linear behaviour of the resistivity is consistent with eq. (8.142), (*ii*) the form of tunnelling conductance is consistent with the linear in ω behaviour of the quasi-particle lifetime close to the Fermi energy, (*iii*) the nuclear relaxation rate has two contributions, one linear in T, coming as usually from particle-hole pairs (Korringa law), and one constant, which may be explained by the form of $\pi(\mathbf{K}, \omega)$ given in eq. (8.141), (*iv*) the existence of frequency and temperature-independent background in the Raman scattering can also be explained by Eq. (8.141), (*v*) The optical conductivity consists of a narrow Drude peak $\sim \omega^{-1}$, and of a broad maximum at higher frequencies. The second feature may be related to the additional absorption due to the excitations responsible for the shape of $\pi(\mathbf{K}, \omega)$ as in eq. (8.141), (vi) finally, the recent ARPES for $Bi_2 Sr_2 Ca Cu_2 O_8$ gave a broad peak close to the Fermi energy, E_F, with the decreasing weight when approaching E_F, and the width proportional to the energetic distance from E_F. This confirms the qualitative prediction which follows from the self energy given by eq. (8.141). Obviously MFL model describes very well the transport properties and optical measurments, but the neutron scattering data and the nuclear magnetic resonance can not be explained satisfactorily.

In connection with the MFL concept explained above, the question of the existence of the Fermi surface in HTSC arises. On one hand, the Fermi surface can well be identified in the photoemission data [218]. On the other hand, there are clear indications that the Fermi liquid close to E_F does not apply and that the entire spectrum is incoherent at E_F, as suggested by the substantial background at higher energies [218].

Crisan and Tataru [219] have made a microscopic analysis of electron–boson interaction using many body methods in the framework of MFL theory. The model Hamiltonian is

$$H_{e-b} = \sum_{p,\sigma} \varepsilon(p) C^+_{p,\sigma} C_{p,\sigma} + \sum_K \Omega(K) b^+_K b_K + \sum_{p,K,\sigma} C^+_{K+p,\sigma} C_{p,\sigma} + (b_K + b^+_{-K}) \quad ...(8.143)$$

where the first term describes the fermionic system, the scond term describes the bosonic system, and the last term describes the electron-boson interaction which has the coupling constant $g(\mathbf{g}-\mathbf{K})$. They have shown that the effective coupling constant between electrons is attractive on the entire energy scale.

The problem of the existence of marginality in the Fermi liquid is still open. However, the wave vector dependence may be an important factor which influences the behaviour of the Fermi liquid.

8.14 SO_2 AND SO_5 MODELS

Wen and Lee [220], Lee et al. [221] and Lee and Wen [222] have discussed the unusual superconducting state of cuprates using the SO(2) formulation of $t-j$ model. The energy scale of the dispersion and a gap feature near the $(\pi, 0)$ point obtained by this theory seems to be consistent with the pseudogap phenomenon. The theory predicts small Fermi surface segments at low doping, which is consistent with the ARPES data [223]. However, the spectral functions in the superconducting state have not been obtained based on this model.

More recently a theory based on SO(5) symmetry has been proposed by Zhang [224] to unify the entire phase diagram encompassing both superconducting and antiferromagnetism, as a function of carrier concentration, i.e. the d-wave superconductivity and antiferromagnetism have a common origin in HTSC cuprates and are manifestations of the superstate, which indicates the AF and superconducting order parameters, in the Hubbard-like Hamiltonian [225]

$$H = -t \sum_{<i,j>\sigma} C^+_{i\sigma} C_{j\sigma} + \frac{J}{2} \sum_{<i,j>} S_i \cdot S_j + U \sum_i n_{i\uparrow} n_{i\downarrow} - \mu \sum_{i\sigma} C^+_{i\sigma} C_{j\sigma} \quad ...(8.144)$$

The electron density $n(\equiv <N>)$ is determined via the chemical potential μ and it is known for $n = 1$ and very U the system is Mott insulator which can exhibit also the AF order. Since at some finite doping $\delta = 1 - n \pm 0$, some cuprates are superconductors, the proponents of the Hubbard model believe that the d-wave superconductivity has its origin also in the Hubbard model. So because the AF order is a regular arrangement of paired electrons with opposite spins on neghbouring sites (some kind of a solid formed of Cooper pairs) and superconductivity is made of moving electrons forming Cooper pairs (superconducting liquid of Cooper pairs). The main assumption of the SO(5) theory is that these two phenomena are manisfestations of some underlying symmetry of Hubbard-like model [224].

We must note that the AF order is characterized by the three component vector n, while d-wave superconductivity is described by the complex (two component) order parameters Δ_d (and Δ_d^+). Because both phenomena are believed to be due to the same origin, it is assumed that the corresponding order parameter form the five-dimensional vector(operator)-superspin state $\psi = (n_1, \mathbf{n}, n_5)$ with

$$n = \Delta_d^+ + \Delta_d, n_s = -i(\Delta_d^+ - \Delta_d)$$

and

$$\Delta_d = \sum Y_d(\mathbf{p}) C_{p\uparrow}^+ C_{-p\downarrow}$$

where $Y_d(\mathbf{p}) = \cos p_x - \cos p_y$. The remaining three components of the superspin-state $n = (n_2, n_3, n_4)$ are the spin components identified with the AF order

$$n = \sum_{p,m,n} C_{p+Q,m}^+ \sigma_{m,n} C_{p,n} \qquad \ldots(8.145)$$

where $\mathbf{Q} = (\pi, \pi)$ is the AF wave vector and σ is the Pauli spin vector, $m, n = 1, 2$.

The central assumption of the model is that this superspin is an irreducible multiplet of the SO(5) group which has 10 symmetry generators (three spin operators S, one charge operator $\mathbf{Q}$ and six π-operators π and π^+) [226].

$$S = \sum_{p,m,n} C_{p,m}^+ \sigma_{mn} C_{p,n}, \quad Q = \frac{1}{2}[N - M],$$

$$\pi^+ = \sum_{p,m,n} Y_d(\vec{p}) C_{p+Q,m}^+ (\sigma, \sigma^y)_{mn} C_{p,n}^+$$

$$\pi = (\pi^+)^+$$

where N is the electron number operator, M is the number of lattice sites, $Y_b(\mathbf{p}) = \cos P_x - \cos P_y$. These operators form generators of the SO(5) group only if one assumes that $Y_d^2(\mathbf{p}) = 1$ which is realized only in a part of the physical space.

The central property is the existence of the π^+ operator especially the one called π_d,

$$\pi_d^+ = \sum_p Y_d(\mathbf{p}) C_{P+Q\uparrow}^+, C_{P\uparrow}^+$$ which also rotates the AF order into the superconducting order.

This SO(5) symmetry concept is valid only if the operator π_d^+ commutes with the Hamiltonian, which is not the case because

$$[H, \pi_d^+] = \omega_\pi \pi_d^+ + ..., \qquad ...(8.146)$$

where
$$\omega_n \approx J/2\,(1-2) + 2(U_{n\downarrow} - \mu) \qquad ...(8.147)$$

In fact eqs. (8.146) and (8.147) exhibit that the SO(5) symmetry is dynamically broken, *i.e.* it is not exact symmetry of the Hamiltonian. However, if the excitation energy ω_n is very small then one deals with very weak dynamically broken symmetry and the whole concept of the AF-superconductivity should work.

Now, the question arises, how big is the energy ω_π? Zhang [224] and Demler and Zhang [225] claim that in the limit $U \to \infty$ and near the half filling ($n = 1$) one has $\mu = UN/2 + \tilde{\mu}$ and $\omega_\pi \approx \frac{J}{2}(1-n) - 2\tilde{\mu}$, where $\tilde{\mu}$ is the chemical potential for the strongly correlated $t-J$ model (where doubly occupancy is not allowed). If one consider this claim to be correct, then the SO(5) symmetry is dynamically broken by the presence of the chemical potential $\tilde{\mu}$, which is the single parameter which triggers the antiferromagnetic transition. For $n = 1$, *i.e.* half filled case $n = 1$ and $\tilde{\mu} = 0$. This means $\omega_\pi = 0$ and the SO(5) symmetry is exact [224].

Zhang and coworkers [224, 226] claim that ω_π is really small, then it should be seen in the neutron scattering in the superconducting state. According to this interpretation the 41 *meV* magnetic resonance peak seen in neutron scattering measurements [227] is just due to this π-excitation, *i.e.* $\omega_\pi = 41$ *meV*. The following reasoning have been provided for this claim :

If π_d^+ is an exact eigen operator, *i.e.* $[H, \pi_d^+] = \omega_\pi \pi_d^+$ then in the $\pi_d^+ - \pi_d^+$ correlation function, the main contribution comes from this π-resonance state, $|\psi_\pi^N\rangle = N_\pi \pi_d^+ |\psi_0^{N-2}\rangle$. Here N_π is the normalization constant and in the half filled case these states are as described above. We have

$$\ll \pi_d^+(-\omega)\,\pi_d(\omega) \gg = -\frac{1}{\pi} \text{Im} \sum_n \frac{|\langle\psi_n^N|\pi_d^+|\psi_0^{N-1}\rangle|^2}{\omega - E_n^N + E_0^N + i\eta}$$

$$= \frac{1}{N_\pi^2}\,\delta(\omega - \omega_\pi) \qquad ...(8.148)$$

Moreover, in the presence of d-wave superconductivity this π-resonance would also manifest itself in the spin-spin correlation function at the wave vector $\mathbf{Q} = (\pi, \pi, \pi)$. We have

$$\chi_Q^+(\omega) = \frac{1}{\pi} Im \sum_n \frac{|\langle\psi_n^N|S_Q^+|\psi_0^N\rangle|^2}{\omega - E_n^N + E_0^N + i\eta} \qquad ...(8.149)$$

where
$$S_Q^+ = \sum_p C_{p+Q,\uparrow}^+ C_{p,\downarrow} \quad \text{and} \quad S_Q^- = (S_Q^+)^+ \qquad ...(8.150)$$

Making use of
$$\frac{1}{2}[\pi_d, S_Q^+] = \pi_d \qquad ...(8.151)$$

One obtains
$$\chi_Q^+(\omega) = -\frac{1}{\pi} N_\pi^2 \operatorname{Im} \frac{|<\psi_0^{N-2}|\pi_d S_Q^+|\psi_0^N>|}{\omega - \omega_\pi \to i\eta} + ...$$
$$= N_\pi^2 |<\psi_0^{N-2}|\Delta_d|\psi_0^N>|^2 \delta(\omega - \omega_\pi) + ... \qquad ...(8.152)$$

The latter result expresses the direct manifestation of the (assumed) SO(5) symmetry in neutron scattering in superconducting state, where the magnetic resonance below T_c is nothing more than the π-resonance, i.e. $\omega_\pi = 41$ meV. Meixer et al. [228] claimed that the π-resonance (in the $\pi_d^+ - \pi_d^+$) correlation function is obtained in calculations (numerical) on small clusters from 10 atoms.

Until now the "41 meV" magnetic resonance seen in the neutron scattering measurements is the only experiment in HTSC cuprates which might be interpreted in terms of SO(5) symmetry, of course, this concept is valuable for these materials.

Recently, Greiter [229] and Baskaran and Anderson [230] have raised serious objections. Greiter [229] claimed that the assumption $\mu \approx UN/2$ in the $U \to \infty$ is simply incorrect, because the chemical potential at half filling ($n = 1$) is discontinuous and therefore $\mu \neq Un/2$. For $n < 1$, the chemical potential jumps to the lower Hubbard band and for $U \to \infty$, it is independent of U, i.e. $\mu \ll U$. On this basis, it is estimated that ω_π is very large, i.e. $\omega_\pi \approx 2$ eV. Thus SO(5) symmetry approach is not a useful concept.

If we consider that the above reasoning is correct then there arises a question how to interpret the numerical results obtained by Meixner et al. [228], i.e. the existence of the low-energy π-resonance in the $\pi_d^+ - \pi_d^+$ correlation function. In that respect Greiter [229] remarked that the theoretical arguments for the π-resonance can run backwards, i.e. instead of going in reasoning from eqns. (8.148) to (8.152), one should follow the opposite way. This means that in the presence of d-wave superconductivity a resonance in the spin-spin correlation function $\chi_Q^+(\omega)$ manifests itself in the $\pi_d^+ - \pi_d^+$ correlation function.

There are also relevant criticisms of the SO(5) model based on the form of the starting and restricted model Hamiltonian in eq. (8.144) [230]. There is a deeper problem in SO(5) approach because the antiferromagnetic order is formed due to the Mott insulating state, while the superconductivity state is formed in the metallic state which is characterized with the Fermi surface [230]. We know that Fermi surface is meaningless in the Mott insulator and hence the physics is dominated by the **discontinuous insulator metal transition**. Baskaran and Anderson [230] remarked that in accordance with the latter fact the antiferromagnetic order and superconductivity cannot be related by the continuous symmetry transformation and therefore the superspin multiplet does not have physical meaning.

Recently Park [231] remarked that SO(5) might not be an approximate symmetry of the t-J model. Eder et al. [232] have claimed that the low energy states of the t-J model from SO(5) multiplets and the half filling ($n = 1$) are obtained from the higher spin states of half filling

($n = 1$) are obtained from the higher-spin states at half-filling through SO(5) rotations. For example, the π-operator π_α ($\alpha = 1, 2, 3$) carries charge -2, a spin triplet and momentum transfer $\mathbf{P}_{transfer} = \mathbf{Q}$ and has d-wave symmetry, while the operator $\hat{\Delta}_d$ carries charge -2, a spin singlet and momentum transfer $\mathbf{P}_{transfer} = 0$ and has d-wave symmetry. However, $\hat{\Delta}_d$ relates the half-filled (hf) ground state which is singlet $| hf, S = 0 >$ and the two hole ($2h$) ground state which is also singlet $| 2h, S = 0 >$, while the π-operator creates the state $| 2h, S = 0 >$ if it is applied at half filling to the state $| hf, S = 1 >$ with the total spin $S = 1$ and the total momentum $\mathbf{P} = \mathbf{Q}$. We must note that the latter transition belongs to the allowed transition because it is believed that $| hf, S = 1 >$ and $| 2h, S = 0 >$ belong to the same SO(5) multiplet. Thus, both operators π and $\hat{\Delta}_d$ can generate the hole-doped ground states. One finds it is interesting to compare their special properties. The spectral function for π-operator is given by eq. (8.148). The form of the spectral function of $\hat{\Delta}_d$ operator has the form

$$<< \hat{\Delta}_d^+(-\omega) \hat{\Delta}_d(\omega) >> = -\frac{1}{\pi} \text{Im} <hf, S = 0 \left| \hat{\Delta}_d^+ \frac{1}{\omega - H + E_{2h} + i\eta} \Delta_d \right| hf, S = 0 >$$

...(8.153)

Park [231] reported that the spectral properties of $\hat{\pi}$ and $\hat{\pi}_d$ operators are very similar but the weight of the π-operator is smaller and therefore the π-operator approximation is not better than the $\hat{\Delta}_d$ operator approximation for the two-hole ground state. This result raises the question about the SO(5) symmetry of the t-J model.

We may note that SO(5) symmetry model omits the electron–phonon interaction. Moreover, the phonon modulation of the chemical potential (due to local charge of the Cu and O levels) can blur the π-resonance.

It seems that SO(5) and other symmetry concepts might be useful and operative in some-small-U exotic metallic systems which exhibit both antiferromagnetic and superconducting order.

Greiter [229], Baskaran and Anderson [233] and Anderson and Baskaran [234] have raised serious criticisms starting from technical points to points of fundamental principles.

8.15 PSEUDOGAP PHENOMENON IN HTSC CUPRATES

We have stated that since the discovery HTSC cuprates by Bednorz and Muller [235], the anomalous normal state properties have been studied from the various points of view. In particular the pseudogap phenomena in underdoped cuprates have been a very important issue. There are a lot of studies for the issue from both experimental and theoretical points of view. However, the complete understanding remains to be obtained.

The pseudogap phenomena mean the suppression of the low frequency spectral weight without any long range order. They are universal phenomena observed in various compounds of under-doped cuprates.

The normal state excitation gap in under-doped cuprates have been indicated by several various experiments. The nuclear magnetic resonance (NMR) experiments have shown the anomalous temperature dependence of the spin lattice relaxation rate $1/T_1$ and the spin susceptibility χ[236-241]. The quantity $1/T_1T$, which increases with decreasing temperature, starts to decrease at the pseudogap onset temperature T^* [236-240]. The spin susceptibility decreases gradually from the rather high temperature, and immediately changes its slope at T^* [241].

The optical conductivity have indicated the suppression of the low frequency spectral weight above the superconducting critical temperature T_c [242]. The suppression is continuous above and below T_c.

The transport coefficients also change their behaviour at T^*. The T-linear in-plane resistivity observed in under-doped cuprates deviates downward [243-245]. The Hall coefficient remarkably deviates downward and decreases with temperature [243-244]. The c-axis resistivity strongly increases in the pseudogap region [246-247]. Stojkovic and Pines [248] and Yanase and Yamada [249] have shown that these behaviours of the transport phenomena are naturally explained by considering the momentum dependent scattering rate owing to antiferromagnetic spin fluctuations.

The angle-resolved photoemission spectrum (ARPES) [250-251] have directly shown in particular, the suppression of the low frequency spectral weight. Moreover, ARPES experiments have shown that the shape of the pseudogap is similar to that of the superconducting gap, and the magnitude does not change at the superconducting critical point [252]. This reveals that the pseudo gap as the d-wave shape and is continuously connected with the superconducting gap.

The onset temperature T^* measured by the above experiments is almost the same as the mean field superconducting critical temperature T_{MF} estimated from the amplitude of the superconducting gap [253].

Renner et al. [254] have shown directly by tunnelling experiments the suppression of the low frequency density of states. In this connection, the gap-like structure which is similar to the normal state pseudogap is observed in vortex cores in the superconducting states under the high magnetic fields [255].

The impurity effects on the reduction of T_c have indicated the suppression of the low frequency density of states [256].

The electronic specific heat is reduced well above T_c and T^* and the step height is extraordinary small in under-doped cuprates [257]. This reveals that the entropy has been already lost at rather high temperature.

The experimental results on pseudogap in HTSC cuprates are reviewed by Timusk and Slatt [258].

Various experiments have indicated the close relationship between the pseudogap phemomena and the superconductivity. In particular, ARPES have directly shown the pseudogap in the one-particle spectral weight [250, 251] and suggested its close relevance and continuity to the superconducting gap [252].

Yanase and Yamada [249] have explained the pseudogap phenomenon as a precursor of the strong coupling superconductivity. They argue that the effective Fermi energy E_F is renormalized by the electron–electron correlation, the ratio T_c/E_F increases in the strongly correlated electron systems. The ratio indicates the strength of the superconducting coupling. Therefore, the strong coupling superconductivity has a general importance for the strongly correlated electron systems. The strong coupling superconductivity necessarily leads to the strong thermal superconducting fluctuations. Such strong fluctuations in the quasi-two dimensional systems have serious effects on the electronic state and gives rise to the pseudogap phenomena.

We have already described earlier that other scenarios have been theoretically proposed for the pseudogap phenomena. In the resonating valence bond (RVB) theory, there are two distinct excitations, **spinon** and **holon**. The pseudogap is described as a spinon pairing (so-called **'spin gap'**). The spinon pairing occurs at T^* (so-called 'spin gap'), and holons condense at T_c. Since 'spin-gap' is based on the singlet pairing, the RVB theory explains the various experiments for under-doped cuprates. However, the continuities from the normal states to the superconducting states and from over-doped to under-doped cuprates are not necessarily obvious, and the physical origin of the spin charge separation is not clear. The magnetic scenarios based on the anti-ferromagnetic or SDW gap formation or their precursor have been proposed by Kamf and Schrieffer [127], Chubukov et al. [128], Pines [129] and Chubukov and Schmalian [259].

The magnetic scenarios based on the anti-ferromagnetic or SDW gap formation or their precursor have been proposed by various authors [127, 130]. In these theories 'hot-spot' which is a part of the Fermi-surface in the vicinity of the anti-ferromagnetic Brillouin zone, exists near $(0, \pi)$ and plays an especial role. Since the quasiparticles at 'hot spot' are strongly scattered by the antiferromagnetic spin fluctuations, the gap structure appears as 'hot spot'. However, considering that the ground state is actually superconductive, we can hardly expect that the magnetic order and the superconductivity are continuously connected with the superconducting transition. Dahm and Tewordt [130] have made calculations by treating the pseudogap phenomena as a precursor of the magnetic instability (anti-ferromagnetism or SDW). We also think it should be possible that the strong spin fluctuations near the magnetic critical point lead to the gap-like structure. In that case, the gap formation first takes place at 'hot spot' because the interaction is strong in the vicinity of $\mathbf{q}(\pi, \pi)$ and $\Omega = 0$. Since the spin fluctuations lead to the transformation of the Fermi surface 'hot spot' comes to exist in the wide region near $(0, \pi)$ [249]. Therefore, the gap structure would have a similar shape to the $d_{x^2 - y^2}$-wave superconducting gap. However, in HTSC cuprates it may not occur. Since the interaction caused by the superconducting fluctuations is strong in the vicinity of $\mathbf{q}(0, 0)$, the important **K**-points for the self energy on the Fermi surface are sure to be on the Fermi surface. On the other hand, in case of the anti-ferromagnetic spin fluctuations, the interaction is strong in the vicinity of $\mathbf{Q} = (\pi, \pi)$. Therefore, the important **K**-points for the quasi-particles with the momentum **K** exists in the vicinity of $\mathbf{K} + \mathbf{Q}$. They are on the Fermi surface only when **K** is on the hot spot [249]. As a result, the quasiparticles slightly apart from 'hot spot' are not directly scattered by the strong interaction. Thus, the pseudogap formation is not impossible but difficult to be attributed to the magnetic interaction on the numerical point of view. Moreover, since the phase transition which really occurs is the superconductivity, the superconducting critical phenomena necessarily take place. It is not obvious how the critical spin fluctuations can exist even in the critical region of the superconductivity. Indeed, the spin-fluctuations have turned out to be reduced in the pseudogap region. Furthermore, the continuity at the superconducting critical point is difficult to be explained naturally by the magnetic scenarios.

The pairing scenarios as a precursor of the superconductivity are classified into several types. The phase fluctuation scenarios have been proposed by Emery and Kivelson [107] and calculated by various authors [260, 261]. The scenario based on the strong coupling superconductivity has been proposed [262, 263] on the basis of the famous Nozieres and Schmitt-Rink formalism [264, 265]. Generally, the strong coupling superconductivity indicates the existence of incoherent Cooper pairs (pre-formed pairs) [264]. The strong attractive interaction produces the pre-formed pairs above the superconducting critical point. The pre-formed pairs condense at the critical point. Yanase and Yamada [266] are of the view that the pseudogap as the gap is brought about by the resonance scattering [267, 268] with the strong superconducting fluctuations. The strong superconducting fluctuations necessarily exist in the case of strong coupling superconductivity in the quasi-two dimensional systems. The strong coupling superconductivity naturally gives rise to the strong fluctuations and the deviations from the BCS mean field descriptions. The strong coupling superconductivity has been discussed from the viewpoint of the crossover from the BCS superconductivity to the Bose-Einstein condensation both in the ground state [265] and at the finite temperature [264]. The concept of Nozieres and Schmitt-Rink formalism [264] takes account of the corrections to the thermodynamic potential and deciding the chemical potential self-consistently. This procedure correspond to counting the number of the bosonic pairs. Their formalism is justified in the low density limit. Several authors have discussed the applicability of the Nozieres and Schmitt-Rink [NSR] formalism to the two dimensional systems [269, 270]. The NSR theory is one of the strong coupling theory describing the crossover from the BCS superconductivity to the Bose–Einstein condensation. There are many works describing the pseudogap such as the cross-over region [262, 271-278]. However, it has been asserted that the NSR theory is justified only in the low density limit and not in HTSC cuprates (nearly half-filled lattice system) which are high density systems [279]. Moreover, the theory based on the NSR theory becomes harder in the strongly correlated systems. In this case, the superconductivity arises from the coherent quasi-particles near the Fermi surface. This situation will be incompatible with the NSR theory in which the chemical potential shifts lower than the bottom of the band.

Recently, the magnetic field effects on the NMR spin-lattice relaxation rate $1/T_1$ have been measured and discussed by several groups to determine the correct scenarios for the pseudogap phenomena [280-283]. The experimental results are interpreted as follows. The magnetic field effects cannot be observed in under-doped cuprates in which the strong pseudogap phenomena occurs in the wide temperature region [280, 281]. In particular, the onset temperature T^* does not vary. On the other hand, the magnetic field effects are visible from optimally-doped to slightly over-doped cuprates in which only the weak pseudogap phenomena are observed in the narrow temperature region [282, 283]. The observed magnetic field dependences are explained by the conventional weak coupling theory [284, 285]. However, for the under-doped cuprates presently no theoretical explanation of the magnetic field effects on the pseudogap phenomena.

A simple possible relationship between pseudogap and self energy can be summarized by the following relationship

$$\sum (\mathbf{K}, i\omega_n) = -E_g^2(\mathbf{K}) G_0(-\mathbf{K}, -i\omega_n) \qquad ...(8.154)$$

where $G_0(-\mathbf{K}, -i\omega_n)$ is the free electron gas Green function (ω_n denotes the usual Matsubara frequency) [286]. Such a form of the self energy was already used to explain the specific form of the spectral function $A(k_F, \omega)$ obtained as a direct result of ARPES experiments [288]. A similar assumption was used as a starting point by Janko et al. [267], Nozieres and Pistolesi [288] and Benfatto et al. [289] in their works related to the physics of HTSC. In the standard case of BCS superconductivity by considering the interaction between the normal state electrons and the fluctuations generated by the superconducting order parameter at temperatures above the transition temperature T_c, this form of the electronic self energy appears naturally [290, 291].

In the past years, the pseudogap phenomena was investigated using a lot of different techniques [258, 292]. However, the correct phase diagram for HTSC is still under question especially when we approach the crossover region between the underdoped and overdoped region, *i.e.* the pseudogap doping dependence is not well known. The systematic measurements of the magnetic field dependences in the various doping rate can be an important verification to determine the origin of the pseudogap in HTSC cuprates.

8.16 CONCLUSION

Since the discovery of HTSC cuprates, a number of mechanisms have been proposed and pursued along different lines. After intensive and extensive theoretical and experimental investigations for about two decades it seems that many scenarios have been screened out. Pairing symmetry in these systems is an important and controversial topic. The recently performed phase-sensitive tests, combined with the refinement of several other symmetry-sensitive techniques, has for the most part settled this controversy in favour of predominantly *d*-wave symmetry for a number of optimally hole- and electron-doped HTSC cuprates.

We now have a qualitative understanding of the 'pseudogap' regime: the gap seems to be due to pairing without long range order. Superfluid properties are not observed in the pseudogap regime because the phase stiffness is so small that thermal fluctuations have destroyed the ability of the material to carry a supercurrent. Most interestingly, the extent of the pseudogap regime in cuprates shows that the pairing temperature grows rapidly as doping is decreased, reaching as high as 300 K. We must consider this as a encouraging sign not to regard ~ 150 K as an upper bound for T_c and to continue to search for materials and mechanisms to achieve superconductivity at room temperature. However, the pseudogap raises the question of whether pairing is a low-energy instability in which quasi-particles are bound into Cooper pairs (as in BCS theory) or is instead a fundamental property of the doped Mott insulating state (as in RVB and other related models).

The scenario for the normal state emerging from the available experimental results poses strong constraints on a theory of HTSC in cuprates. Obviously the pairing mechanism in HTSC cuprates is still, two decades after the discovery, under dispute [293].

Lastly, one may ask what the final theory should predict. First, it should describe the full complex phase diagram. Second, it should reveal the special conditions in the cuprates that lead to this very special behaviour. From this should follow some suggestions for other materials that would show similar behaviour. While it may not be possible to predict T_c accurately—because, for instance, of a lack of precise input parameters—the final theory should give the correct order of magnitude and explain the trends that are observed in the cuprates. These trends include the increase in T_c as we move from single-layer cuprates to those containing two and three copper oxide layers.

REFERENCES

1. A.A. Abrikosov and L.P. Gorkov, *Sov. Phys. JETP* 19, 1243 (1963).
2. K.P. Sinha and S.L. Kakani, *"High Tempreature Superconductivity. Current Results and Novel Mechanisms"*, Nova Science Publishers, New York (1995).
3. S.L. Kakani, *"Superconductivity : Key Problems"*, Arihant Publishers, Jaipur (India) p. 113 (1996).
4. S.L. Kakani, *"Superconductivity : Current Topics"*, Arihant Publishers, Jaipur (India) p. 63.
5. K.P. Sinha and S.L. Kakani, *Proc. Natl. Acad. Sciences, India*, Vol. LXXII, Sec. A, Pt. III, 153 (2002).
6. M. Crisan, *"Theory of Superconductivity"*, World Scientific, Singapore, (1989).
7. J.W. Halley (Ed.) *"Theory of High Temperature Superconductivity"*, Addison-Wesley, Redwood city, CA (1988).
8. S. Lundquist et al. (Eds.), *"Towards the Theoretical Understanding of High T_c Superconductivity"*, Vol. 14 World Scientific, Singapore 1988).
9. S. Fujita and S. Godoy, *"Quantum Statistical Theory of Superconductivity"*, Plenum Press, New York and London (1996).
10. P.W. Anderson, *"The Theory of Superconductivity in the High-T_c Cuprates"*, Princeton University Press, Princeton, New Jersey (1997).
11. S. Maekawa and M. Sato (Eds.), *"Physics of High Temperature Superconductors"*, Springer Verlag, Berlin, Heidelberg New York (1992).
12. A.S. Alexandrov and Sir Nevill Mott, *"Polarons and Bipolarons"*, World Scientific, Singapore (1995).
13. C. Noce et al. (Eds.) *"Superconductivity and Strongly Correlated Electrons,* World Scientific (1994).
14. M. Cyrot and M. Pavuna, *Introduction to Superconductivity and High-T_c Materials,* World Scientific (1992).
15. C.P. Poole, Jr. *"Hand-book of Superconductivity"* Academic Press (2000).
16. F.J. Owens and C.P. Poole, Jr. *'The New superconductors'*, Plenum, New York, (1996).
17. C.P. Poole, Jr. et al., *"Superconductivity"*, Wiley New York (1995).
18. C.C. Tsuei and J.R. Kirtley, *Rev. Mod. Phys.* 72, 969 (2000).
19. E. Dagott, *Rev. Mod. Phys.* 6,763 (1994).
20. A.P. Kampf, *Phys. Rep.* 249, 219 (1994).
21. M.L. Kulic, *Phys. Rep.* 338, 1 (2000).

22. D. Pavuna, in *"The Gap Symmetry and Fluctuations in High-T_c Superconductors"*, Bok et al. (Eds.) Plenum, New York (1998).
23. J.W. Lynn, *High Temperature Superconductivity*, Springer-Verlag (1990).
24. P.W. Anderson, *Science* 235, 1196 (1987).
25. P.W. Anderson, *Science* 256, 1526 (1992).
26. W. Weber, *Phys. Rev. Lett.* 58, 1371 (1987).
27. K.P. Sinha, *Ind. J. Phys.* 35, 434 (1961).
28. C.M. Varma et al., *Phys. Rev. Lett.* 63, 1996 (1989).
29. S.L. Kakani and C. Hemrajani, *Solid State Physics,* Sultan Chand, New Delhi (4th ed., 2005).
30. C. Kittel, *Introduction to Solid State Physics*, 6th ed., (Wiley, New York : 1986).
31. W.E. Pickett et al., *Science* 255, 46 (1992).
32. W.E. Pickett et al., *Phys. Rev. B* 42, 8746 (1990).
33. R.O. Jones and O. Gunnarsson, *Rev. Mod. Phys.* 61, 689 (1989).
34. R. Micnas et al., *Rev. Mod. Phys.* 62, 113 (1990).
35. R.B. Laughlin, *Science* 242, 525 (1988).
36. J. Orenstein and A.J. Mills, *Science* 288, 68 (2000).
37. J. Hubbard, *Proc. R. Soc. London A* 281, 401 (1964).
38. P. W. Anderson, *Mat. Res. Bull. B,* 153 (1973).
39. P. Fazekas and P.W. Anderson, *Phil. Mag.* 30, 432 (1974).
40. G. Kotliar, *Phys. Rev. B* 37, 3664 (1988).
41. D. Vaknin et al., *Phys. Rev. Lett.* 58, 2802 (1987).
42. J.M. Tranquada et al., *Phys. Rev. Lett.* 60, 156 (1988).
43. C. Gros, *Phys. Rev. B* (Preprint).
44. H. Yokoyama and H. Shiba, *J. Phys. Soc. Japan* 56, 3570 (1987).
45. S.A. Kivelson, et al., *Phys. Rev. B* 35, 8865 (1987).
46. Z. Zou and P.W. Anderson, *Phys. Rev. B* 37, 627 (1988).
47. P.W. Anderson, *Phys. Rev. Lett.* 64, 1839 (1990).
48. H.J. Schultz,. *Int. J. Mod. Phys. B* 5, 57 (1991).
49. G. Baskaran et al., *Solid State Commun.* 63, 973 (1987)
50. G. Baskaran and P.W. Anderson, *Phys. Rev. B* 37, 580 (1988).
51. I. Affleck and B. Marston, *Phys. Rev. B* 37, 3774 (1988).
52. G. Baskaran, *Physica Scripta. T* 27, 53 (1989).
53. Z. Zou and P.W. Anderson, *Phys. Rev. B* 37, 627 (1988).
54. L. Ioffe and V.I. Larkin, *Phys. Rev. B* 38, 8988 (1989).
55. P.B. Wiegman, *Phys. Rev. Lett*-60, 821 (1988).
56. P.A. Lee and N. Nagaosa, *Phys. Rev. B* 45, 966 (1992).
57. H. Fukuyama, *Prog. Theor. Phys. (Supplement)* 108, 287 (1992).
58. R.B. Laughlin, *Science* 242, 525 (1988).
59. P.B. Wiegman, *Physica Scripta T* 27, 160 (1989).

60. E. Mele, *Physica Scripta T* 27, 82 (1989).
61. Z.G. Wen and A. Zee, *Phys. Rev. Lett.* 62, 2873 (1989).
62. Z.G. Wen et al., *Phys Rev. B* 39, 11413 (1989).
63. I. Affleck and B.Marston, *Phys. Rev. B* 37, 3774 (1988).
64. S. Chakravarti et al, *Phys Rev. Lett* 60, 1057 (1988).
65. T. Hsu, *Phys. Rev. B 41*, 11379 (1990).
66. B.O. Wells et al., *Phys. Rev. Lett.* 74, 964 (1995).
67. C. Kim et al., *Phys. Rev. Lett.* 80, 4225 (1998).
68. R.B. Laughlin, *Phys. Rev. Lett.* 79, 1726 (1997).
69. G. Baskaran, Invited Talk (preprint, 2000).
70. J.E. Hirsch, et al., *Phys. Rev. Lett.* 60, 1668 (1988).
71. T. Tanamoto et al., *J. Phys. Soc. Jpn.* 63, 2739 (1994).
72. L. Balents et al., *Cond. mat*/980 3036.
73. D. Das and S. Doniach, *Cond. mat/990 2308*
74. G.D. Mahan, 'Many Particle Physics', Plenum Press, (1990).
75. S. Tomonaga, *Prog. Theor. Phys.* 5, 544 (1950).
76. S.M. Luttinger, *Math. Phys.* 4, 1154 (1963).
77. J. Solyom, *Adv. Phys.* 28, 201 (1979).
78. W. Metzner and C. Di Castro, *Phys. Rev. B* 47, 16107 (1993).
79. C. Castellani et al., *Phys. Rev. Lett.* 72, 316 (1994).
80. P.W. Anderson and D. Khveshchenko, *Phys. Rev. B* 52, 16415 (1995).
81. F.C. Zhang and T.M. Rice, *Phys. Rev. B* 37, 3759 (1988).
82. J. Zannen and A.M. Oles., *Phys. Rev. B* 37, 9423 (1998).
83. P. Prelovsek., *Phys. Rev. Lett.* A 37, 9423 (1988).
84. J. Bonca et al., *Euro Phys. Lett.* 10(1) 87 (1989).
85. A. Ramsak, and P. Prelovsek, *Phys. Rev. B* 40, 2239 (1989).
86. H. Eskes et al., *Physica. C* 160, 429 (1989).
87. V.J. Emery and E.A. Kivelson, *Phys. Rev. Lett.* 74, 3253 (1995).
87(a). V.J. Emery and G. Reiter, *Phys. Rev. B* 38, 4547 (1988).
88. R.J. Gooding and V. Elser, *Phys. Rev. B* 41, 2557 (1990).
89. M.C. Gutzwiller, *Phys. Rev.* 137, A 1726 (1965).
90. D. Vollhard, *Rev. Mod. Phys.* 56, 99 (1984).
91. Y. Hasegawa et al., *Phys. Rev. Lett.* 63, 907 (1989).
92. P. Lederer et al., *Phys. Rev. Lett.* 63, 1519 (1989).
93. E. Dagotto et al., *Phys. Rev. B* 10741 (1992).
94. E. Dagotto, et al., *Rev. Mod. Phys.* 66, 703 (1994).
95. M. Grilli and G. Kotlier, *Phys. Rev. Lett.* 64, 1170 (1990).
96. A. Houghton and A. Sudbo, *Phys. Rev. B* 38, 7037 (1988).
97. E.S. Heeb and T.M. Rice, *EuroPhys. Lett.* 27, 673 (1994).

98. E. Dagotto, et al., *Phys. Rev. Lett.* 74, 310 (1995).
99. F. Onufrieva et al., *Phys. Rev. B.* 54, 12464 (1996).
100. N.M. Plakida et al., *Phys. Rev. B.* 55, 1 (1997).
101. R. Zeyher and A. Greco, *Z. Phys. B* 104, 737 (1997).
102. B.O. Wells et al., *Phys. Rev. Lett.* 74, 964 (1995).
103. T. Tohyama et al., *J. Phys. Soc. Japan* 69, 9 (2000).
104. T. Tohyama and S. Maekawa, *Super Cond. Sci. Technol.* 13, R 17 (2000).
105. T. Tohyama et al., *J. Low Temp. Phys.* 117, 211 (1999).
106. S.R. White and D.J. Scalapino, *Phys. Rev. Lett.* 80, 1272 (1998); 81, 3227 (1998) ; 84, 3021 (2000) ; *Phys. Rev. B* 61, 6320 (1998) ; 60, R 752 (1999).
107. V.J. Emery and S.A. Kivelson, *Nature* 374, 434 (1995) ; *Phys. Rev. Lett.* 74, 3253 (1995).
108. G. Seibold, et al., *Phys. Rev. B* 58, 13506 (1998).
109. M. Veillette et al., *Phys. Rev. Lett.* 83, 2413 (1999).
110. E. Jackelmann et al., *Phys. Rev. B* 58, 9492 (1998).
111. S. Daul et al., *Cond. mat* 990 7301 (1999).
112. A. Moreo, *Phys. Rev. B* 45, 5059 (1992).
113. S.C. Zhang, *Science* 275, 1089 (1997).
114. T. Moria et al., *J. Phys. Soc. Jpn.* 59, 2905 (1990); *Physica C* 185-189, 114 (1991).
115. T. Moria, in "Electronic Properties and Mechanisms of High Temperature Superconductors" T. Sukuba Symposium, edited by T. Oguchi et al., (Amsterdam: North Holland) P. 145. (1992).
116. T. Moriya, *J. Phys. Chem. Solids* 53, 1515 (1992).
117. H. Shimahara and S. Takada, *J. Phys. Soc. Jpn.* 57, 1044 (1988).
118. N. Bickers et al., *Phys. Rev. Lett.* 62, 961 (1989); *Int. J. Mod. Phys.* BI. 687 (1987).
119. S. Nakamura et al., *J. Phys. Soc. Jpn.* 65, 4026 (1996).
120. J.R. Schrieffer et al., *Phys. Rev. B* 39, 11663 (1989).
121. D.K. Morr and D. Pines, *Phys. Rev. Lett.* 81, 1086 (1998).
122. K.P. Sinha and N. Kumar, *'Interactions in Magnetically ordered Solids'* (Oxford University Press, Oxford, 1980).
123. P. Monthouse and D. Pines, *Phys. Rev. B.* 47,6069 (1993).
124. J.R. Friedberg et al., *Phys. Lett.* 152 A, 417 (1991).
125. Y. Bar-Yim, *Phys. Rev. B* 43, 359, 2601 (1991).
126. D.M. Eagles, *Physica C* - 211, 319 (1993).
127. A. Kampf and J.R. Schrieffer, *Phys. Rev. B* 41, 6399 (1990).
128. A.V. Chubukov et al., *Philos. Mag. B* 74, 563 (1996).
129. D. Pines, *Z. Phys, B* 103, 129 (1997).
130. T. Dahm and L. Tewordt, *Phys. Rev. B.* 52, 1297 (1995).
131. Y. Yanase and K. Yamada, *J. Phys. Soc. Jpn.* 68, 548 (1999).
132. P.W. Anderson, *Science* 256, 1526 (1992).
133. J.M. Wheatley et al., *Phys. Rev. B* 37, 5897 (1988).
134. J.M. Wheatley et al., *Nature* 333, 121 (1988).

135. T.C. Hsu and P.W. Anderson, *Physica C* 162-164, 1445 (1989).
135(a). P.W. Anderson, *Physica B* 199 and 200, 8 (1994).
136. S. Chakravarty et al., *Science* 261, 337 (1993).
137. T. Xiang and J.M. Wheatley, *Phys. Rev. Lett.* 77, 4632 (1996).
138. S. Chakravarty, *'Josephson Plasma Resonance in High-Temperature Superconductors : A Test of Interlayer Tunneling Theory : Cond-Mat/9703140* (preprint, 1997).
139. G.G.N. Angilella, *Ph. D. Thesis* (Unpublished, 1999).
140. G. Baskaran, *'Mini Workshop on Strongly Correlated Electron Systems'* (July 1998), ICTP, Trieste, Italy.
141. S. Chakravarty et al., *Phys. Rev. Lett.* 82, 2366 (1999).
142. X.G. Wen and P.A. Lee, *Phys. Rev. Lett.* 76, 503 (1996).
143. P.A. Lee et al., *Phys. Rev.* 57, 6003 (1998).
144. N. Nagaosa, *Miniworkshop on Strongly Correlated Electron Systems* (July 1998), ICTP, Trieste, Italy.
145. D.H. Lee, *Cond. mat/9902287* (1999).
146. L.Yin et al., *Phys. Rev. Lett.* 78, 3559 (1997).
147. B. Chattopadhyay and A.N. Das, *Phys. Lett. A* 246, 201 (1998).
148. A.N. Das and B. Chattopadhyay, *Physica C* 308, 226 (1998).
149. Z.X. Shen and J.R. Schrieffer, *Phys, Rev. Lett.* 1771 (1997).
150. M.B. Walker, *Phys Rev. B* 53, 5835 (1986).
151. D. Controzzi, *Physica C* 324, 21 (1999).
152. T. Tešanovic, *Phys. Rev. B* 36, 2364 (1987).
153. U. Hofmann et al., *Solid State Commun.* 70, 335 (1989); *Z. Phys. B* 81, 25 (1990).
154. B.D. Yu et al., *Phys. Rev. B* 45, 8007 (1992).
155. W.A. Raine and W.H. Matthews, *Phys. Rev. B* 47, 422 (1993).
156. A.R. Bishop et al., *Z. Phys. B* 76, 17 (1989).
157. J. Hubbard and K.P. Jain, *J. Phys.* 61, 1650 (1968).
158. Ajay S. Patra and R.S. Tripathi, *Phys. Rev. B* 51, 12658 (1995) ; *Physica C* 274, 73 (1997).
159. S.L. Kakani and A. Kashyapa, *Physica C* (under communication, 2006).
160. K. Yvon and M. Fracois, *Z. Phys. B Condens. Matter* 76, 413 (1989).
161. R.J. Cava et al., *Nature* 345, 602 (1990).
162. G. Baskaran, 'Electron Correlations in Solids' ed. N.H. March, World Scientific, Singapore (2001)
163. P.W. Anderson, *Phys. Rev. B* 30, 1549 (1984).
164. B.K. Chakraverty and C. Schlenker, *J. Phys.* (Paris) 37, C 4-353 (1976).
165. B. K. Chakraverty, *J. Phys. Lett.* (Paris) 40, L99 (1979).
166. S. Lakkis et al., *Phys Rev. B* 14, 1429 (1976).
167. B. Chakraverty et al., *Phys. Rev.* 17, 3781 (1978).
168. A. Alexandrov and J. Ranninger, *Phys. Rev. B* 23, 1796 (1981).
169. A. Alexandrov and J. Ranninger, *Phys. Rev. B* 24, 1164 (1981).
170. A. Alexandrov et al., *Phys. Rev. B* 33, 4526 (1986).

171. I.G. Lang and Yu. A. Firsov, *Sov. Phys. JETP* 60, 856 (1984).
172. A.S. Alexandrov and P.P. Edwards, *Physica C* 331, 97 (2000).
173. A.S. Alexandrov, *Physica C* 341, 107 (2000).
174. A.S. Alexandrov, *Soc. Italiana di Fisica*-Bologna, Italy, 309 (1998).
175. A.S. Alexandrov, *Phil. Mag. B* 81, 1397 (2001).
176. A.S. Alexandrov, *Phys. Rev. Lett.* 82, 2620 (1999).
177. A.S. Alexandrov and V.V. Kabanov, *Phys. Rev. B* 59, 13628 (1999); B 59, 1 (1999-I).
178. J. Ashkenazi in *"Stripes and Related Phenomena"*, Eds. Bianconi and Saini, Kluwer Academic/Plenum Publishers, New York P-27 (2000).
179. J. Ashkenazi, *Cond.-mat/0108383*, 19th Oct., 2001 (J. Phys. and Chem. Solids, Preprint)
180. J. Ashkenazi, in *"High Temperature Superconductivity"*, Eds. S.E. Barnes, et al., American Institute of Physics, 1-56396-880-0/99.
181. S Ranninger and S. Robaskiewicz, *Physica 135 B*, 468 (1985) ; A.S. Alexanderov in Studies in High Temperature Superconductors pp. 1– 69 (2006) (Ed. A.V. Narliker), Nova Science New York.
182. K. Nasu, *Phys. Rev. B* 35, 1748 (1987).
183. A.S. Alexandrov and J. Ranninger, *Physica C* 198, 360 (1992).
184. D. Emin, *Phys. Rev. Lett.* 62, 1544 (1989).
185. D. Emin and M.S. Hillery, *Phys. Rev. B* 39, 6575 (1989).
186. A.S. Alexandrov and J. Ranninger, in *'Proceedings Conf. Lattice Effects in High-T_c Superconductors'* Santa Fl., Jan 13-15 (1992).
187. B.N. Ganguly et al., *Phys. Rev.* 146, 317 (1966).
188. K.P. Sinha, *Solid State Commun.* 79, 735 (1991).
189. K.P. Sinha and M. Singh, *J. Phys. C* 21, L 231 (1988).
190. K.P. Sinha, *Physica B* 163, 664 (1990).
191. J. Ranninger and J.M. Robin, *Physica C* 253, 279 (1995).
192. C.P. Enz, *Phys. Rev. B* 54. 3589 (1996).
193. K.P. Sinha and A. Rastogi, *National Acad. Sci. Lett.* 18, 22 (1995).
194. K.P. Sinha, *Ind. J. Phys.* 66 A, 1 (1992).
195. A.A. Abrikosov et al., *'Methods of Quantum Field theory in Statistical Physics'*, Prentice Hall, Englewood. Cliff. N.J. (1963).
196. J.R. Schrieffer, *'Theory of Superconductivity'*, W.A. Benjamin, Inc. New York (1964).
197. V.Z. Kresin, *Phys. Rev. B* 35, 8716 (1987).
198. J. Bardeen et al., *Phys. Rev.* 106, 162 (1957).
199. Y. Bar-Yam, *Phys. Rev. B* 43, 359, 2601 (1991).
200. H.J. Bornemann et al., *Physica C* 185-189, 1359 (1991).
201. K.P. Sinha and As. Vytheeswarn, *Solid State Commun.* 99, 845 (1996).
202. K.P. Sinha, *'Invited Paper Presented at the Hyderabad Conference of Superconductivity'*, Dec. 15-16 (1997).
203. J.P. Wallington and A.F. Annett, *Phys. Rev. B* 61, D1 (2000).
204. K.A. Muller, *Nature* 377, 133 (1995).
205. H.J. Bornemann et al., *Physica C* 185-189, 1359 (1991).

206. K.P. Sinha and M. Singh, *Phys. Status Solids (b)* 159 : 787 (1990).
207. G.D. Mahan, *Phys. Rev. B* 48, 16557 (1993).
208. J.A. Wilson, *J. Phys. Condens. Matter* 9, 6061 (1997).
209. R.S. Marikiewicz, *J. Phys. Chem. Solids.* 58, 1179 (1997).
210. S. Prakash, *J. Phys. 1 (France)* 7, 611 (1997).
211. M. Singh and K.P. Sinha, *Solid State Commun.* 70, 149 (1989).
212. T. Domanski et al., *Sol. Stat. Commun.* 105, 473 (1998).
213. J. Ranninger and A. Romano, *Phys. Rev. Lett.* 80, 5643 (1988).
214. S. Zhang et al., *Phys. Rev. Lett.* 78, 4486 (1997).
215. B.O. Wells et al., *Phys. Rev. Lett.* 74, 964 (1995).
216. C.M. Varma et al., *Phys Rev. Lett.* 63, 1996 (1989).
217. C.M. Varma, *Phys. Rev B* 55, 14554 (1997).
218. B. Batlogg, et al., (Eds.) in *'Proceedings of the 10th Anniversary of HTSC Workshop on Physics, Materials and Applications'*, World Scientific, Singapore (1996).
219. M. Crisan and L. Tataru, *J. Superconductivity* 8, 341 (1995).
220. X.G. Wen and P.A. Lee, *Phys. Rev Lett.* 76, 503 (1996).
221. P.A. Lee et al., *Phys. Rev. Lett. B* 57, 6003 (1993).
222. P.A. Lee and X.G. Wen, *Phys. Rev. Lett.* 78, 4111 (1997).
223. T. Tohyama and S. Maekawa, *Supercond. Sci. Technol.* 13, R 17 (2000).
224. S.C. Zhang, *Science* 275, 1089 (1997).
225. W. Hanke et al., *Physica B* 280, 184 (2000).
226. E. Demler and E.C. Zhang, *Phys. Rev. Lett.* 75, 4126 (1995).
227. H.F. Fong et al., *Phys, Rev. Lett.* 75, 316 (1995).
228. S. Meixner et al., *Phys. Rev. Lett.* 4902 (1997).
229. M. Geiter, *Cond.-mat./9705049 and Cond.-mat./9705282*.
230. G. Baskaran and P.W. Anderson, *Cond-mat./9706076*.
231. Y. Park, *Cond.-mat./9907437* (Preprint 28th July, 1999).
232. R. Eder et al., *Phys. Rev. B* 57, 13781 (1998).
233. G. Baskaran and P.W. Anderson, *J. Phys. Chem. Sol.* 59, 1780 (1998).
234. P.W. Anderson and G. Baskaran, *Cond-mat./9709195*.
235. J.G. Bednortz and K.A. Muller, *Z. Phys. B* 64, 189 (1986).
236. W.W. Warren et al., *Phys. Rev. Lett.* 62, 1193 (1989).
237. M. Takigawa et al., *Phys. Rev. B* 43, 247 (1991).
238. M.H. Julien et al., *Phys. Rev. Lett.* 76, 4238 (1996).
239. H. Yasuoka et al., *Physica B* 199 and 200, 278 (1994).
240. Y. Itoh et al., *J. Phys. Soc. Jpn.* 67, 312 (1998).
241. K. Ishida et al., *Phys. Rev. B* 58, R 5960 (1998).
242. C.C. Homes et al., *Phys. Rev. Lett.* 71, 1645 (1993).
243. T. Ito et al., *Phys. Rev. Lett.* 70, 3995 (1993).

244. K. Mizuhashi et al., *Phys. Rev. B* 52, R 3884 (1995).
245. M. Oda et al., *Physica. C* 281, 135 (1997).
246. K. Takenaka et al., *Phys. Rev. B* 50, 6534 (1994).
247. Y. Nakamura et al., *Phys. Rev. B* 47, 8369 (1993).
248. B.P. Stojkovic and D. Pines., *Phys. Rev. Lett.* 76, 811 (1996) ; *Phys. Rev. B* 55, 8576 (1997).
249. Y. Yanase and K. Yamada, *J. Phys. Soc. Jpn.* 68, 548 (1999).
250. H. Ding et al., *Nature* 382, 51 (1996).
251. A.G. Loeser et al., *Science* 273, 325 (1997).
252. M.R. Norman et al., *Nature* 392, 157 (1998).
253. M. Oda et al., *Physica C* 281, 135 (1997).
254. Ch. Renner et al., *Phys. Rev. Lett.* 80, 149 (1998).
255. Ch. Renner et al., *Phys. Rev. Lett.* 80, 3606 (1998).
256. J.L. Tallon et al., *Phys. Rev. Lett.* 79, 5294 (1997).
257. J.W. Loram et al., *Physica C* 235-240, 134 (1994).
258. T. Timusk and B. Slatt, *Rep. Prog. Phys.* 62, 61, (1999).
259. A.V. Chubukov and J. Schmalian, *Phys. Rev. B* 57, 11085 (1998).
260. M. Franz and A. J. Millis, *Phys. Rev. B* 58, 14572 (1998).
261. H.J. Kwon and A. T. Dorsey, *Phys. Rev. B* 59, 6438 (1999).
262. M. Randeria: *Preprint (Cond-mat./9710223)*.
263. C.A.R. Sa de Melo et al., *Phys. Rev. Lett.* 71, 3202 (1993).
264. P. Nozieres and S. Schmitt-Rink, *J. Low. Temp. Phys.* 59, 195 (1985).
265. A.J. Leggett, *'Modern Trends in the Theory of Condensed Matter'*, eds. A. Pekalski and R. Praystawa (Spring-Verlag), Berlin (1980).
266. Y. Yanase and K. Yamada, *J. Phys. Soc. Jpn.* 69, 2209 (2000).
267. B. Janko et al., *Phys. Rev. B* 56, 11407. (1997).
268. J. Maly et al., *Preprint. (Condmat/9805018)*.
269. S. Schmitt-Rink et al., *Phys. Rev. Lett.* 63, 445 (1989).
270. A. Tokumitu et al., *Prog. Theor. Phys. Suppl.* 106, 63 (1991).
271. C.A.R. Sa de Melo et al., *Phys. Rev. Lett.* 71, 3202 (1993).
272. M. Randeria *'Bose-Einstein Condensation'*, ed. A. Griffin et al. (Cambridge University Press, Cambridge, 1994).
273. J.R. Engelbrecht et al., *Phys. Rev. B* 55, 15153 (1997).
274. R. Haussmann, *Phys. Rev. B* 49, 12975 (1994).
275. S. Stintzing and W. Zwerger, *Phys. Rev. B* 56, 9004 (1997).
276. J. Maly et al., *Physica C* 321, 113 (1999).
277. S. Koikegami and K. Yamada. *J. Phys. Soc. Jpn.* 67, 1114 (1998).
278. A. Kobayashi et al., *J. Phys. Soc. Jpn.* 67, 2626 (1998).
279. Y. Yanase and K. Yamada, *J. Phys. Soc. Jpn.* 68, 2198 (1999).
280. G.Q. Zheng et al., *Phys. Rev. B* 60, R 9947 (1999).
281. K. Gorny et al., *Phys. Rev. Lett.* 82, 177 (1999).

282. V.F. Mitrovic, *Phys. Rev. Lett.* 82, 2784 (1999).
283. V.F. Mitrovic, et al., *(Preprint), Cond-mat/9901232*.
284. G.Q. Zheng et al., *Preprint (Cond-mat/0006094)*.
285. M. Eschring et al., *Phys. Rev. B.* 59, 12095 (1999).
286. I. Tifrea et al., *Physica C* 371, 104 (2002).
287. O. Tehernyshyov, Ph.D. Thesis, Columbia University (unpublished).
288. P. Zozieres and F. Pistolesi, *Euro Phys. J. B* 10, 649 (1999).
289. L. Benfatto et al., *Euro Phys. J. B.* 17, 95 (2000).
290. E. Abrahams et al., *Phys. Rev. B* 1, 208 (1970).
291. A. Schmid, *Z. Phys.* 231, 324 (1970).
292. J.L. Tallow and J. W. Loram. *Physica* 349, 53 (2001) and references therein.
293. J. Ashkenazi et al. (Eds.) 'New Challenges in Superconductivity: Experimental Advances and Emerging Theories Nato Science Series II (2004).

Emerging Superconductors 9

9.1 INTRODUCTION

The subject of superconductivity has seen a number of new developments in recent years. Although the HTSC cuprates have been the focus of research on superconducting materials, but a number of other noteworthy superconductors have been discovered which cover a large spectrum of chemically different compounds, ranging from alloys to chalcogenides and organic compounds, and crystallize with very different crystal structures. Rare-earth transition metal borocarbides, Heavy fermions, A 15 compounds, Chevrel phases, Fullerides, Charge-transfer salts, Sr_2RuO_4, LiV_2O_4, Quantum spin ladder materials, MgB_2, UGe_2, ruthenate etc. are few examples. In this chapter, we briefly introduce these superconducting materials.

Almost one century has passed since the discovery of superconductivity, and a huge amount of literature on superconductors has been published. For detail study, readers are advised to consult the references [1-19] and references cited in them and also several conference proceedings on superconductivity.

9.2 INTERMETALLIC A 15 SUPERCONDUCTORS

The A 15 phases with superconducting transition temperature (T_c) exceeding 20 K, are the best known family of intermetallic superconductors. The structure type was earlier referred to as β-W, but is now generally called Cr_3Si, sometimes α-UH_3. A 15 compounds are very brittle, the processing to final dimension takes place in more ductile state and mainly in combination with substrate materials for electrical materials. The structure of A 15 is tetrahedrally closed packed, cubic, space group $p_m\overline{3}n$. Structure of A 15 superconductor Nb_3Ge, emphasizing the infinite linear chains of Nb atoms is shown in Fig. 9.1. Transition temperatures for some of A 15 superconductors are given in Table 9.1.

Table 9.1. Transition Temperatures, T_c (K) of Some A 15 superconductors

A-15	Cr_3Si	Mo_3Ir	Nb_3Al	Nb_3Au	Nb_3Ga	Nb_3Ge	Nb_3Sn	Ta_3Sn	Ti_3Ir	V_3Al	V_3Ga	V_3In	V_3Si
T_c (K)	< 1.2	8.5	19.1	11.3	20.7	23.2	18.2	5.8	4.3	9.6	14.8	14	17.1

All metallic materials being superconducting in the range from 14 K to 23 K have the crystal structures of A 15s. Until the so-called HTSC-cuprates were discovered, the record for the maximum T_c has always been held by members of the A 15 family. A 15 members are expected not to be superconducting above 25 K [20], due to the increasing instability of their structure related to the electron-phonon interaction.

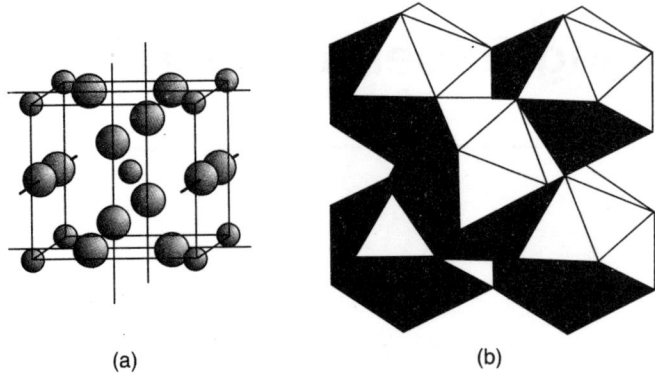

Fig. 9.1 (a) Structure of Nb_3Ge (Cr_3Si type ; A15), emphasizing the linear chains of Nb atoms ; and (b) the framework of $GeNb_{12}$ icosahedra. Small spheres : Ge and large spheres : Nb.

The superconductivity of A 15 compounds can be destroyed with a sufficiently high magnetic field, called the upper critical magnetic field, $B_{c_2}(T)$, which depends on the temperature T and varies from material to material. As the temperature is lowered, $B_{c_2}(T)$ generally increases from T_c to a maximum value, $B_{c_2}(0)$, as the temperature approaches absolute zero (0 K). The A 15 compounds also have very high values of $B_{c_2}(0)$, the highest of which is 44 teslas for a pseudobinary A 15 compound with the composition $Nb_{79}(Al_{73}Ge_{27})_{21}$. Other A 15 compounds with high values of $B_{c_2}(0)$ include Nb_3Al (32 T), Nb_3Ge (39 T), Nb_3Sn (23 T), V_3Ga (21 T), and V_3Si (25 T). For comparison, the highest value of $B_{C_2}(0)$ observed prior to the discovery of high temperature ceramic superconductors was about 60 T for the Chevrel phase compound $Pb\,Mo_6\,S_8$ and now achieved up to > 100 T in bulk materials.

Because of high values of T_c and B_{c_2} and critical current density J_c, the A 15 compounds have a number of important technological applications. Processes have been developed for preparing multifilamentary superconducting wires that consist of numerous filaments of a superconducting A 15 compound, such as Nb_3Sn, embedded in a nonsuper-conducting copper matrix. Superconducting wires can be used in electric power transmission lines and to wind electrically loss-less coils (solenoids) for superconducting electrical machinery (motors and generators) and magnets. Superconducting magnets are employed to produce intense magnetic fields for laboratory research, confinement of high-temperature plasmas in nuclear fusion research, bending beams of charged particles in accelerators, levitation of high-speed trains, mineral separation, and energy storage.

In general, it is possible to describe the physical phenomena relevant for the A 15 s, for example, the high superconducting transition temperature, which is also related to the BCS formula

$$k_B T_c = \hbar\, \omega_D\, exp.\,[-1/N(0)\,|g|\,]$$

by one dimensionality, partially localized states near to or at the Fermi level, elastic softness, and strong electron–phonon interaction.

9.3 CHEVREL PHASES

Chevrel phases were reported by Chevrel et al. [21] in 1971. The compounds have the general formula $M_x Mo_6 X_8$, where M represents any one of a large number (nearly 40) of metallic elements throughout the periodic table, X has values between 1 and 4, depending on the M element; and X is a chalcogen (sulfur, selenium or tellurium). The Chevrel phases are of great interest, largely because of their striking superconducting properties. Chalcogenides show critical temperatures of upto 15.2 K.

Most of the ternary molybdenum chalcogenides crystallize in a structure in which the unit cell, *i.e.*, the repeating unit of the crystal structure, has the overall shape of a rhombohedron with a rhombohedral angle close to 90°. Some of the ternary molybdenum chalcogenides display a slight distortion of the rhombohedral crystal structure at, or below, room temperature to a triclinic in which the three axes of the unit cell, and the three angles between them, are unequal.

The building blocks of Chevrel-phase crystal structure are the M elements and $Mo_6 X_8$ molecular units or clusters. Each $Mo_6 X_8$ unit is a slightly deformed cube with X atoms at the corners and Mo atoms at the face centres. One of these structures, that of $PbMo_6 S_8$ is shown in Fig. 9.2.

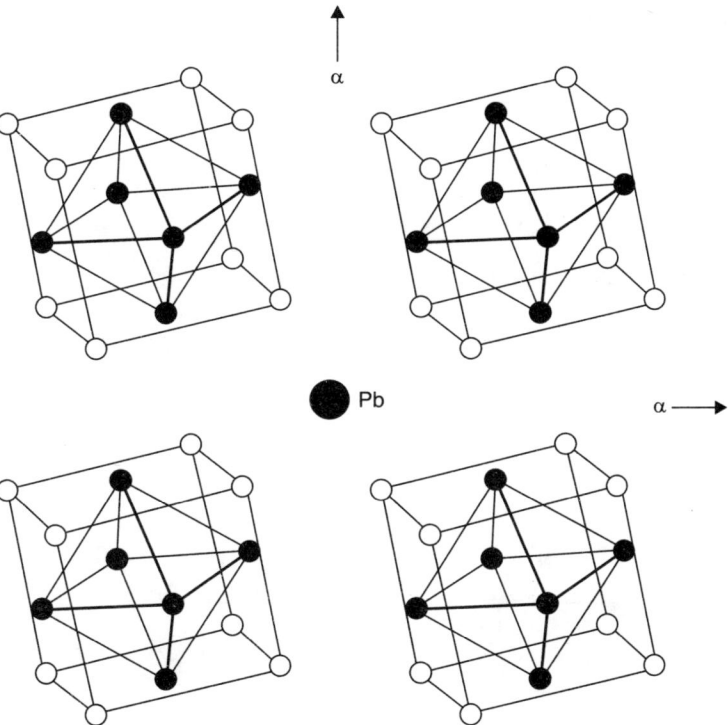

Fig. 9.2 Crystal structure of $PbMo_6 S_8$. Each lead atom is surrounded by eight $Mo_6 S_8$ units, the structure of which is shown in the lower right-hand part of the figure. The rhombohedral angle α is also indicated in the figure.

Chevrel phases are not so much HTSC or LTSC superconductors as high field superconductors, with $PbMo_6S_8$ having an upper critical field of 60 T in bulk samples. In wire form, however, the highest critical fields have been closer to 35 T at 4.2 K, and the wires have low critical currents [22]. Nevertheless, they have a unique application to high-field studies and are under active study by a number of research groups.

Band structure calculations have shown that the Fermi level is situated in a narrow region of bands with mainly Mo-$4d$ character. For $M_x Mo_6 S_8$ a forbidden energy gap is present in the band structure just above the Fermi level and the conduction band is filled for an electron concentration of 24 electrons per Mo_6 cluster. For the corresponding selenides and tellurides this limiting number is lower. As a consequence, the homogeneity ranges of the latter two are systematically shifted toward a lower M content, with respect to the sulfides.

9.3.1 Rhombohedral Ternary Molybdenum Chalcogenides

The rare earth molybdenum sulfides $REMo_6S_8$ and selenides $REMo_6Se_8$ are members of this series whose structure was first characterized by Chevrel et al. [23]. The series of compounds with the general formula $RE_xMo_6S_8$ (x = 1.0 or 1.2) was first reported by Shelton et al. [24]. Nearly all ternary chalcogenides crystallize in a hexagonal-rhombohedral structure with the rhombohedral angle α close to 90°. The crystal lattice structure is similar to as shown in Fig. 9.2 which reveals that the cluster structure of Mo_6S_8 is isolated from the remaining ions (Fig. 9.3).

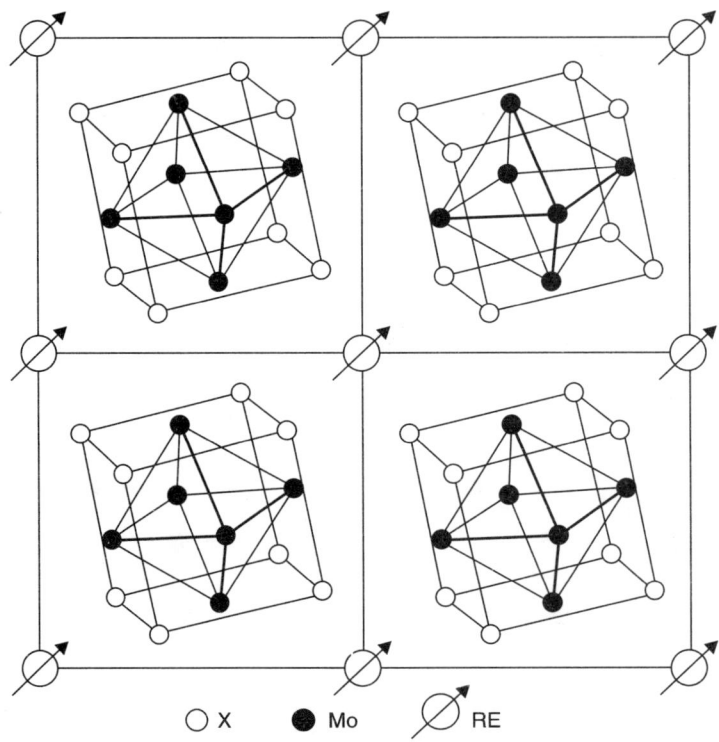

Fig. 9.3 Lattice structure of (RE) Mo_6S_8.

The HoMo$_6$S$_8$ compound belonging to RE$_x$Mo$_6$S$_8$ series exhibit re-entrant superconductivity. This compound first becomes superconducting at a critical temperature T_{c_1} (= 1.82 K) and then on further cooling loses its superconductivity at a second lower critical temperature T_{c_2} (= 0.668 K warming and 0.612 cooling) [25, 26]. The competition between superconductivity and ferromagnetism initially results in a long wavelength (~ 10^2 Å) oscillatory component to the magnetization developing in the superconducting state. This oscillatory magnetic phase exists only in a limited temperature interval, with the superconductivity being destroyed at T_{c_2} and long range ferromagnetic order setting in at sufficiently low temperature [27]. Lynn et al. [28] reported that the compound HoMo$_6$Se$_8$ becomes superconducting at 5.6 K. This does not exhibit re-entrant behaviour in a zero magnetic field down to 40 mK.

Several series of isostructural ternary rare earth compounds have been found that exhibit phase transition from superconducting to antiferromagnetism and true coexistence of superconductivity and antiferromagnetism [29-32]. The rare earth ternary compounds exhibiting superconducting to antiferromagnetic transition are RE chalcogenides RE Mo$_6$ S$_8$ (RE = Nd, Gd, Tb, Dy and Er). The occurrence of antiferromagnetic ordering of rare earth (RE) magnetic moments in the superconducting state has been inferred from a λ-type anomaly in the heat capacity and a cusp in the magnetic susceptibility for RE Mo$_6$S$_8$ compounds [33], and from a feature in the curve of B_{c_2} versus T in REMo$_6$S$_8$ compounds [34]. Neutron diffraction measurements on REMo$_6$S$_8$ compounds [34] and Gd Mo$_6$ S$_8$ [35] confirmed the occurrence of antiferromagnetic ordering associated with Chevrel phase structure. For the antiferromagnetic superconductors ErMo$_6$S$_8$, B_{c_2} increases below T_N, while for others, such as RE Mo$_6$S$_8$ (RE = Tb, Dy and Gd), B_{c_2} decreases rather abruptly below T_N.

9.3.2 Tetragonal Rare Earth Rhodium Borides

The other class of rare earth ternary compounds exhibiting superconductivity is made up of the tetragonal RE rhodium borides. This series of compounds was originally reported by Matthias et al. [36] and has the general formula RERh$_4$B$_4$. Vandenberg and Matthias [37] reported that the crystal structure of these compounds is isomorphic with the corresponding ternary RE cobalt borides reported earlier by Kuz'ma and Bilonizhko [38]. The primitive tetragonal crystal structure of RERh$_4$B$_4$ with the unit cell in dashed outline is shown in Fig. 9.4. [38]. The unit cell contains two formula units.

The compounds for RE = Y, Er, Tm and Lu exhibit superconductivity, whereas the compounds for RE = Gd, Tb, Dy, and Ho are magnetic and never exhibit superconductivity [39]. The compound ErRh$_4$B$_4$, a typical

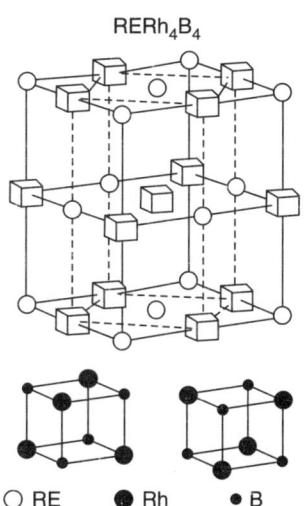

Fig. 9.4 The unit cell for ternary rhodium boride compounds. Dashed lines indicate the tetragonal unit cell. For clarity, the cubes representing the Rh$_4$B$_4$ clusters are not drawn to scale [38].

ferromagnetic superconductor, exhibits *re-entrant superconductivity* and coexistence of superconductivity and long-range ferromagnetic order in a narrow temperature range. Superconductivity and antiferromagnetic order coexist in RE $Rh_4 B_4$ compounds (for RE = Nd, Sm, Tm) [29]. The typical ferromagnetic superconductor Er $Rh_4 B_4$ first becomes superconducting at T_{c_1} = 8.67 K and on further lowering the temperature, it becomes ferromagnetic at T_{c_2} = 0.98 K. Experimental results clearly indicate that superconductivity and long range ferromagnetic order truly coexist in a narrow range of temperature (~ 50 mK). Experiments have shown that the coexistence region is more subtle and complex than a simple pure spiral or vortex intermediate phase separating the superconducting, paramagnetic, and normal ferromagnetic phases.

Coexistence of superconductivity and ferromagnetism has been the most interesting subject in the condensed matter physics for a long time. The early investigations traces back to the original works in the sixties by Clogston [40], Chandrasekhar [41], Abrikosov and Gorkov [42], Fulde and Ferrel [43], and Larkin and Ovchinnikov [44]. These works focused on the spin-singlet superconductivity in a bulk metal with a spin-exchange field, such as produced by ferromagnetically aligned impurities. The presence of the exchange field is unfavourable to the spin-singlet superconductivity.

The first theoretical study on the interplay between superconductivity and magnetic order was published in 1957 by Ginzburg. He showed that coexistence between ferromagnetism and superconducivity is almost impossible. Later, Baltensperger and Strassler pointed out the possibility of the coexistence between superconductivity and antiferromagnetism. They showed that although the time reversal symmetry is broken by the antiferromagnetic order, the superconductivity may coexist with antiferromagnetism in a form of slightly modified pairing system.

Early experiments trying to make these two dissimilar properties—ferromagnetism and superconductivity–demonstrated that doping with a very small amount of rare-earth impurities would completely destroy superconductivity. This is due to interactions between the spins of the electrons and atomic magnetic moments—this interaction attempts to align Cooper pairs, which is mentioned above, would destroy superconductivity. Therefore, forcing superconductivity and ferromagnetism to coexist is a tricky situation. As stated above, in 1977, Matthias and Maple, and Fischer independently discovered that the coexistence between superconductivity and magnetic order is possible in the ternary rare-earth compounds. These compounds present different kinds of order: ferromagnetic, antiferromagnetic and spatial modulated spin order. We have seen that the compound Er Rh_4B_4, which becomes superconductive at 8.7 K, when cooled to 1 K, something interesting happens, *i.e.*, an alternating magnetic structure appears, instead of a typical ferromagnetic ordering. That is, local magnetic moments are aligned, but the amplitude of the magnetization varies sinusoidally. In this intermediary state, the local magnetic moments in a domain interact with those in neighbouring domains, affecting their spin, an attempt to align their magnetic moments. The energy an atom gains by this alignment far exceeds the energy electrons gain by forming Cooper pairs (since domain walls cost energy), so this process snowballs until the material displays a uniform magnetic field. That is, ferromagnetism eventually wins out and superconductivity is destroyed at about ~ 1 K, below which the material is no longer superconductive. However, Er $Rh_4 B_4$ is not truly a ferromagnetic superconductor—the sample displayed either property, depending on its temperature, but not both. In the intermediatory state, the alternating magnetic domains make the material look like an antiferromagnet on a large scale, even though on the atomic scale, it looks ferromagnetic.

It was not until ternary systems such as $HoMo_6S_8$ and $ErRh_4B_4$ were discovered that the coexistence between magnetic order and superconductivity was realized experimentally, and the interaction between these two co-operative phenomena could be explored. We have seen that in these materials, there is a complete sublattice of magnetic ions which is distinct from the sublattice where the superconducting electrons travel. The physical isolation of the magnetic and superconducting electrons yields and exchange energy J which is very small, and leaves the dipole interaction as the dominant coupling between the magnetism and superconductivity. Since there is no chemical disorder to disrupt the magnetic system, the magnetic entropy compels the system to develop long range magnetic order at sufficiently low temperatures. We will discuss the theory of ferro and anti ferromagnetic superconductors in subsequent sections.

9.3.3 Ruthenate Cuprate

Coexistence of ferromagnetism and superconductivity has also been revealed in a high-T_c hybrid ruthenocuprates, $RuSr_2GdCu_2O_8$ with $T_c \leq 40$ K by Tallon et al. [45, 46]. Zero field cooled magnetization and heat capacity measurements show that superconductivity in the system is

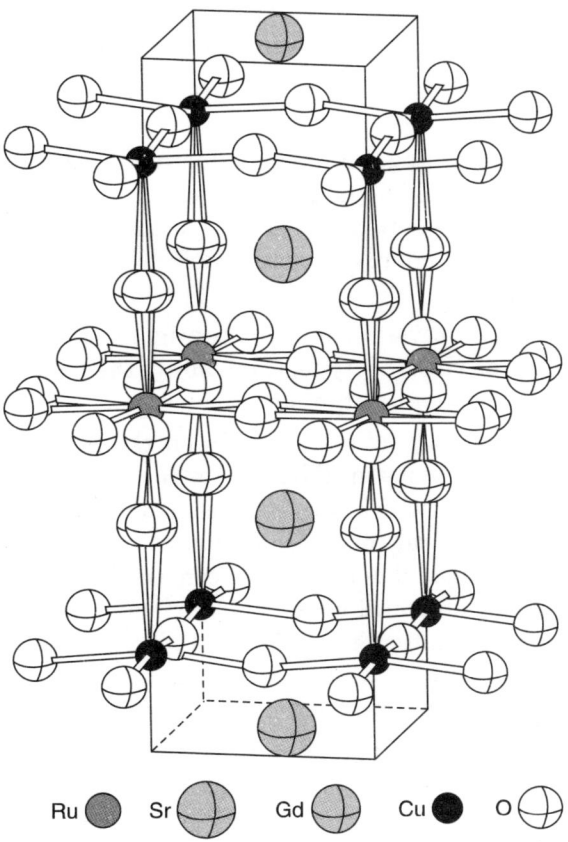

Fig. 9.5 The average crystal structure of $RuSr_2GdCu_2O_8$ showing the disordered rotations of the RuO_6 octahedra.

a bulk property. The SQUID measurements [45, 46] reveal a ferromagnetic transition at a Curie temperature T_M = 132 K. µSR and SQUID measurements show that ferromagnetism is also a bulk, uniform property. Experiments performed on this system [45, 46] indicate that the magnetically ordered state is not significantly modified by the onset of superconductivity. In these systems, the magnetic phase transition to antiferromagnetism with weak ferromagnetism occurs earlier, *i.e.*, at higher temperatures than the superconductivity onset. They can be called "*superconducting ferromagnets*", in contrast to ferromagnetic superconductors, in which the superconducting phase transition precedes to the magnetic one. A number of unusual phenomena and structures have been predicted and observed for superconducting ferromagnets, the spontaneous vortex phase among them. In the spontaneous vortex phase, vortices appear in the ground (equilibrium) state even without an external magnetic field. This produces a drastic effect on the magnetization curve. In the spontaneous phase, there is superconductivity, but without the Meissner effect.

9.3.4 Borocarbides

The R Ni_2B_2C (R = Gd-Lu, Y) series of materials is a recently discovered family of magnetic superconductors [47, 48]. The T_c values for the non-magnetic rare earths Lu and Y quaternary rare earth transition-metal borocarbides (RTBCs) are relatively high, 16.1 K and 15.6 K respectively. The replacement of Y or Lu with a moment-bearing rare earth leads to a suppression of T_c and the advent of antiferromagnetic ordering below the Neel temperature T_N [48]. The T_c and T_N values for this series have a ratio T_c/T_N that ranges from 7.3 for $R = Tm$ to 0.60 for $R = D_y$. $D_yNi_2B_2C$ is of special interest as a rare example of an ordered compound with $T_c < T_N$.

A striking feature distinguishing the superconducting RTBCs from other superconductors is that for certain combinations of elements R and T superconductivity and antiferromagnetic order have been found in RT_2B_2C where the values of the magnetic ordering T_N are comparable to the T_c values, *i.e.* the magnetic energy is comparable to the superconducting condensation energy.

Most of these borocarbides crystallize in the tetragonal $LuNi_2B_2C$ type structure which is an interstitial modification of $ThCr_2Si_2$ type. Contrary to the behaviour of Cu in the cuprates Ni does not carry a magnetic moment in the borocarbides.

Various types of antiferromagnetic structures on the rare-earth sublattice have been found to coexist with superconductivity in RNi_2B_2C for R = *Tm, Er, Ho* and *Dy*. Particularly of interest is the case of $HoNi_2B_2C$ for which three different types of antiferromagnetic structures have been observed: (*i*) a commensurate one with Ho moments aligned ferromagnetically within layers perpendicular to the tetragonal *c*-axis where consecutive layers are aligned in opposite directions, (*ii*) an incommensurate spiral along the *c*-axis and (*iii*) an incommensurate *a*-axis modulated structure with a modulation $\tau \approx (0.55, 0, 0)$. This wave vector emerges in various RNi_2B_2C compounds with magnetic as well as nonmagnetic R elements and is connected with Fermi surface nesting. In an external magnetic field, some of the RNi_2B_2C compounds show metamagnetic transitions combined with a large negative magnetoresistance.

Recently, superconducting $ErNi_2B_2C$ (T_c = 11 K) appeared to develop a net magnetization below 2.3K, well below the onset of long range spin density-wave (SDW) order at T_N = 6 K. This transition does indeed correspond to the development of a net atomic magnetization that coexists with superconductivity, and this results in the spontaneous formation of vortices in the system.

Superconducting Er Ni$_2$B$_2$C orders magnetically at 6 K into a transversely polarized spin density wave (SDW) structure [49, 50], with the modulation wave vector δ along the a-axis and spins along the b-direction. The initial ordering is a simple spin density wave, with an incommensurate modulation of δ approximately equal to 0.55 a^* ($a^* = 2\pi/a$). At 4.2 K additional higher order peaks become observable as the magnetic structure squares up. These peaks are all odd-order harmonics as expected for a square-wave magnetic structure.

Below the weak ferromagnetic transition (T$_{WFM}$) at 2.3 K, the experimental data indicate that a new series of peaks has developed, which are *even* harmonics of the fundamental wave vector. Polarized neutron measurements unambiguously establish that both the odd-order and the even-order peaks are magnetic in origin. The even-order peaks abruptly develops below 2.3 K, concomitant with the development of new (weak) ferromagnetism at T$_{WFM}$. There is a substantial thermal hysteresis associated with the weak ferromagnetic transition, suggesting that this transition is first order in nature.

The present neutron results [48] demonstrate that a net magnetization develops in Er Ni$_2$B$_2$C in the magnetically ordered state at low temperatures, making this the first such "ferromagnetic superconductor" since HoMo$_6$S$_8$, HoMo$_6$Se$_8$, and ErRh$_4$B$_4$. For Er Ni$_2$B$_2$C the net magnetization is much smaller than for these earlier systems, which allows coexistence with superconductivity over an extended T range. This presents the intriguing possibility that in an applied field vortices will form spontaneously when this net atomic magnetization is present. The small angle scattering data of the vortex structure show that the lattice has the expected spacing at lower applied fields. At higher fields as the field begins to penetrate, vortices spontaneously form in addition to those expected from the applied field alone, increasing the vortex density and shifting the vortex peak. The T dependence of this shift makes it clear that this behaviour is directly related to onset of the net magnetization in the system. It is also interesting to note that the vortex pinning is enhanced in both magnetic phases, which may prove useful in high current applications of superconductors.

9.3.5 Ni/Bi Bilayer

There is a unambiguous evidence for the coexistence of superconductivity and ferromagnetism in a simple system - Ni/Bi bilayers [51]. Perhaps this represents one of the first observations of superconductivity and ferromagnetism coexisting in a simple, well defined system. The two states spatially coexist, and the same electrons are apparently responsible for both phenomena.

In many respects, the Ni/Bi system is remarkably novel. Bismuth is an example of an element normally superconducting only under pressure or when quench-condensed onto liquid-helium cooled substrates.

In this case superconductivity and ferromagnetism are separated in space, they strongly interact via magnetic fields.

Interplay of ferromagnetism and superconductivity makes possible interaction of superconducting topological defects (vortices) with ferromagnetic defects (domain walls and Bloch lines). It is expected that this interaction will have a pronounced effect on static and dynamic properties of materials with ferromagnetism–superconductivity coexistence.

9.3.6 Itinerant Ferromagnetic Superconductors

Very recently ferromagnetic superconductivity has been observed in UGe_2 [52], $ZrZn_2$ [53], and URhGe [54]. The superconductivity is confined to the ferromagnetic phase. Ferromagnetism and superconductivity are believed to arise due to the same band electrons. The persistence of ferromagnetic order within the superconducting phase has been ascertained by neutron scattering. The specific heat anomaly associated with the superconducting transition in these materials appears to be absent.

At ambient pressure UGe_2 is an itinerant ferromagnet below the Curie temperature $T_c = 52$ K, with low temperature ordered magnetic moment of $\mu_S = 1.4$ μ_B/U. With increasing pressure the system passes through two successive quantum phase transitions from ferromagnetism to FM superconductivity at $P \sim 10$ K bar, and at higher pressure $P_c \sim 16$ K bar to paramagnetism [52, 55]. At the pressure where the superconducting transition temperature is a maximum $T_{SC} = 0.8$ K, the ferromagnetic state is still stable with $T_c = 32$ K, and the system undergoes a first order metamagnetic transition between two ferromagnetic phases with different ordered moments [56]. The specific heat coefficient $\gamma = C/T$ increases steeply near 11 k bar and retains a large and nearly constant value [57]. The crystal structure of UGe_2 is shown in Fig. 9.6.

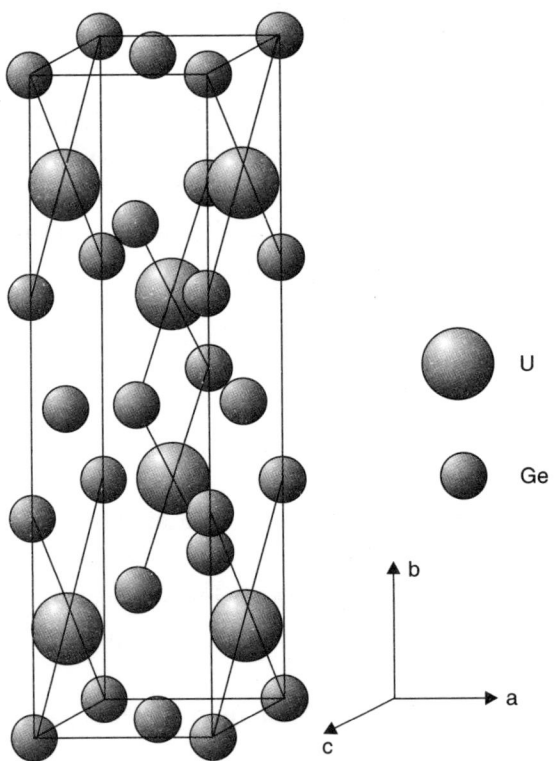

Fig. 9.6 Crystal structure of UGe_2.

The ferromagnets $Zr Zn_2$ and URhGe are superconducting at ambient pressure with superconducting critical temperature $T_{SC} = 0.29$ K, and $T_{SC} = 0.25$ K, respectively. $ZrZn_2$ is ferromagnetic below the Curie temperature $T_c = 28.5$ K with low temperature ordered moment of $\mu_S = 0.17$ μ_B per formula unit, while for URhGe, $T_c = 9.5$ K and $\mu_S = 0.42$ μ_B. The low Curie temperatures (T_c) and small ordered magnetic moments indicate that compounds are close to a ferromagnetic quantum critical point. A large jump in the specific heat, at the temperature where the resistivity becomes zero, is observed in URhGe. At low temperature, the specific heat coefficient γ is twice smaller than in the ferromagnetic phase materials.

Both UGe_2 and $ZrZn_2$ have transition temperatures below 1 K, but these materials and the other like them waiting to be discovered offer theorists and experimenters the chance to investigate the interaction of ferromagnetism with superconductivity in an energetically tranquil low-temperature regime. That opportunity could yield clues to understand another class of materials where superconductivity and magnetism interact: the HTSC cuprates. The crystal structure of $ZrZn_2$ is shown in Fig. 9.7.

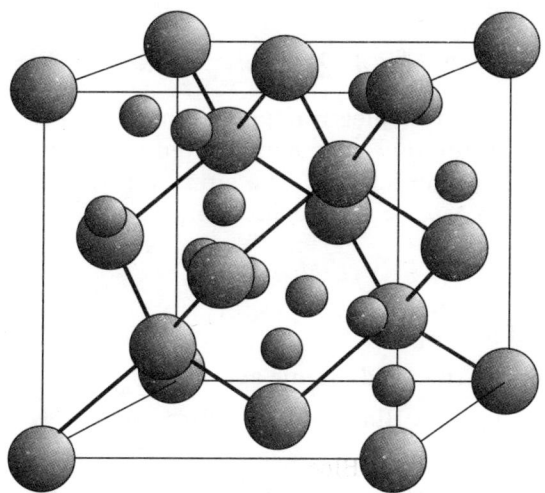

Fig. 9.7 The crystal structure of $ZrZn_2$.

9.3.7 Superconductivity and Ferromagnetism: Mechanism

Superconductivity and ferromagnetism are two states with long-range order that may exist in materials at low temperatures. Both these states compete with each other and it is difficult to understand how they coexist. However their coexistence has been demonstrated experimentally in number of systems, *e.g.* $ErRh_4B_4$, $HoMo_6S_8$, etc. Superconductivity and ferromagnetism coexist microscopically in these systems, but the interplay of these two phenomena transforms the ferromagnetic state with the superconducting regions into oscillatory magnetic state. A simple model could be one with the rare-earth elements forming the ferromagnetic lattice, while the transition metal creates the superconducting state [56, 58].

The Hamiltonian which describes such a model is

$$H = \sum_{K,\sigma} (\varepsilon_K - \mu) C^+_{K\sigma} C_{K\sigma} - \Delta \sum_K (C^+_{K+} C^+_{-K-} + C_{-K-} C_{K+})$$

$$-\frac{J}{N} \sum_{K_1,K_2,f} [-i\mathbf{f} \cdot (\mathbf{K}_1 - \mathbf{K}_2)][C^+_{K_1-} C_{K_2+} S^+_f + C^+_{K_1+} C_{K_2-} S^-_f + C^+_{K_2+} S^z_f - C^+_{K_1-} C_{K_2-} S^z_f] \quad ...(9.1)$$

Here J is the exchange strength of the conduction electrons with periodically localized spins, and is taken to be independent of K, and $<S^z>$ is the spontaneous magnetization per atom, to be determined self consistently from the equation

$$S(S+1) = <S^z>(<S^z>+1) - \frac{1}{N}\sum_K <S^- S^+>_K \qquad ...(9.2)$$

S is a spin operator for the localized moments, which obey the usual spin commutation rules.

The summation over K can be changed into integration using the relation

$$\sum_K = \frac{1}{2\pi}\int d\theta\, 2\pi \sin\theta \int_0^{K_F} dK\, K^2 \qquad ...[9.2(a)]$$

μ is the chemical potential. Δ, the superconducting order parameter is determined self-consistently from the gap equation

$$\Delta = |g|\sum_K <C^+_{K+} - C^+_{-K-}> \qquad ...(9.3)$$

where $|g|$ ($|g|>0$) is the electron–electron, phonon mediated coupling constant, having the dimensions of energy. The summation over K is limited by the wave vector corresponding to the Debye energy $\hbar\omega_D$ at the Fermi surface. The sum over K is restricted to the "interaction shell" $|\epsilon_K| \ll \hbar\omega_D$ and since $\hbar\omega_D \ll E_F$ (Fermi energy), one can put

$$\sum_K = N(0)\int_{\mu-\hbar\omega_D}^{\mu+\hbar\omega_D} \qquad ...[9.3(a)]$$

where $N(0)$ is the density of states.

Here $<...>$ indicates the thermodynamic average. $C^+_{K\sigma}$ and $C_{K\sigma}$ are the fermion creation and annihilation operators for the Bloch states $|K\sigma>$, where $\mathbf{K}$ is the conduction electron wave vector and σ the spin.

Following the equation of motion method and Green's function technique, and retaining terms upto second order in Δ and J, one obtains the following expressions for Δ, $<S^z>$, Curie temperature, T_M and the ac susceptibility χ_{ac} as

$$\frac{1}{|g|N(0)} = A\int_{\mu-\hbar\omega_D}^{\mu+\hbar\omega_D} d\varepsilon_K \left[\frac{B}{2}\left(\frac{1}{\exp\left\{\frac{1}{2}\beta[-B+(B^2+4C)^{1/2}]\right\}+1}\right.\right.$$

$$\left.+\frac{1}{\exp\left\{-\frac{1}{2}\beta[B+(B^2+4C)^{1/2}]\right\}+1}\right) - \frac{B^2+2(C+D)}{2(B^2+4C)^{1/2}}$$

$$\left.\left(\frac{1}{\exp\left\{\frac{1}{2}\beta[-B+(B^2+4C)^{1/2}]\right\}+1} - \frac{1}{\exp\left\{-\frac{1}{2}\beta[B+(B^2+4C)^{1/2}]\right\}}+1\right)\right] \qquad ...(9.4)$$

Here
$$A = \frac{2P + Q <S^z>}{3(\hbar\omega_D)^2 (\hbar\omega_D - 4P)}$$

$$B = 2J <S^z> - 2Q <S^z> \left(1 + \frac{t}{\hbar\omega_D} - \frac{2\Delta^2}{(\hbar\omega_D)^2} - \frac{1}{3}\frac{\Delta^2}{(\hbar\omega_D)^2} t\right)$$

$$C = \frac{\hbar\omega_D}{\hbar\omega_D - 4P}\left[t^2 + \Delta^2\left(1 + \frac{4P}{\hbar\omega_D}\right)\right]$$

$$D = \frac{3(\hbar\omega_D)^2}{2P + Q <S^z>}(\hbar\omega_D + 2P + Q <S^z>) - \Delta^2$$

$$t = \varepsilon_K - \mu$$

$$P = \frac{2J^2}{N} N(0) S(S+1)$$

$$Q = \frac{2J^2}{N} N(0)$$

$$S(S+1) = <S^z>(<S^z> + 1) - \frac{2<S^z>}{N} \qquad \ldots(9.5)$$

$$\times \sum_K \exp\beta \frac{2J^2 <S^z>}{4\pi^2 AN}\left[K_F + \frac{1}{4}\frac{AK^3}{(4\Delta^2 + A^2 K^4)^{1/2}}\right.$$

$$- \frac{(4\Delta^2 + A^2 K^4)^{1/2}}{2AK}\right) \tan^{-1}\frac{2AKK_F}{(4\Delta^2 + A^2 K^4)^{1/2}}$$

$$- \frac{5}{6}\frac{A^2 K^4 K_F}{4\Delta^2 + A^2 K^4 + 4A^2 K K_F^2} \cdot$$

$$\left. + \frac{1}{3}\frac{A^2 K^4 K_F (4\Delta^2 + A^2 K^4)}{(4\Delta^2 + A^2 K^4 + 4A^2 K^2 K_F^2)}\right] - 1\} \qquad \ldots(9.6)$$

Here
$$\Delta = \frac{\hbar^2}{2} m$$

One obtains, Curie temperature

$$T_M(\Delta = 0) = \frac{0.92 J^2 S(S+1) N(0)}{k_B} \qquad \ldots(9.7)$$

$$T_M(\Delta \neq 0) = \frac{0.74 J^2 S(S+1) N(0)}{k_B [1 + 0.23 \Delta^2 N(0)^2]} \qquad \ldots(9.8)$$

and
$$\chi_\omega(\Delta = 0) = 4\mu_B^2 \left(\frac{\hbar\omega_D}{<S^z>} + \frac{m J^2 K_F}{\pi^2 \hbar^2 N}\right)^{-1} \qquad \ldots(9.9)$$

$$\chi_\omega (\Delta \neq 0) = 4\mu_B^2 \left(\frac{\hbar\omega_D}{<S^2>} + \frac{J^2\hbar^2 K_F^5}{40\pi^2 \, N\Delta^2 m} \right)^{-1} \quad ...(9.10)$$

Equations (9.7) and (9.8) clearly reveal that superconductivity and ferromagnetism can truly coexist in the same specimen. The theory explains satisfactorily the experimental results for Er Rh$_4$B$_4$ and HoMo$_6$S$_8$ ferromagnetic superconductors. From eqs. (9.8) and (9.9), it is evident that the usual RKKY interaction is modified, of course by the appearance of the superconducting gap in the quasi electron spectra. Now, on putting $\Delta = 0$ in eq. (9.8), it reduces to eq. (9.7). We may note that a positive term containing $0.23 \, \Delta^2 \, N(0)^2$ appears in the denominator of eq. (9.8). This shows that $T_M (\Delta \neq 0) < T_M (\Delta \neq 0)$, i.e. superconductivity suppress ferromagnetism and also persists below the Curie temperature, T_M.

One obtains the following expressions for electronic specific heat and density of states $N(\omega)$ for the ferromagnetic superconductors

Electronic specific Heat,

$$C_{el}^s = \frac{N(0)}{k_B T^2} \left(\frac{\hbar\omega_D}{\hbar\omega_D - 4P} \right) \int_0^{\hbar\omega_D} dt \, t^2 \, \mathrm{sec} \, h^2 \left\{ \frac{1}{2k_B T} \right.$$

$$\left. \left(\frac{\hbar\omega_D}{\hbar\omega_D - 4P} \right)^{1/2} \left[t^2 + \frac{\Delta^2}{\hbar\omega_D} (\hbar\omega_D + 4P) \right]^{1/2} \right\} \quad ...(9.11)$$

where t and P are given by Equation (9.5). Eq. (9.11) has been obtained using the following relation

$$C_{el}^s = \frac{\partial}{\partial T} \frac{1}{N} \sum_K 2(\varepsilon_K - \mu) <C_{K\uparrow}^+ C_{K\uparrow}> \quad ...(9.12)$$

where $<C_{K\uparrow}^+ C_{K\uparrow}>$ is the correlation function, which can be evaluated using the relation

$$<B(t') A(t)> = \lim_{\varepsilon \to 0} \frac{i}{2\pi} \int_{-\infty}^{\infty} \frac{<<A(t); B(t')>>_{\omega+i\varepsilon} - <<A(t); B(t)>>_{\omega-i\varepsilon}}{e^{\beta\omega} - 1}$$

$$\times \exp[-i\omega(t - t')] \, d\omega \quad ...(9.13)$$

where $<<A(t); B(t')>>$ is the Green function; and employing the identity

$$\lim_{\varepsilon \to 0} \left(\frac{1}{\omega + i\varepsilon - E_K} - \frac{1}{\omega - i\varepsilon - E_K} \right) = 2\pi \, i\delta(\omega - E_K) \quad ...(9.14)$$

Density of States

An important function and the one most susceptible to experimental verification is the density of states as a function of the excitation energy ε_k. The density of states per atom per spin $N(\omega)$ is given by

$$N(\omega) = \lim_{\varepsilon \to 0} \frac{i}{2\pi N} \sum_K [G_{\uparrow\uparrow}(K, \omega + i\varepsilon) - G_{\downarrow\downarrow}(K, \omega - i\varepsilon)] \quad ...(9.15)$$

where
$$G_{\uparrow\uparrow}(K'', K', t) = <<C_{K''\uparrow}\,;\, C^+_{K'\uparrow}>> = -i\theta(t) <[C^+_{K''\uparrow}(t)\,;\, C^+_{K'\uparrow}(0)]>$$
and
$$G_{\downarrow\downarrow}(K'', K', t) = <<C_{K''\downarrow}\,;\, C^+_{K'\downarrow}>> = -i\theta(t) <[C_{K''\uparrow}(t)\,;\, C^+_{K'\downarrow}(0)]>$$
are electron Green's functions.

We obtain

$$\frac{N(\omega)}{N(0)} = \left(\frac{\hbar\omega_D}{\hbar\omega_D - 4p}\right)\left[\frac{1-2P}{\hbar\omega_D} + \frac{2P\Delta^2}{3(\hbar\omega_D)^2}\right] \times \frac{\omega}{\left\{[\hbar\omega_D - 4P]/\hbar\omega_D]\omega^2 - \Delta^2(1 - 4P/\hbar\omega_D)^{1/2}\right\}}$$
...(9.16)

Free Energy

The free energy difference can be calculated by means of the relation [61]

$$(F_S - F_N)/V = \int_0^\Delta d\Delta \frac{d}{d\Delta}(1/[|g|N(0)]^2) \qquad ...(9.17)$$

For lower temperatures, one obtains

$$\frac{F_S - F_N}{V} = N(0)\int_0^{\hbar\omega_D} dt \left[\frac{2}{3}\cdot\frac{P\Delta^2}{(\hbar\omega_D)^3}\right] + \left\{\Delta^2\left(1 - \frac{4P}{\hbar\omega_D}\right)\right/t\right\}$$

$$\tanh\left[\frac{\beta}{2}\left(1 + \frac{2P}{\hbar\omega_D}\right)(t^2 + \Delta^2)^{1/2}\right] - N(0)\int_0^{\hbar\omega_D} dt \int_0^\Delta dt \left[\frac{t^2}{(t^2 + \Delta^2)^{1/2}}\frac{2P}{3(\hbar\omega_D)^3}\right.$$

$$\left. + \frac{1 - 6P/\hbar\omega_D}{(t^2 + \Delta^2)^{1/2}}\right]\tanh\left[\frac{\beta}{2}\left(1 + \frac{2P}{\hbar\omega_D}\right)(t^2+\Delta^2)^{1/2}\right] \qquad ...(9.18)$$

The above result is obtained with the help of partial integration. The first term in Eq. (9.18) is equal to $\Delta_0^2/|g|$. In this way, one obtains:

$$(F_S - F_N)/V = \Delta_0^2/|g| - \frac{2N(0)}{\beta}\left(1 - \frac{8P}{\hbar\omega_D}\right)\exp(-\beta\Delta_0)$$

$$\left(\frac{2\pi\Delta_0}{\beta}\right)^{1/2} - \left(1 - \frac{6P}{\hbar\omega_D}\right)\left\{N(0)\Delta^2 + \frac{\Delta^2}{|g|} + N(0)\Delta^2\ln(\Delta_0/\Delta)\right\}$$

$$+ \frac{1}{3}\left(1 - \frac{6P}{\hbar\omega_D}\right)\frac{N(0)}{\beta^2}\pi^2 - \frac{4}{3}\frac{N(0)}{\beta}\frac{P}{(\hbar\omega_D)^3}$$

$$\int_0^{\hbar\omega_D} dt\, t^2 \ln\left(\frac{\cosh\beta/2(1+2P/\hbar\omega_D)(t^2+\Delta^2)^{1/2}}{\cosh\beta(1+2P/\hbar\omega_D)}\right) \qquad ...(9.19)$$

where Δ_0 is the zero temperature gap, whose value can be determined from

$$\beta\Delta_0 = \pi e^{-\gamma} \simeq 1.76 \qquad ...(9.20)$$

Critical Field

Low temperature critical field can be obtained using

$$F_S(T, 0) = F_N(T, 0) - (8\pi)^{-1} H_c^2$$

and eq. (9.19). One obtains

$$H_c = (8\pi V)^{1/2} \left[\Delta^2/|g| - \frac{2N(0)}{\beta}(1 - 8P/\hbar\omega_D) \right]$$

$$\exp(-\beta\Delta_0)(2\pi\Delta_0/\beta)^{1/2} - (1 - 6P/\hbar\omega_D)$$
$$\{N(0)\Delta^2 + \Delta^2/|g| + N(0)\Delta^2 \ln(\Delta_0/\Delta)\}$$

$$+ \frac{1}{3}(1 - 6P/\hbar\omega_D)\frac{N(0)\pi^2}{\beta^2} - \frac{4}{3}\frac{N(0)}{\beta}\frac{P}{(\hbar\omega_D)^3}$$

$$\int dt\, t^2 \ln\left(\cosh\frac{\beta}{2}\left[1 + \frac{2P}{\hbar\omega_D}\right]\right)(t^2 + \Delta^2)^{1/2}]^{1/2} \qquad ...(9.21)$$

This model explains satisfactorily the interplay of superconductivity and ferromagnetism in $ErRh_4B_4$ and $HoMo_6S_8$.

One can also study the interplay of ferromagnetism and superconductivity in a system using the following model Hamiltonian:

$$H = \sum_{p,\alpha} \varepsilon(p) C^+_{p\alpha} C_{p\alpha} - \Delta \sum_p (C^+_{p\uparrow} C^+_{-p\downarrow} + C_{-p\downarrow} C_{p\uparrow})$$

$$- \frac{I}{2N} \sum_i \sum_{\substack{p,p' \\ \alpha,\beta}} C^+_{p\alpha}\, \sigma_{\alpha\beta} \cdot C_{p\beta} S_i \exp[iR_i(p - p')]$$

$$- \frac{1}{2}\sum_{i \neq j} J_{ij}\, \mathbf{S}_i \mathbf{S}_j \qquad ...(9.22)$$

where the first two terms describes the superconducting state, the third describes the interactions between electrons and spins (with the coupling constant I) and the last term describes the spin-spin interaction (with the coupling constant J) is supposed to be ferromagnetic.

Maekawa and Tachiki [62] obtained the following expression for the critical temperature T_c of the ferromagnetic superconductor as

$$T_c = 1.14\omega_D \exp\left[-1/-(|g|N(0) - \frac{3C_0 I^2}{4p_0^2 a^2} \ln\frac{T_c - T_m + 4p_0^2 \alpha^2}{T_c - T_m}\right] \qquad ...(9.23)$$

where $|g|$ is the BCS electron–electron interaction and $N(0)$ the density of states. The spin-spin correlation has been considered as

$$\chi(q,T) = C_0/(T - T_M + \alpha^2 q^2) \qquad ...(9.24)$$

where T_M is magnetic transition temperature and

$$C_0 = S(S+1)\,T_M \qquad ....(9.24(a))$$

If we consider the interaction between conduction electrons and the local moments in the Born approximation, one can use the results from the theory of dilute alloys to calculate all the physical quantities for the ferromagnetic superconductor [63]. The only difference (which is expected to occur) is that for a ferromagnetic superconductor, one do not perform an "average on the impurities" but a summation on the regular lattice.

The Green's functions of the interest are

$$G^{-1}(p;\omega) = G^{0-1}(p,\omega) - \Sigma(p;\omega) = i\omega - \varepsilon(p)\tau_3 \Delta \tau_1 \sigma_y$$
$$G^{0-1}(p,\omega) = (i\omega - \varepsilon(p)\tau_3 - \Delta\tau_1\sigma_y)^{-1} \qquad ...(9.25)$$

where Σ is taken as

$$\Sigma(p;\omega) = -\frac{1}{2\tau(T)} \frac{(i\omega - \Delta\tau_1\sigma_y)}{(\omega^2 + \Delta^2)^{1/2}} \sin\omega \qquad ...(9.26)$$

Using for the spin-spin correlation the expression (9.24), one obtains the scattering time as

$$\frac{1}{\tau(T)} = \frac{T}{\tau_s} f(T) \qquad ...(9.27)$$

where τ_s is defined as [63]

$$\frac{1}{2\tau_s} = \frac{1}{\tau_1} - \frac{1}{\tau_2} \qquad ...(9.28)$$

where τ_1 and τ_2 are scattering times. The scattering time can be calculated as

$$\frac{1}{\tau(T)} = \frac{T}{\tau_s} f(T) \qquad ...(9.29)$$

where τ_s is defined by eq. (9.28) and

$$f(T) = \frac{1}{(2a\,p_0)} \ln \frac{T - T_M + (2a\,p_0)^2}{T - T_M} \qquad ...(9.30)$$

Equation (9.25) gives

$$\frac{\omega}{\Delta} = u\left(1 - \frac{1}{\tau(T)} \frac{1}{\sqrt{1+u^2}}\right) \qquad ...(9.31)$$

and the equation for the gap

$$\Delta = -|g|\pi T \sum_\omega \int \frac{d^3p}{(2\pi)^3} \frac{1}{4} Tr\,(\tau_1 \sigma_y G(p;\omega)) \qquad ...(9.32)$$

can be expressed in terms of u as

$$\Delta = |g|\,N(0)\,2\pi T \sum_\omega (1+u^2)^{-1/2} \qquad ...(9.33)$$

Near the critical temperature T_c, the eq. (9.33) can be linearized and one obtains

$$\ln \frac{T}{T_{c_0}} = \psi\left(\frac{1}{2}\right) - \psi\left(\frac{1}{2} + \rho_c(T_c)\right) \qquad ...(9.34)$$

where

$$\rho_c(T_c) = \frac{1}{2\pi T_c\,\tau(T_c)} \qquad ...(9.35)$$

The above result has been obtained by Machida and Younger [64] and from Fig. 9.8.

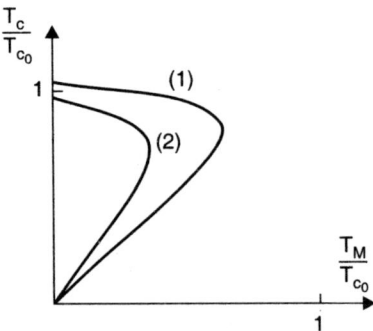

Fig. 9.8 The phase diagram for a ferromagnetic superconductor. The curve (1) corresponds to $(2p_0 a)^2/T_{c_0} = 0.5$ and $N(0) (1/2)^3 S(S + 1)/T_{c_0} = 0.01$. The curve (2) is for the same value of the first parameter but the second parameter is 0.07.

One can easily see that there are always two temperatures which give solution to this equation. The upper critical temperature T_{c_1} which is obtained as

$$T_{c_1} = T_{c_0} - \frac{\pi}{4} \frac{1}{\tau(T_c)} \qquad ...(9.36)$$

can be transformed

$$\frac{T_{c_1}}{T_{c_0}} = 1 - \frac{\pi}{4} \frac{1}{\tau_{\Delta G} (2a\, p_0)^2} \ln \frac{T_{c_0} + (2a\, p_0)^2}{T_{c_0}} - \frac{\pi}{4} \frac{1}{T_{c_0} + (2a\, p_0)^2} \frac{T_M}{T_{c_0}} \qquad ...(9.37)$$

If $T_M \ll T_{c_0}$. The second solution of eq. (9.34) T_{c_2} cannot be calculated analytically, but it is very near to T_M. The upper critical field H_{c_2} has been calculated by the standard method by Ramakrishnan and Varma [65].

The equation for the superconducting order parameter Δ in the presence of the electromagnetic field A is

$$\Delta(x) = g\pi T \sum_\omega \int dx_1\, Q_{\uparrow\downarrow}(x_1 - x, \omega) \left\{ \exp\left[2e \int_x^{x_1} \mathbf{ds} \cdot \mathbf{A}(s)\right] \Delta(x_1) \right\} \qquad ...(9.38)$$

and

$$G_\alpha^o (x, \omega) = \frac{-m}{2\pi |\chi|} \exp\left\{ ip_{0\alpha} |\chi| \sin \omega - \frac{|\tilde\omega|}{V_{0\alpha}} |\chi| \right\} \qquad ...(9.39)$$

where

$$\tilde\omega = \omega + \frac{1}{2} (\tau_0^{-1} + \tau^{-1}(T)) \qquad ...(9.40)$$

In eq. (9.40), τ is the scattering time for the non-magnetic impurities and $P_{0\alpha} = p_0\, \alpha h\, V_0$.

The internal molecular field h acting on the conduction electrons contains a term which describes the interaction of the conduction electrons with the local moments and the Pauli paramagnetic term

$$h = \frac{I\,\chi(T)}{2\mu_0 N} + \mu_0 B \qquad \ldots(9.41)$$

where N is the number of local spins per unit volume and

$$\chi(T) = [(\mu_0 g)^2 N/T] \langle S_q S_{-q}\rangle$$

In the dirty limit $\tau^{-1} \gg T_c$, the upper critical field for a ferromagnetic superconductor is given by

$$\ln\left(\frac{T}{T_{c_0}}\right) + \mathrm{Re}\,\psi\left(\frac{1}{2} + \frac{1}{2\pi T\,\tau(T)} + \frac{ih}{2\pi T} + \frac{DeB}{2\pi T}\right) = \psi\left(\frac{1}{2}\right) \qquad \ldots(9.42)$$

One can generalize eq. (9.42) to include the effect of spin-orbit scattering. If we define the spin orbit scattering time by

$$\tau_{So}^{-1} = C_1 \frac{N(0)}{g}\,4\pi\,|V_{so}|^2 \qquad \ldots(9.43)$$

where V_{so} is the spin-orbit scattering coupling constant. One obtains the equation for the upper critical field $H_{c_2}(T)$ as

$$\ln\frac{T}{T_c} = \frac{1}{2}\left[\left(1 + \frac{b}{\sqrt{b^2 - h^2}}\right)\psi\left(\frac{1}{2} + \rho_-\right) + \left(1 - \frac{b}{\sqrt{b^2 - h^2}}\right)\psi\left(\frac{1}{2} + \rho_+\right)\right] - \psi\left(\frac{1}{2}\right) \qquad \ldots(9.44)$$

where
$$\rho_\pm = \frac{1}{2\pi T}\left[\frac{1}{\tau(T)} + \frac{1}{\tau_{so}} + DeB \pm (b^2 - h^2)^{1/2}\right] \qquad \ldots(9.45)$$

and $b = 1/\tau_{so}$, τ_{so} being the spin-orbit scattering time.

This case becomes important in the high-field type-II superconductors. The molecular field acting on the conduction electrons is linear in I. If the exchange integral J is negative, the spin polarization counteracts the effects the Pauli paramagnetism and increases the upper critical field. This effect is called as *Jaccarino-Peter effect* [66].

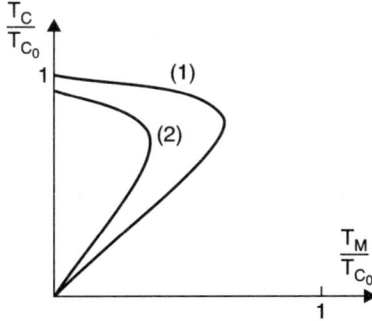

Fig. 9.9 The phase diagram of the ferromagnetic superconductor with dynamical spin-spin correlation. Curve (1) is identical with the curve (1) of Fig. 9.8 and corresponds to $\alpha = 1$, $\beta = 0$. Curve (2) is for $\alpha = 0.5$ and $\beta = 0.5$.

Phase diagram 9.9 reveals that the system always exhibits a re-entrance phenomenon. In this model, there is no region where the two-long range orders coexist, but some ternary compounds exhibit this coexistence. Machida and Yonger [64] have generalized these results by considering the elastic and inelastic scattering of the electrons on the localized spins. These effects can be considered taking for the linearized equation of the order parameter of the form

$$\Delta(T) = gQ(T)\,\Delta(T) \qquad \ldots(9.46)$$

where
$$Q(T) = \pi T \sum_\omega \sum_p \gamma(\omega)\, G(p,\omega) G(p,-\omega) \qquad \ldots(9.47)$$

with the Green function
$$G^{-1}(p,\omega) = i\omega - \varepsilon(p) - \Sigma(\omega) \qquad \ldots(9.48)$$

On performing the momentum integration in eq. (9.47), one obtains

$$Q(T) = \pi T N(0) \sum_\omega \frac{\gamma(\omega)}{|\omega| + \Sigma(\omega)} \qquad \ldots(9.49)$$

where $\gamma(\omega)$ can be calculated from the Dyson equation, one obtains

$$\gamma(\omega) = 1 + \pi T \sum_{p';\omega} T_{\uparrow\downarrow}(\mathbf{p},\omega;\mathbf{p}',\omega')\, G(p,\omega)\, G(-\mathbf{p}'-\omega')\,\gamma(\omega') \qquad \ldots(9.50)$$

where $\Gamma_{\uparrow\downarrow}$ is the irreducible four-point vortex [67]. The salt energy $\Sigma(\omega)$ from eq. (9.49) is given by

$$\Sigma(p;\omega) = \left(\frac{I}{2}\right)^2 \pi T \sum_{p';\omega'} G(p',\omega)\,\chi(\mathbf{p}'-\mathbf{p};\omega-\omega') \qquad \ldots(9.51)$$

where $\chi(\mathbf{q},\omega)$ is the dynamical susceptibility of the localized spin system. Taking for the four-point vortex, the simple expression.

$$\Gamma_{\uparrow\downarrow}(\mathbf{p},\omega;\mathbf{p}',\omega') = -\left(\frac{I}{2}\right)^2 \chi(\mathbf{p}-\mathbf{p}';\omega-\omega') \qquad \ldots(9.52)$$

One obtains the critical temperature as solution of eq. (9.52)

$$\frac{1+\rho_c}{|g|N(0)-\rho_c} = 2\pi T_c \sum_\omega \frac{1}{\omega} \qquad \ldots(9.53)$$

One finds that effective coupling which gives rise to the superconducting state is weakened by the repulsive paramagnon-mediated interaction. Equation (9.53) has two solutions T_{c_1} and T_{c_2}. The first critical temperature can be approximated for $(T_{c_0} - T_{c_1})T_{c_1} \ll 1$ and $\rho_{c_1} \ll 1$ as

$$\frac{T_{c_0} - T_{c_1}}{T_{c_1}} = \rho_{c_1}\left[\frac{1}{(|g|N(0))^2} + \frac{1}{|g|N(0)}\right] \qquad \ldots(9.54)$$

One can calculate the second critical temperature only numerically. The dynamical character of $\chi(\mathbf{q};\omega)$ can be considered taking

$$\chi(\mathbf{q};\omega) = \chi(\mathbf{q})\,[\alpha + \beta\delta_{\omega,0}) \qquad \ldots(9.55)$$

where α and β are adjustable parameters.

One can calculate the critical temperature from the standard relation $1 = gQ(T_c)$, where

$$Q(T) = \frac{N(0)\phi(T)}{1 + \alpha\rho\phi(T)} \qquad ...(9.56)$$

where

$$\phi(T) = 2\pi T \sum_\omega \frac{1}{\omega(1+\alpha\rho) + \beta\rho} \qquad ...(9.57)$$

With these results, one gets two critical temperatures T_{c_1} and T_{c_2} and as stated above, the first one can be approximated for $(T_{c_0} - T_{c_1})/T_{c_0} \ll 1$ and $\rho_{c_1} \ll 1$

as

$$\frac{T_{c_0} - T_{c_1}}{T_{c_0}} = \rho_{c_1}\left\{\alpha\left[\frac{1}{|g|N(0)^2} + \frac{1}{|g|N(0)}\right] + 3\beta\xi(2)\right\} \qquad ...(9.58)$$

The upper critical field H_{c_2} can be calculated in a similar way and the temperature dependence presents the same qualitative behaviour.

This model predicts the occurrence of the re-entrance effect in agreement with the predictions of Machida and Yonger [64] and Kakani and Upadhayaya [58]. Grest et al., [68] calculated the density of states $N_s(\omega)$ for a re-entrant superconductor using the Eliashberg equations for the spin fluctuations. A reasonable agreement with the tunnelling experiments [69] has been obtained, but more accurate calculations taking into account the effect of crystal field splitting have to be performed.

Bulaevski et al. [70] considered the magnetic structures which may appear in the re-entrant superconductors due to the exchange and the electromagnetic dipolar interactions of the localized moments and electrons and the anisotropy. We may note that such a realistic model was necessary to explain all kinds of transitions in the real compounds, because it is impossible to explain all kinds of transitions only by the electromagnetic interaction.

The role of the electromagnetic interaction is, in fact, to give rise to a nonuniform magnetic structure in the superconducting state but the anisotropy could modify the helical structure. In the presence of the magnetic anisotropy, the magnetic structure is dominated by a special phase which has a transverse character but the spins are arranged in the magnetic domains. The domain structure is one dimensional because the energy of the superconducting state depends weakly on the type of the magnetic structure. The transition between the superconducting state to the domain structure may be of the second order or of the first order. In the first case, the domain phase appears by the modification of the helical structures and in the second case, this phase appears by the creation of series of nuclei with alternating opposite direction of magnetization [63].

Machida and Nakanishi [71] have shown that superconductivity, helical order and ferromagnetism really coexist in many ternary systems. They considered the system of conduction electrons as one-dimensional interacting with a localized spin system. They have performed the calculations in the mean field approximation and the self-consistent equations have been solved in terms of the **Weierstrass functions**. The sinusoidal modulated magnetic phase (observed experimentally) is more stable than the helical magnetic phase (observed experimentally) is more stable than the helical structure which appears as a metastable one.

The superconducting state is also spatially modulated, and the solutions of the self-consistent equations revealed the existence of a soliton lattice which has a two-energy gap structure and a spin-density polarization of the conduction electrons. However, the electronic system in this model is considered as one-dimensional, otherwise this explains reasonably the coexistence between the three phases.

9.3.8 Superconductivity and Antiferromagnetism

Several series of isostructural rare earth compounds exhibit true coexistence of superconductivity and antiferromagnetism [5, 6, 8, 14, 17, 58, 59, 72 -75]. In the ternary rare earth systems, the magnetic rare earth ions are distributed periodically throughout the crystal lattice, and each rare earth ion has a partially filled 4f electron shell a corresponding magnetic moment. These compounds are metallic and has provided for the first time experimentally realizable systems for studying the interplay between superconductivity and antiferromagnetism. The rare earth chalcogenides and rare earth rhodium borides are the typical examples of antiferromagnetic superconductors. On cooling, these compounds first become superconducting below the superconducting transition temperature (T_c) and then on further cooling below the Neel temperature (T_N) the system is antiferromagnetically ordered which coexist with superconductivity. Typical examples are $RE_xMo_6S_8$ (for RE = Gd, Tb, Dy and Er) with x = 1.0 or 1.2 and $RERh_4B_4$ (for RE = Nd, Sm and Tm). The most important behaviour about these systems is the anomalous behaviour of the upper critical field H_{c_2} versus temperature T in the vicinity of T_N. The compound $Tb_2Mo_3Si_4$ [76] is an exception in the sense that its T_c ($\approx$ 0.8 K) is much below T_N ($\approx$ 19 K). One of the most interesting and best investigated typical antiferromagnetic superconductors is $Sm\,Rh_4B_4$ (T_c = 2.75 K, T_N = 0.87 K). The system exhibit the anomalous behaviour of upper critical field H_{c_2}. We may note that the two phases, *i.e.* superconductivity and magnetism are competing in these systems, but in an antiferromagnet, the total magnetization is zero, the effect of the superconducting state being more complex. Baltensperger and Strassler [77] were the first who concluded that the two phenomena were not incompatible. Several mechanisms by means of which superconductivity is modified by antiferromagnetic order have been considered such as [5, 58, 78] :

- (*i*) The reduction in pair breaking due to decrease of the mean magnetization and, in turn, the conduction electron spin polarization below T_N ;
- (*ii*) The increase of pair breaking due to magnetic moment fluctuations in the vicinity of T_N ;
- (*iii*) The decrease of the attractive phonon mediated electron-electron pairing interaction by antiferromagnetic magnons ;
- (*iv*) The reduction of available phase space for virtual pair scattering due to the introduction of gaps in one electron excitation spectrum $E(K)$ by the change in lattice periodicity associated with antiferromagnetic order ;
- (*v*) The pairing of electrons with finite momentum.

These studies have provided a basic understanding of the interplay of superconductivity and antiferromagnetism, but most of these studies are phenomenological in nature.

We consider a model situation in which besides the pairing interaction among conduction electrons, there is an effective antiferromagnetic coupling between nearest neighbour localized electrons alongwith a new two body interaction between conduction and localized electrons. The antiferromagnetic ordering appears in this when intersite hopping is taken into account along with intrasite Coulomb repulsion [79]. When the system has both localized and conduction electrons, the mixing between the two states can lead to interesting interactions [80-82]. The above mechanism is relevant for rare earth ternary systems having less than half filled f-atomic shells [83]. The Hamiltonian which describes such an interaction is [84, 85]

$$H = \sum_{K,\sigma} (\varepsilon_K - \mu) C^+_{K\sigma} C_{K\sigma} - g \sum_{K,K'} C^+_{K'\uparrow} C^+_{-K'\downarrow} C_{-K'\downarrow} C_{K\uparrow}$$

$$+ \sum_{i,\sigma} E_l C^+_{l\sigma} C_{l\sigma} + \sum_{\substack{l,m' \\ \sigma,\sigma'}} J_{lm} C^+_{l\sigma} C^+_{m\sigma'} C_{l\sigma'} C_{m\sigma}$$

$$+ \left(\sum_{K,l,m} G^{lm}_K C^+_{l\uparrow} C^+_{m\downarrow} C_{-K\downarrow} C_{K\uparrow} + h.c. \right) \qquad ...(9.59)$$

where the first two terms are the fermion creation and annihilation operators for the Bloch state $|K\sigma\rangle$ with K being the conduction electron wave vector and σ the spin. $(C^+_{l\sigma}, C_{l\sigma})$ and $(C^+_{m\sigma}, C_{m\sigma})$ are the fermion (creation and annihilation) operators for localized electrons. The fermion operators satisfy the usual fermion anticommutation rules.

Following Green's function technique and equation of motion method, we obtain the expression for superconducting order parameter as

$$\Delta = \frac{1}{\beta} \sum_n \int \frac{d^3 K'}{(2\pi)^3} |g| \left\{ \frac{-(\Delta - \Gamma)}{(i\omega_n)^2 - \xi^2_{K'} - (\Delta - \Gamma)^2} \right\} \qquad ...(9.60)$$

where magnetic order parameter $\Gamma = \sum_K G^{ml}_K \langle C_{-K\downarrow} C_{K\uparrow} \rangle$,

$$G^{ml}_K = \langle\langle C^+_{m\downarrow} ; C^+_{l\uparrow} \rangle\rangle,$$

$$\xi_K = \varepsilon_K - \mu$$

where $|g|$ is the electron–electron phonon mediated coupling constant having the dimensions of energy. On using Poisson summation formulas and contour integration (9.60) yields

$$\frac{1}{N(0)|g|} = \int_0^{\hbar\omega_D} d\xi_{K'} \left(\frac{\Delta - \Gamma}{\Delta}\right) \cdot \frac{\tanh\left[\frac{\beta}{2} \sqrt{\xi^2_{K'} + (\Delta - \Gamma)^2}\right]}{[\xi^2_{K'} + (\Delta - \Gamma)^2]^{1/2}} \qquad ...(9.61)$$

where $N(0)$ (or $N(E_F)$) is the density of states for one spin projection at the Fermi surface. On putting $\Gamma = 0$, eq. (9.61) reduces to the standard BCS form

$$\Gamma = \frac{1}{\beta} \sum_{n,l,m} G^{lm}_K \langle\langle C^+_{m\downarrow}, C^+_{l\uparrow} \rangle\rangle \qquad ...(9.62)$$

One obtains

$$\Gamma = \sum_{m,l} G_K^{ml} \frac{\Gamma_{ml} \tanh\left(\frac{\beta}{2}(\tilde{E}_{ml}^2 + \Gamma_m^2)^{1/2}\right)}{2(\tilde{E}_{ml}^2 + \Gamma_m^2)^{1/2}} \quad \ldots(9.62(a))$$

where

$$\Gamma_{ml} = \frac{1}{\beta} \sum_{n,K} G_K^{ml} \ll C_{-K\downarrow}^+ ; C_{K\uparrow}^+ \gg \quad \ldots(9.63)$$

We note that Γ in turn is related to Δ through eq. (9.63). This shows that two phases coexist in some temperature range.

Within the limit $\Delta \to 0$, eq. (9.61) yields the following expression for superconducting transition temperature (T_c)

$$\frac{1}{N(0)|g|} = \int_0^{\hbar\omega_D} d\xi \left[\frac{\beta}{2} \frac{\Gamma^2 \operatorname{sech}^2 \frac{\beta}{2}(\xi^2 + \Gamma^2)^{1/2}}{\xi^2 + \Gamma^2} + \frac{\tanh \frac{\beta}{2}(\xi^2 + \Gamma^2)^{1/2}}{(\xi^2 + \Gamma^2)^{1/2}} \right] \quad \ldots(9.64)$$

Density of States

The density of states per atom per spin $N(\omega)$, is given by

$$N(\omega) = \lim_{\varepsilon \to 0} \frac{i}{2\pi N} \sum_K [G_{\uparrow\uparrow}(K, \omega + i\varepsilon) - G_{\uparrow\uparrow}(K, \omega - i\varepsilon)] \quad \ldots(9.65)$$

The one particle Green's function is

$$G_{\uparrow\uparrow}(K, \omega) = \frac{1}{2}\left[\frac{(\xi_K/\alpha) + 1)}{(i\omega_n - \alpha)} - \frac{((\xi_K/\alpha) - 1)}{(i\omega_n + \alpha)}\right] \quad \ldots(9.66)$$

with

$$\alpha^2 = \xi_K^2 + (\Delta - \Gamma)^2 \quad \ldots(9.66(a))$$

After simplification, one obtains

$$\frac{N(\omega)}{N(0)} = \frac{1}{N} \frac{\omega}{[\omega^2 - (\Delta - \Gamma)^2]^{1/2}} \quad \ldots(9.67)$$

Electronic Specific Heat

The electronic specific heat per atom of a superconductor is determined from

$$C_{es} = \frac{\partial}{\partial T} \frac{1}{N} \sum_K 2\xi_K < C_{K\uparrow}^+ C_{K\uparrow} > \quad \ldots(9.68)$$

where $< C_{K\uparrow}^+ C_{K\uparrow} >$ is correlation function obtained from the following relation

$$< C_{K\uparrow}^+ C_{K\uparrow} > = \lim_{\varepsilon \to 0} \frac{i}{2\pi} \int_{-\infty}^{\infty} \frac{\ll C_{K\uparrow} C_{K\uparrow}^+ \gg_{\omega + i\varepsilon} - \ll C_{K\uparrow} C_{K\uparrow}^+ \gg_{\omega - i\varepsilon}}{e^{\beta\omega} - 1} d\omega \quad \ldots(9.69)$$

$$= \frac{1}{2}\left[\frac{(1 + (\xi_K/\alpha))}{e^{\beta\alpha} - 1} + \frac{(1 - (\xi_K/\alpha))}{e^{-\beta\alpha} - 1}\right] \quad \ldots(9.70)$$

Using (9.70), (9.68) yields

$$C_{es} = \frac{N(0)}{N} \frac{1}{k_B T^2} \int_0^{\hbar\omega_D} d\xi_K \, \xi_K^2 \, \text{sech}^2 \left(\frac{\alpha}{2 k_B T} \right) \qquad ...(9.71)$$

Free Energy

The free energy difference between the superconducting and normal state can be calculated by means of the following relation

$$\frac{F_S - F_N}{V} = \int_0^{\Delta} d\Delta \, \Delta^2 \, \frac{d}{d\Delta}\left(\frac{1}{|g|} \right) \qquad ...(9.72)$$

using (9.61), (9.72) takes the form

$$\frac{F_S - F_N}{V} = \frac{\Delta^2}{|g|} - 2N(0) \int_0^{\hbar\omega_D} d\xi \left[\int_0^{\Delta} d\Delta \, \frac{(\Delta - \Gamma) \tanh(\beta E/2)}{E} \right]$$

where $\quad E = [\xi^2 + (\Delta - \Gamma)^2]^{1/2}$

On evaluating Δ integral, one finally obtains

$$\frac{F_S - F_N}{V} = \frac{\Delta^2}{|g|} - N(0) \left[\frac{1}{2} \Delta (\Delta - 2\Gamma) + (\Delta - \Gamma)^2 \ln\left(\frac{2\hbar\omega_D}{\Delta - \Gamma} \right) - \Gamma^2 \ln\left(\frac{2\hbar\omega_D}{\Gamma} \right) \right]$$
$$+ \frac{1}{3} \pi^2 N(0) (k_B T)^2 - \frac{4N(0)}{\beta} \int_0^{\hbar\omega_D} d\xi \ln(1 + e^{-\beta E})] \qquad ...(9.73)$$

where $\quad E' = [\xi^2 + \Gamma^2]^{1/2} \qquad ...(9.73(a))$

Critical Field (H_c)

The low temperature critical field H_c is given by

$$H_c = [8\pi \, | \, F_S - F_N \, |]^{1/2} \qquad ...(9.74)$$

Using (9.73), one obtains

$$H_c = (8\pi V)^{1/2} \left[\frac{\Delta^2}{|g|} - N(0) \left\{ \frac{1}{2} \Delta (\Delta - 2\Gamma) + (\Delta - \Gamma)^2 \right. \right.$$
$$\left. \ln\left(\frac{2\hbar\omega_D}{\Delta - \Gamma} \right) - \Gamma^2 \ln\left(\frac{2\hbar\omega_D}{\Gamma} \right) \right\} + \frac{1}{3} \pi^2 N(0) (k_B T)^2$$
$$\left. - 4 N(0) k_B T \int_0^{\hbar\omega_D} d\xi \ln(1 + e^{-\beta E}) \right]^{1/2} \qquad ...(9.75)$$

Phase Transition

$$F_S - F_N \neq 0$$

shows that the superconducting to paramagnetic phase transition at T_N is of the first order.

The theory has been applied to the system Sm Rh$_4$B$_4$ and explains satisfactorily the observed experimental results.

The interplay between superconductivity and ternary compounds can also be studied by a model based on the interaction between electrons and sublattice mangetization $<S^z>$ of the antiferromagnetic state. The Hamiltonian describing such an interaction is

$$H_{e-s} = -\sum_{p,Q} H(Q)(C^+_{p\uparrow} C_{p+Q\uparrow} - C^+_{p\downarrow} C_{p+Q\downarrow}) \qquad ...(9.76)$$

where the molecular field $H(Q)$ is

$$H_Q = <S^z> \frac{I}{2} \sum_{p,Q} \delta_{p,Q} \qquad ...(9.77)$$

and is the "d-f" interaction.

In the absence of the interaction which breaks the time reversal symmetry, one finds that the total Hamiltonian commutes with the time-reversal operator T, i.e., $[HT] = 0$. However, for an antiferromagnetic superconductor, the contribution (9.76) does not commute, i.e. $[H_{es}, T] \neq 0$. However, if we consider the operator $Y = TR$ (R being the translation by a vector connecting the two sublattices), one finds $[H, Y] = 0$. This clearly implies that $\psi_{p\alpha}(\mathbf{r})$ is an eigen function of the Hamiltonian H which describes the antiferromagnetic superconductor, then $Y\psi_{p\alpha}(\mathbf{r})$ is also an eigenfunction with the same energy. The pairing in an antiferromagnetic superconductor appears between the states $\psi_{p\alpha}(\mathbf{r})$ and $\exp(i\phi) \psi_{p\alpha}(r)$ where ϕ is an arbitrary phase. Fulde and Keller [86] showed that the one electron spectrum in the presence of the molecular field is

$$E(p) = \frac{\varepsilon(p) + \varepsilon(p+Q)}{2} \pm \left[\frac{1}{2}(\varepsilon(p) - \varepsilon(p+Q))^2 + H^2(Q)\right]^{1/2} \qquad ...(9.78)$$

which has gaps at points $P = \pm Q/2$. The pairing of the electrons in an antiferromagnetic superconductor has been considered in different ways. Machida et al. [87] and Nass et al. [88] considered the following Hamiltonian

$$H = \sum_{p,\alpha} \varepsilon(p) C^+_{p\alpha} C_{p\alpha} + \sum_{p,\alpha} [\alpha H(Q) C^+_{p\alpha} C^+_{p+Q,\alpha} + h.c.]$$

$$+ \Delta \sum_p (C^+_{-p\downarrow} C^+_{p\uparrow} + h.c.) \qquad ...(9.79)$$

where the order parameter Δ describes the pairing between the states $(p_\uparrow, -p_\downarrow)$. Authors neglected the pairing between the states.

Zwicknagl and Fulde [89] used instead, energetically degenerate eigenstates (n, p) of the Hamiltonian

$$H_0 = \sum_n E_n(p) C^+_{np\alpha} C_{np\alpha} \qquad ...(9.80)$$

They obtained the following expression for the critical temperature T_c of an antiferromagnetic superconductor

$$T_c = 1.13 \, \hbar\omega_D \exp[-1/\eta] \qquad ...(9.81)$$

where

$$\eta = g \sum_{i,p} \delta[E_i(p) - \varepsilon_0]\left[1 - \frac{4H^2(p)}{E_{+1}(p) - E_{-1}(p)}\right] \qquad ...(9.82)$$

and $E_{\pm 1}(p)$ corresponds to the two signs from eq. (9.78). From eq. (9.81), one finds that the exchange field may change the effective coupling by a modification of the density of states and by a reduction of the electron–electron interaction due to its influence on the electronic wave function. The latter is more important and leads to a linear decrease of T_c as function of the exchange field.

Model [21] used for the ferromagnetic superconductors can also be applied to the antiferromagnetic superconductor with the mention that last term in eq. (22) gives rise to the antiferromagnetic order. In the Born approximation, Machida [90] obtained the following approximation for the scattering time.

$$\frac{1}{\tau(T)} = 2\pi N(0) \left(\frac{I}{2}\right)^2 \frac{1}{4} \int d\Omega_q \, \chi(q, T) \qquad ...(9.83)$$

where the correlation function for the antiferromagnetic ordered spins has been defined by

$$\chi(q; T) = <S(q) \cdot S(-q)> \qquad ...(9.84)$$

Expression (9.83) shows that the scattering time τ depends on the wave vector $\mathbf{Q}$ which characterizes the antiferromagnetic superlattice structure. If $Q \ll 2p_0$, the superconductivity is affected by the magnetic order because the conduction electrons are effectively scattered on the ordered magnetic ions. For $Q \simeq 2p_0$, the antiferromagnetic order does not affect the superconducting state drastically and the two states may coexist.

In the case $|Q| \ll 2p_0$, the spin correlation is given by

$$\chi_0(T) = \frac{S(S+1)T}{T + T_N} \qquad ...(9.85)$$

where T_N is the Neel temperature of the antiferromagnet eq. (9.83) becomes

$$\frac{1}{\tau(T)} = 2\pi N(0) \left(\frac{I}{2}\right)^2 \chi_0(T) \qquad ...(9.86)$$

The critical temperature T_c can be calculated similar for a ferromagnetic superconductor and one obtains

$$\ln \frac{T_c}{T_{c_0}} = \psi\left(\frac{1}{2}\right) - \psi\left(\frac{1}{2} + \rho(T_c)\right) \qquad ...(9.87)$$

where

$$\rho(T_c) = \frac{1}{2\pi T_S} f(T_c) \qquad ...(9.88)$$

$$f(T_c) = \begin{cases} \dfrac{1}{T + T_N} & ; \; T_c > T_N \\ \dfrac{1}{6 T_N} \left(2 + \dfrac{\chi_{\parallel}(T_c)}{\chi_{\perp}(T_c)}\right) & ; \; T_c < T_N \end{cases} \qquad ...(9.89)$$

$\chi_{\parallel}$ and $\chi_{\perp}$ being the parallel and perpendicular respectively, transverse components of $\chi(T)$ for antiferromagnet. Fig. 9.10.

We may note that the mechanism of the coexistence between the antiferromagnetic order and superconductivity is much more complicated than appear in the simple mechanism presented above. Indeed, the antiferromagnetic order in the spin system gives rise to a change in the density of the electronic states and in the electronic system, there may appear the new condensed state. Charge- Density Waves (CDW) and Spin-Density Waves (SDW). When the triplet electronic states are considered, the last ones give rise to itinerant-electron antiferromagnetism. Due to this particular feature of an antiferromagnetic superconductor, the non-magnetic impurities will have a strong effect on this state.

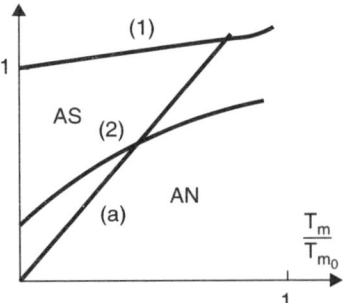

Fig. 9.10 Phase diagram for an antiferromagnetic superconductor. The lines corresponds to $N(0) (I/2)^2 \cdot S(S+1)/T_{co}$ as (1) 0.04 and (2) 0.13. Below the straight line (a) the system is only anitferromagnetic (AN) and below the curves is an antiferromagnetic superconductor (AS).

9.4 HEAVY FERMION AND RUTHENATE SUPERCONDUCTORS

Heavy fermion metals and ruthenate system Sr_2RuO_4 are non-cuprate superconductors have the following properties in common: (a) they are close to magnetic instabilities at low temperature; (b) their superconductivity is unconventional; and (c) unlike the cuprates, they go to superconducting from *Fermi Liquid* normal states (with the exception of UBe_{13} and some quantum critical heavy fermion systems, e.g. $CePd_2Si_2$ [91], $CeRhIn_5$ [92] $CeRhIn_5$ [93], $CeIrIn_5$ [94]). Superconductivity in these systems, which are the typical strongly correlated electron (SCES) materials, has not been explained from the microscopic point of view and commonly believed, but not yet proven, that the pairing mechanism in many of these systems involves magnetic degrees of freedom.

Heavy fermion systems are alloys of $4f$ or $5f$ elements (most often cerium (Ce) or uranium (U)) with non-magnetic elements having complicated band structures. These systems exhibit peculiar metallic properties at sufficiently low temperatures. The word "heavy" reflects the property that in these materials the density of electronic states on the Fermi surface, $N(0) = m^* K_F / \pi^2 \hbar^2$ (where m^* is the effective electronic mass) can be tens or even hundreds of times larger than it is in normal metals. Such electron systems are frequently, called 'heavy-fermion systems' and these heavy fermions are called '*quasi particles*', which are the approximate eigenstates one gets by starting with a non-interacting electron gas and then adiabatically turning on the interactions. The thermodynamics of a 'Fermi liquid' of quasiparticles was developed by Landau, and the Landau–Fermi liquid theory underpins our understanding of metallic behaviour.

Thermodynamic measurements show that the heavy fermion participate in the superconductivity of heavy fermion systems. All heavy fermion superconductors are non-BCS, and both spin singlet and spin-triplet states are believed to have been demonstrated.

An important recent development in heavy fermion superconductivity was the discovery of superconductivity at quantum critical points in magnetic systems. In these cases, the superconductivity is produced by using hydrostatic pressure to suppress the magnetic phase transition to 0 K. In two cases, the superconductivity is confined to a very narrow range of pressures around the critical pressure for the suppression of magnetism-strong evidence that the superconductivity is magnetically mediated.

In 1979, the superconductivity was, however discovered quite unexpectedly in $CeCu_2Si_2$, which is one of typical heavy fermion systems, by Steglich et al. [95]. This system is not nearly ferromagnetic, rather it is nearly **antiferromagnetic**. The discovery of superconductivity in $CeCu_2Si_2$ with remarkable magnetic characters has implied a scenario clearly different from the electron-phonon mechanism in the conventional BCS theory. In fact, physical properties in the superconducting phase at low temperature have shown power-low temperature dependence, different from exponential decay observed in the s-wave superconductors. Thus, $CeCu_2Si_2$ has become the pioneering discovery of unconventional superconductivity in SCES. It is now considered as an even-parity superconductor with line-nodes, probably d-wave pairing symmetry.

Since 1979, many unconventional superconductors have been discovered in heavy fermion systems. Multi-phase diagrams in UPt_3 [96–98] and $U(Be_{1-x}Th_x)_{13}$ [99–102] indicates curious superconductivity with multi-components. In particular, UPt_3 is the odd parity superconductor first discovered in electronic systems [103]. UPd_2Al_3 and UNi_2Al_3 [104, 105] are unconventional heavy fermion superconductors coexisting with antiferromagnetism, and are considered to have even- and odd-parity pairing states, respectively [106–108]. URu_2Si_2 indicates coexistence of unconventional superconductivity and a hidden order [109]. Further more, recent progress in experiments under pressures have promoted discoveries of new superconductors: $CeCu_2Ge_2$ [110–112], $CePd_2Si_2$ [113, 114], $CeRh_2Si_2$ [115], $CeNi_2Ge_2$ [116] and $CeIn_3$ [114, 117]. These systems are AF metals at ambient pressure, while under high pressures, the AF phases abruptly disappear accompanied by superconducting transitions. Except for cubic $CeIn_3$, all other materials have the same $ThCr_2Si_2$-type crystal structure as in $CeCu_2Si_2$ (see Fig. 9.11).

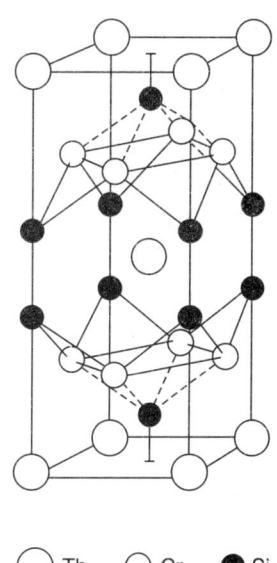

Fig. 9.11 Unit Cell of $ThCr_2Si_2$ structure [118]

Quite recently, several kinds of new heavy-fermion superconductors have been discovered. One is a family of $CeTIn_5$ (T = Co, Rh, and Ir) with $HoCoGa_5$ - type crystals structure [119, 120]. Since the discovery, a variety of experimental investigations have rapidly increased. These heavy fermion compounds possess relatively high transition temperature such as $T_c = 2.3$ K for $CeCoIn_5$, which is the highest among heavy fermion superconductors observed yet. The dominant antiferromagnetic (AF) spin fluctuations have been suggested by the existence of the neighbouring AF phase in the pressure-temperature $(P - T)$ phase diagram, the power law $(T^{1-1.3})$ behaviour in the resistivity, and the magnetic behaviour in the resistivity and the magnetic behaviour observed in the NQR/NMR $1/T_1$. The situation is quite similar to other Ce-based superconductors mentioned above. Another is the coexistence of superconductivity and ferromagnetism in UGe_2 at high pressure and $URhGe$ at ambient pressure [121, 122]. Obviously, heavy fermion compounds are very unusual metals. They are unusual superconductors as well. Heavy-fermion superconductors show a great variety of ground states and offer rich examples to investigate unconventional superconductivity in SCES.

Although many heavy fermion systems have been discovered and investigated, few exhibit superconductivity. It is a persistent mystery why some heavy fermion systems superconduct and other systems do not, and ultimate theory of heavy fermion superconductivity will have to explain this. The number of heavy fermion superconductors has been increased recently by discoveries of superconductivity at magnetic quantum critical points in heavy fermion systems [114, 121, 123, 125].

Heavy fermion metals are very sensitive to impurities and defects, *e.g.* even minute amounts of some impurities can induce large moment antiferromagnetism in paramagnetic heavy fermion systems, while wash out the double superconductivity transition of UPt_3. Obviously, crystal growth plays a central role in this field.

The heavy fermion state exists on the border between magnetism and the non-magnetic ground state (Fig. 9.12). The order on the magnetic side is almost always antiferromagnetic. We may note that heavy fermion behaviour can be seen in both the magnetic and non-magnetic states, but only close to the borderline. Neutron scattering often sees short-range antiferromagnetic correlations in non-magnetic heavy fermion systems, while the ordered moment of magnetic heavy fermion systems is often extremely small, ~ 0.01 μ_B for example (but others show larger moments). Small moment magnetism is still a puzzle.

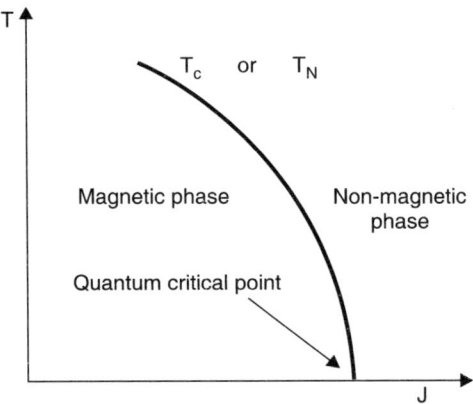

Fig. 9.12 Phase diagram of a heavy fermion metal as a function of the on-site coupling, J, between the *f*-spin and the conduction sea. Heavy fermion behaviour is reported for values of J near the quantum critical point.

The uranium heavy fermion superconducting systems, UBe_{13}, UPt_3, URu_2Si_2, UNi_2Al_3, UPd_2Al_3, with the exception of UBe_{13}, show antiferromagnetic order coexisting with superconductivity. However, the strength of the antiferromagnetic order (as measured by the size of specific heat jump at the antiferromagnetic transition) is much larger in UPu_2Si_2, UPd_2Al_3 and UNi_2Al_3 than it is in UPt_3. In all these heavy fermion superconducting systems, the antiferromagnetic ordering temperature is an order of magnitude larger than the superconducting transition temperature. In $CeCu_2Si_2$ there is phase of unknown order parameter, called the A-phase, which competes with superconductivity [126].

The size of the linear specific heat coefficient, γ, varies from ~ 60 mJ mol^{-1} K^{-2} in URu_2Si_2, to ~ 700 mJ mol^{-1} K^{-2} in UBe_{13} and $CeIrIn_5$. The heavy fermion systems with the largest γ-values have weak or no antiferromagnetic order. This is understandable in that all of the entropy comes from the spin degrees of freedom, and the larger the moment tied up in antiferromagnetic order, the less there is to appear in the entropy at low temperature.

The superconducting transition temperatures are low, the highest T_c being 2 K in UPd_2Al_3. The lower γ, higher T_N systems tend to have higher values of T_c. Although T_c is low, it is a large fraction of the Fermi temperature T_F, which ranges from about 10 to 100 K, so in this sense, the heavy fermion superconductors are high temperature superconductors.

We may note that in all of the heavy fermion superconductors, the specific heat jump at the superconducting transition is of the order of the normal state specific heat, showing that the heavy fermions participate in the superconducting order. For *e.g.*, the specific heat jump for UPt_3 is $[\Delta C/(\gamma(T_c) T_c)]$ ~ 0.8 compared to a BCS value of 1.43. All of the heavy fermion superconductors are strongly type-II superconductors. For example, in UPt_3 at $T = 0 K$, H_{c_1} ~ 3.0 mT and H_{c_2} ~ $3T$. Obviously, this is an enormous value of H_{c_2} for such a low value of T_c. The penetration depth tends to be long while the coherence length is short. In UPt_3 the values are λ ~ 6000 Å, and the coherence length is ξ ~ 100 Å (compared to 0.039 and 380 Å in Nb).

Interestingly, all of the above properties of heavy fermion superconductors are in line with the BCS theory for a metal with heavy electrons. However, there is considerable evidence that the superconductivity in these systems is not BCS phonon-mediated s-wave superconductivity.

The main evidence for unconventional superconductivity in heavy fermion systems is (a) the existence of nodes in the gap, as shown by thermodynamic, transport and spectroscopic measurements such as nuclear magnetic resonance, and (b) the multiple superconducting phases of UPt_3. Fig. 9.13 shows some simple arrangements of nodes for two dimensional representations of the order parameter. We may note that the physics of such nodes is here both more complicated and much richer than in cuprates, because in the cuprates the Fermi surface is cylindrical and so, for example, point nodes at the poles are impossible.

In a conventional s-wave superconductor, the opening of a superconducting energy gap has a dramatic effect on thermodynamic properties, *e.g.* C/T vanishes as $\exp(-\Delta/k_B T)$ as T → 0. In contrast, if there are **nodes** in gap then C/T vanishes much more slowly, as T if there are line nodes and as T^2 if there are point nodes but no line nodes. For example, in Fig. 9.13 it can be seen that $C/T \propto T$ at low temperature, revealing that there are line nodes in the energy gap of UPt_3. Many other measurements are sensitive to nodes, such as thermal conductivity, NMR relaxation rate $1/T_1$, ultrasonic attenuation, etc.

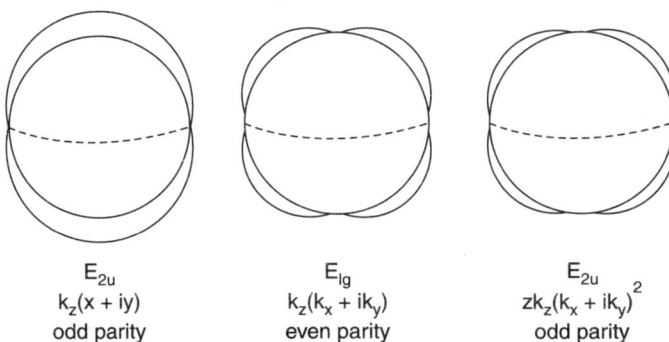

E_{2u}
$k_z(x+iy)$
odd parity

E_{1g}
$k_z(k_x+ik_y)$
even parity

E_{2u}
$zk_z(k_x+ik_y)^2$
odd parity

Fig. 9.13 Arrangement of nodes in the energy gap for some of the possible order parameter basis functions of UPt$_3$, for an elliptical Fermi surface (the actual Fermi surface is much more complicated). All three have a line of nodes around the 'equator', but the first has no nodes on the poles, the second has nodes on the poles with linear dispersion, and the third has nodes on the poles with quadratic dispersion. All these three arrangements will give rise to very different ratios of thermal conductivity along the polar and equatorial directions, κ_c/κ_{ab}: in the first case vanishing as $1/T^2$, in the second vanishing logarithmically and, in the last case, having a finite value in the low temperature limit.

The discovery of multiple superconducting phases in UPt$_3$ (Fig. 9.14) is strong evidence of non-s-wave, multi-component order parameter. There are two possible explanations for the multiple superconducting phases. Either there is an accidental degeneracy between two order parameters with different symmetries, or there is a single double-degenerate order parameter whose degeneracy is lifted by some symmetry breaking field. A natural candidate for the symmetry breaking field is weak antiferromagnetism that appears at 5 K. Pressure experiments have shown that the splitting of the transition scales with the size of the antiferromagnetic moment. In particular, both vanish above at about 4 K [127].

Various experiments shows that order parameter in UPt$_3$ is unconventional, but very few give information about its actual **symmetry.** Recently, low-field NMR has been used to try to measure the spin-susceptibility in the mixed state of UPt$_3$. These measurements reveal in high fields (from 0.4 to 1.6 T) that there is no change in the Knight shift upon entering the superconducting state. At lower fields, there are reductions in the Knight shift for fields applied along c and b, but not along a [128, 129]. These measurements are strong evidence for odd parity superconductivity in UPt$_3$, i.e. either a p-or f-wave orbital state.

Pure UBe$_{13}$ has a superconducting transition at around 1 K. One interesting feature is that the resistivity is very high and does not show any signs of coherence when the superconducting transition occurs. The upper critical field (H_{C_2}) is enormous, nearly 18 T, although this is somewhat sample dependent and strongly temperature dependent as $T \to 0K$ [127]. Upon doping with thorium, a double superconducting transition appears [130]. The normal and superconducting states in this system are far from understood.

In CeCu$_2$Si$_2$ and UPd$_2$Al$_3$, $1/T_1$ measurements are consistent with line nodes [131, 132], while Knight shift measurements [133, 134] see a clear reduction in the spin-susceptibility upon entering the superconducting state; these measurements are obviously consistent with finite-angular momentum spin-singlet (probably d-wave) superconductivity.

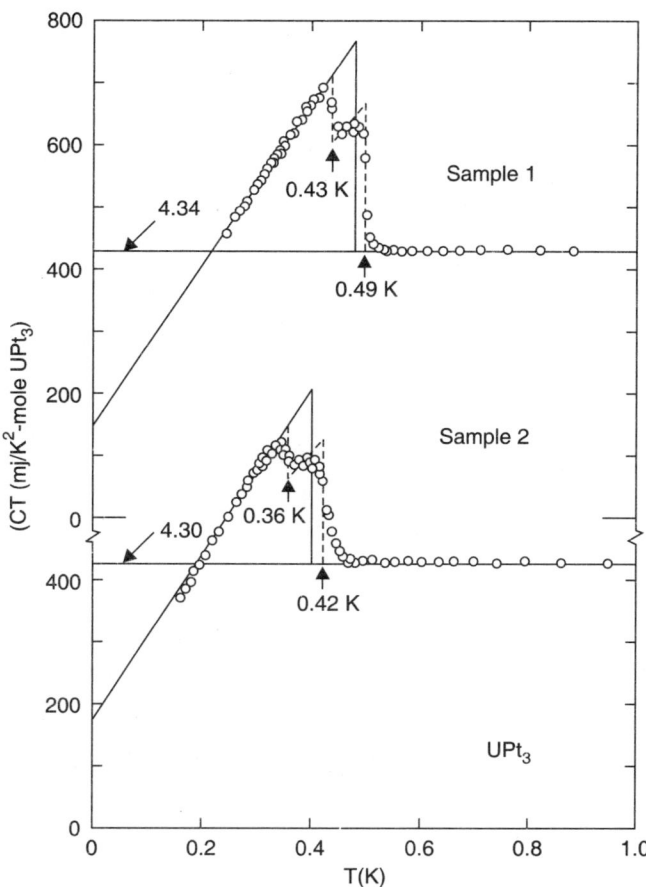

Fig. 9.14 The double transition in specific heat measurements on high quality crystals of UPt_3. We may note that C/T is linear in T below the transition (evidence of line nodes). Moreover, the large constant value of C/T in the normal state is the hallmark of a heavy fermion metal.

Several new superconductors, including neither Ce nor U-atom, have been also discovered and attracted attentions. In the filled skutterudite compound $PrOs_4Sb_{12}$ (T_c = 1.85), the possibility of the double transition has been indicated [135]. In the Pu-based compound $PuCoGa_5$ with the same crystal structure as a family of $CeTIn_5$, a very high transition temperature T_c = 18.5 K has been reported [136].

The series $CeTIn_5$ (T = Co, Rh and Ir) [137-139] possess many similarities with the cuprates and organic superconductors. Obviously, these compounds offer a great opportunity to bridge our understanding in Ce-based heavy-fermion superconductors, HTSC cuprates, and organic superconductors. These compounds are quasi-2D materials with the layered structure which has been confirmed by the deHaas-van Alphen (dHvA) measurements [140-144]. The P-T phase diagram in $CeRhIn_5$ [145] and the phase diagram in the alloy system $CeRh_{1-x}Ir_xIn_5$ [146] indicate that the superconductivity appears in the neighbourhood of the AF phase. The power-law behaviour of the resistivity in the normal phase with $T^{1.3}$ in $Ce Ir In_5$ (T_c = 0.4 K)

[147] and T in CeCoIn$_5$ (T_c = 2.3 K) [148] implies the strong quasi-2D AF spin fluctuation. In the SC phase, no coherence peak just below T_c, the T^3 behaviour at low temperatures in the NMR/NQR $1/T_1$, and the decrease in the Knight shift irrespective of directions indicate the anisotropic even-parity superconductivity, similar to other Ce-based heavy-fermion superconductors [149-154]. In addition, the four fold symmetry in the thermal conductivity in CeCoIn$_5$ [155, 156] strongly suggests the $d_{x^2-y^2}$ wave singlet pairing. Furthermore, the pseudogap phenomenon has been reported in the ^{115}In – NQR measurement in CeRhIn$_5$ in the range of pressures (P = 1.53 – 1.75 G Pa), under which the coexistent phase exist [157]. These similarities, especially the AF spin fluctuation and $d_{x^2-y^2}$ wave pairing, indicate that the underlying physics in the superconductivity in these Ce-based heavy-fermion super-conductors is in common with the cuprate superconductors. This is also inferred from the microscopic electronic state, where one f-electron exists per Ce site and the f-electron band itself is almost half-filled when the hybridization with the conduction electron band is not taken into account.

U-based heavy-fermion superconductors exhibit various kinds of SC phases probably owing to multi f-band structures with two or three f-electrons per U-site.

U-based compounds do not necessarily display clear log T dependence in the resistivity. This is attributed to the multi f-orbitals and relatively large hybridization terms with the conduction electrons. These properties also provide the complicated multi-band structures and many Fermi surfaces with dominant f-character. The Fermi surfaces obtained by the band calculations well explain the dHvA measurements. One of the characteristics of U-based compounds different from Ce-based compounds is the clear coexistence between the SC phase and some kind of magnetic ordered phase. UPt$_3$ is the odd-parity superconductor (T_c = 0.5 K) discovered for the first time in the electronic systems [158, 159]. This superconductor coexists with an unusual AF order below T_N = 5K [160, 161], which is observed in the neutron scattering. We may note that it has not been observed by the static and/or dynamically slow probes. This unusual AF order at $\mathbf{Q}$ = (0.5, 0, 1) becomes the long range order below 20 mK within the resolution of the neutron scattering [162]. In URh$_2$Si$_2$, the unconventional superconductivity below 1.5 K also coexists with a hidden order with a clear jump at 17.5 K in the specific heat [163, 164]. UPd$_2$Al$_3$ is an AF metal with $\mathbf{Q}$ = (0, 0, 0.5) below T_N = 14.5 K [165] and coexists with the anisotropic even-parity superconductivity below T_c = 2 K [166, 167]. The isostructural compound UNi$_2$Al$_3$ is in the SDW state with $\mathbf{Q}$ = (0.5 ± τ, 0, 0.5) and τ = 0.11 ± 0.003 below T_N = 4.6 K [168, 169]. The SDW state coexists with the unconventional superconductivity below T_c = 1.2 K [170] which may be the odd-parity state as indicated by the NMR/NQR measurement [171].

Quantum Critical Superconductivity in Heavy Fermion Systems

The proximity of heavy fermion metals to magnetic instabilities has been exploited recently in studies that use chemical doping or hydrostatic pressure to continuously tune systems through quantum critical points (Fig. 9.12).

Chemical doping introduces disorder, which is strongly pair breaking for unconventional superconductivity, and superconductivity has not been found in such investigations. However, by using hydrostatic pressure superconductivity has found at the quantum critical points of several stoichiometric heavy fermion systems [172].

Emerging Superconductors

Fig. 9.15 shows the magnetic-superconducting phase diagram of $CePd_2Si_2$ [114]. This system is antiferromagnetic at 10 K at ambient pressure. Even in the antiferromagnetic state it shows heavy fermion behaviour as $T \to 0$, having for example a linear specific heat coefficient of $\gamma \sim 250$ mJ/mol/K^2 [173]. However, many quantities are difficult to measure under high pressure. The magnetic transition is signalled by a sharp change in slope of the resistivity versus temperature, while the superconductivity shows up as a drop to zero of the resistivity [114].

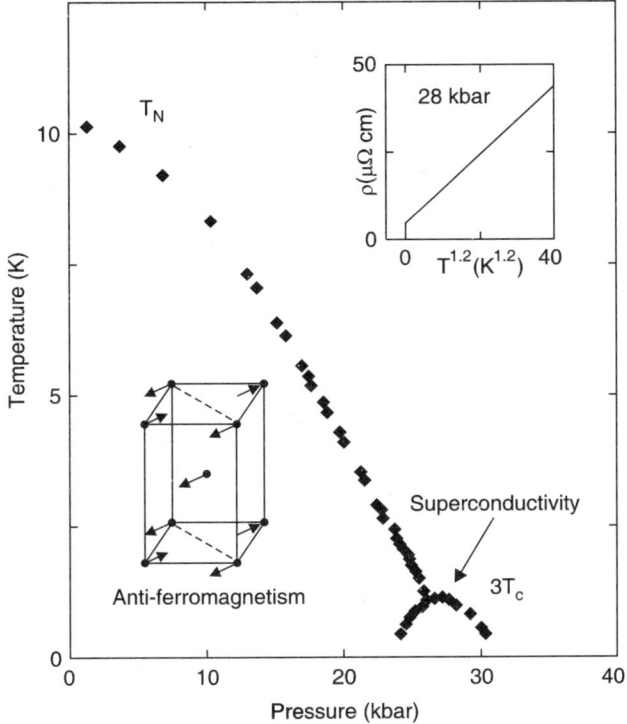

Fig. 9.15 Phase diagram of $CePd_2Si_2$. This is antiferromagnetic at ambient pressure. The anitiferromagnetism is suppressed between 25 and 30 K bar. Around the 'quantum critical point' a superconducting phase is observed. We may note that superconducting transition temperature T_c has been multiplied by three, to make it visible on this plot. The on-set shows the $T^{1.2}$ resistivity in the normal state [114].

Unlike the other heavy fermion systems, the normal state in the vicinity of the quantum critical point is **non-Fermi-liquid** like. The inset of Fig. 9.15 shows that the resistivity of $CePd_2Si_2$ close to its quantum critical pressure follows a $T^{1.2}$ law over a very large temperature range of nearly 40 K. This power law is not in agreement with the usual Fermi liquid behaviour seen in heavy fermion metals—typically T^2 over a small range (up to 1 or 2 K or less) above 0 K.

A similar phase diagram has been found for $CeIn_3$, again with a small superconducting phase existing over a small range of pressures near the quantum critical point, which again lies near 25 k bar [174].

This transition from a non-Fermi liquid normal state to an (presumably) unconventional superconducting state is reminiscent of HTSC cuprates, and indeed there are certain theories of HTSC which assume that there is a quantum critical point that dominates the physics of cuprates [175-177].

The strongest evidence that this superconductivity is mediated by magnetic fluctuations is the way, it dies out so rapidly as the pressure is increased beyond the critical pressure (*i.e.*, the pressure at which the magnetism is destroyed). In contrast, some other heavy fermion materials have been found in which superconductivity and magnetism appear to compete, in that superconductivity is found over a large range of pressures above the critical pressure P_c for destruction of magnetic order, but for $P < P_c$ the superconductivity disappears abruptly. This gives the appearance of a system in which superconductivity is being destroyed by magnetism. In this class may be $CeCu_2Ge_2$, which is antiferromagnetic upto about 50 k bar, above which the magnetism suddenly collapses, to be followed over a broad range of about 150 k bar by a superconducting ground state, and its sister compound $CeCu_2Si_2$, which displays a very similar phase diagram except that everything is displaced to lower pressure by about 50 k bar, placing the magnetic/non-magnetic boundary very close to ambient pressure [178].

The system $CeRh_2Si_2$ is reported to be superconducting near the critical point [179], but it is not clear in which class this falls. The system $CeRhIn_5$ becomes superconducting above about 16 k bar [137], and a companion compound $CeIrIn_5$ [147], which is superconducting at ambient pressure, seems to goes to superconducting from a non-Fermi liquid state.

The accurate measurements under high pressure is a very difficult task and this makes the field a challenging area.

Although such a variety of interesting phenomena have been observed experimentally, there are little theoretical progress from the microscopic point of view. One of the causes is the complicated band structure, since U-based compounds possess several f-electrons per U-site and many Fermi-surface sheets originating from multi f-orbitals. On the one hand, the complicated nature is unfavourable for the theoretical effort, but on the other hand, it is related to a great variety of the ground states. Variety of experiments have indicated that for heavy-fermion compounds, the superconducting transition occurs after the formation of the coherent quasi-particle state with heavy effective mass.

A qualitative understanding of what may happen in these systems is connected with the existence of a kind of Kondo resonance peak at the Fermi energy, and the electrons in this resonance peak are "heavy fermions". This resonance peak is a feature of the f-electron spectrum which forms the "heavy fermions", and thus the f-electrons are paired up in the superconducting state. The pairing presents some particular aspects which show a pronounced difference to usual superconductivity. In the usual superconductors, the Fermi energy ε_F (or the band width of the conduction electrons) is of the order of 10^4 K, the critical temperature T_c is of the order of 10-20 K and the characteristic energy is the Debye temperature $\Theta_D \simeq 10^2$ K, so we have $T_c \ll \Theta_D \ll W$. When the electrons in the Kondo resonance peak undergo the pairing, the situation is totally different ; the width of the resonance peak is of the order of $T_K \simeq 10$ K (T_K is the Kondo temperature) which is only one magnitude order larger than T_c and the pairing mechanism provides for an energy scale θ as large as or even larger than the effective band width $W = T_K$; *i.e*, we have $\theta \geq T_K = W > T_c$.

Several authors have considered periodic Anderson Model (PAM) as a typical model for heavy-fermion systems. This model well describes the dual nature of f-electron systems. For details readers are referred to [5, 180-182] and references cited therein.

Ruthenate system Sr_2RuO_4 is an interesting system for many reasons. This system has the same crystal structure as the parent compound of an important family of HTSC cuprates. The normal state of this system is well understood and almost decided that its superconductivity is unconventional and of spin-triplet kind. Now, we present a brief review about this system.

9.4.1 Ruthenate Superconductor (Sr_2RuO_4)

Maeno et al. [183] discovered superconductivity in Sr_2RuO_4 with T_c as large as 1.5 K in the high purity sample. This compound is layered perovskite oxide metal and possesses the same crystal structure as HTSC $La_{2-x}Sr_xCuO_4$, and it similarly has the quasi two dimensional nature. For example, the resistivity exhibits a large anisotropy; the ratio ρ_c/ρ_{ab} is in the order of several hundreds [183, 184]. The quantum oscillation measurement has also clearly shown the quasi two-dimensional Fermi surfaces [185]. RuO_2 layers are expected to be essential for the metallic behaviour and superconductivity as CuO_2 layers in HTSC cuprates. Despite a wide ranging search for layered perovskite oxide superconductors that do not have copper-oxygen planes, Sr_2RuO_4 is the only one that has been found to date. Both the superconductivity, which is unconventional, and the metallic state, which is not, offer interesting contrasts with cuprates.

The normal state properties of Sr_2RuO_4 have been extremely well characterized by thermodynamic and transport properties. These reveal the normal state to be a highly anisotropic Fermi liquid [184], with a resistivity anisotropy at low temperature of $\rho_c/\rho_{ab} \sim 1400$ [184(a)]. A complete quantum oscillation study [184(b), 184(c)] reveal that the Fermi surface is very simple consisting of three slightly warped tubes of rather square cross-section. Obviously, the normal state properties of this system are better understood than those of either the heavy fermion metals (which have much more complex Fermi surfaces) or the cuprates. Finally, the low superconducting transition temperature ($\sim$ 1.5 K) compared with the Fermi degeneracy temperature of $T_F \sim 25$ K means that a weak coupling theory of superconductivity can be used, as opposed to the more challenging strong coupling theories needed in heavy fermions and HTSC cuprates.

In contrast to HTSC cuprates, Sr_2RuO_4 is considered to be an ideal two-dimensional Fermi-liquid [184], since there is no anomalous behaviour in the normal state. Since T_c is much lower than that of HTSC cuprates, and organic superconductors κ-$(ET)_2 X$, the superconductivity in this system is easily destroyed by the small-perturbation or disorder. For Sr_2RuO_4 it is difficult to find, at least at present a well-defined controlling parameter for the appearance of the superconductivity such as doping in HTSC cuprates.

According to the first-principle band-structure calculations [186, 187], the electronic states near the Fermi level mainly consist of Ru $4d\varepsilon$ orbitals, although the Ru $4d\varepsilon$ and O^{2p} orbitals hybridize with each other. Since it is expected that electrons strongly correlate through coulomb interactions at Ru sites, Sr_2RuO_4 belongs to a class of **'Strongly correlated electron systems'** (SCES). In fact, the Mott insulating state has been found in the related compound $CaRuO_4$ [188], suggesting the importance of strong correlation effect.

Compared to other superconducting compounds, Sr_2RuO_4 has somewhat different nature both in the electronic structure and in the superconducting properties. One interesting issue is that this compound is **multiband system**. The quantum oscillation measurement has shown three quasi-two-dimensional Fermi surfaces, which are defined as α, β and γ sheets [185]. We may note that the observed Fermi surfaces are in good agreement with the first principle band-structure calculation results [186, 187] and the recent ARPES measurement [189]. Another remarkable difference from HTSC cuprates and organic κ-$(ET)_2 X$ is the electron filling. Since the valence of ruthenium ion is Ru^{4+}, four electrons occupy Ru-site on the average. Again according to the quantum oscillation measurement, about 4/3 electrons are included in the γ-band and remained part is included in α and β bands. Thus, it is generally regarded that $Sr_2 Ru O_4$ is far from the half-filling.

The most outstanding and interesting difference is the pairing symmetry, which has recently been suggested to be spin-triplet due to excellent experiments [190]. The most important experimental evidence suggesting the spin-triplet pairing has been obtained by NMR [191, 192]. Ishida et al. [191, 192] have measured the ^{17}O (oxygen)-NMR and Ru Knight shift by applying the magnetic field parallel to the ab-plane, and observed no suppression in the spin susceptibility below T_c. This result excludes the possibility of the spin-singlet pairing, and at the same time, indicates the d-vector along the z-axis. The d-vector is an unusual expression for the internal degree of freedom, which is an interesting topic in the triplet superconductivity [193, 194]. Recent inelastic polarized neutron scattering experiment has suggested the same results [195].

Spin triplet pairing in ruthenate has been experimentally confirmed recently. Slight portion of non-magnetic impurities drastically suppresses the superconductivity [196], in sharp contrast to the impurity effects in conventional s-wave superconductors. The NMR and NQR relaxation rates exhibit no coherence peak just below T_c [197, 198]. The NSR measurement has shown that an internal magnetic field is spontaneously turned on below T_c [199], indicating that the time-reversal symmetry is broken in the SC state of ruthenate. From this result, the chiral state $\hat{d} = (K_x \pm i K_y)\hat{z}$ has been proposed [200]. The temperature dependence of the critical current in $Pb/Sr_2 Ru O_4/Pb$ junction is also consistent with the p-wave pairing state [201].

Theoretically Rice and Sigrist [202] pointed out a possibility of spin-triplet superconductivity in $Sr_2 Ru O_4$, immediately after the discovery of superconductivity in this system. Their insights have been based on the following facts: (i) some Fermi liquid parameters are similar to those of ^3He, which is confirmed to be a spin-triplet p-wave superfluid [194], (ii) the three-dimensional analogous compound $SrRuO_3$ exhibits the ferromagnetism with a Curie temperature $T_c = 160$ K. Rice and Sigrist [202] have considered that the Hund's rule coupling among Ru $4d\varepsilon$ orbitals stabilizes the spin-triplet pairing rather than the spin-singlet one. Although any microscopic justification for the above two insights has not yet been obtained upto now, their excellent prediction itself has obtained a great success.

The theoretical interests on $Sr_2 Ru O_4$ are mainly focused on the two fundamental aspects of superconductivity, namely the **pairing symmetry** and **pairing mechanism.** Concerning the pairing symmetry, the origin of the power-law behaviours is a challenging subject. As mentioned earlier the chiral state without time-reversal symmetry is expected for the internal degree of freedom. Then, assuming the simple momentum dependence of SC gap, for instance, $\Delta(\mathbf{K}) \propto \sin K_x$ [203], the excitation gap opens on the whole Fermi surface. On the contrary, the gap-less power-law behaviours, suggesting the existence of line node, have been observed in common among the several experimental results on the specific heat [204], NMR $1/T_1T$ [150], magnetic field penetration depth [205], thermal conductivity [206, 207] and ultrasonic attenuation rate [208]. We may note that only the point node is derived from the symmetry argument, even if the three-dimensional degree of freedom is taken into account [209, 210]. Thus, if one assumes the Chiral state, the line node should appear only accidentally. Namely, the theoretical proposal on this problem has to rely on somewhat an accidental reason. Among them, the three dimensional f-wave symmetry [162, 163] has been supported by the thermal conductivity measurement [207, 213]. The more improved proposal based on the multi-band effect has been proposed along this line [214]. The essential assumption of this proposal is that the zeros of the order parameter corresponding to the symmetry $\Delta(k) \propto k_x$ is parallel to the plane. When the zeros have a slope, the point node is expected. The pairing state assumed here is generally difficult in view of pairing mechanism, because the Fermi surface of Sr_2RuO_4 is clearly two dimensional. Obviously, this problem should be resolved within the two dimensional model.

Sr_2RuO_4 is perhaps the most interesting unconventional superconductor presently under investigation. Unlike the cuprates, it is superconducting at stoichiometry, so disorder does not strongly affect the normal state properties. Now crystals of exceptional purity have been produced. The melting point is also high. The Fermi surface is very simple and normal state properties of this system are better understood. The effective masses as determined by $dHvA$ are as high as 14.4 m_e [185]. There is strong evidence that Sr_2RuO_4 is a triplet superconductor. It is not obvious why Sr_2RuO_4 should develop line nodes in the gap.

It is quite surprising that Sr_2RuO_4 is a triplet superconductor. Neither neutron scattering nor thermodynamic properties seem to place this system especially close to ferromagnetic ordering, and indeed single band models that assume that the measured magnetic susceptibility provides the pairing interaction predict that Sr_2RuO_4 will be a d-wave superconductor [215]. The scenario that is emerging at the present moment is that the multi-band/ multi-orbital nature of Sr_2RuO_4 is important: On-site Coulomb repulsion between the different d-orbitals has been shown to suppress singlet superconductivity [216] leaving open an avenue for triplet superconductivity to appear. However, Rice et al. [202] have suggested that the superconductivity resides mainly on one sheet of the Fermi surface, which develops a full superconducting gap, while any nodes lie on other sheets of the Fermi surface on which weak superconductivity is induced by a kind of K-space proximity effect. Obviously, Sr_2RuO_4 is the simplest of the unconventional superconductors, but at the same time challenging too.

9.5 ORGANIC SUPERCONDUCTORS

Jerome et al. [217] discovered the first organic superconductor $(TTMSF)_2 PF_6$ in 1980 [T_c = 0.9 K at 12 k bar]. The critical temperature of superconductivity (T_c) in organic materials have risen, exceeding 10 K in 1988 [2]. This superconductor is ambient pressure superconductor based on BEDT-TTF with T_c = 10.4 K. More than 90 organic superconductors have been prepared so far [219, 220]. They are classified into 15 families on the basis of the conducting component molecules, namely TMTSF (7 members with the highest T_c of 3 K at 5 k bar, and 1.4 K at ambient pressure (AP), BEDO-TTF (BO) (2, 1.5 K) BEDT-TSF (BETS) (5, 8 K), BEDS-TTF (1, 7.5 K at 1.5 k bar), DMET (7, 1.9 K) MDT-TTF (1.41 K), S,S-DMBEDT-TTF (1, 2.6 K at 5.8

k bar), DMET-TSF (2, 0.58 K), TMET-STF (1, 3.8 K), ESET-STF (1, 4.8 K at 3.2 k bar), DTEDT (2, 4 K), dmit (8, 6.5 K at 20 k bar and 1.3 K at AP) and C_{60} (3, 26 K).

All these families of organic superconductors are summarized in Tables 9.2 to 9.5.

Table 9.2 Organic superconductors of symmetric donors : TMTSF and TMTTF

Donor	Anion	Symmetry	Ratio	$\sigma_{RT}{}^a$ (S cm^{-1})	$T_{max}{}^b$ (K)	P_c (kbar)	T_c (K)	Characteristics
TMTSF	PF_6	Octahedral	2:1	540	12–15	6.5	1.4	SDW (12 K). FISDW
	AsF_6	Octahedral	2:1	430	12–15	9.5	1.4	SDW (12 K)
	SbF_6	Octahedral	2:1	500	12–17	10.5	0.38	SDW (17 K)
	TaF_6	Octahedral	2:1	300	15	11	1.35	SDW (11 K)
	ClO_4	Tetrahedral	2:1	700		0	1.4	OD(24 K, $a \times 2b \times 2c$). FISDW SDW (5K) by rapid cool
	ReO_4	Tetrahedral	2:1	300	~ 182	9.5	1.2	OD(177 K, $2a \times 2b \times 2c$)
	FSO_3	Tetrahedral-like	2:1	1000	~ 88	5	3	OD(88 K, $2a \times 2b \times 2c$)
TMTTF	Br	Spherical	2:1	260	100	26	0.8	SDW(15 K)

$^a\sigma_{RT}$: Conductivity at room temperature.
$^bT_{max}$: Temperature at maximum conductivity.
^{c}SDW : Spin density wave, OD : order-disorder transition of anion.

Although the total number of the organic superconductors is still less than 1% of the inorganics, the molecular systems have provided many interesting features in their structural and physical properties both in the normal and superconducting phases concerning their dimensionality, electron correlation and superconducting characterstics.

Molecular conductors differ fundamentally from the regular metallic conductors in that they are composed of building blocks (the molecules) that exhibit their own specific properties (*e.g.*, molecular orbitals with their ionization energy or electron affinity, characterstic IR or Raman active modes, and NMR chemical shifts of different atomic sites). In contrast to conventional molecular crystals made up of neutral organic molecules held together by weak van der Waals forces, organic conductors contain molecules with unpaired carriers in π molecular orbitals (open shells) which allow delocalization over all molecular sites in the crystal via strong intermolecular π-overlap. The open-shell character arises from partial oxidation (reduction) of donor (acceptor) molecules with formation of a salt of an inorganic anion (cation).

These molecular conductors are therefore different from the extended conjugated polymers and graphite where the π-system of the extended molecules provides the conducting pathway.

The development of molecular conductors (superconductors) has been stimulated to a large extent by the suggestion made by Little in 1964 [221] that the arrangement of chain

Table 9.3. Organic superconductors of symmetric donors:
BEDT-TTF(ET), BEDO-TTF(BO), BEDT-TSF(BETS), and BEDS-TTF

Donor	Anion[a]	Symmetry	Ratio & Phase	σ_{RT}^{b} (S cm^{-1})	T_{max} (K)	P_c (kbar)	T_c (K)	Characteristics
BEDT-TTF(ET)	I_3	Linear	2:1 β_L	60		0	1.5	SdH, AMRO
	I_3	Linear	2:1 β			0	2.0	dHvA(22–25%, 0.4–0.5 m_e 103%. 4.0 m_e)
	I_3	Linear	2:1 β_H			0	8.1	anneal below 110 K for > 20 h
	I_3	Linear	2:1 α_1			0	~8	press β_L, SdH(51%, 4.65 m_e)
	I_3	Linear	3:2.5 γ			0	2.5	heat α, ε- or ξ-phase
			2:1 θ			0	3.6	charge of donor = – 2.5/3
	I_3	Linear	2:1 κ	20		0	3.6	SdH, AMRO, dHvA(19%, 1.8m_e : 102%, 3.5 m_e) SdH, dHvA(15%, 1.9 m_e : 102%, 3.9 m_e)
	I_3	Linear		30		0		
	IBr_2	Linear	2:1 β	20		0	2.7	$\gamma = 18.9 \pm 1.5$, $\beta = 10.3 \pm \iota$, $\Theta = 218 \pm 7$
	AuI_2	Linear	2:1 β	20		0	4.9	SdH, dHvA(2.0 ± 0.5 m_e, 0.3±0.1 m_e) SdH. AMRO. dHvA (51 – 53% . 4.0 4.2 m_e)
	Cu(CF$_3$)$_4$.TCE	Planar	2:1 κ_L			0	4.0	
	Cu(CF$_3$)$_4$.TCE	Planar	2:1 κ_H			0	9.2	$\gamma = -50$, $\beta = 18$, $\Theta = 203 \pm 10$
	Ag(CF$_3$)$_4$.TCE	Planar	2:1 κ_L			0	2.4	
	Ag(CF$_3$)$_4$.TCE	Planar	2:1 κ_H			0	11.1	
	Au(CF$_3$)$_4$.TCE	Planar	2:1 κ_L			0	2.1	
	Au(CF$_3$)$_4$.TCE	Planar	2:1 κ_H			0	10.5	
	Cu(CF$_3$)$_4$.TBE	Planar	2:1 κ_L			0	5.2	
	Ag(CF$_3$)$_4$.TBE	Planar	2:1 κ_L			0	4.8	
	Ag(CF$_3$)$_4$.TBE	Planar	2:1 κ_H			0	7.2	
	Au(CF$_3$)$_4$.TBE	Planar	2:1 κ_L			0	5.8	

Table 9.3. Continued...

Anion	Geometry	Phase					Notes
Cu(CF$_3$)$_4$·121DBCE	Planar	2:1 κ$_L$			0	5.5	
Ag(CF$_3$)$_4$·121DBCE	Planar	2:1 κ$_L$			0	4.5	
Au(CF$_3$)$_4$·121DBCE	Planar	2:1 κ$_L$			0	5.0	
Cu(CF$_3$)$_4$·121DCBE	Planar	2:1 κ$_L$			0	3.5	
Ag(CF$_3$)$_4$·121DCBE	Planar	2:1 κ$_L$			0	3.8	
Ag(CF$_3$)$_4$·121DCBE	Planar	2:1 κ$_H$			0	7.3	
Au(CF$_3$)$_4$·121DCBE	Planar	2:1 κ$_L$			0	3.2	
Cu(CF$_3$)$_4$·112DCBE	Planar	2:1 κ$_L$			0	4.9	
Ag(CF$_3$)$_4$·112DCBE	Planar	2:1 κ$_L$			0	4.1	
Ag(CF$_3$)$_4$·112DCBE	Planar	2:1 κ$_H$			0	10.2	
Au(CF$_3$)$_4$·112DCBE	Planar	2:1 κ$_L$			0	5.0	
Cl$_2$(H$_2$O)$_2$	Square, Cluster	3:1	500	100	16	2	dianion, charge of donor = + 2/3
Ag(CN)$_2$·H$_2$O	Linear, cluster	2:1 κ	37	150	0	5.0	SdH(17%, 2.7 m_e)
Pd(CN)$_4$·H$_2$O	Square planar, Cluster	4:1 β″	100	70	7	1.2	dianion
Pt(CN)$_4$·H$_2$O	Square planar, Cluster	4:1 β″	280	120	6.5	2	dianion
ReO$_4$	Tetrahedral	2:1	200	81	0.4	2	
Fe(C$_2$O$_4$)$_3$·H$_3$O PhCN	Octahedral	4:1 β″			0	6.5–7.7	AMRO
Cr(C$_2$O$_4$)$_3$·H$_3$O PhCN	Octahedral	4:1 β″			0	6.0	
SF$_5$CH$_2$CF$_2$SO$_3$	None	2:1 β″			0	5.3	SdH(13%, 1.9 m_e). γ = 18.7 ± 1. β = 12.2 ± 1. Θ = 221 ± 7

Table 9.3. *Continued...*

Compound							Remarks
KHg(SCN)$_4$	Polymer	2:1	100		0	0.3	SdH. AMRO (16%. 1.4 m_e 13% 2.53 m_e) γ = 6.4, β = 11.6, Θ = 223
NH$_4$Hg(SCN)$_4$	Polymer	2:1 α		1.2	1.2	0	0.8–1.7	SdH, AMRO(13%, 21 m_e) γ = 25 ~ 26. β = 12.8 ± 0.5, Θ = 221
RbHg(SCN)$_4$	Polymer	2:1 α	380			0	0.5	SdH. AMRO(16.5%, 1.5 m_e) γ = 7.1 β = 11.1, Θ = 226
TlHg(SCN)$_4$	Polymer	2:1 α	6			0	0.1	SdH. AMRO (16%. 1.5 m_e, 10.3%, 4.0 m_e)
Hg$_{2.78}$Cl$_8$	Polymer	4:1 κ	30	25		12	1.8	dianion
Hg$_{2.89}$Br$_8$	Polymer	4:1 κ	5	18		0	4.3	dianion
Cu(NCS)$_2$ Deuterated salt	Polymer	2:1 κ	10–40	90		3.5	6.7	
Cu[N(CN)$_2$]Br Deuterated salt	Polymer	2:1 κ	2–50	50–90		0	10.4	SdH, AMRO.dHvA (15%. 3.5 m_e ; 105%. 6.5 m_e) inverse isotope effect
Cu[N(CN)$_2$]Cl Deuterated salt	Polymer	2:1 κ	2	semi		0	11.2	SdH(4.4%.0.95 m_e at AP. 106%. 6.4 m_e at 9 k bar) might show inverse isotope eflect by slow cooling
Cu[N(CN)$_2$]Cl$_x$Br$_{1-x}$	Polymer	2:1 κ				0.3	11.8	
		2:1				0.3	12.8	
		2:1 κ		semi		0	13.1	SdH, AMRO(15.5%. 102% at 7.7 k bar.)
Cu[N(CN)$_2$]Br$_{0.9}$I$_{0.1}$	Polymer	2:1 κ				0	10	re-entrant super inverse isotope effect
						0	11.5	$x = 0.15$
						0	11.3	$x = 0.25$
						3	5.9	$x = 0.5$

Table 9.3. Continued...

	Cu(CN)[N(CN)₂] Deuterated salt	Polymer	2:1 κ	5—50		0	11.2 inverse isotope effect
						0	12.3
	Cu₂(CN)₃	Polymer	2:1 κ	10	semi	1.5	2.8 SdH, AMRO (3.1% at AP, 96%, 4.5 m_e at 8 k bar)
	Cu$_{2-y}$(CN)$_{3-2y}$ [N(CN)₂]$_y$	Polymer	2:1 κ			0	3–11
BEDO-(BO)	Cu₂(NCS)₃	Polymer	3:1	22		0	1.06 charge of donor = + 1/3
	ReO₄.H₂O	Tetra-hedral	2:1	200	2.5	0	1.5 SdH (0.7%, 1.15 m_e ; 1.5%, 0.9 m_e)
BEDT-TSF (BETS)	GaCl₄	Tetra-hedral	2:1 λ	50		0	8 SdH (100%)
	GaBr₃Cl$_y$	Tetra-hedral	3:1 λ			0	7.8
	GaFCl₃	Tetra-hedral	2:1 λ			0	3.5
	(FeCl₄)$_x$(GaCl₄)$_{1-x}$	Tetra-hedral	2:1 λ			0	3.0–5.4 $x = 0.43$–0.55, re-entrant super conductor
	Cl₂TCNQ		2:1			3.5	0.8
						1.5	1.3
BEDS-TTF	CuN(CN)₂]Br	Polymer	2:1 κ	0.1	25	1.5	7.5

[a]TCE:1,1,2-trichloroethane. TEB:1,1, 2-tribromoethane. 121DBCE:1,2-dibromo-1-chloroethane, 121 DCBE:1,2-dichloro-1-bromoethane. 112 DCBE:1, 1-dichloro-2-bromoethane. PhCN : benzonitrile. C₂O₄: oxalate dianion.
[b]Semi: semiconductor.
[c]SdH: Schubnikov de Haas oscillations, dHvA: de Haas van Alphen oscillations, AMRO : angle dependent magetoresistance oscillations, the numbers in parentheses are the area of the Fermi surface relative to the first Brilloun zone and the cyclotron mass, m_e : electron mass. γ : Sommerfeld coefficient, mJ mol K². β : mJ mol K⁴. Θ: Debye temperature, (K).

Table 9.4. Organic superconductors of unsymmetric donors:
DMET, MDT-TTF, S,S-DMBEDT-TTF, DMET-TSF, TMET-STF, and DTEDT

Donor	Anion	Symmetry	Ratio & phase	σ_{RT} (S cm^{-1})	T_{max} (K)	P_c (k bar)	T_c (K)	Characteristics[c]
DMET	I_3	Linear	2:1	170		0	0.47	
	AuI_2	Linear	2:1	300	40	5	0.55	SDW(20 K)
	IBr_2	Linear	2:1	210		0	0.58	
	$Au(CN)_2$	Linear	2:1	2500	25	2.5	1.1	SDW(25 K)
	$AuBr_2$	Linear	2:1	14	150–180	1.5	1.6	SDW(2.8 K)
	$AuBr_2$	Linear	2:1 κ	200	150	0	1.9	dHvA
	$AuCl_2$	Linear	2:1	230		0	0.83	
MDT–TTF	AuI_2	Linear	2:1 κ	20		0	4.1	$\gamma = 35 + 3.$ $\beta = 16.5 + 1.0$
S,S-DMBED-TTF	ClO_4	Tetrahedral	2:1 κ	0.05	75	5.8	2.6	
DMET-TST	AuI_2	Linear	2:1			0	0.58	
	I_3	Linear	2:1			0	0.4	
TMET-STF	BF_4	Tetrahedral	2:1			0	3.8	
ESET-STF	$Cu[N(CN)_2]Br$	Polymer				3.2	4.8	
DTEDT	$Au(CN)_2$	Linear	3:1	15	4	0	4	
	SbF_6	Octahedral	3:1			0	0.3	

Table 9.5 Organic superconductors of acceptors : M(dmit); and C_{60}

Acceptor	Donor or cation[a]	M	Ratio & phase	σ_{RT} (S cm^{-1})	T_{max} (K)	P_c (k bar)	T_c (K)	Characteristics
M(dmit)$_2$	TTF	Ni	1:2 α	300		7	1.6	
	NMe$_4$	Ni	1:2 β	60	100	7	5	SDW, SdH
	EDT-TTF	Ni	1:1 α	100	12	0	1.3	
	TTF	Pd	1:2 α	800	245	20	6.5	
	TTF	Pd	1:2 α'			22	1.7	
	NMe$_4$	Pd	1:2 β	30	semi	6.5	6.2	
	NMe$_2$Et$_2$	Pd	1:2 α			4	2.4	
	PMe$_2$Et$_2$	Pd	1:2 β'			7	4	
C_{60}	OMTTF·C$_6$H$_6$	K				0	18.8	M was doped into donor C_{60}·C_6H_6
	OMTTF·C$_6$H$_6$	Rb				0	26	
	DBTTF·C$_6$H$_6$	Rb				0	22	

[a]NMe$_4$: tetramethylammonium, NMe$_2$Et$_2$: diethyldimethyammonium, PMe$_2$Et$_2$: diethyldimethylphosphonium, C$_6$H$_6$: benzene; for TTF. EDT-TTF. OMTTF, and DBTTF: (Fig. 9.17).

conductors in a polarizable medium could give rise to superconductivity at higher temperature.

The geometry of the building blocks governs the dimensionality of the transport properties of these particular conductors (between one and three dimensions).

For one-dimensional conductors, the planarity of the molecules is responsible for their particular crystal structure (Fig. 9.16 (a)). All these molecules are modifications of the flat fulvalence skelton. Noticeable intermolecular overlap is allowed when they form stacks. The conduction band derives from π orbitals directed along the stacks and reveals a pronounced 1-D character, i.e., a flat and open Fermi surface [4, 5].

Higher dimensionality conductors are also derived from similar planar molecules that form dimers within each dimer and between neighbouring dimmers (Fig. 9.16 (b)). This crystal structure gives rise to a large conductivity within molecular layers and to about 10^4 times less conduction between sheets. They can be considered as 2-D conductors [224].

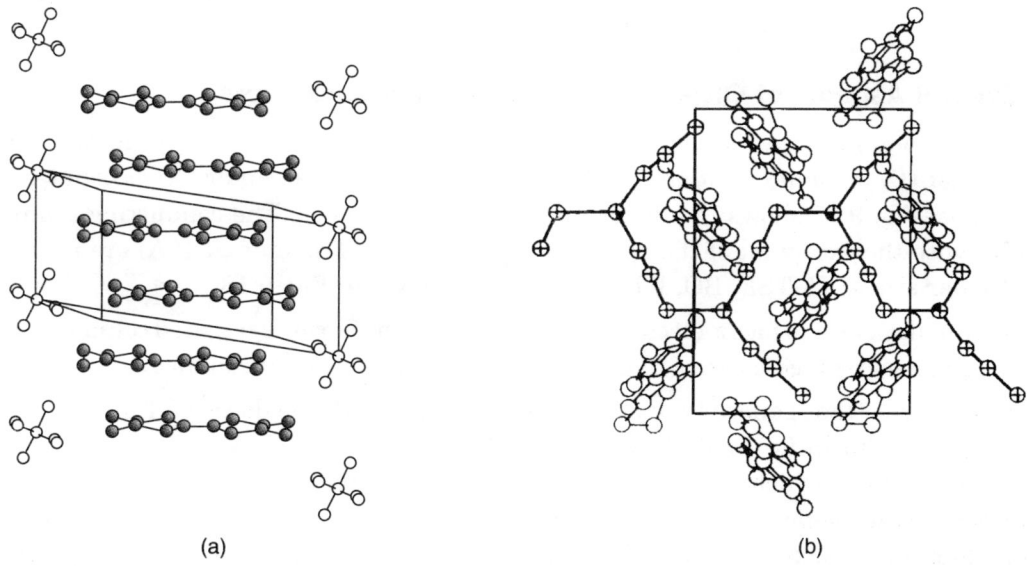

Fig. 9.16 Crystal structure of organic conductors (a) 1-D conductor $(TMTSF)_2 PF_6$ viewed perpendicularly to the high-conductivity axis ; (b) 2-D conductor $(BEDT-TTF)_2 Cu(SCN)_2$ viewed along axis of least conductivity.

The discovery of the spherical C_{60} molecule has opened the way to 3-D isotropic molecular conduction. Three of the four valence electrons of each carbon atom are engaged in sp^2 bonding with the three neighbours. There remains one electron per carbon atom in a p-state which is delocalized over the whole surface of the sphere in π-molecular orbitals. The neutral C_{60} molecule possesses a closed-shell electronic structure, giving rise to an insulating molecular solid. However, populating the highest unoccupied molecular levels (the LUMO) with carriers donated by inorganic cations such as alkali or alkaline-earth metals gives rise to a partly filled conduction band. This band is half-filled in the prototype material A_3C_{60} (A = K, Rb, Cs, etc.). The C_{60} spheres are packed in a face-centered cubic lattice, leaving cavities with octahedral and tetrahedral local symmetry which are all occupied in $A_3 C_{60}$ (Fig. 9.16(c)) [225].

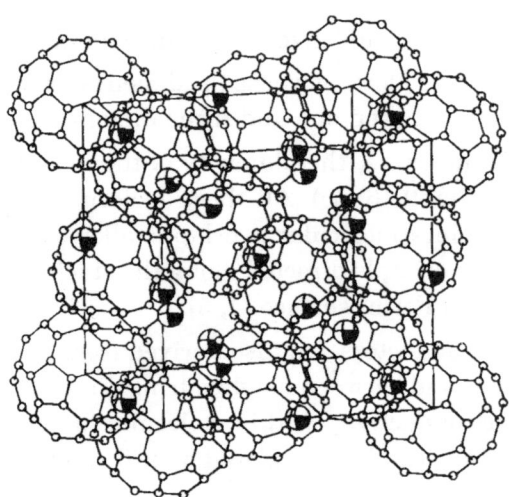

Fig. 9.16 (c) Crystal structure of 3-*D* conductor K_3C_{60}.

9.5.1 General Aspects of Organic Metals and Superconductors

All the known organic superconductors are the charge transfer (CT) complexes consisting of electron donor (D) or counter cation (K) molecules and electron acceptor (A) or counter anion (X) molecules. Fig. 9.17 shows the donor and acceptor molecules. The conducting component molecules, with the exception of the spherical C_{60} molecule (diameter ca 10Å) are nearly flat. The molecular size of TMTSF, BO, ET, and BETS is shown in Fig. 9.18.

They make a face to face π-electron overlap to form a uniform segregated column. Carriers move along the segregated column to give rise to a low-dimensional conductor.

It is reported that generally, in neutral CT complex with the degree CT(γ) less than 0.5, the segregated column is unstable and the complex results in a semiconductor with an alternating column in which *D* and *A* molecules stack alternatively. The organic metals and superconductors are ionic CT complexes with $0.5 \leq \gamma < 1$ and they have uniform segregated columns. However, a non-uniform segregated column does not give metallic complexes. But there is one exception known so far. In the κ-type stacking, molecules form a dimer that is orthogonally aligned to each other to construct a two-dimensional conductor.

The CT complexes are classified as $D^{+\gamma} A^{-\gamma}$ (*DA* type complex), $(D^{+\gamma})_m X_n$ (cation radical salt) and $K_m (A^{-\gamma})_n$ (anion radical salt), where *X* and *K* are closed shell ions. The C_{60} superconductors and alkylammonium [*M*(dmit)$_2$] superconductors belong to the anion radical salts where the radical electrons on molecule A play an essential role in the physical properties of the salts. While the closed shell molecular or atomic ion, K, takes both roles to compensate the charge to neutral and to adjust the distance between the radicals. We may note that only four, *i.e.*, (TTF) [*M*(dmit)$_2$]$_2$ (*M* = Ni, Pd), EDT-TTF [Ni(dmit)$_2$] and (BET)$_2$ Cl$_2$ TCNQ, are of the DA type. In this type of complex, both radical electrons on D and A molecules may participate in the electric conduction provided that the γ is not unity. Since the bandwidth of organic superconductors is comparable or smaller than the on-site Coulomb repulsive energy, the delocalization of electrons is prevented when γ is 1(Mott criterion). Obviously, in the DA type complex, the degree of CT is an important factor to understand the transport properties. Although, the exact γ value has not been determined yet in the above-mentioned DA type superconductors, it was found experimentally that the principal contributor to the superconductivity is the electron on M (dmit)$_2$ C_{60} and BETS molecules. Thus, all the M (dmit)$_2$ and C_{60} superconductors (Table 9.5) are all of the acceptor type (or electron type) superconductors. The other superconductors are cation radical salts with mainly γ = 0.5. Obviously, they are donor type (or hole type) superconductors.

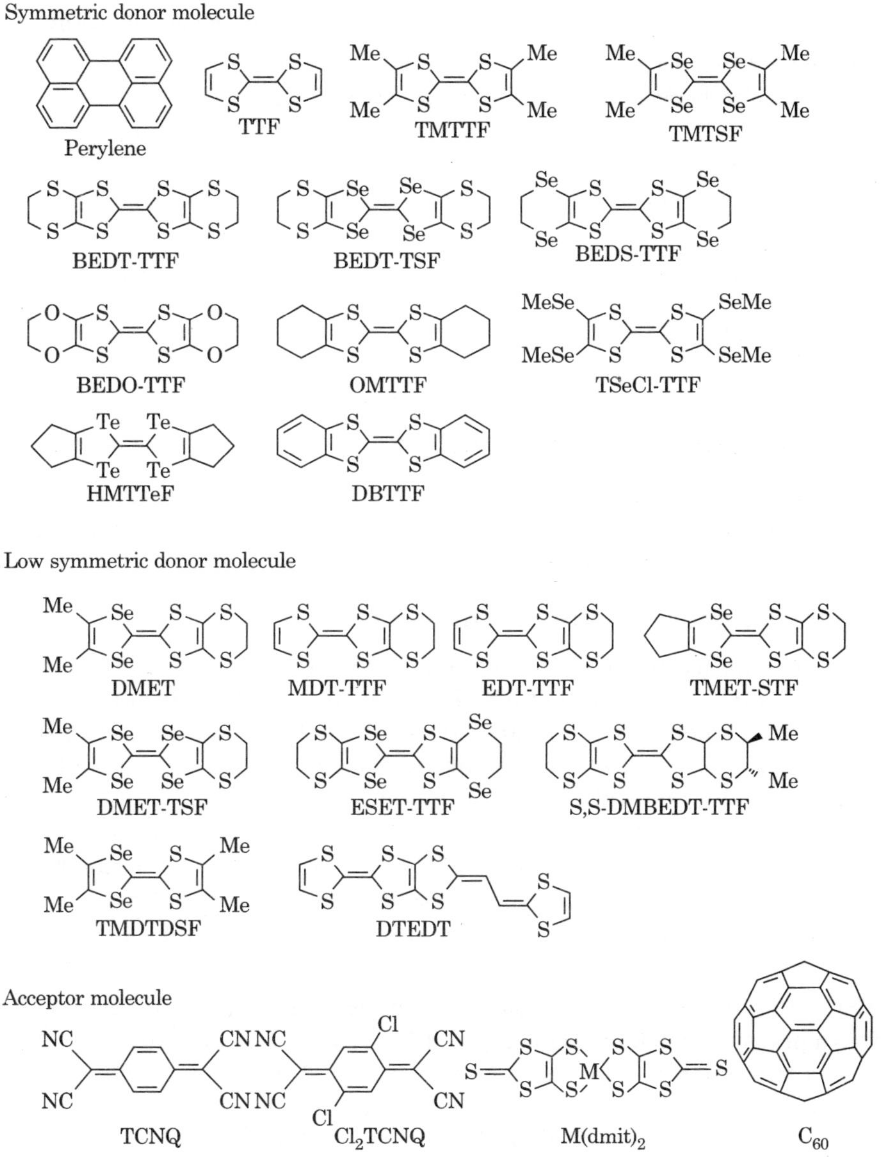

Fig. 9.17 Structures of donor and acceptor molecules.

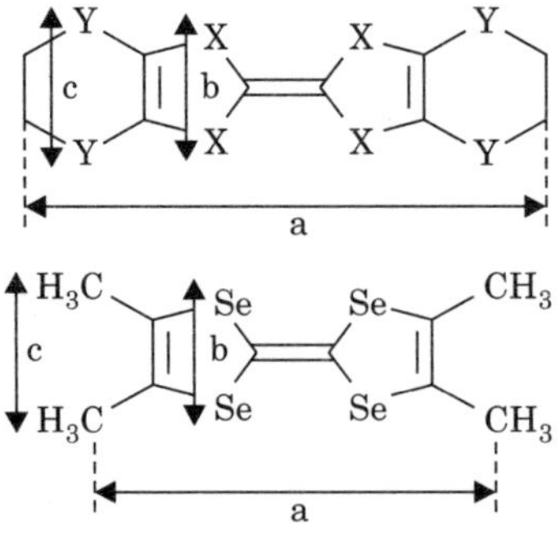

(a) molecular length between the carbon atoms of terminal ethylene (or methyl) groups
(b) molecular width between the inner chalcogen atoms
(c) molecular width between the outer chalcogen atoms (or carbon atoms of methyl groups)

	a (Å)	b (Å)	c (Å)
X = S Y = O BEDO-TTF(BO)	10.1	2.99	2.90
X = S Y = S BEDT-TTF(ET)	11.8	2.96	3.54
X = Se Y = S BEDT-TSF(BETS)	12.1	3.19	3.52
TMTSF	9.15	3.15	3.12

Fig. 9.18 Molecular geometry of BEDO-TTF(BO), BEDT-TTF(ET), BEDT-TSF(BETS) and TMTSF.

The organic superconductors prepared so far have been developed on the basis of the concept of increased diamensionality by chemical method with or without physical method (pressure), and their T_c's are advanced by the concept of the density of states.

9.5.2 Electronic Structure and Superconductivity

The salts of the tetramethyl tetraselenafulvalene (TMTSF) molecule are regarded as prototype materials for 1–D molecular conductors. The overlap between organic stacks along the transverse b-direction amounts to $t_{\perp b} \approx 20 - 30$ meV, which is about one tenth the value for the intrastack overlap ($t_\parallel \approx 200 - 300$ meV), making the conduction band about 1 eV wide. The overlap is negligible along the c-direction. Such a marked anisotropy causes the Fermi surface to be nearly planar, intersecting the $k_\parallel$ axis at wave vectors $\pm k_F$, where the Fermi wave vector for a 1–D conductor is related to the density of carriers per unit cell ρ by the relation $2k_F = \rho\pi/a$ [226, 227]. The stabilization of superconductivity below a certain critical temperature T_c requires overcoming the strong divergence that develops at low temperature in a 1–D electron gas towards an insulating behaviour.

(TMTSF)$_2$ PF$_6$, the organic compound in which superconductivity was first observed [228], requires a pressure of 9 k bar. However, (TMTSF)$_2$ ClO$_4$ is actually the only 1–D material that undergoes a transition to superconductivity at T_c = 1.2 K under ambient pressure. These temperature values are rather low, but higher T_c values are reached in materials where the shape of the Fermi surface is less likely to lead to the stabilization of insulating ground states (the 1–D and 2–D molecular superconductors, Fig. 9.19).

In 2–D conductors, the Fermi surface is tubular with $t_{(a,b)} = 100 - 150$ meV, $t_{\perp c} \approx 0.5$ meV, and consequently T_c rises upto 8.1 K in $(BEDT-TTF)_2 I_3$ [bis (ethylenodithio) tetrathiafulvalene and even upto 10 – 12 K for anions such as $Cu(SCN)_2^-$.

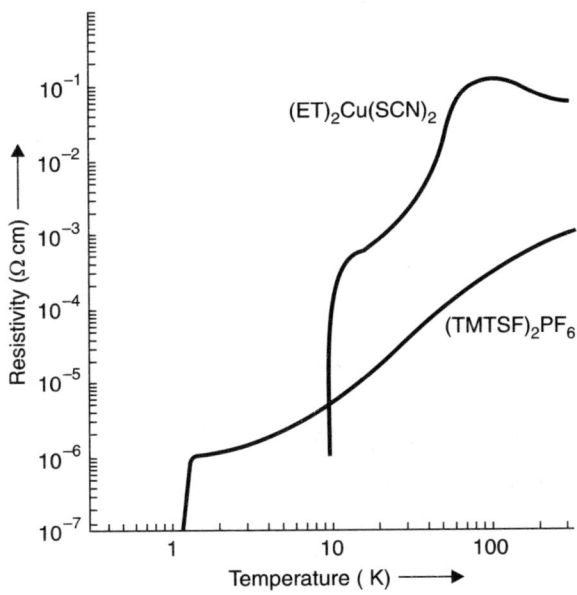

Fig. 9.19 The superconducting transition of 1 – D (under pressure) and 2–D organic conductors.

The discovery of organic superconductor $(TMTSF)_2 PF_6$ [228] has made a great impact on the field of superconductivity. The intensive exploitation in this field has revealed a universality of superconducting phenomena and attached much interests on the physical aspects of organic materials [219]. Among them, one of the most interesting superconductors is κ-type $(BEDT-TTF)_2 X$. We will express as $\kappa (ET)_2 X$ for an abbreviation. These compounds have quasi-two-dimensional electronic structures, which have been confirmed by the Shubrikov-de Hass experiments [229] and by the strong anisotropy in the electronic transport [230].

$\kappa(ET)_2 X$ compounds are one of the central attraction in this area because of their typical features as SCES. A typical phase diagram is shown in Fig. 9.20. We have two ordered phases by tuning the pressure P. One is the AF insulating state in the lower pressure region and the other is superconducting phase in the higher pressure region [231, 232]. We may note that the superconductivity occurs when the AF order disappears. The transition temperature decreases with increasing P. No doubt, this phase diagram includes similar aspects to that of HTSC cuprates, but there are several differences between these two systems. First, the phase transition from the AF to the SC state is the first order. Second, the carrier number in the conduction band is always half-filling per dimer, independent of the control parameter P. The pressure widens the bandwidth W and thus, it controls the parameter U/W. The decrease in U/W by the pressure has been confirmed by many experimental facts. For instance, the resistivity becomes smaller together with the superconducting T_c [231-236]. Therefore, the Mott transition in these compounds is regarded to be "bandwidth controlled", which is contrasted with the "filling controlled" Mott transition, as observed in HTSC cuprates [237].

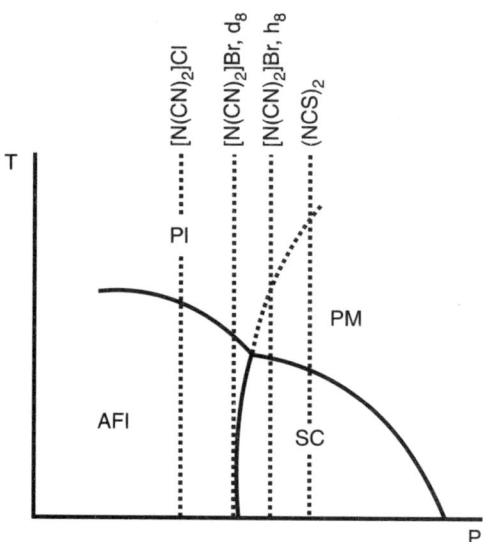

Fig. 9.20 Schematic phase diagram of $\kappa - (ET)_2 X$ organic compounds.

A series of these compounds have different properties at ambient pressure due to the kinds of anions X: $\kappa - (ET)_2$ Cu [N(CN)$_2$] Cl is the AF insulator [238]. The deuterated $\kappa - (ET)_2$ Cu [N(CN)$_2$] Br is located on the boundary between the two phases [239]. $\kappa - (ET)_2$ Cu [N(CN)$_2$] Br and $\kappa - (ET)_2$ Cu(NCS)$_2$ show the superconductivity at the ambient pressure. This systematic change due to X is regarded as an effect of the chemical pressure, as described in the phase diagram of Fig. 9.20.

The highest T_c among $\kappa - (ET)_2 X$ compounds is about 13 K. From the theoretical point of view, this value is the same order as the HTSC cuprates, when T_c is scaled by the bandwidth. In general, organic materials have smaller bandwidth by an order, because they are constructed from the molecular orbitals. The scaling between T_c and W have suggested a similar pairing mechanism to the HTSC cuprates.

The similarity between organic and HTSC cuprate superconductors can be extended to the anomalous properties in the normal state. In particular, the pseudogap has been also observed in the NMR $1/T_1 T$ [240, 241] with $T^* \sim 50$ K. However, it is an important difference that T^* is much higher than T_c and the electronic state is almost incoherent above T^*. Thus, the different nature around T^* is expected. However, the similar properties in the electronic state indicate the manifestation of the pseudogap with the same origin. Recently, Matsumoto and Kanoda [242] have given a clear understanding on this problem by measuring the magnetic field dependence of the NMR $1/T_1 T$; the SC fluctuation appears from the new crossover temperature T_c^* which is between T_c and T^*. The electronic state below T_c^* is regarded as the pseudogap state induced.

1–D and 2–D organic superconducting materials display the general features which are common to superconductors, namely a state with zero resistance and the expulsion of magnetic flux below T_c. They are type-II superconductors. There exist a broad field range between B_{c_1} and B_{c_2} in which the magnetic field is not fully expelled from the material (the vortex state).

Emerging Superconductors

For all organic conductors, parameters describing the superconducting phase are strongly anisotropic (Table 9.6).

Table 9.6 Properties of organic superconductors

Property*	Compound (dimensionality)		
	$TM_2 X$ (1-D)	χ-ET_2 X (2-D)	A_3C_{60} (3-D)
T_c, K	1.2 (X = ClO_4)	9(X = $Cu(SCN)_2$); 12(X = $Cu[N(CN)_2]Br$)	19(A = K); 33(A = Rb_2Cs)
H_{c_1} (0 K), 10^{-4} T	0.2(a); 1(b); 10(c)	1(a, b); 45(c)	
H_{c_2} (0 K), T	2.8(a); 2.1(b); 0.16(c)	30(a, b); 2(c)	49
ξ (0 K), Å	700(a); 335(b); 20(c)	174(a, b); 7(c)	26
λ (0 K), Å		5000 – 10000	2400 – 4800

*T_c = Critical temperature; H_{c_1}, H_{c_2} = Critical fields ; ξ = Coherence length ; λ = Penetration depth.

9.5.3 Other Salient Features of 1-D Superconductors

TM_2 X Phase Diagram

A direct consequence of the 1-D character of the $TM_2 X$ series is the existence of a wide variety of ground states which can be observed in various members of the family depending on parameters such as the chemical composition of the organic molecule or of the inorganic counterion and also on hydrostatic pressure.

Fig. 9.21 shows the variety of ground states; it is a generic phase diagram for the $TM_2 X$ family.

On the left of the phase diagram, $TM_2 X$ organic compounds exhibit a metal-like behaviour at high temperature, but below the temperature T_ϱ, where the resistivity passes through a broad minimum, a loss of the charge degree of freedom is observed, leaving the spins localized and behaving like those of a chain quantum antiferromagnet. Coulombic repulsions are responsible for such localization (Wigner-like localization) [243]. In the same temperature range the electron-phonon interaction gives rise to a 1-D fluctuation with a doubled periodicity which is detected by X-ray experiments. Below 20 K, these 1-D lattice fluctuations order themselves three-dimensionally and the system undergoes a phase transition with loss of the spin degrees of freedom. The ground state is a spin-peierls (SP) phase in which the lattice is tetramerized and the spins are coupled in a singlet state of total spin zero.

As we move towards the right hand side of the phase diagram, the role of electron-phonon interaction becomes less dominant but electron-electron interactions remain dominating, as shown by the temperature dependence of the spin susceptibility, which is still strongly reminiscent of the 1-D antiferromagnet. In this region of the phase diagram, the new order that becomes established below 19–12 K is characterized by a modulation of the spin density with a wave vector **Q** which provides the best nesting of the Q-1-D Fermi surface [266]. It is the presence of this additional periodicity in the exchange potential which is responsible for the opening of a gap 2Δ at the Fermi level accompanied by a metal insulator transition.

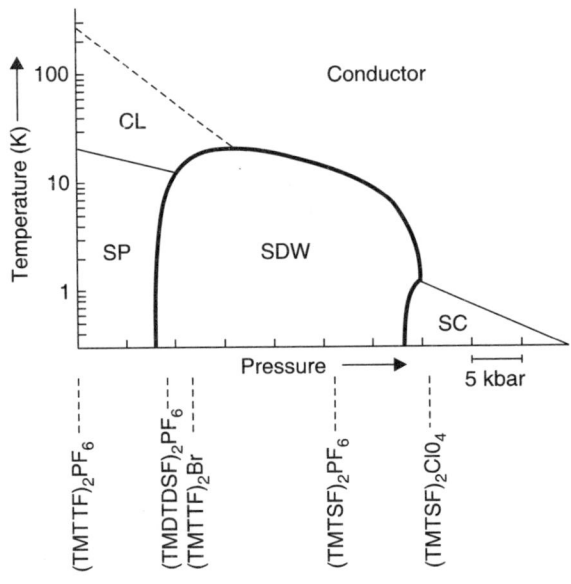

Fig. 9.21 The generalized phase diagram for the TM_2X series
TMTTF   = tetramethyltetrathiafulvalene;
TMDTDSF   = tetramethyldithiadiselenafulvalene;
TMTSF   = tetramethyltetraselenafulvalene
SP   = Spin-Peierls;
SDW   = Spin density wave;
SC   = Superconducting ground states;
CL   = Region where charges are localized by Coulombic forces.

Finally, on the very right of the phase diagram, the spin density wave (SDW) ground state itself is suppressed by an enhanced transverse coupling and superconductivity becomes stable instead.

As far as the conducting phase is concerned, one cannot regard the electron gas as an ordinary electron gas. Once more the one-dimensionality and the existence of strong electron-electron Coulombic repulsions make this metallic state remarkable. In spite of a fairly high conductivity-reaching 10^6 (Ω cm^{-1}) at low temperature for the compounds in the $(TMTSF)_2 X$ series the susceptibility of the conducting state develops a strong tendency toward antiferromagnetism at low temperature. NMR experiments have shown that the antiferromagnetic response becomes prominent below 150 K and displays a power low divergence $\chi_s \approx (T/E_F)^{-\gamma}$ with $\gamma \approx 0.85$ as expected from the theory of the 1-D correlated electron gas. On the other hand, the uniform spin susceptibility measured by a Faraday technique is temperature dependent, very much like that of a 1-D quantum antiferromagnet. Clearly, the superconducting instability of $(TMTSF)_2 X$ conductors arises out of a background of antiferromagnetic fluctuations.

9.5.4 Spin Density Wave Collective Conduction

The antiferromagnetic state of phase diagram 9.21 also exhibits various novel physical properties. It is the nesting of the Fermi surface with the wave vector $\mathbf{Q}$ which drives the transition from a metal at $T > 12$ K in $(TMTSF)_2 PF_6$ to an insulator exhibiting a magnetic modulation or spin density wave at the same wave vector $\mathbf{Q}$.

The broken symmetry of the SDW ground state, characterized by a magnetic modulation at wave vector $\mathbf{Q}$ in $(TMTSF)_2 X$, $X = PF_6$, NO_3 etc., is responsible for the activated behaviour of the conductivity through the single particle gap 2Δ. Moreover, two types of collective excitation arise in the SDW condensate. These are the magnon excitations, which manifests themselves in the antiferromagnetic resonance, and the phase mode of the condensate which is gapless provided the spin modulation is non-commensurate with the underlying lattice. In $(TMTSF)_2 PF_6$, the wave vector corresponding to the optimum nesting of the Fermi surface is incommensurate, as shown by various measurements sensitive to local magnetic fields. Therefore, as long as $\mathbf{Q}$ is incommensurate with the underlying lattice, the electrons can slide freely without any cost in energy, almost like in a superconductor. However, in real materials the interaction between as SDW (or the associated CDW) and impurities or crystal defects provides of finite pinning energy and a threshold electric field E_T must be reached before the condensate can contribute to the conduction at zero frequency [Fig. 9.22]. In all SDW phases of the $(TMTSF)_2 X$ series E_T is restricted to the range of a few mV/cm [244]. Moreover, the linear increase of E_T versus the density of lattice defects suggests that the phase of the spin modulation is locally adjusted at each defect site to provide the largest pinning energy (strong pinning limit).

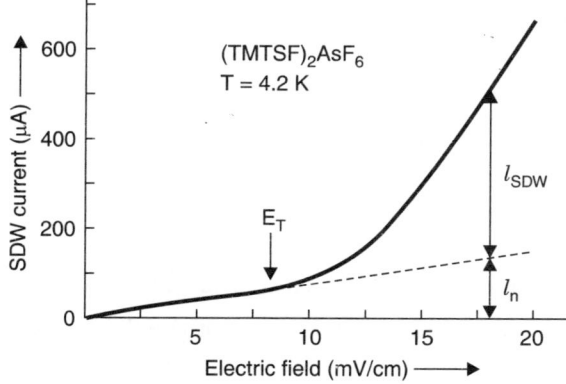

Fig. 9.22 Current-voltage characteristics in the SDW state showing the onset of nonlinear conduction due to the collective motion of the entire electron gas above a threshold field E_T.

Direct evidence for spin density wave transport is the detection of a current oscillating at a frequency proportional to the d.c. current carried collectively. The recent observation of such oscillations, the harmonic and sub-harmonic locking of this oscillation to an external a.c. source and a motional narrowing of NMR spectrum in the sliding SDW state have provided firm evidence for the existence of a novel collective transport in the SDW condensate [245].

9.5.5 QUASI 1-D Electrons in High Magnetic Fields

The response of the Q-1-D electron gas to the application of a strong magnetic field has provided new and unexpected results, namely the transition from a Q-1-D conductor to a quasi-2-D antiferromagnetic semimetal induced by a magnetic field at low temperature. The instability of the Q-1-D electron gas at high field can be **inferred** from the observation of magnetoresistance oscillations about a certain threshold field in $(TMTSF)_2 PF_6$ under pressure. Moreover, NMR data have shown that the ground state, which is stabilized by the magnetic field, is characterized by the existence of a spin density wave (SDW).

These phenomena are fairly well explained by a model in which, crudely speaking, the field restores enough one-dimensional character in the energy dispersion for the SDW dispersion to become stable at a finite temperature [246]. We may note that this is also equivalent to the effect of negative pressure. However, the experimental data of $(TMTSF)_2 PF_6$ show that the effect of the magnetic field is slightly more sophisticated. A sequence of phases must be crossed before the so called $N = 0$ phase (which is believed to correspond to the insulating SDW ground state, which is stable at ambient pressure) can be reached above 18 T (Fig. 9.23(a)) [247]. The integer N is related to the deviation of $\mathbf{Q}_x$ from $2k_F$, and denotes a phase containing N fully occupied Landau levels, quantized by the magnetic field. The situation $N = 0$ corresponds to zero density of states at the Fermi level, i.e. a gap which is opened over the whole Fermi surface. Phases with $N = 1, 2, 3, ...$, which are observed below 18 T correspond to a sequence of semimetallic phases with a very low density of carriers (10^{-2}/unit cell). In these subphases 1, 2, or 3, ... Landau levels are completely filled. The situation where the Fermi level falls between the completely filled and completely empty levels minimizes the diamagnetic energy of the 2-D carriers. This situation prevails in a finite range of magnetic fields, and the wave vector of the magnetic modulation within a given subphase can vary linearly with the field. First-order phase transitions are expected between various sabphases, in agreement with the observation of hysteresis (Fig. 9.23). The quantization of the Hall resistance at $Q_{xy} = h/2Ne^2$ (factor 2 is appearing from the spin degeneracy) is a direct consequence of the field dependence of the $\mathbf{Q}$ vector. The $(TMTSF)_2 PF_6$ Hall data (Fig. 9.23(b)) agree very well with the quantized nesting theory [248].

The quantization of the Hall resistance in the field induced SDW (FISDW) phases is indeed very reminiscent of the quantum Hall effect of the 2-D electron gas in semiconducting heterostructures. There is, however, an important difference between these two phenomena. In both cases the quantization requires a reservoir of nonconducting electronic states. This reservoir is provided by localized states in the gap between conducting Landau levels or the electron-hole (spin modulation) condensate for the 2-D electron gas and the FISDW, respectively.

Emerging Superconductors

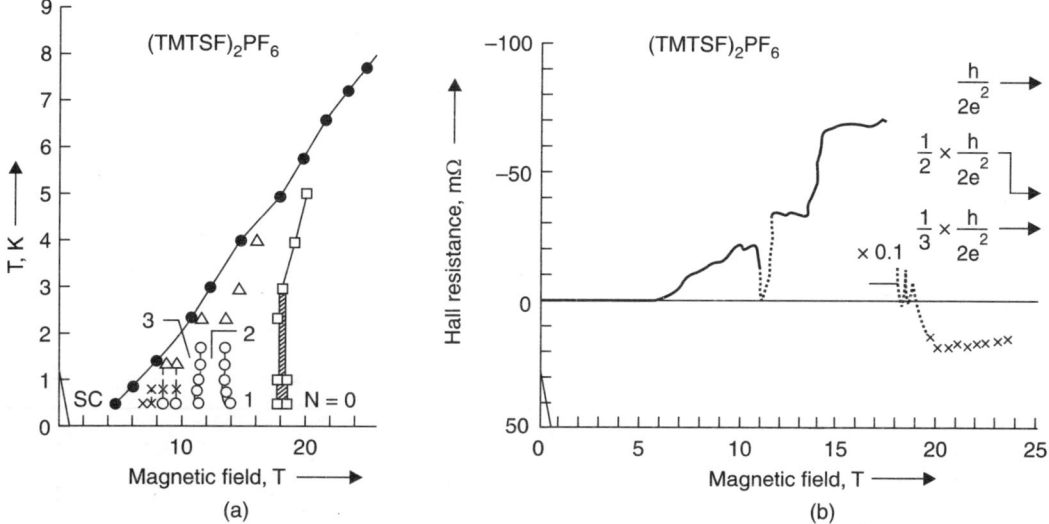

Fig. 9.23 Temperature versus magnetic phase diagram (a) and quantized Hall resistance (b) for $(TMTSF)_2PF_6$ under pressure.

As far as 1-D conductors are concerned the effectiveness of the on-site Coulombic repulsion is enhanced by the nesting properties of the Fermi surface. The resulting competition between the Cooper and Peierls channels is particularly damaging for superconductivity which cannot be stabilized above 1 K in 1-D conductors derived from the TTF skeleton.

This competition is, however, less severe in 2-D materials where consequently T_c can be raised upto 12 K. Attempts to increase T_c by further decreasing the intermolecular overlap [and thus increasing density of states $N(E_F)$] using more bulky inorganic ions have failed to provide metal-like materials, but compounds with interesting magnetic properties have been obtained.

9.5.6 Superconducting Characteristics

The organic superconducting system TMTSF is non-ideal type II superconductor with low carrier density ($\sim 10^{21}$ cm^{-3}). The following are the superconducting characteristics for the most extensively studied $X = ClO_4$ salt among $(TMTSF)_2 X$ (See Table 9.6). The lower critical fields (B_{c_1}) are 0.02 mT (a-axis) 0.10 mT (b) and 1.0 mT (c) at 0.5 K, and the upper critical fields (B_{c_2}) are 2.8 T (a), 2.1 T (b) and 0.16 T (c) at 0 K (extrapolated). The extrapolated B_{c_2} (a) at 0K is a little larger than the Pauli limit (B_{Pauli}) given by

$$B_{Pauli} (k\ O_e) = 18.4\ T_c = \Delta_0/\sqrt{2}\ \mu_B \qquad \ldots(9.90)$$

where Δ_0 is the gap parameter at 0 K ($T_c = 1.4$ K and the BCS relation $2\Delta_0 = 3.53\ k_B\ T_c$ gave $B_{Pauli} = 2.6$ T). The Ginzburg-Landau (GL) coherence length (ξ) are $\xi_a = 706 - 873$ Å, $\xi_b(0) = 335 - 385$ Å, and $\xi_c(0) = 20.3 - 22.7$ Å, those are longer than those of the lattice parameters. Critical current J_c is extremely small; J_c (0.5 K) = 0.1 ± 0.05 A – cm^{-2} along the c-axis and one order of magnitude higher along the a-axis. The specific heat C above T_c is represented as

Table 9.6. Superconducting characteristics of selected TMTSF and BEDT-TTF(ET)

Super-conductor	B_{c_1} (mt)	$B_{c_2}(T)$	$\xi(0)$(Å)	J_c(A cm^{-2})	dT_c/dP (K k bar^{-1})	γ (mJ mol^{-1} K^{-2})	β (mJ mol^{-1} K^{-4})	$\Theta(K)$	ΔC (mJ mol^{-1} K^{-1})	Δ_0(meV)
(TMTSF)$_2$ClO$_4$	0.02(a, 0.5K) 0.10(b, 0.5K) 1.0(c, 0.5K)	2.8(a, 0K) 2.1(b, 0K) 0.16(c, 0 K)	706–837 335–385 22.7	−1(a,0.5 K) 20.3 0.1 ± 0.05 (c, 0.5 K)	−0.08	10.5	11.4	213	214	0.44, 3.68
β_L-(ET)$_2$I$_3$	0.005 (a, 0.1 K) 0.009 (b', 0.1 K) 0.036 (c^*, 0.1 K)	2.09(a, 0K) 2.48(b',0K) 0.81 (c^*, 0K)	633(a) 608(b') 29(c^*)		−0.3	24 ± 3	19	197 ± 5		1
β_H-(ET)$_2$I$_3$		25(//,0K, 1.6kbar) 2.7(⊥, 0K, 1.6 k bar)	127(//) 10(⊥)		−1.4					
β-(ET)$_2$IBr$_2$	0.39(a, 0K) 1.6(b, 0K)	3.36(a, 0K) 3.60(b, 0K) 1.5(c, 0K)	463(a) 444(b) 18.5(c)		−0.7					
β-(ET)$_2$AuI$_2$	0.4(a, 0K) 2.05(b, 0K)	6.63(a, 0K) 0.51(c, 0K)	249(a) 19.2(c)		−1.0					
κ-(ET)$_2$Cu(NCS)$_2$	0.07(///5K) 20(⊥,1.5 K)	24.5(//,0.5K) 5.5(⊥, 0.5 K)	29 ± 5(//) 3.1 ± 0.(⊥)	1 − 1.3 × 10 (5K, 50G). 10^2(// b) H //c, 5K. 0(G)	−3.0 − 3.5	34 25 ± 3	10 11.2	223 215 ± 10	7.3 × 10^2 5.3 × 10^2	0.8, 2.1, 4.2.2 − 9.8
κ-(ET)$_2$Cu[N(CN)$_2$]Br		30.6 (//,1.5K) 7.4 (⊥,1.5 K)	23 ± 4 5.8 ± 1.0(⊥)		−2.4	22 ± 3 0.06 ± 0.43	12.8 12.0	210 − 15 212	D-salt	

$$C = \gamma T + \beta T^3 \qquad ...(9.91)$$

with $\gamma = 10.5$ mJ mol^{-1} K^{-2} and $\beta = 11.4$ mJ mol^{-1} K^{-4} and shows a jump due to superconductivity at 1.22 K with the increment of C; $\Delta C = 21.4$ mJ/mol/K. These values yield $\Delta C/\gamma T_c = 1.67$, which is close BCS value of 1.43.

The superconducting gap measurements are not unanimously decided. A Schottky barrier formed by evaporating Ga Sb onto the $X = ClO_4$ salt gave an extremely large gap of 3.68 meV, which corresponds to $T_c = 12$ K. A (TMTSF)$_2$ ClO$_4$ per amorphous Si/Pb junction gave a gap of 0.44 meV, which is close to that expected from the BCS theory (eq. 9.92), $2\Delta_0 = 0.42$ meV at $T_c = 1.4$ K,

$$2\Delta_0 = 3.53 \, k_B T_c \qquad ...(9.92)$$

Obviously, some superconducting characteristics of the $X = ClO_4$ salt are consistent with the simple BCS theory. However, a microscopic investigation by proton (^{1}H) NMR pointed out the unconventional nature of the superconductivity of this salt. The temperature dependence of the relaxation rate (T_1) of protons at zero magnetic field indicates that $1/T_1$ did not exhibit the Hebel-Slichter coherent perk just below T_c characteristic to simple BCS isotropic gap. Below it, $1/T_1$ shows T^3 dependence suggesting that $X = ClO_4$ salt has an anisotropic order parameter having lines of zero on the Fermi surface.

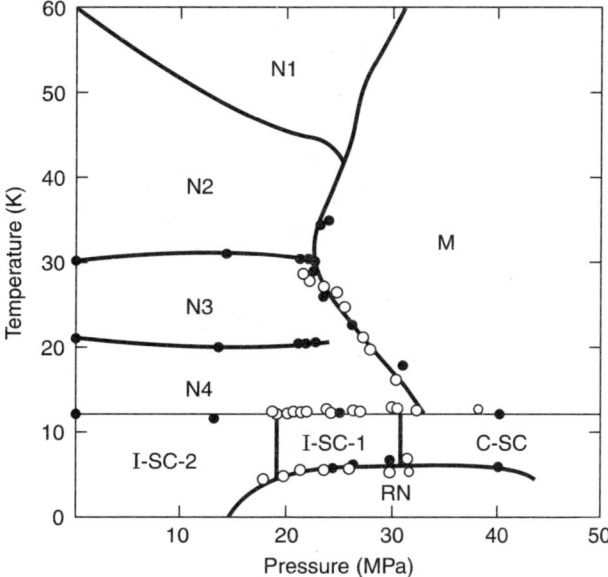

Fig. 9.24 Pressure temperature phase diagram of κ-(ET)$_2$ Cu[N(CN)$_2$] Cl (M : metallic phase, N1 – N4 : non-metallic phase, RN – re-entrant metallic phase, C-SC : Complete superconducting phase, I-SC-1, I-SC-2 : Incomplete superconducting phase. N1 may correspond to 'fuzzy metallic phase'. Spins on ET molecules have one-dimensional antiferromagnetic interactions (N2) which grow antiferromagnetic interactions (N3) and exhibit weak ferromagnetism (N4).

For organic superconductors κ-(ET)$_2$ Cu(CN)[N(CN)$_2$], κ-(ET)$_2$ Cu(NCS)$_2$ and κ-(ET)$_2$ Cu[N(CN)$_2$] Br, T_c was suppressed rapidly by pressure ($dT_c/dP = -2.4$ to -3.5 bar^{-1}). While for the organic superconductor κ-(ET)$_2$ Cu[N(CN)$_2$] Cl T_c increased with increasing pressure at low pressures (12.8 K at 0.25 – 0.3 k bar) then was suppressed at higher pressures ($dT_c/dP = -3.5$ K k bar^{-1}). Below the superconducting phase appeared a new non-metallic phase, which was suppressed above 0.55 k bar but reappeared by the magnetic field cycling (superconductor-to-insulator reentrant). Very complicated T–P phase diagrams were proposed (Fig. 9.24), where the superconductivity resides next to the magnetic field [249].

The D-salts exhibited almost the same temperature dependence of resistivity to those of the corresponding H-salts of above-mentioned four organic systems. Sometimes different temperature dependence of resistivity from those shown in Fig. 9.25 was observed in the D-salt of κ-(ET)$_2$ Cu [N(CN)$_2$] Br, which is thought to reside closest to the Mott insulating regime among I–IV. The D-salts of I ($T_c = 12.3$ K), II ($T_c = 11.2$ K) and IV ($T_c = 13.1$ K at 0.3 k bar) showed higher T_c than the corresponding H-salts (inverse isotope effect). In the case of III, normal isotope effect was observed occasionally ($T_c = 11.2$ K). However, it has been claimed that the slow cooling through a phase transition at around 80 K of III induces an inverse isotope effect on this salt too. However, the increase of T_c is observable when outer ethylene groups are replaced by heavier isotope atoms even though the electron densities on these ethylene groups are negligible.

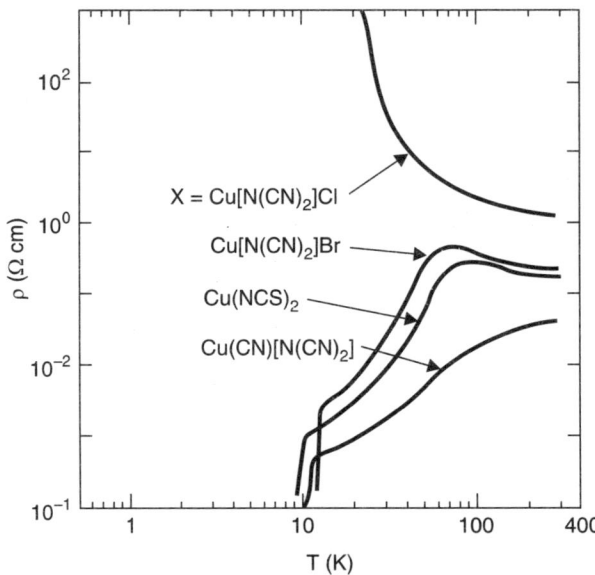

Fig. 9.25 Temperature dependence of the resistivity of κ-(ET)$_2$ Cu(CN) [N(CN)$_2$] (I), κ-(ET)$_2$ Cu(NCS)$_2$ (II), κ-(ET)$_2$ Cu[N(CN)$_2$] Br (III) and κ-(ET)$_2$ Cu[N(CN)$_2$] Cl (IV).

The isotope effect is just the opposite of what one expects from the weak coupling simple BCS theory. In a system with strong electron correlation, the index α in equation

$$T_c \sim \theta_D \sim \text{(isotope mass)}^\alpha, \quad \alpha = -0.5 \qquad \ldots(9.93)$$

approaches zero, but does not exceed zero. In the above salts, the α values are 4.6 for I, 3.6 for II, and 1.1 for IV.

9.5.7 Other Organic Superconductors

Among the superconductors in Tables 9.2 to 9.5, the coexistence of superconducting and magnetic phases was observed in κ-$(ET)_2$ $Cu[N(CN)_2]$ Cl, β''-$(ET)_4$ $[M(C_2O_4)_3]$ H_3 O Ph CN (M = Fe, Cr, C_2O_4) is oxalate dianion [250, 251] and $\lambda - (BETS)_2$ $[Fe_x Ga_{1-x} Cl_4]$ [252]. In the first case ET radical spins play both roles of superconductivity and magnetism, and they compete by themselves. For the latter two cases, the organic donor molecules take the part of superconductivity and Fe or Cr ions the magnetic part. The superconductivity of the second material seems to have no interaction with the magnetic spins on Fe^{3+} or Cr^{3+}, which exhibit paramagnetic behaviour. In the last case, x was estimated as 0.43 and 0.55. The superconductivity appeared at 4.0 – 4.5 K for x = 0.43 and an insulating state emerged below it; 3.0 – 3.5 K at AP. For x = 0.55, the superconductor-to-insulator re-entrant phenomenon was reported at 1 k bar. The superconductivity originated from the BETS electrons. The antiferromagnetism due to Fe^{+3} spins competes with each other.

The highest T_c in Tables 9.2–9.5 was reported in the Rb doped OMTTF C_{60} benzene neutral CT complex [253]. The complex consists of a two-dimensional layer of C_{60} molecules, which is sandwiched by the layers and consisting of OMTTF and benzene. The doping was achieved at 64 – 67°C with Rb and 50 – 55°C with K. No practical superconductivity was detected in the Rb doping to pristine C_{60} and C_{60} CT complexes of ET, hydroquinone, TSe C_1-TTF, EDT-TTF and HMTTeF under the condition. The T_c of 26 K was reported on Rb doped OMTTF C_{60} benzene by SQUID and the volume fraction was 10%. The $C = C A_g$ mode of C_{60} by the Raman scattering experiments on pristine, K-doped and Rb-doped OMTTF C_{60} benzene appeared at 1467, 1446 and 1453 cm^{-1}, respectively, suggesting that the C_{60} molecules in the Rb - doped sample have a lower reduction state than in the K-doped one.

9.5.8 Theory of Organic Superconductivity

The study of new class of synthetic conductors in which superconductivity has been discovered has contributed to a better experimental and theoretical understanding of the physics in low-dimensional electron gases (one or two dimensional), although there is still no firm explanation for the driving mechanism leading to the attractive pairing in organic superconductors. The interplay between antiferromagnetic and superconducting ground states in the TM_2 X series, the role of Coulombic interactions in the magnetic properties of the low dimensional electron gas above T_c, and the strong sensitivity to non-magnetic impurities in both one and two-dimensional organic superconductors are all factors playing in favour of a non-conventional superconducting mechanism. However, it is still premature to establish a close connection between the mechanism of superconductivity in existing organic superconductors and the model proposed by Little in 1964 [254]. Moreover, the stabilization of spin density wave (SDW) phases at high magnetic field and coherent motion of a SDW state under high electric field are remarkable products of the developments of these new compounds.

In organic superconducting systems that are quasi two-dimensional, charge density waves (CDW), spin density waves (SDW), and superconductivity may occur. The T_c of the materials that present superconductivity and CDW or SDW is low, but the transition from the normal state to a state of superconductivity and CDW or SDW present a great interest [255, 256]. In comparison to ternary and heavy fermion superconductors [5, 59], the situation of interplay of superconductivity and magnetism in organic superconductors is quite different. In ternary superconductors, the antiferromagnetism can coexist with superconductivity because spin fluctuations are averaged out and are ineffective in destroying the superconducting states. This has been observed in $D_y\,Mo_6\,Se_8$ [257]. The ferromagnetic long-range order leads to the re-entrant phenomenon, *e.g.* in $Er\,Rh_4\,B_4$ [258]. For UPt_3, the superconductivity, appears below antiferromagnetic order. However, for organic superconductors, *e.g.* BEDT-TTF salts, it is generally agreed that the counter anions of the salts contribute to neither conduction nor localized magnetism. The origin of the magnetism is sought in the Fermi surface nesting, and hence the superconductivity and the magnetism in these salts are not spatially separated, even if they stem from a different part of the Fermi surface. Obviously, the competition between superconductivity and magnetism in organic superconductors is much more delicate [259].

Several attempts have been made to explain the interplay of superconductivity and magnetism and the microscopic origin of superconducting phase transition in the organic metals [260]: Various models have been proposed (*e.g.*, electron-electron interaction via intermolecular vibrations, libron-electron coupling, magnetic coupling, electron-electron interaction, spin-fluctuation mechanism, excitonic model, two band mechanism, EMV interaction, *g*-ology mechanism, etc.) [59]. However, none of these mechanisms are capable of explaining all the observed features in these systems. However, arguments on the mechanism are making steady progress and theory and experiments have been interacting rather directly in many phases. Experimental results favour an anisotropic metal model based on the concept of partial electron dielectrization [261].

As stated above, in organic superconductors, which are quasi one or two-dimensional systems, CDW, SDW, and superconductivity may occur. The T_c of these systems that present superconductivity and CDW or SDW is low, but the transition from the normal state to one of these states presents a great interest. This interest is due to the fact that it is known that in the systems with low dimensions ($d = 1, d = 2$), the fluctuations of the order parameter suppress the ordered phases. However, these condensed phases exist, and this can be explained only by the fact that a real system does not have a low dimensionality, but in fact have three dimensionality and present a high anisotropy. In this way, one can treat the system as a quasi-one-dimensional or quasi-bi-dimensional system [180, 262].

The simplest model for the study of this problem proposed by Bilbro and McMillan [263] consists of dividing the momentum space into two regions, (1) and (2). Kakani and coworkers [264] assumed that the Fermi surface satisfies certain nesting conditions in the region 2, which allows CDW or SDW gap formation. The nesting wave vector $\mathbf{Q}$, which characterizes the SDW state, is assumed to be $2\mathbf{Q} = \mathbf{G}$, with $\mathbf{G}$ as the reciprocal lattice vector of the original lattice. The remaining region of the Fermi surface is denoted as region 1, where only the superconducting energy gap is allowed to open [265].

Kakani and coworkers [264] consider the model Hamiltonian of a superconductor with nesting surface sections, where the formation of the CDW or SDW state is possible [266] as

$$H = H_0 + H_S + H_{AM}$$

$$= \sum_{ip}^{\alpha\beta} [\xi_i(p)\delta_{\alpha\beta} - \mu_B^* \sigma_z^{\alpha\beta} H] C_{ip\alpha}^+ C_{ip\alpha}$$

$$+ \frac{1}{2} \sum_{ijlm} V_{ijlm} \sum_{pp'q}^{\alpha\beta} C_{i(p+q)\alpha}^+ C_{j(p'-q)\beta}^+ C_{mp'\beta} C_{ip\alpha}$$

$$- \mu_B^* \sum_{ij}^{\alpha\beta} H_{ij}(Q) \sigma_z^{\alpha\beta} \sum C_{i(p+q)\alpha}^+ C_{ip\beta} \qquad ...(9.94)$$

where H_0 describes the electronic conduction band in the normal phase. The normal term, H_S, is the familiar phonon-mediated electron–electron interaction responsible for superconductivity. The third term, H_{AM}, represents anitferromagnetic ordering of ions.

In eq. (9.94) a system of Bloch electrons whose eigen energy with reference to the Fermi level is represented by $\xi_i(P)$, effective Bohr magneton is abbreviated μ_B^*, Pauli spin matrices as $\sigma_z^{\alpha\beta}$, H is external magnetic field, $C_{ip\alpha}^+$ and $C_{ip\alpha}$ are, respectively, creation and annihilation electron operators of Bloch states $|p\alpha\rangle$ with band index i, crystal momentum p, and spin α. In the second term of eq. (9.94), V_{ijlm} is the matrix element of contact-effective electron-electron interaction, which satisfies certain symmetry requirements, provided strong mixing of various branches of the nonreconstructed elementary excitation spectrum takes place. Also in second term of model Hamiltonian, the sum of all wave vectors extend in the first Brillouin zone, and the subscripts **p + q** and **p–q** define modulo reciprocal lattice vectors. In the third term, $H_{ij}(\mathbf{Q})$ are matrix elements of the antiferromagnetic molecular field.

In this model, the electron spectrum decomposes into two parts—a nondegenerate one with branch $\xi(\mathbf{p})$, and a degenerate one with two branches related by the condition

$$\xi_1(\mathbf{p}) = -\xi_2(\mathbf{p} + \mathbf{Q}) \qquad ...(9.95)$$

with **Q** as wave vector of the CDW or SDW state. Here, the Fermion operator C satisfied the usual anticommutation relationship :

$$[C_{ip\alpha}, C_{jp'\alpha'}^+] = C_{ip\alpha} C_{jp'\alpha'} - C_{jp'\alpha'}^+ = \delta_{ij} \delta_{pp'} \delta_{\alpha\alpha'}$$

and

$$[C_{ip\alpha}, C_{jp'\alpha'}] = [C_{ip\alpha}^+, C_{jp'\alpha'}^+] = 0 \qquad ...(9.96)$$

The microscopic theory of organic superconductors can be developed using the Green's function technique and equation of motion method. Green's function equation are obtained as

$$[i\omega_n - \xi_n(k) + h] G_{nl}^{\mu\sigma} - D_{nj}^{\mu\beta} G_{jl}^{\beta\sigma} + \Delta_{nl}^{\mu\beta} F_{jl}^{\mu\beta} = \delta_{nl}^{\mu\sigma}$$

$$[i\omega_n - \xi_n(k) + h] F_{nl}^{\mu\sigma} - D_{nj}^{\mu\beta} F_{jl}^{\beta\sigma} - \Delta_{nj}^{\mu\beta} = 0 \qquad ...(9.97)$$

Here, $D_{nj}^{\mu\beta}$ and $\Delta_{nj}^{\mu\beta}$ are self-energy parts and satisfy the following self-consistent conditions:

$$D_{nj}^{\mu\beta} = T\sum \{V_{njmi} - SpV_{nijm}\, G_{mi}^{\alpha\alpha}\} + H_{nj}\sigma_z^{\alpha\beta}$$

$$\Delta_{nj}^{\mu\beta} = T\sum \{V_{njmi}\, F_{mi}^{\mu\beta}\}$$

with $\qquad \omega_n = (2n+1)\pi T,\ \mu_B^* H = h$ ...(9.98)

and $\qquad \sigma_z^{\alpha\beta} = \begin{bmatrix} 1 & 0 \\ 0 & -1 \end{bmatrix}$

Here, H is external magnetic field. Assuming

$$\Delta_{11} = \Delta_{22} = \Delta \text{ and } D_{21} = D_{12} = D\,[\sigma_0(1-\alpha) + \alpha\sigma_3],$$

where $\alpha = 1$ for SDW and $\alpha = 0$ for CDW state, one obtains the following equations:

$$G_{11} = [i\omega_n + \xi_K + h][(i\omega_n + h)^2 - \xi_K^2 - (1-2\alpha)D^2 - \Delta^2]/D^r \quad\text{...(9.99)}$$

$$G_{21} = -\Delta[(i\omega_n + h)^2 - \xi_K^2 - D^2 - (1-2\alpha)\Delta^2]/D^r \quad\text{...(9.100)}$$

$$F_{11}^+ = -\Delta^+[(i\omega_n + h)^2 - \xi_K^2 - (1-2\alpha)D^2 - \Delta^2]/D^r \quad\text{...(9.101)}$$

where $\qquad D^r = [(i\omega_n + h)^2 - \xi_K^2 + \delta_+^2][(i\omega_n + h)^2 - \xi_K^2 - \delta_-^2]$ ...(9.102)

with $\qquad \delta\pm = \delta \pm \alpha D$ ...(9.103)

$\qquad \delta = [\Delta^2 + (1-\alpha)D^2]^{1/2}$ ...(9.104)

From the equations (9.99) to (9.101), we obtain the self-consistent equation for the gas as

(i) In the **presence of CDW state** (i.e., $\alpha = 0$),

$$1 = |g|\,N_d(0)\int d\xi_K\,[\tanh\beta\{h + (\xi_K^2 + \Delta^2 + D^2)^{1/2}\}/2$$
$$- \tanh\beta\{h - (\xi_K^2 + \Delta^2 + D^2)^{1/2}/2\}]/2(\xi_K^2 + \Delta^2 + D^2)^{1/2}$$
$$+ |g|\,N_{nd}(0)\int d\xi_K\,[\tanh\beta\{h + (\xi_K^2 + \Delta^2)^{1/2}\}/2$$
$$- \tanh\beta\{h - (\xi_K^2 + \Delta^2)^{1/2}\}/2]/2(\xi_K^2 + \Delta^2)^{1/2} \quad\text{...(9.105)}$$

(ii) In the **presence of SDW states** (i.e., $\alpha = 1$), Δ becomes

$$1 = |g|\,N_d(0)\int d\xi_K\,[\{\Delta + D\}\{\tanh\beta(h + [\xi_K^2 + (\Delta+D)^2]^{1/2}/2$$
$$- \tanh\beta(h - [\xi_K^2 + (\Delta+D)^2]^{1/2})/2/[\xi_K^2 + (\Delta+D)^2]^{1/2}$$
$$+ \{\Delta - D\}\{\tanh\beta(h + [\xi_K^2 + (\Delta-D)^2]^{1/2})/2$$
$$- \tanh\beta(h - [\xi_K^2 + (\Delta-D)^2]^{1/2}\}/[\xi_K^2 + (\Delta-D)^2]^{1/2}]$$
$$/4\Delta + |g|\,N_{nd}(0)\int d\xi_K\,\tan h\beta\{h + (\xi_K^2 + \Delta^2)^{1/2}\}/2$$
$$- \tan h\beta\{h - (\xi_K^2 + \Delta^2)^{1/2}/2\}/2(\xi_K^2 + \Delta^2)^{1/2} \quad\text{...(9.106)}$$

Emerging Superconductors

where $N_d(0)$ and $N_{nd}(0)$ are density of states for the insulting and conducting cases at the Fermi surface respectively. The total density of states at the Fermi surface is

$$N(0) = N_d(0) + N_{nd}(0) \qquad \ldots(9.107)$$

The **insulating order parameter** (D) is given by

$$D = I \sum_p <C_p^+ C_{p+Q}>$$

This parameter is limited within the region of insulator. One obtains

(i) For $\alpha = 0$ (i.e., insulator order parameter for the CDW case)

$$1 = IN_d(0) \int d\xi_K [\tanh \beta \{h + (\xi_K^2 + \Delta^2 + D^2)^{1/2}\}/2$$
$$- \tanh \beta \{h - [(\xi_K^2 + \Delta^2 + D^2)^{1/2})/2]/2(\xi_K^2 + \Delta^2 + D^2)^{1/2} \qquad \ldots(9.108)$$

(ii) For $\alpha = 1$ (i.e., insulator order parameter for the SDW case)

$$1 = IN_d(0) \int d\xi_K \{[D+\Delta][\tan h \beta \{h + (\xi_K^2 + (D+\Delta)^2)^{1/2}\}/2$$
$$- \tanh \beta \{h - (\xi_K^2 + (D+\Delta]^2)^{1/2}\}/2]/(\xi_K^2 + [D+\Delta]^2)^{1/2}$$
$$+ [D-\Delta]^2)^{1/2} - (D-\Delta)[\tanh \beta \{h + (\xi_K^2 + [D-\Delta]^2)^{1/2}/2$$
$$- \tanh \beta\{h - (\xi_K^2 + [D-\Delta]^2)^{1/2}\}/(\xi_K^2 + [D-\Delta]^2)^{1/2}\}/4D \qquad \ldots(9.109)$$

Order Parameter at Zero Temperature

Let Δ_0 is the SC order at zero temperature in the absence of CDW or SDW, in the presence of CDW or SDW as $\Delta(0)$ and the insulating order parameter in the absence of superconductivity as D_0. One obtains:

(i) $\alpha = 0$, **CDW case**

For $h = 0$ (in the absence of magnetic field) and $D = 0$ (absence of insulator), one obtains

$$\Delta_0 = 2\hbar\omega_D \exp[-1/|g|N(0)] \qquad \ldots(9.110)$$

For $\Delta(0)$, one obtains

$$1 = |g|N_d(0) \int d\xi_K /[\xi_K^2 + \Delta^2(0) + D^2]^{1/2} + |g|N_{nd}(0) \int d\xi_K /[\xi_K^2 + \Delta^2(0)]^{1/2}$$
$$\ldots(9.111)$$

and $\quad D_0 = 2E_B \exp[-1/N_d(0)] \qquad \ldots(9.112)$

(ii) $\alpha = 1$, **SDW case**

One obtains

$$1 = |g|N_d(0) \int d\xi_K [\{\Delta(0)+D\}/\{\xi_K^2 + [\Delta(0)+D]^2\}^{1/2}$$
$$+ \{\Delta(0)-D\}/\{\xi_K^2 + [\Delta(0)-D]^2\}^{1/2}]/2\Delta(0)$$
$$+ |g|N_{nd}(0) \int d\xi_K /[\xi_K^2 + \Delta^2(0)]^{1/2} \qquad \ldots(9.113)$$

and $\quad |\Delta^2(0) - D^2|^{1/2} [\Delta(0)+D]/[\Delta(0)-D]|^{D/2\Delta(0)} = \Delta_0 \qquad \ldots(9.114)$

With two order parameters, the following situations can arrive:

(i) $\Delta = 0, D = 0$ (i.e., normal state)

(ii) $\Delta \neq 0, D = 0$ (i.e., superconducting state without anisotropy with T_{c_0} as onset temperature of superconducting state in the absence of CDW/SDW state.

(iii) $\Delta = 0, D \neq 0$ (i.e., anisotrophic state without superconductivity T_{w_0} (T_{s_0}) as on set temperature of CDW/SDW state without superconductivity, respectively.

(iv) $\Delta \neq 0, D \neq 0$ (i.e., superconducting state coexisting with anisotropy; i.e., CDW/SDW state).

Kakani and coworkers [264] have made a detailed study of last situation. They obtained equations for CDW and SDW states separately by taking Fermi wave vector $k_F = \sim n^{1/2}$ and density of states constant, where n is the number of atoms per unit length of the system [59]. The theory clearly reveals that superconductivity and SDW state coexist in $k - (BEDT - TTF)_2$ $Cu(NCS)_2$ salt with a layered crystal structure. The display of two gap structures is attributed to a short coherence length of organic metals. Tunnelling data are in full agreement of the two gap structure along with the anisotropy of each group. The theory also explains satisfactorily the behaviour of density of states, specific heat, free energy etc. [264].

The study of the new class of synthetic conductors in which superconductivity has been discovered has contributed to a better experimental and theoretical understanding of the physics in low-dimensional electron-gases (one- or two dimensional), although there is no firm explanation for the driving mechanism leading to the attractive pairing in organic superconductors. Since the phonon excitation is generally strong in the organic conductors, a class of materials should be s-wave superconductors due to the electron-phonon mechanism. However, some class of the organic superconductor can be categorized into strongly correlated electron systems and the unconventional superconductivity is expected, e.g., (TMTSF) PF_6, where the possibility of not only the d-wave pairing, but also the triplet pairing has been presented [267]. The microscopic calculations [268-271] seem to indicate the predominance of the d-wave superconductivity. No doubt, the development of the satisfactory microscopic theory of organic superconductors is one of the interesting issues.

9.6 FULLERENE SUPERCONDUCTORS

The twin discovery of molecular [1] and bulk C_{60} [2] have laid the foundation for a vast new field of study in condensed matter physics. Much of the interest in this new form of carbon originates from the subsequent discovery of superconductivity when bulk C_{60} is doped with alkali metal [274]. The similarity of the C_{60} structure of the geodesic domes of R. Buckminster Fuller [275] has given rise to the name Fullerene and Buckyball to the C_{60} molecule in particular. Early references to the shape of the C_{60} molecule led to some use of the terminology foot ballene (European) or Soccerballene (USA). These have mostly given way, initially to the unwieldy buckminster fullerene and now very widely to simply fullerene to denote the C_{60} molecule and in fact the class of closed C_m molecules with m-60 or greater. The crystalline C_{60} solid is called fullerite, while the crystalline compounds $A_n C_{60}$ (and offshoots) are called Fullerides [276]. There are several surprising aspects of the phenomenon in C_{60} that have attracted attention. These are [18]:

(i) The molecular nature of C_{60}, in particular the collective motion and residual disorder among the molecules.

(ii) The three dimensionality of the molecule and the crystal lattice.

(iii) The relatively high superconducting transition temperature of alkali doped C_{60} that is only surpassed by cuprates.

9.6.1 Metal-Doped Alkali Fullerene Superconductors

The discovery of superconductivity in potassium-doped C_{60} at 18 K by Hebard et al [277] was an unexpected observation that has attracted a great deal of attention in the scientific community, and has resulted in a tremendous increase in the number of experimental and theoretical papers. In the past few years, the field of fullerene superconductivity has undergone a remarkable development with the identification of the superconducting phase in (K, Rb, Cs)$_3$C$_{60}$ [277-281], NH$_3$Na$_2$CsC$_{60}$ [282], Ca$_5$C$_{60}$ [283], Ba$_6$C$_{60}$ [284], RbCs$_2$C$_{60}$ [281].

The fullerene molecule contains 60 carbon atoms forming 12 pentagons and 20 hexagons, fused into a pseudo sphere. From a structural point of view the "balls" behave like single metal atoms and adopt closed-packed arrangements, preferentially a face-centered cubic arrangement, similar to the one found in elementary Cu. At room temperature the molecules are orientationally disordered, but below 249 K a preferred orientation is adopted and the space group is lowered from $Fm\bar{3}m$ to $Pa\bar{3}$.

Upto about 10 metal atoms per fullerene unit can be inserted into this basic structure, which contains one octahedral voids per C_{60} molecule. At the composition A_3C_{60} both the octahedral and the tetrahedral sides are occupied by single atoms. The resulting arrangement of metal atoms and fullerene molecules for K_3C_{60} is shown in Fig. 9.26.

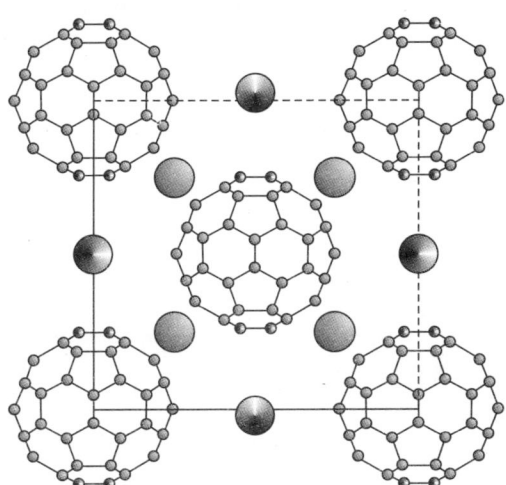

Fig. 9.26 Structure of K_3C_{60} in a partial projection along [001]. Dark shading and fullerene molecule : $z = 0$; light shading : $z = 1/4$.

In substituted fullerides such as Na_2CsC_{60}, the larger metal atoms preferentially occupy the octahedral sites, whereas in Na_2KC_{60} a disordered arrangement of cations was found. In "stuffed" $Na_2Cs(NH_3)_4C_{60}$ part of the Na atoms are surrounded by four ammonium molecules forming a tetrahedron. These complex cations occupy the octahedrol sites, whereas the remaining Na and Cs atoms are randomly distributed over the tetradhedral sites.

The alkaline metal atoms fully donate their electrons to the C_{60} unit. Superconductivity is observed for a metal-to-fullerene ratio close to 3 : 1, with critical temperatures near 30 K measured for $Na_2Cs(NH_3)_4C_{60}$, Rb_3C_{60}, Rb_2CsC_{60}, and Cs_2RbC_{60}. $T_c = 45$ K is reported for nominal $Rb_{2.7}Tl_{2.2}C_{60}$. For the alkaline earth metals, the charge transfer is not complete and superconductivity is found for a higher metal-to-fullerence ratio, e.g. Ca_5C_{60} (8.4 K), Sr_6C_{60} (4K) and Ba_6C_{60} (7K). T_c increases monotonically with increasing cell parameter for both the *fcc* and *bcc* packing.

Following these pioneering discoveries, extensive work has been carried out both on physics and chemistry of fullerene superconductors. These fullerene superconductors now represent the highest T_c molecular superconductors. The great interest in these superconductors is in particular due to these systems being a completely new class of superconductors, the large value of their T_c, and the question of whether or not such a large value of T_c can be caused by the coupling to phonons alone [6, 59]. There has therefore been a great effort over the past few years to characterise and understand both the normal state and superconducting properties of these systems. The most fundamental understanding of the mechanism of superconductivity is the correlation observed between T_c and the interfullerene spacings [285 – 287], which was established in the early stages, i.e., that the T_c of fullerene supercondcutors increase with the interfullerene spacings. The chemical substitution of alkali metal sites and the application of high pressure yielded the same relation between T_c and interfullerene spacing within the error of 10%.

Fullerenes have other unique features as building blocks for constructing solids. The first thing to be pointed out is the spherical shape of the molecule. C_{60} in particular, has an exceptionally high symmetry, which is referred to as icosahedral (I_h) symmetry. The molecular structure is identical to a soccer ball (Fig. 9.27). The I_h symmetry of the C_{60} molecule is directly reflected in the molecular electronic structures.

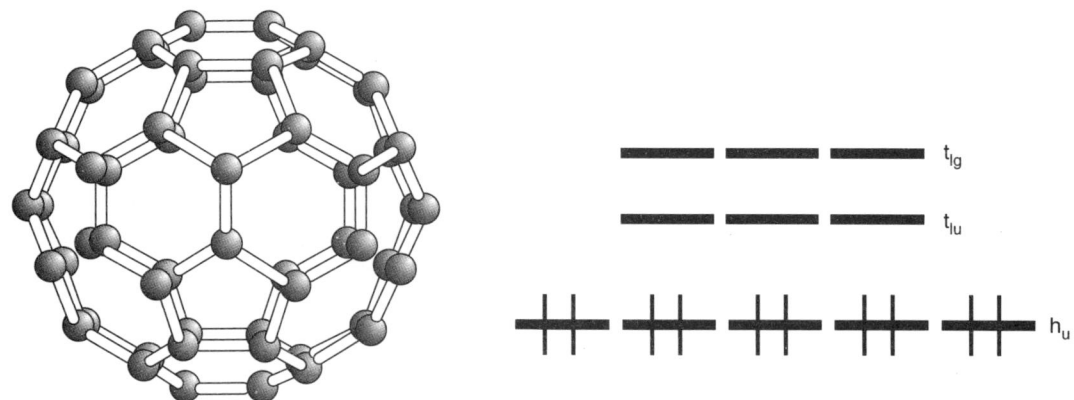

Fig. 9.27 The molecular structure and electronic energy levels of C_{60}. Doped electrons are introduced into the LUMO (t_{1u}) and, subsequently, to the LUMO + 1 (t_{1g}) states.

Fig. 9.27 also shows the molecular energy levels near the Fermi level obtained from the Huckel molecular orbital calculation [288]. For example, the highest occupied molecular orbital (HOMO) is five fold degenerate, while the lowest unoccupied molecular orbital (LUMO) and LUMO + 1 state are both triply degenerate. The degeneracy of the latter two levels is quite important, since electrons are introduced into LUMO and LUMO + 1 bands upon doping of alkali metals or alkaline earth metals. We call these two levels as t_{1u} and t_{1g} states, respectively, following the symmetry group representation.

The second feature of fullerene molecule is the capability of intercalation and, at the same time, a large capacity of electrons. Alkali metals and alkaline earth metals can be doped up to six per C_{60} molecule, forming for example, Ba_6C_{60}. The nominal counting of the valence of the C_{60} molecule in this compound is $n = 12$, assuming that the Ba ion is divalent. Moreover, the valence n can be controlled from 0 to 12 per molecule, by changing the number and species of intercalated elements.

Combining the two characteristics, the physical interest in the research of C_{60} intercalation compounds is that one is able to tune the electron filling of two triply degenerate electronic levels over a wide range. The most fundamental question related to superconductivity is the correlation between the number of electrons transferred to C_{60} (molecular valence n) and the occurrence of superconductivity. According to the simple band picture, when the formal electron counting of C_{60} molecule is 0, 6 and 12, the materials should be band insulators, while they are metallic at the other electronic states where the LUMO (t_{1u}) or LUMO + 1(t_{1g}) states partially filled.

A current conclusion on alkali intercalated compounds is that superconductivity appears only at the half-filled states in the t_{1u}. Superconductivity at different valence states is strictly prohibited for the alkali doped systems.

In sharp contrast, the criteria for superconductivity are significantly different in the t_{1g} band. As for the alkaline earth binary materials, Ba_4C_{60} and Ca_5C_{60} are superconductors with onset temperatures 6.5 and 8.4 K, respectively. Though the crystal structure of Ca_5C_{60} is not fully solved yet, it is believed to be a simple cubic structure with the same molecular arrangement as fcc [289]. A schematic illustration of the structural sequence of K_3C_{60} and Ba_xC_{60} with $3 \leq x \leq 6$ is shown in Fig. 9.28. The compounds $K_3Ba_3C_{60}$ corresponds to the half-filling of the t_{1g} band having formal valence of C_{60} in $K_3Ba_3C_{60}$ as $n = 9$. The position of K and Ba ions are disordered and, hence $K_3Ba_3C_{60}$ is regarded as a 1:1 solid solution of K_6C_{60} and Ba_6C_{60}. This monovalent compound was reported to be a superconductor at $T_c = 5.6$ K. Isostructural $Rb_3Ba_3C_{60}$ was also found to be a superconductor at $T_c = 2K$ [290]. Optimization of the synthesis route may produce a rich variety of ternary compounds, in which one is able to tune the t_{1g} band filling continuously, keeping the crystal structure unchanged.

Fig. 9.29 summarizes various fullerene superconductors in terms of the relation between the T_c and the molecular valence (band filling). Although there exists a vast variety of fullerene intercalation compounds, Fig. 9.29 provides a simple insight into the chemical criteria for fullerene superconductivity : (a) when C_{60} is intercalated only with alkali metals, superconductivity occurs only at the trivalent state with the fcc structure. When superconductivity is realized despite this strict constraint, T_c reaches above 30 K, (b) when C_{60} is intercalated with alkaline earth or rare earth metals, superconductivity occurs irrespective of molecular valence and crystal structure. Particularly, as far as the t_{1g} band is concerned, superconductivity always appears once the partial band filling is achieved. However, T_c stays lower than 10 K.

Obviously, fullerene superconductors are classified into two groups : alkali doped systems (t_{1u} systems) and alkaline earth doped systems (t_{1g} systems). However, the current understanding of rare earth doped systems is still poor.

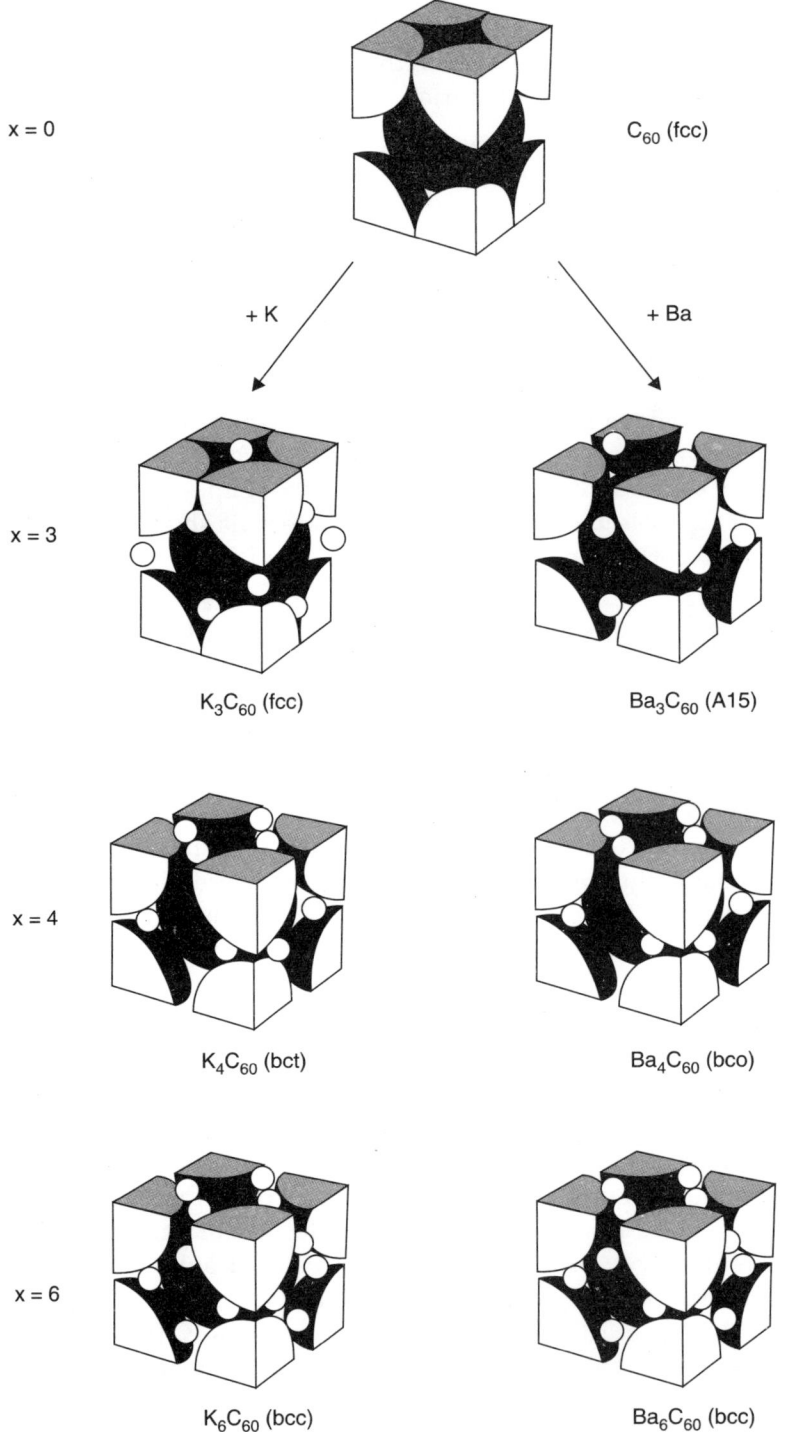

Fig. 9.28 A schematic representation of the structural sequence of K_xC_{60} and Ba_xC_{60} with $3 \leq x \leq 6$. The large and small spheres represent the C_{60} molecule and intercalated metal ion, respectively.

Emerging Superconductors

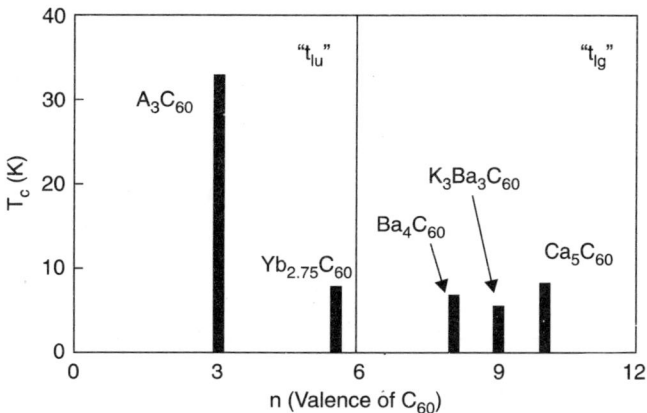

Fig. 9.29 T_c versus band filling correlations in fulleride compounds.

Clearly, superconductivity in aklali metal doped fullerene intercalation compounds which provide partially filled t_{1u} bands occurs in the fcc structure and exclusively at the half-filled state. Once these conditions are violated, T_c rapidly decreases and finally disappears. The symmetry of the molecule and crystal structure seems quite important for this group of materials. On the other hand, superconductivity in the alkaline earth metal doped C_{60} is insensitive to crystal structure and band filling. It seems likely that superconductivity appears once the t_{1g} state is partially filled. However, T_c does not exceed 10 K, in sharp contrast to the t_{1u} systems. These features are possibly explained by the broadened conduction band due to orbital mixing. The larger bandwidth suppresses the instability to the insulators and stabilizes the metallic states and thus superconductivity, however, $N(E_F)$ is lower reducing T_c.

At this stage one may ask a serious question about the superconductivity of larger fullerenes. All the bulk superconductors identified so far are C_{60}-based compounds. At present, nobody knows whether other higher fullerenes with lower symmetry than C_{60} could afford superconductors or not. Of particular interest is that noncubic Ba_4C_{60} becomes superconducting, suggesting that the degeneracy of the molecular orbital is not a necessary condition. We can say that intercalation of higher fullerenes is an exciting area to be explored in the near future.

Theory

A number of theoretical explanations have been advanced to account for the superconducting phase transition in the A_3C_{60} compounds [291-301]. The usual weak coupling BCS theory is cast in terms of phonons. Other pairing-mediating excitations could be involved and several alternative pairing schemes have been advanced [291-301]. However, these models partially explain the observed features in these systems.

Recently, the "fullerenes" C_N which have the hollow cage structures of carbon have been intensively investigated [6, 18,19]. There are several experimental indications that the doped fulleride system show polaronic properties due to Jahn-Teller distortion, for e.g. (i) The electron spin resonance study on the radical anion of C_{60} by Kato et al. [302] has revealed the small g factor, $\gamma = 1.9991$, and this is associated with the residual orbital angular momentum due to

Jahn-Teller distortion; (*ii*) photoemission studies of C_{60} and C_{70} doped with alkali metals by Takahashi et al. [303] and Chen et al. [304] have shown peak structures, which cannot be explained by a simple band-filling picture; (*iii*) Morita et al. [305, 306] have shown that when poly (3-alkylthiophene) is doped with C_{60} interband absorption of the polymer is remakably suppressed and the new absorption peak evolves in the low energy range. The Jahn-Teller splitting of LUMO in C_{60} state and/or the Coulomb attraction of positively charged polaron to C_{60} might occur; (*iv*) Matus et al. [306] found two peaks around 1.5 and 1.7 eV below the gap energy 1.9 eV while measuring luminescence of neutral C_{60}. They have interpreted this as the effect of the polaron excitor; (*v*) The experiments on the dynamics of photoexcited states by Lane et al. [307] have shown the interesting roles of polaron.

The low temperature properties of many polaron system have been studied by Alexandrov and Ranninger [308, 309]. They have shown that the ground state of electrons, strongly coupled to phonons, appears to be charged Bose-liquid, consisting of singlet or triplet pairs (small bipolarons with charge $2e$ and an enhanced band mass). They have further shown that the most pronounced characteristic feature of small polarons is a strong mixing of electron and phonon degrees of freedom.

Following Alexandrov and Ranninger [310], Kakani and coworkers [311] have developed a theory of doped fulleride superconductors by generalizing the Holstein [312] and Lang-Firsov [313] and Gorkov-Nambu [314, 315] formalisms in order to evaluate the Green's functions for electrons coupled to phonons and considering the range of coupling which corresponds to small polaron formalism ($\lambda \geq 1$).

The model Hamiltonian for the system reads as

$$H = H_e + H_{ph} + H_{e-ph} + H_{ee} \quad \ldots(9.115)$$

where
$$H_e = \sum_{K,S} E(K) C^+_{KS} C_{KS} \quad \ldots(9.116)$$

is the kinetic electron energy in the initial Bloch band. Where K and S denotes the state with quasi momentum and spin respectively. $E(K)$ is bare band energy. H_{ph} is the vibration energy of the lattice,

$$H_{ph} = \sum_q \omega(q) d^+_q d_q \quad \ldots(9.117)$$

where $\omega(q, \nu)$ is the phonon dispersion d_q are phonon operators $q = (q, \nu)$, with ν the type of the vibrational mode.

The electron-phonon interaction is described by the Frohlich Hamiltonian :

$$H_{e-ph} = \sum_{K,q,S} \{\omega(q)\gamma(q)\frac{1}{\sqrt{2N}} C^+_{K+q,S} C_{K,S} d_q + h.c.\} \quad \ldots(9.118)$$

in which $\omega(q)$ and $\gamma(q)$ are the phonon frequency and the interaction matrix element in a parent crystal without charge carriers respectively. Correspondingly, one obtains

$$\gamma_q^2 = \frac{4\pi e^2}{q^2 \Omega \omega_0}\left(\frac{1}{\varepsilon_\infty} - \frac{1}{\varepsilon_0}\right) \quad \ldots(9.119)$$

In the case of optical longitudinal phonon with frequency $\omega(q) = \omega_0$ and $\varepsilon_0, \varepsilon_\infty$ are the dielectric constants of the crystal with and without taking account of the ionic part. Ω is the volume of the unit cell and N is their number. For acoustic phonons, one obtains

$$\gamma^2(q) = E_D^2 \left[\frac{q}{uM}\right], \qquad \omega(q) = \omega_0 \qquad \ldots(9.120)$$

where E_D is the deformation potential, u is the sound velocity and M is the mass of the elementary cell. For intermolecular phonons.

$$\gamma^2(q) = \gamma_0^2$$

Now, the combined Hamiltonian reads as

$$H = \sum_{K,S} E(K) C_{KS}^+ C_{KS} + \sum_{KqS} \frac{1}{\sqrt{2N}} \omega(q)\, \gamma(q)\, dq$$

$$C_{K+q,S}^+ C_{K,S} + h.c.] + \sum_q \omega(q)\, d_q^+ d_q + V_c \qquad \ldots(9.121)$$

where V_c is the Coulomb repulsion. Hamiltonian (9.121) includes electron-phonon and electron-electron correlations. Using the site representations, the Hamiltonian is diagonalized and one obtains.

$$\tilde{H} = \sum_{\substack{m,n,s \\ m \neq n}} t(m-n) C_{ms}^+ C_{ns} \exp\left[\sum_q d_q \left\{\frac{1}{\sqrt{2N}} \gamma(q)^{iqm}\right.\right.$$

$$+ \sum_q \omega_q \left(d_q^+ d_q + \frac{1}{2}\right) + \sum_{mns} \left\{V_c(m-n) - \sum_q \omega(q)\left(\frac{1}{\sqrt{2N}}\right)^2\right.$$

$$\left. \gamma^2(q)\, e^{iq(m-n)} \right\} C_{ms}^+ C_{ns}^+ C_{ns} C_{ms} \qquad \ldots(9.122)$$

We define the following one particle temperature electron normal (G) and anomalous (F) Green's functions :

$$G(K, \omega_n) = -\frac{1}{2}\sum_m \int_{-\beta}^{+\beta} dl\, e^{i\omega_n l + iKm} \ll l_l\, C_{0\sigma}(l)\, C_{m\sigma}^+(0) \gg \qquad \ldots(9.123(a))$$

$$F(K, \omega_n) = \frac{1}{2}\sum_m \int_{-\beta}^{\beta} dl\, e^{i\omega_n l + iKm} \ll l_l\, C_{0\sigma}(l)\, C_{m\sigma}(0) \gg \qquad \ldots(9.123(b))$$

Applying the Lang and Firsov decoupling transformation [313] and neglecting the polaron-polaron coupling and after simplification, one obtains

$$G(K, \omega_n) = e^{-g^2} \left[\frac{u_K^2}{i\omega_n - \varepsilon_n} + \frac{v_K^2}{i\omega_n + \varepsilon_n} + \frac{1}{N}\sum_{l=1}^\infty \frac{g^2}{l!}\right.$$

$$\sum_{K'} \frac{u_{K'}^2(1-n_{K'})}{i\omega_n - l\omega_0 - \varepsilon_{K'}} + \frac{v_{K'}^2\, n_{K'}}{i\omega_n - l\omega_0 + \varepsilon_{K'}} + \frac{u_{K'}^2\, n_{K'}}{i\omega_n + l\omega_0 - \varepsilon_{K'}} + \frac{v_{K'}^2(1-n_K)}{i\omega_n + l\omega_0 + \varepsilon_{K'}}\right\}\right] \qquad \ldots(9.124)$$

$$F(K, w_n) = e^{-g^2}\left[u_K v_K\left(\frac{1}{i\omega_n - \varepsilon_K} - \frac{1}{i\omega_n + \varepsilon_K}\right) + \frac{1}{N}\sum_{l=1}^{\infty}\frac{(-1)^l g^{2l}}{l!}\right.$$

$$\sum_{K'} u_{K'} v_{K'}\left\{\frac{1-n_{K'}}{i\omega_n - l\omega_0 - \varepsilon_{K'}} - \frac{n_{K'}}{i\omega_n - l\omega_0 + \varepsilon_{K'}}\right.$$

$$\left.\left.+ \frac{n_K}{i\omega_n + l\omega_0 - \varepsilon_{K'}} - \frac{1-n_{K'}}{i\omega_n + l\omega_0 + \varepsilon_{K'}}\right\}\right] \quad ...(9.125)$$

where

$$u_K^2 = \frac{1}{2}\left(1 + \frac{\xi_K}{\varepsilon_K}\right), \quad v_K^2 = \frac{1}{2}\left(1 - \frac{\xi_K}{\varepsilon_K}\right)$$

$$u_K v_K = -\frac{\Delta}{2\varepsilon_K} \quad \text{and} \quad n_K = n(\varepsilon_K)$$

with

$$n(x) = \left[\exp\left(\frac{x}{k_B T}\right) + 1\right]^{-1}$$

$$\varepsilon_K = \sqrt{\xi_K^2 + \Delta^2(K)}$$

$$\Delta_K = -\frac{1}{2}\sum_{K'} V(K - K')\frac{\Delta(K')}{\varepsilon_{K'}} \tanh\frac{\varepsilon_{K'}}{2k_B T}$$

and

$$V(K) = \frac{1}{N}\sum_m V(m)\exp(iKm)$$

The energy dispersion for the polaronic band is given by

$$\xi_K = \left\{\sum_m \sigma(m, 0)\tau(m) e^{iKm} - \mu\right\}$$

having a narrow band half width $W \ll D$ (where $D = ZT(m)$)

Using

$$\Delta = |g|\sum_K <C_K^+ C_{-K}^+> \quad ...(9.126)$$

and

$$\sum_K = N(0)\int_0^{\hbar\omega_D} d\xi_K \quad ...(9.127)$$

where $<C_K^+ C_{-K}^+>$ correlation function, obtained from eq. (9.125) as

$$<C_K^+ C_{-K}^+> = \left[\frac{1}{2} + \frac{1}{2}\frac{\xi_K}{\varepsilon_K}\tanh\frac{\beta\varepsilon_K}{2}\right] + \frac{1}{N}\sum_{l=1}^{\infty}\frac{g^{2l}}{l!}$$

$$\left[\frac{1}{2}+\frac{1}{2}\frac{\zeta_{K'}}{\varepsilon_{K'}}\tanh\frac{\beta(l\omega_0-\varepsilon_{K'})}{2}-\frac{n_{K'}\,\varepsilon_{K'}}{2\varepsilon_{K'}}\left(\tanh\frac{\beta(l\omega_0+\varepsilon_{K'})}{2}-\tanh\frac{\beta(l\omega_0-\varepsilon_{K'})}{2}\right)\right]$$
...(9.128)

Using (9.126) to (9.128), the gap equation becomes

$$\Delta = |g|\,N(0)\int_0^{\hbar\omega_D} d\xi_K \left[\frac{-\Delta(k)}{2\sqrt{\xi_K^2+\Delta^2(K)}}\tanh\frac{\sqrt{\xi_K^2+\Delta^2(K)}}{2k_BT}\right]$$

$$\frac{-|g|}{N}\sum_{l=1}^{\infty}\frac{(-1)^l}{l!}g_1^{2l}\,N(0)\int_0^{\hbar\omega_D}\sum_{K'}\xi_K\frac{\Delta(K')}{2\sqrt{\xi_{K'}^2+\Delta^2(K')}}$$

$$\times\left\{\tanh\frac{1}{2k_BT}\left(l\omega_0+\sqrt{\xi_{K'}^2+\Delta^2(K')}\right)-\frac{1}{\exp\left\{\frac{\xi_{K'}^2+\Delta^2(K')}{k_BT}\right\}+1}\right.$$

$$\left.\times\left(\tanh\frac{1}{2k_BT}\left(l\omega_0+\sqrt{\xi_{K'}^2+\Delta^2(K')}\right)+\tanh\frac{1}{2k_BT}\left(l\omega_0-\sqrt{\xi_{K'}^2+\Delta^2(K')}\right)\right)\right\}$$ ...(9.129)

Right hand side of equ. (9.129) has two terms which are quite independent. First term varies with K whereas second term varies with K', hence one can define two superconducting order parameters for the system. The two independent terms finally give the two equations as,

$$\frac{1}{|g|\,N(0)}=\int_0^{\hbar w_D}d\xi_K\left[\frac{1}{2\sqrt{\xi_K^2+\Delta^2(K)}}\tanh\left\{\frac{\sqrt{\xi_K^2+\Delta^2(K)}}{2k_BT}\right\}\right]$$ ...(9.130)

and with $l = 1$, the other equation is

$$\frac{1}{|g|\,N(0)|g_1^2|}=\int_0^{\hbar\omega_D}\frac{d\xi_{K'}}{2\sqrt{\xi_{K'}^2+\Delta^2(K')}}\left\{\tanh\frac{1}{2k_BT}\right.$$

$$\left(l\omega_0+\sqrt{\xi_{K'}^2+\Delta^2(K')}\right)\frac{1}{\exp\sqrt{\left\{\frac{\xi_{K'}^2+\Delta^2(K')}{k_BT}\right\}}+1}$$ ...(9.131)

With the help of eqs. (9.130) and (1.131), one can study the behaviour of superconducting order parameter with temperature.

Electronic specific heat

One can determine the electronic specific heat per atom of a superconductor from the following relation.

$$C_{es}=\frac{\partial}{\partial T}\left[\frac{1}{N}\sum_K 2\xi_K<C_K^+\,C_K^+>\right]$$ ...(9.132)

Using (9.128), eq. (9.132) gives

$$C_{es} = \frac{N(0)}{N} \int_0^{\hbar\omega_D} \frac{2\xi_K}{2k_B T^2} d\xi_K \left\{ -\frac{1}{2} \xi_K \operatorname{sech}^2 \left(\frac{\xi_K}{2k_B T} \right) \right.$$

$$+ \frac{n_{K'}}{2} \frac{\xi_{K'}}{\xi_{K'}} \left[(l\omega_0 + \varepsilon_{K'}) \operatorname{sech}^2 \left(\frac{l\omega_0 + \varepsilon_{K'}}{2k_B T} \right) \right.$$

$$\left. \left. + \frac{n_{K'}}{2} \frac{\xi_{K'}}{\xi_{K'}} \left((l\omega_0 - \varepsilon_{K'}) \operatorname{sech}^2 \frac{(l\omega_0 + \varepsilon_{K'})}{2k_B T} \right) \right\} \right\} \quad ...(9.133)$$

Density of States Function

For $\omega > 0$, the density of states function is defined as

$$N(\omega) = \lim \frac{i}{2\pi} [G_{11}(K, \omega + i\eta) - G_{11}(K, \omega - i\eta)] \quad ...(9.134)$$

Using the following identity

$$\lim_{\eta \to 0} \frac{1}{2\pi} \left[\frac{1}{\omega + i\eta - \omega_n^+} - \frac{1}{\omega - i\eta - \omega_n^+} \right] = i\delta(\omega - \omega_n^+)$$

After simple calculations, one obtains

$$\frac{N(\omega)}{N(0)} = \frac{1}{N} \sum_{l=1}^{\infty} \frac{g^{2l}}{l} \sum_{K'} \left\{ \frac{(i\omega_n - l\omega_0)}{\sqrt{i\omega_n - l\omega_0)^2 - \Delta^2}} - \frac{(i\omega_n + l\omega_0)}{\sqrt{(i\omega_n + l\omega_0)^2 - \Delta^2}} \right\} \quad ...(9.135)$$

The theory has been applied to K_3C_{60} and Rb_3C_{60} and explains the observed experimental features satisfactorily.

9.7 MgB$_2$

The discovery of non-cuprate "high temperature" superconductivity in Mg B$_2$ has stimulated new excitement in condensed matter physics [316]. MgB$_2$ possesses the highest $T_c \simeq 39$ K among intermetallic compounds and the strong isotope effect on T_c. MgB$_2$ introduced a new binary inter-metallic superconductor with three atoms per unit cell (see Fig. 9.30).

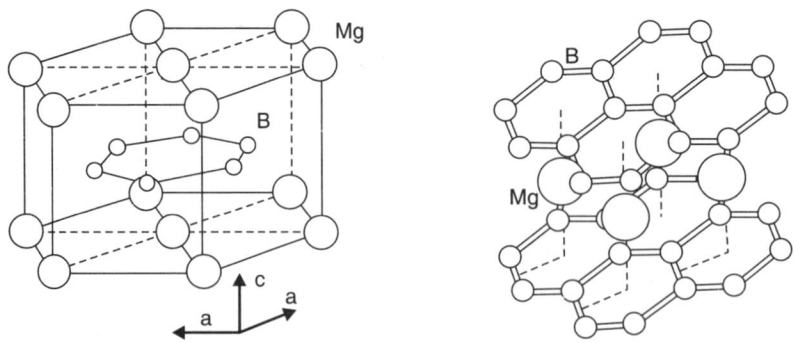

Fig. 9.30 Magnesium diboride (MgB$_2$) belongs to the AlB$_2$ (space group p6/mmm) family of structures. Graphite type B layers separated by hexagonal close-packed layers of Mg [317].

MgB$_2$ exhibits a strong anisotropy in the B-B lengths similar to graphite. The distance of boron planes is much larger than the in-plane B-B distance. The values of unit cell parameters differ from those of Nagamatsu et al. [316], where the reported values of the cell parameters are $a = 3.086$ Å and $c = 3.524$ Å. The electronic states at the Fermi level, which are the highest occupied electronic states, are mainly either $\sigma-$ or π-bonding boron orbitals (Fig. 9.31 b-d). The σ-bonding states are confined in the boron planes. Thus, MgB$_2$ may be unique with partially occupied σ-bonding states in a layer structure. Because the charge distribution of the σ-bonding states is not symmetrical with respect to the in-plane positions of boron atoms, the σ-bonding state couple vary strongly to the in-plane vibration of boron atoms (Fig. 9.31 e). Approximately 30% of the density of states at the Fermi energy is due to planar boron $p_{x,y}$ states (σ bonds) that have little dispersion in the z-direction, giving rise to nearly cylindrical hole Fermi surfaces of 2D character. The remaining 70% of states in boron p_z states (π bonds) that are strongly hybridized with the Mg s-p orbitals, have three-dimensional (3D) character and give rise to mostly electron like Fermi surfaces. No d-electrons exist in either Mg or B.

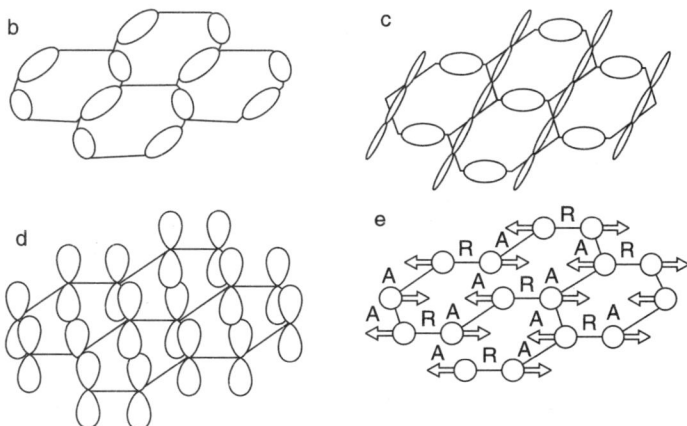

Fig. 9.31   Boron atoms form honeycomb planes, and magnesium atoms occupy the centres of the hexagons in between boron planes (b and c), σ-bonding states at the Fermi level derived from boron $p_{x,y}$ orbitals. (d), A π-bonding state at the Fermi level derived from boron p_z orbitals. (e), A vibrational mode of boron atoms that couples strongly to σ-bonding electronic states at the Fermi level. As boron atom move in the arrow directions, shortened bonds, marked with 'A', become attractive to electrons, whereas elongated bonds, marked with 'R', become repulsive. The σ-bonding states (b, c) couple strongly to the vibrational mode because they are mainly located in either the attractive or the repulsive bondings of the mode. The π-bonding states (d) do not couple strongly to this mode.

9.7.1 Properties of MgB$_2$

(*i*) **Critical Temperature:** Superconducting onset and end point transition temperatures as observed by Nagamatsu et al. [316] from resistivity measurement are, respectively, 39 K and 38 K as shown in Fig. 9.32. It is found that critical temperature, T_c, decreases with the increase of pressure (Fig. 9.33) as observed both from the pressure dependent resistivity and magnetic susceptibility [319] measurements. Doping effect on both the Mg and B sites of MgB_2 by suitable elements revealed that T_c decrease at various rates with the doping of different elements though the doping of few elements showed either a small increase of it or remains unchanged [319]. Poddar et al [320] have fitted resistivity (ρ) data on polycrystalline MgB_2 samples following the Bloch-Gruneisen (BG) formula [321] given by

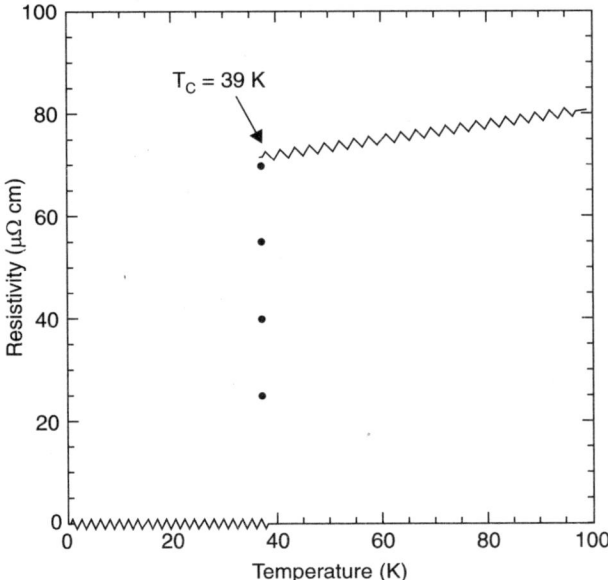

Fig. 9.32 Variation of resistivity of MgB_2 under zero magnetic field [316].

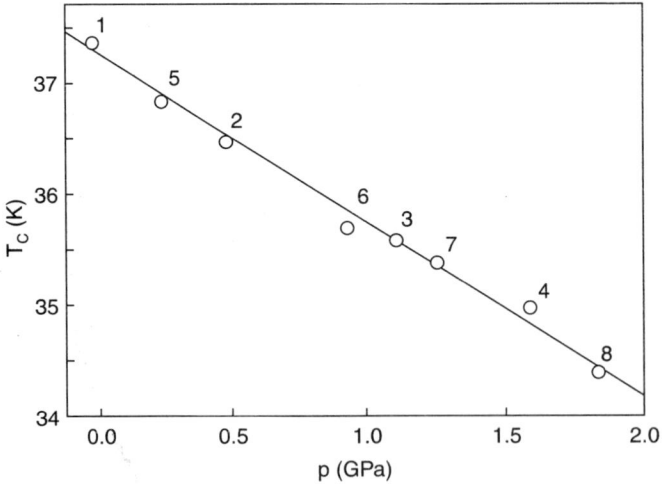

Fig. 9.33 Pressure dependence of T_c from ac susceptibility measurement. The numbers present the sequential order of the experimental runs [318].

Emerging Superconductors

$$\rho(T) = \rho_0 + (m-1)\rho'\theta \left(\frac{T}{\theta}\right)^m J_m\left(\frac{T}{\theta}\right) \qquad ...(9.136)$$

where
$$J_m\left(\frac{T}{\theta}\right) = \int_0^{\frac{\theta}{T}} \frac{x^m e^x \, dx}{(e^x-1)^2} \qquad ...(9.137)$$

Here ρ_0 is the sum of the residual resistivity of the grains and the resistance of the links between the grains, θ is Debye temperature and ρ' is the temperature coefficient of resistivity which can be expressed as

$$\rho' = \frac{6\pi k_B \lambda_{tr}}{\hbar e^2 \, 2N(0) <v_F^2>} \qquad ...(9.138)$$

where $N(0)$ is the density of state at the Fermi level and $<v_F^2>$ is the mean square electron velocity at the Fermi surface.

9.7.2 Critical Fields

The lower critical field $B_{c_1}(0)$ of a well-shaped cylindrical dense sample was found to be 176 Oe [322]. The B_{c_2} of polycrystalline MgB_2 at 10 K was found to be 14 T [323]. But B_{c_2} of polycrystalline MgB_2 was enhanced and the magnetic-field induced broadening was significantly reduced with increasing grain size [324].

Single crystalline MgB_2 exhibited [324] remarkable anisotropy of $B_{c_2}(T)$ and irreversibility field $B^*(T)$ with anisotropy ratio $\gamma \approx 3 \pm 0.2$.

9.7.3 Critical Current Density

The nanocrystalline samples [325] revealed $J_c = 10^5$ A cm^{-2} at 20 K with 1 T magnetic field [Fig. 9.34]

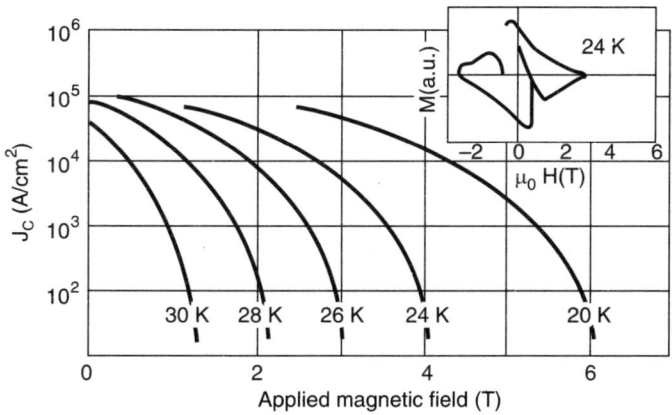

Fig. 9.34 Critical current density (J_c) calculated from magnetization data [325]. The inset shows the magnetization loop at 24 K (SQUID data) with two jumps at low magnetic fields.

9.7.4 Thermoelectric Power (TEP)

The variation of thermoelectric power, (S) with temperature is shown in Fig. 9.35 [326]. The TEP of MgB_2 is positive and decreases with the decrease of temperature [318]. The TEP with a positive sign showed [327] a linear increase up to ~ 150 K but indicated a saturated behaviour at high temperatures. The positive sign of TEP shows that the carriers of this inter-metallic superconductor are hole-type. However, the saturated behaviour at higher tempera-tures suggest the two types of charge carriers, *i.e.* holes as well as electrons. The results showed that the metallic hole carriers from σ bands are important to explain the temperature dependence of TEP. The decrease of T_c and increase of TEP with pressure [326] were explained by the two metallic bonds (hole and electron bands) model. Considering the linear $S \propto T$ dependence below 150 K, one can interpret the low temperature part of the curve as the diffusion theremopower, S_d, of hole type metals. For metals, Mott has given an expression [328] for the diffusion thermopower as

$$S_d(T) = -(\pi^2 k_B^2 T/3e)\, [\delta \ln\rho(E)/\delta E]_{E=E_F} \qquad ...(9.139)$$

where k_B is the Boltzmann constant and e is the charge of the carriers. Assuming that the conductivity is proportional to the energy and that there is a T-independent mean free path for the carriers, the expression (9.139) becomes

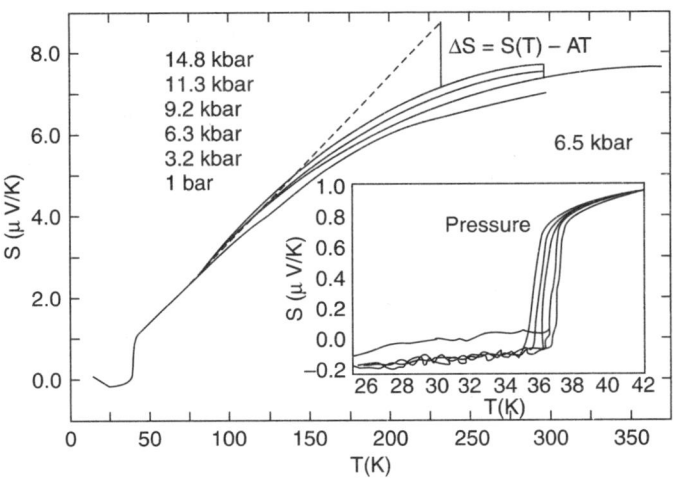

Fig. 9.35 Pressure dependent thermoelectric power (S) of MgB_2. A is the slope of the linear $-T$ dependence curve between T_c and 120 K [326]. The inset shows the magnified low temperature region.

$$S_d(T) \approx \pi^2 k_B^2 T/3eE_F \qquad ...(9.140)$$

where E_F is Fermi surface. We may note that eq. (9.140) is derived considering a spherical Fermi surface together with a T- independent relaxation. This approximation limits the validity of eq. (9.140) at low temperatures. Consequently, one cannot explain the deviation of S from linearity at T_0 and its saturation close to room temperature cannot be explained by eq. (9.140). This feature seem to be related to the complexity of the Fermi surface and the existence of the electron type sheets. Band structure calculations show that the Fermi surface of MgB_2 is far from spherical. Poddar et al. [320] have estimated $E_F = 0.8$ eV from the shape of $S(T)$ at low temperature region and using eq. (9.140). This value corresponds to $v_F \approx 5.3 \times 10^7$ cm/s.

9.7.5 Thermal Conductivity

Thermal conductivity (k) of single-crystalline hexagonal MgB_2 in normal state, superconducting state and in mixed state have been elaborately studied by Sologubenko et al. [329] both in presence and absence of magnetic fields. The ab-plane thermal conductivity (k) as a function of magnetic field B with orientations both parallel and perpendicular to the c-axis and at temperatures between 0.5 and 300 K has been investigated [329]. In the mixed state, k (B) at constant temperatures, revealed features that were not typical for common type-II superconductors. A non-linear temperature dependence of the electronic thermal conductivity was observed in the field-induced normal state at low temperatures. This behaviour was at variance with the Wiedemann and Franz law, and suggested an unexpected instability of the electronic subsystem. The thermal conductivity showed (Fig. 9.36) suppression below T_c without a clear anomaly at T_c which is considered to be as a result of the normal charge carrier concentration below T_c [327]. k is reported 49.2 mW/cmK at 29.6 K, increases with the increase of temperature and reaches a value of 100 mW/cm-K at 200 K. Above 200 K, instead of monotonous increase with temperature many groups observed a distinct maximum. As stated above, the thermal conductivity differs in value and behaviour from those of simple metal. In fact, the observed values of k are rather low with respect to the high carrier concentration of MgB_2 (two carriers per unit cell) as well the occurrence of the maximum of k at high temperature compared to other metals together with the lack of anomaly at the superconducting transition. The low value of k can be related to strong e-phonon coupling which reduces both phonon and electron mean free path. The most puzzling aspect is the absence of anomaly in k at T_c. We may remember that in low temperature superconductors where quasi-particles are the main heat carriers, k decreases below T_c owing to their condensation. In order to identify the main heat carriers (electrons or phonons) in MgB_2 in the normal state, Poddar et al. [320] have made the following calculations.

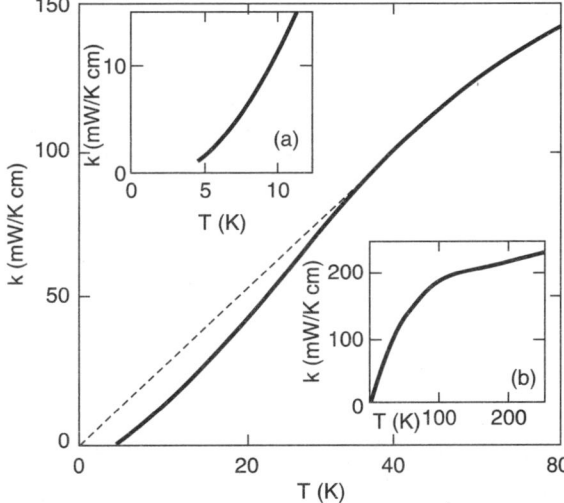

Fig. 9.36 Temperature dependence of thermal conductivity (k) of MgB_2 as a function of temperature below 60 K [327]. The dashed line is guide to eye. The insets (a) and (b) show those upto 10 K and 250 K respectively.

The thermal conductivity in a metal is given by the sum of the electron thermal conductivity, k_e, and the phonon thermal conductivity, k_p and may be written as

$$k = k_e + k_p \qquad ...(9.141)$$

According to the Mathiessen's rule

$$k_e = (W_e^p + W_e^i)^{-1} \text{ and } k_p = (W_e^p)^{-1} \qquad ...(9.142)$$

where, W_e^p and W_e^i are the thermal resistivities for scattering with phonons and impurities, respectively. Following Berman [331], one can write

$$W_e^i = \rho_0/L_0 T \qquad ...(9.143)$$

and

$$W_e^p = \frac{R}{L_0 T}\left(\frac{T}{\theta}\right)^5 J_5\left(\frac{\theta}{T}\right)\left[1 + \frac{3}{\pi^2} n_a^{2/3}\left(\frac{\theta}{T}\right)^2\right] - \frac{1}{2\pi^2}\frac{J_7\left(\frac{\theta}{T}\right)}{J_5\left(\frac{\theta}{T}\right)} \qquad ...(9.144)$$

and

$$W_p^e = \frac{R}{L_0 T}\left(\frac{\theta}{T}\right) J_5\left(\frac{\theta}{T}\right)\frac{\pi^2 n_a^2}{27\left\{J_5\left(\frac{\theta}{T}\right)\right\}^2} \qquad ...(9.145)$$

where $R = 2\rho'\theta$, $L_0 = 2.44 \times 10^{-8}$ V^{2}K^{-2}
and n_a is the number of electrons per atom. Poddar et al. [320] found that the thermal conductivity estimated from resistivity analysis is considerably less than the experimental data. Similar discrepancy has also been reported by Putti et al. [332]. It is worthwhile to note that the factor by which the experimental data differs from the theoretical values depends on the sample preparation condition which determines the porosity, grain boundary, impurity, etc. These factors affected resistivity of the samples and in turn the thermal conductivity results.

The reported values for $\gamma = 2.4$ mJ/mol K^2 and $\chi_0 = 16.2 \times 10^{-6}$ emu/mol obtained from the heat capacity measurements and susceptibility measurements respectively [333]. The value of density of states (DOS) estimated from γ and χ_0 is found to be 0.083 – 0.086 states/eV atom-spin. This is close to the value (0.12 states/eV-atom spin) obtained from band structure calculations [334].

9.7.6 Hall Effect

The mixed-state Hall resistivity ρ_{xy} and longitudinal resistivity ρ_{xx} in superconductivity, MgB$_2$ thin films have been measured as a function of the magnetic field over a wide range of current densities from 10^2 to 10^4 A cm^{-2} [335]. A universal Hall scaling behaviour was observed [335] with a constant exponent β of 2.0 ± 0.1 in $\rho_{xy} = A\rho_{xx}^\beta$, which was independent of the magnetic field, the temperature and the current density. Jin et al. [336] have investigated the temperature and magnetic field dependence of Hall coefficient (R_H) of two well-characterized superconducting states. Their result showed that the normal state R_H value was positive (Fig. 9.37) and increased with increasing temperature, independent of the applied magnetic field. They also reported $R_H^{-1} \propto T$ (40 – 300 K) and the Hall angle θ_H followed Cot $\theta_H \propto T^2$ (100-300 K). As the sample was cooled below T_c (H), R_H decreased rapidly with temperature and changed its sign before it reached zero. The position and magnitude at which R_H showed a minimum depending on applied field [336].

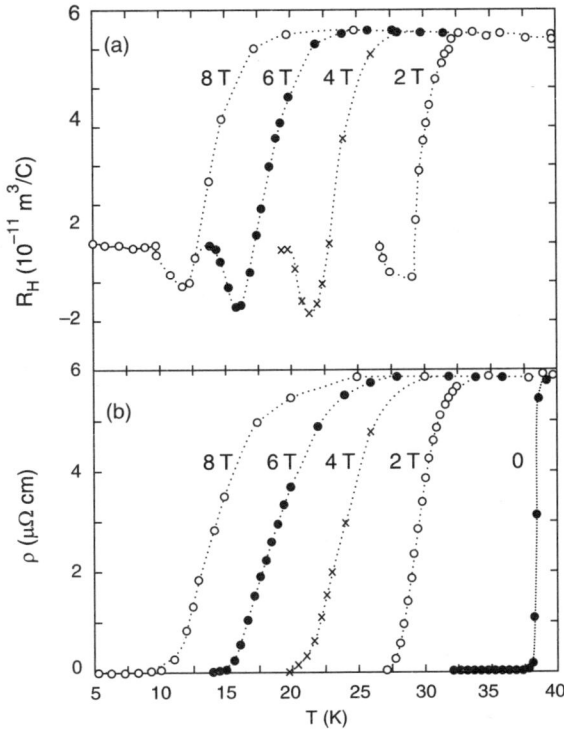

Fig. 9.37 Temperature dependence of Hall coefficient (a) R_H and (b) longitudinal resistivity of MgB$_2$ thin films [336] at different magnetic fields.

9.7.7 Specific Heat

Yang et al. [337] have measured the low temperature specific heat $C(T)$ for polycrystalline MgB$_2$. Fig. 9.38 shows the C vs. T and $\frac{C}{T}$ vs. T curves for polycrystalline MgB$_2$ [337] at $H = 0$ and 8 T. Fig. 9.39 shows the $\frac{\Delta C}{T}$ vs. temperature for polycrystalline MgB$_2$ [337] where $\Delta C = C(H) - C(8\ T)$. Together with the small specific heat jump $\Delta C/\gamma T_c = 1.09$, γ being the anisotropic parameter stated as the anisotropic s-wave or multicomponent order parameter. Yang et al. [337] observed that the magnetic field dependence of anisotropic parameter $\gamma(H)$ was neither linear for a fully gapped s-wave superconductor nor $H^{1/2}$ dependent for nodal order parameter. Walti et al. [338] Based on a modified Debye-Einstein model, Walti et al. [338], have achieved a rather accurate account of the lattice contribution to the specific heat, separating the electronic contribution from the total measured specific heat. Walti et al. [338] estimated the electron-phonon coupling constant λ from their result for electronic specific heat, to be of the order of 2, significantly enhanced compared to common weak-coupling values ≤ 0.4.

Soon after the discovery of superconductivity at 40K in MgB_2, it was proposed that this finding is expected within two fundamentally different theoretical frameworks : the BCS electron-phonon theory [339, 340] and the theory of hole superconductivity [341]. Both the views have claimed to be consistent with various experimental observations. However, the situation is still not clear and more decisive experiments are needed.

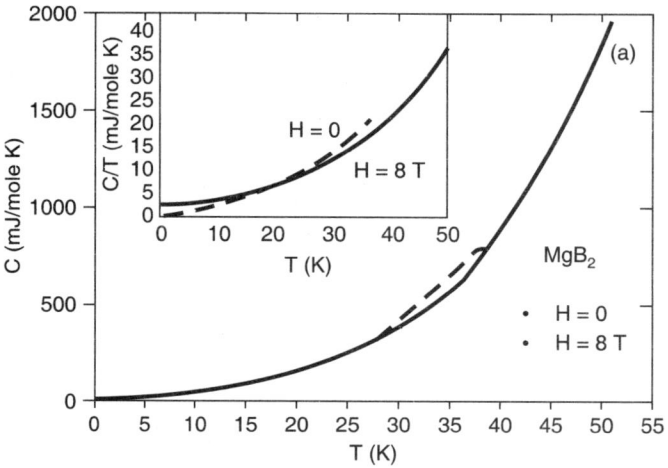

Fig. 9.38 Temperature variation of specific heat of MgB_2 at H = 0 and 8 T [337]. The anomaly around 39 K manifests the bulk superconductivity. The inset shows $\frac{C}{T}$ vs. T.

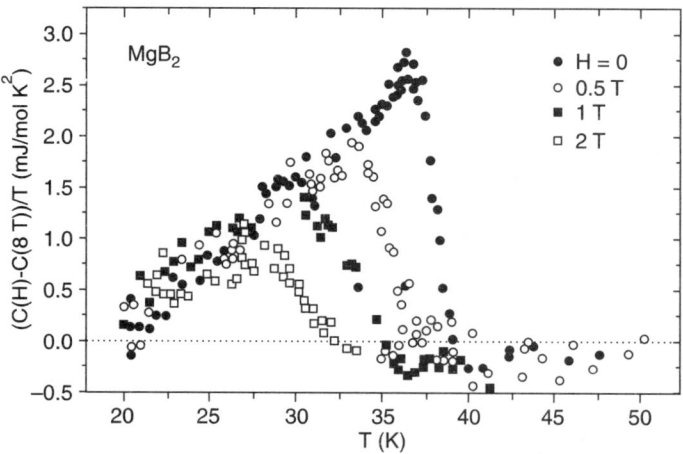

Fig. 9.39 Variation of $\frac{\Delta C}{T} = (C(H) - C(8\ T))/T$ with temperature at different magnetic fields (0 − 2 T) for polycrystalline MgB_2 [337].

9.8 PHOTO-INDUCED SUPERCONDUCTORS

Photoconductivity has a long history and occurs in a variety of insulating and semiconducting materials [342]. The existence of a metal insulator (M-I) transition in HTSC make these materials attractive systems for photoexcitation experiments. Raman scattering measurements indicate the appearance of normally forbidden modes in fully oxygenated photoexcited samples [343]. Transient photoinduced charges of more than ten orders of magnitude in the surface resistivity of $YBa_2Cu_3O_x$ (YBCO) single crystals have been reported earlier [344]. Later on, persistent superconductivity in insulating YBCO films was observed [345, 346]. It was shown that laser illumination induces a systematic decrease with long relaxation times in the electrical resistivity $\rho_{xx}(T)$ of oxygen deficient YBCO films. These experimental findings opened up the possibility for the existence of photoinduced superconductivity. Nieva et al. [347] have unambiguously shown that the decrease in resistivity in superconducting YBCO films is accompanied by a simultaneous increase in T_c. This confirmed the expectations of a photo induced transition to the metallic state and of photoinduced superconductivity in HTSC cuprates. Although Nieva et al [348] reported a photoinduced change in the Hall coefficient at room temperature in conjuction with the T_c enhancement, the relation between the density of photoinduced holes and both the normal-state and superconducting properties has not been clarified yet. The microscopic mechanism underlying these phenomena is also still controversial [347, 349].

Photoexcitation offers an attractive alternative method to vary the concentration of charge carriers with a number of advantages; for *e.g.*, n_c can be increased without the added complication of a change in chemical composition and crystal structure, and n_c can be increased in a transient manner with sub-nanosecond resolution using pulsed-laser techniques. The former is important for it allows one to study the system as a function of n_c without simultaneously changing the degree of disorder; the latter is important for it provides one with the opportunity to probe the dynamics of the electronic system at short times and well away from equilibrium. For details of photoinduced superconductivity from theoretical prediction to experimental discoveries, readers are advised to consult recent review by K.P. Sinha [350].

9.9 NEW DISCOVERIES SINCE MgB_2

Two serious claims about new superconductors, made by two different groups [351, 352]. The first consisted in the sample preparation of a carbon-sulphur composite with a indication of superconductivity at 35 K [351]. The second claim of existence of superconductivity at 1K in pyrochlore structure of $Cd_2Re_2O_7$ [352]. The structure of $Cd_2Re_2O_7$ is cubic, with a space group Fm 3 m and unit cell of 10.27 Å [Figs. 9.40, 9.41]. A unit cell contains eight chemical formulae, while the asymmetric unit contains one Cd in the 16-fold position $\left(\frac{1}{2}, \frac{1}{2}, \frac{1}{2}\right)$, one Re cation in the 8-fold position (000) and two oxygen atoms, 01 and 02 in the 48 fold $\left(0.31, \frac{1}{8}, \frac{1}{8}\right)$ and 8 fold $\left(\frac{3}{8}, \frac{3}{8}, \frac{3}{8}\right)$ positions respectively [353].

A superconductor made of potassium, osmium and oxygen having formula KOs_2O_6 with T_c = 9.6 K is reported in the literature by the Japanese physicists at the University of Tokyo. Another superconductor reported is boron doped diamond with T_c = 4 K [354]. Diamond is usually an electrical insulator and well known for being exceptionally hard. It also conducts heat well and can withstand strong electric fields. These properties make it attractive for electronic applications especially when doped with charge carriers, such as boron.

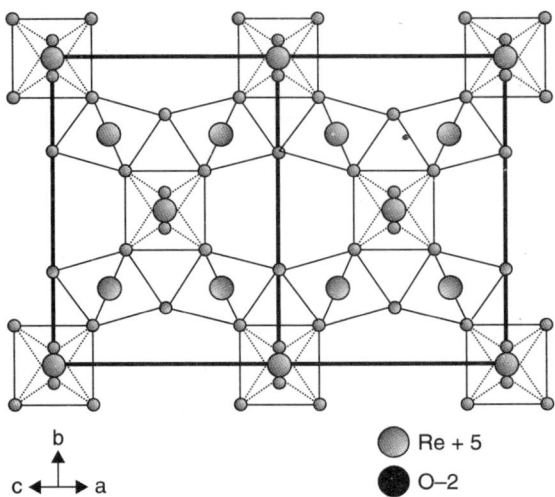

Fig. 9.40 The cubic unit cell of $Cd_2Re_2O_7$ viewed along the [110] axis. Only the Re cations and the 0-1 atoms are shown. These atoms form a corner-sharing three-dimensional array of octahedra centred on the Re cations. The strings of octahedra run along the [10 – 1] axis. The octahedra appearing as single belong to octahedral strings running along the [110] axis. The Cd tetrahedra centred on the 0-2 atoms are located in the hexagon-shaped channel generated by the octahedral array.

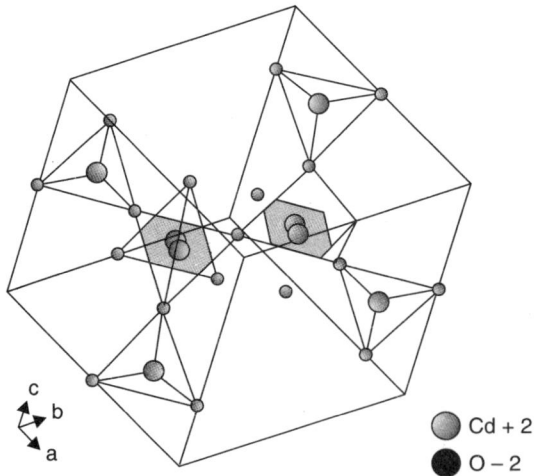

Fig. 9.41 The cubic unit cell of $Cd_2Re_2O_7$ is viewed almost along the [111] axis. Only the Cd cations forming the three-dimensional tetrahedral framework and the 0-2 atoms at the centre of the tetrahedra are shown.

9.10 FUTURE HIGH-T_c SUPERCONDUCTORS

The search for novel superconducting materials has always been an important integral part of superconductivity research. The efforts are going on and we hope to find such high temperature superconductors which may overcome the existing problems and expedite commercialization of HTSC devices. As room temperature superconductors are attained, new challenges for scientists and new promises for technologists will undoubtedly created.

REFERENCES

1. B.T. Matthias et al., *Rev. Mod. Phys.* 35, 1 – 22 (1963).
2. B.W. Roberts, *J. Phys. Chem. Ref. Data* 5, 581 (1976).
3. S.V. Vonsovsky et al., *'Superconductivity of Transition Matals'*, Springer Series in Solid-State Sciences, Vol. 27, Springer-Verlag, Berlin (1982).
4. E.M. Savitsky et al., *'Superconductivity in Alloys of Nobel Metals* (In Russian), Metallurgiya, Moscow (1985).
5. K.P. Sinha and S.L. Kakani, *'Magnetic Superconductors: Recent Developments'*, Nova Science Publishers, NewYork (1989).
6. S.L. Kakani, *Superconductivity: Current Topics*, Bookman Associates, Jaipur, India (2001).
7. L.I. Berger and R.W. Roberts, in *CRC Handbook of Chemistry and Physics* 78th ed. (D.R. Lide, Ed.), pp. 12-60 to 12-86, CRC Press, Boca Raton (1987).
8. C.R. Poole Jr. *Handbook of Superconductivity,* Academic Press, San Diego (USA) (2000).
9. T. Ishiguro et al. *'Organic Superconductors'*, 2nd ed. Springer, Berlin (1998).
10. A.A. Varlamov et al., *Adv. Phys.* 48, 655 (1999).
11. A.I. Larkin et al., (eds.), *Handbook on Superconductivity : Conventional, Unconventional Superconductors,* Springer, Berlin (2002).
12. F. Steglich in T. Kasuya and T. Sato (eds.): *Theory of Heavy Fermions and valence Fluctuations* Springer, Berlin (1985).
13. P.G. de Gennes, Phys. Kondens, Materie 3, 79 (1964) *Superconductivity in Metals and Alloys* (Benjamin New York, 1969).
14. H. Suhl and M.B. Maple, (eds.) *'Superconductivity in d- and f-band Metals'* (Academic Press, New York, 1980).
15. D. Jerome and L.G. Caron (eds.), *Low-Dimensional Conductors and Superconductors,* Plenum Press, New York, (1987).
16. A.F. Hebard, *Physics Today* (Nov. 1992), 26.
17. M.B. Maple and φ. Fischer (eds.), *'Superconductivity in Ternary Compounds'*, Springer, Berlin, New York (1982).
18. A.P. Ramirez, *Superconductivity Review*, 1, 1 (1994).
19. O. Gunnarsson, *Rev. Mod. Phys.* 89, 575 (1997).
20. W. Klose, *Sommershule fur Supraleitung, Pegnite, Germany : DPG,* 1970, p. 14.
21. R. Chevrel et al., *J. Solid State Chem.* 3, 515 (1971).
22. N. Cheggour et al., *J. Appl. Phys.* 84, 2181 (1998).
23. R. Chevrel et al., *Res. Bull.* 10,1169 (1975).

24. R.N. Shelton et al., *Phys. Rev. B* 34, 199, (1986).
25. L.D. Woolf et al., *J. Low Temp. Phys.* 35, 651 (1979).
26. L.D. Woolf et al., *Phys. Lett.* 71 A, 137 (1979); ibid, *Phys. Lett.* 71 A, 363 (1979).
27. J.W. Lynn et al., *Phys. Rev. B* 31, 5756 (1985).
28. J.W. Lynn et al., *Phys. Rev. Lett.* 52, 133 (1984).
29. H.C. Hamakar et al., *Solid State Commun.* 31, 139 (1979); *ibid,* 32, 289 (1979).
30. M. Ishikawa et al., *J. Phys. (Paris)* 39, C 6 – 1379, (1978).
31. D.E. Moncton et al., *Phys. Rev. B* 16, 801 (1977).
32. D.E. Moncton et al., *Phys. Rev. Lett.* 41, 1133 (1978).
33. R.W. McCallum et al., *Solid State Commun.* 24, 391 (1977).
34. M. Ishikawa and O. Fischer, *Solid State Commun.* 23, 37 (1977); 24, 747 (1977).
35. M.B. Maple et al., *Phys. Lett.* 77 A, 487 (1980).
36. B.T. Matthias et al., *Proc. Nat. Acad. Sci,* (USA) 74, 1334 (1977).
37. J.M. Vandenberg and B.T. Matthias, *Proc. Natl., Acad. Sci.* USA 74, 1336 (1977).
38. Yu B. Kuz'ma and N.S. Bilonizhko, *Sov. Phys. Crystallog.* 16, 697 (1972).
39. A.J. Freeman et al., *J. Magn. Magn. Mater.* 7, 296 (1978).
40. M.A. Clogston, *Phys. Rev. Lett.* 9, 266 (1962).
41. B.S. Chandrasekhar, *Appl. Phys. Lett.* 1, 7 (1962).
42. A.A. Abrikosov, and L.P. Gorkov, *Zh. Eksp, Theo. Fiz.* 39, 1781 (1960) [Sov. Phys. JETP 12, 1243 (1961)].
43. P. Fulde and A. Ferrel, *Phys. Rev.* 135, A 550 (1964).
44. A. Larkin and Y. Ovchinnikov, *Sov. Phys. JETP* 20, 762 (1965).
45. J.L. Tallon et al., IEET Trans., *Appl. Sup.* 9, 1051 (1999).
46. C. Bernhard et al., *Phys. Rev. B* 59, 14099 (1999).
47. Ch. Mazumdar et al., *Physica B* 223 and 4, 102 (1996).
48. R. Nagarajan et al., *Phys. Rev. Lett.* 72, 274 (1994).
49. K.H. Muller and V.N. Narozhnyi, *Rep. Prog. Phys.* 64, 943 (2001).
50. J.W. Lynn and S. Skanthakumar, *"Handbook on Physics and Chemistry of Rare Earths,"* Ch. 19, Vol. 31, Eds. K.A. Gschneidner et al., North Holland, Amsterdam (2001) p. 313.
51. J.W. Lynn et al., *Phys. Rev. B* 55, 6584 (1997).
52. J.W. Lynn et al., *Phys. Rev. Lett.* 94, 37006 (2005).
53. S. Saxena et al., *Nature* (London) 406, 587 (2000).
54. C. Pfleiderer et al., *Nature* (London) 412, 58 (2001).
55. D. Akoi et al., *Nature* (London) 413, 613 (2001).
56. A. Huxley et al., *Phys. Rev. B* 63, 144519 (2001).
57. C. Pfleiderer and A.D. Huxley, *Phys. Rev. Lett.* 89, 147005 (2002).
58. N. Taleiva et al., *J. Phys. Condens. Matter* 13, L 17 (2001).
59. S.L. Kakani and U.N. Upadhyaya, *J. Low Temp. Phys.* 53, 221 (1983); *Phys. Stat. Sol. (b)* 125, 861 (1984); *Phys. Stat. Sol. (b)* 135, 235 (1986); *Phys. Stat. Sol. (a)* 99, 15 (1987); *J. Low Temp. Phys.* 70, 5 (1988).
60. S.L. Kakani and Madhu Kakani, *'Superconductivity: key Problems'*, Bookman Associates, Jaipur, India (1996).

61. S.L. Kakani and C. Hemrajani, 'Recent Advances in Superconductivity', Today and Tomorrow, New Delhi (1990).
62. A.L. Fetter and J.D. Walecka, 'Quantum Theory of Many Particle Systems' (McGraw Hill, 1971).
63. S. Mackawa and M. Tachiki, *Phys. Rev. B* 18, 4688 (1978).
64. M. Crisan, 'Theory of Superconductivity', World Scientific, Singapore (1989) (p. 114).
65. K. Machida and D. Younger, *J. Low Temp. Phys.* 35, 449 (1979); *ibid* 35, 561 (1979).
66. T.V. Ramakrishnan C.M. Varma, *Phys. Rev. B* 24, 137 (1981).
67. V. Jaccarino and M. Peter, *Phys. Rev. Lett.* 9, 290 (1962).
68. T. Matsuura et al., *Prog. Theor. Phys.* 57, 713 (1977).
69. G.S. Grest et al., *J. Mag. Mag. Mat.* 31 – 34, 501 (1983).
70. U. Poppe et al., *Physica 108 B*, 805 (1981).
71. L. Bulaevski et al., *Phys. Rev. B* 28, 1370 (1983).
72. K. Machida and H. Nakanishi, *Phys. Rev. B* 30, 122 (1984).
73. S.V. Vonsovsky et al., 'Superconductivity of transition metals, their alloys and compounds', Springer Verlag (1982).
74. K. Machida, *Appl. Phys. A* 35, 193 (1984).
75. A.I. Buzdin and L.H. Bulaevskii, *Usp. Fiz. Nauk.* 149, 45 (1986) [Sov. Phys. Usp. 29 95), 412 (1986)].
76. C. Rossel, Il. Iuovo Cimento, 2D, 1834 (1983).
77. F.G. Aliev et al., *Pisma Zh. Eksp. Teor. Fiz.* 43, 143 (1986) [JETP 43, 182 (1986)].
78. W. Baltensperger and S. Strassler, *Phys. Kondens Mater*-1, 20 (1963).
79. K.N. Shrivastava and K.P. Sinha, *Phys. Reports,* 115, 93 (1984).
80. P.W. Anderson, in *Magnetism, eds.* G.T. Rado and H. Suhl, Academic Press, New York (1963) Vol. 1.
81. P.W. Anderson, *Phys. Rev.* 126, 41 (1961).
82. J.R. Schrieffer and P.A. Wolff, *Phys. Rev.* 149, 491 (1966).
83. B. Coqblin and J.R. Schrieffer, *Phys. Rev.* 185, 847 (1969).
84. J. Hubbard, *Proc. R. Soc.* (London), A 276, 238 (1963).
85. K.P. Sinha, *Solid State Commun.* 31, 201 (1979).
86. T.C. Loya and S.L. Kakani, *Pramana-J. Phys.* 43, 41 (1994).
87. P. Fulde and J. Keller in "Superconductivity in Ternary compounds," Eds. O. Fischer and M.B. Maple (Springer, Heidelberg, 1982) p. 249.
88. K. Machida et al., *Phys. Rev.* 822, 2307 (1980).
89. M.J. Nass et al., *Phys. Rev. Lett.* 46, 614 (1981).
90. G. Zwicknagl and P. Fulde, *Z. Phys.* 43, 23 (1981).
91. K. Machida, *J. Low Temp. Phys.* 37, 583 (1989).
92. N.D. Mathur et al., *Nature* 394, 39 (1998).
93. R. Movshovich et al., *Phys. Rev. B* 53, 8241 (1996).
94. H. Hegger et al., *Phys. Rev. Lett.* 84, 4986 (2000).
95. C. Petrovic et al., *Euro phys. Lett.* 53, 354 (2001).
96. F. Steglich et al., *Phys. Rev. Lett.* 43, 1892 (1979).

97. G.R. Stewart et al., *Phys. Rev. Lett.* 52, 679 (1984).
98. G.R. Stewart, *Rev. Mod. Phys.* 56, 755 (1984).
99. K. Hasselbach et al., *Phys. Rev. Lett.* 50, 1595 (1989).
100. M. Sigrist and K. Udea, *Rev. Mod. Phys.* 63, 239 (1991).
101. H.R. Ott et al., *Phys. Rev. Lett.* 63, 93 (1989).
102. H.R. Ott in : D.F. Brewer (Ed.), *Progress in Low Temperature Physics,* XI, North Holland, Amsterdam, 1987, p. 215; Phys. Acta 60, 62 (1987).
103. Y. Kuramoto and Y. Kitaoka, in : J. Birman et al., (Eds.), *Dynamics of Heavy Electrons,* Oxford, New York, 2000.
104. H. Tou et al., *Phys. Rev. Lett.* 77, 1374 (1996).
105. C. Geibel et al., *Z. Phys. B* 84, 1 (1991).
106. C. Geibel et al., *Z. Phys. B.* 83, 305 (1991).
107. M. Kyogaku et al., *J. Phys. Soc. Japan* 62, 4016 (1993).
108. H. Tou et al., *Physica B* 230 – 232, 360 (1997).
109. K. Ishida et al., *Phys. Rev. Lett.* 89, 03700 (2002).
110. T.T.M. Palstra et al., *Phys. Rev. Lett.* 55, 2727 (1985).
111. D. Jaccard et al., *Phys. Lett. A* 163, 475 (1992).
112. D. Jaccard et al., *Rev. High Pressure Sci. Technol.* 7, 412 (1998).
113. D. Jaccard et al., *Physica B* 259 – 269, 1, (1999).
114. F.M. Grosche et al., *Physica B* 223 – 224, 50 (1996).
115. N.D. Mathur, *Nature,* 394, 39 (1998).
116. R. Movshovich et al., *Phys. Rev. B* 53, 8241 (1996).
117. S.J.S. Lister et al., *Z. Phys. B* 103, 263 (1997).
118. I.R. Walker et al., *Physica C* 282 – 287, 303 (1997).
119. E.V. Sampath Kumaran, et al., 'Valence Fluctuations in Solids*, L.M. Falicov, et al., (ed.) North-Holland (1981), p. 193.
120. H. Hegger et al., *Phys. Rev. Lett.* 84, 4986 (2000).
121. C. Petrovic et al., *Europhys. Lett.* 53, 354 (2001).
122. S.S. Saxena et al., *Nature* 406, 587 (2000).
123. D. Aoki et al., *Nature* 413, 613 (2001).
124. E. Vargoz, and D. Jaccard, *J. Magn.Magn Mater.* 177, 294 (1998).
125. H. Hegger et al., *Phys. Rev. Lett.* 84, 4986 (2000).
126. C. Petrovic, *Europhys. Lett.* 53, 354 (2001).
127. G. Bruls et al., *Phys. Rev. Lett.* 72, 1755 (1994).
128. J.P. Brison et al., *Physica B* 280, 165 (2000).
129. H. Tou Kitaoka et al., *Phys. Rev. Lett.* 80, 3129 (1998).
130. H. Tou Kitaoka et al., *Phys. Rev. Lett.* 77, 1374 (1996).
131. H.R. Ott et al., *Phys. Rev. B* 33, 126 (1986).
132. Y. Kitaoka et al., *J. Magn. Magn. Mater* 52, 341 (1985).
133. H. Tou Kitaoka et al., *J. Phys. Soc. Jpn.* 64, 725 (1995).
134. Y. Kitaoka et al., *Jpn. Appl. Phys. Suppl.* 26, 1221 (1987).

135. M. Kyogaku, *Physica B* 186 – 188, 285 (1993).
136. E. D. Bauer et al., *J. Phys. Soc. Japan Suppl.* 71, 23 (2002).
137. J.L. Sarro et al., *Nature* 420, 297 (2002).
138. H. Hegger et al., *Phys. Rev. Lett.* 84, 486 (2000).
139. C. Petrovic et al., *Europhys. Lett.* 53, 354 (2001).
140. C. Petrovic et al., *J. Condens. Matter* 13, L 337 (2001).
141. Y. Haga et al., *Phys. Rev. B* 63, 060503 (2001).
142. R. Settai et al., *J. Phys. Condens. Matter* 13, L 627 (2001).
143. D. Hall et al., *Phys. Rev. B* 64, 212508 (2001).
144. H. Shishido. et al., *J. Phys. Soc. Japan* 71, 162 (2002).
145. Y. Onuki et al., *Physica B* 312 – 313, 13 (2002).
146. Y. Kitaoka et al., *J. Phys. Chem. Solids* 63, 1141 (2002).
147. P.G. Pagliuso et al., Z. Fisk. *Phys. Rev. B* 64, 100503 (2001).
148. C. Petrovic et al., *Euorphys. Lett.* 53, 354 (2001).
149. C. Petrovic et al., *J. Phys. Condens. Matter* 13, L 337 (2001).
150. Y. Kohori et al., *Eur. Phys. J.B.* 18, 601 (2000).
151. G.q. Zheng et al., *Phys. Rev. B* 86, 4664 (2001).
152. Y. Kohiri et al., *Phys. Rev. B* 64, 1345 26 (2001).
153. N.J. Curro et al., *Phys. Rev. B* 64, 180514 (2001).
154. Y. Kohori et al., *Physica B* 312 – 313, 126 (2002).
155. T. Mito et al., *Physica B* 312 – 313, 16 (2002).
156. K. Izawa et al., *Phys. Rev. Lett.* 87, 057 002 (2001).
157. K. Izawa et al., *J. Phys. Chem. Solids* 63, 1055 (2002).
158. S. Kawasaki et al., *Phys. Rev. B* 66, 054521 (2002).
159. H. Tou et al., *Phys. Rev. Lett.* 77, 1374 (1996).
160. H. Tou et al., *Phys. Rev. Lett.* 80, 1329 (1998).
161. G. Aeppli et al., *Phys. Rev. Lett.* 60, 615 (1988).
162. G. Aeppli et al., *Phys. Rev. Lett.* 63, 676 (1989).
163. Y. Koike et al., *J. Phys. Soc. Japan* 67, 1142 (1998).
164. M.B. Maple, *Phys. Rev. Lett.* 56, 185 (1986).
165. K. Matsuda et al., *Phys. Rev. Lett.* 87, 087203 (2001).
166. A. Krimmel et al., *Z. Phys. B* 82, 161 (1992).
167. C. Geibel et al., *Z. Phys. B* 84, 1 (1994).
168. M. Kyogaku et al., *Physica B* 230 – 232, 360 (1997).
169. A. Schröder et al., *Phys. Rev. Lett.* 72, 136 (1993).
170. J.G. Lussier et al., *Phys. Rev. B* 56, 11749 (1997).
171. C. Geibel, *Z. Phys. B* 83, 305 (1991).
172. K. Ishida et al., *Phys. Rev. Lett.* 89, 037002 (2002).
173. S.R. Julian et al., *J. Magn. Magn. Mater.* 177, 265 (1998).
174. R.A. Steeman et al., *Solid State Commun.* 66, 103 (1988).

175. I.R. Walker et al., *Physica C 282,* 303 (1997).
176. S. Sachdev and J. Ye, *Phys. Rev. Lett.* 69, 2411 (1992).
177. G. Keppli et al., *Science* 278, 1432 (1997).
178. C. Castellani et al., *Z. Phys. B-Conden. Matter* 103, 137 (1997).
179. E. Vargoz and D. Jaccard, *J. Magn. Magn. Mates.* 177, 294 (1998).
180. R. Movshovich et al., *Phys. Rev. B* 53, 8241 (1996).
181. M. Crisan, *Theory of Superconductivity,* Ch. 7, p. 248, World Scientific (1989).
182. K. Yamada and K. Yosida, *Prog. Theor. Phys,.* 76, 621 (1986).
183. S. Doniach and E.H. Sondheimer, *'Green's Functions for Solid State Physicists',* Ch. 11, p. 235, Imperial College Press (1998).
184. Y. Maeno et al., *Nature* 372, 532 (1994).
185. Y. Maeno et al., *J. Phys. Soc. Japan* 66, 1405 (1997).
185. (a) N.E. Hussey et al., *Phys. Rev. B* 57, 5505 (1998).
185. (b) A.P. Mackenzie et al., *Phys. Rev. Lett.* 76, 3786 (1996).
185. (c) C. Bergemann et al., *Phys. Rev. Lett.* 84, 2662 (2000).
186. A.P. Mackenzie, et al., *Phys. Rev. Lett.* 76, 3786 (1996).
187. T. Oguchi, *Phys. Rev, B* 51, 1385 (1995).
188. D.J. Singh, *Phys Rev. B* 52, 1358 (1995).
189. S. Wakarsuzi and Y. Maeno, *Phys. Rev. Lett.* 84, 2666 (2000).
190. A. Damascelli et al., *J. Electron Spectrosc. Rerat. Phenom.* 114, 641 (2001).
191. Y. Maeno et al., *Phys. Today* 54, 42 (2001).
192. K. Ishida et al., *Nature* 396, 658 (1998).
193. K. Ishida et al., *Phys. Rev. B* 63, 060507 (2001).
194. M. Sigrist et al., *Rev. Mod. Phys.* 63, 239 (1991).
195. A.J. Leggett, *Rev. Mod. Phys.* 47, 331 (1975).
196. J.A. Duffy et al., *Phys. Rev. Lett.* 85, 5412 (2000).
197. A.P. Mackenzie et al., *Phys. Rev. Lett.* 80, 161 (1998).
198. K. Ishida et al., *Phys. Rev. B* 56, R 505 (1997).
199. K. Ishida et al., *Phys. Rev. Lett.* 84, 5387 (2000).
200. G.M. Luke, *Nature* 394, 558 (1998).
201. M. Sigrist et al., *Physica C* 317 – 318, 134 (1999).
202. R. Jin et al., *Prog. Theor. Phys.* 100, 53 (1998).
203. T.M. Rice and M. Sigrist, *J. Phys. : Condens Matter* 7, L 643 (1995).
204. K. Miyake and O. Narikiyo, *Phys. Rev. Lett.* 83, 1423 (1999).
205. S. Nishizaki et al., *J. Phys. Soc. Japan* 69, 572 (2000).
206. I. Bonalde et al., *Phys. Rev. Lett.* 85, 4775 (2000).
207. M. A. Tanatar et al., *Phys. Rev. B* 63, 064505 (2001).
208. K. Izawa et al., *Phys. Rev. Lett.* 86, 2653 (2001).
209. C. Lupien et al., *Phys. Rev. Lett.* 86, 5986 (2001).
210. H. Kusunose and M. Sigrist et al., *Euro. Phys. Lett.* 60, 281 (2002).
211. E. I. Blount, *Phys. Rev. B* 32, 2935 (1985).

212. Y. Hasegawa et al., *J. Phys. Soc. Japan* 69, 336 (2000).
213. H. Won and K. Maki, *Europhys. Lett.* 52, 427 (2000).
214. M. A. Tanatar et al., *Phys. Rev. Lett.* 86, 2649 (2001).
215. M.E. Zhitomirski and T.M. Rice, *Phys. Rev. Lett.* 87, 057001 (2001).
216. P. Monthoux (Preprint).
217. T. Takimoto, *Phys. Rev. B* 62, R 14641 (2000).
218. D. Jerome et al., *J. Phys. Lett.* 41, L 95 (1980).
219. H. Urayama et al., *Chem. Lett.* 55 (1988).
220. T. Ishiguro et. al., *Organic Superconductors,* 2nd edn. (Berlin – Springer) 1998.
221. J.M. Williams et al., *'Organic Superconductors (Including Fullerenes)* (Englewood Cliffs, NJ : Prentice Hall), 1992.
222. W.A. Little, *Phys. Rev. A* 134, 1416 (1964).
223. K. Bechgaard et al., *Solid State Comm.* 33, 1119 (1980).
224. D. Jerome, *Science* 252, 150924 (1991).
225. J.M. Williams et al., *Science* 252, 1501 (1992).
226. R.C. Haddon, Acc. *Chem. Res.* 25, 127 (1992).
227. D. Jerome and H.J. Schulz, *Adv. Phys.* 31, 299 (1982).
228. D. Jerome and J.P. Farges (eds.) : *'Organic Conductors',* Marcel Dekker, NewYork (1994), p. 405
229. D. Jerome et al., *J. Phys. Lett.* 41, L – 95 (1980).
230. K. Oshima et al., *Phys. Rev. B* 38, 938 (1988).
231. L. I. Buravov et al., *J. Phys. I* (France) 2, 1257 (1992).
232. K. Kanoda, *Physica C* 282 – 287, 299 (1997).
233. K. Kanoda, *Hyperfine Interactions* 104, 235 (1997), and references therein.
234. Yu. V. Sushko et al., *Physica C* 185 – 189, 2681 (1991).
235. Yu. V. Sushko et al., *Physica C* 185 – 189, 2683 (1991).
236. D.M. Watkins and G.A. Yaconi, *Phys. Rev. B* 44, 4666 (1991).
237. Y.V. Sushko and K. Andres, *Phys. Rev. B* 47, 330 (1993).
238. M. Imada et al., *Rev. Mod. Phys.* 70, 1039 (1998).
239. K. Miyagawa et al., *Phys. Rev. Lett.* 75, 1174 (1995).
240. A. Kawamoto et al., *Phys. Rev. B* 55, 14140 (1997).
241. H. Mayaffre et al., *Europhys. Lett.* 28, 205 (1994).
242. A. Kawamoto et al., *Phys. Rev. Lett.* 74, 3455 (1995).
243. M. Matsumoto and K. Kanoda (Preprint).
244. V. Emery et al., *Phys. Rev. Lett.* 48, 1039 (1982).
245. S. Tomic et al., *Phys. Rev. Lett.* 62, 462 (1989).
246. G. Kriza et al., *Phys. Rev. Lett.* 66, 1922 (1991).
247. L.P. Gor'Kov and A.G. Lebed', *J. Phys. Lett.* 45, L 433 (1984).
248. J.R. Cooper et al., *Phys. Rev. Lett.* 63, 1984 (1989).
249. H. Heritier et al., *J. Phys. Lett.* 45, L 943 (1984).
250. H. Ito et al., *J. Phys. Soc. Jpn.* 65, 2987 (1997).
251. M. Kurmoo et al., *J. Am. Chem. Soc.* 117, 12209 (1995).

252. L. Martin et al., *J. Chem. Soc. Chem. Commun.* 1997, 1367 (1997).
253. H. Kobayashi et al., *J. Am. Chem. Soc.* 119, 12392 (1997).
254. A. Otsuka et al., *Mat Res. Soc. Symp. Proc.* 488, 495 (1998).
255. W.A. Little, *Phys. Rev. A* 134, 1416 (1964).
256. Yu. V. Sushko and K. Andres, *J. Supercond.* 7, 937 (1994).
257. T. Ishiguro et al., *J. Supercond.* 7, 657 (1994).
258. D.E. Moncton, *Phys. Rev. Lett.* 60, 615 (1988).
259. G.W. Crabtree, *Phys. Rev. Lett.* 49, 1342 (1982).
260. T. Ishiguro and K. Yamaji (eds.), *Organic Superconductors* (Springer – Berlin), 1990.
261. V.Z. Kresin and W.A. Little (eds.), *Organic Superconductivity* (Plenum, NewYork) 1990.
262. G. Saito and S. Kagoshima, eds. *'The Physics and Chemistry of Organic Superconductors'* (Springer – Berlin) 1990).
263. M. Kohmoto and Y. Takoda, *J. Phys. Soc. Jpn.* 59, 154 (1990).
264. G. Bilbro and W.L. McMillan, *Phys. Rev.* 14, 5 (1976).
265. Anil Surana, R.K. Paliwal and S.L. Kakani, *J. Sup.* 13, 335 (2000).
266. S. Mazumdar, *Solid State Commun.* 66, 4, 427 (1988).
267. A.M. Gabovich et al., *Phys. State. Sol.* 141, 557 (1987).
268. I. J. Lee et al., *Phys. Rev. B 62, R* 14669 (2000).
269. H. Shimahara, *J. Phys. Soc. Japan* 58, 1735 (1989).
270. H. Kino and H. Kotani, et al., *J. Phys. Soc., Japan* 68, 1481 (1999).
271. H. Kuroki et al., *Phys. Rev. B* 63, 094509 (2001).
272. T. Nomura and K. Yamada et al., *J. Phys. Soc. Japan* 70, 2694 (2001).
273. H.W. Kroto et al., *Nature* 318, 162 (1965).
274. L.D. Kratschmer et al., *Nature* 347, 354 (1998).
275. C. Haddon et al., *Nature* 350, 320 (1991).
276. R.B. Fuller, *'Invention – The Presented Works of Buckminster Fuller'*, St. Martins, NewYork (1983).
277. W.E. Pickett, *Solid State Phys.* 48, 225 (1994) (Academic Press, NewYork).
278. A.F. Hebbard et al., *Nature* 350, 600 (1991).
279. K. Holccer et al., *Science* 252, 1154 (1991).
280. M.J. Rosseinsky et al., *Phys. Rev. Lett.* 66, 2830 (1991).
281. C.C. Chen et al., *Science* 253, 888 (1991).
282. K.K. Tamgaki et al., *Nature* 352, 222 (1991).
283. O. Zhou et al., *Nature* 362, 433 (1993).
284. A.R. Kortan et al., *Nature* 355, 529 (1992).
285. T.T.M. Palstra et al., *Solid State Commu.* 93, 327 (1995).
286. R.M. Fleming et al., *Nature* 352, 787 (1991).
287. O. Zhou, *Science* 255, 833 (1991).
288. J. Diederichs et al., *J. Phys. Chem. Solids* 58, 123 (1997).
289. R.C. Haddon et al., *Chem. Phys. Lett.* 125, 459 (1986).
290. A.R. Kortan et al., *Nature* 355, 529 (1992).
291. Y. Iwasa et al., *Phys. Rev. B* 57, 13395 (1998).

292. G. Baskaran and E. Tossati, *Curr. Science* 61, 33 (1991).
293. S. Chakravarty et al., *Science* 254, 970 (1991).
294. L. Pietronero and S. Strassler in *"Fullerenes: Status and Prospectus"*, C. Talani, et al., (eds.) P. 225 (World Scientific, 1992).
295. K. P. Sinha, *Solid State Commun.* 83, 291 (1992).
296. Y. Takada, *Physica C* 185 – 189, 419 (1991).
297. K. Harigave, *Prog. Theor. Phys.* (Preprint).
298. H. Zheng and K.H. Bennemann, *Phys. Rev. B* 46, 11993 (1992) – II).
299. K. Harigave, *Phys. Rev. B* 45, 13676 (1992 – I).
300. A.S. Alexandrov, *JETP Lett.* 55, 189 (1992).
301. M. Schluter et al., *Phys. Rev. Lett.* 68, 526 (1992).
302. V.P. Antropov, et al., *Phys. Rev. B* 48, 7651 (1993 – II).
303. K. Kato et al., *Chem. Phys. Lett.* 180, 446 (1991).
304. T. Takahashi et al., *Phys. Rev. Lett.* 68, 1232 (1992).
305. C.T. Chen et al., *Nature* 352, 603 (1991).
306. S. Morita et al., *J. Appl. Phys.* 31, 2822 (1992).
307. M. Matus et al., *Phys. Rev. Lett.* 68, 2822 (1992).
308. P.A. Lane et al., *Phys. Rev. Lett.* 68, 887 (1992).
309. A.S. Alexandrov and J. Ranninger, *Phys. Rev. B* 23, 1976, 1981, ibid 24, 1164 (1981).
310. A.S. Alexandrov et al., *Phys. Rev. B* 33, 1164 (1986).
311. A.S. Alexandrov and J. Ranninger, *Physica C* 198, 360 (1992).
312. S.C. Tiwari, R.S. Paliwal and S.L. Kakani [Under Communication, 2006].
313. T. Holstein, *Ann. Phys.* (N.Y.) 8, 325 (1959).
314. L.G. Lang and Yu. A. Firsov, *Zh. Eksp. Teor. Fiz.* 43, 1843 (1962) [Sov. Phys. JETP 16, 1301 (1963)].
315. G.D. Mahan, *Many Particle Physics* (Plenum, New York 1982).
316. D.N. Zubaroev, Upsekhi, *Fiz. Nauk. Sov. Phys.* 70, 5 (1960).
317. J. Nagamatsu et al., *Nature* 410, 63 (2001).
318. C. Buzea and T. Yamashita, *Supercond. Sci. Technol.* 14 R 115 (2001).
319. B. Lorenz et al., *Phys. Rev. B* 64, 012507 (2001).
320. S. Mollah et al., *Indian. J. Phys.* 77 A (1), 9 (2003).
321. A. Poddar et al., *Physica C* 390, 191 (2003).
322. Y. Kong et al., *Cond-mat/0102 499*, (2001).
323. S.L. Li et al., *Phys. Rev. B.* 64, 094 522 (2001).
324. D.C. Larbalestier et al., *Nature* 410, 186 (2001).
325. A.K. Pradhan et al., *Phys. Rev. B* 64, 212509 (2001).
326. A. Gumbel et al., *Appl. Phys. Lett.* 80, 2725 (2002).
327. E. S. Choi et al., *Cond. Mat/0104454* (2002).
328. T. Muranaka et al., *Phys. Rev. B 64,* 020505 (R) (2001).
329. N.F. Mott and E.A. Davis, *Electronic Processes in Non-crystalline Materials,* Clarendron, Oxford (1979).

330. A.V. Sologubenko et al., *Phys. Rev. B* 66, 014504 (2002).
331. P.G. Klemens, in : R.P. Type (Ed.), 'Thermal Conductivity', Vol. I, Academic Press, London (1969) (Chapter 1).
332. R. Berman, 'Thermal Conduction in Solids, Clarendron Press, Oxford, 1976, p. 133.
333. M. Putti et al., in A.V. Narlikar (Ed.) *Studies of High Temperature Superconductors* Vol. 38, Nova Science Publishers, NewYork.
334. Y. Wang et al., *Physica C* 355, 179 (2001).
335. A.Y. Liu et al., *Cond-mat/0103570* (2001).
336. W.N. Kang et al., *Phys. Rev. B* 65, 184520 (2002).
337. R. Jin et al., *Phys. Rev. B* 64, 220506 (R) (2001).
338. H.D. Yang et al., *Phys. Rev. Lett.* 87, 167003 (2001).
339. Ch. Walti et al., *Phys. Rev. B* 64 172515 (2001).
340. J. Kortus et al., *Phys. Rev. Lett.* 86, 4656 (2001).
341. J.M. An and W.E. Pickett, *Phys. Rev. Lett.* 86, 4366 (2001).
342. J.E. Hirsch, *Cond-mat/0102115, Phys. Lett. A* 282, 392 (2001).
343. A. Rose, in *Interscience Tracts on Physics and Astronomy,* Vol. 19, edited by R.E. Marshak, (Interscience, NewYork, 1963).
344. D.R. Wake et al., *Phys. Rev. Lett.* 67, 3728 (1991).
345. G. Yu et al., *Solid State Commun.* 72, 345 (1989).
346. V.I. Kudinov et al., *Phys. Rev. Lett. A* 151, 358 (1990).
347. V.I. Kudinov et al., *Phys. Lett.* A-157, 290 (1991).
348. G. Nieva et al., *Appl. Phys. Lett.* 60, 2159 (1992).
349. G. Nieva et al., *Phys. Rev. B* 46, 14249 (1992).
350. V.I. Kudinov et al., Phys. Rev. B 47, 9017 (1993).
351. K.P. Sinha, 'Phenomenon of photoinduced Superconductivity; From theoretical prediction to experimental discoveries' *Studies of High Temperature Superconductors,* Vol. 16, Ed. A.V. Narlikar, Nova Science Publishers, New York (1996).
352. R. Ricardo da Silva et al., *Phys. Rev. Lett.* 87, 147 001 (2001)
353. M. Hanawa et al., *Phys. Rev. Lett.* 87, 187001 (2001).
354. P.C. Donohue et al., *Inorg. Chem.* 4, 1152 (1965).
355. E. A. Ekimov et al., *Nature* 428, 542 (2004).

Applications of Superconductivity   10

10.1 INTRODUCTION

Thoughts concerning the technical applications of superconductivity are as old as superconductivity itself. Today superconductors can be found in applications as commonplace as MRI (magnetic resonance imaging) systems installed in almost every modern hospital and cellular-telephone base station, as well as monumental applications such as particle accelerators. The most spectacular applications are in the high current regime. By high currents, we usually mean at least tens, typically hundreds and possibly mainly thousands of amperes. Obviously, the zero resistance exhibited by a superconductor should be of immense benefit when using these high currents. The main commercial significance of superconductors is in power transmission. The power transmission is an important factor in energy distribution essential for the developing as well as the developed countries. In high temperature superconductivity lies a promise for the future. A promise of reduction of transmission losses by high temperature superconductors are expected to be useful for numerous power appliances, including transmission cables, transformers, current limiters, motors and generators, etc. High efficiency, high power density, and improved materials utilization, are some of the benefits associated with HTSC.

Superconducting magnets of every desired size and geometry have now been planned and manufactured. They are employed not just in scientific research, in high energy and solid state physics, as was the case in 1960s, they are already in planning and use for controlled fusion, for superconducting motors with output of several thousand kilowatts, for energy storage facilities and for magnetic levitation of trains *i.e.*, MAGLEV.

The application of superconductors in measurement technology is no less revolutionary; here it opens up for us the opportunity to increase the sensitivity of many determinations by orders of magnitude over what could be achieved with normal conducting circuitry. Superconducting circuit and memory elements could bring decisive improvements to large computers, under certain conditions.

New HTSC oxides provide us with superconductors which can be employed at the temperature of liquid nitrogen (T_B = 77 K) or above. At present, bulk samples, thin and thick films, wires and tapes based HTSC materials are fabricated. Small and medium scale models of HTSC electric power devices have been designed, constructed and tested. Some power projects have also begun to use superconductors based transformers and fault limiters. Compact, quiet, light weight and superefficient HTSC transformers are environment friendly and oil free. A current controller is designed to react to and reduce unanticipated power disturbances in the utility grid, preventing loss of power to consumers or damage to utility grid equipment. The primary application of HTSC current controller will be for the reduction of fault current. A new generation of superconducting fault limiters is being called upon due to their ability to respond within a few milliseconds to limit thousand of amperes of current. HTSC motors are cryogenically cooled, super efficient synchronous machines with HTSC field windings. HTSC motors may increase machine efficiency beyond 98% reducing losses by as much as 50% compared to conventional motors, producing energy savings, reduced pollution per unit energy produced and lower lifecycle cost.

The main requirement for commercial electric power applications is that, wires should be flexible and strong enough to carry large current in a magnetic field. The HTSC wire fabrication technology still needs improvement to meet cost and performance requirements for widespread applications. Overcoming this technology barrier is a main focus of the second generation wire development effort. [1–3 and references therein] This chapter is devoted to brief discussion of existing applications and future prospects of superconductors.

10.2 SUPERCONDUCTING MAGNETS

The first successful superconducting magnet that used a type-II superconductor was built in 1954 by George Yntema of the University of Illinois. The magnet, which used iron pole pieces with niobium (Nb) wire windings produced a field of 0.7 T. In the course of developing his magnet, Yntema made the important observation that the transport current of his wire depends on the amount of metallurgical cold work introduced into the wire during the drawing process. Thus, unlike the critical temperature and fields thermodynamic variables that describe the superconducting state the critical current in a type-II superconductor is, to a large extent, a metallurgical variable that depends on processing methods.

In the early 1960s, the discovery by John Kunzler of the high superconducting density in niobium-tin ($Nb_3 Sn$) and the almost simultaneous invention by John Hulm of a method to produce niobium zirconium (Nb Zr) wire in long lengths led to a flurry of coil construction activities. By 1963, General electric Co. has produced a Nb_3 Sn tape superconducting magnet with a field of greater than 10 T. About a decade late, multifilament Nb_3 Sn wire was developed. We may mention here that stable superconducting wires and tapes are basic requirements for a superconducting magnet. The multifilament configuration allows for improved magnet stability and increased current densities: almost all high-field research magnets used today are based on multifilament Nb_3 Sn technology.

The discovery of HTSC superconductivity in 1986 brought about a resurgence in magnet technology. HTSC allow for magnets operating well above liquid helium temperature, 4.2 K. But those materials also have very high critical fields, which suggests that they can also be exploited at low temperatures, at least as high-field inserts to Nb_3 Sn based superconducting magnets that operate in liquid helium. Researchers are actively pursuing HTSC-insert magnet technology, primarily for application in NMR spectrometers that probe protons with resonance frequencies above 1 GHz. That frequency corresponds to a magnetic field of 23.5 T [4].

10.2.1 Superconducting Magnet Technology

Engineers who design high-field superconducting magnets seek to minimize the volume that contains the magnetic field and also the potential energy stored in the magnet. The reason that the volume needs to be minimized is simple: stored magnetic energy is proportional to the volume that contains the magnetic field, and the cost and difficulty of building a magnet increases with increasing stored energy. The potential energy in a magnet needs to be as low as possible so that the coil can be protected in case the superconductor experiences a transition to the normal state, an event known as **quench**.

A superconductor can experience local fluctuations in the temperature or magnetic field. If the superconductor is operating near its critical condition, (*i.e.*, near the temperature-dependent field that separates the superconducting and normal states), then such fluctuations will lead to a decrease in the superconductor's critical current. That decrease forces flux motion in the filament, which heats up the filament and further decreases the critical current. If unchecked, this process can lead to thermal runaway and quench. During the runaway, the magnet's potential energy is deposited as heat into both the coil structure that is in thermal contact with the magnet. That undesirable event wastes time and liquid helium, and can damage the coil.

To minimize the occurrence of quenches, superconducting magnets use conductors made from a composite of fine superconductor filaments with diameters of the order of microns in a conductive **matrix**. The matrix, which also aids in the production of long lengths of ductile wire, is usually copper (Cu) for Nb alloys or silver (Ag) in the case of HTSCs. The increased surface-to-volume ratio of the filamentary superconductor enhances its ability to transfer current and heat to the surrounding conductive matrix in the event of momentary normalization. The positive feedback process that can lead to quenches is thus damped, and thermal runaway does not occur. Final filaments have an additional advantage: they reduce so called AC losses, flux flow losses due to changing magnetic fields within the superconductor. Such losses are particularly common during charging and discharging, or during the production of a time varying magnetic field.

By virtue of its upper critical magnetic flux density B_{c_2} Nb Ti at 4.2 K is used in magnetic fields upto 10 T, and at 1.8 K (superfluid He), upto 12 T. The predominant superconductor for the range of 12-20 T is Nb_3 Sn. Other candidates for use in high fields are Nb_3 Al and V_3 Ga.

In general metallic superconductors are fabricated by co-extrusion (extrusion, drawing, rolling) in the shape of multi-filament or multi-core conductors containing a great number (upto tens of thousands) of thin superconducting filaments embedded in matrix of normally conducting material. Nb Ti conductors for *d.c.* applications are made with a matrix of high-conductivity copper (Fig. 10.1(*a*)), which can be processed directly with the ductile Nb Ti. In the case of the more brittle Nb_3 Sn, a complicated fabrication method (bronze technique) leads to Nb_3 Sn filaments in a high-resistance matrix of Cu Sn [5]. For *d.c.* applications, this is combined with a core or jacket of high conductivity Cu, isolated from the Cu Sn by a tantalum diffusion barrier (Fig. 10.1(*b*)). In both cases, the high conductivity copper stabilizes the conductors in case of malfunctions (*e.g.*, a sudden transition to the normally conducting state, known as quenching), by acting as a short term current and heat bypass. The ratio of superconductor to stabilizing metal can vary between 1 : 1 and 1 : 20, depending on the application.

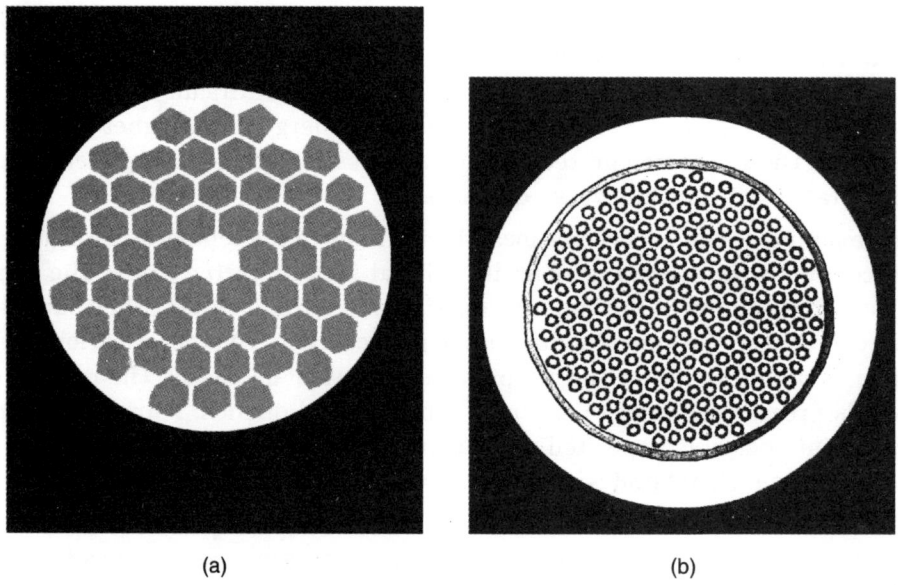

Fig. 10.1 Metallic multifilament superconductors (*a*) 27594 Tb Ti filaments, diameter 5 µm, in Cu matrix. (*b*) 23000 Nb_3 Sn filaments diameter 4 µm, in Cu Sn matrix. The filaments are assembled into bundles (After Vacuumschmelze Hanau).

The diameters of the superconductor filaments lie in the range of 1-50 µm for *d.c.* applications. The small diameters are required to confer stability against changes in the magnetic flux, which might initiate normal conduction. The criterian for adiabatic stability is given approximately by [6] :

$$J_c D_{SC} < \sqrt{\gamma\, C_V(T)(T_c - T_0)/\mu_0} \qquad ...(10.1)$$

where J_c is the critical current density of the superconductor, D_{SC} the thickness of the superconductor, γ the mass density, $C_v(T)$ the specific heat, T_c the critical temperature of the superconductor, and $\mu_0 = 4\pi \times 10^{-7}$ H/m the permeability of the free space.

With regard to time-variant magnetic fields, the filaments must be electrically decoupled from one another, which is achieved by twisting the conductors. For power applications, conductors rated at between 100 A and 100 kA are required. These capacities are attained by fabricating several conductors (Fig. 10.1) into transposed stranded wires or Robel bars; the individual conductors must be electrically decoupled for *a.c.* application. Fig. 10.2 shows a Robel bar used for the field winding of a 400 MVA test generator in development.

If metallic multifilament superconductors are subjected to periodically varying (*a.c.*) fields at line frequency (50 or 60 Hz) and significant amplitude (*e.g.*, in generators, transformers, and motors), the filament diameters must be lowered to 0.1 µm to minimise hysteresis losses. The hysteresis loss per unit volume is proportional to the filament diameter [8]

$$P_H = \frac{1}{2} f D_{SC}\, \Delta B\, J_c\, V_{SC} \qquad ...(10.2)$$

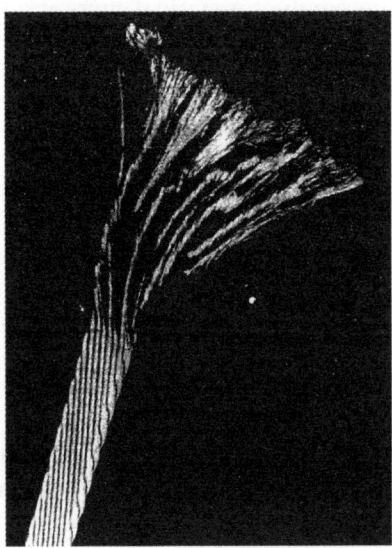

Fig. 10.2 Transposed high-current conductor (Robel bar) made of Nb Ti multifilament conductors, current 10 K at 5 T (After Vacuumschmelze Hanau).

where f is the frequency, D_{SC} the diameter of the superconducting filaments ΔB is the field amplitude (= $2\sqrt{2}\ B_{eff}$, perpendicular to filament thickness), J_c is the critical current density, and V_{SC} is the volume of the superconductor.

At lower field amplitudes, such as those in high-power transmission cables, the hysteresis losses are independent of superconductor thicknesses:

$$P_H = \frac{2}{3} \mu_0^2 f\, \Delta B^3\, A_{SC}/J_c \qquad ...(10.3)$$

where A_{SC} is the area of the superconductor exposed to the field.

To minimise coupling losses (eddy current losses) in the metal matrix, high-resistance alloys such as Cu Ni are employed.

The local superconducting-to-normal fluctuations that do lead to quenches must rapidly propagate through the coil and the current should decay quickly enough so that no specific hot spot exceeds a temperature T_{loc} of order 100 K. The maximum decay time τ_D consistent with this local propection requirement can be expressed as a balance between Joule heating and the thermal current of the coil material:

$$J_0^2\, \tau_D = \int_{T_0}^{T_{loc}} (\delta/\rho)\, C\, dT \qquad ...(10.4)$$

In the balance formula, J_0 and T_0 are the current density and temperature during magnet operation, δ is the magnetic material density, ρ is the temperature-dependent resistivity of the magnet, and C is the magnet's heat capacity. Typical decay times are of the order of seconds.

There are a number of methods for enhancing normal current decay and thus preventing hot spots. They include heating the coil to enhance normal zone propagation and inductive coupling to resistive secondary windings. In most high-field magnets, the local protection requirement, rather than the superconductor critical current density, determines the maximum current density allowed.

10.2.2 Global Protection

When a quench occurs, all the stored magnetic energy is converted to heat in the coil and its cold structure. The combined mass of the coil and its cold structure m must be able to absorb the stored energy in a manner that will allow the coil to be recooled with liquid helium and recharged. The global protection requirement means the coil cannot be heated above a maximum temperature T_{max}. As with the local protection requirement, T_{max} is determined by energy balance

$$E = \frac{1}{2} LI^2 = \int_{T_0}^{T_{max}} mC\, dT \qquad ...(10.5)$$

In the global energy balance relation, L is the coil's magnetic induction and I is its current. To minimise the possibility of mechanical damage due to stress induced by thermal expansion T_{max} is usally kept below about 100 K.

The total voltage across a coil during quench may be just a few volts, but turn-to-turn voltages within the winding can often substantial on the order of kV. The higher voltage is possible because, as the current changes during a quench, the inductive and resistive voltages can nearly cancel each other across the coil, but not over a single coil winding. It is crucial that a superconducting magnet have an electrical insulation system that allows the magnet to withstand the high turn-to-turn voltages without damage to the coil. Superconductors for magnets approaching 20 T can present special problems because the superconducting phase is often produced with a heat treatment after the coil is wound. For such superconductors, the insulation system usually consists of a fibreglass braid or cloth that be impregnated with epoxy to form a solid block winding after the heat treatment is complete. Designers of insulation systems must keep in mind that the systems often occupy a significant fraction of winding cross-section. Thus, insulation can reduce the number of ampere-turns per unit length, a particularly undesirable penalty in the high-field region of the coil.

The conductors within a high-field superconducting magnet are subject to enormous pressures when the magnet is operating. At 20 T, the magnetic pressure is 320 megapascals (3200 atmospheres), enough pressure to cause inelastic deformation in copper. In small bore, moderate-field magnets, the conductor itself can withstand the pressure. But as the magnetic field increases, additional structural material must be incorporated within the winding or on the outside of the coil. Many high-field superconducting magnets are subdivided into multiple nested coils to minimize accumulated stress.

According to the Virial theorem, the minimum structural mass is directly proportional to the magnetic energy stored

$$M_{min} = \delta_S E/\sigma \qquad ...(10.6)$$

where σ is the allowable stress in the structural material, and δ_S is the material's density. To be structurally stable, solenoids typically require several times the virial theorem mass. Superconductor properties can be degraded by strain, which adds another constraint to the structural material.

10.2.3 Conventional Superconducting Materials

Superconducting magnets are subject to thermal instability, leading to the loss of superconductivity, known as **quench**, in which the critical values of field, temperature, and current density are exceeded and fail to recover. This phenomenon generally begins in a localized region of the coil, then spreads to the rest of the magnet or magnet system with a "quench propagation velocity". All magnets use some form of composite superconducting wire, in which superconducting filaments carry current in parallel with normal conductor, known as the stabilizer. The stabilizer has the dual purpose of preventing quench in the face of disturbances and of protecting the magnet from excessive temperatures and pressures when unwanted quench occur. Following a quench, current rapidly transfers from the superconducting material to the stabilizer, since composites are always designed so that the resistance of the stabilizer is much less than that of the superconducting material in its normal state.

Quench protection, adequate thermal insulation, and ability to withstand high pressure are features of all superconducting magnets. So is the availability of long lengths of superconducting wire, frequently in excess of 1 km [4]. Superconducting wires can be fabricated from many different materials, and each has its advantages and disadvantages.

Superconducting magnets that produce fields below about 10 T use filamentary niobium titanium (Nb Ti) composite conductors. The alloy is ductile, easily fabricated, and relatively inexpensive. The maximum field that can be produced by the composite, as with all superconducting materials, is determined by the critical field B_{C_2}. The exact value of $\mu_0 H_{c_2}$ for Nb Ti depends on the specific alloy composition. It ranges from 10 to 11 T at 4.2 K and increases to 14-15 T at 0 K.

The critical current density J_c in a superconductor is another key parameter that determines superconducting magnetic performance. In practice, superconducting magnet builders typically refer to the *engineering critical current density J_E*, which is normalized to the composite wire cross-section. Fig. 10.3 shows that J_E decreases with increasing magnetic field. (It also decreases with increasing temperature). The ampere-turns needed to generate a magnetic field are determined by the superconducting current available, and it is rarely economical to use a particular material at fields in which its J_E is less than 100 A/mm^2. Fig. 10.3 shows that this economy criterion limits Nb Ti to magnets with fields below about 10 T at 4.2 K.

Magnets that generate fields above 10 T require a superconducting material with a higher B_{C_2} than Nb Ti can offer. In more than 90% of such magnets, that material is filamentary $Nb_3 Sn$ composite wire. As mentioned earlier, the many filaments provide increased electromagnetic and thermal stability along with reduced AC losses. They also allow for higher overall current densities. Furthermore, lengths of conductor can be connected with superconducting joints, which allows engineers to build persistent current magnets, necessary for applications needing highly stable magnetic fields.

Although $Nb_3 Sn$ forms an intermetallic compound with a well-defined crystal structure, the $Nb_3 Sn$ composite depending on just how it is made, can have a range of Sn content. Mechanical processing of the brittle compound is impossible, so composite wire fabrication is done by coreducting the Nb and Sn as separate components. The most robust large scale method, the so-called bronze process, incorporates Nb filaments in a bronze (Cu Sn) matrix. This method, however, limits the Sn content in the composite to about 14% of the bronze by weight. Increasing the Sn fraction, which increases B_{C_2} and J_E for a given field, is done by preparing pure or low alloy Sn along with the Nb in a Cu matrix, the so-called Sn process. For both methods, the post process wire contains no Nb_3 Sn : Magnets wound with post-process wire are heat treated near 1000 K to convert the components to the superconducting Nb_3 Sn compound. At 4.2 K, Nb_3 Sn can have $\mu_0 H_{C_2}$ as high as 25 T. At the same temperature, useful values of J_E persists to fields of 19-21 T, depending on the fabrication method.

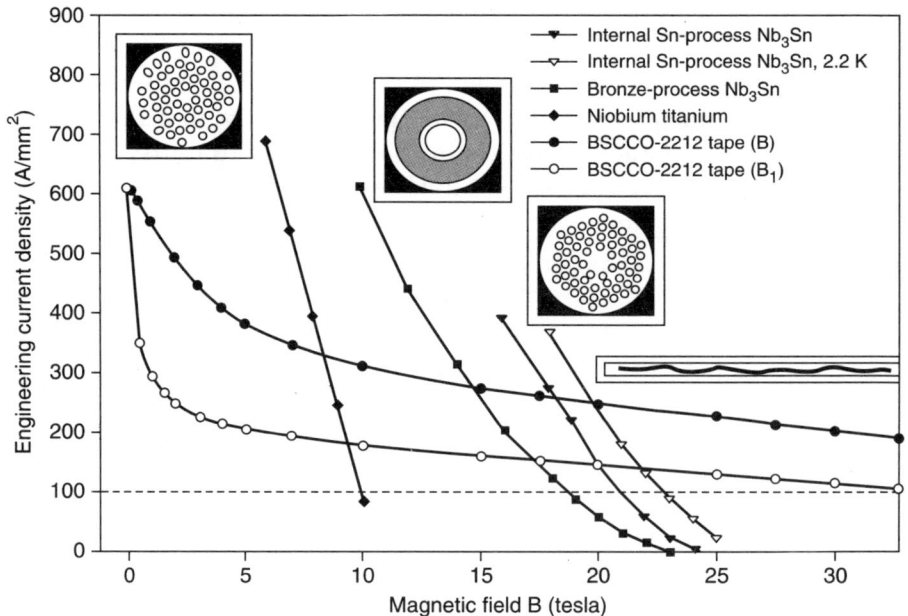

Fig. 10.3 Engineering current density J_E decreases as the applied magnetic field B increases. The plot shows the falloff for several technologically important superconductors. The dashed line at 100 A/mm² represents the minimum current density for economically practical magnet development. With the single exception noted in the legend, all measurements are made at 4.2 K. Data are given for both internal tin-process and bronze-process niobium tin (Nb_3 Sn) wires and for field applied both parallel ($B_\parallel$) and perpendicular ($B_\perp$) to ceramic BSCCO-2212 tape. The insets are cross-sectional photographs of multifilament wires and the ceramic tape.

Both magnet technology (primarily stress control quench management, and reduced temperature operation) and conductor development have contributed to the increases in Nb_3Sn magnet field strength from the 10 T levels of the early 1960s to the present state-of-the-art strengths of 22-23 T. For example Fig. 10.3 shows that operating an internal Sn process Nb_3Sn magnet at 2.2 K gains an additional 2 T compared to a 4.2 K operation.

Over the years, J_E at high fields has increased steadily in Nb_3 Sn compounds, primarily as a result of increased B_{C_2}. Incorporating more Sn was one route that enabled the increase of critical field ; another was doping the Nb_3 Sn layer by alloying either the precusor Nb filaments, the bronze matrix or the pure Sn filaments. Ginzburg-Landau-Abrikosov-Gorkov (GLAG) formulation of the BCS theroy shows that B_{C_2} is proportional to both the normal state resistivity and the critical temperature in the superconductor. Obviously, any addition to the Nb_3Sn that increases the resistivity without decreasing T_c will push up B_{C_2}.

Instead of pure Nb, today's high field composites typically use niobium tantalium (Nb Ta) filaments, titanium additions or both in the Cu Sn or Sn. After more than 20 years of effort, it appears that superconducting magnet builders working with Nb_3Sn have reached the limit of what they can accomplish with alloying additions and increased Sn content. They are shifting their attention to other materials in their never ending quest for higher fields.

10.2.4 Ceramic HTSC Materials

Ceramic HTSC have been developed primarily for applications well above 4.2 K and indeed, a number of low-field magnets currently operate above 20 K. But HTSC superconducting materials also possess properties that make them candidates for use in high field magnets that operate at liquid helium temperatue. In particular, tapes fabricated from the BSCCO-2212 compound material ($Bi_2 Sr_2 Ca Cu_2 O_x$) have a very slow decline in J_E with magnetic fields above 20 T as shown in Fig. 10.3. Fig. 10.3 also shows that there is a substantial difference in J_E, depending on whether the field is applied parallel or perpendicular to the tape face. Even in the perpendicular orientation, which leads to the smallest J_E, the 100 A/mm^2 standard occurs above 30 T. That makes BSCCO-2212 a strong candidate for high field laboratory magnets. Fig. 10.4 displays a coil made with an Ag/BSCCO-2212 conductor that is designed to be inserted as part of a larger magnet. Presently, the world record of about 23.5 T for an allsuperconducting magnet is held by the Tsukuba Magnet Laboratory at the National Institute for Materials Science in Japan. That magnet contains an inner most BSCCO-2212 coil surrounded by Nb_3Sn and Nb Ti coils. Inclusion of the BSCCO-2212 coil added 5.4 T to the total field.

The BSCCO materials (there are useful compounds other than the 2212 formula) are generally fabricated as tapes because these materials have a significant anisotropy in crystal structure and transport properties. The large aspect ratio typically used for tapes makes magnet engineering difficult. Round wire BSCCO-2212 was fabricated as early as 1989, but until about five years ago its J_E was substantially lower than that in tapes. Engineers continue to work toward further improving round wire-BSCCO-2212.

A serious challenge much more so for HTSC ceramic superconducting materials than for low-temperature ones is understanding and controlling dissipation mechanisms such as flux motion. Dissipation control is crucial for persistent current magnets used, for *e.g.* in NMR spectroscopy. An additional challenge is to develop persistent current joints for HTSC superconducting materials that will allow very steady fields. However, till now, joints in these materials are not upto NMR standards.

To address problems of anisotropy, dissipation control and joint technology, magnet builders are considering alternate high-field superconductors such as niobium aluminium (Nb_3Al) and tantatium tin (Ta_3Sn). Both, however, have been difficult to fabricate in the long, consistent lengths required for magnet applications.

Fig. 10.4 HTSC coils can be inserted into magnets to boost the generated magnetic field. This coil was developed through a collaboration between Oxford Instruments Superconducting Technology and the National High Magnetic Silver/BSCCO-2212 Superconductor, sol-gel insulation and epoxy impregnation. The coil is wound with more than a kilometer of conductor and has boosted a background 19-tesla field by an additional 3 tesla.

We may note that at temperatures lower than 40 K or so, HTSC ceramics also performs pretty well. This significant improvement at ~ 40 T is because of the **irreversibility line**. This means that at the higher temperatures, however HTSC cannot be regarded as a high field material. Nevertheless HTSC can have a role in achieving higher fields than the present limit of ~ 21 T for superconducting magnets provided they are used at lower temperatures. At temperatures higher than the irreversibility line, HTSC is still superconducting, of course, but is only useful for low field applications. This can include, however, being used for lossless ampere turns in order to provide the MMF for iron cored magnets.

$Y Ba_2 Cu_3 O_7$ and the lead-doped bismuth compound $(Bi Pb)_2 Sr_2 Ca_2 Cu_3 O_{10}$ (Bi-2223), with a high degree texture are also the principal compounds for magnet. The state of the art is less advanced than in metallic superconductors, particularly as to the critical current density at 77 K and the control of the fabrication process for long conductor lengths. However, major progress has already been made.

The powder in silver tube process [11] is employed to make wires and thin tapes of Bi-2223 in lengths of upto 1000 m. Critical current densities are 10^4 A/cm^2 in zero magnetic field. Current densities of 7×10^4 A/cm^2 have been achieved in short laboratory samples. Because its critical current density at 77 K is highly sensitive to magnetic fields, Bi-2223 at 77 K appears suitable only for low field applications such as cables, current limiters and possibly transformers.

The preferred configuration is also the multifilament conductor (Fig. 10.5). Such conductors can be stretched upto 0.3% without degradation of the critical current density. The filament diameter can be made at least an order of magnitude larger than with metallic superconductors, chiefly because of the higher heat capacity at 77 K (adiabatic stability), and in view of hysteresis losses at 50-60 Hz, the higher efficiency of liquid nitrogen cooling systems.

Two approaches have been taken in the development of highly textured conductors of $YBa_2Cu_3O_7$, which appears better suited than Bi-2223 for application involving moderate and strong magnetic fields. The first is melt-texturing growth (MTG), while the second involves thin-film techniques, laser ablation, and chemical vapour deposition (CVD) for the production of thick layers deposited on metal tapes (Hastelloy) with intermediate buffer films (ZrO_2). The MTG process is particularly suitable for the fabrication of massive superconducting parts such as those used in passive magnetic bearings, superconducting permanent magnets, or magnetic shieldings. Critical current densities J_c at 77 K of 4×10^4 A/cm^2 have been attained in magnetic fields of 1.5 T [12]. The other approach has led to laboratory specimens with YBCO a few μm thick and having critical current densities at 77 K : 6×10^5 A/cm^2 (zero field), 8×10^4 A/cm^2 [13].

Fig. 10.5 High-temperature multifilament superconducting wire made of Ag matrix with 37 filaments 100 μm in diameter.

Summarizing, we can say that there are very roughly two distinct regions for HTSC ceramics : (*i*) the high temperature, low field region (HTLF) and (*ii*) the low temperature, high field region (LTHF).

10.2.5 Applications of Superconducting Magnets

A large number of small superconducting magnets are already produced commercially as standard products with field strengths upto 8 T. They find many applications, particularly in solid state physics.

The majority of high-field superconducting magnets are produced for **high-frequency nuclear magnetic resonance (NMR) spectrometers.**

10.2.5.1 NMR Investigations

NMR investigations are an important method of structure determination in organic chemistry. Here the substance being investigated is brought into a magnetic field. Because of directional quantization the nuclei of hydrogen atoms only able to take up two orientations in the magnetic field, namely parallel and antiparallel. These two orientations differ in their potential energies. The parallel orientation has lower potential energy *i.e.*, work has to be performed to displace the moments from their parallel orientation. The energy difference between the two orientations is proportional to the magnetic field

$$\Delta E = 2\ \mu B$$

where μ is the component of the magnetic moment in the direction B and B is the magnetic field.

The hydrogen atoms of an organic molecule are distributed between these two orientations. The energetically more favourable energy level is somewhat more heavily occupied. Fig. 10.6 illustrates this distribution schematically. It this system is now irradiated with electromagnetic radiation with quantum energy $\Delta E = h\nu$, then this radiation will excite transitions between the levels with energies between E_1 and E_2. Since the lower level is more densely occupied more transitions will take place from the lower to the upper level than in the reverse direction, *i.e.*, radiation will be absorbed. If the magnetic field is maintained constant and the frequency ν of the radiation varied then a typical absorption signal is obtained at the resonance frequency : $\nu_R = \Delta E/h$ (Fig. 10.6(*a*)).

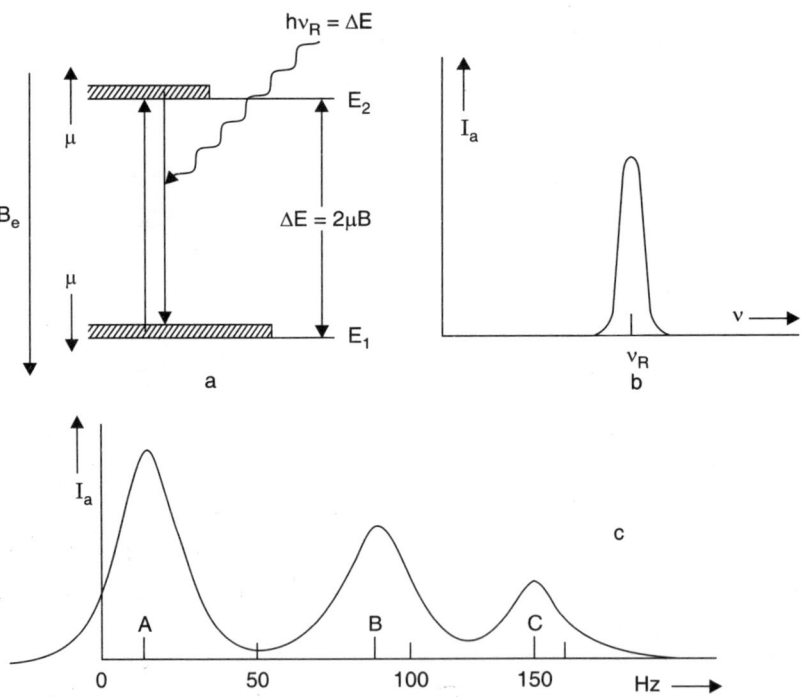

Fig. 10.6 Schematic representation of nuclear spin resonance. (*a*) term scheme (*b*) absorption signal, (*c*) spectrum of ethyl alcohol C_2H_6O. Lines *A*, *B* and *C* represent the protons of the CH_3, CH_2 and OH groups of the C_2H_6O molecule. The splitting shown was obtained with a resolution of 10^6.

The crucial factor is that the hydrogen nucleus does not just the external field B_e but also its environment. This causes the resonance frequencies of the hydrogen nucleus in different chemical environments e.g., as CH_3 or CH_3O groups of an organic molecule, to differ some what. The frequency displacement is known as the "chemical shift". Thus it is possible to identify the environments of hydrogen nuclei using NMR methods. This method is of great importance for the determination of the structures of organic molecules.

Now the resolution of the lines is difficult or impossible when the breadth of the lines is greater than their distance apart on the frequency axis. All that is then observed is a large, unresolved signal. Apart from the instrumental constants (e.g., the homogeneity of the magnetic field) the width of the lines is determined by the interaction of the nucleus being observed with more distant groups in the molecules. This interaction leads to the absorption line for a particular group of protons being split up into a number of narrower lines and, hence having a certain half width.

High magnetic fields can bring about an appreciable improvement in the resolution here. The chemical shift is proportional to the external magnetic field, i.e., the lines of various proton groups move further apart as the magnetic field increases. The splitting as a result of interaction with neighbouring groups is independent of the external field B_e, i.e., the half-width associated with these interactions remains unaffected even at high field and to determine their chemical shifts.

The high fields offered by superconducting magnets, therefore, offer advantages. However, it is necessary in the case of high resolution nuclear magnetic resonance spectrometers to maintain the magnetic field very homogeneous in the region of the sample. Every field inhomogeneity increases the half width of the line. Superconducting magnets for 7.5 T has field inhomogeneity in the sample volume (a sphere of 1 cm diameter) is less than 2×10^{-7}. With these magnets it is possible to resolve nuclear resonance lines whose frequencies only differ by $\nu = 0.05$ s^{-1} with a measurement frequency of 2.7×10^8 s^{-1}. This results in the enormous resolution of 5×10^9. The fact that it is now possible to construct such magnets so that they consume less than 1 litre liquid helium daily for cooling purposes means that they have virtually completely displaced conventional iron magnets in the field of high resolution spectroscopy.

NMR has found a very promising application in the field of medicine in recent years in the form of NMR tomography. Here the nuclear spin resonance described in Fig. 10.5 (a) is employed for the investigation of human tissue. This provides entirely new diagnostic evidence.

While the high resolution nuclear spin resonance basically exploits the shift in the resonance frequencies of protons (other nuclei can also be investigated) as a result of their differing chemical environments (chemical shift (Fig. 10.5 (b)), in NMR topography it is the strengths of the resonance signals that are employed as a measurement of the density of the protons. In addition, it is possible to choose the relaxation times of the nuclear spins as the parameter to be measured. The relaxation times are determined by the environment of the protons and can be employed diagnostically.

In order to localize the nuclear spin signals i.e., to produce a map of the signal height or the relaxation time, the constant magnetic field B_e is overlaid with a locally varying magnetic field, e.g. a field that increases in the X-direction. Depending on the resonance frequency it is then possible to localize the position of the nuclei being detected on the x-axis. Appropriate variation of the direction of the axis allows compilation of "pictures" of the proton density and the relaxation times. A considerable computational effort is naturally required as it is in X-ray tomography.

Superconducting coils are employed for the creation of the required magnetic fields. This has opened up an entirely new field of application for superconducting magnets. Instruments are being built today which can investigate the whole body of a patient. NMR tomography is of particular importance for the investigation of pathological changes in the brain.

Summarizing, we can say that the majority of high-field superconducting magnets are produced for high frequency NMR spectrometers. In addition of NMR, high-field superconducting magnets are also used in ion cyclotron resonance (ICR), electron magnetic resonance (EMR), and ultra-low temperature physics experiments.

NMR spectroscopy is a very active field of research, particularly in chemistry and biology with their stringent requirements for field homogenity and temporal stability. High magnetic fields offer numerous advantages, including improved spectral senstivity and resolution. In many applications, increasing the magnetic field yields multiple benefits; a small boost in field strength leads to a significant improvement in capability.

The spectroscopy of aligned samples of membrane proteins is one such application. Spectroscopists who use high fields enjoy the standard (and significant) improvements in resolution and sensitivity achieved with such fields. But an increase in field strength also increases relaxation times, which leads to sharper peaks for the associated resonance transitions. Moreover higher fields enhance a sample's alignment which further reduces line width. Because of the improved alignment, one can use few glass plates in preparing a sample and get more sample exposed to the magnetic field. A larger sample, better alignment, and increased relaxation times lead to multiplicative improvement in NMR performance with field strength.

NMR imaging at fields around 1T is a routine medical diagnostic. At higher fields, NMR imaging can give extraordinary images of micro organisms and other microstructures. Fig. 10.7 for example, shows an NMR image of a single neutron, along with an accompanying hydrogen spectrum. The ability to obtain such single-cell NMR images is a recent development, possible only with high field, high frequency instruments.

Fourier transform ICR is the highest resolution mass spectroscopy technique available. It can resolve the difference in mass between isotopically identical molecules a difference of less than the mass of one electron. In addition to opening exciting fundamental research opportunities, ICR also has broad technological applications. For example, one can use the technique to resolve and identify the more than 20,000 compounds found in crude oil and to identify the "fingerprint" of accelerants used in arson fires. The main commercial interest in FT ICR is for drug development. Principal applications include identifying the best drug targets from vast combinatorial libraries and identifying the site and nature of post translational modifications in proteins. This fall, the NHMFL expects to install a 15-T superconducting magnet.

EMR is complimentary to NMR and X-ray crystallography. Broadly applicable in chemistry, biology and material science, it is particularly valuable for analyzing large heterogeneous structures. As with NMR and ICR, the resolution of EMR increases with magnetic fields. Higher fields produce higher resonant frequencies, which result in increased spectral dispersion and increased sensitivity.

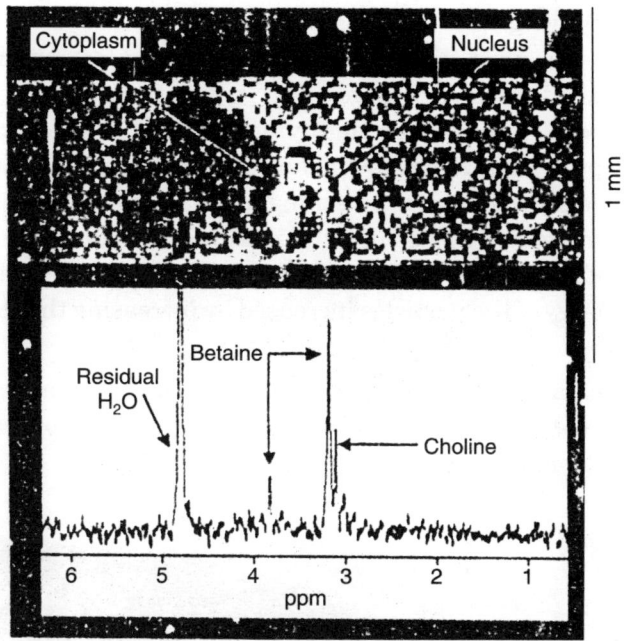

Fig. 10.7 A single neutron isolated from the abdominal ganglion of the sea hare '**Aplysia California**'. The neutron was imaged at 17.6 T, corresponding to a proton resonance frequency of 750 MHz. Such images can be used to define a region for localized proton spectroscopy. The hydrogen spectrum accompanying the neutron image displays resonance peaks for the cell's principal intracellular metabolites. The high magnetic field, in addition to narrowing the spectral line widths, allows images and spectra to be obtained reasonably quickly. The image of the neutron was acquired in 20 minutes and the spectrum was obtained from a 15.6 nanoliter sample in 25 minutes.

Superconducting magnets also find application in solid state research. The best methods for the determination of the quantum states of electrons in metals and semiconductors employ magnetic field. In these fields, the electrons rotate in circular paths as a result of the Lorentz force measurement of the frequency of this movement provides information concerning characteristic parameters of the electronic system e.g. concerning the effective mass of electron[1]. However, this rotational frequency can only be measured if, put in a simplified manner, the electrons make at least one rotation within a mean free path, in other words, in the time between two collisions. Since the circular paths are large in small fields, it is only possible to carry out such experiments on very pure samples with large mean free paths. This also makes

---

[1] A quantum condition for the motion of the electrons in a magnetic field B allows only certain paths and also leads to characteristic oscillations of the magnetic susceptibility and electrical resistance as a function of $1/B$. These oscillations are known as the De Hassvan Alphen effect and the Shubnikov-De Hass effect.

it possible now to introduce scattering centers deliberately and then to study their effects on the electron system. Superconducting magnets are being increasingly employed for this purpose too.

10.2.5.2 Superconducting Magnets for High Energy Physics

High energy physics opens up a wide field of applications for magnets made of metallic superconductors, in particular for very large ones. In this field, particles are now accelerated to several GeV (1 GeV = 10^9 eV). These highly energetic charged particles *e.g.*, protons or deutrons, have to be kept in their tracks by means of suitable magnetic fields. To simplify the matter somewhat, this is all the easier the stronger the magnetic field which is available. In particular, increasing magnetic fields make it possible to reduce the diameter of circular accelerators while maintaining the energy constant or on the other hand, the diameter can be kept constant and the energy of the particles increased by increasing the strength of the magnetic field.

Many gigantic *d.c.* magnets have been built for elementary particle detectors as well as beam guidance magnets outside the accelerator ring, and have been operated without problems for years in many laboratories worldwide.

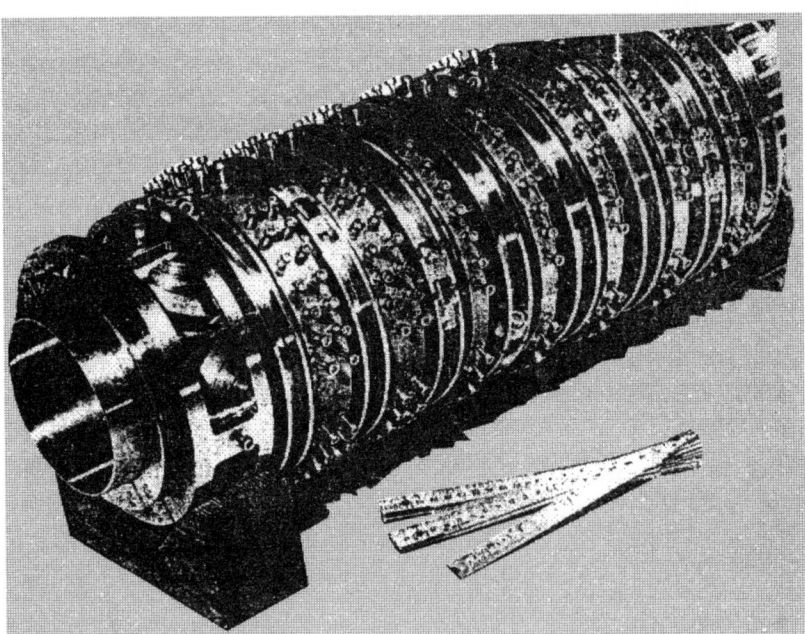

Fig. 10.8 Superconducting quadrupole magnet with Nb_3 Sn winding. Length 1 m, warm diameter 12 cm, maximum field 3.5 T, field gradient 0.37 T/cm.

Fig. 10.8 shows a superconducting magnet which can also be employed to guide the particles after they have left the accelerator. In this case, it is not necessary to change the field with time. Magnets of this type are known as quadrupole magnets; in them two extended pairs of coils (1 m long) generate cross magnetic fields. The maximum field is 3.5 T which achieves the magnetic field gradient of 0.37 T/cm required to focus the particles. The coils are immersed in a cryostat which has a thermal opening of 12 cm diameter for the steel tube containing the beam. This magnet illustrates very vividly that superconducting magnets are built today even for special purposes. These magnets have yielded appreciable savings in electrical energy consumption and operating costs. Their higher magnetic flux densities (a factor of two to five greater than conventional iron core magnets) permit substantial reductions in size.

These advantages have led to all modern ring-type particle accelerators being built with superconducting magnets *e.g.* 1000 GeV proton accelerator rings (Tevatron) at Fermilab in USA and the Hera accelerator at DESY, Hamburg, which have 1000 and 630 dipole and quadrupole magnets, respectively installed in their 6 km circumference rings. Larger machines of the same type has also been planned and constructed.

Replacing metallic superconductors with HT superconductors has little effect on installation costs but the electrical energy costs for running these accelerators can be lowered approximately by a factor of four.

10.2.5.3 Magnetic Separation and Purification

In a magnetic separation process, ferromagnetic or paramagnetic particles are extracted from a stream of nonmagnetic material by the action of magnetic forces. The forces acting on the magnetic particles is given by

$$F \sim \chi V \mu_0 H \nabla H \qquad \ldots(10.6)$$

where χ is the susceptibility of particles, V the particle volume, H the magnetizing field, and ∇H is the field gradient acting on the particle.

The use of superconducting magnets makes it possible, above all, to greatly increase the magnetic field strength and field gradient over the values in conventional magnetic separators ($\mu_0 H \leq 1T$). This allows magnetic separation of even weakly ferromagnetic or paramagnetic particles of small size, for which expensive and environmentally-unfriendly chemical flotation methods were formerly required. Other potential applications of separators with superconducting magnets are the removal of magnetic impurities from minerals, water treatment, desulfurization, recovery of radioactive materials, and several other chemical and biological processes. This type of magnetic separator has the further advantage of a low energy consumption, roughly an order of magnitude less than for equal-capacity systems using water-cooled windings.

Several pilot plants with liquid-helium-cooled Nb Ti windings have been tested in the past with success [14]. A favoured separator type is the high gradient magnetic separator (HGMS) shown in Fig. 10.9. The magnetic field is generated by a superconducting solenoid; the field gradient is established by a stack of fine magnetic steel meshes (matrix) in which the magnetic particles are trapped while nonmagnetic particles can pass without hindrance. The magnetic field is then turned off, and the magnetic material is washed out with water, after which the process begins a new.

Making use of HTSC ceramic materials cheaper and more robust liquid nitrogen cooling can be employed, which will greatly improve the market prospects for superconducting separators.

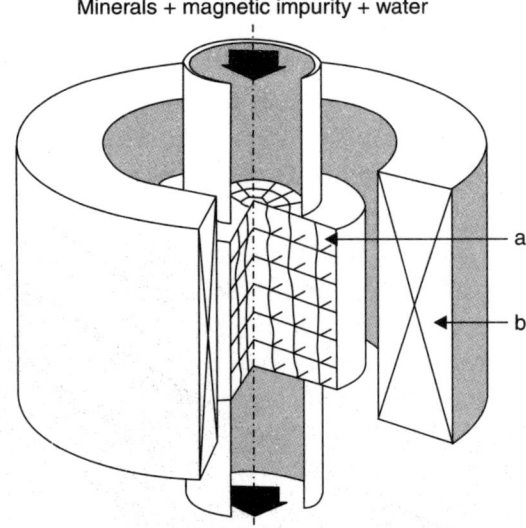

Fig. 10.9 High-gradient magnetic separator (HGMS). $a \rightarrow$ magnetic steel matrix, $b \rightarrow$ superconducting magnetic field coil.

10.2.5.4 Magnetic Levitation Vehicles

Because of the increasing demand for inter-city business and personal travel and the highway and air traffic congestion, high-speed ground transporation has been reexamined. The two principal technology options are the very high speed steel-wheel-on-rail systems (such as Japanese bullet trains and French TGV) and the magnetically levitated (MAGLEV) systems. The former is limited to a maximum speed of about 300 km/h because of the loss of the traction beyond this speed. On the contrary, maglev systems have demonstrated speeds of about 500 km/h. While the investment costs of steel-wheel systems would be lower than those of maglev systems, their maintainence is expected to be higher due to the close track tolerance that must be maintained.

There are two maglev systems: **repulsive** and **attractive**. The Japanese system utilizes the former, while the Germans the latter. The attractive system uses only conventional technology and was once considered to be more economical, than the repulsive one, because no refrigeration is required. If superconductors could be maintained at 77 K using liquid nitrogen, this advantage disappears. The repulsive system has the further advantage (over the attractive one) of a large air gap (6 in. vs. 0.5 in.) which relaxes the design requirements and is thus dynamically stable.

Magnetic levitation using superconducting magnets was first suggested in 1963 by Powell [15], soon after the discovery of type-II superconductors with their implications for carrying large currents. In 1967, Powell and Danby proposed a system [16] using a less expensive conducting guideway at room temperature. Later, they conceived the novel idea of a null-flux suspension system that would minimize the drag force and thus require much propulsion power [17]. Since then maglev systems became objects of considerable study in several countries, most notably Japan, Germany and the UK.

Applications of Superconductivity

As stated above, two different kinds of maglev are competing today's prototype phase of development. The difference is in the way the vehicle attaches to the guideway either attractive levitation forces (electromagnetic suspension) or repulsive forces (electrodynamic levitation). Beginning in the 1970s, research in Germany explored both systems, but more recently only electromagnetic systems are in use [18]. In Japan, both systems were tried and electrodynamic system was preferred [19, 20].

(*i*) **Electromagnetic Suspension.** Electromagnetic systems (EMS) depend on the attractive forces between electromagnets and a ferromagnetic (steel) guideway, as shown in Fig. 10.10 (*a*). Because the force of attraction increases with decreasing distance, such systems

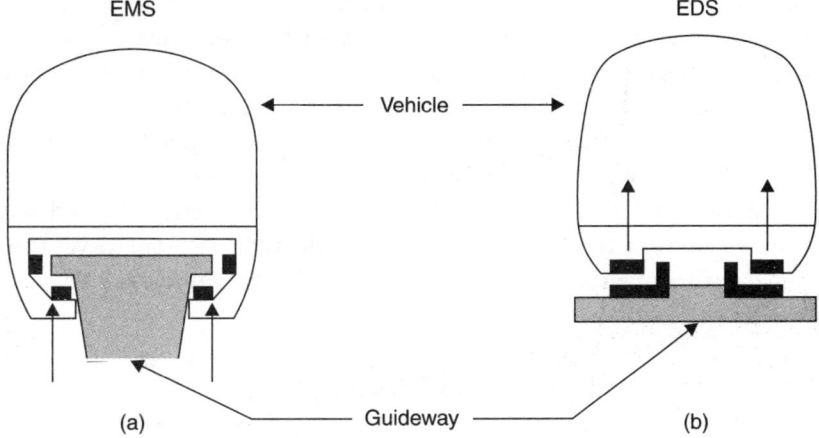

Fig. 10.10 Alternate maglev systems : (*a*) In the **Electro Magnetic System** (EMS), the attractive force between the magnets on the vehicle and guideway draw the vehicle upward and lift it slightly above the guideway. Other magnets are used to achieve later positioning and stability : (*b*) In the **Electro Dynamic System** (EDS), repulsive forces keep the vehicle away from the guideway.

Fig. 10.11 Magnetically levitated train being tested at the Emsland (Germany) test facility.

are inherently unstable and the magnet currents must be carefully controlled to maintain the desired suspension height. Furthermore, the magnet to guideway spacing needs to be small (only a few centimeters at most). On the other hand, it is possible to maintain magnetic suspension even when the vehicle is standing still, which is not true for electrodynamic (repulsive force) systems. In the system of Fig. 10.10 (*a*) a separate set of electromagnets provides horizontal guidance force, but the levitation magnets, acted on by a moving magnetic field from the guideway provide the propulsion force. The German Transrapid TR-07 (Fig. 10.11) is designed to carry 200 passengers at a maximum speed of 500 km/hr. The levitation height is 8 mm and power consumption is estimated to be 43 MW at 400 km/hr.

(*ii*) **Electrodynamic Levitation.** In the EDS system (Fig. 10.10 (*b*)), short-circuited superconducting coils carrying a persistent current are mounted on the either side of the chassis of the passenger vehicle (Fig. 10.12). Stationary room-temperature copper suspension and guidance coils are located on either side of the roadway; when the vehicle moves relative to these, the magnetic flux of the superconducting coils (Ca 3 T) induces currents in the stationary coils, which exert a repulsive electromagnetic force that stably supports and guides the vehicle. An ironless linear synchronous motor with active roadway provides the propulsive force. Stationary copper drive coils embedded in the roadway are sectionally fed with a variable-frequency three phase current to produce a travelling electromagnetic wave; the interaction of this wave with the *d.c.* flux of superconducting magnets in the vehicle generates the propelling force. EDS system has the following advantages:

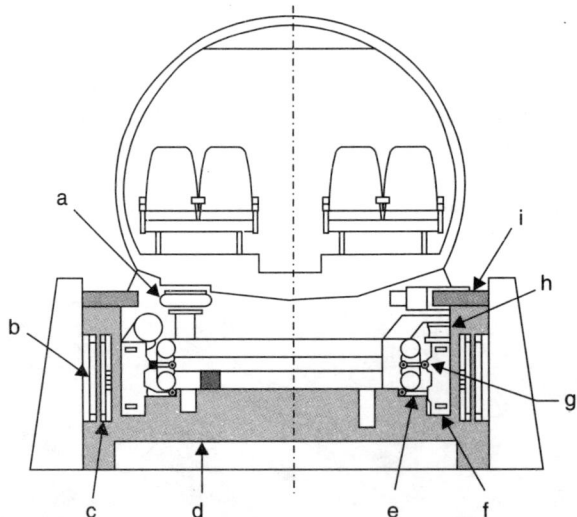

Fig. 10.12 EDS : Cross-section through the Japanese MLU 002 experimental vehicle : (*a*) Air-spring-suspension; (*b*) Propulsion coils; (*c*) Suspension and guidance coils; (*d*) Auxiliary support wheel; (*e*) Chassis; (*f*) Superconducting magnet; (*g*) Magnet fixing; (*h*) Air-spring suspension; (*i*) Auxillary guide wheel. In the EDS, the repulsive forces keep the vehicle away from the guideway.

(*i*) Wide air gap ($\geq$ 100 mm) for suspension, guidance and propulsion.

(*ii*) No need for air-gap regulation.

(*iii*) Problem–free transmission of high power to the high-speed vehicle through contactless propulsion.

(*iv*) No need for magnet power supply on board the vehicle.

The repulsive levitation force is inherently stable with distance, and comparatively large levitation heights (20-30 cm) are attainable by using superconducting magnets. Three different configurations are shown in Fig. 10.13—a flat horizontal conductor, a split L-shape conductor, and an array of short-circuit coils on the sidewalls. Each has its own advantages and disadvantages. The proposed Japanese high-speed maglev system uses interconnected figure-eight (null-flux) coils on the sidewalls as shown in Fig. 10.13 (*c*). The null-flux arrangement tends to reduce the magnetic drag force and thus the propulsion power needed.

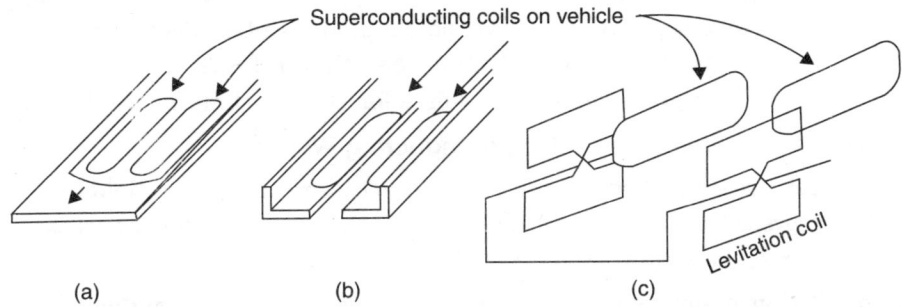

Fig. 10.13 Three different arrangements for EDS (repulsive force) levitation systems : (*a*) Superconducting coils over a flat conducting guideway; (*b*) Superconducting coils with a split L-shape guideway; (*c*) Superconducting coils with short-circuit figure 8 coils connected in null flux arrangement. The proposed Japanese system is similar to arrangement (*c*).

Obviously there are reasons for and against each type of system, and which is best may be determined by operating costs and marketplace factors such as passenger comfort. The high-pitched screech of metal against metal at 300 mph may be so unnerving that no efficiency of operation is sufficient to outweight it. Until real units are built and refined, all such speculations are idle.

Special Considerations for HTSCs

Prototypes of early 1990s are based on LTSCs or conventional electromagnets. The superconducting magnets use NbTi, but subsequent changeover to HTSCs is planned. The role of HTSCs, if any lies in the future. Nevertheless it is not too early to consider certain design features that are implied by HTSCs. In any future application of HTSCs, it will be necessary to give thought to the mechanical aspects of the superconducting medium. Magnetic forces on superconductors must act through the HTSC itself, and therefore these materials must have substantial mechanical strength. Bonding to steel or aluminium frames will of course be done, but it must be remembered that the forces at first on the HTSC, and are then transmitted to the supporting structure. Obviously, it is necessary to give considerable attention to the stresses and strains that will take place in various HTSC applications.

For example, in maglev the train ride may seem very smooth to the passengers because it is gliding above the rails, but there are repeated pulses of magnetic force acting between the car and the guideway. Each pulse places a stress on the HTSC material. Over a sustained period of operation, if tiny cracks develop and the supercurrent consequently decreases, the pounding will get worse, leading to early failure of HTSC components. Similar scenarios can be imagined for any HTSC application in which cycling of a component takes place.

One worry associated with HTSCs is the operating cost attributable to AC losses. The smooth ride experienced by passengers is only possible if the undulations in the guideway are compensated by slightly varying the stand off distance between vehicle and guideway, all at 300 mph. That will show up at the superconducting magnets as an AC variation in the magnetic field of the permanent magnets, and AC fluctuations dissipate energy. We may note that the maglev system is designed to run entirely on DC. Experience at test tracks in Japan and Germany teach us how to design around this problem using Nb-based superconductors, but we have almost no data on AC losses in YBCO or the other HTSCs so far; only preliminary indications that the AC loss problems will be worse for HTSCs than for the LTSCs.

10.2.5.5 Generators and Motors with Superconducting Windings

Coils with iron cores are employed to generate the necessary magnetic fields in conventional motors and generators. This limits the practical maximum field strength than can be obtained from a machine of given volume.

Superconducting magnets allow the generation of very much higher magnetic fields. This means that superconducting machines can be made much smaller for the same output. This can be of crucial importance for certain applications. The cryoscopic complication must naturally be taken into account in every estimate of economy. However, careful calculations have revealed that the low temperatures necessary do not have a serious effects on the competitiveness of rotating machines even today.

The power of an electrical machine with a cylindrical air gap is described by the proportionality

$$P \approx A_{eff} B_r D^2 L^n \qquad ...(10.7)$$

where D is the diameter of the air gap between the rotor and stator windings, L the length of the active part of the machine, n the rotation speed of the rotor, A_{eff} the effective linear current density at the periphery of the air gap (stator winding), and B_r the radial magnetic flux density in the air gap.

Superconducting windings without magnetic iron give far higher magnetic flux densities than in conventional machines ($B_r \approx 1$ T). Omitting of iron also makes it possible to increase the effective linear current density A_{eff}. Consequently, superconducting machines can be built much smaller and lighter than conventional machines with the same power rating. The low electrical losses of superconducting windings (all losses in d.c. windings are thermal) and the low rotor friction losses (reduced size) lead to marked gains in efficiency, especially for higher power gains. A comparison of parameters between superconducting generator and conventional one is given in Table 10.1.

Applications of Superconductivity

Table 10.1 Comparison of prameters between superconducting generator and conventional one

	Superconducting generator	Conventional generator
Rating capacity (MW)	200	$\simeq$ 200
Rating voltage (kV)	20	$\simeq$ 15
Rating power factor	0.9	$\simeq$ 0.9
Rating field current (A)	3000	$\simeq$ 2500
Flux density at the center of stator (T)	$\simeq$ 1.1	$\simeq$ 0.85
Maximum flux density of field windings (T)	5.2	$\simeq$ 2.0
Synchronous reactance ($p.u.$)	0.3	$\simeq$ 1.7
Open-circuit time constant (s)	650	$\simeq$ 6
Generator weight (tons)		
Rotor	15	$\simeq$ 40
Stator	121	$\simeq$ 220
Total	136	$\simeq$ 260
Efficiency (%)	99.35	$\simeq$ 98.3

Preferred types of machines are *d.c.* and synchronous devices, because superconducting *d.c.* windings (without hysteresis losses) can be used here. An outline of the system is shown schematically in Fig. 10.14. It consists of turbines driving a rotor, a motor turning the rotor at

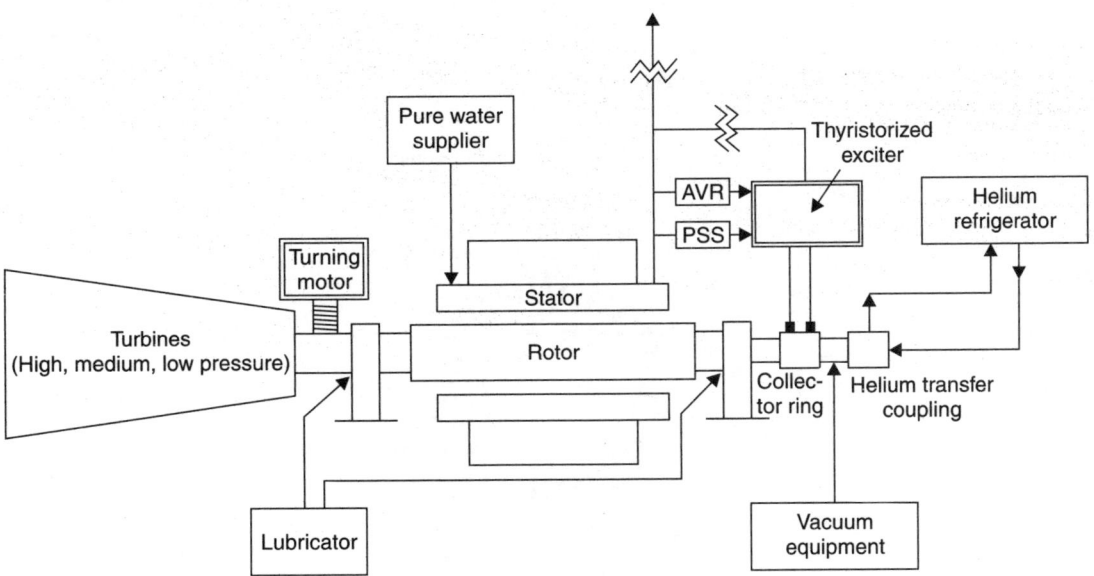

Fig. 10.14 Superconducting generator system (outline).

low speed (~ 80 rpm) while the field windings are cooled down or warmed up, pure water equipment for cooling the armature windings, a helium refrigerator, a vacuum pump for evacuation of the rotor, and so on. The latter two pieces of equipment are unique to a superconducting generator. In addition, static excitation equipment for charging the field windings through the collector rings is used. As a stator, air gap windings are adopted to allow a high magnetic flux density.

The typical structure of a superconducting generator is shown in Fig. 10.15. One of the features of a superconducting generator is that a multicylindrical rotor with thermal insulation by vacuum enables cooling down to 4.2 K. Superconducting field windings are connected with a rotating shaft through a torque tube. A cylinder for radiation shielding is mounted between the field windings and the damper. The damper serves as protection against rotor vibration and shielding against higher harmonics arising from armature reaction. Liquid helium is supplied through a helium transfer coupling.

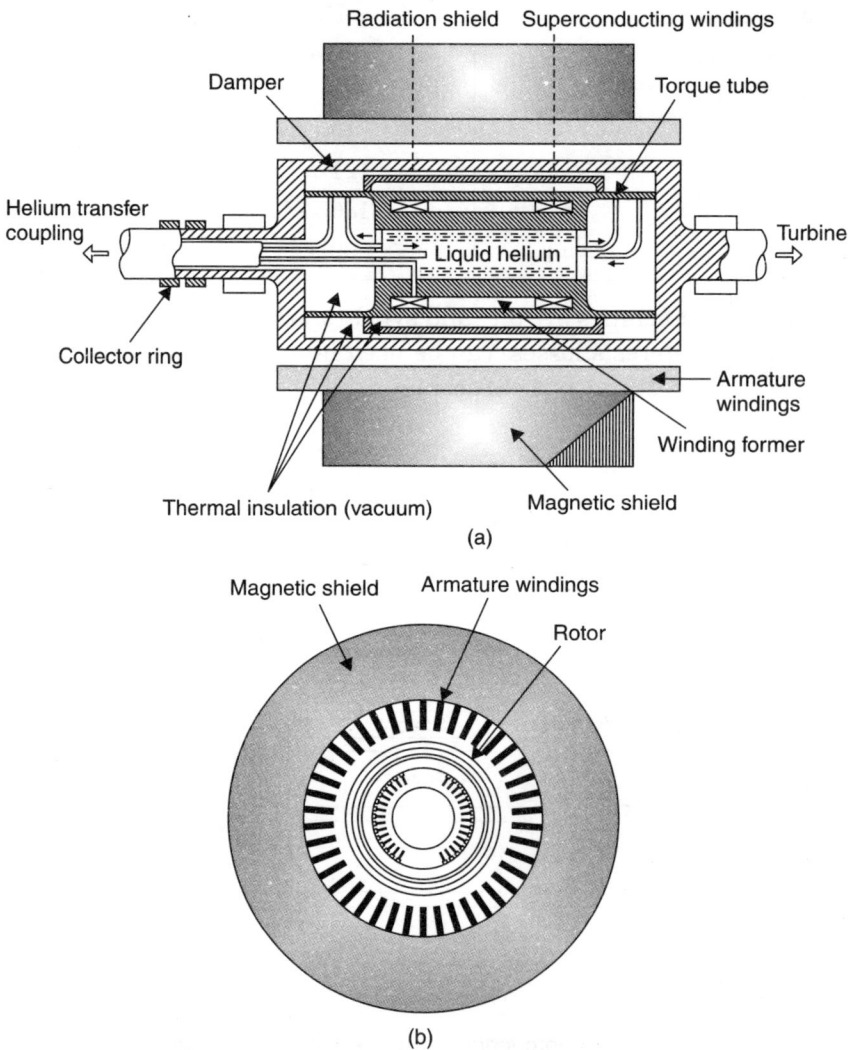

Fig. 10.15 (*a*) Conceptual structure of a superconducting turbine generator, and (*b*) schematic illustration of a generator cross-section.

Preferred types of machines are *d.c.* and synchronous devices, because superconducting *d.c.* windings (without hysteresis losses) can be used here. Efforts are being made worldwide to create large superconducting synchronous generators for use in power plants. The present design has only the rotating field winding made of superconducting material, with the stationary three phase winding built of water cooled stranded conductors that contain fine copper wires. In comparison with conventional generators, the volume and weight are reduced by about a factor of two for large machines (500 – 1000 MVA), and the efficiency is increased from 98.85% to 99.5%. The economic breakeven point for NbTi conductors is in the range of 600 – 800 MVA.

Fig. 10.16 shows the concept for a rotor with a two-pole superconducting field winding, whose conductors are laid in the grooves of a nonmagnetic steel cylinder. Conductors for field windings are generally wound inside nonmagnetic slots and supported wedges to minimize wire motions due to centrifugal and electromagnetic force as shown in Fig. 10.17. The conductors are cooled down with liquid helium. Low density helium heated up with ac losses, and so on, flows into helium vessel on the inner radius side of the rotor by a large centrifugal force, while low temperature high-density liquid helium flows into the slots and subsequently, flow of liquid helium will occur in slots. Therefore, it is important to design appropriately cooling channels inside the slots so that the conductors may be sufficiently cooled. The most important project currently is the Japanese Super GM program, in which Japanese industry and utility companies are developing three rotors with various superconducting field windings for a model generator of the 70 MW class [21].

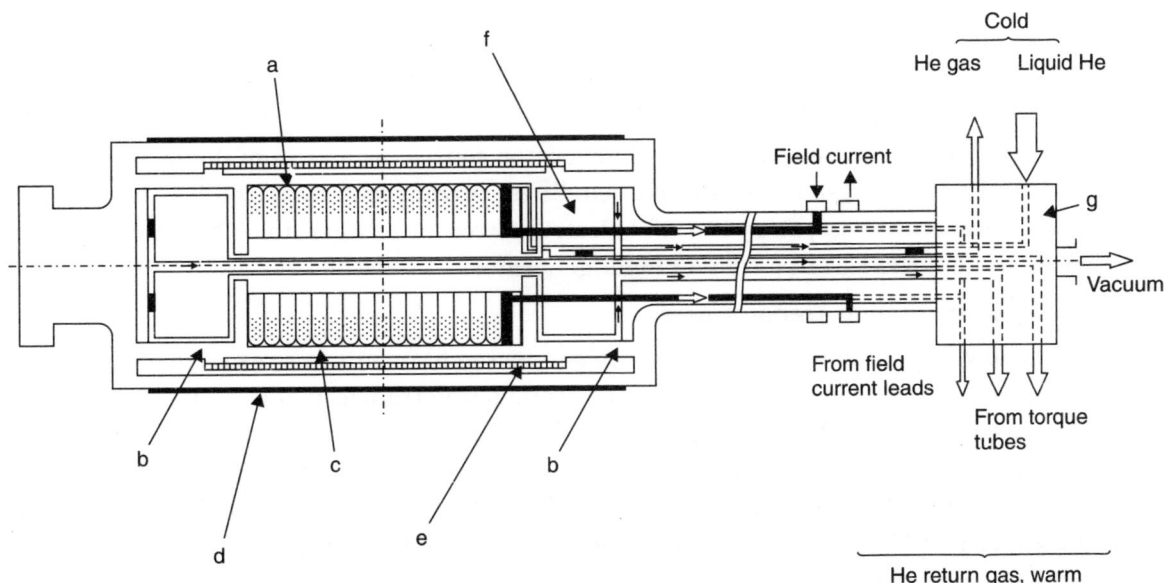

Fig. 10.16 Diagram of rotor with field winding $a \rightarrow$ superconducting field winding; $b \rightarrow$ torque tube extension; $c \rightarrow$ torque tube; $d \rightarrow$ warm damper shield; $e \rightarrow$ thermal radiation shield; $f \rightarrow$ vacuum; $g \rightarrow$ helium transfer unit.

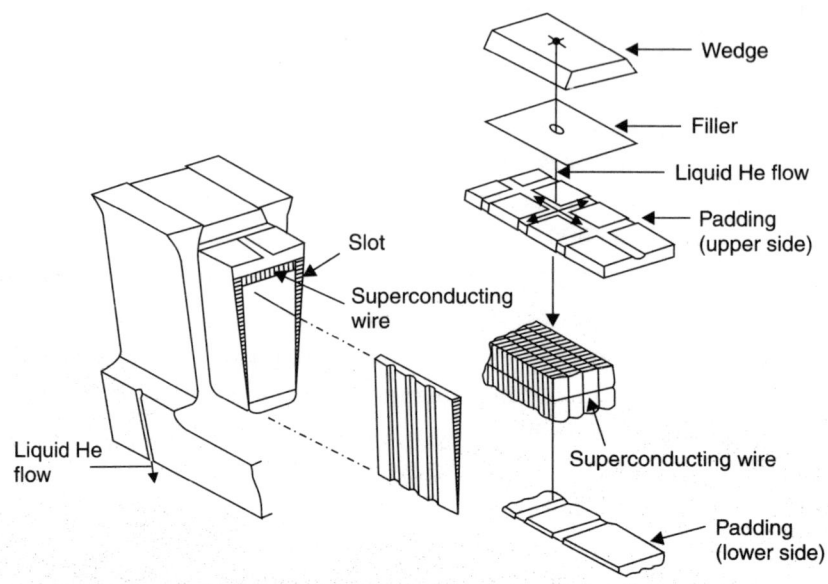

Fig. 10.17 Structure inside a slot.

A field winding with HTSCs and liquid nitrogen cooling will shift the economic breakeven point into the 200 – 400 MVA power range. Furthermore, a completely superconducting machine (with superconducting three phase winding) would also be conceivable with HTSCs.

In large *d.c.* machines, the use of superconducting windings is limited by problems in transmitting current from the stationary to rotating parts (commutator in heteropolar machines, very high currents in homopolar machines).

Energy losses in a superconducting generator consist primarily of Joule loss (copper loss) in armature windings, ac loss in field windings, current lead loss, iron loss in magnetic shield body, mechanical loss, stray loss and electric power for refrigerator. The total losses are approximately half the loss in a conventional generator.

10.2.5.6 Superconducting Transformers

Advances in low-loss *a.c.* superconductors with submicron filaments in a resistive Cu Ni matrix have given rise to increased interest in the superconducting transformer (Fig. 10.18). For more efficient removal of hysteresis losses, the iron core is held at room temperature, cooled with a liquid such as oil. The high-voltage and low-voltage windings are placed in a thermally insulated tank that surrounds the iron core; it is made of glass-fibre-reinforced plastic in order to block eddy current losses. The windings are cooled in a bath of liquid coolant, which is also to perform part of the function of the dielectric and so must be subcooled to prevent the formation of gas bubbles, which would result in undesirable partial discharge. Other vital components are the low-voltage and high-voltage bushings [22].

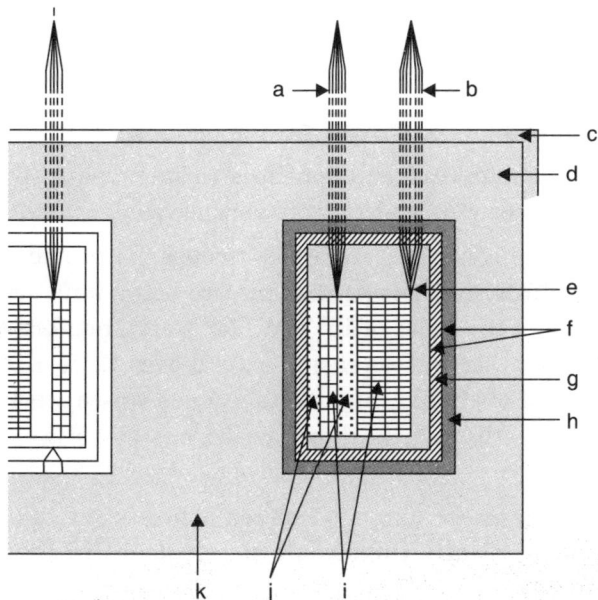

Fig. 10.18 One phase superconducting transformers. $a \to$ low-voltage connection; $b \to$ high-voltage connection; $c \to$ oil tank; $d \to$ iron yoke; $e \to$ liquid coolant + electrical insulation; $f \to$ coolant tank (glass-fibre-reinforced plastic); $g \to$ thermal insulation; $h \to$ superconducting low-voltage and high-voltage windings; $j \to$ electrical insulation + coolant liquid; $k \to$ iron core.

The advantages of superconducting transformers are :

(*i*) Marked reduction in 50 Hz winding losses, resulting in higher efficiency.

(*ii*) Enhanced magnetic flux due to low winding losses permitting a reduction in the cross-section of flux-guiding iron and thus a substantial decrease in the weight and hysteresis losses of the iron core.

(*iii*) Short-circuit current limiting through the transition of superconducting windings to the resistive normally conducting state resulting in simpler mechanical design of the winding (milder forces) and thus leading to higher current densities and lighter weight windings.

Recently a number of small single-phase model transformers rated at 100 kVA have been built by using submicron Nb Ti filament conductors and have demonstrated the advantages of superconducting technology.

The benefits of superconductivity become much greater if the helium cooled windings of metallic superconductors are replaced by liquid-nitrogen-cooled HTSC-windings. In a study, the following values have been obtained for a 1000 MVA power transformer with windings made of a multicore HTSC wire with 5 μm filaments (critical current density 10^5 A/cm^2) in a matrix of Ag-Al alloy (conventional figures are given in parentheses for comparison).

Weight of iron core	91 t (280 t)
Weight of windings	5 t (58 t)

Iron losses	100 kW (300 kW)
"Copper" losses	200 kW (2000 kW)
(converted to 300 K)	

10.2.5.7 Superconducting Cables for Power Transmissions

The energy consumption of mankind will continue to increase in future. Great efforts are being made to find new sources of power for this very reason.

As the amount of power consumed increases so does the problem of power transmission. In order to obtain higher efficiencies power stations are being built with ever greater outputs per unit. This power output of the order of 10^3 MW (10^9 watt) then has to be transmitted to the consumer. Here the interconnection often has to extend over large distances. This is achieved today largely with the use of overhead, high-voltage lines which work at a voltage of 380 kV and more. The subsidiary disribution is then carried out at voltages from 110 kV down to 20 kV.

The increasing consumption naturally makes it necessary to extend the transmission grid. Here it is becoming virtually impossible particularly in the great conurbations to continually add overhead power lines. The effect on the landscape and particularly on land use of high voltage transmission lines makes it appear ever more necessary to transfer power transmission to underground cables. A great deal of development work and investment will be required in this field for the future.

Naturally one thinks too of the exploitation of superconductivity for this purpose. A conductor with no ohmic resistance appears at first glance, to be virtually ideal for power transmission. There is no doubt that superconductivity basically offers a new and attractive possibility.

High power superconducting cables are of interest for underground power transmission near densely populated areas. Because transmission distances are relatively short (10 – 100 km), three phase cable is preferred. Fig. 10.19 is a schematic of such a cable with HTSC cores and liquid nitrogen cooling. The flexible, hollow cable cores are of coaxial design. The actual phase conductors, made of superconductors (wires, tapes) in helical form are enclosed in electrical insulation (wound tapes of polypropylene or polyethylene) and surrounded by a return conductor in which the *a.c.* current flows in the opposite direction to the phase conductors. As a result no *a.c.* electromagnetic fields are produced outside the cores. A rigid or flexible double-walled thermal insulation layer surrounds the flexible cable cores; liquid coolant flows through the inner tube, while the outer one is at ambient temperature.

The practicability of superconducting cables has already been demonstrated in a number of experimental cables with helium cooled metallic superconductors. The largest project was a two phase model cable of 115 m length with a power capacity of 1000 MVA at a voltage rating of 138 kV [23]. However, cables with helium cooling become economically competitive with conventional water-cooled oil cables only when carrying 4000 MVA and up. A schematic representation of the cross-section of helium cooled superconducting cable is shown in Fig. 10.20. Estimates indicate that the economic breakeven point for nitrogen cooled HTSC cables may be as low as 1000 MVA.

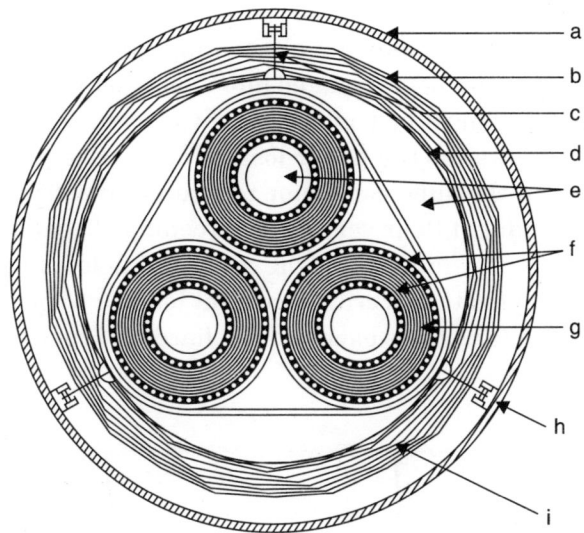

Fig. 10.19 Liquid nitrogen cooled three-phase cable with HTSCs
$a \to$ outer steel pipe; $b \to$ super insulation; $c \to$ suspension; $d \to$ nitrogen tube; $e \to$ nitrogen coolant; $f \to$ superconductors; $g \to$ electrical wrapped-tape insulation; $h \to$ suspension track; $i \to$ vacuum.

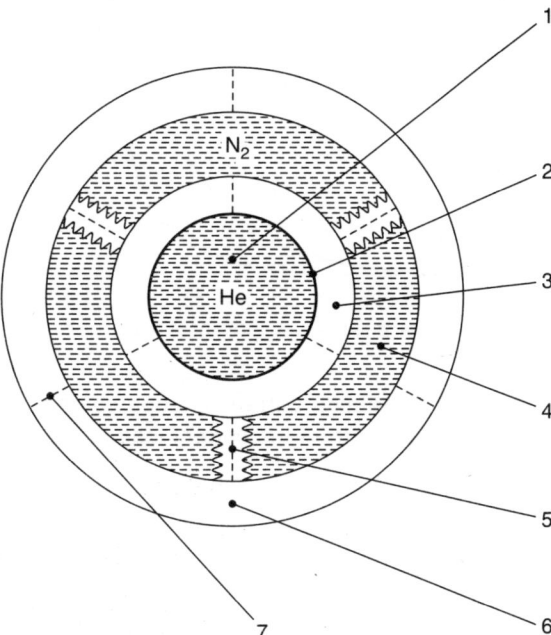

Fig. 10.20 Schematic representation of the cross section of a superconducting cable. 1 liquid He; 2 Nb conductor on a carrier tube; 3 vacuum insulation; 4 liquid nitrogen; 5 electrically insulating supports; 6 vacuum insulation; 7 support construction with low thermal conductivity.

Alternating Current Cables

Certain residual losses have to be accepted for alternating current cables. They result from the interaction of the electrons that are not correlated into Cooper pairs with the alternating electric field. The heat that is produced must be transported away by the coolant at 4 K. This requires 500 times as much electric power as the power loss involved. Every effort must, therefore, be made to keep these losses in the superconductor to a minimum. Type-I superconductors are very suitable for this on account of their very large screening effects. Fig. 10.21 illustrates the power losses per m² surface as a function of the magnetic field at the surface. If maximum losses of 0.025 W/m² are accepted then a critical current load for a conductor 10 cm in diameter is 5500 A for Pb and 16000 A for Nb. As expected the hard (or type-II) superconductors do not offer any particular advantages.

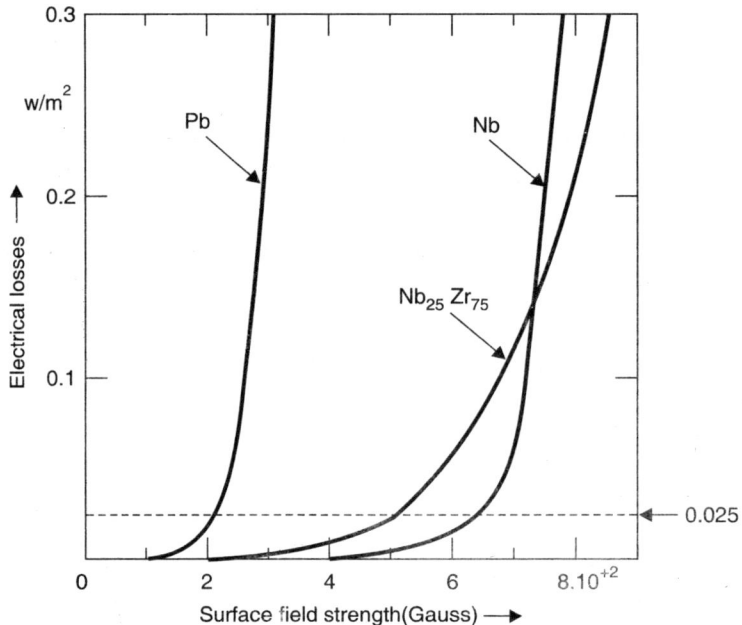

Fig. 10.21 Electrical losses for alternating current in superconductors as a function of the surface magnetic field strength of the current load.

Many constructional suggestions involve the use of rigid tubes, which can only be manufactured in relatively short lengths and which have to be joined in vacuum-tight manner with good contact between the superconductors. Klaudy [24] suggested ring corrugated tubes instead of rigid tubes, which avoids these constructional difficulties. Such corrugated tubes can be constructed in lengths upto 1500 m and wound onto cable drums. Distancing pieces of poor thermally conducting materials guarantee the necessary centering (Fig. 10.22).

The innermost tube through which liquid helium flows contains many three phase conductors suitably arranged. The space between tubes 2 and 3 is cooled with liquid nitrogen. The space between tubes 1 and 2 and tubes 3 and 4 are for thermal insulation.

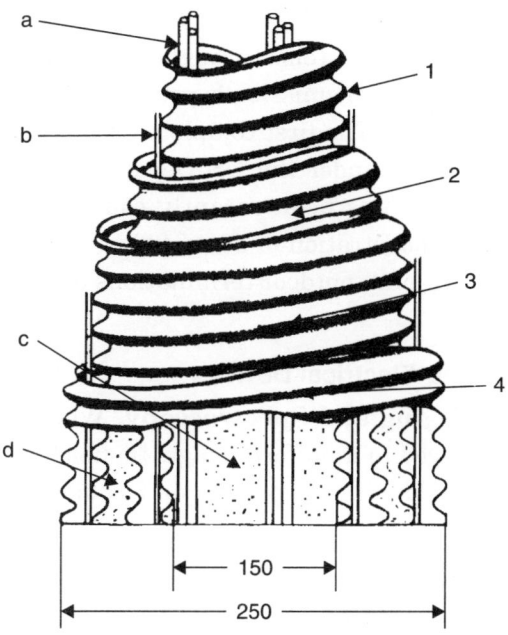

Fig. 10.22 A superconducting cable of corrugated tubes. 1, 2, 3 and 4 corrugated tubes; $a \to$ three-phase conductor, superconductor Nb or Pb; $b \to$ distance pieces of poorly thermally conducting materials; $c \to$ liquid helium $d \to$ liquid nitrogen.

Summarizing, there are four main components of a superconducting cable system:

(i) **Electrical core.** Which contains the current carrying conductor, the dielectric insulation and the grounded shield;

(ii) **Cryogenic envelope.** Which isolates the 'ambient' external surroundings from the low temperature superconducting electric core;

(iii) **Terminations.** Which provide a thermal, current and voltage transition from the cold core to the outside power system at the cable end;

(iv) **Refrigeration equipment.** Which conditions the cryogenic fluid used to cool the cable.

There are relevant differences between the electrical cores and terminations of the dc and ac cables, however, the cryogenic envelope and the cooling equipment are similar for both types, at least in their general design.

Superconducting transmission cables can provide two to five times the current transmission capability of a conventional cable with the same cable diameter.

Applications that benefit from the increased capacity are retrofitting into existing cable pipe or ducts; reducing the 'trenching' effort required for a new cable run; one-to-one relocation of overhead lines to underground and reduced transformation steps in urban feeders (while providing the same power). It is anticipated that the transmision efficiency can be improved significantly over conventional lines at comparable loading.

Since the present transmission systems throughout the world are mostly ac, the current research on superconducting power transmission is primarly focused on ac applications with a number of HTSC ac cable programs. The first complete prototype cable system, developed by Pirelli Cavi and Sistemi and EPRI under the US Department of Energy Superconductivity Partnership Initiative (SPI) [25] has been recently completed and tested. This activity culminated in the successful demonstration of 50 m 400 MVA 115 kV pipe-type HTSC cable system (comprising HTSC cable, joint, outdoor terminations and LN_2 cooling plant) capable of carrying 2 k A_{rms} ac.

10.2.5.8 Novel Superconducting Electrical Devices

(i) Magnetic Resonance Imaging and Spectroscopy. When a nucleus with a finite spin number, such as a proton is placed in a uniform magnetic field the nuclear magnetic moment precesses about the magnetic field with the precession frequency (Fig. 10.23 (a))

$$v_0 = \frac{\gamma B_0}{2\pi} \quad \ldots(10.8)$$

where γ is the magnetogyric ratio and B_0 is the magnetic field strength. When the transverse components of the moment reaches the equilibrium state, they are canceled out and the total magnetization vector M is directed along the magnetic field.

To tilt the vector M away from the equilibrium state, an oscillating transverse magnetic field B_1 with a frequency v_0 is applied to the magnetic moment. If the moment is observed from a coordinate system revolving at the frequency of v_0, the moment rotates about the x'-axis (Fig. 10.23 (b)). The B_1 field is applied as a radio-frequency (RF) electromagnetic field pulse. The pulse that tips the magnetization to the Y' axis is called a 90° pulse. On the other hand that which inverts the magnetization to the z'-axis is called a 180° pulse.

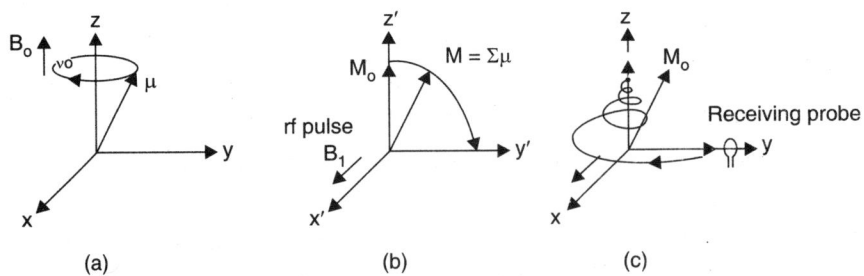

Fig. 10.23 Nuclear magnetic resonance : (*a*) precessional motion of a magnetic moment in a magnetic field; (*b*) mutation of the magnetic moment by the RF pulse (motion is observed in rotating coordinate system); (*c*) free induction decay after the RF pulse.

After the RF pulse, the induced magnetization decays to the initial state, the recovery process is termed **free induction decay** (FID). The FID is characterized by two relaxation times : a longitudinal time, T_1 and a transverse relaxation time, T_2. T_1 is the recovery time for the induced magnetization M_z to return to its equilibrium value as a result of spin-lattice interactions. T_2 is the decay time for M_x and M_y caused by spin-spin interactions. During the FID, an oscillatory voltage is induced in a probe on the X- or Y-axis (Fig. 10.23 (c)).

In MRI, a small magnetic field gradient is generated by a set of gradient coils, i.e., G_x, G_y and G_z. The FID signal in the presence of field gradients gives a cross-sectional distribution of the Fourier transformed data of the magnetization of the human body. Thus, if the FID signals are observed sequentially with a stepped change in the field gradient, the body image can be constructed in Fourier space. (Whole-body imaging employs magnetic fields ranging from 0.15 to 1.5 T. Image quality improves with increasing magnetic flux density). The set of FIDs is then inversely Fourier transformed to obtain a cross-sectional image of the body. Two types of gradient field sequences are possible [26].

(i) **Multiple-angle Projected Method.** The angle between the field gradient and the X-axis is changed successively (Fig. 10.23 (a)). The FID signal is obtained for each field gradient.

(ii) **Fourier Transform Method.** The gradient G_y is applied for a time t_y, which is then changed to the X-direction by the G_x gradient. The FID is observed under G_x slope. A set of FID signals, obtained for different t_y values gives a magnetization image in the Fourier space (Fig. 10.23 (b)). The data acquisition time of the FID signals ranges from a few seconds to a few minutes.

The magnetic moment, T_1 or T_2 can be obtained separately by appropriate pulse sequences consisting of 90° and 180° pulses. The sequence is chosen depending on the information required for medical diagnosis. The standard magnetic imaging matrix consists of 256 × 256 elements, while that for high resolution mode is 512 × 512. MRI for medical applications employs a 32-bit mini computer, which enables reconstruction of the real part of the image of the human body in 2 to 3s. A cross-sectional image of a human head obtained by using an MRI system is shown in Fig. 10.24.

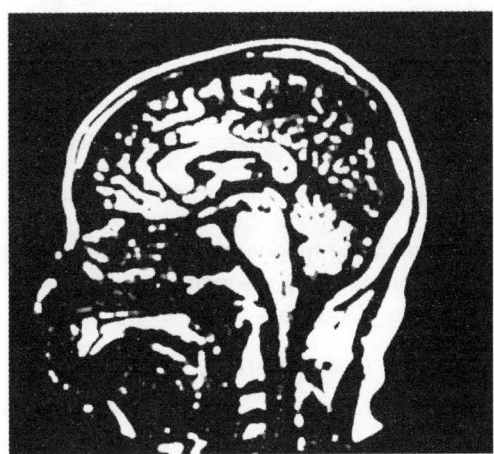

Fig. 10.24 Cross-sectional imaging of the human head.

A cross-sectional view of a superconducting magnet and a cryostat for MRI is shown in Fig. 10.25.

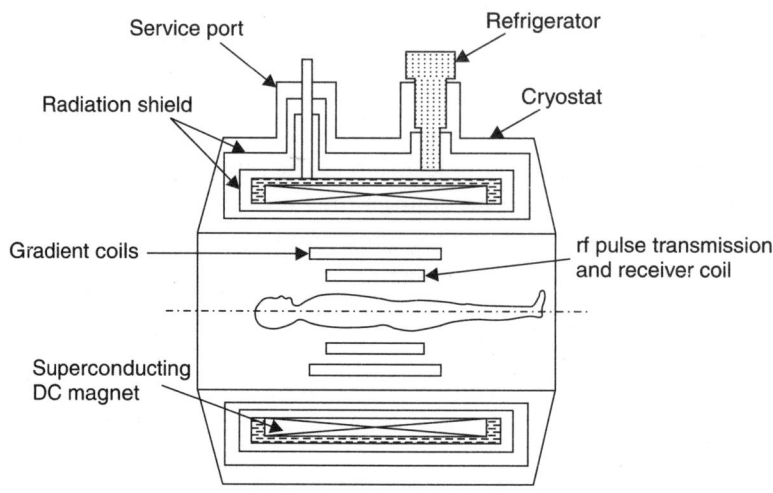

Fig. 10.25 Cross-sectional view of a superconducting magnet and a cryostat for MRI.

A *dc* superconducting magnet is often used because it can generate a high magnetic field, such as 0.35 to 2.0. Image quality improves with increasing magnetic flux density. For this reason, some 85% of currently installed MRI systems (Ca. 18000) are equipped with superconducting magnets operating at flux densities between 0.5 and 2 T. The magnets are of cylindrical design (solenoids), 2 – 3 m long with an internal bore about one meter in diameter in which the patient is placed (Fig. 10.26).

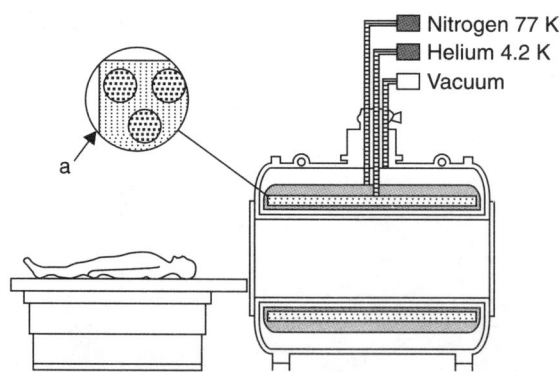

Fig. 10.26 Superconducting magnet providing magnetic flux densities ≥ 0.5 T for magnetic resonance imaging (MRI)
(*a*) Windings of Cu-NbTi multifilament conductors.

Detailed specifications for the magnet are :

(*i*) **Magnet Field Strength.** A 0.35 to 2.0 T central field is required to obtain a high signal-to-noise ratio. Such a field is obtained with a superconducting *DC* magnet as stated above.

(ii) **Field Homogeneity.** Field inhomogeneity is less than 20 ppm in a 0.5 m diameter spherical volume (DSV). The high spatial field homogeneity better than ± 10 ppm in a sphere 0.5 m in diameter required for these applications is achieved by using multicoil configurations and shim coils that surround the patient, like the high frequency and gradient coil systems.

(iii) **Field Stability.** Field decay rate is as low as 0.1 ppm/h, which means that the field does not change considerably during data acquisition. The magnets exhibit high field stability over time ($\Delta B/B < 10^{-7}$ h^{-1}) due to the persistent-current operation that is possible with superconducting coils [27].

At present, all these superconducting magnets are constructed with NbTi metallic multicore conductors. A changeover to HTSCs would lower capital cost by about 15%. In addition, cooling costs would be cut by $100000 to $200000 for a five-year service life.

(iv) **Helium-Consumption.** The helium consumption rate is as low as 0.1 to 0.4 L/h. Liquid helium lasts for several months in the cryostat.

The specifications for proton MR magnet are summarized in Table 10.2.

Table 10.2 Specification of a magnet for proton MRI

Property	Value
Field strength	0.35 – 2.0 T
Field homogeneity	20 ppm/0.5m DSV
Field stability	0.1 ppm/h
Room temperature bore	1.0 m
Helium consumption	0.1 – 0.4 L/h
Weight	7.5 – 30 t

In the proton MRI, high resolution, precise, and high-speed imaging are being sought. In near future the following systems are foreseen (Table 10.3).

Table 10.3 NMR properties for some nuclei

Nuclei	Resonance frequency at 1T (MHz)	Natural adundance(%)	Relative sensitivity
^1H	42.6	99.98	100
^{13}C	10.7	1.1	0.02
^{31}P	17.2	100	8.3

From table 10.3, it is obvious that other interesting nuclei for medical applications are ^{31}P and ^{13}C. We can see from the table that their concentration in living tissues and their NMR sensitivities are much lower than the concentration of proton nuclei. Therefore, a higher field, such as 2 to 4 T, is necessary to obtain a good signal-to-noise ratio. Several 4 T whole-body MRI systems have already been constructed which aim at obtaining magnetic resonance spectroscopy (MRS) of ^{31}P in the human body [28]. Relative concentrations of the adenosine triphosphate (ATP), phosphorcreatine (PCr), and inorganic phosphate (Pi) give a patient's energy metabolism condition.

Since the natural abundance of ^{13}C is only 1.1%, a labeled compound must be used for imaging. The ^{13}C nucleus is useful for tracing metabolic products and to obtain the metabolic pathway of a particular carbon atom in a human body.

10.2.5.9 Magnetic Systems for Magnetic Confinement Fusion Reactors

Very great efforts are being put into controlled nuclear fusion *i.e.*, the fusion of 4 hydrogen nuclei to a helium nucleus, such as takes place spontaneously in the hydrogen bomb, so that it takes place in a controlled manner. The energy involved in this process is very large since hydrogen is available on earth in virtually unlimited quantities. This thermonuclear process could be a future energy source. At the moment the investigations of fusion are all in the field of basic research. The difficulties that must be overcome to achieve controlled fusion are enormous. In order to set up a process that will yield excess energy it is necessary to heat the hydrogen gas to at least a few ten million degrees. This hot plasma, that practically only consists of hydrogen nuclei (the isotope tritium with a nucleus made up of a proton and two neutrons *i.e.*, $_1H^3$ is employed) and electrons can naturally not be enclosed in material vessels. However, since they are charged particles their paths can be deflected in a magnetic field. For this reason it is possible, in spite of their high velocity, to hold the particles in a reaction volume by the use of sufficiently high magnetic fields of suitable geometry. The magnetic fields required are so high that it is certain that they can only be generated economically with superconducting magnets. Obviously, magnetic plasma confinement is a promising route to the economic utilization of thermonuclear fusion energy. Two toroidal magnetic field configurations, the Tokmak (Fig. 10.27 (*a*)) and the Stellarator (Fig. 10.27 (*b*)) appear especially suitable. Commercial fusion reactors will require Superconducting magnet systems to achieve the required power-plant efficiency.

The coil layouts of both Tokmak and Stellarator are shown in Fig. 10.27. In the axially symmetrical Tokmak configurations the magnetic surfaces which as surfaces of constant pressure contain the helical field lines, are established by superposing the toroidal main field and a poloidal magnetic field. The poloidal field is generated chiefly by an electric current induced in toroidal plasma and also by annular internal transformer coils and external poloidal field coils. In the stellarator, the magnetic surfaces are generated solely by currents in external coils so that a purely modular design is possible for the coil system. Magnetic flux densities of 5 – 6 T are required on the ring-shaped plasma axis in the tokmak configuration corresponding to maximum magnets fields of 12 T at the superconducting coils which can be attained only with Nb_3Sn conductors. The stellarator makes less stringent requirements on the magnetic field, which can be satisfied with NbTi conductors.

The technical feasibility of large magnets for fusion has been demonstrated in a number of large superconducting tokmak systems, *e.g.* LCT project at Oak Ridge National Laboratory

Applications of Superconductivity

in USA a toroidal configuration comprising six D-shaped magnets; and the TORE-SUPRA at the Centre d' Etudes Nucleaires Cadarache in France [29]. The largest international project is the development of a magnet system for International Thermonuclear Experimental Reactor (ITER). This will be a tokamak-type with D-shaped toroidal field coils measuring 18×11 m and poloidal field coils upto 25 m in diameter; the maximum flux density will be around 13 T. Other magnet configurations include the stellarator project (Wendelstein $W > -X$) at the Max-Planck Institute for Plasma Physics, Garching, Germany and Large Helical Device (LHD) at the National Institute for Fusion Science (NIFS) in Japan.

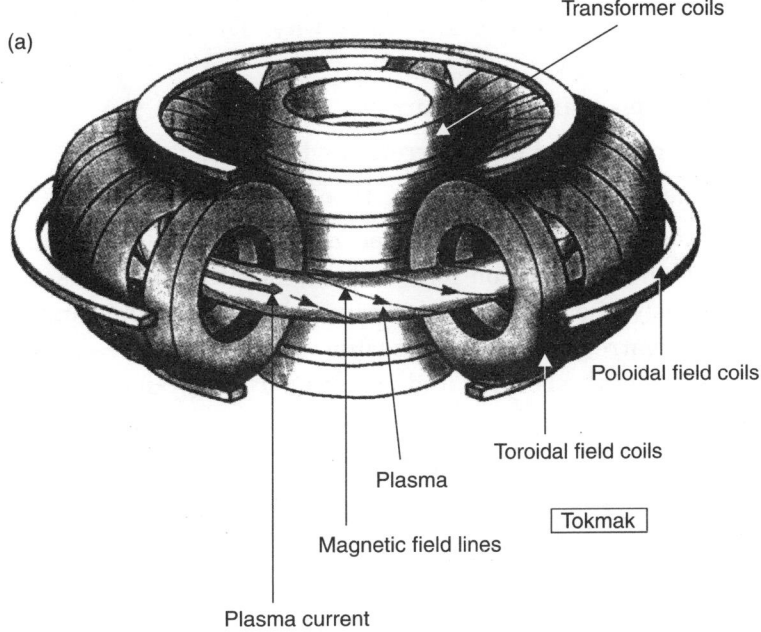

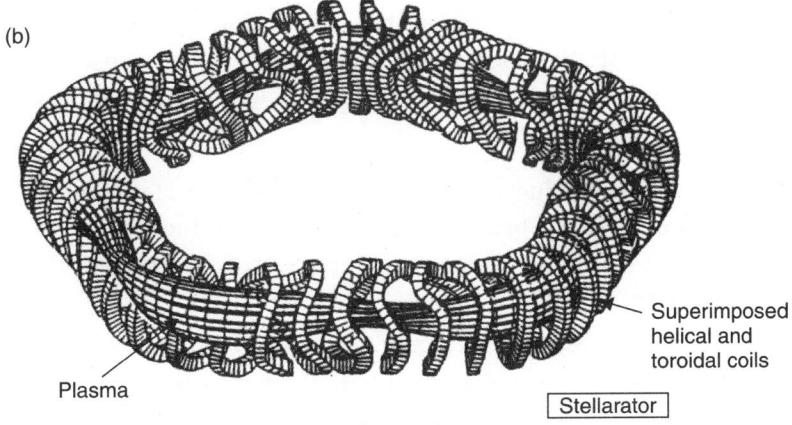

Fig. 10.27 Principal magnet configurations for magnetic-confinement fusion reactors (*A*) Tokmak; (*B*) Stellarator.

The state of the art is based exclusively on metallic superconductors. There would be a variety of benefits in changing over to HTSCs. First, the overall efficiency of a fusion power plant could be increased by 2 – 4%. Simple cooling would result in higher reliability and availability as well. However, this application would impose the most severe requirements (magnetic fields, forces) on the HTSCs.

10.2.5.10 Bearings

A frictionless bearing is every mechanical designer's dream, and a bearing in which the two surfaces never make contact is close to that ideal. Magnetic bearings achieve mechanical separation and when other dissipative factors are minimized, they are extremely good bearings, allowing rotational speeds not attained any other way. Three different types of magnetic bearings are possible :

(i) Coil magnet systems using direct current,

(ii) Electrodynamic or induced eddy-current devices, and

(iii) Passive Meissner-effect bearings using type-II superconductors.

A simple magnetic bearing consists of a permanent magnet rotor suspended between two superconductors Magnetic repulsion forces cause the rotor to be suspended in mid air, when it is able to spin freely; the only friction is caused by aerodynamic and magnetic drag. If the rotor tries to drift off centre, a restoring force due to flux pinning restores it. This is known as the **magnetic stiffness** of the bearing and is an important design parameter.

To increase the magnetic stiffness in practical bearing, a superconducting cylinder that acts as a radial bearing might be added. However, this increases the magnetic drag on the rotor. Superconducting bearings need to be controlled in five directions : up, down, side-to-side pitch and yaw. Efforts to increase magnetic stiffness through clever design and improved materials, are underway at several laboratories. Although a low-stiffness bearing is more forgiving of an out-of-balance rotor, most applications require a larger stiffness than is presently available.

Lift force is not generally considered to be a problem. Actually magnetic pressure is a better parameter for characterizing bearings, because it does not depend on the area of the magnet of the superconductor. Pressure of 10^5 N/m^2 have been reported with a permanent magnet and a superconductor, which is probably large enough for magnetic bearings.

Hybrid Bearings

At the Texas centre for superconductivity at the University of Houston (TCSUH), advances have been made toward using hybrid bearings made of superconductors and permanent magnets [30]. This concept exploits magnet-to-magnet repulsion to **support** the load and use a superconductor to **stabilize** it.

The repulsive force between two like magnets is akin to a ball sitting atop the crown of a hill : any slight perturbation in any direction will start it rolling. Earnshaw's theorem [31] shows that a system composed only of static forces with an inverse square law between the system components, such as in a system with permanent magnets and paramagnets, is unstable.

Applications of Superconductivity

Denoting positive constants by α and β, the force between two dipole magnets M_1 and M_2 is $\alpha M_1 M_2$, and the stability parameter is $-\beta M_1 M_2$; the negative sign indicates that it is unstable. Superconductors, by contrast, are very good diamagnets and are therefore to provide a negative value for one M_1, thus yielding a positive stability parameter.

The TCSUH configuration is called '**Hybrid Superconductor-Magnet Bearing**' (HSMB) and is conceptually illustrated in Fig. 10.28. The "crown of a hill" potential surface has a dimple in it (left) because a superconductor is placed between the mutually repelling south poles of two permanent magnets (right). This combination provides the necessary stability.

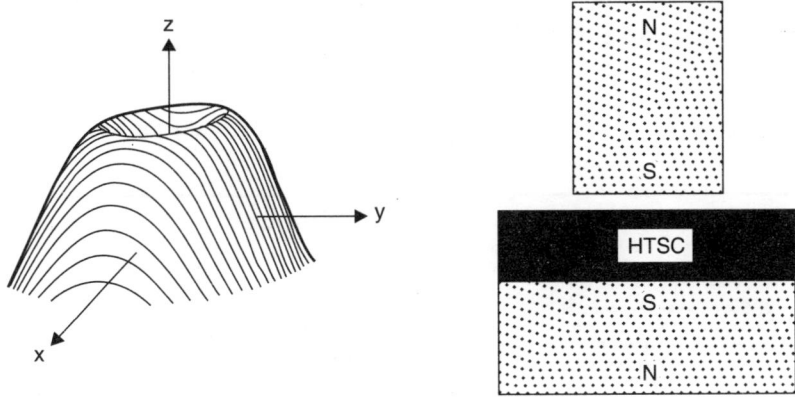

Fig. 10.28 Force potential surface associated with hybrid-superconductor magnet.

In actual practice, several different bearings can be constructed this way. Fig. 10.29 shows some of them: at right is a thrust bearing based on magnetic repulsion, in the centre is a journal bearing; and at left is a pair of attracting magnets to assist the trust and journal bearings.

The next step is to tailor the materials processing methods (for materials used in bearings) toward meeting specific device parameters. This may mean sacrificing some lifting power in order to gain mechanical strength or resistance to damage from vibration. In earlier embodiments, superconducting levitation bearings had limited applications because they simply could not lift the load required in many situations. However, in the TCSUH configuration, the superconductor need only provide the **stability**, whereas the **lifting** is provided by regular magnets. This is very similar to the concept employed since about 1900 in centrifuge bearings in which a set of permanent magnets provides a carefully balanced lift force and a small mechanical pivot bearing provides stability. Using this concept opens the door to a much wider range of practical applications. As in any potential use of HTSCs, the trade-off to be made between the cost of refrigeration and the benefit of better overall system performance in this case meaning lower frictional losses in the bearings.

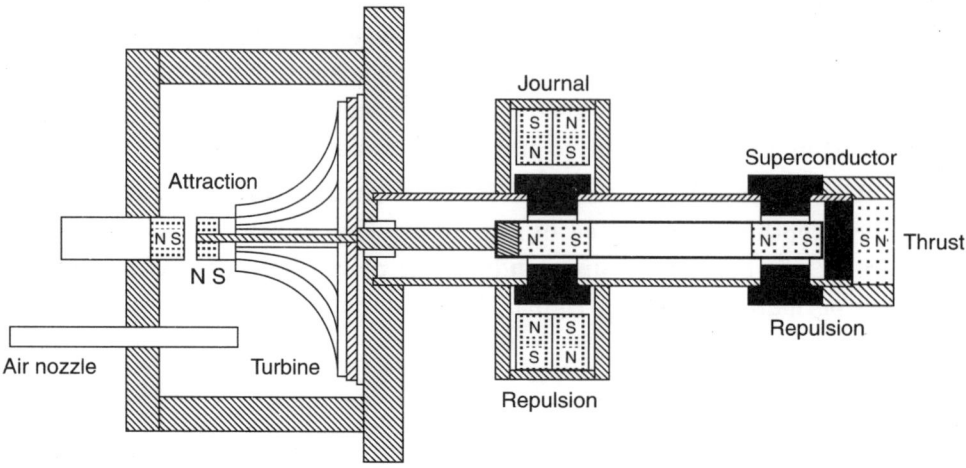

Fig. 10.29 Alternative forms of the hybrid superconductor bearing.

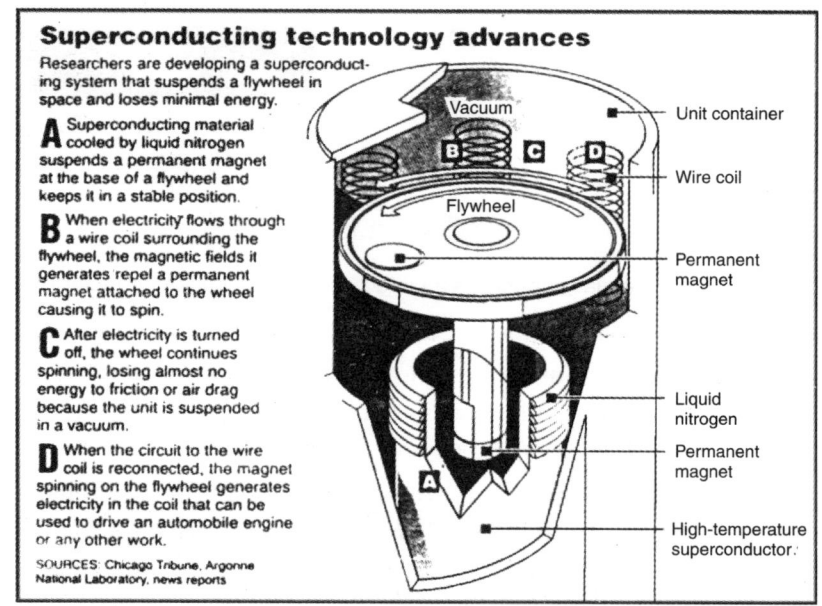

Fig. 10.30 Conceptual drawing of flywheel energy storage.

Flywheel Energy Storage. One of the foremost applications of superconducting bearings is in energy storage via flywheels. The concept is illustrated in Fig. 10.30. The entire apparatus is housed in a vacuum chamber to eliminate losses due to air drag. Only the HTSC bearing need to be cooled to LN_2 temperatures; that bearing acts on a permanent magnet attached to the flywheel, both to suspend it and to keep it stable.

Energy is transferred into and out of a flywheel as follows: A permanent magnet is mounted in the flywheel. Current flows through a coil adjacent to the flywheel, and that repels the magnet whose motion causes the flywheel to spin. Proper synchronization between the current pulses and the angular position of the rotating magnet is required to accelerate the spinning flywheel. Once the flywheel reaches top speed, the electricity is turned off, and the wheel continues spinning, losing almost no energy to friction or air drag (because of vacuum). Later when the circuit is reconnected the magnet spinning on the flywheel generates electricity in the coil, which can be used to run an external load.

Presently, this type of storage is not being done. For an electrical utility wishing to store power for 12 hours or so, flywheels have not been considered competitive (compared to batteries, pumped hydro, etc.) because of friction losses in the bearings. However, they could be competitive if their bearings were to have very low losses [32].

For electrical energy storage, HTSC bearings suggest the hope of reaching very high round-trip efficiencies for power stored in a flywheel over 12 hours. Laboratory models have demonstrated very low energy loss rates, but scaling up to a full size flywheel without introducing new loss mechanisms will be a serious engineering challenge.

10.2.5.11 Magnetohydrodynamic (MHD) Energy Conversion

MHD generators permit the direct conversion of thermal enthalpy to $d.c.$ electricity. Strongly heated, ionized gas (natural gas) flow at high velocity perpendicular to an imposed magnetic field between electrodes on either side of the flow channel. If such a generator is installed as the first stage upstream of a conventional steam-turbine loop, the overall efficiency of the combination power plant can be enhanced to 50%. For this, however, at least 20% of the enthalpy must be extracted from the hot gas, which requires magnetic field strengths of at least 6 T. Due to these high values and the huge dimensions of actual commercial systems, dipole magnets 25 m long with a 5 m free bore are required. Therefore, only superconducting magnets can be considered on account of efficiency. The properties of NbTi conductors are adequate.

Although a number of smaller model systems with superconducting magnets have been tested successfully, work ceased worldwide in the early 1980s due to aging problems with the channel wall and electrodes. A new MHD programme was begun in Italy in 1989 [33].

As the electrical energy consumed by superconducting magnets is only 1% of the energy generated, a change one to HTSCs would not result in any major savings in operating costs or gains in efficiency for an MHD system. The same applies to installation costs.

10.2.5.12 Superconducting Magnetic Energy Storage (SMES)

The use of energy storage to supplement base-load capacity for an electric power system has long been recognised as an attractive method to maximize the generation or transmission capacity of the utility. During off-peak periods, electric power can be stored and then recovered during high demand conditions to offset the need for larger generation or expanded transmission capacity. In addition to this load-levelling application, energy storage, with the right features, can provide system benefits to improve voltage stability, frequency control, reactive power consumption and provide rapid response power during momentary faults or complete power interruptions. A wide range of energy storage technologies such as hydroelectric, pumped gas storage, batteries etc., can be considered depending on geographic, environmental and overall cost factors.

The energy storage technologies that directly involve superconductivity can be divided into two distinct categories: superconducting magnetic energy storage (SMES) and flywheel energy storage (FES) or electromechanical storage.

The ability of superconductors to carry direct current without losses means that a persistent can be maintained in a short-circuited superconducting coil with no external current supply. This makes superconducting coils very attractive as a means of storing electrical energy directly i.e., without conversion to other forms of energy. The stored energy is proportional to the square of the magnetic flux density and to the volume of the coil. A network-controlled rectifier (rectifier/inverter) connects the storage coil with the line. The operating state of the storage device (charging, discharging, storage) is set by controlling the firing angle of the thyristors (Fig. 10.31). In comparison with other energy storage means, SMES has the advantage of very short response times (< 100 ms), while larger storage devices have the further advantage of high efficiency (> 90%). In addition, they can be used for several tasks in the grid simultaneously such as peak handling, grid stabilization, and hot standby.

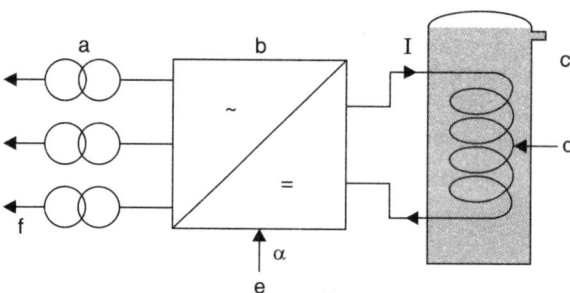

Fig. 10.31 Operation of a superconducting magnetic energy storage (SMES) device (*a*) Transformer; (*b*) Rectifier/inverter; (*c*) Liquid nitrogen; (*d*) Superconducting coil; (*e*) Control; (*f*) To three-phase grid.

Increased use of SMES in future grids may reduce the need for new plant construction and improve the utilization of existing fossil-fuel plants. This would yield an appreciable cut in fossil-fuel consumption and in the associated CO_2 emissions and other problems.

In 1991, development of a storage unit (100 kWh, 20 MW) for grid stabilization began in Japan. An Engineering Test Model (ETM) with a storage capacity of 20 MWh is under discussion in the USA. "Micro" SMES units (1 MW) are already being used to protect sensitive manfacturing equipment against voltage spikes.

The designs of large-scale storage units and experimental systems are based on metallic superconductors at 77 K would have little effect on the capital costs of a storage unit but would improve the efficiency, even in large units (+ 3%). For small and medium sized storage plants (≤ 1 MWh) HTSC are an essential precondition for acceptable efficiency levels [34].

10.2.5.13 Superconducting Current Limiters

Increasing short-circuit currents in power grids impose steadily growing dynamic loads integral components such as generators, transformers, and switch gear. The notion that is being explored currently is to use the fast (≤ 1 ms) transition of superconductors from the superconducting to the normally conducting state (from $R \approx 0$ to very high R) to limit fault currents to acceptable values. The transition can be initiated spontaneously (when critical values are exceeded) or controlled externally (with pulsed supplemental current or supplemental magnetic field).

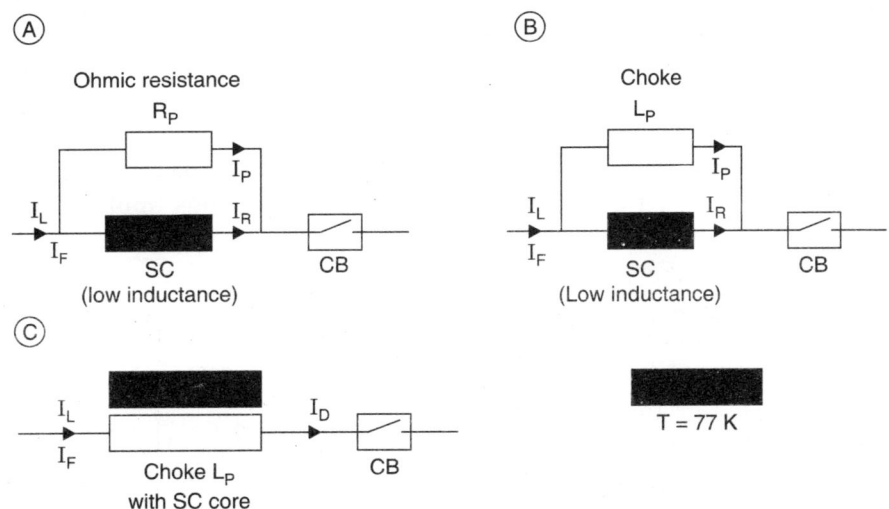

Fig. 10.32 Main types of superconducting current limiters (one phase shown)
(A) Resistive; (B) Hybrid; (C) Inductive
SC = Superconductor; CB = Conventional breaker.

Fig. 10.32 shows the three basic current limiter concepts now being studied for three-phase grids. In figures 10.32 (A), and (B), the superconductor is in the form of a wire or tape of well defined length in a low-inductance configuration (e.g., a bifilar winding). In the resistive limiter (Fig. 10.32 (A)), the superconductor is connected in parallel with an ohmic resistance R_p kept at room temperature. When the superconducting winding (trigger coil) becomes normally conducting ($R_S \gg R_p$) the current switches over to the parallel resistance R_p, which limits it to predetermined value. After 50 ms, the limited current is interrupted by a conventional breaker. The resistive limiter is also well suited to limiting direct currents. The hybrid type (Fig. 10.32 (B)) also has a low-inductance superconducting trigger coil, but here the current is limited by a choke (limiter coil), a high-inductance copper or superconductor winding connected in parallel with the trigger.

The inductive class (Fig. 10.32 (C)) has a number of variants such as a normally conducting choke whose windings are shielded from its iron core by a cylinder made of massive superconducting material. In normal operation, the cylinder is superconducting and the inductance of the coil is very low. If a fault current causes the cylinder to become normally conducting, it loses its shielding action and the iron core of the choke become fully effective ; the inductance of the coil increases abruptly and limits the current.

The feasibility of superconducting current limiters at medium voltages has been demonstrated with a number of small, single-phase experimental devices using metallic superconductors. In the most advanced project, the Japanese industry is collaborating with a large electric utility to develop a three-phase, 6 kV/2kA limiter [35]. The use of HTSCs is also being tested, because the simplicity of cooling with liquid nitrogen instead of helium should yield a decrease of 70% in the operating losses and savings of 60% for the cooling plant.

Superconducting Fault-Current Limiters (SCFCL)

A fault-current limiter (FCL), whether it employs a superconductor or not, is basically a variable impedance that is installed a series with a circuit breaker. During a fault the impedance of the FCL increases rapidly to a value (typically a fraction of an ohm on a distribution sytem and several ohms on a transmission system) at which the fault current is reduced to a level that the circuit breaker can easily interrupt. In the case of a typical superconducting FCL (SCFCL), a fault causes the device to switch from a superconducting, low impedance state to a normal, high impedance state. Use of FCLs is being considered for various applications within the power sector where the potential for large fault current exists. These applications include : protection of generator and transformer circuits, use with a main breaker on transmission or distribution systems, use with a breaker joining transmission or distribution subsystems or together with feeder breakers or individual subsystems. A typical application of a FCL is illustrated in Fig. 10.33 Because a FCl will likely be called upon to operate

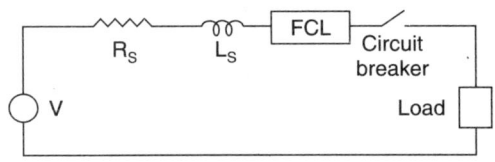

Fig. 10.33 A typical FCL application where R_S and L_S are source resistance and inductance.

very rarely, its performance and losses under normal operation are extremely important. Ideally, an FCL should have zero impedance at normal load and negligible auxillary losses under normal operation. To effectively limit fault currents, a FCL must 'instantly' sense the fault and switch impedance within, a fraction of cycle (several ms) and must be able to maintain this state until the circuit breaker operates (fraction of a second). Since many faults are due to temporary occurrences such as lightning strikes or animals, tree branches etc., shorting out lines, many power systems automatically reclose the circuit breaker within a matter of 0.1 s. Thus, a FCL must be able to return to its low impedance state prior to this first reclosure to provide protection from a continuing fault. In addition, a FCL should have high reliability, low capital cost and a long lifetime that is comparable to other power system components.

Most **SCFCL** concepts involve the switching of a superconducting element from the superconducting to normal state. A switching speed of several milliseconds would be fast enough to prevent a build up of large fault currents but slow enough to prevent large voltage spikes due to stray inductance. Several low temperature superconductor (LTSC) and HTSC FCLs have demonstrated acceptable switching characteristics. Prior to operation of the circuit breaker, the superconducting element must dissipate large amounts of energy (100s of MJ for transmission systems and almost 1 MJ for distribution systems). This dissipation causes the superconducting elements to heat up, making recovery to the superconducting state prior to reclosure difficult. The bridge type FCL does not suffer from this problem since the superconducting inductor does not normalize during a fault condition.

A number of SCFCL concepts have been designed and tested in recent years throughout the world, and several more are under development. One of the most simplistic approaches for a SCFCL uses HTSC thick films which show high resistance in their normal state and have essentially zero resistance in the SC state. A resistive SCFCL offers nearly ideal system performance, since the device in the passive SC mode shows zero resistance and, thus, presents no impedance or losses to the system. A 'resistive' SCFCL is basically self-triggering and designed to rapidly normalize during a current transient to produce a sufficient increase in resistance to limit the current rise. A major requirement for a resistive SCFCL is the ability to completely recover to the SC state during the time the circuit breakers are clearing the fault.

To date, no single SCFCL design concept has been established as a clear winner. Each design has its relative advantages and disadvantages. The outstanding issue for thin film, HTSC FCLs is demonstrating the ability to develop adequate resistance while carrying the current required for power circuits. The outstanding issue of LTSC FCLs is reducing the level of ac loses to provide acceptable operating losses. The outstanding issue for the HTSC-based, bridge type FCL is the present high cost of HTSC wire to justify use of a superconducting rather than a conventional inductor. So far, no design has demonstrated acceptable cost, performance and reliability for widespread use on an electric power system.

10.2.5.14 Trapped Flux Devices for Manufacturing

Magnets and electromagnetic forming devices such as **dent pullers** have been used for a long time in a variety of manufacturing applications. This includes the use of permanent magnets to attract and hold ferromagnetic objects, to clamp objects and assemblies during manufacturing where in some circumstances, permanent magnets may constitute the only practical way of clamping objects in confined spaces during the manufacturing process. In addition to these permanent magnets, larger electromagnetic clamps and chucks also find extensive application in the manufacturing industries. Both permanent magnets and electromagnets have significant limitations although they are useful in a wide variety of applications. Electromagnets pose a hazard due to very high currents and voltages that are required to generate magnetic fields of sufficient strength to be useful in industrial applications. Moreover, due to their bulk and necessary power leads, electromagnets are frequently not well suited for use in confined areas. On the other hand, permanent magnets generally have a limited clamping force due to their low strength magnetic fields. In addition, their magnetic fields cannot be shut off, or easily redirected in a portable device. The replacement of conventional magnetic technology with trapped-flux superconductors in devices such as magnetic clamps and dent pullers offer great advantages over conventional technologies in aerospace applications. The trapped-flux superconductors are based on $YBa_2Cu_3O_{7-x}$ (YBCO) material with $T_c = 92$ K.

Bulk trapped-flux superconductors, offer a practical solution to a flexible tooling system where superconductor can produce a powerful but controllable magnetic clamp are durable can survive autoclave environments and unlike permanent magnets, can be easily turned off and on (Fig. 10.34). Magnetic clamps can be constructed from solid superconducting materials in a disc or ring shapes. Such magnets can be energized by subjecting the superconductor to a changing magnetic field (ΔB). The superconductor generates currents in response to the change. Various routes may be taken to arrive at a charged trapped flux magnet : (a) field cooling (FC), in which a superconductor is cooled in a strong external field before changing that field to zero (thus supplying the required changing magnetic field); (b) zero field cooling (ZFC), where the material is cooled in 'zero field', after which the field is ramped up, held until the system reaches equilibrium, and then brought down again to zero; and (c) pulsed field changing (PFC), which is a rapid non-equilibrium version of zero field cooling.

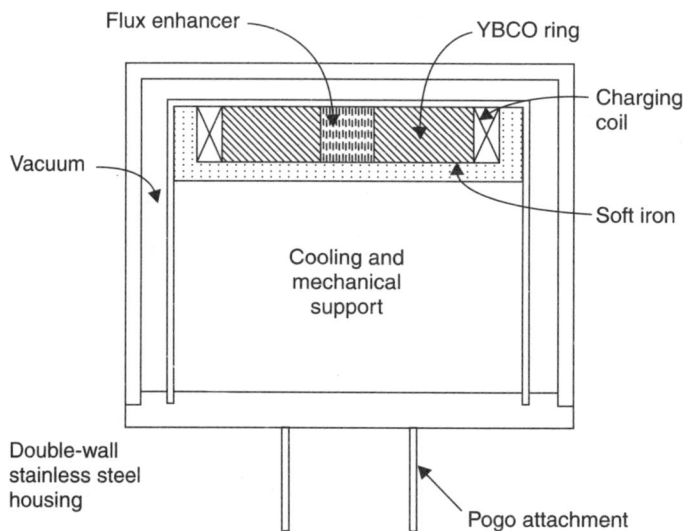

Fig. 10.34 Trapped-flux superconducting magnetic clamp

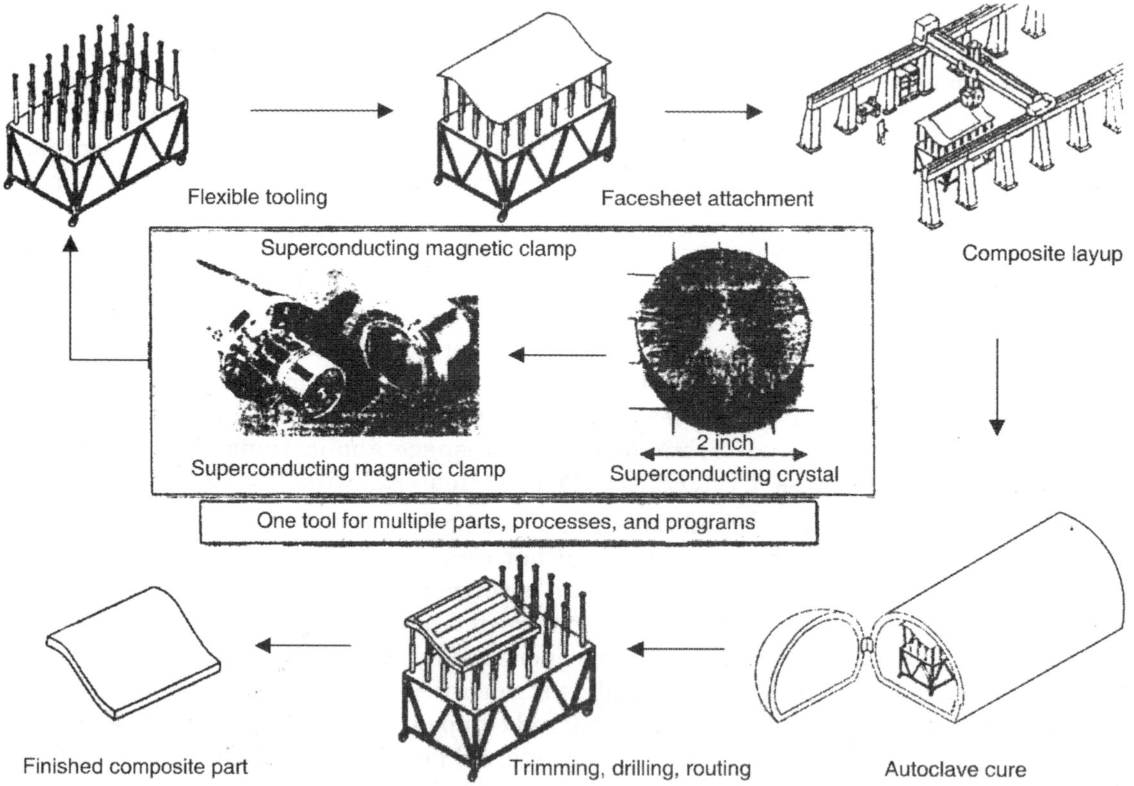

Fig. 10.35 Flexible tooling concept with trapped-flux superconducting clamps for composite processing.

Applications of Superconductivity

Pulsed field charging has recently been studied at Boeing and elsewhere as an efficient means of charging flux-trap magnets. PFC is attractive because of practical considerations in the external field source. In trapped-flux magnetic clamp, YBCO superconductors has been energized with pulsed fields exceeding 20 kG using a solenoid wound with 0.7 mm copper wire. Total weight of the coil reported is just 220 gm. Obviously, pulsed field charging is thus a virtual necessity for a wide range of practical applications.

By properly orienting and careful design of the magnetic circuit return path (with soft iron, etc.) powerful clamping devices can be configured. The trapped magnetic field in the HTSC can be atleast 1.2 T at 77 K in 25 mm diameter single grain. The trapped field increases with increasing current density. The current density increases with decreasing temperature thus greatly increasing the trapped-flux capability and the resulting clamping force (Fig. 10.34). The flexible tooling concepts where superconducting clamps replace the traditional vacuum clamps is shown in Fig. 10.35. The operational advantage of superconductors is that the magnetic force can be turned off when desired, by warming it upto a temperature above its transition point.

Superconducting Dent Puller

In the manufacture of commercial and military air planes, electromagnetic dent pullers are used for removing inadvertent damage during manufacture. Very significant time and cost savings could be realized with the availability of safe, portable dent pullers. A unique application of the dent puller is for removal of dents from structures that have no access to the backside, such as honeycomb airplane control surface assemblies, double walled engine nacelles, and tubes.

The basic concept employed in an electromagnetic dent puller, originally developed at Boeing involves producing a high negative magnetic field gradient across the facesheet of a metallic (or electrically conductive) structure. By using capacitors and selectively cancelling the exterior field, the interior field is then rapidly collapsed back towards the flux concentrator, thus producing the outward force on the facesheet of the structure (Fig. 10.36). By varying the magnetic field level and quenching rate within the structure, the pulling force on the facesheet can be regulated and precisely controlled.

When subjected to a pulsed magnetic field, the basic phenomena occurring in a metal sheet is the diffusion of the magnetic field into and through that sheet. Specific pulse frequencies are required to first allow the magnetic field to readily penetrate the metal sheet (push); and to then enable the same sheet to shield or prevent the rapidly collapsing field from easily diffusing back through the sheet (effective pull). The tension force is achieved by first establishing a magnetic field on both sides of the metal sheet with a field generating coil on the outside of the metal sheet. The coil is kept energized until the low frequency field has completely penetrated, as illustrated by "Slow Bank Action" in Fig. 10.36. If the magnetic field at the outsde is then rapidly reduces to zero, the magnetic pressure on the inside will push towards the outside resulting in the effective pulling force, as illustrated by a "Fast Bank Action" in Fig. 10.36.

Although conventional electromagnetic dent pullers are used in the aircraft industry they are not utilized very widely due to several critical limitations. They are very large, heavy systems and very dangerous in operation (495 V and 30,000 A). They require large cables connecting the capacitors and hand-held unit, making the accessibility of the dent puller to the areas where they need to be used very impractical or sometimes impossible.

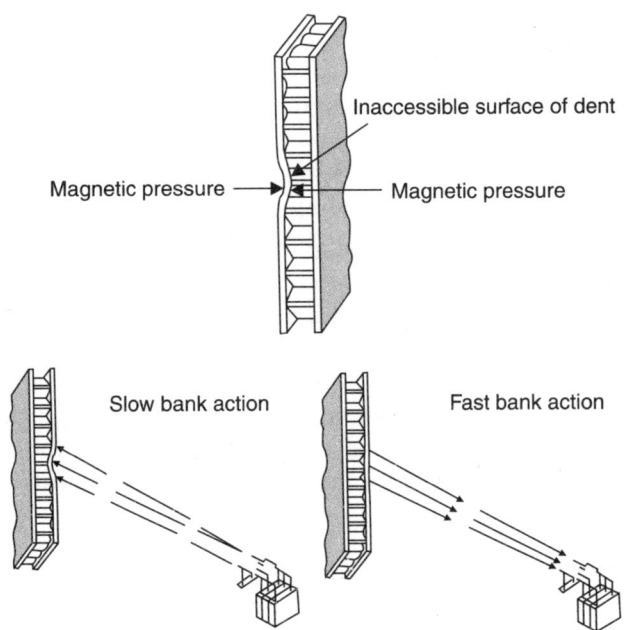

Fig. 10.36 Electromagnetic dent puller operating sequence.

In the superconducting dent puller, the capacitor can be replaced by magnetic energy stored in trapped field magnets. Advantages occured with the superconducting system include safety (during quench, the developing resistive and inductive voltages cancel), portability and simple operation under hazardous and remote conditions. A schematic flow diagram for a superconducting dent puller is shown in Fig. 10.37. The device operational principle include : (a) magnetic field trapping; (b) magnetic flux pumping; and (c) magnetic field quenching. Theoretical design calculations show that field strengths of 1 – 2 T are required in a working superconducting dent puller. The magnetic field quench times required in a dent puller will vary, but are generally in the 10 – 100 μs range. Pulling force is a function of the magnetic field change. Trapped-flux superconducting magnets with sufficient strength are currently available. The most challenging step required in the operation of the superconducting dent puller is to non-destructively quench the trapped field in the superconductor within the required time frame. One way to achieve a quench is by thermally inducing a flux-jump by inserting a small heater into a cored trapped-field magnet. However, due to the large heat capacity and low thermal conductivity of YBCO, this technique works only in relatively small samples. Other techniques that may be required for large samples are under intensive investigation. These techniques include AC and RF heating methods that can deliver energy directly into the interior of the bulk superconductor, once it is in the normal state, so that we would not be limited by the low thermal conductivity.

The effectiveness and ultimate clamping and pulling strength of trapped devices such as clamps and dent pullers is directly proportional to the strength of the trapped magnetic field of a bulk superconductor, or surface strength of a permanent magnet. These devices can greatly benefit from increases in current density capability of superconducting bulks and from larger single superconducting domain sizes. As the trapped magnetic field capability in a superconductor increases, the mechanical strength of the bulk superconductors needs to be improved to accomodate the resulting hoop stresses associated with high-trapped fields and from mechanical forces exerted on the superconductor by the clamping or pulling operations.

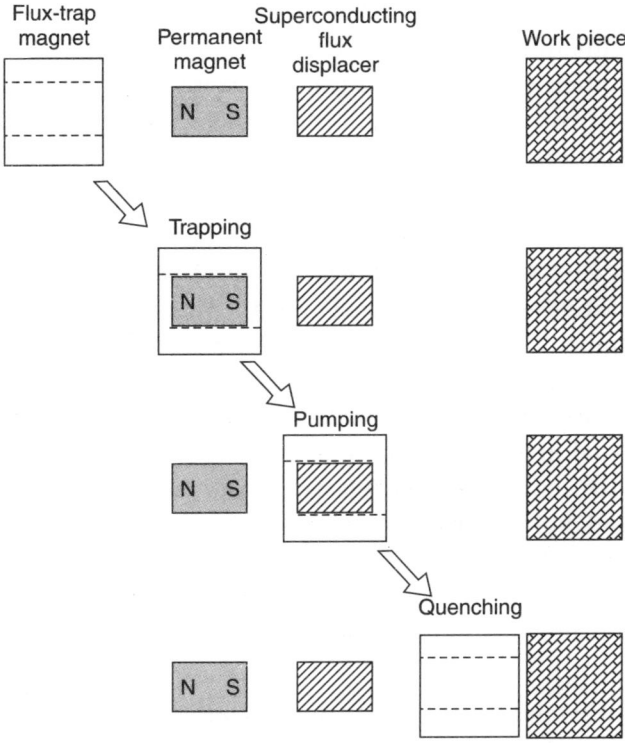

Fig. 10.37 A conceptual flow diagram for a superconducting dent puller.

10.3 ELECTRONIC APPLICATIONS OF SUPERCONDUCTIVITY

Superconducting electronics [36 – 39] (SCE) have become an indispensable resource for physical measurement and detection, for ultrafast "super logic chips", for many high frequency applications, and for certain calibration purposes due to special advantages of superconductor electronics over the conventional counterparts in terms of higher speed, less noise, lower power consumption and much higher frequency limit [40 – 43]. Superconducting electronic circuits are used where very low noise, very high sensitivity, minimal losses, short switching times, and stability of voltage calibration are crucial, *e.g.*, communication, high precision and high frequency electronics magnetic field measurements, superfast computers, etc. Advantages over competing semiconductor devices include, the macroscopic quantum effects of superconductivity (Josephson effects), unparalleled low noise levels, and novel effects such as quantum interference and voltage-frequency conversion governed solely by physical constants. In such applications, superconducting components are employed in the most critical areas of physical instrumentation, ultrafast logic circuits, and high frequency engineering and also find use as voltage standards. Because cooling is always desirable when minimal noise is desired, superconducting electronics represents the technology of choice under extreme requirements, particularly in research and in military hardware. Potentiality of several superconductor electronic devices have already been established using conventional LTSCs. Helium-Cooled SQUID systems have become firmly established in laboratories, where they serve as extrasensitive magnetometers and susceptometers and as instruments for measuring derived quantities.

The need for cooling to temperatures below T_c was long considered a major drawback and prevented the adoption of superconducting electronics in conventional hardware. Even where superconducting devices surpassed all competition (or where there was no competition in the first place), as in biomagnetic medical diagnostics using SQUID arrays, the cost of cooling and shielding has presented a major obstacle to the commercial success of superconducting electronics.

Important milestones in the commercialization of superconducting electronics include the transition to competitive electronics based on HTSC cuprates and the rapid progress made recently in the development of cooling equipment for electronics, especially the pulse-tube cooler. High T_c and also higher energy gap of HTSC cuprates extends the capability of 'superconductor electronics' quite considerably [44]. Rapid advancement in the synthesis of thin HTSC thin films and artificial grain boundary HTSC Josephson junctions have excited considerable interest in the development of electronic devices, e.g. Quantum Interference Device (SQUID), microwave and digital devices which are being found to be very promising for future applications. Some of the HTSC devices have already been commercialized [45]. The field sensitivity achieved in recently developed HTSC-SQUIDS are sufficiently high for several applications such as in boimagnetism measurements, non-destructive evaluation, geophysical measurements, etc.

The use of HTSCs in microwave passive devices has an advantage over normal conductors such as copper and silver in terms of low insertion loss and high gain due to their lower surface resistance. In past few years, HTSC based filters and subsystems have been considered for application in mobile communication as well as for satellite and some specific radio astronomy applications [46 – 52]. The use of HTSC filters in cellular base station is being investigated to achieve better sensitivity. About 1000 HTSC filter subsystems have been deployed worldwide with millions of hour of cumulative operators [48]. Several HTSC digital circuits such as A/D converter, shift resistors, rapid single flux quantum (RSFQ) circuit etc., have been developed successfully [53]. The use of HTSC in RSFQ digital circuit is expected to raise the operating temperature and frequency limit due to higher energy gap.

10.3.1 Superconducting Effects Important for Electronic Applications

(*i*) **Pure Inductances.** In a superconductor *d.c.* current transport remains virtually lossless upto frequencies in the megahertz range [losses comparable to copper are observed only at 100 GHz in niobium at 4 K or in Y BCO at 77 K], provided two conditions are met :

- The magnetic field amplitudes must be sufficiently small so that no flux penetrates into the superconductor.
- Flux "frozen" in by the cooling process must not be able to migrate (*i.e.*, it must be pinned).

Applications of Superconductivity

These requirements can be satisfied in electronic applications. Two important applications of the effect are therefore possible: shielding of (cyro) electronic circuits and lossless "transformation" of magnetic flux from zero frequency up into the megahertz range. In the design of these applications, the superconducting structure can be treated as a simple equivalent circuit composed of ideal inductances and capacitances with no resistive components. If the capacitances are negligible, the integrated Faraday induction law

$$n_1 \Delta\phi_1 = - I \Sigma L_i \qquad \ldots(10.9)$$

describes a "flux transformer" comprising a closed series circuit of inductances L_i; when the flux in L_1, having n_1 turns, changes by $\Delta\phi_1$, a current I is induced in the circuit. This circuit I acts to oppose the flux in L_1, and if the L_i have been properly selected it will displace flux from L_1 to the L_i as desired. The L_i can, of course, be any extended structures, not just coils.

Within the stated constraints, this flux transformation is independent of frequency. The effect has proven extremely useful for transmitting magnetic flux sensor. It is also important in the construction of flux concentrators/transformers for SQUID applications ranging from simple susceptometers to gravity-wave detectors and monopole detectors [36, pp 178 – 190; 39, pp 388 – 404].

If there is no inductive load ($L_i = 0$ for $i > 1$), L_i provides ideal shielding of external flux changes. This principle can be employed for creating volumes practically free of magnetic fields by "inflating" an initially folded superconducting pouch [39; pp 354 – 359].

Small High-Frequency Losses

In superconducting materials, electrical losses vanish if the current densities $J < J_c$ are constant over time. The losses at small high-frequency power increase as a function of f^2 but still remain very low over a wide frequency range. For the best epitaxial YBCO films, the surface resistance R_S (77 K) at 100 GHz is always an order of magnitude smaller than in cooled copper. Fig. 10.38 illustrates this point, comparing R_S (f) for high-quality YBCO films with the surface resistance

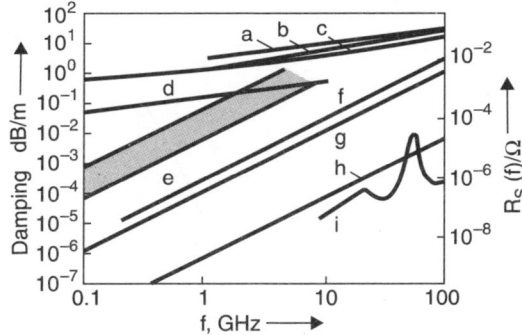

Fig. 10.38 Measured surface resistance together with calculated and measured strip-line losses for YBCO and copper at 77 K (dielectric losses not included)— a → copper strip line at 300 K; b → copper strip line at 70 K; c → copper; d → RG-9 coaxial cable; e → copper strip line at 77 K (measured); f → epitaxial YBCO; g → YBCO strip line at 77 K (theoretical); h → YBCO strip line at 27 K (theoretical); i → atmospheric propagation at sea level.

of copper. At a given frequency, losses increase rapidly above a high-frequency power threshold, corresponding to a magnetic field amplitude of 0.03 T, beyond which flux penetrates into the superconductor. Signal propagation in superconductors is nearly dispersionless below $f/\text{GHz} = 73\ T_c/\text{K}$, so that this effect can be utilized for high-rate information transmission.

Superconducting high–frequency resonators (cavities) and low-loss delay lines and filters for frequencies upto 100 GHz used to be important only in military hardware (special radar techniques). HTSCs are now permitting use in civilian areas: remote sensing of the earth, mobile radio and civilian radar.

Energy Gap Effects

Superconductor insulator-superconductor (SIS) tunnel junctions are used for extremely low noise direct ("video") or heterodyne detection of millimeter radiation in radio astronomy (Figs. 10.39 and 10.40). These devices are opted at temperatures well below T_c. Josephson effects are purposely suppressed by a static magnetic field. The devices utilize two effects :

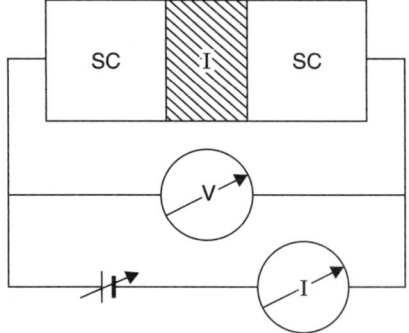

Fig. 10.39 Circuit for measuring *I* versus *V* characteristic of a tunnel diode.

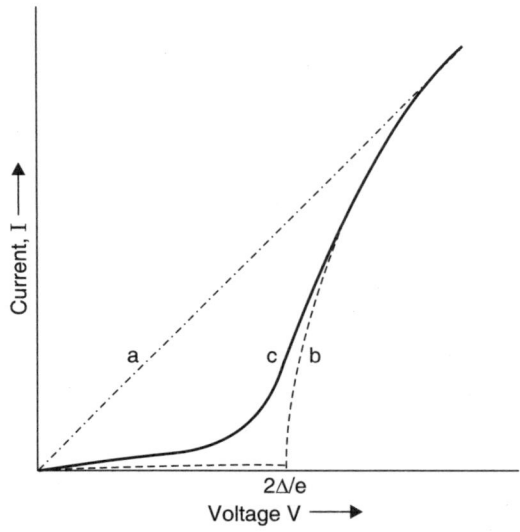

Fig. 10.40 Qualitative curve of I versus *V* for the circuit 39 *a* → Ohmic curve, *b* → Curve for $T = 0$, *c* → Curve for $0 < T < T_c$.

First, the extreme nonlinearity of the current-voltage characteristic for SIS junctions made with LTSC (abrupt change in tunnel current at the total energy gap);

Second, the quantum effect of phonon-assisted tunneling. With helium-cooled Pb/Bi/In oxide-Pb/Bi tunnel junctions, the quantum limit (cutoff) of the mixer noise temperature $T_M = \hbar\omega/2k_B$ has been reached for detection in the 100 GHz range; this makes the technique indispensable in radio astronomy. [36, pp . 259 – 284]. The state of the art is an integrated submillimeter LTSC detector circuit [54]. It does not appear desirable to replace LTSC with HTSC in this case, since as yet no SIS tunnel junction has been devised on the basis of HTSCs.

Quantum Interference Effects

Cooper pairs "overlap" within the coherence length and in this way, the coherence property is propagated. Macroscopic structures of superconductors exhibit macroscopic interference phenomena. The magnetic flux enclosed by a macroscopic (*e.g.*, on the scale of millimeters) superconducting ring is quantized; this can be understood as the constructive interference of Cooper-pair waves with a multiple of 2π phase difference along the ring, so that the magnetic flux enclosed by the superconducting ring is a corresponding multiple of the magnetic fluxoid quantum $\phi_0 = h/2q = 2.0679 \times 10^{-15}$ T m^2. The Josephson junctions make it possible to introduce magnetic flux into a superconducting ring in continuous fashion and to utilize the macroscopic quantum interference in the ring for measurement purposes (the basis of SQUID devices), *i.e.*, the strong magnetic field dependence of the Josephson current phenonenon (*i.e.*, magnetic field controlled quantum interference of a Josephson junction) is employed in the design of extremely sensitive magnetic field sensors called SQUIDS. In accordance with the ratio of areas, external magnetic fields act much more strongly on quantum interference in the macroscopic ring (area 1000 μm^2) than on quantum interference in a single Josephson junction enclosing less than 1 μm^2 of the field. For applications with magnetic control, Josephson junction optimized with respect to noise level and switching time are purposely made small ("point" contacts) and are linked with well-defined superconducting interferometer ring structures. Josephson junctions epitomize one of the most profound macroscopic quantum aspects of superconductivity, while at the same time enabling a wide range of applications than any other aspect of superconducting electronics.

All of the devices which employ Josephson effects rely on combining this macroscopic quantum property with another fundamental quantum principle exemplified in superconductors : *i.e.*, **magnetic flux quantization**. The combination of these two effects has led to two rather different type of device for applications : (*i*) superconducting logic gates which have been realized in a number of different fronts, the front runner for widespread application is **rapid single flux quantum** (RSFQ) logic. The essential features of RSFQ logic are that the energy involved in writing or sorting a bit of information or operating a single gate is orders of magnitude smaller than that corresponding to any semiconductor device. This results in very low heat dissipation. In addition the switching speed is very high (~ few picoseconds), the devices have an essentially quantized nature, arising from the fact that a voltage pulse which is propagated through the system maintains its product of its amplitude and duration constant at a quantized value of 2×10^{-15} Vs (= 2×10^{-15} Wb, corresponding to a single quantum of magnetic flux). The conversion of analog to digital signals and vice versa is the other unique niche applications where Josephson junction based superconducting electronics have already demonstrated a clear advantage over corresponding semiconductor devices. The unique properties of Josephson junctions allow a number of unusual architectures for A to D conversion, (ii) the second major application device which combines both Josephson effects and flux quantization is the SQUID. Most SQUIDS are applied as analog measuring device where in general the speed of the device is not high (although broadband picovoltmeters with a bandwidth as high as 15 MHz have been demonstrated). The most important application of SQUIDS envisages their use to detect and image biomagnetic activity arising from the human body. SQUID systems (including those based on HTSC cuprates) are being used for non-destructive evaluation in various industrial enviornments e.g., inspection for cracks in aircraft frames or wheels.

Josephson junctions are also used as mixers and electromagnetic radiation detectors. This junction relies on the extreme nonlinearity which is present in I-V characteristic of a Josephson or quasi-particle tunnel junction. The presence of a superconducting energy gap is responsible for generating a sharp jump in the conductance at voltages which exceed the superconducting energy gap. This extreme nonlinearity, coupled with the ability of Josephson junctions (of sufficiently small capacitance) to respond upto a frequency f corresponding to the gap $2\Delta/e (f - \Delta/h)$, for the HTSC cuprates can be as high as 30 T Hz. In this section, some of these applications are described.

10.3.2 Josephson Junctions, Tunnel Junctions and Weak Links

The Josephson junctions is essentially nonlinear component in superconducting electronics. Fig. 10.41 illustrates the basic current voltage (I/V) characteristics of Josephson junctions. Such an element can be modeled by a parallel network of a Josephson junction, an ohmic resistance or SIS tunnel characteristic, and a capacitance, where I and V values of a Josephson junction are related to each other and to time by the Josephson equation

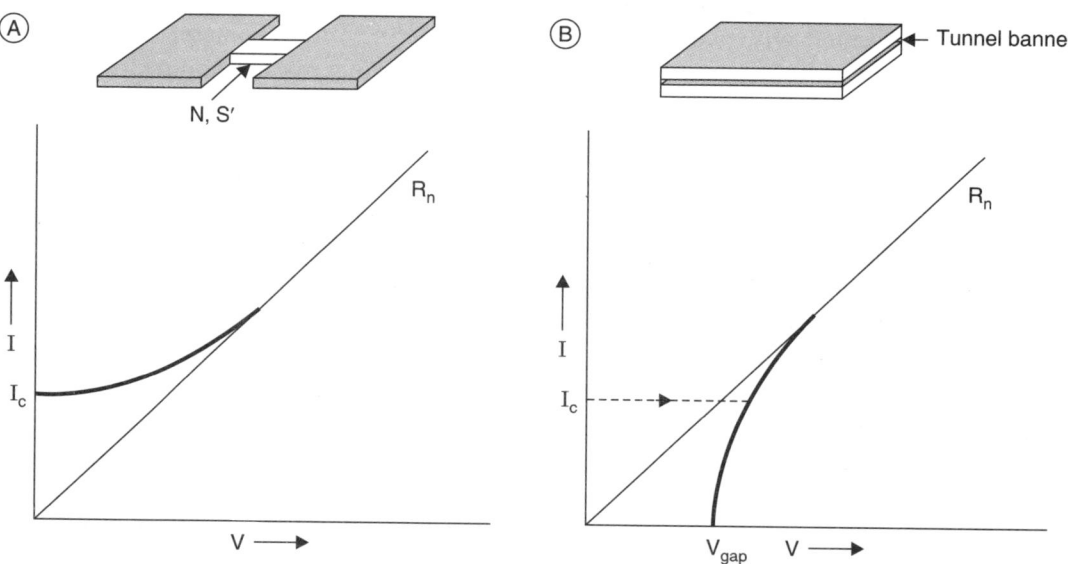

Fig. 10.41 Ideal current-voltage characteristics of the Josephson junction : (A) → Steady behaviour; (B) → Switching behaviour. N → normal conducor; S', S → Superconductor.

$$I = I_c (\sin \delta_0 + 2e\, VL/h)$$

where I_c is the critical current, δ_0 the phase offset related to the "frozen in" flux.

The characteristic of Fig. 10.41 (A) is time average and is steady; that of Fig. 41 (B) exhibits switching behaviour. The transition between Josephson current and quasiparticle current, which can take place on a picosecond time scale, long formed the basis for all major logic applications of superconductivity, with the Josephson threshold value switch as active element and with I_c as threshold value. The "weak link" characteristic (Fig. 10.41 (A)) is well-suited to sensor (SQUID) applications and new logic circuits of RSFQ type. Characteristic Fig. 10.41 (A) can (but need not) be derived from Fig. 10.41 (B) by bridging with a resistance; Fig. 10.41 (A) presupposes a weak coupling link between the superconductors S, not the existence of a tunnel junction. Characteristic shown in Fig. 10.41 A may arise if the two superconductors S are connected by a normally conducting bridge or a bridge made up of a "weakened" superconductor. For quantum coherence to be preserved between the superconductors S, the bridges must have dimensions comparable with the superconducting coherence length in the well material. For well-characterized tunnel junctions-the noise properties can be theoretically described in an acceptable manner (to within a factor of two to three); for bridges deviations of upto two orders of magnitude can occur, so that large scale integrated circuits have not been feasible so far [55]. It is not easy to prepare S-I-S Josephson junction in HTSC as it is usually done in LTSC due to short coherence length of HTSC cuprates. It has been found that natural grain boundaries in HTSC materials behave as Josephson junctions [56, 57]. The weak link nature of grain boundaries was more clearly established by growing YBCO HTSC epitaxial films on a $Sr\,TiO_3$ bicrystal substrate [58]. The bicrystal substrate is fabricated by fusing two single crystal substrates. When HTSC film is grown epitaxially on the bicrystal substrate, a single grain boundary is realized. Dimos et al. [58] found that the critical current across the grain boundary is a function of misorientation angle between the two crystals. A 45° misorientation angle formed by rotation about the C-axis can reduce the critical current density by four orders of magnitudes than that of the film. The understanding of weak link nature of grain boundary in HTSCs has necessitated the development of single crystal film technology for electronics application.

Detailed studies of these grain boundaries have led to develop several techniques for realizing artifical grain boundaries and junctions such as edge junctions, multilayer ramp junctions and bicrystal junctions whose behaviour are similar to LTSC Josephson junctions. Grain boundaries in HTSCs are of central importance for numerous applications ranging from the electronic circuits and sensors such as SQUIDS. We may note that Josephson junctions made of HTSC cuprates, which can be operated successfully at elevated temperature (70 K), have all been of type shown in Fig. 10.41 (A).

It is interesting to note that the development of SCE devices based on the Josephson effect proceeded on three levels : basic physics, device and circuit innovation, and materials science and processing development. No doubt, materials processing technology has paced the implementation of SCE devices from their discovery to today. The first Josephson junctions were thin film structures, whose fabrication was primitive by today's standards. The pioneering research by Jaklevic et al. [59] used two junction thin film SQUIDs. Shortly after the initial discoveries, fabrication of many Josephson-effect devices and SQUIDs changed in two ways : from thin film tunnel junctions to point contact [60] and other weak link structures [61]; and from two-junction or dc SQUIDs to one-junction or rf SQUIDs [62, 63].

Fabrication of thin film tunnel junctions was not easy, were time consuming, required access to and expertise with vacuum systems and frequently failed. Zimmermann [60] and other workers soon demonstrated mechanical methods of producing Josephson junctions generally referred to as point contacts. There are basically two principal electrical distinctions between Josephson tunnel junctions and other Josephson junctions : (i) the physical structure. Tunnel junctions usually have an insulating barrier separating two superconductors. Other types of junctions usually have a conducting barrier, either normal metal or weak superconductor (ii) the difference in their electrical characteristics is shown in Fig. 10.42. Tunnel junctions are typical under-damped, and hence have multi-valued, hysteretic current-voltage characteristics. The others are usually over-damped and have single-valued non-hysteretic voltages.

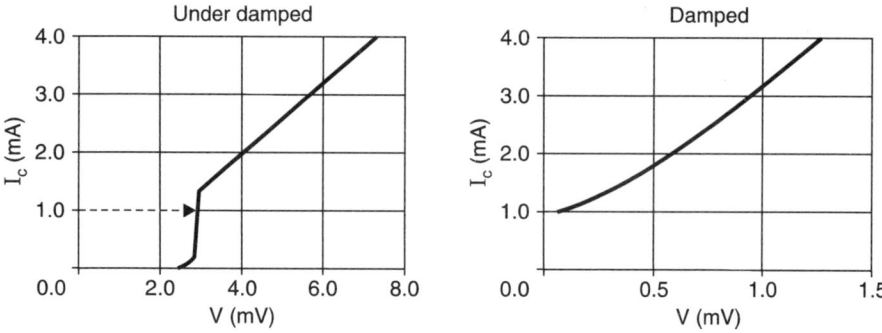

Fig. 10.42 dc current voltage characteristics for under-damped and damped-Josephson junctions. The critical current is 1 mA in both cases. Calculated for Nb junctions. Note that we expanded the voltage scale for the damped junction shown at the right by more than a factor of 5.

Since the discovery of HTSC cuprates, there has been a worldwide effort to produce Josephson junctions, SQUIDs and other thin film devices and circuits that operate as high as 77 K. Modest success has been achieved for thin film Josephson-effect devices suitable for SQUIDs [64]. So far, Josephson junctions made of HTSC cuprates, which can be operated successfully at elevated temperature, i.e., 70 K, have all been of type shown in Fig. 10.41 (A). Various types of Josephson junctions made with HTSC superconductors are shown in Fig. 10.43 (A) to (G).

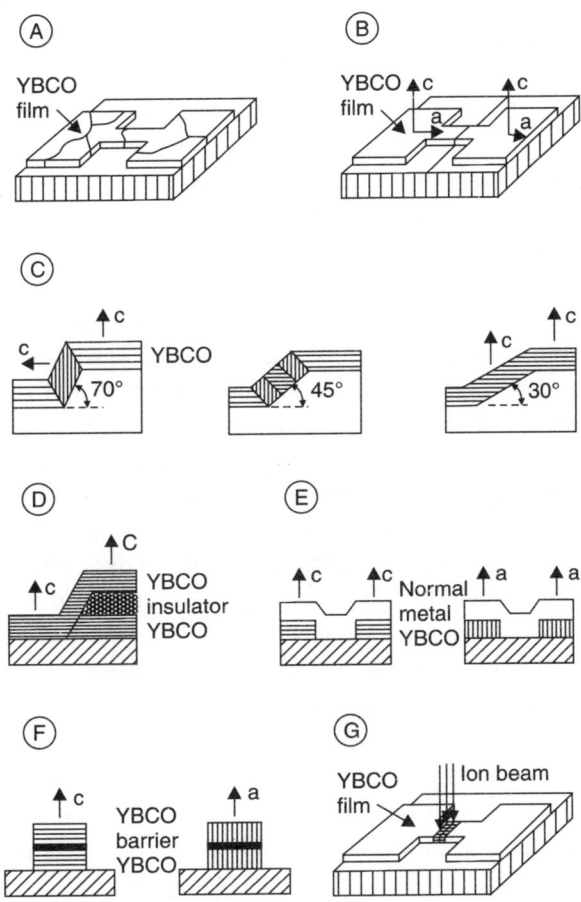

Fig. 10.43 A–G. Types of Josephson junctions made with high-temperature superconductors. (A) Grain-boundary junction (polycrystalline film); (B) Grain-boundary junction (bicrystalline substrate or bicrystalline buffer); (C) Step-edge junction (two different junctions connected in series); (D) Edge junction with epitaxial barrier; (E) Normally conducting bridge as in situ step junction; (F) Planar junctions (currently not for 77 K operation); (G) Weakened bridge.

10.3.3 Josephson Circuits, Digital Circuits, Digital Signal Processing and Voltage Standards

Circuits containing Josephson junctions and superconducting paths forming closed structures are interferometers of the SQUID type and are discussed in next section. As current-driven switches, they have found some use in logic circuits and A/D converters as well. Important switching elements for Josephson circuits, besides the Josephson junctions proper, are thin film resistors, microstrip lines (select lines, bias lines), and connections formed by superconducting current paths on a substrate coated with a dielectric film. Various types of latching logic gates, ranging from a relatively large-area interferometer to the very compact gates in four junction logic, and a variety of gate families and memory cells have been devised, these have been developed to LSI (large-scale integrated) complexity. The crowning achievement in this line is a functioning four-bit microprocessor running at 770 MHz. The state of art comprises superconducting structures of niobium or NbN with tunnel barriers of aluminium oxide. The mean power consumption of a logic gate in this technology is 10 µW at 4 K. Despite advanced niobium technology, there is yet no satisfactory solution for superconducting fast-access mass storage, and indeed no solution is in prospect. This defect might be remedied with a superconducting shift register as working memory, with hybrid combinations employing cooled CMOS (complementary metal oxide silicon) semiconductor mass memory ; such an approach is pursued in USA and Japan. A more ambitious and promising technique is being explored in USA : the non-latching rapid single flux quantum (RSFQ) logic family employing niobium superconductor [38 pp. 1 – 49]. Processor frequencies well over 100 GHz surpassing the semiconductor competition by more than two orders of magnitude appear feasible. Special processors, especially for ultrafast digital signal processing, are of interest here. In this logic family, the bit is the presence of a switching pulse leasing a few picoseconds in a time window corresponding to a 2π phase change in the Josephson relation. Josephson junction of the type shown in Fig. 41 (A) are used. The functionality of the ultrafast RSFQ basic circuits has already been demonstrated with niobium. One sensational result is the demonstration that a 32-bit RSFQ shift register at 77 K can be operated at a frequency of over 100 GHz [76]. The experiment involved HTSC nanobridges which however, exhibit an unacceptable error rate (one wrong bit every 10 min), but the convertibility of RSFQ circuits to the still-unperfected HTSC technology has now been proved feasible at least for fault tolerant circuits.

The basis of the Josephson voltage standard is that a Josephson junction irradiated with high frequency radiation (frequency f) exhibits current steps at multiples of the voltage $V = h/2ef$. Modern Josephson voltage standards for 1 – 10 V consist of thousands of hysteretic niobium aluminium oxide-niobium Josephson junctions connected in series, which are integrated into a superconducting microstrip conductor in such a way that all junctions are excited at the same high frequencies (70 GHz) [37 pp. 21 – 36]. It appears that new design will make it possible to lower the operating frequency to 12 GHz [77]. In this way, the cost of HF system can be cut by about an order of magnitude, which is conducive to adoption in industry. We will discuss them in details in subsequent sections.

Josephson Junction

Fabrication of integrated circuit devices based on LTSC Josephson junctions has advanced to large scale integration. Nb Josephson junctions (JJ) based process technology has achieved 10 k gates/cm^2 circuits [65 – 67]. Nb N-based IC process technology has been used to demonstrate circuits upto 2 k gates/cm^2 operating at 10 K. Efforts are in process to develop high performance NbN-based circuits because they have higher gap voltage and higher temperature of operation. This eases the cryocooler requirements, particularly for mobile platforms and air and spacecraft. Fig. 10.43 depicts the cross-section of a typical Nb-based IC process [68]. In addition to the junction trilayer, most processes use three or four Nb interconnecting (wire) layers and silicon oxide insulation layers (either SiO_x or SiO_2). The oxide layers serve as electrical isolation between the metal layers that form the active circuits.

Advanced processing techniques, with the goals increasing circuit speed and solving critical issues such as minimizing I_c spreads as junction sizes decrease, have been developed to produce smaller features [69, 70].

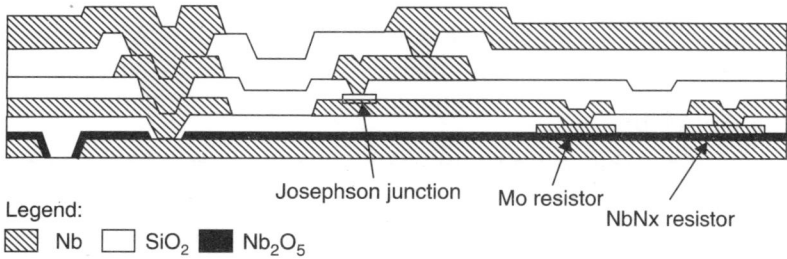

Fig. 10.44 Cross-section of a typical Nb-based superconductor IC on a silicon wafer substrate.

IC fabrication consists of three process modules: thin film deposition, lithographic patterning which transfers the design into a photosensitive material on the wafer, and etching, which transfers the circuit pattern into the thin film itself. These three steps are repeated a number of times, alternating metal and oxide film layers.

HTSC Josephson Junction IC Devices

SCE devices have a unique requirement :

Cryocooling. After the discovery of HTSC cuprates, the prospects for reducing this requirement seemed near. Although HTSC still needs cooling by at least 223 K below ambient for operation at 77 K, the required cooling would be considerably less but at the cost of significant complexity in manufacturability.

Although single-layer YBCO junction processes meet many requirements for SQUID magnetometry, a reliable low-noise multi layer circuit would improve even this application. High-margin digital circuits need junctions with low parasitic inductance. To achieve low parasitic inductance, one requires high layer-to-layer alignments and the fabrication of junctions directly adjacent to a superconducting ground plane.

HTSC Bicrystal Junction

Josephson junctions are invariably formed at grain boundaries in HTSC films. Although the bicrystal junction is the most common approach for single SQUID applications, extending this technique to digital integrated circuits present many difficulties (Fig. 10.45). Regardless

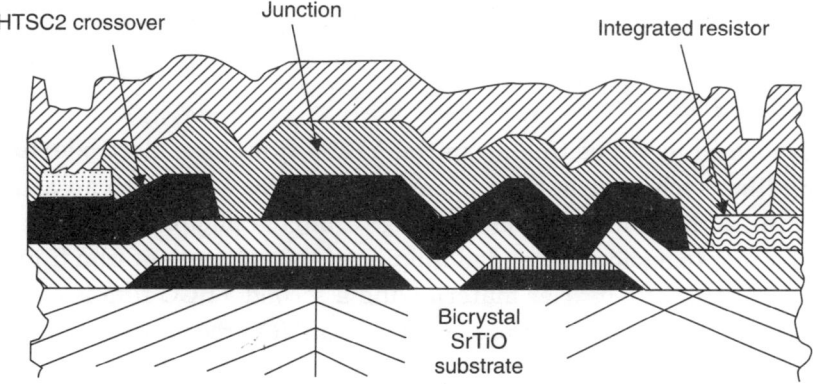

Fig. 10.45 Bicrystal junction in multi-layer structure.

of whether a ground plane is positioned above or below the junction, this approach produces a weak link directly adjacent to the junction that can alter inductance. Also, interconnects crossing the bicrystal boundary will suffer from a junction that reduces its current.

HTSC Step-edge Junctions

Like the bicrystal junction, the step edge junction uses the natural boundary formed between misoriented grains of a YBCO film, where the bicrystal junction is parallel to the substrate the step-edge junction is orthogonal to the substrate surface. Fig. 10.46 shows the cross-section for an *IC* process based on step-edge junctions. Since the first film deposited on a single crystal substrate often produces higher critical current material than subsequent films, this has proven to be a good junction type for use in magnetometry which needs low-noise HTSC films for best operation. The step-edge technique is not limited to a single substrate grain boundary. Junctions can be placed at the designer's discretion as close as photolithography allows in order to reduce parasitic inductance.

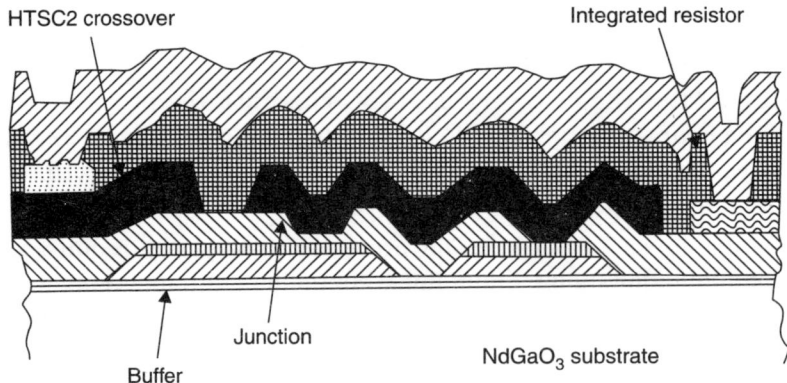

Fig. 10.46 Step-edge junction in multi-layer process.

Step edge and bicrystal junctions both processes result in an exposed grain boundary that should be passivated to provide greater stability against environmental and aging effects. They also have the flexibility of post-fabrication targeting of I_c by reversible oxygen annealing. Though these junctions have advantages, it is difficult in practice to control the natural formation of grain boundaries necessary to consistently reduce junction critical current variations below 30% [71].

HTSC Ramp-edge SNS

This junction is formed by depositing an initial YBCO film (base electrode) and overlaying it with a dielectric film, such as strontium titanate. A shallow edge, with an angle less than 30° from the substrate, is then formed by ion milling through a photolithographic mask. The edge angle is small in order to avoid unintended flipping of the subsequent film from *c*-axis to *a*-axis [72]. A bilayer consisting of a barrier material and a second YBCO film is then deposited over the edge. Fig. 10.47 illustrates the cross section of an *IC* process using ramp-edge junction. The $I_c R_n$ product of these junctions can vary from tens of microvolts to millivolts, depending on the cleaning of the base electrode. Although junctions with the higher values of $I_c R_n$ are generally desired for higher high-speed circuit applications, these are often achieved by a less controllable, hence less reproducible interface resistance.

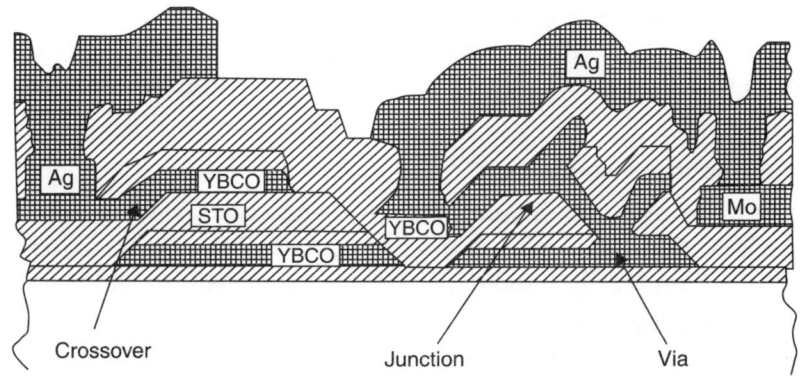

Fig. 10.47 Ramp-edge junction in multi-layer process.

A promising variation of the SNS ramp-edge junction is the interface–engineered junction. Instead of depositing a non-superconducting layer, one creates the barrier by transforming the edge surface by a sequence of plasma treatments and anneals [73]. Such approaches can span a large parameter space using ion beam etching, annealing at elevated temperatures, plasma exposure and chemical treatments. They have been successful to a varying extent, including achieving junction I_c variation as low as 8% $(1 - \sigma)$ [74].

Ion-beam Damaged Junction Process

There are numerous research efforts to use high-energy focused electron and/or ion beams to fabricate junctions by localized damage [75]. Simple digital circuits have been demonstrated with these junctions. Fig. 10.48 illustrates a multi-layer process incorporating an ion beam junction. Ion beam processes have the flexibility to target the junction I_c by repeating the ion beam exposure at the junction to reduce the I_c or to anneal out damage to increase I_c.

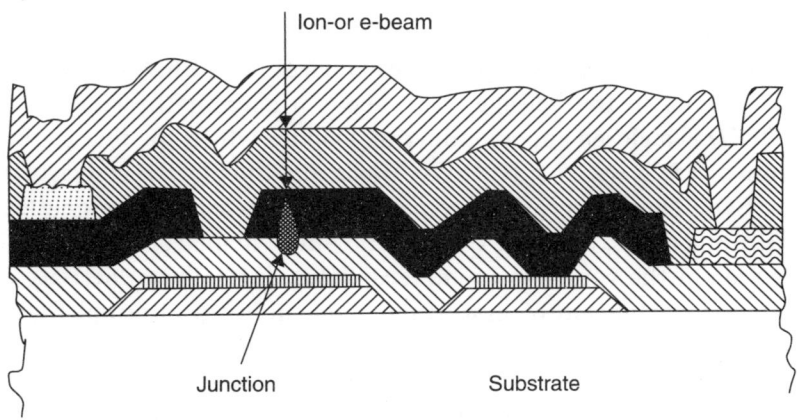

Fig. 10.48 Ion beam junction in multi-layer process.

Trilayer HTSC Junctions

Vertical trilayer junctions between a-axis YBCO films offer a low inductance geometry that would be direct parallel to Nb circuit (see Fig. 10.44). However, depositing uniform a-axis films, which allows use of the longer coherence length orientation has proved much more difficult than c-axis deposition. Since a-axis growth proceeds more rapidly than c-axis films, any c-axis inclusion would create a large variation in topology that would compromise junction parameters.

Vias and Interconnect Cross-overs

Control of film epitaxy and stoichiometry are essential to growing smooth, defect free, superconducting YBCO films. One must optimize and control the lattice match of the substrate to the film deposition temperature, and deposition gas composition and pressure.

In order to fabricate vias and interconnect cross-overs, the basic requirements for defect-free films remain the same but accomplishing them is considerably more difficult.

Photolithography and Etching

Photolithographic processes in use for Nb circuit processing may not be desirable for YBCO because YBCO reacts with the trace chemicals dissolved in water. Contact of the YBCO film with water is to be avoided or minimized. Passivating YBCO with a $Sr\,TiO_3$ film is one method of limiting exposure of YBCO surfaces to potentially degrading chemical reactions. Unfortunately edges exposed during etching are more difficult to passivate. One will have to take great care to remove contamination by chemicals employed in photolithography before subsequent epitaxial deposition. An innovative approach to avoiding such contamination is by multi-layer masking techniques that avoid wet processing.

10.3.4 SQUIDs

Superconducting Quantum Interference Devices (SQUIDs) are extremely sensitive detectors of magnetic flux and magnetic fields, electrical current, or any other physical quantity that can be converted into magnetic flux. SQUID is essentially a flux to voltage transducer. Basically, the SQUID converts threading its loop into a quantity that can be detected by a subsequent electronic stage. SQUID exploits two fundamental effects of superconductivity : the Josephson effect and the single-valuedness of the quantum phase in a superconducting ring. The first key element is the Josephson junction which is characterized by a limited critical current at zero voltage and by switching to the voltage state above a current threshold. The second basic element of the SQUID is a closed superconducting loop for which the flux is quantized in units of the flux quantum $\phi_0 = h/2_e = 2.07 \times 10^{-15}$ Wb. Basically, the SQUID converts magnetic flux to electrical voltage. SQUIDs can be used for instance, for measuring magnetization, magnetic susceptibility, magnetic fields, current, voltage and small displacements. SQUID systems are used for the detection of the very weak magnetic fields due to brain activity and for the non destructive testing or evaluation of materials. Even quantum-limited SQUIDs are under development for the detection of gravitational waves passing a resonance mass antenna [78–80]. At very low temperatures, the intrinsic lower limit of energy resolution of a SQUID per hertz bandwidth is given by the quantum noise, $\hbar \approx 10^{-34}$ J/Hz, as has been confirmed in laboratory specimens. For practical devices with efficient flux coupling, good helium-cooled

SQUIDs achieve energy resolutions around 30 $\hbar$ and field resolutions of 10^{-15} T/Hz$^{1/2}$ at frequencies down to < 1 Hz. This high sensitivity has made SQUIDs vital for many important experiments and makes possible the measurement of extremely weak biomagnetic signals from human organs such as the heart and brain, in the range of $10^2 - 10^5$ fT/Hz$^{1/2}$ (corresponding to 10^{-8} to 10^{-5} times the earth's magnetic field). Measurement of biomagnetism can be regarded as the pioneering application of superconductivity in normal technology outside the physics laboratory. SQUIDs also find applications in the field of geophysics and nondestructing testing [80, 81].

There are two main versions of the SQUID : *dc* SQUID and *rf* SQUID. The *dc* SQUID consists of a superconducting loop interrupted by two resistively shunted Josephson junctions and the voltage across the parallel junctions represents the output signal. If a *dc* current passed through the *dc* SQUID, i.e., d.c. SQUID is biased with a *dc* current $I_B \approx I_c$, where I_c is the critical current of the Josephson junction, the voltage U across it depends in strongly nonlinear way on the magnetic flux threading the SQUID loop. The prefix '*dc*' implies that it is biased with direct current. Both Josephson junctions in the *dc* SQUID have identical characteristic. Critical current of the *dc*-SQUID is an oscillatory function of magnetic flux with periodicity of one flux quantum ϕ_0. The value of one flux quantum, $\phi_0 (= h/2e)$ is 2×10^{-7} G/cm^2. The *rf* SQUID has only one junction in the loop and the read out is realized by a resonant LC circuit coupled to the loop. The average voltage across the resonant circuit is the output signal. Fig. 10.49 shows the two basic SQUID configurations $A \rightarrow dc$ SQUID; $B \rightarrow rf$ SQUID; C and $D \rightarrow$ Extreme positions of the I versus V curve on variation of the magnetic flux; $E \rightarrow$ At a constant operating point I_B, the voltage varies with period ϕ_0. In Fig. 10.49, if magnetic flux ϕ enters the superconducting ring, (with inductance L_s) through the two identical Josephson junctions, the current voltage characteristic of the ring will vary between two extreme positions (Fig. 10.49 (C)) corresponding to fluxes $n\,\phi_0$ and $\left(n + \dfrac{1}{2}\right)\phi_0$, where $\phi_0 = 2.0679 \times 10^{-15}$ Wb is the magnetic fluxoid quantum and n is an integer. If a fixed bias current $I_b > 2 I_c$ is passed through the ring (where I_c is the critical current of the individual Josephson junction), the voltage drop V across the ring varies periodically in the flux with period ϕ_0 (Fig. 10.49 (E)).

The *rf* SQUID is biased with an *rf* current applied to the SQUID through inductively coupled tank circuit. Here also for an appropriate biasing *rf* voltage across the SQUID is an oscillatory function of magnetic flux with the periodicity of ϕ_0 i.e., varies with resonance frequency ω_T between two extreme positions, again corresponding to flux values of $n\,\phi_0$ and $\left(n + \dfrac{1}{2}\right)\phi_0$ in the ring as shown in Fig. 10.49 (D). If the amplitude of the high frequency current is set equal to a current step of the I_{rf} versus V_{rf} curve and the flux ϕ in the ring is varied, then V_{rf} also varies periodically with period ϕ_0 (Fig. 10.49 (F)).

To use SQUID as a magnetometer, it is operated in 'flux-locked-loop', mode in which the SQUID output varies the applied flux [82]. The magnetic field sensitivity of a bare SQUID is not very large due to small SQUID loop. In order to increase the magnetic field sensitivity of the SQUID, a flux transformer with a large pick up loop is used. The secondary of the transformer is coupled to SQUID. In order to eliminate the effect of background magnetic field such as the earth field, a gradiometer which consists of two counter wound pick up coil is used.

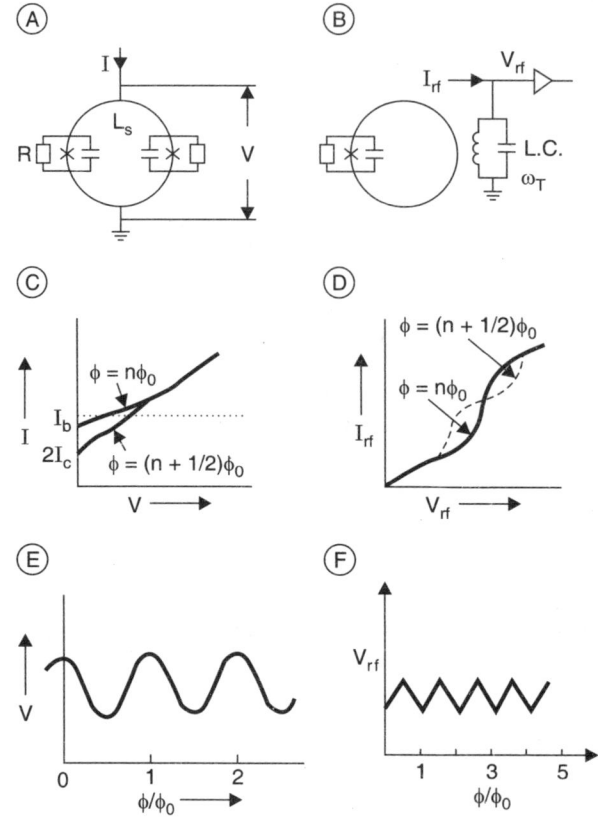

Fig. 10.49 The two basic SQUID configurations (A) d.c. SQUID; (B) RF SQUID; (C), (D) Extreme positions of the I versus V curve on variation of the magnetic flux; (E) At a current operating point I_B, the voltage varies with period ϕ_0; (F) For an appropriate biasing, rf voltage across the SQUID is an oscillatory function of magnetic flux with the periodicity of ϕ_0.

Apart from the two above mentioned basic configurations for the SQUID, there are also some modified versions, *e.g.*, the **relaxation oscillation SQUID [83, 84]**.

SQUIDs have been realized in low-T_c and HTSC cuprate materials. In general low-T_c-SQUIDs are based on niobium–aluminium technology whereas in the case of HTSC cuprate SQUIDs YBaCuO is the most frequently used material. Prior to the development of HTSC SQUIDs, low-T_c SQUIDs were well developed and in use for laboratory measurements, and, several other applications [81] widespread application of low-T_c SQUID could not pick up due to the requirement of liquid helium for its operation which is not easily available in remote area besides being costly and difficult to manage. The development of HTSC SQUIDs operating at 77 K hold promise for wider applications for the SQUIDs in remote area.

Almost immediately following the discovery of HTSC cuprates, SQUIDs using natural grain boundary junctions (NGBJ) have been demonstrated which show sufficient magnetic field sensitivity for several applications [82]. There were problems in HTSC NGBJ SQUIDs such as junction critical current could not be well controlled and the $L I_c$ product was suboptimal, but working devices were demonstrated at 77 K. It was relatively easy to get one natural grain boundary junction of reasonable characteristic for fabricating *rf*-SQUID than two identical NGBJ for *dc* SQUIDs. Fig. 10.50 (*a*) shows schematic of a NGBJ *rf*-SQUID which have been fabricated on HTSC polycrystalline film. The microbridge is made so small that it contains only one grain boundary junction. Fig. 10.50 (*b*) shows voltage-flux characteristics of Hg (Tl)-Ba-Ca-Cu-O *rf*-SQUID at two operating temperatures 77 K and 113 K. The $V - \phi$ modulations have been observed upto 121 K which is the highest operating temperature for the SQUID reported so far [85]. Artifical HTSC junctions such as edge junction, bicrystal junction, multi layer ramp junction, biepitaxial junction have been employed to fabricate *rf* and *dc* SQUIDs [82]. Fig. 10.51 (*a*) shows schematic of a high-T_c SQUID fabricated on a Y BCO film deposited on a Sr TiO$_3$ bicrystal substrate. Fig. 10.51 (*b*) shows the voltage–flux characteristics of the bicrystal junction *dc*-SQUID at different biasing current at 77 K [82].

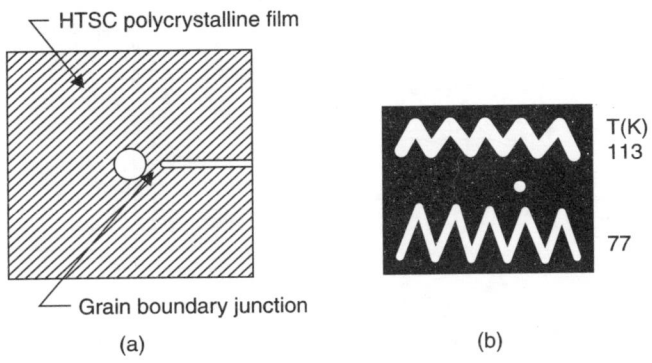

Fig. 10.50 (*a*) Schematic of NGBJ *rf*-SQUID on a polycrystalline high-T_c film, (*b*) voltage-flux characteristics of Hg (Tl)-Ba-Ca-Cu-O film *rf*-SQUID at two operating temperatures.

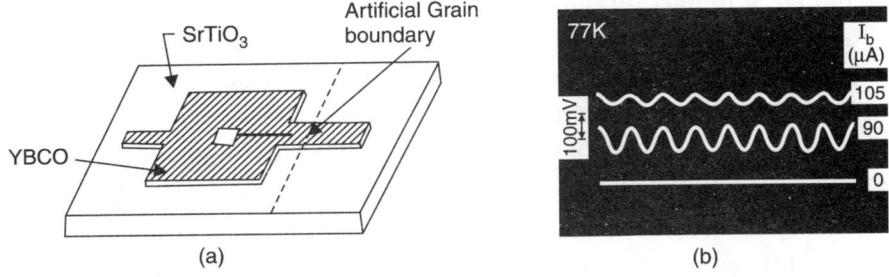

Fig. 10.51 (*a*) Schematic of bicrystal junction *dc*-SQUID, (*b*) voltage-flux characteristics of YBCO bicrystal junction *dc*-SQUID at different biasing current [82].

The readout electronics of a sensitive SQUID magnetometer operates as a null detector by strong reverse feedback with the aid of a compensating coil, which compensates a change in external flux ϕ with a flux resolution down to 10^{-6} ϕ_0/Hz^2. Important factors in the optimal functioning of a SQUID with characteristics shown in Fig. 10.49 (*C*) and (*D*) are the selection of parameters such that $2L_s I_c \approx \phi_0$ (quantization condition) and $I_c \sim k_B T/\phi_0$ is large enough that thermal flux noise at the SQUID operating temperature is not by itself of the order of $\phi_0/2$ (maximum driving of SQUID). This constant limits SQUID inductances to less than 10^{-9} H at 4 K and less than 10^{-10} H at 77 K. As a result, stringent standards apply to microfabrication of

HTSC SQUIDs to concentrate magnetic flux in approximately SQUID rings a few micrometer in diameter. The flux concentration is achieved by means of flux transformers which are fabricated as thin film structures (Fig. 10.52 (A)).

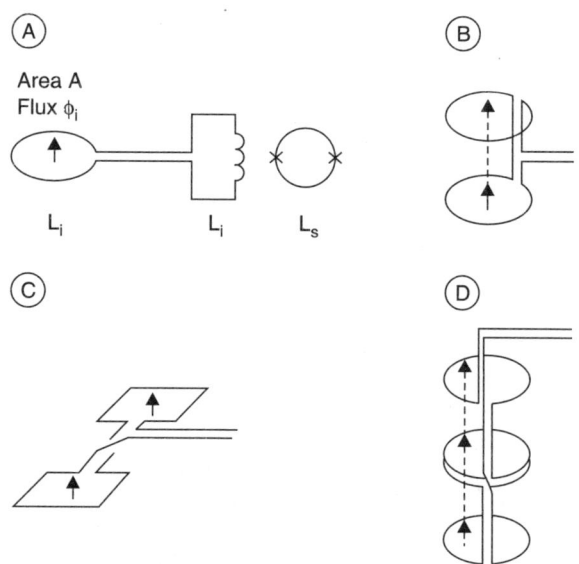

Fig. 10.52 Flux transformer
(A) Magnetometer; (B) First-order axial gradiometer;
(C) Second-order axial gradiometer; (D) First-order planar gradiometer.

During the past decade, considerable progress in the performance of SQUID readout electronics was achieved. Many novel readout concepts were developed since the early 1990s which were stimulated in part by the need to simplify the SQUID electronics for biomagnetic multichannel systems. For magnetically unshielded systems and high-frequency applications, the dynamic performance was increased considerably upto the theoretical limits. Improved surface-mount electronic components enabled strong miniaturization and opened new possibilities. In short, SQUID readout electronics followed the general trend in electronic equipment extensive computer control, increase in flexibility and speed, and reduction in power and size.

Different approaches have been used for coupling flux transformers to develop high-T_c SQUID magnetometer. These are large washer magnetometer, direct coupled magnetometer, transformer coupled and multilevel magnetometer.

Flux concentrators without thin-film coils, in the form of slotted single-layer film structures, are less effective as transformers but are easy to manufacture.

The sensitivity of a SQUID magnetometer at low frequencies is dominated by $1/f$ noise. In addition to junction noise, another major source of $1/f$ noise in SQUIDs is the random motion of stray flux mainly in the pick-up loop and weak spots in the film crossovers. In the past few years YBCO film quality and crossover design have improved, thus reducing the flux-motion noise by some eight orders of magnitude since the earliest HTSC days.

Flux concentrations without thin-film coils, in the form of slotted single-layer film structures are less effective as transformers but are easier to manufacture.

The magnetic noise of the HTSC SQUID magnetometers lies in the range of 10 – 50 fT/Hz$^{1/2}$ in the whole noise region (1 fT = 10^{-15} T). It is entirely adequate for several applications such as biomagnetic measurements (magneto cardiography), non-destructive evaluation, magnetic microscopy and geophysical measurements [86].

Given the great sensitivity of SQUIDs, interference suppression is a key problem. The first thing is to maintain the function of the SQUID that is experiencing interference, then to isolate the signal for measurement from the noise. In the presence of external fields that vary only slightly with position, very small local field gradients can be isolated and measured if the pick up coil is made in the form of a gradiometer of first or higher order (Fig. 10.52 A – C). Strong interference fields are generally compensated with passive and active shielding (magnetically shielded compartments, large coils). The more open the system, the sooner it will run into dynamic limiting. If this limiting is not overcome, electronic and "software" gradiometers may be used. In these approaches, magnetometer signals are corrected in either analog or digital mode. The problem of suppressing interference is identical for SQUIDs based on low- and high-temperature superconductors. The problem with LTSC-SQUIDs is not so much their field sensitivity ($\approx$ 1 ft/Hz$^{1/2}$ at $\geq$ 1 Hz) as the performance and cost/benefit ratio of the system. A SQUID magnetometer optimized in this respect is shown in Fig. 10.53. This represents the state of art for LT SQUIDs in large shielded systems.

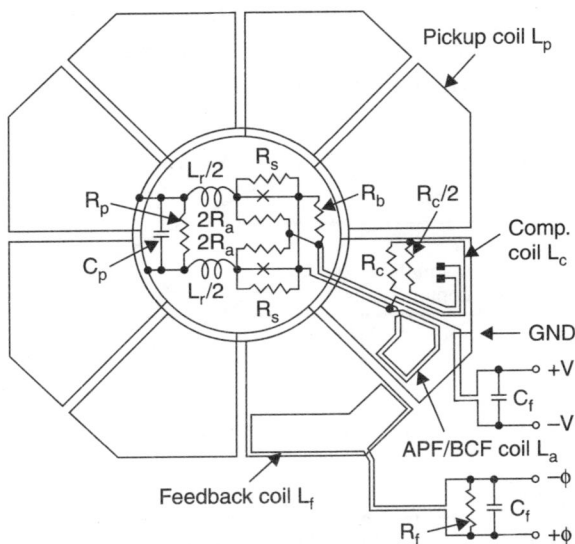

Fig. 10.53 Layout and circuit diagram of a SQUID magnetometer designed by PTB, Berlin (chip size 7.2 × 7.2 mm^2).

SQUID Applications

Low-T_c and also HTSC-SQUIDs are used in large variety of applications. The applications of *rf* SQUIDs in practice are, however, rather limited due to their somewhat inferior properties as compared to *dc* SQUIDs.

LTSC dc SQUIDs have been commercially available since the 1980s. The reliable and robust sensors are based on Nb/Al technology and equipped in such a way that a variety of standard characteristics of materials, like susceptibility and magnetization can be determined in a straigthforward method. More complicated, double stage SQUID configurations where a dc SQUID signal is readout with a series array of SQUIDs to improve the SQUID output level are also available from industry. Apart from separate sensors also whole SQUID-based instruments are offered by industry, $e.g.$, rock magnetometers, multichannel biomagnetometers, and apparatus for metrology.

During the past few years HTSC-SQUIDs have also been offered by some companies. These sensors, based on YBaCuO, have not yet reached the reliable status of the low-T_c devices. The materials problems produce HTSC junctions with well predicted properties and to arrive at an integrated device are still considerable. Research is carried in many laboratories worldwide on improving SQUID characteristics with respect to sensitivity and low frequency noise.

Strong features of a SQUID are its extreme sensitivity, the frequency-independent response (apart from the $1/f$-noise at low frequency) and the ability to measure changes on a background value with unchanged sensitivity. In each SQUID application one or more of these features play a dominant role [78, 87 – 89].

Biomagnetism is one of the main areas for the application of SQUIDs. The various biomagnetic signals are in the frequency band from about 1 to 100 Hz which is very appropriate for SQUIDs. Frequency spectrum of biomagnetic signals compared with interference (noise) fields in an urban environment is shown in Fig. 10.54. In particular, magnetoencephalography and cardiography have been studied with multichannel systems. For example, over 200 channels are used for mapping the magnetic field pattern around the head with low-T_c SQUID sensors

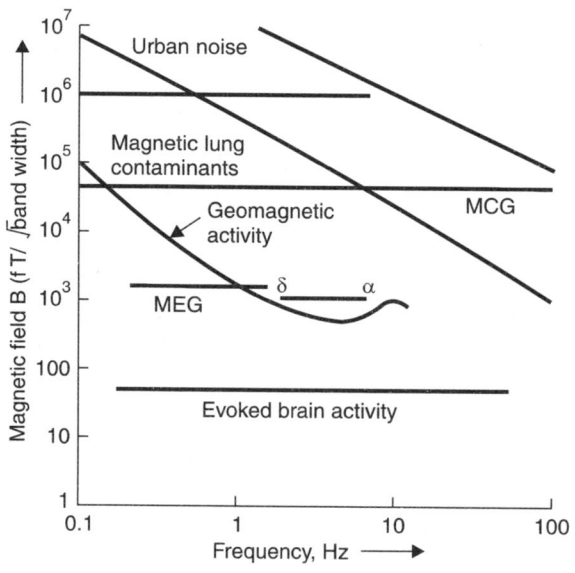

Fig. 10.54 Frequency spectrum of biomagnetic signals compared with interference (noise) fields in an urban environment.

placed in a helmet-like cryostat. In such large multichannel systems, special attention has to be paid to the elimination of cross talk between the various channels and to the reduction of the number of connecting wires to room temperature in order to keep the helium consumption at an accepted level. Magnetic field of human heart is 10^{-10} T and this can be measured using HTSC SQUID coupled with a HTSC gradiometer [86, 90, 91]. Fig. 10.55 shows result of a magnetocardiogram (MCG) measurement using a HTSC-SQUID.

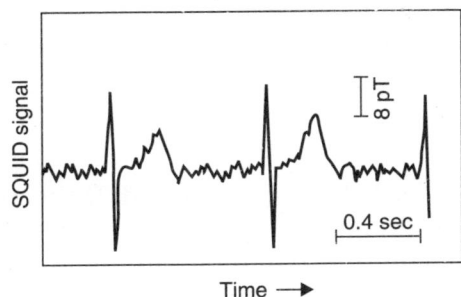

Fig. 10.55 Magnetocardiogram (MCG) as recorded using high-T_c SQUID operating at 77 K [91].

The widespread application of SQUIDs is in the field of non-destructive testing and evaluation. Cracks, corroded areas or areas with (magnetic) impurities in materials can be detected with SQUID systems by measuring the magnetic flux changes due to the disturbances in conductivity or magnetization. The sample is at room temperature and it is a challenge to minimize the distance between the cooled sensor and the material under investigation. The property that a SQUID measures–flux changes– is now of great importance for it allows the measurement in static backgrounds leading to improved lateral resolution. HTSC systems are coming up quickly. SQUID systems are used for aircraft wheels testing.

SQUID systems play an important role in determining the magnetic properties of the earth [81]. This concerns for instance the characterization of specific earth samples with a rock magnetometer. Samples with a length of about 1 m are taken from a bore hole or from different rock positions and the magnetic stratification is studied, giving information on possible natural resources or the magnetic history of the earth as is investigated in paleomagnetism. Another approach is the mapping of earth's magnetic field and electromagnetic impedance. Both low-T_c and high-T_c SQUIDs have established their value in this kind of experiment. The latter have been demonstrated by Foley et al. [92] in airborne geomagnetic prospecting applications.

SQUID systems are used for measuring gravitational forces. In this case, the SQUID operates as a displacement sensor. One can also make global observation for mineral surveying. An application, which needs a SQUID with sensitivity near the quantum limit, is the detection of gravitational waves with a resonant mass antenna [93]. The passage of a gravitational wave gives a very small distortion of a deeply cooled mass which is a bar or sphere cooled to the mK temperature range. A displacement sensitivity of the order of 10^{-23} m/Hz$^{1/2}$ can be obtained which is expected to be adequate for the direct detection of the gravitational waves. For these applications, only low-T_c SQUIDs can fulfil the desired criteria.

SQUIDs are used as a readout sensor for other detectors. A significant application is the read-out of a bolometer for particle detection [94]. Very fast SQUIDs are needed and digital SQUID concepts in low-T_c materials may open the broad field of particle research.

10.3.5 Biomagnetism

This is a modern method of medical investigation where the magnetic field originating from the human body is measured and interpreted in a way to lead to diagnostic findings [36]. The magnetic fields under investigation in biomagnetism may result from magnetized particles or ion currents related to body functions and are thus very weak. Signal amplitudes typically measured range from 10^{-9} to 10^{-14} T in a frequency range from dc to 1 kHz. This requires very sensitive magnetic field sensors, $i.e.$, a domain of SQUIDs and very efficient methods to suppress electric and magnetic interference.

Depending on the particular field of application, different SQUID systems are required. Modern biomagnetic systems are very complex. With only a few exceptions they are operated in a magnetically shielded room and contain upto 300 low T_c SQUID-sensors maintained at liquid helium temperatures in non-magnetic fibreglass reinforced epoxy dewars. The data acquisition, computation and visualization of results require state of the art technology. How high T_c SQUIDs and liquid nitrogen cooling will reduce system complexity and thus would reduce costs remains to be seen.

We may note that for all electric diagnostic techniques $i.e.$, electroencephalography (EEG) and electrocardiology (ECG), corresponding magnetic counterparts exist, $i.e.$, magnetoencephalography (MEG) and magnetocardiography (MCG).

There are occurrences of practical phenomena which cannot be detected by electric measurements but are highly visible in magnetic data [37]. Experimental data clearly reveal that magnetic measurements contain information that their electrical counterparts do not disclose. In addition, magnetic methods need no electrodes and derive the data in a faster and reproducible manner. The resolution of ECG and EEG is limited by the noise of the measurement system. At present the best signal-to-noise ratios of both methods are comparable. Since SQUID systems are still under continuous improvement, it follows that the magnetic modalities may become more sensitive than the electrical ones in the near future.

The advantages of biomagnetism are also evident for the special application of fetal MCG [38], where the magnetic method allows a detection of the fetal heart signal over a long period during pregnancy than the electrical method.

Obviously, biomagnetism is a facinating application of superconductivity. Very sophisticated LTS-SQUID systems are commercially available and employed in different fields with growing success. Technological developments leading to unshielded operation and/or utilization of HTSC-technology are highly desirable as they should trigger an appreciably wider distribution of these promising methods.

10.3.6 Non-destructive Evaluation (NDE)

Both low T_c and high T_c-SQUIDs are used as sensors for the magnetic detection of a variety of surface breaking and deep lying flaws in metals, including iron, steel, aluminium and titanium and in certain other materials such as concrete. The methods employed are first **magnetometry**, in which the defect is polarized by a magnetic field, and the induced dipole moment is detected by a SQUID and secondly **galvanometry** in which the flaws distort the flow of eddy currents excited in the specimen and the SQUID measures the subsequent distortion of the magnetic field generated by these currents. Each of these methods was in use for NDE before SQUID sensors became involved and in fact, other sensors such as fluxgates have displayed adequate sensitivity for most purposes. The reasons for incorporating SQUIDs do not lie in their extreme electromagnetic senstivity, but rather in other features, that they offer: these include, (*i*) the use of gradiometric spatial differentiation techniques to improve spatial resolution, (*ii*) the high dynamic range of devices which allow the detection of small changes in large polarizing fields and (*iii*) the use of frequencies far lower (and skin depths larger) than those common in conventional eddy current techniques, allowing detection of faults at greater depths.

In NDE, the aim is to measure interruptions of currents by cracks or corrosion in metals such as around rivets in airplane winds. Low-T_c SQUID operating at 4.2 K has already been successfully used to demonstrate the potentiality and uniqueness in SQUID-based NDE studies. In low-T_c-SQUID-NDE system, standoff distance between sensor at 4.2 K and surface of the test object at room temperature is relatively large which affects the spatial resolution. In case of high-T_c SQUIDs, the standoff distance can be easily reduced with acceptable boil-off rate of liquid nitrogen (77 K). This will increase the spatial resolution as well as overall sensitivity of the system [98, 99]. These new developments have allowed NDE to be extended to a number of problems of concern, including the detection of deep lying defects in engineering structures such as pipelines and reactor vessels, and in aircraft structures and components.

A recent addition among SQUID instruments is the **SQUID microscope**, which although not applicable to conventional NDE problems such as crack and void detection. A typical SQUID microscope operates with SQUIDs and pick-up coils, but a strong emphasis is put on high spatial resolution using tiny detection coils with the smallest possible separation between the coils and the room temperature specimen under examination. Very ingenious techniques have led to cryostat windows so thin that 20 μm standoff distances have been achieved with 20 μm coil diameters. Microscopes have been used to make magnetic maps of the ink-dots on bank notes, to study flux distribution in superconducting films where they can resolve individual flux vortices and to study the behaviour of magnetotactic bacteria.

10.3.7 Microwave Devices

The low surface resistance of HTSCs directly translates into extremely high Q^* values for resonant microwave devices. The orders of magnitude increase of Q-value is very attractive to the designers of resonators and filters. Besides resonators, several HTSC based microwave devices, *e.g.* filters, multiplexer, receivers, delay line, antenna, phase shifter, etc. with superior performances have been demonstrated [100 – 113].

According to the structures, resonators can be classified into three categories, *e.g.* one dimensional, *i.e.*, transmission line type; two dimensional, *i.e.*, planar type; three dimensional *i.e.*, cavity or bulk type. Filters can also be classified in the same way as the resonators used as their building blocks.

---

*Q-value is defined as the ratio of storage energy versus dissipated energy within a *rf* period.

In terms of surface resistance and power handling capability, the epitaxially grown thin film HTSC materials are superior compared with bulk or thick film HTSC materials. The majority of HTSC resonators and filters are made of thin film HTSC materials on lattice matching substrates such as La AlO_3, MgO or R-cut sapphire.

Fig. 10.56 (a) shows a half-wavelength microstrip line (microstrip line is a transmission line comprising a line and a ground plate on the front side and back side of substrate respectively) HTSC resonator with a Q-value of 13000 measured at 5.5 GHz and 80 K [114]. Fig. 10.56 (b) shows a 5-pole (a pole corresponds to a resonacne) microstrip line filter with parallel-coupled have wavelength resonators as its building blocks [115]. The main advantage of HTSC filter is to make very narrow band filters with many poles for steep skirts and high off-band rejection without suffering from high in-band loss. This is specially important for telecommunication applications, as these applications demand the most efficient utilization of an assigned frequency bandwidth.

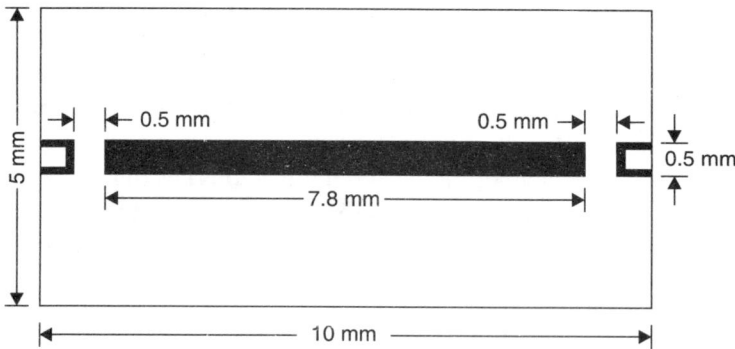

Fig. 10.56 (a) The design of a 5 GHz HTSC microstrip line resonator on a 20-mm thick $LaAlO_3$ substrate.

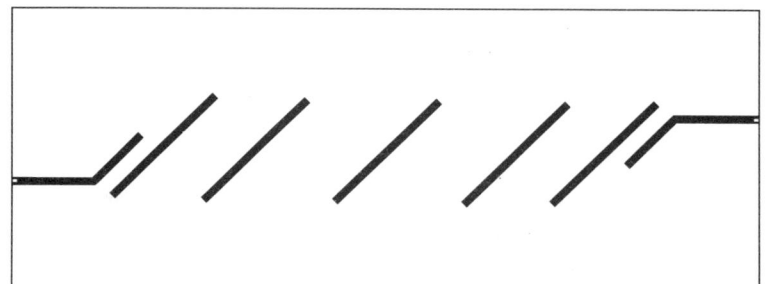

Fig. 10.56 (b) The design of an X-band 5-pole HTSC microstrip line bandpass filter on 20-mm thick $LaAlO_3$ substrate.

The power handling capability of HTSC devices is limited by the critical current density, J_c of HTSC materials. For one dimensional resonators and filters with narrow line width typical transmitting power level is in milliwatts. To raise the critical current density J_c of the HTSC materials, a two dimensional configuration for the HTSC resonators and filters can be used. A 4-pole HTSC filter is shown in Fig. 10.56 (a). This square resonator can be treated as a half-wave length transmission line resonator with a very broad line width equal to its length. The widening of HTSC line width does help to increase the power handling capability. However, the problem of current crowding at edges still exists. Therefore, the power handling capability improvement is limited.

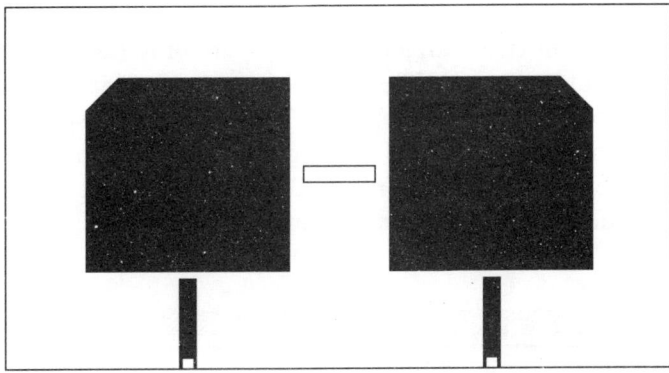

Fig. 10.57 (*a*) A 4-pole dual-mode HTSC bandpass filter.

To solve the current crowding problem at edges, a new type of resonator for high power HTSC filters was invented by Shen et al. [116] as shown in Fig. 10.57 (*b*). This is a 6 GHz 3-pole HTSC filter with TM_{01} mode octagon-shaped planer resonators. Since in TM_{01} mode, the current flows in the circular direction and hence there is no current at the edge due to the fact that the magnetic flux 'wrap around' problem is eliminated.

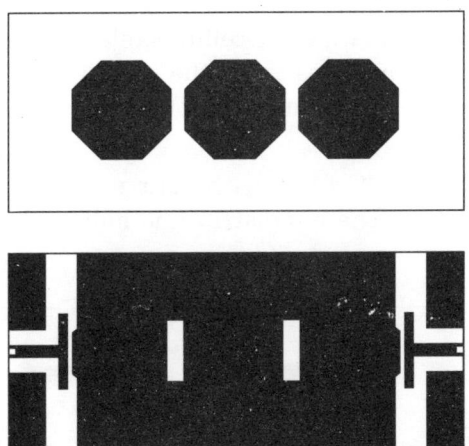

Fig. 10.57 (*b*) The layout of a 3-pole 6.04 GHz HTSC bandpass filter with TM_{01} mode octagonal resonators on a 20-mm thick $LaAlO_3$ substrate. The upper and lower figures show the front-side and back-side of the filter circuit respectively.

The power handling capability of HTSC filters is also limited by their coupling circuits. The coupling mechanisms can be put into two categories: **gas coupling** and **direct contact coupling**. However, none is suitable for high power HTSC filters.

Three Dimensional Resonators and Filters

One can form a three-dimensional HTSC resonator in a different ways : (*a*) **cavity type** : the structure is the same as the conventional resonant cavity made of metal such as copper, (*b*) **HTSC dielectric hybrid type** : Fig. 10.58 shows a typical example, which is a TM_{011} mode HTSC sapphire resonator [117]. This resonator consists of a cylindrical sapphire rod sandwiched by two HTSC films. Due to extremely low surface resistance of the HTSC films and extremely low loss tangent (loss tangent is a measure of *rf* loss in dielectric materials) of the sapphire material, Q-value of such HTSC sapphire resonator is extremely high *i.e.*, 10^6 at 80 K and 10^7 at 20 K. Since the *rf* current in the HTSC-sapphire is spread over a large area of the HTSC film, the power handling capability of resonator is very high. Therefore, this resonator is ideal for serving as the block of high power HTSC filter handling tens of kilowatts. Moreover, the number of poles for such a filter can be very large without sacrifice of its in-band insertion loss due to the extremely high Q-value.

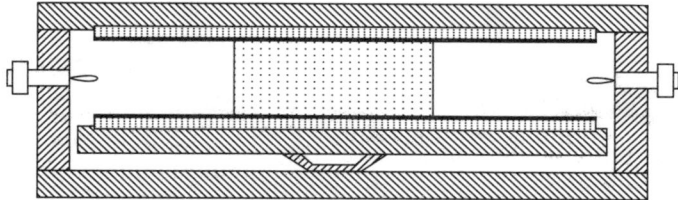

Fig. 10.58 A cross-sectional view of a TM_{011} mode HTSC sapphire resonator in a copper case with input and output coupling circuits.

New Development and Application

HTSC filters have very attractive market for cellular telephone and personal communication system (PCS). Their operating frequencies are in the sub-giga-Hz and low-giga-Hz ranges. The conventional filter configurations are not suitable for such low frequency HTSC filters because of their large size.

The configuration of an HTSC filter with 'hair pin' resonators as its building blocks [118, 119] is shown in Fig. 10.59. The microstrip line half-wavelength resonators are folded into a 'hair-pin' shape.

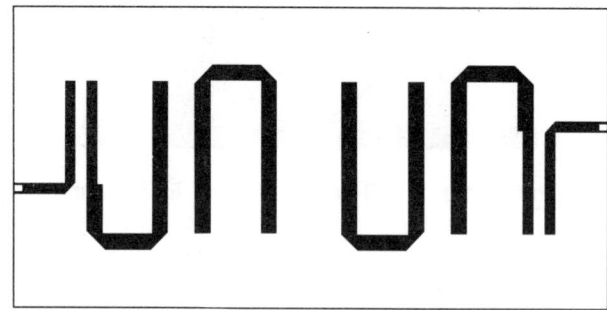

Fig. 10.59 A 4-pole HTSC microstrip line bandpass filter with the 'hair-pin' type resonators.

An HTSC filter with the **'staggered resonator array'** design [119] is shown in Fig. 10.60. This design starts with an array of parallel microstrip lines with equal spacing to form a section of periodic structure. Periodic structure has alternated passing bands and stop bands. The overall size of this filter is smaller than the conventional half wavelength HTSC filters.

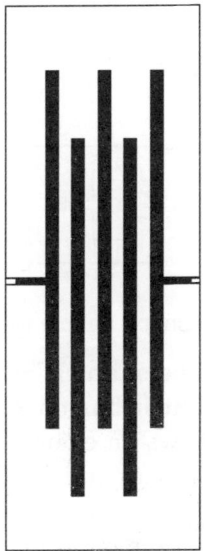

Fig. 10.60 A staggered five-resonator-array HTSC filter in the strip line form with two 20-mm thick LaAlO$_3$ substrates.

To further reduce the size of HTSC filter some novel approaches are needed, *e.g.,* use of slow-wave transmission lines; using 'lumped elements' such as spiral inductor and inter-digital capacitor as the building blocks of the resonant circuit [120, 121].

There has been intensive research to reduce the size of miniature HTSC filters. It seems that efforts in this direction will lead to some new types of very small HTSC filters with good performance.

For wireless communication, the assigned frequency band is limited and very expensive. Therefore, how to utilize the available frequency band more efficiently becomes an important issue. In order to put more channels into a given frequency band, one requires the filter having very steep skirt at the edges of passing band. There are two ways to increase the slope of the skirt : (*i*) to increase the number of poles in the filter, *i.e.*, to add resonators. To utilize the substrate area more efficiently, a modified hairpin type resonator is used as a building block, which is called *J*-shape resonator due to its shape (Fig. 10.61). The 16-pole filter and 32-pole filter are fabricated on a half and a whole 3″ diameter substrate respectively.

Further increasing the number of resonators in an HTSC filter is not practical : (*i*) the size of substrate increases proportional to the number of resonators. The large substrate makes the HTSC filter bulky, and more difficult to cool down to cryogenic temperatures; (*ii*) the signal group delay (group delay is the time for signal passing through a circuit) in a filter is approximately proportional to the number of resonators in the filter. Large group delay is undesirable for digital communication, because it increases the variation of delay time, causing signal distortion.

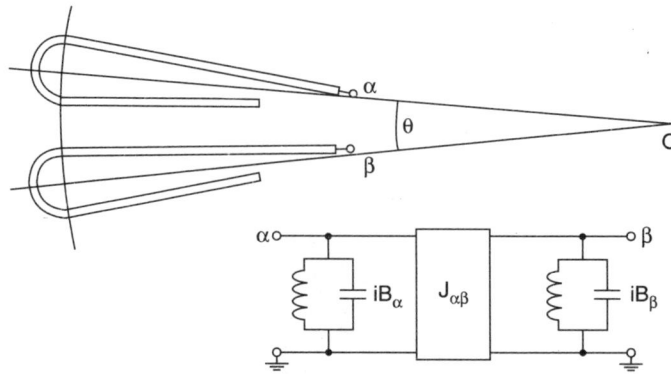

Fig. 10.61 Layout of two coupled J-shape hairpin resonators and the equivalent circuit.

The other way to increase the slope of an HTSC band-pass filter is to introduce 'zeros' (zero also corresponds to a resonance but it has opposite effect of the pole) into its complex frequency response plane. Fig. 10.62 shows an example, in which thin transmission lines are added into the HTSC band-pass filter circuit to introduce additional coupling between non-adjacent resonators. These lines provide an alternate signal path to cancel out the signal at frequencies close to the bottom of skirts to make them steeper.

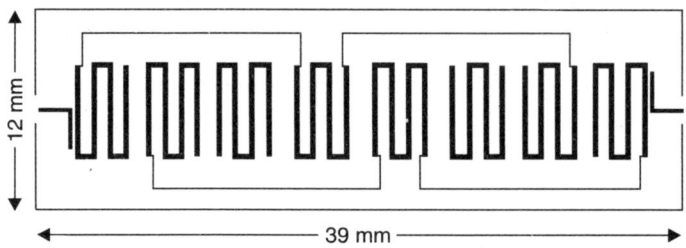

Fig. 10.62 Layout of the 8-pole filter on 0.5-mm thick La AlO$_3$ substrate.

For digital communication, it often requires uniform group delay time over the signal frequency band to avoid signal distortion caused by different delay times for different frequency components. For conventional Chebyshev design (filter with transmitting frequency response as Chebyshev function is called Chebyshev design), the group delay time as a function of frequency has two sharp frequencies, which make the group delay time non-uniform over the passing band of the filter. Fig. 10.63 shows an example. It uses four dual-mode resonators to form an 8-pole filter and two cross couplings, M-14 and M 58, between resonators 1 – 4 and 5 – 8, respectively.

Characteristic of some passive microwave devices are summarized in Table 4 [122].

In past few years, there has been a lot of interest in developing HTSC based subsystems for wireless and communication, satellite applications. Several research projects are in progress for the development of HTSC microwave components and subsystems for use in the transreceiver station of digital mobile communication systems of the second generation (GSM, DAM PS/IS – 136, PDC etc.) and for the forthcoming third generation (W-CDMA, CDMA 2000). Narrow band and compact HTSC receiver band pass filters have been developed for mobile communication applications. The low loss and high selectivity of the planar HTSC multipole filters can induce an enhancement of the base station sensitivity and capacity. However, the major challenge in the development of HTSC based subsystems are not just HTSC components alone but also the associated cryo-packaging and crycooler integration of the subsystems.

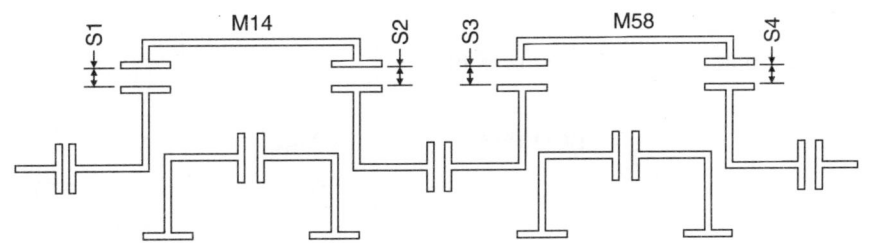

Fig. 10.63 The layout of an 8-pole dual-mode filter.

Table 10.4 Characteristics of some passive microwave devices
[Q ≡ quality factor, f_0 ≡ resonance frequency, IL ≡ insertion loss, BW ≡ bandwidth, η ≡ radiation efficiency of antenna]

Device	High-T_c superconductor	Substrate	Characteristics
Cylindrical cavity resonator	YBCO thick film	YSZ	f_0 = 5.66 GHz, Q = 7.15 × 10^5 (77 K)
Dielectric resonator (sapphire)	TBCCO thin film	LaAlO$_3$	f_0 = 5.55 GHz, Q = 3 × 10^6 (80 K)
Microstrip Resonator	YBCO thin film	LaAlO$_3$	f_0 = 5 GHz, Q = 1.6 × 10^4 (77 K)
Planar disk Resonator	YBCO thin film	LaAlO$_3$	f_0 = 1 GHz, Q = 5 × 10^5 (77 K)
11-pole band pass filter	YBCO thin film	LaAlO$_3$	f_0 = 1.778 GHz, BW = 0.6%, IL = 0.6 dB (65 K)
Cross-coupled band pass filter	YBCO thin film	MgO	f_0 = 5 GHz, BW = 4%, IL = 0.3 dB (77 K)
Microstrip band stop filter	YBCO thin film	LaAlO$_3$	f_0 = 1.623 GHz, BW = 0.32%, IL = 1 dB (63 K)
Lumped element band pass filter	YBCO thin film	MgO	f_0 = 1.777 GHz, BW = 0.84%

(Contd.)...

3-pole lumped element band pass filter	YBCO thin film	LaAlO$_3$	f_0 = 2 GHz, BW = 2%, IL = 0.1 dB (77 K)
Delay line	YBCO thin film	LaAlO$_3$	Delay = 22.5 ns. f_0 = 20 GHz, IL = 5 dB (77 K)
Phaseshifter	YBCO thin film	LaAlO$_3$	Phase shift = 1000°/dB (77 K)
Magnetically tunable 3-pole band pass filter	YBCO thin film	LaAlO$_3$	f_0 = 10 GHz, BW = 1%, tuning $\geq$ 13% (77 K)
Electrically tunable 3-pole band pass filter	YBCO thin film	LaAlO$_3$	f_0 = 2.5 GHz, BW < 2%, tuning $\geq$ 15% (77 K)
Patch antenna	YBCO thin film	MgO	f_0 = 20 GHz, BW = 6.7%,
Loop antenna	YBCO thin film	YSZ	f_0 = 410 MHz, BW = 1.4%, η = 80% (77 K)

10.3.8 Transmission Lines

Superconducting transmission lines can be of two types : (*i*) superconducting coaxial cables ; and (*ii*) thin film microwave miroelectronic components. The superconducting coaxial cable is used as a basis for low loss delay lines having a low frequency dispersion and is considered as a communication line in combination with a high power cable operating at cryogenic temperature. Much more widespread in modern electronics are the superconducting microwave microelectronic components. They are applied as interconnections in microwave integrated circuits (MIC), as a fundamental part of resonators and filters based on microstrip lines or coplanar waveguides (CPW), and as feeders in microwave planar antennas.

Presently, the most attention is being paid to development and applications of thin film transmission lines based on MTSCs (YBCO, TBCCO). The fundamental characteristics of thin film transmission lines are the propagation parameters : phase velocity, attenuation constant and characteristic impedance. The propagation parameters are determined by geometry of the line, dielectric permittivity of the substrate materials, current distribution in the cross-section of the transmission line and the surface impedance of the superconducting film. The surface impedance of the superconducting film depends on the temperature, the film thickness and the structural quality of the film. We may note the microwave power handling capability of the superconducting transmission line, which is extremely sensitive to the defect structure of the superconducting film.

The specific area of application for transmission lines is delay lines. The use of superconducting materials makes it possible to realise a delay line with extremely low attenuation and extremely low frequency dispersion of the delay time. The impressive parameters of such delay lines were obtained with the delay lines on a niobium coaxial cable operating at 4.2 K. Thin film delay lines based on LTSCs and HTSCs are used in systems for microwave signal processing, *e.g.* in so-called chirp radar.

10.3.9 Antennae

Antenna plays an important role in transmitting and receiving signals through free space. Numerous technical applications rely on "wireless" propagation through the atmosphere and through ground. Mobile communications, wireless local area networks (W-LANs), terrestrial and spaceborne point-to-point links as well as broadcasting systems may serve as examples from the wide field of communications. Prominent examples for remote sensing and navigational applications are radar techniques, GPS, direction finding, radio astronomy, microwave radiometry and geophysical subsurface probing. Examples for further sensor applications of electromagnetic waves are gas spectroscopy and moisture content measurements. The electromagnetic spectrum employed in these various applications reaches from Hz (*e.g.* long range navigation) to the THz (sub-millimeter) regime (*e.g.* gas spectroscopy).

The technical systems for these applications comprise radio frequency (RF) circuits where RF signals are generated, frequency converted, modulated and amplified and where analog and digital signal processing is applied in order to gain the desired information. In these circuits electromagnetic (*em*) waves are propagating in transmission lines and wave guide structures. The function of an antenna is to provide a transition from the guided waves in the RF circuits to the waves in open space and vice versa. Antenna structures possess one or several ports which are connected to transmission lines and/or directly connected to active elements (*e.g.* transistors). In the transit mode of an antenna, guided electromagnetic waves are incident at the antenna port(s) and the function of the antenna is to radiate the incident RF power into the surrounding open structure. This has to be done in a pre-specified manner with respect to the directional dependence of the radiation density, polarization and phase ('radiation pattern'). Moreover, the transition from guided waves to waves in the open structure is associated with certain 'imperfections' which cannot be avoided due to physical restrictions. This includes the fact that the radiated power is lower than the incident power due to reflections at the antenna ports and dissipation in the antenna structure (conductor losses and dielectric losses). As a consequence of the unavoidable storage of reactive electromagnetic field energy in the antenna structure and its close vicinity ('near field zone') the worth of the frequency range within which the input reflection factor can be kept sufficiently low is restricted (limiting 'matching bandwidth').

In the receive mode of antenna, *em* waves from different directions are impinging upon the antenna. The function of the antenna is to convert these waves into guided waves available at the antenna port(s). This, conversion is associated with a pre-specified 'spatial filtering according to the radiation pattern. Therefore, the ratio of the power available at the port(s) to the radiation density of the incident wave becomes a function of the direction of incidence and polarization. Through reciprocity antenna properties in the receive mode are uniquely related to properties in the transit mode. This applies to the radiation pattern as well as to the matching at the ports. If in the transit mode radiated power is reduced due to reflections at an antenna port, the same antenna will exhibit a reduced power delivered to the matched load in the receive mode.

Desired properties of the antenna depend on the specific system application and can therefore be very different for different applications. Antenna may be required to possess a highly directive, moderately directive and omni-directional radiation pattern.

Highly directive antenna with rotationally symmetric pencil beams or with fan beams are often realized as reflector antennae with one parabolic reflector, illuminated by a moderate directive feed antenna or with two reflectors (*e.g.* parabolic main reflector and hyperbolic subreflector). Alternatively, arrays composed of a large number of antenna elements with low directivity and a mutual spacing of about 0.5 – 1.5 wavelengths may be used for highly directive antennae.

There are other applications where an antenna with wide angular coverage is required, *e.g.* the antenna for a handheld station for mobile communication. These antennae can typically be realized as three-dimensional dipole or monopole antennae, or as printed antenna (*e.g.* micro-strip patch antenna).

Antennae with multiple ports allow the radiation pattern to be changed electronically. This 'beam forming' is based on a controlled phase and amplitude 'weighing' of the waves received at the ports (receive mode) or fed into the ports (transit mode). This weighing can be performed with analog control elements, *e.g.* in antenna arrays with electronic beam steering. Alternatively, weighing can be realized in the digital part, leading to the concept of simultaneous adaptive digital beam-forming ('smart antennae').

The required (matching) frequency bandwidth of antennae varies significantly as a function of the system application. Antennae for conventional mobile communication applications (*e.g.* GSM standard) require a fractional frequency bandwidth of about 2 – 3%. Envisaged ultra-wideband systems (UWS) for communication applications are using picosecond pulses in order to spread the electromagnetic power over an extremely wide frequency band (several GHz), resulting in a spectral power density which is far below the power density of co-existing systems ('spread-spectrum concept). Consequently, a bandwidth of more than one octave is required for UWS antennae. This also applies to antennae for bore-hole radar applications where multiple-octave bandwidth pulses are used to achieve a sufficient range resolution.

The application of HTSC technology to antenna techniques is justified only in those cases where a considerable system benefit can be achieved, *e.g.* low FR conductor losses. There are several applications related with antenna such as superdirective antenna arrays or small antenna in use of HTSC material in place of normal conductor leads to an increase of efficiency due to smaller value of surface resistance. Loss in matching network, which connects the active device to antenna is particularly important for short dipole antenna and this may be significantly reduced by use of HTSCs with low surface resistance. The first HTSC antenna was demonstrated by the Birmingham group in 1988 [123]. This antenna consisted of a short dipole (20 mm) made out of bulk polycrystalline YBCO material. The antenna was operated at 550 MHz and enhancement in the gain was 12 dB as compared to the copper antenna operating at room temperature. Since then, several other groups have successfully fabricated different type of HTSC antenna [124] such as microstrip antenna and mender line antenna. For a 2.4 GHz antenna, the conventional patch would have a length of 42 mm on RT-duroid ($\varepsilon_r = 2.2$) whereas the YBCO '*H*' antenna on La AlO$_3$ substrate ($\varepsilon_r = 24$) is of 6 mm in length. HighT_c superconductor 4-element patch array antenna has been developed for satellite communication that showed 3.6 dB higher gain as compared to copper array [125]. For a 16-element YBCO HTSC superdirective array, a supergain of 4.2 has been reported [126].

Provided that conductor losses limit the system relevant performance of antenna and matching network structures, implementation of HTSC technology should be considered. However, technological restrictions as well as cost have to be taken into account.

10.3.10 Refrigerations for Cryoelectronics

Although some superconductors now operate at temperatures well above liquid nitrogen temperature (77 K), by everyday standards they still need to be very cold. In order to achieve this, refrigeration, using cryoengineering at various levels of sophistication depending on the application, is needed. The availability of reliable, long lived, low vibration, compact, economical refrigerators offering capacities of 1 W at 80 K is very important for the wider adoption of cryoelectronics in industry. Liquid nitrogen dewars are bulky and need to be refillled with liquid nitrogen after a certain duration. This poses a serious problem for widespread application of HTSC electronic devices. For more extensive and controlled temperature ranges, mechanical refrigerators, usually known as **cryocoolers**, are necessary. These have to access the range approximately 120 K down to 4 K or lower. There are different type of cryocoolers and broadly speaking two types of cycles recuperative cycles, such as Joule-Thompson and Brayton and regenerative cycles, *e.g.* Stirling, Gifford, McMahon and pulse tubes [127]. All of these rely on gas compression and expansion. The differing requirements of various superconducting devices often lead to different cooling methods being employed. The recuperative systems are usually used for large-scale open cycle application such as liquefiers or sometimes on very large superconducting magnets that need to operate below 4 K. The refrigeration power required for HTSC electronic devices is usually less than a few Watts at temperatures between 60 and 80 K.

One of the most challenging problems in cooling superconductors with cryocoolers is that of reducing the associated vibration and electromagnetic interference (EMI) caused by the motor and other moving parts. The problem is most serious with SQUID devices because of their extreme sensitivity to magnetic fields and to vibration in the earth's magnetic field. Excessive vibration of superconductor-insulator-superconductor (SIS) junctions for microwave receivers can lead to distortions of the signature.

To reduce noise in a superconducting device caused by the cryocooler, we must take care of : (*i*) selection of cryocooler type, (*ii*) selection of materials, (*iii*) distance between cryocooler and superconductor, (*iv*) mounting platforms, (*v*) shielding, and (*vi*) thermal damping, and (*vii*) signal processing.

Interference problems are less severe when using cryocooler with other superconducting devices, *e.g.* microwave filters for wireless communication or in satellite stations [121]. An integrated receiver front and unit with high-T_c noise reduction filter and low noise amplifier (LNA) has been developed by several research groups [130 – 132]. They have made use of high Q high-T_c superconducting filters, low noise cryogenic amplifier and highly reliable cooler. A 60 channel superconducting input multiplexer integrated with pulse tube cryocooler [133] and miniaturized cooling system for HTSC antenna [134] have also been reported in the literature.

Major-progress has been made in Stirling and pulse-tube coolers. A Stirling cooler made by TRW offers cooling power of 250 mW at 65 K; the cooler weigh 1.4 kg and the electronics 0.75 kg. The compressor consumes 12 W, and the electronics 5 W, part of which is used to suppress vibration. The lifetime is claimed to be 10 years [128]. Pulse-tube coolers have the further advantage that there are no moving parts in the vicinity of the cooler and the compressor can be located many meters away. Toshiba, Nihon University and the Institute of Space and Astronautical science [135] have developed a cooler with a 278 W linear drive, providing 10 W of cooling power at 80 K and a minimum temperature of 35 K.

In general, it is the regenerative type of cryocooler that finds most application areas. The last category of regenerative cryocooler dealt with in the pulse tube, which is relatively recent development, deals explicitly with advances in pulse tubes that have made them serious candidates for many applications. A comparative study of various cryocoolers shows that a particular type of pulse tube can be the most efficient of all cryocoolers.

The simplified cooling that can be employed with HTSCs, coupled with substantial progress in cryoelectronic HTSC devices, leading to performance figures similar to those achieved with LTSCs, have extended the range of electronic applications of superconductivity to include nonlaboratory settings.

Cost always play an important role when attempting to market a superconducting device. In terms of any marketable product, the superconducting device and the cryocooler must be considered an inseparable pair. Often the cost of the superconductor/cryocooler system is dominated by the cost of cryocooler. Typically the user in the high-T_c device field is interested in, Mean Time to Failure (MTIF) of at least 3 years and usually 5 years. Some maintainance during this time may be tolerated, but it should be minimal (a few hours at the most) and usually not more than once in a year. Coil moving parts must be oil free and the need for such long lifetimes under such conditions generally requires non-contact bearings. Such special bearings can greatly enhance the cost. The development of cryocoolers for space applications has led to cryocooler lifetimes of about 10 years with no maintainence but costs are usually very high. Reducing these costs for commercial applications, but still maintaining a comparable reliability is the key issue of research on cryocoolers.

Summarizing we can say that for power applications, efficiency cost and reliability are the main considerations whereas with superconducting devices, such as SQUIDs, considerations like vibrations EMI and oscillation are important.

10.3.11 High-T_c Superconductor Radiation Detectors

Highly sensitive high-T_c superconductor bolometers for measuring thermal radiation can be created simply by operating superconducting strip in the temperature range containing the very steep $R(T)$ transition to normal conduction, then measuring the voltage drop across the strip, which varies as a function of the incident radiation. Above 77 K, infrared (IR) sensitivity close to the theoretical limit of $D^* = 2 \times 10^{10}$ cm Hz$^{1/2}$ W^{-1} can be achieved for 0.2 Hz to 2 Hz which is about one order better than the commercially available pyroelectric or thermopile IR detector. Suitable design will allow response times in the microsecond range and shorter.

Broadly speaking, high-T_c superconductor IR detector can be placed under two main categories : (i) thermal and (ii) non-thermal or quantum. In the thermal approach, incoming radiant power causes a rise in temperature of the HTSC lattice. This rise in temperature, then modulate some HTSC property such as resistance. The thermal detector is potentially broad band, limited by spectral properties of the absorber. However, it is potentially slow especially due to the thermal inertia of the substrate.

An alternate approach is the IR or quantum detector. In IR or quantum detector, the power directly interacts with the HTSC material with a quantum efficiency. However, the basic idea is not to excite the phonons, but rather to directly influence the Cooper pairs. The incoming signal pulse directly leads to a reduction in the Cooper pair density, without influencing the phonons. Reduction in the Cooper pair density modifies HTSC property e.g. kinetic inductance or critical current. The quantum detector is potentially very fast, but at the expense of not being quite as broadband as the thermal one. Requiring Cooper pairs, quantum detectors operate at or below T_c whereas thermal detectors can operate below, at, or above T_c.

Prototypes of bolometers for IR and far IR, which can be used in spectrometry, radio astronomy, and space applications have also been reported. High-T_c bolometers will be favoured as compared to LTSC bolometer in astronomy and atmospheric physics where the higher operating temperature provide distinct advantage. One such detector is also intended for deployment in the HTSSE experiment. The state of the art is as follows : sensitivity upto 1750 V/W, noise down in 7 ± 10^{-10} W/Hz$^{1/2}$ at 10 Hz, signal risetimes < 0.5 ms, chopper modulation at 100 Hz for 45 nm thick YBCO on a thinned Si substrate with buffer [136].

For fabricating high-T_c film IR detector, high-T_c films needs to be prepared on a substrate having good thermal conductance and excellent-IR performances. It is reported that silicon and sapphire work well. However, a buffer layer is deposited before depositing high-T_c films [137].

10.3.12 Superconducting Transistors

Interest in superconducting transistors is due to unique potential benefits of these devices. A superconducting, and thus virtually loss-free, on state of the drain source channel (DS channel) is the most important example of such an advantage, the ability to sustain high current densities is another. In addition, with regard to fabrication and operation, superconducting transistors are compatible with other superconducting components, and may in some cases even be required to interface with semiconducting electronics. There are superconducting base transistors [138, 139], superconducting FET [140] and others [141]. However, no gain was observed with LTSC transistors, because the superconducting energy gap is much smaller than that of a semiconductor. When the limitations of Josephson junctions became apparent research on superconducting transistors was stepped up and intensified even more when HTSC cuprates was discovered [142].

HTSCs have lower carrier concentration as compared to LTSCs. Thus, the penetration depth (λ) of HTSCs is larger. This provides the possibility that an applied electric field could penetrate deep enough to produce a measurable change in the supercurrent and thus gives rise to field effect transistor (FET). The grain boundaries are generally have lower carrier density and the applied electric field can modify the Josephson current across the grain boundary embedded in a superconducting drain-source channel in a similar way as gate voltage modifies the current through the semiconductor FET. For a bicrystal grain boundary junction, the application of electric field has been found to produce drastic change in critical current of the grain boundary. Two other approaches have also been explored to produce high-T_c three terminal device using grain boundaries. These are using quasiparticle injection in the grain boundary and modulating grain boundary's resistance by local magnetic field. Based on these ideas, several high-T_c FET have been reported [142]. However, still high-T_c transistors are at research stage and these devices can succeed if their performance guarantees a substantial advantage over that of cooled semiconducting electronics at the time of introduction to the market. This can only be achieved if they exploit basic superconducting properties of HTSC cuprates. Obviously, loss-less-gain and high-frequency performance will be crucial parameters. Presently our understanding of such basic aspects as noise, speed or power handling capabilities is still so limited that it is fairly impossible to predict accurately the ultimate performance of these high-T_c transistors. Research in this field may lead to devices that can be used for exciting science experiments and hopefully also to devices having significant commercial applications.

10.3.13 Digital Circuits and Electronics

Superconducting digital electronics is a promising field and offers the possibility of high speed operation at extremely low power. The low power allows high packing density. Beyond semiconductors, it is the most promising technology. Presently, the developments of Rapid-Single Flux Quantum (RSFQ) logic circuits are the most exciting field in superconducting electronics. RSFQ has ultra-fast operating speed of several hundred GHz and very small power dissipation of the order of 10 nW. The basic principle of the RSFQ device is very simple. The quantum flux stored in a superconducting loop in used as an individual bit of information. However, the quality of Josephson junction (JJ) included in the loop is most important factor to construct integrated RSFQ circuit.

The shortest switching time per gate [143] is 2.5 ps for a Nb Josephson junction operated at 4.2 K. The operating voltage V_0 is proportional to T_c of the material used. For Nb with T_c = 9 K, $V_0 \sim 3$ mV. Hence, we expect the operating voltage of ~ 30 mV for a 90 K HTSC. In order to compensate thermal noise at high temperature, a high current is required. For a low temperature Josephson switching device, the operating current I_0 is 1 – 0.5 mA. Accordingly, at 77 K, we have $I_0 \sim 2 - 10$ mA. This results in an increase of power consumption by a factor of 200 operating at 77 K with HTSC. This consumption, giving rise to heat, makes a higher density of Josephson devices at 77 K difficult.

The switching time for a Josephson gate is determined by the RC time constant, where C is the junction capacitance and R the impedance of the strip line. When HTSC are used, we expect the junction dimensions to be approximately the same as that of the LTSC junction. If the junction barriers are the same, C should be about the same for a junction using HTSC. Roughly $R \simeq V_0/I_0$, and is expected to be about the same for the switches both at 4.2 and 77 K. As a result, the switching time for a 77 K Josephson junction is about the same as that for a low-temperature junction operated at 4.2 K. Theoretically, the limiting switching time is determined by the energy gap E_g. The theoretical limiting switching time τ is then $\tau = \hbar/E_g$. For Nb with $E_g \sim 3$ mV, τ is 0.2 ps, which is 10 times shorter that observed presently. In order to achieve such a short switching time, C has to be reduced. The dimension of the junction has thus to be reduced to ~ 0.1 µm. For HTSC, the dimension should accordingly to be reduced to ~ 0.1 µm, which is too small for the current technology to be accomplished. However, most modern superconductor logic work concentrates on RSFQ. The RSFQ family offers a unique combination of high speed, low power and the demonstrated capability to design and make working circuits of LSI complexity [144, 145]. Some of the major applications simply make use of these properties of speed and less power. Others derive additional benefits from the quantum nature of underlying physics, which can provide great precision in analog circuits.

HTSC digital circuits are more suitable for use in single flux quantum (SFQ) circuits than LTSC ones, because the junctions in SFQ circuits must be overdamped. HTSC Josephson junctions (JJs) are naturally overdamped, which means that their $I - V$ curves do not show hysteresis. As explained above the $I_c R_n$ product Df HTSC junctions expected to be larger than that of LTSC junctions due to its dependence on the gap voltage of the superconductor.

Several HTSC digital circuits have been fabricated *e.g.* RSFQ circuits consisting of *dc* SQF converters, Josephson transmission line, RS-flip flop, read out SQUID and Analog to Digital (A/D) converter, comparators, voltage divider, sigma-delta modulator, sampler, etc [146 – 152]. These HTSC digital circuits are summarized in Table 10.5.

Table 10.5. HTSC superconductors digital circuits

HTSC ditital circuits	Kind of JJs	Number of JJs	Number of HTSC layers	Maximum operating temperature (K)
3 bit shift resistor	Bicrystal	26	1	50
Sampler	Ramp-edge	5	3	50
Ring oscillator	FEBI	15	1	39
T-FF	Ramp edge	14	3	65
Comparator	Ramp edge	8	3	30
Σ-Δ Modulator	Bicrystal	10	2	33
Voltage divider	Ramp edge	11	2	27

To develop large HTSC digital circuits, the spread of critical current of HTSC Josephson junction (JJ) needs to very small. Presently, for a 100 ramp edge HTSC junction, a spread of 8% is reported. It is expected that with 1000 HTSC junctions, 6% deviations will be achieved soon. This reveals that the sigma-delta (Σ–Δ) **Analog to Digital (A/D) converter** will be made in the yield of 50% if suitable designs are prepared. As stated earlier, the problem of thermal noise and parasitic inductance will have to be handled. LTSC flash converters have demonstrated 20 GSs^{-1} operation. This is the fastest single converter [153] made to date in any technology. Sigma delta (Σ–Δ) [154] and delta converters [155, 156] offer high dynamic range at speeds to say 10 – 200 MS s^{-1}. The performance is comparable to the semiconductor state of the art, and is expected to exceed it soon.

Time to digital converters [158] and digital auto-correlators [158] have been made for applications in high energy physics and radio astronomy.

10.4 SUPERCONDUCTING MIXERS

There are essentially two very successful concepts of superconducting mixers : (*i*) superconducting-insulator-superconductor (SIS) tunnel junction mixer, and (*ii*) the superconducting hot electron bolometer (HEB) mixer. HTSC Josephson mixers may become competitive to Schottky mixers or *e.g.* space-borne applications with passive radiation cooling. Presently, SIS and HEB mixers seem to be the preferred choice for ultimate performance, at least in the range 100 GHz upto a couple of terahertz if cooling down to 4 – 10 K can be provided. In semiconductor technology, High Electron Mobility Transistor (HEMT) amplifiers (which do not need a LO oscillator) are approaching the performance of SIS mixers upto about 100 GHz and Schottky mixers have been operated beyond 2T Hz. Excellent noise properties and possibilities to produce net conversion gain are the advantages with superconducting mixers. Superconducting mixers also require extremely low LO power. Presently, SIS mixer give the lowest receiver noise temperatures between 100 GHz and 1 THz and HEB mixers upto 5 T HZ. In some applications, these mixers may have fairly low saturation power, but this may be improved by using arrays of mixing elements. The another disadvantage of these mixers is the requirement of cooling to cryogenic temperatures. SIS mixers are the best elements with quantum limited noise levels upto twice their gap frequency, typically beyond 1 T Hz. Above this frequency, superconducting HEB mixers seem to be the successor to the SIS technology [162 – 165].

10.5 OTHER SUPERCONDUCTING DEVICES

(*i*) **Particle and Photon Detection.** Particle detection can be made more sensitive than any other method by using devices based on various aspects of superconductivity *e.g.* the extremely rapid change of resistance with temperature, which occurs as a pure superconductor thin film passes through its superconducting transition, provides the basis for an extremely sensitive low thermal mass thermometer. By combining this with a mechanism for absorbing particles such as photons, phonons or massive particles (including atoms, molecules, ions or electrons) the small temperature rise produced by absorption of a single particle may be readily detected. Depending on the nature of the absorbed particle and the transfer of the absorbed energy to bolometer such a detector may also be used as a spectrometer *i.e.*, a detector with absorbed particle energy determination.

Normal Metal-Insulator-Superconductor (NIS) and Transition Edge Sensor (TES) calorimeters have been developed. Both these devices have high sensitivity and can have very good very resolution which depends upon the operating temperature and size of the absorber. An energy resolution $\delta E = 20$ eV for 6 keV X-rays with $\tau = 15$ µs has been achieved with a NIS detector operating at 80 mK and having 0.5 µm thick Au absorber with an area of 100×100 µm^2 [166].

More extensive developments of TES devices have been pursued. In most cases the temperature of operation is adjusted to be a particular value in the range roughly 50 – 300 mK by using a bilayer such as Al-Ag, Ti-Au, or Mo-Cu and using the proximity effect to determine the T_c of the bilayer film [166]. Careful methods of fabrication lead to reproducibility and to transition widths of less than 1 mK. Excellent energy resolutions for X-rays have been achieved with TES devices e.g. 2 eV at 1.5 keV [168] and 4.5 eV at 6 keV [168]. TES devices have also been used to measure the energies of single infrared, optical and UV photons [169], and arrays of devices have been developed [170]. Although these successes are remarkable, there remains the limitation on the detector volume due to the electronic specific heat of the N-metal for the NIS junction and of supercoductor at T_c for NIS.

The development of particle and photon detectors based on superconductivity has matured to that stage where detector devices are being used almost routinely in basic science and in some technological applications. Interestingly these detectors do not just improve upon conventional techniques, but are being used to carry out measurements which cannot be done with any conventional techniques. Developments have reached to that stage where device fabrication is being carried out by industry. We expect that the area of particle and photon detection and spectrometry will soon become a very significant part of the field of applied superconductivity.

(*ii*) **Optical Sensors.** Optical sensors are evolving applications uses very small superconducting devices. Although the energy of a visible photon is greater than that of the binding energy of a cuprate electron pair (so that cooper pairs, *i.e.* superconductivity can be disrupted when they are incident on a superconductor), nevertheless it has been demonstrated that thin microbridges of superconductor can show a fast response to incident near infrared or visible radiation. Interestingly, some features of this optical response persist into the normal state of cuprate thin films.

The development of superconducting single-flux-quantum (SFQ) logical devices, based on resistively shunted Josephson tunnel junctions requires ultrafast and ultrasensitive detectors for fibre optic interconnects between the outside world and the SFQ processor, which must be able to work at cryogenic temperatures and be technologically compatible with SFQ integrated circuits. The superconducting optical-to-electrical transducers can transform the input information coded in the form of a train of ultrafast optical pulses to the electrical domain and, subsequently, feed it into the ultrafast superconducting processor. Supeconducting optical sensors also find applications in traditional areas of optoelectronics and infrared (IR) imaging. Mid-IR optical radiation spectrum is especially important, since 3 – 5 µm and 10 – 13 µm bands correspond to the transmittance windows in the earth atmosphere (minimal H_2O absorption), thus, they are crucial for *e.g.*, effective satellite communication and sensing. Optical fibres or the few µm radiation spectrum with **ultra-low** losses can also be fabricated and form the basis for future advanced telecommunication systems. Contrary to current semiconductor optoelectronic compounds that lack adequately low band-gap values, superconducting photodetectors have essentially no band-gap limitations and are known to be very effective mixers in the terahertz frequency area as well as sensitive X-ray radiation detectors. NbN hot electron superconducting photodetectors (NbN HESPs) are able to detect single quanta of not only visible but also IR radiation and successfully compete with other designs (such as photo multipliers and superconducting-insulator tunnelling structures [171]) as single-photon detector

[172]. Besides very high sensitivity and picosecond response time, the other unique advantage of the ultrathin NbN-hot electron device is that it operates at approximately 10 K and this is an appropriate operating temperature for the NbN-digital circuitry. Thus, the entire sensor/processor unit can be jointly cooled to 10 K, and this temperature is much more readily attained than the 4 K required for Nb-based superconducting devices. Very recently, new photoresponse experiments with femtosecond optical excitations and optoelectronic registration of transient waveforms have been developed and provided the most direct information on nonequilibrium processes in HTSC [173 – 175].

HTSC HESPs are ideally suited for digital and communication applications because of their high absorption coefficient at essentially any wavelength and the ultrafast response. Simple YBCO microbridges exhibit single-picosecond response times, making them not only the cheapest, but also one of the fastest optoelectronic switches.

The discovery of HTSCs has also prompted interest in the development of HTSC bolometric photodetectors. The best reported value for detectivity D^* for YBCO bolometers are of the order of 10^{10}, which is well above 10^8 reported for best uncooled devices. However, since D^* generally increases with the decrease of temperature and hence LTSC bolometers operating at helium temperatures or below are consistently better than YBCO detectors at 77 K. Uncooled IR YBCO detectors are presently, the most attractive HTSC-based devices for mass commercial applications.

(*iii*) **Metrology.** Metrology is the science of precision measurement of physical quantities and has developed a close relationship with superconductivity and superconducting materials.

The main source of close relationship between metrology and superconductivity lies with the extreme accuracy of *ac* Josephson relationship and the fact that a SQUID can be converted into a highly sensitive detector of almost any physical quantity by choice of a suitable input transducer. But other areas of overlap include the provision of precise and stable homogeneous magnetic fields as well as the implementation of low loss microwave resonators for clocks and frequency standards. Novel direct current comparator devices have also had great impact.

In this chapter, we have presented in brief a few applications of superconductivity. The field is very vast and cannot be covered in one chapter.

REFERENCES

1. P.J. Lee (Ed.); *Engineering Superconductivity,* Wiley Interscience (2001).
2. D.A. Cardwell and David S. Ginley, *Handbook of Superconducting Materials* : Volume 11 Characterization, Applications and Cryogenics', Institute of Physics Publishing (2003).
3. S.L. Kakani, '*Superconductivity : Current Topics*', Bookman Associates, Jaipur (2001).
4. S.W. Van Sciver and K.R. Marken, *Physics Today,* August 37 (2002).
5. M. Suenaga and A.F. Clark (Eds.) : '*Filamentary A 15 Superconductors*', Plenum Press, NewYork (1980), pp. 17 – 34.
6. M.N. Wilson (Ed.) : '*Superconducting Magnets*' Clarendon Press, Oxford (1983).
7. L. Intichar, and D. Lambrecht :'*Superconductivity in Energy Technologies*', VLI-Verlag Düsseldorf (1990), pp. 98 – 140.

8. H. Brechna (Ed.) : *'Superconducting Magnet Systems*. Springer Verlag, Berlin (1973).
9. H. Kolm et al. (Eds.), *High Magnetic Fields* (Proc. conf.), MIT Press, Cambridge, Mass; and J. Wiley, NewYork (1962).
10. J.E. Kunzler, *IEEE Trans. Tagn.* 23, 396 (1987).
11. K. Sato et al., *Proc. Int. Symp. Superconductivity,* 2nd, 1990, 335 – 340.
12. D.F. Lee et al., *Physica C* 202, 83 (1992).
13. R.P. Reade et al., *Appl. Phys. Lett.* 61, 2231 (1992).
14. P.A. Cheremnykh, *IEEE Trans. Magn.* MAG- 28, 651 (1992).
15. J.R. Powell, *ASME Railroad Conference*, Paper 63-RR-4 (April 23 – 25, 1963).
16. J.R. Powell and G.R. Danby, *Mech. Eng.* 89, 30 (1967).
17. J.R. Powell and G.R. Danby, in *'Recent Advances in Engineering Science'*, ed. A.C. Eringen (Gordon and Breach, NewYork, 1970).
18. W. Menden et al., *Proc. Maglev '89 Intl. Conf.,* pp 11 – 18 (IEE, Japan).
19. Y. Kyotani, *IEEE Trans. Magn.* 24, 804 (1988).
20. T. Nagaika and H. Takatsuka, *Proc. Maglev' 89 Intl. Conf.,* pp. 29 – 35 (IEE, Japan).
21. S. Fuchino et al., *IEEE Trans. Magn.* MAG-28, 279 – 282 (1992).
22. G. Bogner, *'Superconductivity in Energy Technologies,'* VDI-Verlag, Dusseldorf, pp. 123 – 139 (1990).
23. R.F. Giese, *'Superconducting Transmission cables,* Argonne National Laboratory 9700 South Class Avenue, Argonne 111, 60439, Feb. 17, 1993.
24. P.A. Klaudy, *Advances in Cryogenic Eng.* 2, 684 (1966).
25. A. Bolza, *IEEE Trans. Appl. Supercond.* 7, 339 (1997).
26. K. Roth, *Proc. 11th Int. Conf. on Cryogenic Engineering*, Butterworth, Woburn, Mass. p. 48 (1986).
27. H. Siebold; *"Physics and Materials Science of High Temperature Superconductors"*, NATO ASI Ser. Ser. E 81, 665 – 680 (1990).
28. J. Vetter et al., *IEEE Trans. Magn.* MAG-24 1285 (1988).
29. B. Turck, IEEE Trans. Magn. MAG-25, 1473 – 1480 (1989).
30. C.K. McMichael et al., *Appl. Phys. Lett.* 60, 1893 (1992).
31. S. Earnshaw, *Trans. Cambridge Phil. Soc.* 7, 97 (1842).
32. B.R. Weinberger et al., *Appl. Phys. Lett.* 59, 1132 (1991).
33. F. Negrini et al., *IEEE Trans. Magn.* MAG-28, 390 – 393 (1992).
34. P. Komarek, *Proc. Int. Symp. Superconductivity*, 5th 1199 – 1204 (1992).
35. D. Ito et al., *IEEE Trans. Magn.* MAG-28, 438 – 441 (1992).
36. H. Weinstock and M. Nisenoffs (eds.), *'Superconducting Electronics'*: Springer Verlag, Berlin (1988).
37. G. Costabile et al. (eds.): *'Nonlinear Superconducting Electronics and Josephson Devices,* Plenum Press, NewYork (1990).
38. T. Ruggiero and D.A. Rudman: *'Superconducting Devices',* Academic Press, Boston (1990).
39. A. Barone (ed.): *'Principles and Applications of Superconducting Quantum Interference Devices',* World Scientific, Singapore (1992).
40. A.H. Silver, *IEEE Trans. Appl. Supercond.* 7 : 69 (1997).

41. T.V. Duzer, *IEEE Trans. Trans. Appl. Supercond.* 7, 98 (1997).
42. J.M. Rowell, *IEEE Trans. Appl. Supercond.* 9, 2837 (1990).
43. K.K. Likharev, *Supercond. Sci. Technol.* 3, 325 (1990).
44. S. Tanaka, *Physica C* 341 – 348, 31 (2000).
45. A.I. Braginski, *IEEE Trans. Appl. Supercond.* 9, 2825 (1999).
46. R.R. Mansour, *IEEE Trans Microwave Theory and Techniques* 50, 750 (2001).
47. B.A. Willemsen, *IEEE Trans. Appl. Superconductivity* 11, 60 (2001).
48. R.B. Greed et al., *IEEE Trans. Appl. Superconductivity* 9, 4002 (1999).
49. D. Jedamzik et al., *IEEE Trans. Appl. Superconductivity* 9, 4022 (1999).
50. M. Klauda et al., *IEEE Trans. Microwave Theory and Techniques* 48, 1227 (2000).
51. E.R. Soares, et al., *IEEE Trans. Microwave Theory and Techniques* 48, 1190 (2000).
52. E.M. Saenz et al., *IEEE Trans. Appl. Superconductivity* 11, 395 (2001).
53. J. Mannhart and P. Chaudhari, *Phys. Today* 58, 48 (2001).
54. Y.M. Zhang et al., *'4th Int. Superconductive Electronics Conf. ISEC'* 93, August 11 – 14, Boulder, CO, pp. 2c – 4, 9 (1993).
55. J.S. Martens et al., in [54] pp. 2d – 4, 17.
56. A.K. Gupta et al., *Pramana-J. Phys.* 28, L 705 (1987).
57. P. Chaudhari et al., *Phys. Rev. Lett.* 60, 1653 (1988).
58. D. Dimos et al., *Phys. Rev. B* 41, 4038 (1990).
59. R.C. Jaklevic et al., *Phys. Rev.* 140, A, 1628 (1965).
60. J.E. Zimmerman and A.H. Silver, *Phys. Rev.* 141, 367 (1966).
61. J. Lambe et al., *Phys. Lett.* 11, 16 (1964).
62. A.H. Silver and J.E. Zimmerman, *Phys. Rev. Lett.* 15, 888 (1965).
63. A.H. Silver and J.E. Zimmermann, *Phys. Rev.* 157, 317 (1967).
64. D. Koelle et al., *Rev. Mod. Phys.* 71, 631 (1999).
65. Y. Wada, *Proc. IEEE* 77, 1194 (1989).
66. S. Yano et al., *IEEE Trans. Magn.* 27, 26 18 (1991).
67. S. Kotani et al., *IEEE J. Solid-State Circuits* 26, 612 (1991).
68. L.A. Abelson, *'Superconductive Electronics Process Technologies'*, Ext. Abstr. 5th Intl. Superconductive Elec. Conf. 1, 1 (1997). Physikalisch-Technische Bundesanstalt (Braunschuweig, Germany).
69. M. Aoyagi et al., *IEEE Trans. Appl. Supercond.* 7, 2644 (1997).
70. H. Numata, *IEEE Trans. Appl. Supercond.* 7, 2282 (1997).
71. J.M. Murduck et al., *Ext. Abstr. 4th Int. Supercond. Elec. Conf.*, 242 Centennial Conference (Boulder, CO).
72. C.L. Jia et al., *Physica C* 175, 545 (1991).
73. B.H. Moeckly and K. Char, *Appl-Phys. Lett.* 71, 17 (1997).
74. T. Satoh et al., IEEE *Trans. Appl. Supercond.* 9, 3141 (1999).
75. A. Pauza et al., *J. Appl. Phys.* 5612 (1997).
76. J.S. Martens et al., *4th Int. Superconductive Electronics Conf. ISEC'*, 93 August 11 – 14, Boulder, CO, pp. 2d – 4, 17.

77. H.G. Meyer et al., in [76] pp. 3b – 1, 128.
78. D. Drung, *Supercond. Sci. Technol.* 16, 1320 (2003).
79. R. Fischer et al., *Biomagnetism : Clinical Aspects*, eds. M. Hoke et al., Elsevier, Amsterdam, pp. 585.
80. J.P. Wikswo Jr., *IEEE Trans. Appl. Supercond.* 5, 74 (1995).
81. J. Clarke, *IEEE Trans. MAG*-19, 288 (1983) and *IEEE Trans. MAG*-61, 8 (1973).
82. A.K. Gupta and N. Khare in *"Studies of High Temperature Superconductors"*, ed. A.V. Narlikar, Nova Science Publishers, New York, Vol. 12, p. 43 – 94 (1994).
83. C. Baumgartner et al., (eds) in *'Studies in Applied Electromagnetics and Mechanics'*, Vol-7 (1995) (IOP Press, Amsterdam).
84. O.V. Lounasma et al., *Proc. Natt-Acad. Sci.*, 93, 8809 (1996).
85. N. Khare et al., *Supercond. Sci Technol.* 11, 517 (1998).
86. K. Enpuku and T. Minotani, *IEICE Trans. Electron E* 83-C, 34 (2000).
87. J. Clarke, *'SQUID fundamentals SQUID sensors : Fundamentals, Fabrication and Applications* ed. H. Weinstock (Kluwer : Dordecht) p. 1 – 62 (1995).
88. G.B. Donaldson et al., *'The New Superconducting Electronics*, ed. H. Weinstock and R.W. Ralston (Kluwer : Deventer) pp. 181 – 220 (1993).
89. J.C. Gallop, *'SQUIDs, the Josephson Effects and Superconducting Electronics'* (Hilger, Bristol).
90. P.J. Bosch et al., *Cryogenics* 37, 139 (1997).
91. Y. Tavrin et al., *Appl. Phys. Lett.* 62, 1824 (1993).
92. C.P. Foley et al., *IEEE Trans. Appl. Supercond.* 9, 3786 (1999).
93. D.G. Blair, *'The Detection of Gravitational Waves,* (Cambridge University Press, Cambridge).
94. K.D. Irwin et al., *Appl. Phys. Lett.* 64, 3497 (1996).
95. H. Weinstock (ed.), *"SQUID Sensors : fundamentals, fabrications and applications"*, (NATO ASI Series E : Appl. Sci Vol. 329) Amsterdam : Kluwer Academic Publishers.
96. K. Brockmeier et al., *'Magnetocardiography and 32-lead potential mapping : Repolarization in normal subjects during pharmacologically induced stress*, J. Cardiovasc. Electrophys. 8, 615 – 626 (1997).
97. C. Baumgartner et al. (eds.), 'Biomagnetism: fundamental research and clinical applications studies in applied electromagnetics and mechanics Vol-7 (Amsterdam : IOP Press). (1995).
98. C. Carr et al., *Cryogenics* 36, 691 (1996).
99. A. K. Gupta et al., *Measurement Science Technology* 8, 111 (1997).
100. M. Mishra et al., *IEEE Trans. Appl. Supercond.* 11, 4128 (2001).
101. T.W. Button and N.M. Alford, *Appl. Phys Lett.* 60, 1378 (1992).
102. Z.Y. Shen et al., *IEEE Trans. Microwave Theory and Techniques* 40, 2424 (1992).
103. K.H. Yong et al., *Appl. Phys. Lett.* 58, 1789 (1991).
104. H.T. Kim et al., *IEEE Trans. Appl. Supercond.* 9, 3909 (1999).
105. S. Ohshima et al., *IEEE Trans. Electron E* 83 C 2 (2001).
106. H.T. Su et al., *IEEE Trans. Appl. Supercond.* 11, 349 (2001).
107. M. Reppel et al., *Electronics Letters* 34, 929 (1998).
108. S.H. Talisa et al., *IEEE Trans. Appl. Supercond.* 5, 2291 (1991).

109. G.F. Dionne, *IEEE Trans. Microwave Theory and Techniques* 44 1361 (1996).
110. D.E. Oates and G.F. Dionne, *IEEE Appl. Supercond.* 9, 4170 (1999).
111. A.T. Findikoglu et al., *Appl. Phys. Lett.* 66, 3674 (1995).
112. D.C. Chung, *IEEE Trans. Appl. Supercond.* 11, 107 (2001).
113. L.P. Ivrissimtzis et al., *IEEE Trans. Applied Supercond.* 41, 33 (1994).
114. C. Walker et al., *IEEE Trans. Microwave Theory Tech.* 39, 1462 (1991).
115. Z.Y. Shen, *'High-Temperature Superconducting Microwave Circuits'*, Artech House, Boston [1994].
116. Z.Y. Shen et al., *IEEE Trans. Appl. Supercond.* 7, 2446 (1997).
117. Z.Y. Shen et al., *IEEE Trans. Microwave Theory Tech.* 40, 2424 (1992).
118. W.G. Lyons et al., *IEEE Trans. Magn.* 27, 2537 (1991).
119. G.L. Matthaei et al., *IEEE MIT-S Int. Microwave Symp. Digest* 2, 457 (1996).
120. G.L. Hey-Shipton, *IEEE MTT-S Int. Microwave Symp. Digest* 3, 1493 (1996).
121. Q. Huang et al., *IEEE MITS-S Int. Microwave Symp. Digest* 1, 371 (1998).
122. N. Khare, *Science Letts.* 25, 239 (2002).
123. S.K. Khamas et al., *Electronics Letters* 24, 460 (1988).
124. M.J. Lancaster, *'Passive Microwave Devices, Applications of Superconductors,* Cambridge University Press, Cambridge (1997).
125. M.I. Ali et al., *IEEE Trans. Appl. Supercond.* 9, 3077 (1999).
126. L. P. Ivissindtzis et al., *Electron Lett.* 30, 92 (1994).
127. R. Radebaugh, *'Advances in Cryocoolers',* Proc. ICEC 16/ICMC, Elsevier Science, Oxford (1997), pp. 33.
128. C.M. Jackson et al., *4th Int. Superconductive Electronics Conf. ISEC' 93,* August 11 – 14, Boulder, Co. pp. 6a – 4, 341 (1993).
129. B.A. Willemsen, *IEEE Trans. Appl. Superconductivity* 11, 60 (2001).
130. M. Klauda et al., *IEEE Trans Microwave Theory and Techniques* 48, 1227 (2000).
131. J.S. Hong et al., *IEEE Trans. Microwave Theory and Techniques* 48, 1240 (2000).
132. K. Satoh et al., *Physica C* 357 – 360, 1495 (2001).
133. R.R. Mansour et al., *IEEE Trans. Microwave and Theory Techniques* 48, 1171 (2000).
134. K. Ehata et al., *IEEE Trans. Appl. Superconductivity* 11, 111 (2001).
135. Y. Ohtani et al., *CEC/ICMC* (1993).
136. Q. Li et al., *Appl. Phys. Lett.* 62, 2428 (1993).
137. L.R. Vale and R.H. Ono, *IEEE Trans. Appl. Superconductor* 11, 762 (2001).
138. T. Tamura et al., *Jpn. J. Appl. Phys.* 24, L 709 (1985); *J. Appl. Phys.* 60, 711 (1986).
139. D.J. Frank et al., *IEEE Trans. Magn.* 21, 721 (1985).
140. T.D. Clarke et al., *J. Appl. Phys.* 51, 2736 (1980).
141. T. Kawakami and H. Takayanagi, *Jpn. J. Appl. Phys.* 26, Suppl. 26 – 3 (1987).
142. J. Mannhart, *Supercond. Sci. Technol.* 9, 49 (1996).
143. S. Kotani et al., *Tech. Digest IEOM,* 1987, p. 865.
144. K. Likharev and V. Semenov, IEEE Trans. Appl. Supercond. 1, 3 (1991).
145. K. Likharev, Physics World 10, 39 (1997).

146. B. Oelze et al., Appl. Phys. Lett. 70, 658 (1997).
147. M. Hidaka et al., *IEEE Trans. on Appl. Supercond.* 9, 4081 (1999).
148. B. Ruck et al., *Appl. Phys. Lett.* 72, 2328 (1998).
149. A.G. Sun et al., *IEEE Trans on Appl. Supercond,.* 9, 3825 (1999).
150. A.H. Sonnenberg et al., *Physica C* 326 - 327, 12 (1999).
151. B. Ruck et al., *Physica C* 326 - 327, 170 (1999).
152. K. Saito et al., *Appl. Phys. Lett.* 76, 2606 (2000).
153. S.B. Kaplan et al., *IEEE Trans. Appl. Supercond.* 9, 3020 (1999).
154. D.L. Miller et al., *IEEE Trans. Appl. Supercond.* 9, 4026 (1999).
155. S.V. Rylov et al., *IEEE Trans. Appl. Supercond.* 9, 3016 (1999).
156. V.K. Semenov et al., *IEEE Trans. Appl. Supercond.* 9, 3026 (1999).
157. O.A. Mukhanov et al., *IEEE Trans. Appl. Supercond.* 9, 3619 (1999).
158. A.V. Rylyakov et al., *IEEE Trans. Appl. Supercond.* 9, 3623 (1999).
159. G. Gao et al. (preprint).
160. M. Dorozevets et al., *IEEE Appl. Supercond.* 9, 3606 (1999).
161. B.S. Krasik et al., *J. Appl. Phys.* 81, 1581 (1997).
162. M. Krough et al., *IEEE Trans. Appl.* Supercond. 11, 962 (2001).
163. A. Barone and G. Paterno, *'Physics and Applications of the Josephson Effect'* (Wiley, NewYork), 1982.
164. T. Van Duzer and C.W. Turner, *'Principles of Superconductive Devices and Circuits* 2nd ed. (Englewood Cliffs, NJ : Prentice Hall).
165. M. Nahum and J.M. Martinis, *Appl. Phys. Lett.* 66, 3203 (1995).
166. J.M. Martinis et al., *Nucl. Instrum. Methods A* 444, 23 (2000).
167. D.A. Wollman et al., *Nucl. Instrum. Methods A* 444, 145 (2000).
168. K.D. Irwin et al., *Nucl-Instrum. Methods A* 444, 184 (2000).
169. B. Cabrera et al. *Appl. Phys. Lett.* 73, 735 (1998).
170. A.J. Miller et al., *Nucl. Instrum. Methods A* 444, 445 (2000).
171. P. Verhoeve et al., *IEEE Trans. Appl. Supercond.* 7, 3359 (1997).
172. G.N. Goltsman et al., *Appl. Phys. Lett.* 79, 705 (2001).
173. F.A. Hegmann et al., *Appl. Phys. Lett.* 67, 285 (1995).
174. C. Jackel et al., *Phys. Rev. B* 54, R6889 (1996).
175. M. Hangyo et al., *Appl. Phys. Lett.* 69, 2122 (1996).

Glossary

- **Almost-localized Fermi liquid.** Metallic system which under a relatively small change of external parameter such as temperature, pressure, or composition undergoes a transition to the Mott insulating state. In such a metal, electrons have large effective mass. At low temperature the system may order antiferromagnetically or undergoes a transition to the superconducting state. Both the nonstoichiometric oxides (such as $La_{2-x}Sr_xCuO_4$), and heavy fermion systems are regarded as almost localized Fermi liquids.

- **Anneal.** To heat and then slowly cool a material to reduce brittleness. Annealing of HTSC cuprates usually follows sintering and is done in an oxygen-rich atmosphere to restore oxygen lost during calcination. The oxygen content of a ceramic superconductor is critical. For example, YBCO with 6.4 atoms of oxygen will not superconduct. But YBCO with 6.5 atoms will.

- **Anti-ferromagnetism.** A state of matter where adjacent ions in a material are aligned in opposite or "anti-parallel" arrays. Such materials display almost no response to an external magnetic field at low temperatures and only a weak attraction at high temperatures. Some researchers are of the opinion that anti-ferromagnetism in cuprates plays a role in the formation of Cooper pairs and, thus, in facilitating a superconductive state in some copper oxide compounds.

- **Bardeen, Cooper and Schrieffer (BCS) theory.** This is the first widely accepted microscopic theory to explain superconductivity. BCS theory asserts that, as electrons pass through a crystal lattice, the lattice deforms inward towards the electrons generating sound packets known as "phonons". These phonons produce a trough of positive charge in the area of deformation that assists subsequent electrons in passing through the same region in a process known as phonon-mediated coupling, *i.e.* electrons with opposite spins and momenta mediated by lattice deformation (phonon) forms Cooper pairs and the resultant attractive interaction overcomes their mutual repulsion. At a critical temperature (T_c), the electron system undergoes a transition to a condensed state of pairs which is characterized by zero *dc* electrical resistance and a strong diamagnetism.

- **Borocarbides.** Superconducting borocarbides are compounds containing both boron and carbon in combination with rare-earth and transition elements; some of which exhibit the unusual ability to return to a normal, non-superconductive state at temperatures below the critical temperature, T_c.

- **BSCCO.** This is an acronym for a superconductor system containing the elements Bismuth, Strontium, Calcium, Copper and Oxygen. A small amount of lead is also included in these compounds to promote the highest possible T_c. BSCCO has probably found the widest acceptance among HTSC cuprates applications due to its unique properties. These compounds exhibit both an intrinsic **Josephson effect** and anisotropic (directional) behaviour.

- **Ceramics.** Ceramic superconductors are inorganic compounds formed by reacting a metal with oxygen, nitrogen, carbon or silicon. The best-known of these are the copper-perovskites. Ceramics are typically hard, brittle, heat resistant materials formed by a process known as solid-state reaction.

- **Charge Reservoirs.** In superconductors, charge reservoirs are the layers that may control the oxidation state of adjacent superconducting planes (even though themselves are not superconducting). In the layered cuprates, these consist of copper-oxide chains.

- **Chevrel (Phases).** A class of molybdenum chalcogenides (compounds containing Group VI elements S, Se or Te along with molybdenum and a positively charged metal ion). With the help of a novel fabrication technique B_{C_2} (upper critical field) in the Chevrel Pb $Mo_6 S_8$ from 50 T increases to > 100 T in bulk materials.

- **Coherence Length.** The size of a Cooper pair-representing the shortest distance over which superconductivity can be established in a material. Coherence length is typically of the order of 1000 Å ; although it can be as small as 30 Å in the HTSC copper oxides.

- **Cooper Pair.** Two electrons with opposite spins and momenta form a pair in accordance with BCS theory is called a Cooper pair. Below the superconducting transition temperature, T_c, Cooper pairs form a condensate – a macroscopically occupied quantum state which flows without resistance. However, since only a small fraction of the electrons are paired, the bulk does not qualify as being a "Bose-Einstein condensate".

- **Correlated Electrons.** Electrons with their kinetic (or band) energy comparable to or smaller than the magnitude U of electron-electron repulsion. This situation is described by the condition $U \gtrsim W$, where W is the width of a starting (bare) energy band. Strictly speaking, we distinguish between the limits of almost-localized Fermi liquid for which $U \lesssim W$, and the limit of strongly correlated electrons (the spin-liquid) in which $U \gg W$. The term **correlated electrons** means that the motion of a single electron is correlated with that of others in the system.

- **Critical Field** (B_c). This is the maximum magnetic field that a superconductor can endure before it is "quenched" and returns to a non-superconducting state. A higher T_c also brings a higher B_c.

- **Critical Current Density** (J_c). This is the maximum current that a superconductor can carry. As the current flowing through a superconductor increases, the T_c will usually decrease.

- **Critical Temperature** (T_c). This represents the critical transition temperature below which a material begins to superconduct. The sudden loss of resistance in a superconductive medium may occur across a range as small as 20 millionths of a degree or, in the case of some stoichiometrically imperfect compounds, tens of degrees.

- **Diamagnetism.** The ability of a material to repel a magnetic field. Many naturally-occurring substances (like water, wood and paraffin, and many of the elements) exhibit **'weak'** diamagnetism. Superconductors exhibit **strong** diamagnetism below T_c. In a few rare compounds, a material may become superconductive at a higher temperature than the point at which diamagnetism appears. However, as a rule, the onset of strong (perfect) diamagnetism is one of the most reliable ways to ascertain when a material has become superconductive.

- **D-wave.** A form of electron pairing in which the electrons travel together in orbits resembling a four-leaf cover. Wave functions help theoreticians describe (and predict) electron behaviour. Recently, the d-wave models of superconductivity have gained substantial support over s-wave pairing as mechanism by which HTSC cuprates might be explained.

- **Energy Gap.** This is the energy required to break up a pair of electrons. According to BCS theory, the energy gap (in meV), $E_g = 3.5\, k_B T_c$, where k_B is Boltzmann's constant. Since electron-pairing is universally agreed to be the method by which superconductivity occurs, E_g is the amount of energy required to disrupt the superconducting state.

- **Electron Spin Resonance** (ESR) (Also EPR : Electron Paramagnetic Resonance). This is another mechanism by which superconductivity might be explained in some materials. ESR is the response of electrons to electromagnetic radiation or magnetic fields at discrete frequencies. Electrons, as they move, create tiny magnetic moments. Nearby electrons are influenced either beneficially or adversely. When the moments are complimentary, the electrons become paired and can help each other move through a crystal.

- **Exchange Interaction.** Part of the Coulomb interaction between electrons which depends on their resultant spin state. If the spin-singlet configuration is favoured in the ground state, then the interaction is called antiferromagnetism. The exchange interaction provides a mechanism of magnetic ordering in Mott insulators; it may also correlate electrons into singlet or triplet pairs in a metallic state, particularly when the pair-exchange coupling J of electron pair is comparable to the kinetic energy of each of its constituents.

- **Fermi Liquid.** Term describing the state of interacting electrons in a metal. Equilibrium properties of such systems are modeled by a gas of free electrons with renormalized characteristics such as the effective mass. The properties at low temperatures are determined mainly by the electrons near the Fermi surface. The electron-electron interactions lead to specific contributions to the transport properties of such a system.

- **Ferrite.** These are ceramics with magnetic properties. Many of the same elements used in ferrites (*e.g.*, Ba, Sr, Tm, O) are also key constituents in ceramic superconductors. This may be an important clue in understanding HTSC in cuprates.

- **Ferro-magnetism.** A state wherein a material exhibits magnetization through the alignment of internal ions (neighbouring magnetic moments). This contrasts with paramagnetism, which is temporary, much weaker and results from unpaired electrons.

- **Flux-Lattice.** A configuration created when flux lines from a strong magnetic field try to penetrate the surface of Type-II superconductor. The tiny magnetic moments within each resulting vortex repel each other and a periodic lattice results as they array themselves in an ordered fashion.

- **Fluxon.** This is the smallest magnetic flux (flux quantum) that exists in nature. Just as electrons are a quantized flux. The term is used in association with vortices, which result from magnetic fields penetrating Type-II superconductors in single fluxon quanta.
- **Flux-Pinning.** The phenomenon where a magnet's lines of force (called flux) become trapped or "pinned" inside a superconducting material. This pinning binds the superconductor to the magnet at a fixed distance. Flux pinning is only possible when there are defects in the crystalline structure of the superconductor (usually resulting from grain boundaries or impurities). Flux pinning is desirable in HTSC cuprates in order to prevent "flux creep", which can create a pseudo-resistance and depress J_c and B_c.
- **Four-Point Probe.** This is the most common method of determining the T_c of a superconductor. Wires are attached to a material at four points with a conductive adhesive. Through two of these points a voltage is applied and, if the material is conductive, a current will flow. Then, if any resistance exits in the material, a voltage will appear across the other two points in accordance with Ohm's law ($V = IR$, voltage equals current times resistance). When the material enters a superconductive state, its resistance drops to zero and no voltage appears across the second set of points. By using the four-point method, instead of just two points, resistance in the adhesive and wires can be ignored ; as the second set of points do not themselves conduct any current and can, therefore, only reflect what voltage exists across the body of the material.
- **Gapless Superconductivity.** It is well known that the superconducting order parameter and energy gap are two different concepts. The order parameter describes the number of Cooper pairs in the condensate and determines the basic features of superconductors: the Meissner effect and the absence of electric resistance; while the energy gap is the minimum binding energy of the Cooper pairs, which manifests itself in the low-temperature heat capacity, thermal conductivity, absorption of electromagnetic radiation, and ultrasound, etc. A typical example of the distinction between them is the gapless superconductivity observed in superconductors containing magnetic impurities where there is a finite superconducting order parameter but a vanishing energy gap. In this case, the Bose condensate of the Cooper pairs does not contain all the pairs due to the pair-breaking effect of magnetic impurities, the order parameter describes the coherent wave function of the condensate pairs, while the energy gap may vanish. In reality, the magnetic impurities do not provide the only pair-breaking mechanism. The proximity effect in an FM/SC junction leads to gapless superconductivity.
- **Grain Boundaries.** These are the surfaces separating individual crystals, or grains, in a crystalline solid. These represent planar defects in an otherwise perfect crystalline material.
- **Green's Functions** (or propagator as it is sometimes called). Green's Function (G) is a well established tool for dealing with problems of many particle assemblies in condensed matter physics. The main advantage of the Green's function method over other methods is that a reasonable value of G can be obtained by using perturbation theory to determine self energy.

- **Heavy Fermions.** These are the compounds containing the elements cerium, ytterbium, or uranium; whose (inner shell) conduction electrons often have effective masses (called **quasi particle** masses) several hundred times as great as that of a "free" (normal) electron mass. This gives them what is known as a low Fermi "energy" and makes them unlikely and unusual-superconductors. Cooper pairing in these systems seems to arise from the magnetic interactions of the electron spins.

- **Hole.** Hole is a positively-charged vacancy within a crystal lattice resulting from the shortage of electron in that region. Holes are typically induced by doping a material with an impurity. However, they can also be synthesized electronically with devices like the field-effect transistor (FET). Modern electronic devices rely heavily on holes (as p-type semi-conductors) to function. There is evidence that the holes of hypocharged oxygen in charge-reservoirs are, in fact, what makes possible HTSC in the layered cuprates.

- **HTSC.** This is an acronym for "High-Temperature Superconductor" (or superconductivity). There is no widely-accepted temperature that separates HTSC from LTSC (Low temperature superconductors). All the superconductors known prior to the discovery of superconducting oxocuprates are usually classified as LTSC. The barium-lanthanum-cuprate fabricated by Muller and Bednorz, with a T_c of 30 K, is considered to be the first HTSC material. Certainly, any compound that will superconduct above the boiling point of liquid nitrogen (77 K) would be HTSC.

- **Hubbard Subband.** A term describing each of the two parts of an energy band in which it splits when the electron-electron repulsion energy is comparable to their kinetic energy. The Hubbard splitting of the original band induced by the interaction explains in a natural way the existence of the Mott insulating state in the case of an odd number of electrons per atom (*i.e.,* when the atomic shells composing a band state are only half-filled).

- **Hysteresis (loop).** Hysteresis, as it applied to a superconductor, relates to the dynamic response of a superconductor to a strong magnetic field impinged upon it. As the strength of a nearby magnetic field (B) increases, the critical transition temperature (T_c) of a superconductor will decrease. And, at some point superconductivity will completely disappear, as it becomes "quenched". However, as the magnetic field is gradually withdrawn, the superconductor may NOT-immediately return to a superconductive state. Here in lies the hysteresis. The graph of B-vs-T_c is different retreating than it is advancing (creeping a "loop" shape). This fact is to be weighed carefully in high-current applications where the superconductor B_c may, even briefly, be exceeded ; as significant power losses can result.

- **Infinite Layer.** Infinite layer compounds have no clear separation between molecules. Rather than electrostatic bonding between discrete molecules to form a bulk crystalline aggregate, all the atoms are bound together by covalent or co-ionic bonding to form the equivalent of one huge molecule. $(Ba, Sr) CuO_2$ and $Na_2 Ba_6 Si_{46}$ are examples of "infinite layer" or "infinite network" superconductor compounds.

- **Isotope Effect.** The isotope effect in superconductors is the influence atomic mass contributes to the critical temperature (T_c) of a superconductor. For example, $^{203.4}Hg$ has a T_c of 4.126 K, while ^{198}Hg has a T_c of 4.177 K. Since both forms of Hg have the

Glossary

same lattice structure, this difference in T_c can be attributed solely to the difference in mass. Isotope effect in HTSC cuprates is almost negligibly small.

- **Josephson Effect :**

  (i) **DC Josephson Effect.** A phenomenon predicted by Brian Josephson that electrons would "tunnel" through a narrow (< 10 Å) non-superconducting region, even in the absence of an external voltage. In a normal conductor, electrical current only flows when there is a voltage differential and contiguous electrical connection. It has been shown that the Josephson effect arises from the incoherent phase relationships between superconducting electrons in the two (separated) superconductors.

  (i) **AC Josephson Effect.** A.C. Josephson effect is where the current flow oscillates as an external magnetic field impinged upon it increases beyond a critical value (at a frequency of $2eV/h$, where e is the electron charge, V is the voltage that appears and h is Planck's constant). [This oscillation frequency has, in fact, resulted in an upward revision of Planck's constant from 6.62559×10^{-34} to 6.626196×10^{-34}]

- **Josephson Junction.** A thin layer of insulating material sandwiched between two superconducting layers. Electrons "tunnel" through this non-superconducting region in what is known as the "Josephson effect". (The standard volt is now defined as the voltage required to produce a frequency of 483597.9 GHz in a Josephson junction.

- **Magnon.** Quanta of spin waves excitations of magnetically ordered systems. In the ground states the magnetic moments are aligned due to the exchange interaction. When spin deviation occurs due to temperature of external perturbation, the deviation propagate like waves, giving rise to spin waves. Magnons obey Bose statistics.

- **Meissner Effect.** Exhibiting diamagnetic properties to the **total** exclusion of all magnetic fields. This is a classic hallmark of superconductivity and can actually be used to levitate a strong rare-earth magnet.

- **Mott Insulator.** Insulator containing atoms with partially field $3d$ or $4f$ shells. These systems order magnetically (usually antiferromagnetically) when the temperature is lowered. Thus, they differ from ordinary (Bloch-Wilson) or band insulators which are weakly diamagnetic and are characterized by filled atomic shells, separated from empty states by a gap.

- **Mott Transition.** The Mott transition is the shift from an insulating to a metallic state in a material. HTSC cuprates are composed of CuO_2 planes that are separated from each other by ionic "blocking layers". Although it has one conduction electron (or hole) per Cu site, each CuO_2 plane is originally insulating because of the large electron correlation. That behaviour is typical of the Mott insulator state, in which all the conduction electrons are tied to the atomic sites. The superconducting state emerges when holes from the blocking layers dope the CuO_2 layers in a way that alters the number of conduction electrons and triggers the Mott transition. There are views that the strong antiferromagnetic correlation, which originates in the Mott-insulating CuO_2 sheets and persists into the metallic state, could be a possible mechanism of HTSC - cuprates.

- **Organics.** Organic superconductors are sub-class of organic conductors that include molecular salts, polymers and pure carbon systems (including carbon nanotubes and C_{60} compounds). They may also be referred to as "molecular superconductors". They are typically large, carbon-based molecules of 20 or more atoms, consisting of a planar organic molecule and a non-organic anion.

- **Penetration Depth** (also London Penetration Depth). This term relates to how deeply a magnetic field will penetrate the surface of a superconductor. An external magnetic field impinged upon a Type-II superconductor will decay exponentially into the surface based on the paired electron density within the superconductor (only a small fraction of the electrons are in a superconductive state).

- **Perovskites.** A large family of crystalline ceramics that derive their name from a mineral known as a perovskite. They are the abundant minerals on earth and have a metal to oxygen ratio of approximately 2 to 3. Copper oxide superconductors are layered perovskites.

- **Phase Slip** (also Quantum Phase-Slip). A point where a material in a superconductive state spontaneously changes from one state to another, generating a topological defect. This defect causes paired electrons to become "out of step" with each other, producing a voltage and, ergo, non-zero electrical resistance. This phenomenon has been observed in ultra-thin wires less than a few tens of nanometers of diameter. Though bulk superconductivity may persist ($T < T_c$), one consequence of phase slip is a lower current-carrying state. A similar phenomenon occurs in Josephson junctions.

- **Phonon.** This is a quanta of lattice waves, *i.e.*, a quantum of sound; the smallest unit of energy of a vibration corresponding to a sound wave. Phonon is analogous to a photon or light quantum.

- **Polaron.** The object that results when an electron in the conduction band of a crystalline insulator or semiconductor polarizes or otherwise deforms the lattice in its vicinity. The polaron comprises electron plus its surrounding lattice deformation. (Polarons can also be formed from holes valence band). If the deformation extends over many lattice sites, the polaron is "large", and the lattice can be treated as a continuum. Charge carriers inducing strongly localized lattice distortions form 'small' polarons.

- **Proximity Effect.** The phenomenon where a thin film of non-superconductive material in close proximity with a superconductor takes on superconducting properties. The Josephson junction is a device that takes advantage of this phenomenon. The inverse proximity effect is where just the opposite occurs. A non-superconductive metal can enhance the T_c of an adjacent superconductor. This inverse effect has been observed with silver and lead.

- **P-wave.** A rare form of electron pairing in which two electrons travel together in spherical orbits, with both having the same direction of rotation.

- **Quasiparticle.** A bare particle that is "dressed" or clothed by a cloud of other surrounding particles. Quasiparticles behave similarly to bare (normal) particles, but usually have a larger effective mass due to this cloud moderating interactions with other particles.

- **Quench.** The phenomenon where superconductivity in a material is suppressed; usually by exceeding the maximum current the material can conduct (J_c) or the maximum magnetic field it can withstand (B_c).

- **Re-entrant** (behaviour). A condition where a material retreats from its superconductive state and then re-enters it. This can be caused by a strong external magnetic field that dynamically exceeds the B_c of the material and/or is misaligned (in the case of some organic superconductors), a discordant temperature below T_c (in the case of some borocarbides) or by J_c hysteresis (momentarily exceeding the critical current density, causing the T_c to shift downward).

- **Resistance.** The opposition of a material to the flow of electrical current through it. Energy lost due to resistance is a result of vibrations at the molecular level and manifests itself as heat in proportion to the square of the current flow. In a superconductor all resistance disappears below a certain temperature. However, this applies only to direct current (DC) electricity. Other types of losses result when transporting alternating current (AC). Examples of this include hysteresis, reactive-coupling and radiational losses. In the HTSC cuprates, the power loss in applications like transmission lines is inversely proportional to the critical current density for the low magnetic field applications. This limitation can be compensated to some degree by increasing the ratio of voltage to current. In Type-II superconductors carrying high-frequency alternating current, "skin effect" losses also result as the energy tends to migrate to the surface where the conductive medium is incontiguous, producing a pseudo-resistance. In some materials the amount of resistance may also depend on the direction of current flow (anisotropic resistivity) and/or presence of an external magnetic field (Hall effect).

- **Room Temperature Superconductor.** There are no confirmed room temperature superconductors (as was once reported for lithium-beryllium-hydride and for lead-silver-carbonate). However, it has been theorized that a metallic form of hydrogen **might** be a room temperature superconductor. In 1996, physicists at Lawrence Livermore Laboratory were able to **briefly create metallic hydrogen**. But its existence was fleeting and no measurement of the Meissner effect was possible. Zero resistance has been observed at room temperatures in **ballistic quantum wire**. However, having one-dimensional geometry, this wire does not exhibit the Meissner effect, except when configured as a closed loop.

- **Sinter.** The process of heating a material to just below its melting point. An extended period of sintering is the method by which the constituent components of a ceramic superconductor are inherently brittle, sintering helps promote intergranular bonding and hardness.

- **SQUID.** A superconducting loop interrupted in two places by Josephson junctions. When sufficient electrical current is conducted across the squid body, a voltage is generated proportional to the strength of any nearby magnetic field. The SQUID, an acronym for Superconducting Quantum Interference Device, is the most sensitive detector known to science.

- **Stripes.** These are microscopic rivers of charge that flow across the surface of a Type-II superconductor. It is theorized that stripes encourage "holes" to pair up and, as such, may play a role in facilitating charge transfer. Recently, it was shown that there exists a critical value of micro-strain that must be exerted upon the CuO_2 planes for strips to form.

- **Superconductor.** An element, inter-metallic alloy, or compound that will conduct electricity without resistance below a certain temperature. However, this applies only to direct current (DC) electricity and to finite amounts of current. All known superconductors are solids, none are gases or liquids. And all require extreme cold to enter a superconductive state. Once set in motion, current will flow forever in a closed loop of superconducting material making it the closest thing to perpetual motion in nature. Superconductivity is a "macroscopic quantum phenomenon". In addition to being classified Type-I and Type-II, superconductors can be categorized further by their dimensionality. Most are 3-D. But some compounds, like surface-doped Na WO_3 and some organic superconductors are 2D. $Li_2 CuO_2$ and single-walled carbon nano-tubes have shown rare 1-D superconductivity. In addition to repelling magnetic fields, enhanced thermal conductivity, higher optical reflectivity and reduced surface friction are also properties of superconductors. The term "superconductor" is also used in some instances to refer to materials that have near infinite thermal conductivity such as carbon nano-tubes. Very recently Edwards et al. (Chem Phys Chem 7, 2015 (2006)) argue that superconducting charged Bose liquid may be found in a true liquid state of condensed matter at ambient pressure. They proposed that fluid–ammonia solution may be stabilized and observed as a high–T_c superconducting liquid (230 K) or at least a vitreous superconductor in the corresponding quenched solutions (160 K).

- **Susceptibility.** A measure of relative amount of induced magnetism in a material. Magnetic susceptibility is often used in lieu of resistance measurements to determine the transition temperature of a superconductor. Although, on occasion, the two techniques produce very different T_c's. In a typical superconductor, the (arbitrary) value of susceptibility will change from zero to a negative number as the temperature drops through. However, in some materials it changes from positive to negative, as paramagnetism yields to diamagnetism.

- **S-wave.** A form of electron pairing in which the electrons travel together in spherical orbits, but in opposite directions.

- **Thin Film (deposition).** A method of fabricating ceramic superconductors to more precisely control the growth of crystalline structure to eliminate grain boundaries and achieve a desired T_c. This can involve Pulsed-Laser Deposition (PLD) or Pulsed-Electron Deposition (PED) of the material. A variation of this technique can be used to **increase** the T_c of a superconductor by growing it on a supporting material with a smaller inter atomic spacing. The supporting material acts as a molecular "gridle" to compress the atomic lattice of the superconductor, thereby raising its transition temperature. Superconductive tape is made using thin film deposition technology.

- **Translational Symmetry.** As it applies to superconductivity, translational symmetry is where the process of charge transfer is **repeated exactly** as the charge carriers (paired electrons) traverse the solid. In a normal conductor, latent heat continuously vibrates the atomic lattice, deflecting mobile free electrons and preventing "perfect" translational symmetry. In a superconductor this scattering tendency is overcome.

Glossary

- **Tungsten-Bronze.** A nebuluous term used to describe alikali metal tungstenates, vanadates, molybdates, titanates and niobates. The term was originally coined to describe $Na_x WO_3$ compounds ; the crystals of which look much like the copper-tin alloy known as bronze. There have been reports of superconductivity as high as 91 K for a surface-doped sodium tungsten-bronze. Perhaps this may be the first HTSC material discovered that does not contain any copper.

- **Ultraconductor.** Materials known as ultraconductors™ display room-temperature resistance many orders of magnitude **lower** than the best metallic conductors. Examples of these materials include oxidized polypropylene (OAPP) and other polymers. Since superconductor™ is a colloquial term, these materials might better be described as "hyperconductors". The Meissner effect cannot be confirmed in them, but strong (giant) diamagnetism is in evidence. Some of them may actually find acceptance in high-current applications ahead of superconductors as a result of their low losses at ambient temperatures and pressures.

- **Undressing.** The process by which a **quasiparticle** becomes more like a bare (normal) particle. It is theorized that this may be a driving force behind superconductivity, as undressed electrons are significantly lighter and can, thus, conduct current more readily.

- **Unit Cell.** A unit cell is the smallest assemblage of atoms, ions, or molecules in a solid, beyond which the structure repeats to form the 3-dimensional crystal lattice.

- **Vortices.** Swirling tubes of electrical current induced by an external magnetic field into the surface of a superconducting material that represents a topological singularity in the wave function. These are particularly evident in Type-II superconductors. Recent research suggests that flux vortices may NOT possess quantum values (equal to multiples $h/2e$). But may instead have but a tiny fraction of the basic unit of magnetism. The movement of vortices may produce a **pseudo resistance** and, as such, is undesirable. While superconductivity is a "macroscopic" phenomenon, vortices are a "mesoscopic" phenomenon.

- **YBCO.** An acronym for a well known ceramic superconductor composed of Yttrium, Barium, Copper, and Oxygen. This was the first truly "high temperature" ceramic superconductor discovered; having a transition temperature well above the boiling point of liquid nitrogen – a commonly available coolant. Its actual molecular formula is $Y Ba_2 Cu_3 O_7$, making it a "1-2-3" superconductor. YBCO compounds exhibit d-wave electron pairing.

Appendix 1

DEFINITE INTEGRALS

The gamma function $\Gamma(z)$ and the Riemann zeta function $\zeta(z)$ are defined as follows:

$$\Gamma(z) \equiv \int_0^\infty dt\, e^{-1}\, t^{z-1} \qquad \text{Re } z > 0$$

$$\zeta(z) \equiv \sum_{p=1}^\infty p^{-z} \qquad \text{Re } z > 1$$

These definitions converge only in the specified regions of the complex z plane, but the functions can be analytically continued with the general relations

$$\Gamma(z)\,\Gamma(1-z) = \frac{\pi}{\sin \pi z}$$

$$2^{1-z}\,\Gamma(z)\,\zeta(z)\cos\left(\frac{1}{2}\pi z\right) = \pi^z\,\zeta(1-z)$$

The gamma function also satisfies the functional equation

$$\Gamma(z+1) = z\Gamma(z)$$

and it therefore reduces to the factorial function for integral values

$$\Gamma(n+1) = n! \qquad n \text{ integral}$$

The digamma function $\psi(z)$ is the logarithmic derivative of the gamma function

$$\psi(z) = \frac{d}{dz}\log \Gamma(z) = \frac{1}{\Gamma(z)}\frac{d\Gamma(z)}{dz}$$

At $z = 1$, it reduces to

$$\psi(1) = \int_0^\infty dt\, e^{-t}\ln t \equiv -\gamma$$

where γ is Euler's constant. Useful numerical values are listed below

$$\zeta(0) = -\frac{1}{2} \qquad\qquad \Gamma\left(\frac{1}{2}\right) = \sqrt{\pi} \approx 1.772$$

$$\zeta\left(\frac{3}{2}\right) \approx 2.612 \qquad\qquad \Gamma(1) = 0! = 1$$

Appendix 1

$$\zeta(2) = \pi^2/6 \approx 1.645 \qquad \Gamma\left(\frac{5}{4}\right) \approx 0.9064$$

$$\zeta\left(\frac{5}{2}\right) \approx 1.341 \qquad \Gamma\left(\frac{3}{2}\right) = \frac{1}{2}\sqrt{\pi} \approx 0.8862$$

$$\zeta(3) \approx 1.202 \qquad \Gamma\left(\frac{7}{4}\right) \approx 0.9191$$

$$\zeta(4) = \pi^4/90 \approx 1.082 \qquad \gamma \approx 0.5772$$

$$\zeta'(0) = \left(\frac{d\zeta(z)}{dz}\right)_{z=0} = -\frac{1}{2}\ln(2\pi) \qquad e^\gamma \approx 1.781$$

Numerous series and definite integrals may be expressed in terms of the preceding functions:

$$\sum_{p=0}^{\infty} \frac{1}{(2p+1)^n} = (1 - 2^{-n})\,\zeta(n) \qquad n > 1$$

$$\int_0^\infty dx \, \frac{x^{n-1}}{e^x - 1} = \Gamma(n)\,\zeta(n) \qquad n > 1$$

$$\int_0^\infty dx \, \frac{x^{n-1}}{e^x + 1} = (1 - 2^{1-n})\,\Gamma(n)\,\zeta(n) \qquad n > 0$$

$$\int_0^\infty dx \, x^{n-1} \operatorname{cosech} x = \int_0^\infty dx \, \frac{x^{n-1}}{\sinh x}$$

$$= 2(1 - 2^{-n})\,\Gamma(n)\,\zeta(n) \qquad n > 1$$

$$\int_0^\infty dx \, x^{n-1} \operatorname{sech} x = \int_0^\infty dx \, \frac{x^{n-1}}{\cosh x}$$

$$= 2\Gamma(n) \sum_{p=0}^{\infty} (-1)^p (2p+1)^{-n} \qquad n > 0$$

$$\int_0^\infty dx \, x^{n-1} \operatorname{sech}^2 x = \int_0^\infty dx \, \frac{x^{n-1}}{\cosh^2 x}$$

$$= 2^{2-n}(1 - 2^{2-n})\,\Gamma(n)\,\zeta(n-1) \qquad n > 0$$

$$\int_0^1 dx \, x^{m-1}(1-x)^{n-1} = \int_0^\infty dx \, \frac{x^{m-1}}{(1+x)^{m+n}} = \frac{\Gamma(m)\,\Gamma(n)}{\Gamma(m+n)}$$

$$\int_0^\infty dx \, \ln x \, \operatorname{sech}^2 x = \left[\frac{\partial}{\partial n} \int_0^\infty dx \, x^{n-1} \operatorname{sech}^2 x \right]_{n=1} = -\ln \frac{4e^\gamma}{\pi}$$

The Bessel function $K_0(x)$ may be defined by the integral representation

$$K_0(x) \equiv \int_0^\infty dt\, e^{-x \cosh t}$$

and we find

$$\int_0^\infty dx\, K_0(x) = \int_0^\infty dt\, \text{sech}\, t$$

$$= 2\left(1 - \frac{1}{3} + \frac{1}{5} - \cdots\right) = 2 \arctan 1 = \frac{\pi}{2}$$

Subject Index

A

Abrikosov lattice 113
Accoustic attenuation 58, 162
A.C. Josephson effect 68, 69
A.C. resistivity 39
A.C. susceptibility 93
Alkali metal doped C_{60} fullerence superconductors 14
Alternating current cables 584
Anderson lattice model 387
Anderson lattice model Hamiltonian 331
Angle resolved photoelectron spectroscopy 397
Angular correlation annihilation radiation 397
Anisotropic type-II superconductors 114
Antenna 633
Antiferromagnetic 487
Average pinning force density 118

B

Band structure of $Bi_2Sr_2CuO_6$ 336
Band structure of $Bi_2Sr_2CaCu_2O_8$ 336
Band structure of $YBa_2Cu_3O_6$ 333
Band structure of $YBa_2Cu_3O_7$ 331
BCS model 149
Bean model 111
Bearings 592
Binary alloys and intermetallic compounds 19
Biomagnetism 622, 624
Bipolaronic model 424
Bloch oscillations 75
Bloch Theorem 142
Bloembergen-Rowland interaction 389
Borocarbides 466
Bose-Einstein statistics 40
Bosons 40

C

Ceramic HTSC materials 563
Charge carriers in cuprates 183
Charge segregation 428
Charge-transfer insulators 384
Chemical potential 150
Chevrel phases 20, 461
Clogston-Chandra Sekhar limit 170
Coexistence of antiferromagnetism and superconductivity 387
Coherence length 38, 49, 60, 101
Columnar defects 261
Conventional superconducting materials 561
Cooper pairs 38, 143
Correlated volume 126
Critical current 47
Critical current density 250, 537, 559
Critical field 424, 473, 483, 537
Critical field B_3 105
Critical fields 83, 160, 177
Critical magnetic field 43
Critical state 121
Critical structures 190
Critical temperatures 190
Cuprates 179

D

D C Josephson effect 68
Debye frequency 146
de Hass-van Alphen resonance 397
Density of states 472, 482
Density of states function 423
Dent pullers 599
Depairing critical current 117
Dielectric constant of a gas of electrons 144
Digital circuits and Electronics 638

Dimmers 505
Discontinuous insulator metal transition 444

E

Effects of oxygen vacancies 329
Elasticity of flux line lattice 125
Elementary pinning force 126
Electrodynamic levitation 574
Electromagnetic suspension 573
Electrical resistivity 29
Electron mean free path 35
Electron lattice interaction 164
Electron like 396
Electronic coupling mechanisms 394
Electronic states of the HTSC couprates 323
Electronic structure of Bismuth and Thallium HTSC 335
Electronic structure of Thallium Compounds 336
Electronic specific heat 472, 482, 533
Electronic structure and superconductivity 508
Electronic applications of superconductivity 603
Electromagnetic suspension 573
Eliashberg equation 164
Eliashberg theory 165
Ellipsoids in magnetic fields 90
Energy gap 56
Energy gap effects 606
Engineering critical current density 561
Entropy 50, 160
Ettings hausen effect 289
Eular's constant 154
Exchange interaction and real space pairing 374
Extended wave 377
Exotic superconductivity mechanisms 394
Extrapolation length 106

F

Fermi liquid 396, 407
Fermi liquid theory 405
Fermions 40
Fermi surface 174
Fermi surface nesting 327
Field homogeneity 589
Field stability 589

Fluoxid 47, 63
Flux creep 48, 127
Flux flow 119
Flux line lattice 103
Flux pinning 47, 246
Flux quantum 393
Flux quantization 62
Fluxon 63
Flux state 404
Flywheel energy storage 595
Fourier transform method 587
Free energy 53, 160, 423, 473, 484
Free induction decay 586
Fulde-Ferrell state in superconductors 173, 175
Fullerence superconductors 524

G

Gapless superconductivity 139, 176
Gauge theory 404
General aspects of organic metals and super conductors 506
Generators and motors with superconducting windings 576
Global protection 560
Giant flux creep 128
Ginzburg-Landau parameter 49, 61, 101, 217
Ginzurg-Landaw theory 99
GL theory and BCS theory 162
Grain boundary 245
Grain boundary weak links 234
Granularity 219
Gutzwiller projection operator 401
Gutzwiller projector 401
Gyromagnetic ratio 223

H

Hall coefficient 184
Hall effect 540
Hartree-Fock energy 159
Heat capacity or specific heat 51
Heavy fermions 389
Heavy fermion and Ruthenate superconductors 486
Heavy superconductors 16, 488
Heitler-London valence bond state 399

Subject Index

High frequency electromagnetic properties 59
High frequency nuclear magnetic resonance spectrometers 565
High temperature superconductors 188
High-T_c superconductor radiation detectors 636
HTSC bicrystal junction 613
HTSC Josephson junction IC devices 613
HTSC josephson junction 78
HTSC ramp-edge SNS 614
HTSC step-edge junctions 614
Hole like 396
Holes 33
Holon 378, 403, 447
Hubbard modes 355, 387
Hubbard subbands and hole states 359
Hund's rule exchange 389
Hybrid bearings 593
Hybrid superconductor magnet bearing 593
Hybridized systems 371, 380
Hysteresis cycle 120

I

Individual vortex motion 87
Insulating order parameter 523
Interaction between vortices 111
Interlayer and intralayer effects in HTSC cuprates 422
Interlayer pair tunneling mechanism 416
Intermetallic A-15 superconductors 459
Intragrain critical current densities 232
Inverse ac josephson effect 70
Ion beam damaged junction process 615
Ionic radius 213
Irreversibility line 130, 249, 564
Irreversibility temperature 249
Isotope effect 58, 138
Itinerant electron ferromagnetic superconductors 17
Itinerant ferromagnetic superconductors 468

J

Jaccarino-Peter effect 477
Josephson circuit 611
Josephson current 393
Josephson effect 68
Josephson junction 74, 225, 608, 612
Josephson-like tunneling 417

K

Kinetic exchange 389
Kinetic exchange interaction and magnetic phases 365
Kondo effect 17, 153
Kondo interaction 382
Kramers-Kronig relations 167

M

Macroscopic quantum interference effects 70
Magnetic coupling mechanisms 394
Magnetic decoration techniques 133
Magnetic flux quantization 607
Magnetic interactions, polarons and pairings 384
Magnetic levitation 64
Magnetic levitation vehicles 572
Magnetic phase diagram 89
Magnetic polaron 366
Magnetometry 624
Magnetic resonance imaging and spectroscopy 586
Magnetic separation and purification 571
Magnetic systems for magnetic confinement of fusion reactors 590
Magnetohydro dynamic energy conversion 595
Magnon pairing mechanism 394
MAGLEV 555
Marginal Fermi liquid 397
Marginal Fermi liquid model 440
Mechanical effects 58
Meissner effect 7
Meissner-Ochsenfeld effect 42
Hg-Ba-Ca-Cu-O systems 321
Metal doped fullerene superconductors 525
Metal-insulator transitions 361
Metastable states 109
$Mg\ B_2$ 534
Microbridge 76
Microscopic (BCS) theory 148
Microwave devices 625
Microwave properties 168

Mixed state 46, 109, 110
Mixed s-d phases 377
Mixed valent systems 384
Modified photon mechanisms 394
Momentum space 38, 395
Mott-Hubbard insulators 384
Mott-like metal insulator 398
Mott insulator 398
Mott transition 353
Multiple angle projected method 587
Multiband system 496
Muon spin rotation experiments 223

N

Narrow band systems 353, 374
Neel state 399
Nernst effect 290
Ni/Bi bilayer 467
NMR investigations 566
Nodal Fermi liquids 405
Nodes 489
Non-destructive evaluation 624
Non-Fermi liquid 566
Normal electrons 39
Normal metal 140
Normal state instability 142
Novel mechanisms of electron pairing 374

O

One dimensional conductors 505
Optical sensors 641
Optimal 287
Order parameter at zero temperature 523
Organic superconductor 23, 497
Overdoped 287

P

Pancake shaped vortices 86
Pairing mechanism 496
Pairing symmetry 496
Pairing symmetry in Lochon model 435
Pair rearrangement 417
Paramagnetic effects in superconductors 170

Particle and photon detection 640
Pauli exclusion principle 40
Peltier effect 289
Penetration depth 59, 161
Perfect diamagnetism 41
Phase diagram 187
Phase diagrams for one layer $Cu\,O_2$ layer 310
Phase diagrams for two $Cu\,O_2$ layers 311
Phase diagrams for three $Cu\,O_2$ layers 317
Phase transition 483
Photoemission 397
Photo-hole spectral function 421
Photo induced superconductors 543
Photolithography and etching 616
Pinning centres 257
Pinning force 104
Pinning of flux vortices 123
Polaron like stripons 428
Power applications 264
Power law dependence 251
Primary crystallization effect of the 2212 phase 314
Proximity effects 107
Properties of correlated electrons 383
Pseudo Coulomb potential 164
Pseudo gap 187, 291, 372, 392, 402
Pseudo gap phenomenon in HTSC cuprates 445

Q

Quantum critical superconductivity in heavy fermion systems 492
Quantum interference effects 607
Quantum-spin ladder superconductors 15
Quantum spin liquid 400
Quantum spin-liquid state 325
Quasi-1D electrons in high magnetic fields 514
Quasi particle peak 392
Quasi particles 373
Quench 557, 561

R

Rare earth transition metal borocarbides 12
Rapid single flux quantum 607
Real space 38
Relaxation effects 249

Subject Index

Relaxation oscillation SQUID 618
Refrigerations for cryoelectronics 635
Representative phase diagrams 295
Resonating valence bond theory 397
Resonance valence bond 394, 400
Reverse critical states and hysteresis 136
Rhombohedral ternary molybdenum chalcogenides 462
Right-Leduc effect 289
Ring supercurrent 43
Ruthenate cuprate 465
Ruthenate superconductor 495
Rutger's formula 52

S

Screened Coulomb pseudo potential 164
Seaback effect 288
Shubnikov phase 109
Silsbee effect 131
Solve bosons 378
Slave bosons language 382
Slave-boson theories 404
Small polaron excitations 428
Soliton 365
SO_2 and SO_5 models 441
Spatial homogeneous superconductor 171
Spin fermion model 413
Spin fluctuation mechanism 411
Specific heat 161, 282, 423, 541
Spin density wave collective conduction 513
Spin gap 447
Spin liquid 368, 379
Spinons 378, 403, 447
SQUID applications 613
SQUID microscope 625
SQUIDs 25, 610, 616
Stripe 405, 428
Strongly correlated electrons 365
Strongly correlated electrons systems 495
Supercell model 330
Superconducting cables and power transmission 582
Superconducting current limiters 597
Superconducting electrons 40

Superconductors 39
Superconductivity and antiferromagnetism 480
Superconductivity 1
Superconductivity and ferromagnetism : mechanism 469
Superconducting properties of HTSC cuprates 217
Superconducting fault current limiters 598
Superconducting dent puller 601
Superconducting magnets 556
Superconducting magnets for high energy physics 570
Superconducting magnetic energy storage 595
Superconducting mixers 640
Superconducting transformers 580
Superconducting transistors 637
Symmetry 490

T

Tl-Ba-Ca-Cu-O systems 319
Ternary rare earth rhodium borides 22
Tetragonal rare earth rhodium borides 463
Tetragonal-ortho rhombic transition 329
Theory of organic superconductivity 519
Thermal conductivity 55, 539
Thermal conductivity of HTSC cuprates 273
Thermally activated flux creep 127, 129
Thermodynamic critical field 100
Thermodynamic functions 156
Thermoelectric and thermomagnetic efects 286
Thermoelectric power 538
Thomas Fermi length 143
Thomas Fermi theory 144
Three dimensional resonators and filters 628
T-J model 407
$T M_2 X$-phase diagram 511
Tomographic Luttinger liquid model 407
Tomonga Luttinger model 405
Transmission lines 632
Transport current in a magnetic field 88
Trapped flux devices for manufacturing 599
Trilayer HTSC junctions 616
t-t' - t'' - J model 410
Tunnel effects 66
Tunnel junction 608
Two-fluid model 95

Type-I superconductors 44
Type-I and type-II superconductors 102
Type-II superconductors 45

V

Valence bond coupling 397
Vias and interconnect cross-overs 616
Voltage standards 79
Vortex anisotropies 85
Vortex pinning 47
Vortices 61, 84, 110

W

Weak links 608
Weierstrass functions 479

Z

Zero current barrier height 250